MW00679230

# The Monograph

The Monographs within the *Tall Buildings and Urban Environment* series correspond to the Council's group and committee structure. The basic objective of the Council's Monograph is to document the most recent developments in the state of the art in the field of tall buildings and their role in the urban habitat. The present listing, a modification of what appeared in the original Monograph, includes all current topical committees. Some are collaborating to produce volumes together, and Groups DM and BSS plan, with only a few exceptions, to combine all topics into one volume.

**PLANNING AND ENVIRONMENTAL CRITERIA (PC)**
Philosophy of Tall Buildings
History of Tall Buildings
Architecture
Rehabilitation, Renovation, Repair
Urban Planning and Design
External Transportation
Parking
Social Effects of the Environment
Socio-Political Influences
Design for the Disabled and Elderly
Interior Design
Landscape Architecture

**DEVELOPMENT AND MANAGEMENT (DM)**
Economics
Ownership and Maintenance
Project Management
Tall Buildings in Developing Countries
Decision-Making Parameters
Development and Investment
Legal Aspects

**SYSTEMS AND CONCEPTS (SC)**
Cladding
Partitions, Walls, and Ceilings
Structural Systems
Foundation Design
Construction Systems
High-Rise Housing
Prefabricated Tall Buildings
Tall Buildings Using Local Technology
Robots and Tall Buildings
Application of Systems Methodology
Industrial Tall Buildings

**BUILDING SERVICE SYSTEMS (BSS)**
HVAC/Energy Conservation
Plumbing and Fire Protection
Electrical Systems
High-Tech Buildings
Vertical & Horizontal Transportation
Environmental Design
Urban Infrastructure
Indoor Air Quality

**CRITERIA AND LOADING (CL)**
Gravity Loads and Temperature Effects
Earthquake Loading and Response
Wind Loading and Wind Effects
Fire
Accidental Loading
Safety and Quality Assurance
Motion Perception and Tolerance

**TALL STEEL BUILDINGS (SB)**
Commentary on Structural Standards
Methods of Analysis and Design
Stability
Design Methods Based on Stiffness
Fatigue Assessment & Ductility Assurance
Connections
Cold-Formed Steel
Load and Resistance Factor Design (Limit States Design)
Mixed Construction

**TALL CONCRETE AND MASONRY BUILDINGS (CB)**
Commentary on Structural Standards
Limit States Design
Selection of Structural Systems
Optimization
Elastic Analysis
Nonlinear Analysis and Limit Design
Stability
Stiffness and Crack Control
Precast Panel Structures
Creep, Shrinkage, & Temperature Effects
Cast-in-Place Concrete
Precast-Prestressed Concrete
Masonry Structures

The original five-volume Monograph on the Planning and Design of Tall Buildings was published between 1978 and 1981 and updated with several volumes. They can be ordered through the Council.

**Planning and Design of Tall Buildings**, 5 volumes (1978-1981 by American Society of Civil Engineers)

**High-Rise Buildings: Recent Progress** (1986, Council on Tall Buildings)

**Second Century of the Skyscraper,** (1988, Van Nostrand Reinhold Company)

**Tall Buildings: 2000 and Beyond** (1990, 1991, Council on Tall Buildings)

Council Headquarters
**Lehigh University,**
13 East Packer Avenue
Bethlehem, Pennsylvania 18015, USA

# Architecture of Tall Buildings

**Library of Congress Cataloging-in-Publication Data**

Architecture of tall buildings / Council on Tall Buildings
and Urban Habitat, Committee 30 (Architecture) : contributors, Mir M. Ali...
[et al.] : editorial group, Mir M. Ali, chairman, Paul J. Armstrong, editor.
p. cm. — (Tall buildings and urban environment series)
Includes bibliographical references and index.
ISBN 0-07-012540-6
1. Tall buildings. 2. Architecture—Environmental aspects.
I. Ali, Mir M., date. II. Armstrong, Paul J. III. Council on
Tall Buildings and Urban Habitat. Committee 30 (Architecture)
IV. Series.
NA6230.A72 1995
720′.483—dc20 94-47110
CIP



2 3 4 5 6 7 8 9 0 DOC/DOC 9 0 0 9 8 7 6 5

ISBN 0-07-012540-6

*For the Council on Tall Buildings, Lynn S. Beedle is the Editor-in-Chief and Dolores B. Rice is the Managing Editor.*

*For McGraw-Hill, the sponsoring editor was Joel Stein, the editing supervisor was David E. Fogarty, and the production supervisor was Pamela A. Pelton. This book was set in Times Roman by Cynthia L. Lewis of McGraw-Hill's Professional Book Group composition unit.*

*Printed and bound by R. R. Donnelley & Sons Company.*

This book is printed on acid-free paper.

# Council on Tall Buildings and Urban Habitat

# Council on Tall Buildings and Urban Habitat

**Sponsoring Societies**
International Association for Bridge and Structural Engineering (IABSE)
American Society of Civil Engineers (ASCE)
American Institute of Architects (AIA)
American Planning Association (APA)
International Union of Architects (UIA)
American Society of Interior Designers (ASID)
Japan Structural Consultants Association (JSCA)
Urban Land Institute (ULI)
International Federation of Interior Designers (IFI)

The following identifies those firms and organizations who provide for the Council's financial support.

**Patrons**
Al Rayes Group, Kuwait
Consolidated Contractors International, Athens
Dar Al-Handasah "Shair" & Partners, Amman
DLF Universal Limited, New Delhi
Zuhair Fayez & Associates, Jeddah
Jaros, Baum & Bolles, New York
Kuwait Foundation for the Advancement of Sciences, Kuwait
National Center for Earthquake Engineering Research, Buffalo
National Science Foundation, Washington, D.C.
Shimizu Corporation, Tokyo
The Turner Corporation, New York

**Sponsors**
Birmann S/A Comercio e Empreedimentos, São Paulo
Europrofile Tecom, Luxembourg
George A. Fuller Co., New York
T.R. Hamzah & Yeang Sdn. Bhd., Selangor
HL-Technik, A.G., Munich
Hochtief, Essen
Hong Kong Land Group Ltd., Hong Kong
Kone International (Europe) SA, Helsinki
John A. Martin & Assoc., Inc., Los Angeles
Walter P. Moore & Associates, Inc., Houston
Nippon Steel, Tokyo
Otis Elevator Co., Farmington
Ove Arup Partnership, London
Leslie E. Robertson Associates, New York
Samsung Engineering & Construction Co. Ltd., Seoul
Saud Consult, Riyadh
Schindler Elevator Corp., Morristown
Siecor Corporation, Hickory
Takenaka Corporation, Tokyo
Tishman Construction Corporation of New York, New York
Tishman Speyer Properties, New York
Wing Tai Construction & Engineering, Hong Kong
Wong & Ouyang (HK) Ltd., Hong Kong
Nabih Youssef and Assoc., Los Angeles

**Donors**
American Iron and Steel Institute, Washington, D.C.
W.R. Grace & Company, Cambridge
Haseko Corporation, Tokyo
Hollandsche Beton Maatschappij BV, Rijswijk
Hong Kong Housing Authority, Hong Kong
Iffland Kavanagh Waterbury, P.C., New York
Leigh & Orange Ltd., Hong Kong
O'Brien-Kreitzberg & Associates, Inc., Pennsauken
RTKL Associates, Inc., Baltimore
Skidmore, Owings & Merrill, Chicago
Steen Consultants Pty. Ltd., Singapore
Syska & Hennesy, Inc., New York
Thornton-Tomasetti/Engineers, New York
Werner Voss & Partners, Braunschweig
Wong Hobach Lau Consulting Engineers, Los Angeles

**Contributors**

American Institute of Steel Construction, Chicago
H.K. Cheng & Partners Ltd, Hong Kong
Hart Consultant Group, Santa Monica
Institute Sultan Iskandar, Johor
INTEMAC, Madrid
JHS Construcao e Planejamento Ltd., São Paulo
Kuala Lumpur City Centre Berhad, Kuala Lumpur
The Kling-Lindquist Partnership, Inc. Philadelphia
LeMessurier Consultants Inc., Cambridge
Lim Consultants, Inc., Cambridge
Meinhardt Australia Pty. Ltd., Melbourne
Meinhardt (HK) Ltd., Hong Kong
Middlebrook & Louie, San Francisco
Mitchell McFarlane Brentnall & Partners Int'l. Ltd., Hong Kong
Mueser Rutledge Consulting Engineers, New York
Obayashi Corporation, Tokyo
Charles Pankow Builders, Inc., Altadena
Projest SA Empreendimentos e Servicos Tecnicos, Rio de Janeiro
PSM International, Chicago
Skilling Ward Magnusson Barkshire Inc., Seattle
Teng & Associates, Inc., Chicago

**Contributing Participants**

Advanced Structural Concepts, Denver
Architectural Services Dept., Hong Kong
Australian Institute of Steel Construction, Milsons Point
Lynn S. Beedle, Hellertown
W.S. Bellows Construction Corp., Houston
Alfred Benesch & Co., Chicago
Bolsa de Imoveis Est. São Paulo, S.A., São Paulo
Boundary Layer Wind Tunnel Laboratory (U. Western Ontario), London
Boris Limited, London
Brandow & Johnston Associates, Los Angeles
Brooke Hillier Parker, Hong Kong
Buildings & Data, S.A., Brussels
Callison Architecture, Inc., Seattle
CBM Engineers Inc., Houston
Cermak Peterka Petersen, Inc., Fort Collins
CMA Architects & Engineers, San Juan
Construction Consulting Laboratory, Dallas
Crane Fulview Door Co., Lake Bluff
Crone & Associates Pty. Ltd., Sydney
Davis Langdon & Everett, London
DeSimone, Chaplin & Dobryn Inc., New York
Dodd Pacific Engineering, Inc., Seattle
Flack & Kurtz, New York
Sir Norman Foster and Partners, London
Haines Lundberg Waehler International, New York
Healthy Buildings International Inc., Fairfax
Hellmuth, Obata & Kassabaum, Inc., San Francisco
Ing. Muller Marl Gmbh, Marl
International Iron & Steel Institute, Brussels
Irwin Johnston and Partners, Sydney
Raul J. Izquierdo, Mexico
J.A. Jones Construction Co., Charlotte
DMJM Keating, Los Angeles
Lend Lease Design Group Ltd., Sydney
Lerch Bates & Associates Ltd., Surrey
Stanley D. Lindsey & Assoc., Nashville
Marcq & Roba, Brussels
Enrique Martinez-Romero, S.A., Mexico
Maunsell Consultants (Singapore) Pte. Ltd., Singapore
Mitsubishi Estate Co., Ltd., Tokyo
Moh and Associates, Inc., Taipei
Morse Diesel International, New York
Multiplex Constructions (NSW) Pty. Ltd., Sydney
Nihon Sekkei, U.S.A., Ltd., Los Angeles
Nikken Sekkei, Ltd., Tokyo
Norman Disney & Young, Brisbane
N. V. Besix SA, Brussels
Park Tower Group, New York
Peddle Thorp Australia Pty. Ltd., Brisbane
Cesar Pelli & Associates, New York
Rahulan Zain Associates, Kuala Lumpur
RFB Consulting Architects, Johannesburg
Rosenwasser Grossman Cons. Engrs., PC, New York
Emery Roth & Sons Intl. Inc., New York
Rowan Williams Davies & Irwin Inc., Guelph
Samyn and Partners, Architects and Engineers, Brussels
Sato & Boppana, Los Angeles
Sepakat Setia Perunding (Sdn.) Bhd., Kuala Lumpur
Sergen Servicos Gerais de Engenharia S.A., Rio de Janeiro
SOBRENCO, S.A., Rio de Janeiro
South African Institute of Steel Construction, Johannesburg
Steel Reinforcement Institute of Australia, Sydney
STS Consultants Ltd., Northbrook
Studio Finzi, Nova E Castellani, Milano
Taylor Thomson Whitting Pty. Ltd., St. Leonards
Tooley & Company, Los Angeles
Tractebel Development SA, Brussels
B.A. Vavaroutas & Associates, Athens
VIPAC Engineers & Scientists Ltd., Melbourne
Wan Hin & Company Ltd., Hong Kong
Weidlinger Associates, New York
Woodward-Clyde Consultants, New York
Yolles Partnership, Inc., Toronto

## Other Books in the Tall Buildings and Urban Environment Series

*Cast-in-Place Concrete in Tall Building Design and Construction*
*Cladding*
*Building Design for Handicapped and Aged Persons*
*Semi-Rigid Connections in Steel Frames*
*Fire Safety in Tall Buildings*
*Cold-Formed Steel in Tall Buildings*
*Structural Systems for Tall Buildings*

Planning and Environmental Criteria

# Architecture of Tall Buildings

## Council on Tall Buildings and Urban Habitat

## Committee 30 (Architecture)


**CONTRIBUTORS**

*Mir M. Ali*
*Kathryn H. Anthony*
*Paul J. Armstrong*
*Eli Attia*
*David P. Billington*
*Sarah L. Billington*
*Roger Brockenbrough*
*Pao-Chi Chang*
*Massimo Pica Ciamarra*
*Raymond J. Clark*
*Henry J. Cowan*
*Carolyn Dry*
*Mahjoub Elnimeiri*
*S. K. Ghosh*
*Myron Goldsmith*
*Lawrence W. Haines*
*Nestor Iwankiw*
*Douglas Stephen Jones*
*Michael K. Kim*
*Scott Melnick*
*Clare Monk*
*Tanwir Nawaz*
*Gyo Obata*
*Inigo Ortiz*
*Robert I. Selby*
*Bratish Sengupta*
*David C. Sharpe*
*Alfred Swenson*
*Kenzo Tange*
*Varkie Thomas*
*Ken Yeang*
*James P. Warfield*

**Editorial Group**

**Mir M. Ali,** Chairman
**Paul J. Armstrong,** Editor



**McGraw-Hill, Inc.**

New York San Francisco Washington, D.C. Auckland Bogotá
Caracas Lisbon London Madrid Mexico City Milan
Montreal New Delhi San Juan Singapore
Sydney Tokyo Toronto

## ACKNOWLEDGMENT OF CONTRIBUTIONS

This Monograph was prepared by Committee 30 (Architecture) of the Council on Tall Buildings and Urban Habitat as part of the *Tall Buildings and Urban Environment* series. The editorial group was composed of Mir M. Ali, Chairman; and Paul J. Armstrong, Editor.

Special acknowledgment is due those individuals whose papers formed the major contribution to the chapters in this volume. These individuals and the chapters or sections to which they contributed are:

Eli Attia, Sections 1.2–1.5
Inigo Ortiz, Sections 1.6, 7.12
Massimo Pica Ciamarra, Section 1.7
Kenzo Tange, Section 1.8
Gyo Obata, Section 1.9
Mahjoub Elnimeiri, Chapter 2
Myron Goldsmith, Chapter 2
David C. Sharpe, Chapter 2
Ken Yeang, Section 3.1
Robert I. Selby, Sections 3.2, 3.4
James P. Warfield, Sections 3.2, 3.4
Tanwir Nawaz, Section 3.3
David P. Billington, Chapter 4
Sarah L. Billington, Chapter 4
Mir M. Ali, Sections 1.10, 2.4, 3.5, 4.5, 5.1–5.3, 5.5, 8.3, 11.1–11.5, 11.8, 11.9, 12.1, 12.2, 12.5
Pao-Chi Chang, Sections 5.4, 9.7
Alfred Swenson, Sections, 5.4, 9.7
Kathryn H. Anthony, Chapter 6
Paul J. Armstrong, Chapter 7, Sections 1.1, 1.10, 3.5, 5.4, 6.5, 9.8, 10.1, 11.1, 12.1
Varkie Thomas, Sections 8.1, 8.2, 9.3–9.6
Lawrence W. Haines, Section 8.4
Douglas Stephen Jones, Section 8.4
Henry J. Cowan, Section 9.1
Michael K. Kim, Secton 9.2
Raymond J. Clark, Sections 9.3–9.6
Roger Brockenbrough, Sections 10.2–10.6
Nestor Iwankiw, Sections 10.7–10.10
Scott Melnick, Sections 10.7–10.10
S. K. Ghosh, Section 11.6
Carolyn Dry, Section 11.7
Bratish Sengupta, Section 11.10
Clare Monk, Sections 12.1, 12.3, 12.4, 12.5

Additional contributions were received from the following: Aparna Bapu, Patrick Dolan, C. R. Dutta, David Eilken, Jill Eyres, Mary Ellen Gibbon, Mark Gray, Robert S. Jordan, Chi-Hwa Kao, Marilyn Kelly, Anne McDermott, Lina Nath, Ripal Patel, Robert Pennypacker, Jennifer S. Salen, Douglas W. Shoemaker, Leszek J. Slowk, Robert J. Smith, and Albert Wong. The Committee Chairman and Editor were given quite complete latitude in the editing process. Frequently, length limitations precluded the inclusion of much valuable material. The Bibliography contains all contributions.

## COMMITTEE MEMBERS

Mir M. Ali, Samuel Aroni, L. Barbiano Belgioyoso, Massimo Pica Ciamarra, Frank L. Codella, Henry J. Cowan, Norman Edwards, Fred L. Foote, Yona Friedman, Myron Goldsmith, Bruce Graham, Robert Gutman, F. Hart, Jeffrey Heller, Walter Henn, Michel Holley, Takekuni Ikeda, Kay Vierk Janis, Akel Ismail Kahera, Frederick Krimgold, A. Lamela, Matthys P. Levy, G. Makarevich, Rafael Obregon, Sergio Edio Pellegrini, M. V. Posokhin, H. Rehman, K. Schwanzer, Alfred Swenson, Richard H. Tracy, Ken Yeang, Mufit Yorulmaz, Markku Viitasald.

## GROUP LEADERS

The committee on Architecture is part of Group PC of the Council, *Planning and Environmental Criteria*. The leaders are:

Bill P. Lim, Chairman
Yona Friedman, Vice-Chairman
David Thompson, Vice-Chairman
Werner Voss, Editor

# Foreword

This volume is one of a series of Monographs prepared under the aegis of the Council on Tall Buildings and Urban Habitat, a series that is aimed at documenting the state of the art of the planning, design, construction, and operation of tall buildings as well as their interaction with the urban environment.

The present series is built upon an original set of five Monographs published by the American Society of Civil Engineers (1978–1981) as follows:

*Volume PC: Planning and Environmental Criteria for Tall Buildings*

*Volume SC: Tall Building Systems and Concepts*

*Volume CL: Tall building Criteria and Loading*

*Volume SB: Structural Design of Tall Steel Buildings*

*Volume CB: Structural Design of Tall Concrete and Masonry Buildings*

Following the publication of a number of updates to these volumes, it was decided by the Steering Group of the Council to develop a new series. It would be based on the original effort but would focus more strongly on the individual topical committees rather than the groups. This would do two things. It would free the Council committees from restraints as to length. Also, it would permit material on a given topic to reach the public more quickly.

The result is the *Tall Buildings and Urban Environment* series, being published by McGraw-Hill, Inc., New York. This Monograph joins seven others, which were released beginning in 1992:

*Cast-in-Place Concrete in Tall Building Design and Construction*

*Cladding*

*Building Design for Handicapped and Aged Persons*

*Fire Safety in Tall Buildings*

*Semi-Rigid Connections in Steel Frames*

*Cold-Formed Steel in Tall Buildings*

*Structural Systems for Tall Buildings*

This particular Monograph was prepared by the Council's Committee 30, Architecture. The original chapter on this topic was a part of Volume PC. It dealt with the many issues relating to tall building architecture when it was published in 1981. The committee decided, however, that there was a need to produce a Monograph that would encompass the issue of architecture in a holistic and comprehensive manner.

The scope of this Monograph, while broad, focuses on the specific issues that affect and shape the design of tall buildings in today's complex and changing world. The committee has endeavored to preserve the in-depth nature of the subjects they have chosen to explore in this Monograph. It provides contributions of the study of tall building architecture from the perspective of 50 leading professionals in Australia, Canada, India, Italy, Japan, Malaysia, Spain, and the United States. This unique international work examines the many issues that must be taken into account when designing tall buildings as part of the urban habitat.

## The Monograph Concept

The Monograph series is prepared for those who plan, design, construct, or operate tall buildings, and who need the latest information as a basis for judgment decisions. It includes a summary and condensation of recent developments for design use; it provides a major reference source to recent literature, and it identifies needed research.

The Monograph is not intended to serve as a primer. Its function is to communicate to all knowledgeable persons in the various fields of expertise the state of the art and the most advanced knowledge in those fields. Our message has more to do with setting policies and general approaches than with detailed applications. It aims to provide adequate information for experienced general practitioners confronted with their first high rise, as well as opening new vistas to those who have been involved with them in the past. It aims at an international scope and interdisciplinary treatment.

The Monograph series was not designed to cover topics that apply to all buildings in general. However, if a subject has application to *all* buildings, but also is particularly important for a *tall* building, then the objective has been to treat that topic.

Direct contributions to this Monograph have come from many sources. Much of the material has been prepared by those in actual practice as well as by those in the academic sector. The Council has seen considerable benefit accrue from the mix of professions, and this is no less true in the Monograph series itself.

## Tall Buildings

A tall building is not defined by its height or number of stories. The important criterion is whether or not the design is influenced by some aspect of "tallness." It is a building in which tallness strongly influences planning, design, construction, and use. It is a building whose height creates conditions different from those that exist in "common" buildings of a certain region and period.

## The Council

The Council is an international group sponsored by engineering, architectural, construction, and planning professionals throughout the world, an organization that was established to study and report on all aspects of the planning, design, construction, and operation of tall buildings.

The sponsoring societies of the Council are the American Institute of Architects (AIA), American Society of Civil Engineers (ASCE), American Planning Association (APA), American Society of Interior Designers (ASID), International Association for Bridge and Structural Engineering (IABSE), International Union of Architects (UIA), Japan Structural Consultants Association (JSCA), Urban Land Institute (ULI), and International Federation of Interior Designers (IFI).

The Council is concerned not only with the building itself but also with the role of tall buildings in the urban environment and their impact thereon. Such a concern also involves a systematic study of the whole problem of providing adequate space for life and work, considering not only technological factors, but social and cultural aspects as well.

## Nomenclature and Other Details

The general guideline for units is to use SI units first, followed by American units in parentheses, and also metric when necessary. A conversion table for units is supplied at the end of the volume. A glossary of terms also appears at the end of the volume.

The spelling was agreed at the outset to be "American" English.

Relevant references and bibliography will be found at the end of each chapter. A composite list of all references appears at the end of the volume.

An appendix is included which identifies the lead professional firms: developer, architect, structural engineer, and services engineer. It will provide a handy reference and obviates the incorporation of such information throughout the text.

From the start, the Tall Building Monograph series has been the prime focus of the Council's activity, and it is intended that its periodic revision and the implementation of its ideas and recommendations should be a continuing activity on both the national and the international levels. Readers who find that a particular topic needs further treatment are invited to bring it to our attention.

## Acknowledgments

This work would not have been possible but for the early financial support of the National Science Foundation, which supported the early program out of which this Monograph developed. More recently the major financial support has been from the organizational members of the Council, identified in earlier pages of this Monograph, in particular Nikken Sekkei Ltd. of Tokyo, Japan, which provided special funding for this volume. The confidence of these firms as well as the individual members of the Council is appreciated.

All those who had a role in the authorship of the volume are identified in the acknowledgment page that follows the title page. Especially important are the contributors whose papers formed the essential first drafts—the starting point. Their names are indicated in the title page.

The primary conceptual and editing work was in the hands of the leaders of the Council's Committee 30, Architecture. The Chairman is Mir M. Ali of the University of Illinois at Urbana-Champaign, Illinois, United States. Comprehensive editing was the responsibility of Paul Armstrong, also of the University of Illinois.

Overall guidance was provided by the Group Leaders: Bill B-P Lim of Queensland University of Technology, Brisbane, Australia; David Thompson of RTKL Associates, Inc., Baltimore, Maryland, United States, Yona Friedman of Paris, France; and Werner Voss of Werner Voss und Partner, Braunschweig, Germany.

*Lynn S. Beedle*
*Editor-in-Chief*

*Lehigh University*
*Bethlehem, Pennsylvania*
*1995*

*Dolores B. Rice*
*Managing Editor*

# Preface

The architecture of tall buildings changes continuously with the times. A review of the literature has revealed the need to address and inquire into the many complex facets of tall building architecture. The goal of Committee 30 was to compile a Monograph on the architecture of tall buildings that would be comprehensive in scope. Thus, the original chapter on architecture in the previously published Monograph (CTBUH, Group PC, 1981) has been substantially updated and expanded. Further, the Committee wanted the Monograph to be international in scope to incorporate tall building architecture as related to the industrialized cities of the United States, Europe, and Asia, as well as to include examples from other parts of the world where tall building technology must meet the needs of rapidly expanding populations and constrained physical economic resources.

The architecture profession is also changing to keep up with rapid technological progress. Because of the client's need to complete projects in a timely and efficient manner, architects have to meet a number of pressing requirements that demand an understanding of the various interdisciplinary forces interacting with one another during the design process.

According to the Council, a tall building is not defined by its height or number of stories. The Foreword of the Monograph *Planning and Environmental Criteria of Tall Buildings* states that it "is a building in which 'tallness' strongly influences planning, design, and use. It is a building whose height creates different conditions in the design, construction, and operation from those that exist in 'common' buildings of a certain region and period" (CTBUH, Group PC, 1981).

The tall building design and construction process is collaborative, requiring the input of architects, engineers, urban planners and designers, economists, sociologists, technical specialists, and other consultants. The efforts of all professionals must be supported by the most up-to-date technology, research, and innovative thinking regarding tall building architecture in this complex endeavor. Case studies are provided in several chapters where detailed building examples are incorporated. Current applications of materials, technology, and design are presented in addition to anticipated future needs.

Chapters 1 to 5 examine the genesis of tall buildings, their urban and political influence, and the social role they play through their planning, structural expression, form, and aesthetic expression. Chapter 1 addresses the historical development of tall buildings and general issues and criteria of architectural design. Case study examples provide a comprehensive view of the design process from an international perspective. Chapter 2 focuses on the interaction between academic research and professional practice. Using case studies developed in the design studio, it demonstrates their applications to engineering and design and the significance of the role of research to innovative thinking in structural systems. Urban design and development are examined in Chapter 3. A case study examines the introduction of tall buildings into West Philadelphia, and their projected impact on the image, vitality, and urban coherence of a large American city are included in this chapter.

The role of engineering, while always present and instrumental in the design process, takes on even greater importance and emphasis in the design of tall structures. Chapter 4 focuses on the notion of structuralist architecture and the significant contributions of the structural engineer. Chapter 5 examines the broader and more subjective issues of aesthetics and form. The psychological and emotional responses to aesthetics and form, while elusive, are manifested through historical styles and cultural *milieux*.

Although the physical and aesthetic impact of tall buildings on their environment is significant, the designer must also recognize the psychological and behavioral impact, which is described in Chapter 6. Chapter 7 examines the integral roles that architectural image, technology, and materials play as determinants of the form and aesthetics of tall building facades. The nature of the wall, building structure, and cladding are presented from an architectural point of view.

Chapters 8 and 9 present building system automation and integration, environmental systems, and their impact on the design, comfort, safety, environmental impact, and physical maintenance of tall buildings. Chapter 8 also includes a case study of "smart" or "intelligent" building design, providing a glimpse into the future. Automation is also playing a greater role in the design of tall buildings now that architects have incorporated computers into the architectural office. A future Monograph on computer-aided design of tall buildings and the application of automated building systems is expected to address these issues in greater depth.

Advances in structural engineering, materials, and construction techniques have enabled architects to develop new forms of expression and to build to unprecedented heights. Chapters 10 to 12 examine the roles that materials contribute to the design and form of tall buildings.

The Editorial Group of Committee 30 extends its gratitude to the Council on Tall Buildings and Urban Habitat for its cooperation throughout this project. The continuous support and guidance by Dr. Lynn S. Beedle and the collaboration of Mrs. Dolores B. Rice are sincerely appreciated. Dr. Bill Lim and Mr. Leslie Robertson deserve special thanks for their assistance in establishing contacts. We are also indebted to the University of Illinois Research Board, which has provided financial support for this project. Nikken Sekkei Ltd. of Tokyo, Japan, also provided a special grant for the production of this volume. Further, we appreciate the contribution made by the School of Architecture at the University of Illinois at Urbana-Champaign both for its resources and for the staff, who have assisted in the production of this Monograph.

The Editorial Group also extends its appreciation to the tireless efforts of its Research Assistants Jennifer Salen, Douglas Shoemaker, Aparna Bapu, and David Eilken. In particular, Jennifer Salen and Douglas Shoemaker made valuable contributions to Sections 8.4 and 9.8, respectively. Without their assistance and continuous effort the compilation of this Monograph would not have been possible. Finally, appreciation is extended to all contributors for providing the subject matter of this document and to the other members of Committee 30 who have provided valuable input.

*Mir M. Ali,*
*Chairman*

*Paul J. Armstrong,*
*Editor*

# Contents

# 1

# Architectural Design: Issues and Criteria

Tall buildings impact the balance of an urban environment. Focus on the tall building's responsiveness to the quality of its surroundings, its urban and environmental impact, as well as its own functional program is essential to the designer. Within the immense range of qualities and intertwined factors that determine the urban habitat as a place for living, form is critical in tall building architecture and the cityscape. This chapter seeks to address the broad issues related to the architectural design process, to examine form as a design issue, and to assess the impact of tall buildings on the urban areas in which they are to serve.

It does not, however, provide the reader with a clear articulation or a listing of specific design criteria that must be considered in the design process of tall buildings. (See Subsection 2 of Section 1.10.) No attempt has been made herein to critically review the few case study examples presented in this chapter against a set of specialized design criteria. The case studies exemplify various aspects of the design process in a general manner and are expected to provide the reader with valuable information about a diverse range of issues and criteria unique to tall buildings.

Section 1.1 addresses the history of the tall building and its development as a unique building type. Sections 1.2 and 1.3 focus on the genesis of tall buildings, the political urban influence, and the role of the tall building in society. Section 1.4 looks at the supertall buildings, their specific qualities, and the inseparable cohesiveness of their form and structure. Sections 1.5 through 1.9 present a series of studies examining the architectural design issues of tall buildings. Section 1.5 introduces four tall buildings with focus on each building's architectural form and its impact on the built environment. Section 1.6 looks at a European tall building design approach for the Mapfre Tower, a tall building within the Olympic village in Barcelona, Spain. Italian tall buildings and their development are the focus of Section 1.7. Section 1.8 addresses architectural design issues for an urban project in Tokyo, Japan. Section 1.9 presents the project descriptions and unique design aspects of seven other tall building projects, five of which have already been completed. These projects represent a spectrum of varied design concepts and challenges with which an architect must deal during the design process. Finally, Section 1.10 discusses future directions in design and research needs.

## 1.1 HISTORICAL DEVELOPMENT OF TALL BUILDINGS

### 1 Monuments and Towers

In terms of their formal language, tall buildings are often categorized historically with towers and monuments where their landmark status carries specific attributes of imagery and place. Prescott (1962) writes:

> The tall building began as an isolated monument, and can be formally considered as a monolith in a visual world having a closed system of values. The symbolic column can be traced historically beginning with the free-standing Egyptian obelisk. The Romans developed their own version with Trajan's Column (A.D. 98) which became the prototype for Western monuments such as Nelson's Column (A.D. 1843). Today the tall building is symbolically devalued and placed in a related system of architectonic forms.

Building height is evident in certain building forms from ancient times in the various civilizations of the East and the West (Gregory, 1962). Its earliest manifestations can be found in the stupa and the pyramid forms with their significant base widths and depths. Other ancient tall structures are the Tower of Babel, the Colossus of Rhodes, the Pyramids of Egypt, and the Mayan temples of Mexico. But the true evolution of the high-rise building perhaps begins with the emergence, in the sixth century A.D., of the tower-like pagodas of Japan derived from the Chinese (Gregory, 1962). In the Middle East, the minaret of Muslim architecture, from the square towers of the Great Mosque of Damascus (707 A.D.) to the slender spires of Turkey and India (for example, the Kutub Minar in Delhi), exerted a significant influence on the Gothic and Renaissance architecture of the West. The classical and Gothic revivalists of the nineteenth century exploited the syntactic and symbolic power of tall buildings. The tower of the medieval town hall, for example, held a significance of civic dignity to the people of the time, and climaxed in England with the towers of Westminster Palace (1840–1867 A.D.). With the Industrial Revolution a tradition of tall buildings in industrial architecture, from the tall brick chimney to the reinforced concrete cooling tower, has evolved steadily, with church towers and spires continuing to reflect religious symbolism. In the latter half of the nineteenth century a new architectural type, the commercial tall building, was made possible by the introduction of the steel frame and the invention of the elevator. Its rational "gridiron" expression was largely the result of design innovations initiated by William Le Baron Jenney, Louis Sullivan, John Wellborn Root, Daniel Burnham, and the other Chicago Loop architects and further developed by the modern architects, including Mies van der Rohe, Walter Gropius, and later Skidmore, Owings & Merrill (Hart et al., 1985; Goldberger, 1989). Refer to Chapters 5 and 7 of this Monograph for additional information on Sullivan and the first Chicago school as well as the influence of modernism. For insight into the development of the steel frame, refer to Chapter 10.

### 2 Tall Buildings: Form and Structure

Prescott (1962) suggests that the development of the inhabitable high-rise tower is actually a double movement. First, the honest expression of function and of materials in commercial building is not regarded as architecture per se, but is significant in the development of architectural expression. The main evolutionary change of the tall building is in its function, from a campanile watchtower or minaret to a spacious office building (Gregory, 1962). This was followed by a simulation of the new forms in old materials which became aesthetically acceptable to commercial architecture. Eventually this was superseded by the renewed strength of an architecture of eclectic historicism springing largely, but not exclusively, from the classic school in England, and the strong traditions of the Ecole des Beaux-Arts in the United States promoted by architects such as McKim, Mead, and White; Graham, Probst, and Anderson; and Daniel Burnham.

The aesthetic function of the tall building has not changed. Its *raison d'être* in the past was to emphasize a focal point, a dominant feature, or the importance of certain aspects of life (Gregory, 1962). The church tower, the campanile (Fig. 1.1), and the minaret were built as reminders of God's omnipresence, made visible to all within the town and out in the fields. The profile of towers when viewed at a distance against the sky creates a stimulating outline or silhouette, providing contrast and interest to the context. According to Gregory (1962), when the contextual idea of the tall building is forgotten and the tall building becomes an end in itself, its townscaping quality is compromised.

Ada Louise Huxtable (1984) points out that the tall building is "the landmark of our age." The tall building represents this century's most stunning architectural phenomenon and its most overwhelming architectural presence. The question of how to design tall buildings, a problem that challenged Sullivan, has never really been resolved. The earliest examples of high-rise buildings were based on historical models which, upon cursory examination, may seem appropriate and even exemplary. But resorting to past vocabularies for a revolutionary building type seemed out of character with the spirit of the new age of modernism (Goldberger, 1989; Huxtable, 1984).

**Fig. 1.1 The campanile was a reminder of God's omnipresence, made visible to all within the town and out in the fields.** (*Courtesy: Paul Armstrong.*)

The development of the tall building has been the apotheosis of modern building technology and aesthetic form. Its evolution represents the synthesis of structure, materials, zoning and code requirements, energy, aesthetics, and social and cultural prestige. In this sense, the tall building is a truly "holistic" building type. Ultimately, the tall building continues to be an art form, and the search for a "skyscraper style" has been the consuming goal of every radical advancement or conservative retrenchment (Huxtable, 1984).

To a significant degree, the tall building was made possible through innovations in steel-frame technology and the invention of the elevator by Elisha Graves Otis in 1853. The tall building's historical origin can be traced to Chicago and New York and was the result of a fortuitous confluence of technology, zoning, and timing. The American office building was the result of a unique combination of industrialization, business, and real estate during a period of unparalleled physical and economic growth and expansion in American cities. Similar growth and development are now occurring in the cities of the Asian Pacific region.

***Four "Skyscraper Ages."*** Cesar Pelli (1982) and Ada Louise Huxtable (1984) identify four "skyscraper ages," beginning with the innovative engineering accomplishments in Chicago (see Chapter 7). The *first skyscraper age* is characterized by its functional orientation and reflects advances in structural engineering and the expression of the steel frame in the elevation. Essentially, the early tall building was the servant of engineering and, as such, was a phenomenon through which innovation was driven by business requirements and economy.

William Le Baron Jenney's Home Insurance building of 1884–1885 initiated the innovative use of the structural steel frame which would characterize much of subsequent tall building design. Chapter 10 presents a historic outline on cast-iron and steel-frame construction. However, the full potential of the steel frame as an architectural expression of the building's facade would not be fully realized until the next century (see Chapter 7). The evolution toward pure structural expression was initially hampered by the desire of architects to relate building facades to historic models and an uncertainty about the ultimate form or appearance that might be achieved.

The *second skyscraper age,* or eclectic phase, sought aesthetic solutions through the application of historical models. Fueled by the ascendancy and popularity of the French Academy and the Beaux-Arts in this country, it continued well into the twentieth century, until construction was halted by the Great Depression. Bruce Price's classical tripartite solution for the American Surety building of 1894–1896 in New York is considered the epitome of Beaux-Arts–inspired high-rise design.

Huxtable classifies the *third skyscraper age* as the modern phase. This third and influential phase was ushered in by the European school represented by Gropius, Mies, and Le Corbusier and is loosely united under the somewhat misleading rubric of the *international style.* Modern architecture was largely a reaction against the Beaux-Arts eclecticism and historically based architectural models. Its revolutionary approach sought to redefine architecture from the ground up. These architects rejected ornamentation and historical reference and, instead, embraced a more technological and rational expression of building form.

Modernism rejected both history and eclecticism and sought an architectural style free of the encumbrances of the past. The international style buildings, which emerged between the world wars, are distinguished by simple, stereometric shapes; unitary volumes wrapped in a thin, weightless skin of glass, plaster, or similar material; and no ornamental details (Norberg-Schulz, 1974). Characterized by its functionalism and direct expression of materials and structure, the international style tall building has become a corporate icon throughout the world. International style buildings are invariably monolithic "boxes," which are functionally defined by the economies of leasable floor areas, building structure, and core-to-exterior-wall relationships. The tripartite classical definitions of base, shaft, and crown are all but absent. Typically, the entire building is raised on columns (*pilotis*). The top of the building is always flat. The stereometric shape of the building, clad with a glass curtain wall, expresses its vertical continuity from base to shaft to top with little articulation or variation.

The *fourth skyscraper age* is perhaps the most prolific of the four skyscraper ages and has radically changed the American skyline in particular. Huxtable calls this phase the post-modern phase, but unlike modernism, which sought stylistic clarity, this phase is pluralistic and includes many architectural styles including post-modernism and late modernism (Jencks, 1988). This period is also marked by some extraordinary advances in technology, such as new concepts in structural systems, advances in curtain-wall design incorporating specialized glazing and photoactive solar devices, and increasingly thinner and lighter stone veneer claddings.

Jencks (1988) distinguishes between post-modern and late-modern architecture based on differences in historical continuity and design methodology:

> The Post-Modernists claim positively that their buildings are rooted in place and history...and that they bring back the full repertoire of architectural expression: ornament, symbolism, humor, and urban context....Relying on modern constructional means and historical memory, elite and popular meanings, [their] buildings are doubly-coded....
>
> Late-Modern architects, by contrast, disdain all historical imagery but that of their immediate forebears. They concentrate on the perennial abstractions of building—space, geometry, and light—and generally refuse to discuss stylistic issues at all....Instead, they approach building as a series of technical problems to be solved by teamwork...or abstract propositions of geometry and organization.

According to Jencks (1988) a post-modern building incorporates both modern design influences (for example, abstraction, preference for stereometric forms, and the direct and straightforward expression of materials and construction) with vernacular, historical, regional, metaphorical, and contextual design influences. Post-modern architecture is inherently complex since it does not ascribe to a single philosophy or even a single notion of style. Instead, post-modern architecture tends to be eclectic, often freely combining historically derived architectural styles with contextual references. Post-modern high-rise buildings are likely to have clearly defined divisions of base, shaft, and crown. The tops of the buildings tend to be distinctive points, spires, or stepped forms. While utilizing modern materials and construction techniques, the shafts are often articulated with "ornamental" detail and polychromatic colors derived from the component materials themselves, or through the application and layering of other materials. The base of post-modern buildings may be monumental, or they may be scaled down to conform to the context of the streetscape and the scale of the pedestrian.

Late-modern architecture, on the other hand, may eschew beauty altogether in deference to expression and function (Jencks, 1988). Buildings such as the Hong Kong and Shanghai Bank headquarters (1980–1986) use high-tech imagery to express their structural and mechanical systems, and to create a machine aesthetic. Chapter 7 discusses the Hong Kong and Shanghai Bank in more detail.

The *fourth skyscraper age* reflects a personal appropriation of eclectically derived historical forms and symbols as well as a return to neotraditional urban planning principles (Huxtable, 1984). Architects and architectural firms are currently designing multifunctional tall buildings that derive their formal and stylistic vocabularies from the classic skyscrapers of the twenties and thirties. Asymmetrical or stepped towers loom over low-rise and midrise volumes and are designed to integrate with the urban context.

***Urban Concern in Design.*** In the late 1970s and early 1980s, architects began to shift their attention from the making of abstract shapes toward a recognition, long overdue, of urbanistic values. This post-modern tendency in architecture was underscored by a return to historical reference, abandoning the abstract severity of the rationalist box and the dream of an international style. But the redefinition of tall buildings in terms of historical precedent was not exclusively at the hands of the architect or the engineer. Goldberger (1989) writes that the "impetus came from the receptiveness of clients toward alternatives to the ubiquitousness of the austere glass box which had characterized modern architecture."

By the end of the 1970s, the success of a few notable skyscrapers of the previous decade such as the IDS Center in Minneapolis and the John Hancock Tower in Boston changed public opinion about tall building design and aesthetics. At the end of 1980 most of the large skyscrapers under construction in New York contained at least some sort of public atrium, retail mall, or "galleria" (Goldberger, 1989), distinguishing these new buildings from their predecessors and signaling a more humanistic and urban concern in tall building design.

The Procter & Gamble headquarters in Cincinnati, for example, is noteworthy not only for its references to the skyscraper vocabulary of the 1920s but, most significantly, for its masterful urban design and scale (Boles and Murphy, 1985). The Procter & Gamble headquarters is comprised of two massive towers linked by glass bridges which form a visual gateway image to the downtown. Completing a one-block composition with the original 1956 Ekstut Tower headquarters building, the new headquarters' L-shaped configuration encloses a garden on two sides and completes the urban fabric. Refer to Chapter 3 for a more in-depth discussion of urban design and tall buildings.

## 3 Recent Trends in Tall Building Design

***High Tech.*** Recent developments in tall building design include high-tech buildings that exploit and humanize technology, function, and ideology (Jencks, 1988). High-tech architects do not consider high-tech architecture as a style, but instead view it as an approach to design which promotes structural and mechanical expression, transparency and movement, and bright coloring to distinguish building services and structure (Jencks, 1988). The notion of structural expression through high-tech architecture is questioned by others (see Chapter 5). High-rise buildings such as the Hong Kong and Shanghai Bank headquarters, Lloyds of London, and L'Institut du Monde Arabe in Paris fall into this category (see Chapter 7).

***"Classical Eclecticism."*** Huxtable (1984) asserts that "'classical eclecticism' is the most radical of the current new directions in skyscraper architecture, not because it espouses the classical sources that the modernists so emphatically rejected, but because it rejects the continuing imperative of structure as a chief generator of architectural style." Huxtable's argument is problematic because it raises the possibility that an architectural aesthetic need no longer reinforce or refer to the physical and structural realities of the building. This is a total departure from the virtually unbroken line of structural determinism that has been a basic tenet of skyscraper design.

***Climatic Response and Sustainability.*** Other directions in tall building design seek to address issues of context and climate, such as the United Gulf Bank in Manama, Bahrain. The adaptation of massive concrete exterior walls with deeply recessed windows acts as solar screens and thermal regulators in Bahrain's hot, desert climate.

Sustainability and "green architecture" (see Chapter 9) have become important architectural issues today as concerns about depletion of natural resources, reuse of natural and synthetic materials, as well as conservation of nonrenewable energy resources take on global proportions. The proposed Warsaw Tower, for example, promotes the use of recycled materials for its exterior cladding. Energy conservation in tall building design has become a determinant for the building's form, exterior cladding materials, amount of exterior glazing, and even the location and the size of its atrium.

The character of today's high-rise cities lies in the good design of each and every building, that each may be sustained on its own merits, but gaining from contrast with equals. The international style buildings erected ubiquitously throughout the world without regard to context, culture, or history cancel each other out, nullifying even their prestige component. For additional information, refer to Chapters 5 and 7, and also to Chapters PC-2 and PC-6 of *Planning and Environmental Criteria* (CTBUH, Group PC, 1981).

## 1.2 FORM AS A CRITICAL ISSUE

Tall building design may be approached philosophically in a number of ways. For example, Kavanagh (1972) views the tall building "as a basic cultural form representing the spiritual and creative aspirations of man...." Codella and McArthur (1973), on the other hand, feel that "our buildings today reflect our civilization...a direct result of the importance of time and money."

The best approach to designing tall buildings is still an open subject of professional debate. The role of the architect in designing them is a point of contention, particularly from the point of view of form giving. There is a classical view of the architect as a generalist, who must have a working knowledge and familiarity with multiple disciplines, because, according to Syrovy (1973), although "technical criteria limit the architectural design to a certain extent...the fact that the majority of technical problems are solved for every building individually provides great possibilities of expression and great resources of inspiration." Thus, despite the fact that form could be determined from technological considerations, it must be closely correlated to the architect's design instinct and aspirations.

### 1 Scale and Context

The appearance of the first tall building, just over a century ago, marked the beginning of what has emerged as the most radical change that cities across the world have ever experienced. Within a fraction of the entire history of urban civilization, the evolution of cities has so fundamentally shifted that this condition cannot yet be comprehended entirely. The unprecedented proliferation of tall buildings of constantly increasing scale (within existing urban fabrics, or within new fabrics that have been molded according to the old fabric) has caused the cumulative deterioration of environmental conditions, thus becoming detrimental to the quality of urban life (Mumford, 1952):

> Actually the skyscraper, from first to last, has been largely an obstacle to intelligent city planning or architectural progress....This gives to most of the discussions of proper form of the skyscraper, even to Louis Sullivan's, an air of studied irrelevance....Aesthetically, Sullivan reduced the problem of the expression of the skyscraper to that of a conventional base for shops and services, a shaft for offices, and a capital to screen the elevator machinery....In his preoccupation with the face of the skyscraper, Sullivan contributed little to the plan....

So the lesson of collective order [as exemplified by the development of the entire block as in the case of the Monadnock Building] was never learned; and the precedent [the Monadnock Building] set out was not even feebly recovered until Rockefeller Center was laid out.

The unique significance of tall buildings lies both in their positive attributes and in their deficiencies. Their attributes develop from the aspirations of humankind to create soaring structures that reach great heights. Their deficiencies lie in the fact that their physique is overwhelming and might establish unfavorable conditions. Moreover, the production of tall buildings is extraordinary because of the complex interactions and limitations set by economics, zoning ordinances, and prevailing technologies, not solely by architectural considerations (Prescott, 1962; Barnett, 1986).

The following discussion will focus on this most critical factor of form. At the urban scale, form is one of the most fundamental architectural design issues of tall buildings and inseparable from the quality of the urban habitat (Cullen, 1961; Rowe and Koetter, 1978).

## 2 The Tall Building Phenomenon

*Tall building* cannot be defined in terms of its height or number of floors since the outward appearance of tallness is a relative matter. As such, there is no consensus on what constitutes a tall building or at what height, number of floors, or proportions a building can be described as tall. Beedle (1971) defines the tall building by stating: "The multistory building is not generally defined by its overall-height or by the number of stories, but only by the necessity of additional operation and technical measures due to the actual height of the building." In the Foreword of the Monograph *Structural Design of Tall Concrete and Masonry Buildings* (CTBUH, Group CB, 1978), Beedle also states that the typology is "a building in which 'tallness' strongly influences planning, design and use," or "a building whose height creates different conditions in the design, construction, and use than those that exist in 'common' buildings of a certain region and period." Tall buildings are the product of need, not whim. They are synonymous with the twentieth century and result from its major forces. Their complex genesis is found in demographics, location, and land values, resulting in society's need for growth. The need for growth, due to increased densities in urban centers around the world, persists and has been increasing exponentially since the beginning of urbanization. There is no reason to expect this trend to change. No doubt, buildings will grow even taller. However, there is no clear indication that a denser environment is inferior to a less dense one. Density alone does not make the urban quality of space (Aregger and Glaus, 1975; Giedion, 1967). For further information about social effects of environments see Chapters 3 and 6 and Chapter PC-3 (CTBUH, Group PC, 1981).

Tall buildings have neither historical nor regional architectural precedents. They did not evolve, as in the past, from their predecessors, the walk-ups, but were the product of a conceptual breakthrough. From the beginnings of urban organization, the height of buildings was limited by the physical ability for a person to comfortably climb stairs and the need for egress during a fire. For centuries buildings reached their optimal height of four or five stories. In response to this limitation, and the ever growing densities, an idea emerged which searched for ways to accommodate the new urban condition. Included in the solution, vertical stacking evolved for the purpose of better utilizing scarce urban land. While tall buildings gained acceptance for a wide range of uses, their other beneficial aspects were also realized: greater flexibility for their users in expanding or contracting of operational areas, adjacency of complementary uses, shared facilities, and better utilization of energy, to name but a few. Although never the generator of tall buildings, technology has always responded by providing means that made their creation viable. Today's technology of tall buildings allows their adaptability to all climatic, geographic, and physical conditions. Indeed, their global proliferation expresses their universality (Scuri, 1990).

## *3 The Political Urban Factor*

In addition to the customary participation of business, finance, and government, a new active partner is now involved in the process of creating tall buildings—the local community. In New York this organized effort first appeared in the form of politically appointed community boards, followed by the self-appointed civic groups, scores of them in Manhattan alone. These potent civic groups have proven to be extremely influential participants in designing the legislative processes. They have affected virtually every aspect of the architectural design of tall buildings.

For example, in large metropolitan cities such as New York, Chicago, and Los Angeles, large-scale buildings are often cited by local, community-based environmental interest groups for their inappropriateness to context and a number of other social and economic problems. Although well-intentioned, these local "aesthetic" boards, however, sometimes fail to look beyond architecture as the cause of more complex social, economic, and cultural problems. In their endeavor to effectively address these complex issues, some cities have brought together community groups—civic leaders, architects, planners, engineers, sociologists, and other professional and nonprofessional groups—into effective partnerships.

By governing building envelopes, zoning regulations carry the potential of becoming the formal unifying denominator of an urban environment. Zoning can become a potent control mechanism. The New York City 1916 Zoning Resolution, the first of its kind and most illuminating ever, was brought about in reaction to deep urban canyons that were created on the narrow streets of lower Manhattan. These metropolitan ravines deprived the city's inhabitants of daylight, sunlight, and natural ventilation. The zoning regulation combined urban concerns with ecoenvironmental safeguards to create a direct, simple, and extremely effective set of governing rules. As a result, urban coherence was achieved as a function of street width. Streets were lined up with buildings of uniform height; taller ones on wide streets, shorter ones on narrow streets, with slender, elegant towers rising upward above them (Broadbent, 1990):

> The city was divided into commercial and residential zones where the heights, and the volumes, of buildings were prescribed. On any given street facades could rise vertically to a prescribed height above which buildings had to be set back. Setbacks were determined, for a given site, by drawing an imaginary plane from the center line of the street to the cornice or parapet line, as specified, of the facade and beyond. Anything higher then had to be set back behind the pyramidal envelope formed as these imaginary planes from the different sides of the buildings intersected. Towers could rise out of the pyramid provided that they covered no more than 25% of the site.

Even with tall handsome towers emerging throughout the city, daylight reached street level, with only diminutive and negligible adverse ecological effects detected. The towers were never viewed as being excessively high or detrimental; rather the contrary was true. The early skyscrapers were the basis of pride for New York and a source of inspiration for the rest of the world. Buildings such as the Woolworth, Chrysler, and

Empire State are all products of the 1916 legislation and have become landmark tall buildings, exemplary of responsible urban architecture.

The clarity of purpose of this brief resolution and its urban consequences are astonishing when compared with the 1961 Zoning Resolution that followed, and are still in effect, albeit in different form. Driven predominantly by responses to lobbying interests, and averaging about one a day, more than 8000 amendments have been added since its enactment. The present complex document has lost its clarity of purpose, even becoming incomprehensible, by invoking whatever was gained in 60 years of cumulative legislation and knowledge in one sweep. This legislative instrument provided the politically appointed group with power to "negotiate" zoning as befitting their agendas. For example, buildings such as the IBM and AT&T headquarters in midtown Manhattan were products of this strategy. These and many other buildings violated the most sacred rules that zoning was meant to guarantee: a livable urban habitat. By allowing these buildings to rise straight up from the street line to their full height [over 182.9 m (600 ft)] with no setbacks, New York returned to its pre-1916 zoning condition, on a much larger scale. The canyons again emerged (Codella, 1983).

It is very clear that the quality of the urban environment, specifically in view of its ever increasing complexity, cannot depend on bureaucratically prescribed instructions alone. Instead, scientifically based, learned performance parameters must be employed as urban-control standards along with the active participation and support of informed local civic interest groups. Further information about the sociopolitical influences of tall buildings may be found in Chapters 2 and 3, and in Chapter PC-4 (CTBUH, Group PC, 1981).

## 4 The Dual Role of Tall Buildings

Tall buildings should respond to two primary criteria: first, to the smaller circle of its affected users, and second, to the larger urban environment. In regard to the first criterion, the building itself must be gauged relative to its purpose, how it lives up to its expectations. The second criterion must be evaluated in its function as an element in the immediate urban setting. The degree to which tall buildings add to or detract from the quality of their urban surroundings is dramatic, affecting not only the immediate users but, because of their size and scale, the context of the entire city now and in the future.

However, as this unique typology emerged, two problem areas appeared as well. The first was associated with the tall building's facades and ornamentation, since the basic rectilinear shape of buildings was determined by economics and engineering, driven by cost, building footprint limitations, and other interests. The second was related to the form of tall buildings, their considerable mass and scale. As early as the turn of the century, tall buildings were recognized to be adversely affecting their surroundings. In *The Chicago Period in Retrospect,* published in 1924, Sullivan claims that the Chicago architects of the late nineteenth century understood the high-rise building and its urban impact better than their New York counterparts (Mumford, 1952):

> The architects of Chicago welcomed the steel frame and did something with it. The architects of the East were appalled by it and could make no contribution to it. In fact, the tall office buildings fronting the narrow streets and lanes of lower New York were provincialisms, gross departures from the law of common sense. For the tall office building loses its validity when the surroundings are uncongenial to its nature; and when such buildings are crowded together upon narrow streets or lanes they become mutually destructive.

Yet this was never acknowledged as a relevant architectural design issue. Instead, tall buildings were perceived as planning concerns for which zoning ordinances were administratively created and imposed as controls. [Further, the architect might have been more concerned with a major commission and the client's desire (ego or economic drives) than the collective good of the community.]

These two problem areas have continuously been treated in isolation. Rarely, if ever, have they been combined into a singular design issue. Therefore, an enigma exists today in the fact that there is little consensus regarding these two issues, especially noting that even aesthetically pleasing buildings are detrimental to the urban environment, including those praised by distinguished architectural critics. The issue of urban and social characteristics of tall buildings has been discussed in detail in Chapters 3 and 6, and in Chapters PC-7 and PC-8 (CTBUH, Group PC, 1981).

## 5 The Stylistic Debate

With the arrival of the tall building came distinct schools of thought which attempted to find an appropriate expression for this new typology. The Chicago School architects advocated pragmatic, simple, direct, and structurally honest approaches to engage the new type. Recognizing the lack of formal precedent, they searched for the structural essence of tall buildings, leading to great inventiveness. Architects of the New York School, on the other hand, adopted historicism as a direction for designing tall buildings. Classical and Gothic models were adapted and applied to the form, ornamentation, and articulation of tall buildings. By arbitrarily applying given styles to a new scale, this school of thought looked backward toward the past in searching to define a new category. More specifically, in Chicago architects were struggling with the new form and its components. For example, curtain-wall design was isolated and highlighted as an innovative design feature, whereas in New York, curtain walls were made to look like traditional masonry buildings. The academic technique founded in the New York approach led to stylistic theories put into practice, curtailing the emerging pragmatism (Mumford, 1952). (See Chapter 5 for a comprehensive discussion on this topic.)

## 6 Need for a New Design Approach

Throughout the history of tall buildings, constructive far-reaching thoughts were present. However, the manner in which these thoughts were manipulated, ignored, or sometimes misunderstood helped create prevailing inconsistencies and confusion within the practice. Louis Sullivan wrote and preached extensively on the criteria and design of tall buildings, yet all aspects of his work have still not been fully realized architecturally. The current lack of comprehensive knowledge of the complex and pragmatic urban, social, and functional concerns intrinsic to this building type must be addressed. A comprehensive design process is needed to identify all essential principles which unify them into a coherent design based on formal strategies of urbanism.

Responsible architecture, in the framework of this discussion, is based on the recognition that no single building can be looked at as standing alone, but rather within a larger context. As opposed to architectural arbitrariness, this approach addresses specific site conditions, both urban and environmental, resulting in buildings that are the products of their own particular set of criteria. In this method, superficial architectural styles are not to be applied. Rather, formal consequences are to be investigated. By being emphatic to the specific urban context within the design process, buildings are more inclined to create a responsive urbanism that is morphologically coherent. Architecture should be comprehensive and expand beyond form and ornamentation in order to engage essential urban issues (Huxtable, 1984; Lynch, 1990).

## 1.3 TALL BUILDINGS IN THE URBAN ENVIRONMENT

### 1 The Responsive Process

The design of tall buildings expands from the pragmatic to the inspirational, organizing a wide array of programmatic items into a holistic physical reality. Its underlying purpose is to provide the appropriate physical environment for a specific utilitarian function and urban context, commensurate with human needs and aspirations. However, urban context alone has no particular meaning. Its purpose lies in understanding that urban areas are to serve society. Only then can a designer respond faithfully to the public and the site (Polshek, 1988).

The introduction of a building into a city results in an intervention within the urban context. For designers of tall buildings the city consists of other buildings and open spaces within the urban environment (Soleri, 1973). The insertion of large structures into the urban environment alters the preexisting urban condition. Although a difficult phenomenon to describe, urban equilibrium refers to parts of cities we have all experienced as integrated and balanced. Urban equilibrium results from a considered balance between builtform and open space, and it is an ideal condition that may never be reached. However, responsibly designed buildings move our urban environment toward the goal. Since a balanced environment provides the optimal conditions for the livelihood of humankind, jeopardizing its integrity must be kept to an absolute minimum (Zelanski, 1987).

Tall buildings greatly impact the scale and context of the urban environment due to their proportional mass and height (see Chapter 3). Whether standing alone or blending into the urban context, the larger the building's mass, the greater its impact. The intervention of tall buildings in urban environments affects daylighting, sunlight, shadows, and even air movements, creating downdrafts and unexpected gusts. The form of the building's mass is the most crucial part in determining the quality of the disturbances and their magnitude. The design process of a tall building should attempt to keep environmental disturbances to a minimum by the orchestration of its form. Beyond this environmentally responsible approach to tall building form, the pure exhilaration of human experience that only tall buildings can provide must also be addressed (Barnett, 1973).

In their design approach, some architects juxtapose formal geometry against a functional, program-driven system, producing complex results. The mass of a building is encountered as pliable, and its form responsive to both active and passive forces from its urban setting and from within the building itself. These forces are applied as "vectors" of varying magnitude and directions, acting on the building's mass in direct proportion to values assigned based on hierarchy. The form of the building's mass is then manipulated to arrive at an equilibrium that responds to these vectors by contradicting, invalidating, or harmonizing with them. This morphological transformational process tends to be a dynamic, effective, and practical tool for arriving at a physical equilibrium between the tall building and its environment (Jencks, 1980).

### 2 The Tall Building and Its Physical Entity

The design process of tall buildings must include a physical analysis of their elements and components. The act of creating an urban building appropriates a portion of the open, free space of the environment—a public domain—and transforms it into an enclosed, private, solid mass. Buildings cover portions of the atmospheric sky dome (the diffuser of sunlight that emits ambient daylighting), reducing the availability of this natural resource, which is fundamental for the well-being of people. Indiscriminately, buildings also cast shadows and change air movement patterns around adjacent public areas. These adverse effects are caused by the tall building's mass and configuration, not its architectural styles, surface decorations, or materials.

In principle, tall buildings are made of three basic elements:

1. Vertical stacks of functional areas (that is, the *floors*)
2. Vertical service areas that act as an umbilical cord to all the functional areas (that is, the *core*)
3. An outside enclosure that protects these parts from the elements (that is, the *skin* or *building envelope*)

Tall buildings combine a multitude of well-coordinated assemblies of complex systems. They are the true "machines for living" of our times. Their high level of complexity mandates that the principles on which these systems are arranged be highly logical, simple, clear, and direct for their effectiveness. These systems are manifested in a variety of components, ranging from the permanent to the semipermanent to the transitory, all to be adaptable to different levels of modification.

The permanent component is the structure. Tall buildings' structures are designed in accordance with stringent structural code requirements so that they can withstand large gravitational and lateral forces. The structure is protected by the skin from deterioration caused by environmental effects. As such, they could theoretically last forever, or as long as their dimensional and load-bearing properties maintain their validity. The structure should be designed within given magnitudes, to readily accept a range of modifications such as increased local load-bearing capacity and penetrations for the passages of services.

The semipermanent components are the "skin," the "core" (except for its structural components), the mechanical systems (environmental and other), the public and material circulation systems, and the public amenities such as bathrooms. Over time, wear, obsolescence, and changing technologies require the replacement or the modification of many building systems. Therefore the semipermanent systems and components must be designed to accommodate modifications by allowing required capacities and accessibility.

In terms of the monetary value invested in tall buildings, the service core commands relatively the highest. This suggests that the core must be arranged with the greatest level of compactness and efficiency possible, resulting in a quite uncompromising organization. On the other hand, the functional areas, commanding relatively the least monetary value, are malleable. Provided that their expected operational performance is maintained, they can assume a variety of configurations. Furthermore, the repetitive nature of the components and the sophisticated technology of the nonbearing curtain wall used in all tall buildings offer numerous levels of flexibility needed for achieving complex building shapes. This pliability of both the functional areas and their enclosures becomes the most valuable asset for shaping the form of tall buildings.

Transitory components are the installations made to accommodate particular users. Anticipating the inevitable transformations, whether caused by changing needs or use, functional areas in all tall buildings must be designed as multipurpose, anonymous spaces. For this reason, optimal dimensioning and structural and mechanical element standards have evolved over the years for these areas, along with an understanding of the limits within which deviation can occur. Understanding this allows for the utilization of a wide range of plan forms, which can be used without sacrificing the effectiveness of the functional areas. Here lies the potential for configurations which impact the building's own performance within its urban context (Schmertz, 1975).

### 3 The Tall Building as an Urban "Building Block"

Tall buildings are the major building blocks of the urban environment, creating a city's spectacular skyline (Posokhin, 1973). They define public spaces that create urban theaters providing the backdrop for human activity in cities. In respect to their effect on the environment and their visual perception, all tall buildings should be composed of three distinct sections. The designer can use this information as a basis for an analysis of tall building design for an effective, pragmatic methodology that could bring society closer toward more responsible architecture (Jencks, 1980). In the following, the three main sections of the tall building viewed vertically are discussed. The classical viewpoint on this issue was outlined by Louis Sullivan in 1896, not long after the advent of the earliest skyscrapers [see Chapter PC-6 (CTBUH, Group PC, 1981)]. For an additional viewpoint, see Chapter 5.

***Base.*** The building's base, the part that is seen from street level, is contained within the 40° cone of vision. Depending on the depth of open space in front of the building, this section of the building usually rises to a height of five to eight stories. Interfacing with the urban setting, this section of the building is a crucial determinant of the building's contextual quality (*Architectural Record,* 1974). It anchors the building into the metropolitan fabric, defines the street wall and its texture, and contains the building's public-oriented uses. Being relatively low in height, the base of the building has little effect on the urban ecology. However, it has significant impact on the scale and definition of the street and the "humanizing" effect of tall buildings. Geometrically, it derives its planform and dimensional properties from the configurations of the adjoining public areas and the urban context.

***Shaft.*** This section extends from the building's base upward and becomes the prominent form signifier of a tall building. It is most critical in altering the quality of interaction between the building and the ecoenvironmental conditions. This section of a tall building is potentially detrimental for the reason that it covers a major portion of the atmospheric sky dome, and it alters the patterns of air movements in its surroundings. The negative consequences of air patterns, such as downdrafts, updrafts, and turbulence, must be considered in the design process (*Architectural Record,* 1974). The configuration of the midsection is critical in determining whether the building's scale is perceived as imposing or considerate, overwhelming or accommodating. This section of the building is ideally formed through a careful analysis of the site, the context, and the environmental conditions of the building's site. Here the promise for an appropriate and exciting expression of verticality is sought as a means of dynamically tying the earth with the sky.

***Top.*** The top of the building generally has a reduced footprint compared to its midsection. As a result, it rarely affects the ecoenvironmental conditions. In the rare occasion where the imprint of this section of the building is the same size or larger than that of the midsection, it does affect ecological circumstances and must be analyzed accordingly. Its architectural significance lies in the formal transformation from the midsection and its delineated silhouette upon the sky. This section of the building accentuates the building's own identity, and is ideally formed by formal influences of both the lower section, the midsection, and the city's skyline (MacMillan and Metzstein, 1974).

The key for generating architectural coherence from these three sections lies in the morphological transformations from one to another. The geometric cues must be drawn from what already exists, the context, and the urban ecoenvironmental condition. The more the contextual considerations are applied, the more responsive the resulting building is to its urban settings, and the greater is their mutual integrity. Although the range of these concerns is vast, it can be reduced for practical purposes to only those factors that affect physical urbanity. At best they will expand to include the nonpragmatic values drawn from applicable humanistic culture. A more comprehensive discussion of the aesthetics of tall buildings is presented in Chapter 5. Narrative accounts about the tall building as an urban component may also be found in Chapters 3 and 6 as well as in Chapter PC-8 (CTBUH, Group PC, 1981).

## 1.4 THE SUPERTALL BUILDING

A discussion of the form of tall buildings would be incomplete nowadays if not expanded to include the supertall building (STB). Reasons similar to those that brought about the tall building at the turn of the century are prevailing again for this new building type. The STB (more than about 70 or 80 stories high) is currently receiving increased attention in several places around the world. This growing interest and expanding need is indicative of its impending commercial appearance, not too far in the future. Notwithstanding the many resemblances between the tall building and the STB, it would be, however, a gross error to apply design solutions indiscriminately from one to the other. Due to their different economies of scale, two particular factors stand out as fundamentally distinctive—structure and form.

### 1 The Structure

Unlike the current generation of tall buildings, the cost-effectiveness of a STB's structure becomes the requisite for its existence. This most critical factor sets it apart from all other building types. Remaining components of the STB can be translated linearly from the current governing tall building terminology, yet only within a certain range of magnitude represented by levels of adaptation. For example, this range can be exemplified by the STB's conveyance systems of both people and materials. Here intensive adaptation is required. The other extreme can be illustrated by the interior fittings that require little or no adaptation.

Wind velocity grows exponentially with the increase of altitude. Because of their enormous height, STBs are subject to lateral forces of immense magnitudes. The effect of these forces results in acute reactional behavior throughout the building. Even the most sophisticated structural principles of conventional tall buildings do not completely satisfy the performance expected from the STB, within its affordable economical parameters (Guise, 1985).

As structural engineers search to comprehend this new building type, only recently have they begun to realize that continued linear improvements of familiar principles is inefficient. The introduction of a new, more effective approach for designing STB structural solutions is imperative for their implementation (Hubka, 1973). These solutions must be sought via a thorough understanding of the aerodynamic and gravitational forces affecting the building, and the structure's reaction to these forces. The objective is to achieve a basic form which is least resistant. In addition, spatial arrangements of its structural components must function in a synchronized synergic reaction, resulting in a reduction of the structure's material and mass.

The profile of this structural form tends to be wide at its base, for greater stability and improved slenderness ratio (Robertson, 1973). The uppermost zone tends to be narrow and near circular at its top, for a reduction of lateral forces caused by the high-velocity winds. The mass of this structure is concentrated at the periphery of its form, so that its own weight counteracts the uplift caused by lateral forces, avoiding the need for expensive anchoring technics that prevent it from toppling (Siegel, 1962).

### 2 The Form

The form and the structure of an STB are absolutely inseparable, and must complement each other for the effective performance of their assembly. Since the structure of the STB is of such crucial significance, the building's form must submit to and be in conformance with its structural principles. In this context, form is defined exclusively by the building's enclosure—its interface with the environment (Zalcik and Franco, 1974). As such, it determines the aerodynamic behavior of the building consequential to the natural forces acting on it, such as drag coefficient, drift, vortex shedding, and oscillation. The form also greatly determines the environmental conditions of the STB's urban context, such as cumulative downdraft wind resulting from the building's great height, and also issues of daylighting and shadows (see case study 1.5.2).

The inevitability of intertwining the structure and form of an STB allows for a unique opportunity in the investigation of its design. It seems reasonable that the most efficient structure emerges as one that is most environmentally responsive to its location. The slender conical structure and form described provide the most light and air to its context while minimally interfering with the urban sector. These two factors, structure and form, merge into one response that presents the better structure as the proper form. Discussions developing urban design, structural expression, aesthetics, facades, and the economics of tall buildings are presented in Chapters 3, 4, 5, 7, and 10 to 12, as well as in Chapters PC-5 to PC-8 (CTBUH, Group PC, 1981).

## 1.5 ARCHITECTURAL DESIGN PERSPECTIVES: FOUR CASE STUDIES

In this section four case studies are presented to illustrate the architectural design process from different perspectives, as discussed in the foregoing.

### 1 101 Park Avenue, New York (Figs. 1.2 to 1.5)

The program called for an effective and efficient investment-type office building. The site's frontage on Park Avenue, between 40th and 41st Streets, faces a big blank wall created by a rising viaduct, and a descending narrow street that remained after the viaduct split the magnificent 140-ft-wide avenue into two. This wall, rising to a height of 6.1 m (20 ft), in addition to the steeply inclined road, is not conducive to urban activity. Furthermore, 40th and 41st Streets both descend sharply from Park Avenue eastward, leaving only one corner of the site with "presence" on the avenue: the corner at 40th Street.

This most significant corner point became the morphogenesis of the project. The design of the building evolved through this diagonal main axis, passing through to become the prime generator of the form. In respect to the corner and the city block geometry, a 45° angle was selected for the simplicity of integration of the orthogonal grid (Fig. 1.6).

**Fig. 1.2 Office tower, 101 Park Avenue, New York.** (*Courtesy: Eli Attia Architects.*)

**Fig. 1.3 Office tower, 101 Park Avenue, New York.** (*Courtesy: Eli Attia Architects.*)

**Fig. 1.4 Office tower, 101 Park Avenue, New York.** (*Courtesy: Eli Attia Architects.*)

**Fig. 1.5 Office tower, 101 Park Avenue, New York.** (*Courtesy: Eli Attia Architects.*)

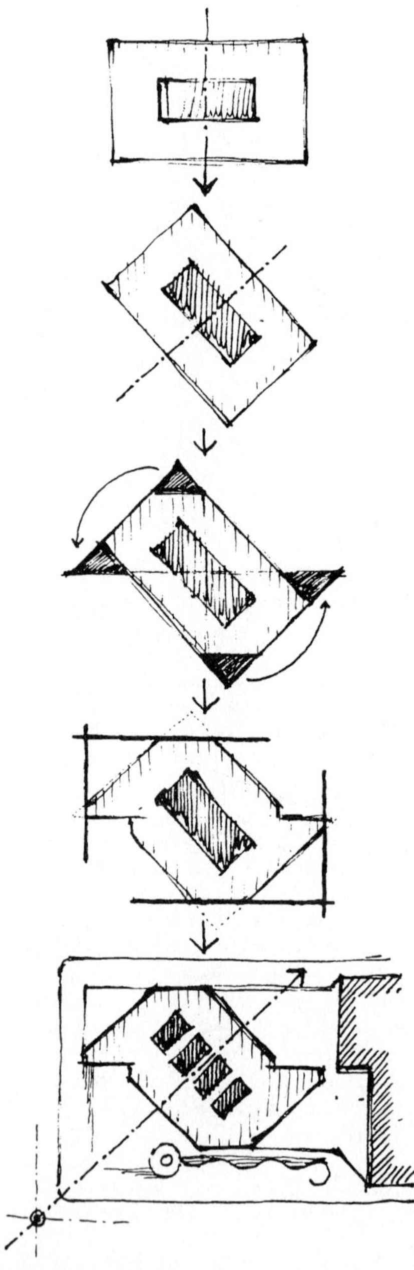

**Fig. 1.6** **The 45° angle was selected for the simplicity of integration of the orthogonal grid. 101 Park Avenue, New York.**

A 1858-$m^2$ (20,000-$ft^2$) rectangular floor plan was used as a basis for clarity, effectiveness, and economy. By evaluating the site and its unique conditions, the plan was then manipulated accordingly. The orthogonal truncations of the rotated rectangular plan form brought the composition back to the city grid and created alignments at the street with its six-story base. This lower portion of the building, which falls within the 40° cone of pedestrian vision, maintains the urban fabric. The form of the tower portion responds to urban criteria emanating from zones above, allowing more daylight and air to reach the public areas at street level. In addition, the new plan configuration offers eight corner offices per floor, two of which have wide 315° panoramic views. This geometric arrangement of the plan reduces the horizontal dimensions of surface planes, causing a greater number of facets to be seen from any viewing angle. By accentuating the verticality of the overall form, the building appears less imposing and apparently airy within the context.

A public plaza was included to gain the zoning bonus which allowed maximum floor area. Welcoming this inclusion, both building and urban context greatly benefited. The plaza provides a link of transition between outdoors and indoors as well as desired relief in an area of the city that has no public open space other than sidewalks. In addition, it would act as a dissipating medium for the many thousands of people that access the building daily. Consequently, an urban breathing space is provided commensurate with, and complementary to, the adjacent large structures. The formal portion of the plaza, its vertex congruent with the "most significant point" and centered on the main axis of the building, is in line with the Park Avenue lobby. The informal part of the plaza, on the south side, benefiting from the largest amount of sunlight on the building's site, is a large seating and strolling area contained and protected from winds by the building. With tall buildings it is important, even in relatively quiet climatic circumstances, to consider severe air movements that can be produced around their shafts (Aynsley, 1973; Maki, 1972).

To demonstrate how the form of the building and its placement on the site affects daylighting and the perception of openness at street level, a comparison between 101 Park Avenue and 535 Madison Avenue (the AT&T headquarters building) may be made. Both buildings are about the same height, over 182.88 m (600 ft), both with a tower portion of about 1858 $m^2$ (20,000 $ft^2$) per floor. The total area above grade of 101 Park Avenue is greater by about 50% than that of the AT&T. (Motivated by company "pride," the AT&T headquarters building has been stretched to exceed in height the much larger adjacent IBM headquarters.) Both buildings were built about the same time and were subject to the same bulk regulations of the New York City Zoning Resolution. Both buildings, for architectural reasons, were built to the avenue street line through an approved special permit form the City Planning Commission. Environmental influences such as sun and shadow are also discussed in Chapter PC-6 (CTBUH, Group PC, 1981).

What is evident from the sky-dome diagram in Fig. 1.7 is that the AT&T covers 99% of its own site's sky dome, whereas 101 Park Avenue covers 57% of its corresponding sky dome. This means that 101 Park Avenue allows 43 times more sky-dome-produced daylighting (4300%) than the AT&T building. Similar observations apply to air and the perception of openness relative to both buildings.

101 Park Avenue is considered in real estate circles to be one of the most lucrative buildings in New York City. Since its completion in 1982 it has been fully occupied, commanding some of the city's highest rents. The success of this building, in addition to the usual prerequisites, is attributable to its form and interface with its context.

### *2 10 Columbus Circle, New York (Fig. 1.8)*

The origin of this project's form, program, and scale derived from the significance of its site. Columbus Circle is probably the most suitable location for a major urban statement of civic significance and pride in New York. It is also one of the city's major subway hubs. Broadway, Central Park South, Central Park West, Eighth Avenue, 58th and 60th Streets, all meet at this unique and prominent urban junction. The proposed building would be a 301,925-m$^2$ (3.25 million ft$^2$) 137-story mixed-use tower, the tallest in the world. Located at Manhattan's geographic center, where major commercial, residential, hotel, recreational, and performing arts centers merge, the building would materialize at the confluence of several major thoroughfares.

The form of the tower develops from decagonal prisms, varying in diameter and height. Their dimensional relationships are governed by $\phi = 1.618$ (the golden section), its rate of change and derivatives. The tower's "growth" follows morphological rules similar to many growth forms in nature. The decagonal form and the governing dimensional order are derived from the tower's structural system. This system is based on a

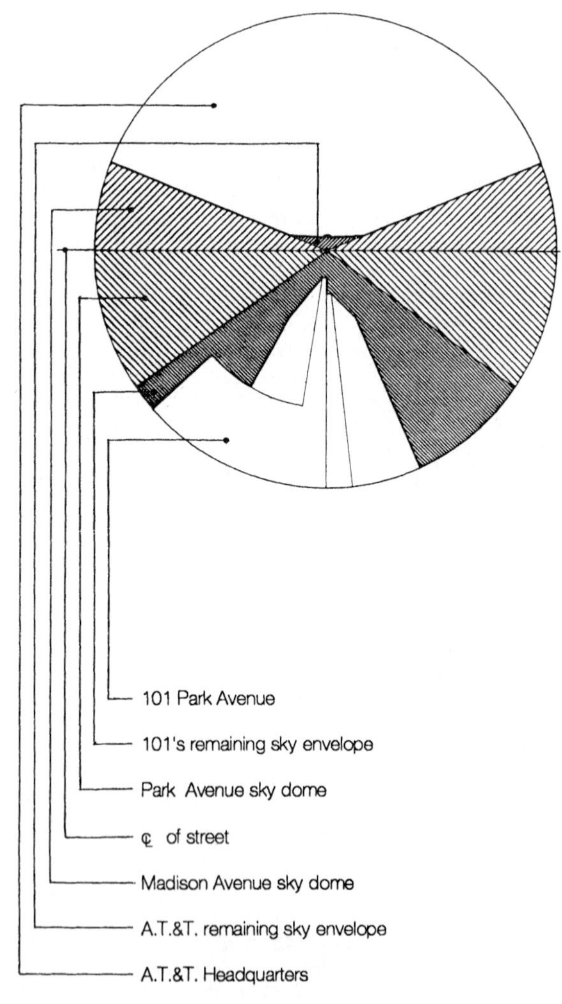


**Fig. 1.7 Sky-dome diagram, 101 Park Avenue, New York.**

pentagonal antiprism as its module, forming a three-dimensional space truss which possesses multiple golden-section relationships.

The distinctive form of this building provides manifold results. Maximum amounts of light and air would be provided to its surroundings, better than any existing building in New York. Typical of tall buildings in this city, the sunlight surrounding these buildings is typified by narrow, mobile shadows (Lamela, 1973). The dimensional changes and high prevalence of setbacks would dissipate the wind force before reaching street level, eliminating downdrafts. By being the tallest, the building would cast the longest shadow; however, its effect would be unimposing. Using the criteria given by the sun's movement and the tower's narrow profile, the casted shadow would stay at one spot for no more than eight minutes over the course of a day (Figs. 1.9 and 1.10).

The building is shaped narrowest at the top, where wind load is the greatest, and widest at the bottom, where wind load is least and the greatest stability is required structurally. Furthermore, wind pressure load and wind drag are further reduced, about 50% compared to a rectangular building, due to the nearly cylindrical shape of the structure (Fig. 1.11). Because of the aerodynamic characteristics of the building's form, the tower

**Fig. 1.8 Model of supertall building, 10 Columbus Circle, New York.** (*Courtesy: Eli Attia Architects.*)

is expected to be comfortable for its users without the utilization of artificial devices such as tuned mass dampers. The tower's peripheral antiprism geometry brings its structural loads straight down to the ground at five points along its perimeter. Its own weight eliminates uplift, thereby greatly reducing the complexity of the tower's foundation (Fig. 1.12). The environmental impact of wind on tall buildings is discussed in further detail in Chapter PC-6 (CTBUH, Group PC, 1981).

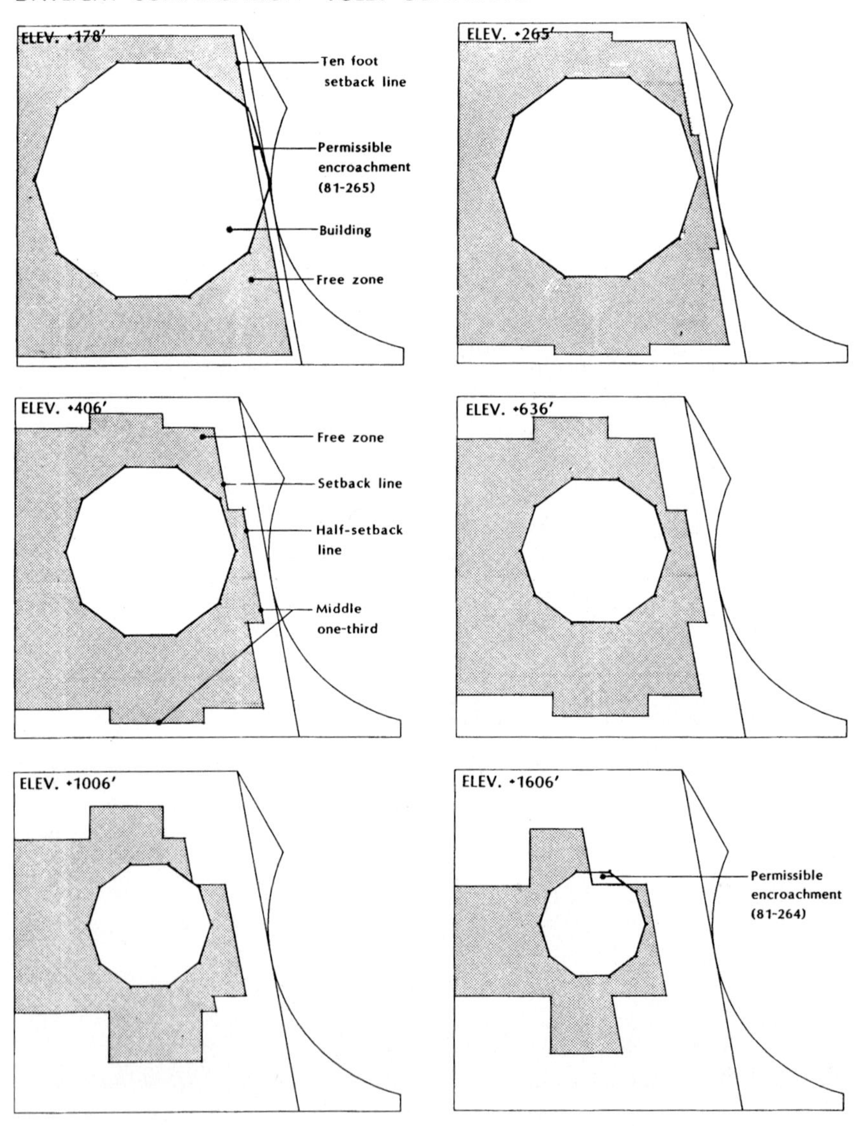


**Fig. 1.9 Daylight compensation, 10 Columbus Circle, New York.**

### 3 *Kostabi World Tower, New York*

The tallest structures have always inspired humankind. The tall building as a signifier of priorities within society has historically changed to conform with its sponsorship: from political to religious to commercial. Art emerging as a global economic factor calls for a large-scale financial approach for providing accommodations for its needs.

A single unity rises, or emerges, from the ground up, gradually at first, then soars with ever increasing intensity upward—an urban "crescendo." None is more appropriate

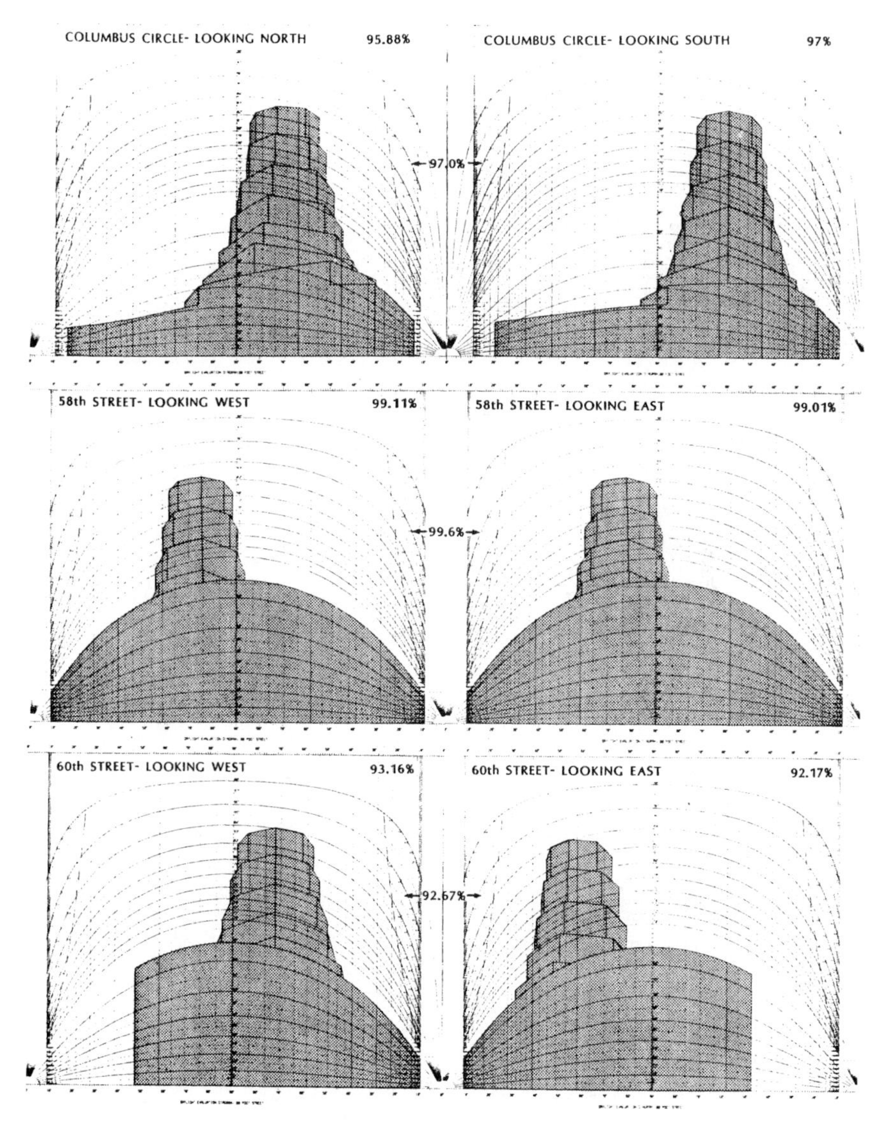


**Fig. 1.10 Waldrum diagram daylight score, 10 Columbus Circle, New York.**

today for such an urban gesture as to signify a domicile for the arts. The unprecedented mixed-use structure is to be exclusively devoted to art. It is to comprise housing for artists, studios, workshops, galleries, museums, schools, theaters, offices, libraries, hotels, bookstores, and restaurants; basically a self-sufficient vertical art city (Fig. 1.13).

At certain moments in the evolution of a city some urban locations become ripe for accommodating urban events of great significance. Located in a fittingly underdeveloped area of New York City, the Greenpoint, Brooklyn, building is to rise to a height of 605.6 m (2000 ft) with an initial floor area of 90,290 m$^2$ (1,000,000 ft$^2$). As the need arises, the flexibility of the design of this proposed building allows its floor area to be

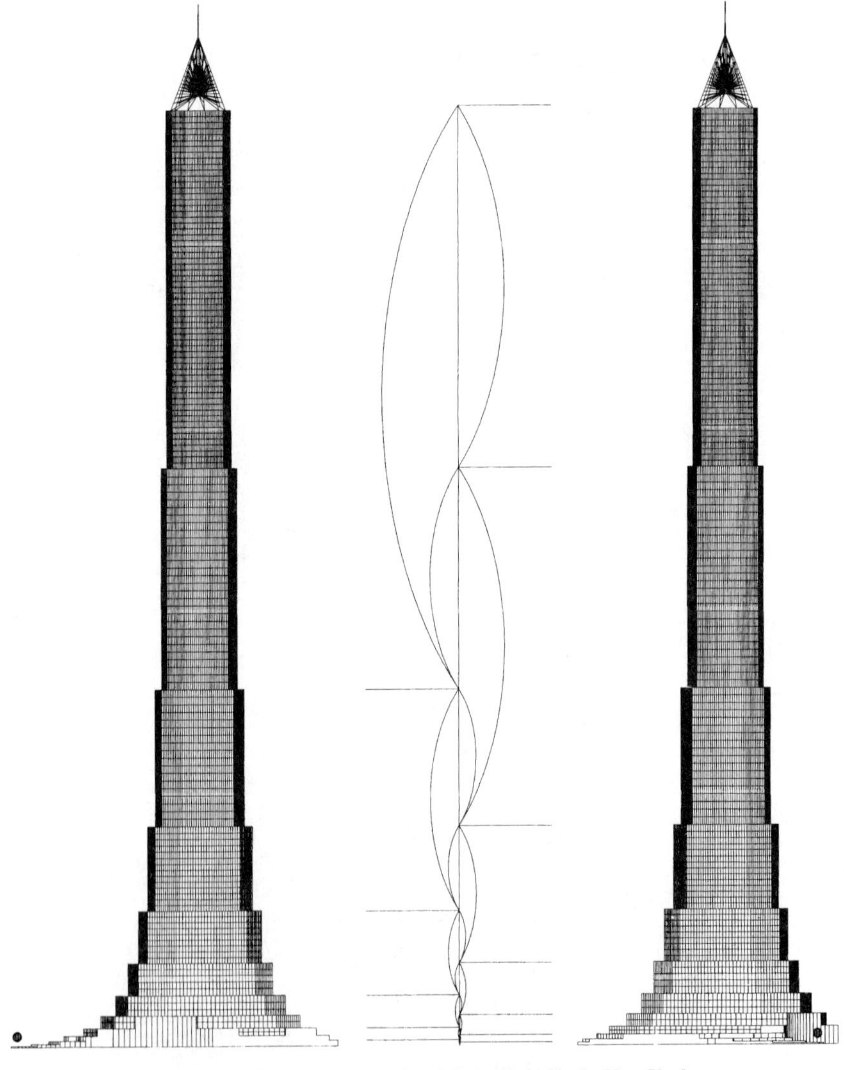

**Fig. 1.11 Wind-load analysis, 10 Columbus Circle, New York.**

increased to up to 280,000 $m^2$ (3,000,000 $ft^2$) and can accommodate a wide variety of spatial changes. Areas can be added or removed freely without affecting the permanent structure. The lower areas of the tapering structural cage are proposed to incorporate multilevel, open-air gardens and parks, above which a complex of museums and galleries would be found near ground level. As the skyscraper rises to express purity of structure and form, the landscaping around the building would also emerge from an ecology of coequal purity. Part wetland and brook, part forest and meadow, the surroundings would reflect the ecosystem once present, and the one most probable to reemerge as global warming takes place.

The structure is round, wide at its base and narrow at its top, having a continuity of no vertical columns (Figs. 1.14 and 1.15). The system is complete and well balanced. Each and every one of its parts has a purpose, equal in validity to all other parts within the system. All the fragmented parts are indispensable and act in harmony with the whole. It is a system within which all sizes, ratios, and proportions are determined and governed by strict geometric disciplines of growth, as in nature. The greater the axial forces, the shorter the structural members are, for reduced buckling. Every external action is carried throughout the tower, causing all other members within the system to react. All exterior forces are translated into greater stability of the system itself; its

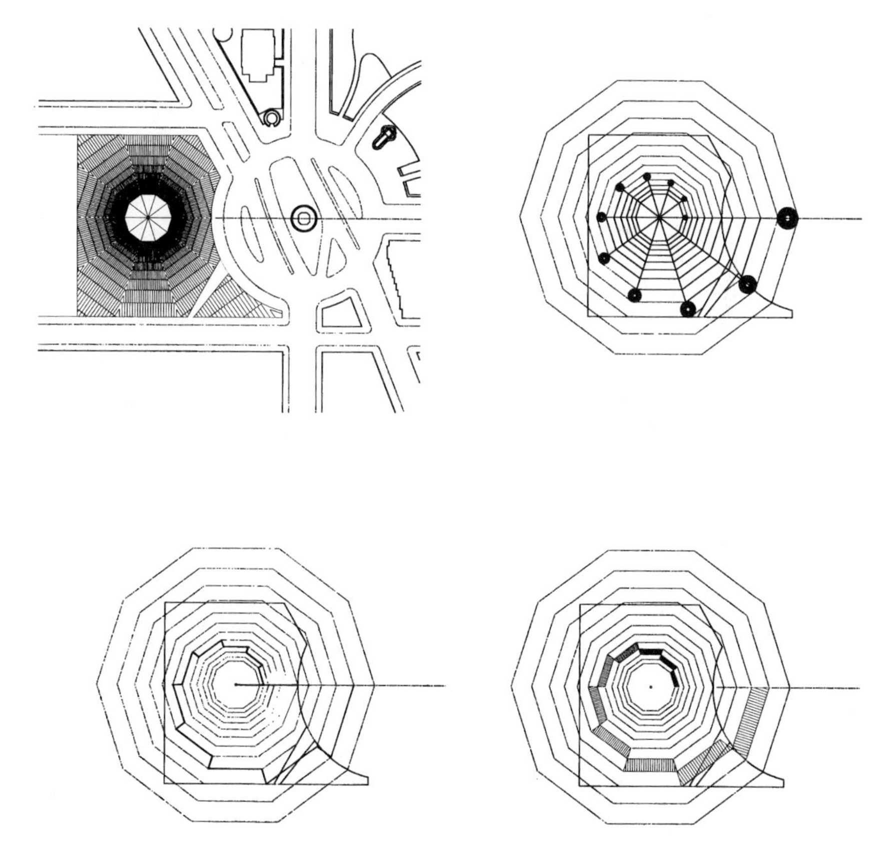

**Fig. 1.12 Plan views showing tower foundation, 10 Columbus Circle, New York.**

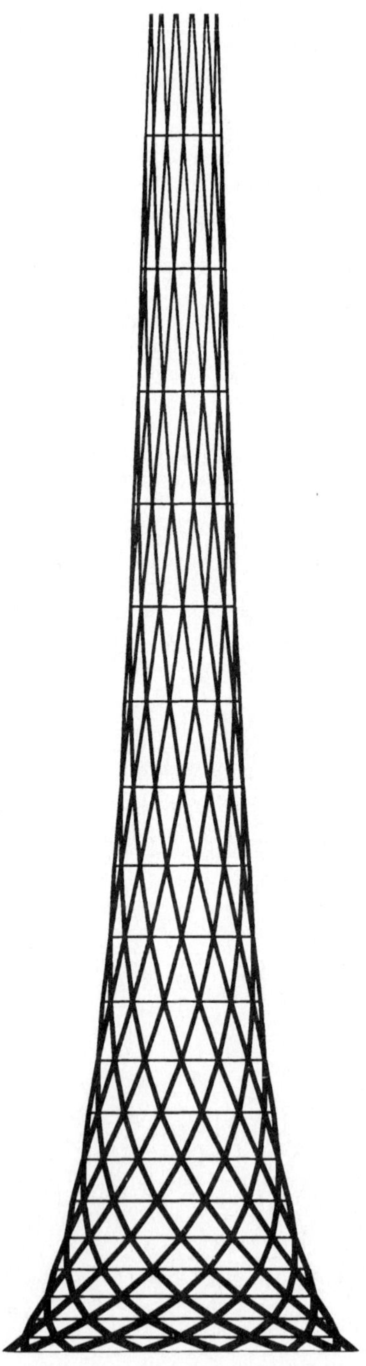

Fig. 1.13 Urban "crescendo," Kostabi World Tower, New York.

geometry translates its own gravity into lateral stability. And so, the tower is a system devised by humans to effectively accommodate human needs and desires.

### *4 The Shalom Center, Tel Aviv (Fig. 1.16)*

The Shalom Center is strategically located at the confluence of major national road networks. It highlights a place not yet highlighted, the present equivalent of the crossroads where ancient trading routes met. Physically and symbolically, the center is a project of regional, national, and international significance. Located on an edge site that is neither city nor suburb, the Shalom Center mediates between these two districts—the heart of the city and the outlying areas. From the unique qualities of the site and program a new typology emerges that is as much a bridge between the different districts of the city as it is a true urban destination. This bridge becomes an oasis floating above the network of roads. The center's design takes cues from this context, embracing it and learning from it, rather than denying or ignoring it.

The Shalom Center is a mixed-use project containing approximately 110,000 $m^2$ (1,200,000 $ft^2$) of office space, 14,000 $m^2$ (150,000 $ft^2$) of apartments, 32,000 $m^2$ (350,000 $ft^2$) of retail stores and theaters, and 21,000 $m^2$ (226,000 $ft^2$) of public spaces and amenities, including a grand roof garden and a 3000-car parking garage. The project as a whole encompasses some 261,000 $m^2$ (2,800,000 $ft^2$). The proposal is conceived as a grouping of buildings that form a community. The three towers are shaped into the three most basic and pure forms: the square, the circle, and the equilateral triangle (Fig. 1.17). These shapes, as different and as far apart as any formal entities can be, are brought together to create an informal whole, an independent, harmonious, and tranquil

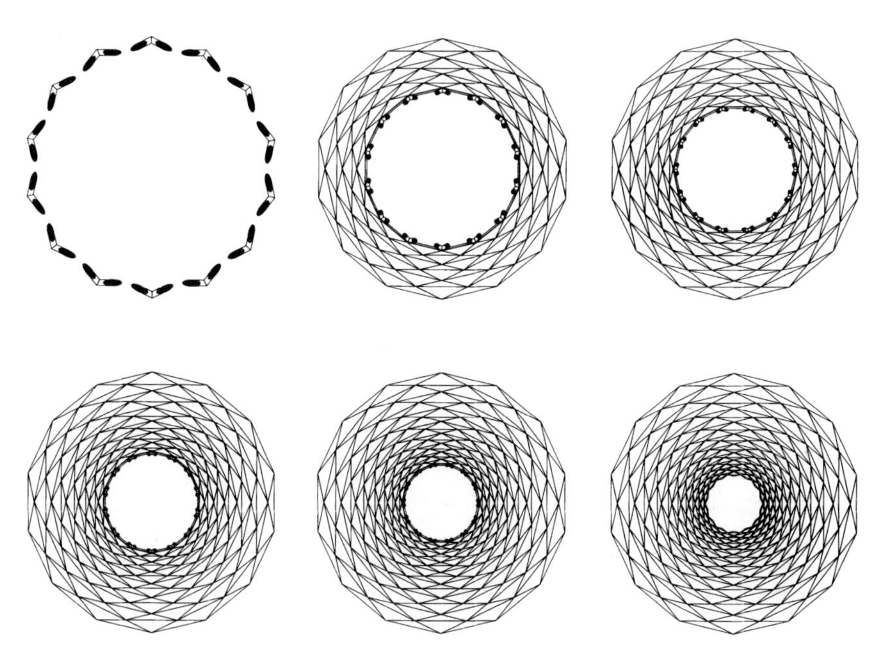

**Fig. 1.14 Structural diagram in plan, Kostabi World Tower, New York.**

grouping. By using varying heights and different floor layouts the users are provided with a wide range of choices and flexibility. The three towers embrace the low, Y-shaped public-oriented building. Its three branches between them extend outward to three distant points: Jerusalem, old Jaffa, and true North. This ensemble of forms, like a gathering of different people into a unified whole, symbolizes the center's name: Shalom (Fig. 1.18).

One tower creates a marker or a statue, but a gathering or a group of towers creates a center and a focus. The center is organized around a vertical circulation area at the geometric center of the complex. This vertical court slices through all public areas and is conceived as an urban space. Acting as a well of light and the project's central focus, it begins from the lowest parking level through the shopping level, up to the roof gar-

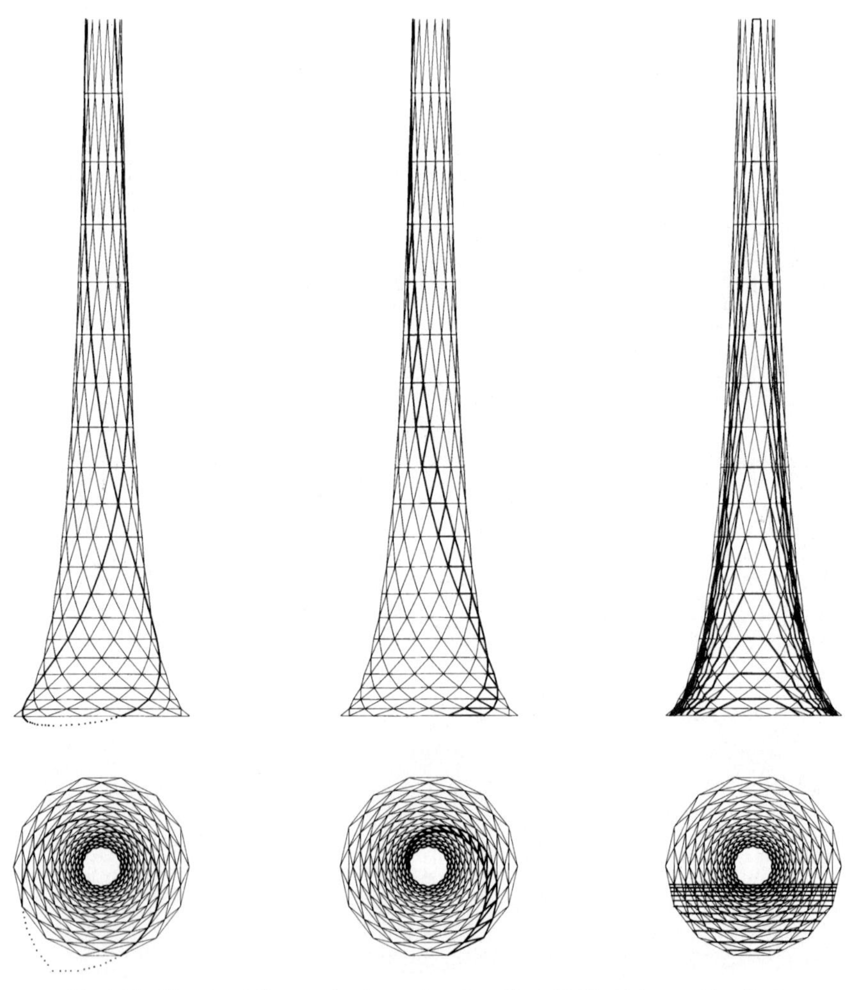

**Fig. 1.15 Structural diagram in plan and section, Kostabi World Tower, New York.**

den. There is no single dominant facade. Every vantage point modifies the relationships created between the three forms. This quality appeals equally to all parts of the metropolis, so that the center appears simultaneously as a gateway, a destination, and a transitional element from one part of the city to the next.

The genesis of the center's hexagonal plan geometry is derived from the three towers and their positioning. The dimensional modularity of this hexagonal geometry was determined by planning and structural considerations, so that it will effectively accommodate the various dimensional requirements of the different use groups: retail stores, public spaces, parking. This geometry is carried through the structural arrangements of the low-rise portion, from its lowest parking floor up to the public roof garden. Structurally, the towers consist of an inner central core, which provides lateral stiffness and carries gravity loads, and of peripheral columns appropriately spaced to accommodate planning needs. The usable office areas are therefore column-free for maximum flexibility.

The towers are envisioned as forms that react with the sky and moving sunlight, creating qualities of light and shadow that resonate and reflect the shapes of the buildings.

**Fig. 1.16 Model of Shalom Center, Tel Aviv, Israel.** (*Courtesy: Eli Attia Architects.*)

The solidity, the whiteness, and the fine-textured concrete surface of the towers are sensitive to variations of light and accentuate the verticality and purity of form. In addition, the concrete visually gestures to the "white city"—the international style tradition that Tel Aviv has immortalized. The glass panels are of light sky-blue reflective glass; they also respond to the more subtle nuances of the changing sky. The size and spacing of the vertical members are varied, creating wavy patterns unique to each form, emphasizing their sculptural qualities and causing them to shimmer in the light. Accordingly, the Shalom Center is perceived as a group of solid forms and a mirage.

The residential portions of the center are located on the top floors, affording privacy and the most dramatic views from the apartments, the duplex penthouses, and the rooftop amenities. In keeping with Tel Aviv's modernist traditions, the Shalom Center is raised on pilotis, creating a shaded arrival hall at its entry level. This space is open and airy with careful landscaping throughout, serving as a natural transition from the

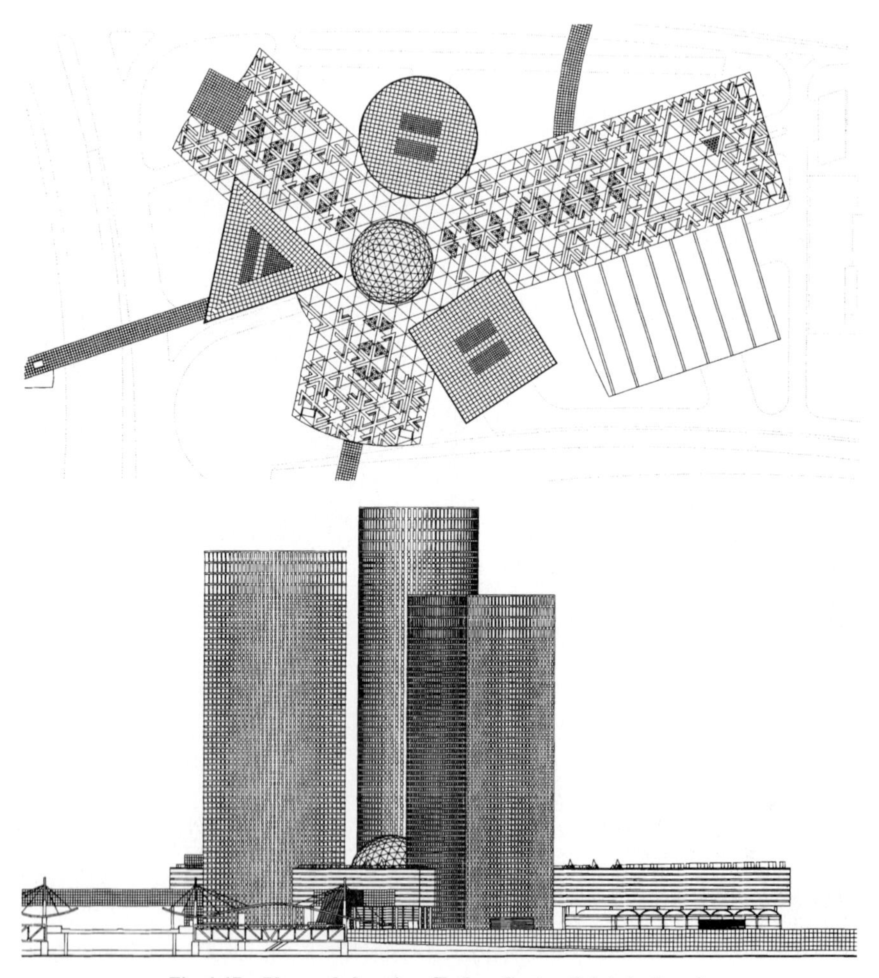

**Fig. 1.17 Plan and elevation, Shalom Center, Tel Aviv, Israel.**

highway to the more intimate scale of the center. At the southwest corner, facing Kaplan Street, a glass cube welcomes pedestrians, bringing them up to the shopping mall and the public areas on the roof. The pavilion is positioned within the center of a large, landscaped sculpture garden, signifying the urban importance of this corner and accentuating the different elements of the entry floor. The vertical court is conceived as an urban place that forms the core of the center as well as the pedestrian network. From this central light-filled atrium radiate three directional two-story streets, illuminated by overhead skylights. The large roof over the shopping mall, above which the pure forms of the three towers emerge, is envisioned as an oasis separated from the highways. This roof garden is an indoor/outdoor facility for public use, and an environment for a wide range of cultural activities. Protruding from its surface are functional, volumetric, sculptural glass forms consisting of the spherical dome, the glass cube, and triangular pyramidal skylights, all derived from, and echoing, the three basic forms of the towers. A maze composed of hedges emanates radially outward from the central glass dome on the three axes of the center. Groves of palm trees will grow from within this maze, capturing the breeze as they did for generations in this ancient land.

**Fig. 1.18 Model of Shalom Center, Tel Aviv, Israel.** (*Courtesy: Eli Attia Architects.*)

The Shalom Center becomes an urban event that accepts and celebrates the various conditions of the site and the complex program. With maximum visibility, the project will achieve a distinctive identity on the skyline of Tel Aviv and become integrated with the city's system of scattered urban centers.

## 1.6 ARCHITECTURAL DESIGN ISSUES: SPAIN

High-rise architecture in Spain is a relatively recent phenomenon. Spain's political position after World War II isolated it from the exponential growth experienced by the European democracies. Other factors, such as limited economic resources, prohibitive development costs, and limited technology, encouraged low- and mid-rise building development. Recently Spain has begun to develop high-rise building projects. Although Spain has brought in the expertise of American and European architects and engineers for many high-rise building projects, it continues to train and encourage its own architects and engineers to design and build tall buildings that reflect Spain's unique heritage and cultural identities.

The 1992 Olympic Games in Barcelona created an opportunity for major architectural design projects throughout the city by a host of internationally renowned architects. These carefully considered civic improvements are often placed directly at the service of the urban design and planning issues. The Olympic village, designed to house 15,000 people, is strategically located within, rather than outside, the city walls. Barcelona planners brought it within the existing city to replace obsolete factories and open the entire city to the seafront.

The following case study focuses on the Mapfre Office Tower, a tall building located at the Olympic village in Barcelona. At an urban knuckle between the Olympic village and the new Olympic port, the Mapfre Office Tower forms a gateway between the city and the water with a second high-rise building, a structurally exhibitionistic hotel caged in a cross-braced exoskeleton.

While discussing a particular aspect of design for this project, specific references are made to other chapters in this Monograph and to other Monographs for general information on the subject.

### 1 Architectural Programming

The celebration of the 1992 Olympic Games in Barcelona brought about a series of urban changes which have substantially modified the city's image. The Mapfre Office Tower is the most important symbolic element of the Olympic village and has opened the city to the sea. The building is located on a site between the marina and a group of 2500 homes which served as the athletes' residences during the Olympic Games (Fig. 1.19, right foreground). The site plan for the area defines the site as a low-rise area of two floors for use as a shopping center, a zone of four floors of offices, and a third zone containing the tower, 33 by 33 m (108 by 108 ft) in plan with 43 floors and 43,000 $m^2$ (462,850 $ft^2$) of usable space. Including the easements, the total built-up area is 75,000 $m^2$ (807,300 $ft^2$).

***Shopping Center.*** The two-story shopping center is located in a large esplanade facing the marina, landscaped with a large number of palm trees which emphasize its recreational nature. A large concrete canopy, projecting from the level of the ceiling of the ground floor, attracts the existing pedestrian traffic and serves as a terrace for the next floor, overlooking the marina. Its location within the group is denoted by a large glass drum which establishes a formal relationship with the lower office building as it matches the finishes on the glass of the office tower and is aligned with the axis of Juan de Austria Street. It also establishes an interesting visual relationship with the marina due to the landscaping of the entire sloping roof of the main shopping center, on which the tower's body appears to be supported.

***Office Building.*** The four-story office building has been designed around the definition of a circular-section volume, the main facade of which perfectly fits the snaking line of the Avenida Litoral, offering a perfectly regular facade of rectangular prefabricated concrete windows. The texture and pattern of the facade are inspired by Catalan modernism (Fig. 1.20). The rear facade is the backdrop, which enables the shopping center building to express itself fully and forcefully against a neutral background of prefabricated concrete with a clean texture and quiet pattern on its first three floors. Panoramic windows provide a view of the sea over the shopping center and add variation to the top floor. Its structural modularity is in 7.20-m (23.6-ft) units, allowing the provision of 3.60-m (11.55-ft) offices, each with a window and an air-conditioning module, making optimal use of the arrangement of offices within the building. The 13-m (42.65-ft)-long circular section of this building is cut by a volume of stainless steel which contains the entry hall, elevators, stairs, and services, and appears from the Avenida Litoral like the keel of a boat—a clear reference to its maritime environment.

**Fig. 1.19 Mapfre Tower (right in foreground), Barcelona, Spain.** (*Courtesy: Ortiz Leon Arquitectos.*)

**Fig. 1.20 Mapfre Tower, Barcelona, Spain.** (*Courtesy: Ortiz Leon Arquitectos.*)

***Office Tower.*** The office tower is the symbolic building, not only of this group but also of the entire Olympic village. Its volumetric definition was created on the basis of its location, perimeter, and height specified in the site plan. The following criteria were applied to these parameters:

1. A strong differentiation between its volume and the lower buildings to give the building the presence and importance required by its urban intention.
2. A design for the lower areas that takes advantage of the other buildings to define an empty volume four stories high, the main access vestibule to the building, opening out toward the existing rotunda of the Avenida Carlos I via a monumental cylinder framed with a glass wall 4 m (13 ft) high, and closed off to the Avenida Litoral by a solid wall of stainless steel which also contains the space occupied by the emergency stairs opening to the exterior on each floor.
3. Given that the building is designated as solely for office use, the design of a facade module meeting the following requirements:

   *a* When seen from afar, the building should appear clear, quiet, and of understandable scale. This was achieved with a continuous horizontal window and cleaning gallery on each floor which perfectly describe the facade and denote the number of floors contained in its height of 153.5 m (492 ft).
   
   *b* When viewed in proximity, the facade should have more detail than the simple definition of horizontal lines. The windows therefore have been inclined outward so that the facade vibrates with the reflections of the sea and the nearby buildings and so that the amount of glass observed in the facade increases as one approaches the building.
   
   *c* All the energy and environmental comfort issues in a building of this type should be perfectly resolved. The use of high-quality materials and solar screening devices, even in the cleaning gallery, enhances the aesthetics and physical comfort of the office areas.
   
   *d* Stainless steel should be chosen for reasons of quality—for durability in an aggressive atmosphere, for style because the building is located next to a marina, and for image as it can be seen from almost everywhere in the city.

The 1100-$m^2$ (11,850-$ft^2$) floor space can be divided into four independent offices per floor. Its most important elements are:

1. Concrete core with metallic pillars spaced 3.60 m (12 ft) apart in the facade, defining the modularity of the offices and installations (Figs. 1.21 and 1.22)
2. Four emergency staircases per floor, grouped in pairs and open to the exterior, so that each of the four offices per floor has its own independent emergency exit
3. Four independent groups of toilets per floor
4. Elevators and lobby areas of high-quality materials such as white dolomite marble and sycamore wood
5. Large, open office spaces with lighting between the pillars and a space of 11 m (36 ft) between them and the concrete core, with impressive views over the city, enhanced by the sloping perimeter glass, eliminating interior reflections
6. The perimeter cleaning galleries

A business center is planned for the lower floors of the tower, with meeting rooms, a fully equipped conference room with a capacity for 168 persons, secretarial services, an exhibition room, and offices. All these spaces surround the building's large four-story lobby.

**Fig. 1.21 Mapfre Tower, Barcelona, Spain.** (*Courtesy: Ortiz Leon Arquitectos.*)

## 2 Building Systems, Components, and Materials

The various building systems and components were integrated in the design process. These and other related issues are discussed in the following.

***Foundations.*** The complexity of the foundations is due to the location of the building next to the sea, on sand that has accumulated over the years, to its height, and to the need to create two basement levels below the sea level. All of this has been achieved using concrete cutoff walls and specially designed slabs to create an underground "inverted swimming pool," using the well-point method for controlling the water-related problems during construction. The tower's core has its foundations in a concrete block 2.60 m (8.5 ft) high by 20 m (65.5 ft) square resting on 49 piles of 1.50-m (5-ft) diameter and 18-m (59-ft) depth. Although complex, these foundations caused no delays in the construction since the theoretical calculations, based on geotechnical studies, proved to match the real conditions found in the subsoil during the work.

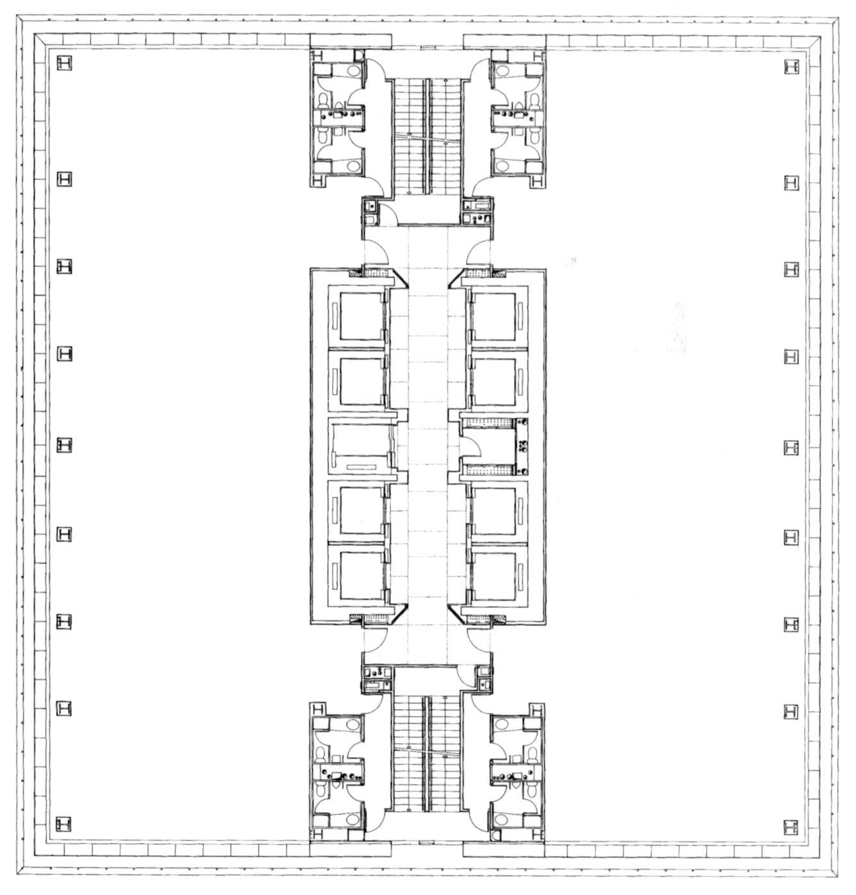

**Fig. 1.22 Office floor plan, Mapfre Tower, Barcelona, Spain.**

***Structure.*** The tower's structure is a mixture of a reinforced concrete core built with the sliding coffer system and with high-strength metallic perimeter pillars (Fig. 1.23). The location of these pillars at 11 m (36 ft) from the core creates large open office spaces of 11 by 33 m (36 by 108 ft). The pillars are 3 m (10 ft) apart, creating the ideal spacing for office modules.

All the metallic parts of the structure are bolted together and fireproofed. The concrete core absorbs the horizontal forces whereas the pillars absorb the vertical forces. This type of structural system facilitated rapid assembly, provided stair carriers, and

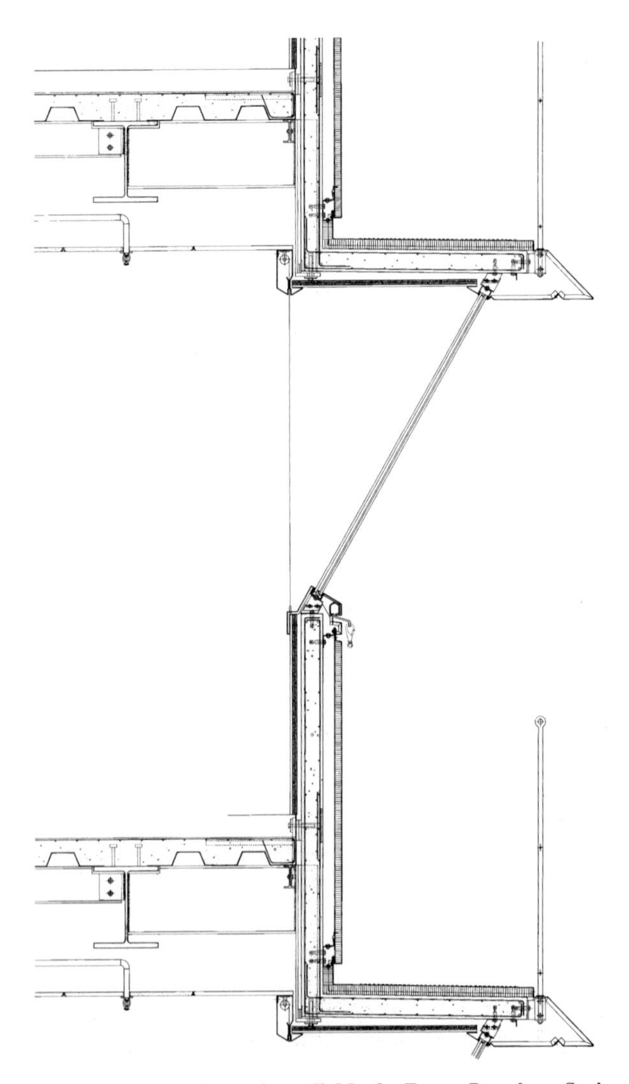

**Fig. 1.23 Section through exterior wall, Mapfre Tower, Barcelona, Spain.**

used prefabricated concrete facade breastwork. General information about structural systems and materials for tall buildings can be found in Chapters 10, 11, and 12.

***Facade.*** The tower's external skin in this case is stainless steel and sloping blue glass with galleries and eaves on each floor, which make its silhouette attractive and unique with respect to other nearby office buildings (Fig. 1.24). The choice of stainless steel as the principal material for the facade was due to the following characteristics:

1. Its great durability as a material
2. Its brightness and widespread maritime use, bringing the building into harmony with its location next to the sea
3. The attribute that stainless steel is a recyclable material manufactured from stainless-steel scrap, from which expensive nickel and chromium, giving it its rust-proof properties, can be recovered
4. The fact that it allows the creation of exclusive designs using folds, pressings, and shapings, in turn allowing the entire facade to be bolted into place, facilitating speed of installation and avoiding difficult-to-control welding in the field
5. The ease with which the entire facade is bolted to the stainless-steel pieces embedded in prefabricated concrete parts
6. Its resistance to the effects of atmospheric pollution and acid rain with low cleaning and maintenance requirements

After various studies, the problem of creating a facade that protects the building from the environment and especially from direct sunlight gave rise to a solution with the following design outcome:

1. Inspection, cleaning, and maintenance galleries on the facade of each floor are indispensable in an aggressive environment. There are two cupboards on each floor for cleaning tools, with a water tank and tap.
2. Sunlight control canopies to prevent direct sunlight from entering the office areas.
3. Windows which slope outward to prevent interior reflections, guaranteeing a perfect view of the city and the sea and increasing the available volume of each floor. Furthermore, they attract much less dirt.
4. The use of different finishes of stainless steel to achieve maximum visual quality.
5. Good overall thermal insulation in the building, giving it great thermal inertia.
6. The glass comprises three sheets, two of which are bonded together, giving the glazed area great safety and preventing glass from falling to the street in the case of breakage.
7. Important blind breastwork between the floors which, together with the corbels, helps prevent the exterior propagation of fire between floors.

A narrative account concerning the imagery and composition of facades in general may be found in Chapter 7. Further, a list of attributes characterizing the quality and performance of facades is presented in the Appendix of Chapter 7.

***Prefabricated Services.*** To optimize building time and to help eliminate the need for on-site craftspersons, all of the building's services were designed as prefabricated concrete modules, coming from the workshops completely finished with sanitary ware, tiles, mirrors, plumbing, doors, and the like. The only work required on site was to place

**Fig. 1.24 Mapfre Tower, Barcelona, Spain.** (*Courtesy: Ortiz Leon Arquitectos.*)

them in their definitive positions and connect the installations. This resulted in a high level of finish and standardization, which is indispensable for this type of work.

These modules came equipped with all the necessary service elements: waste bin, cupboard for cleaning tools, odorless hangings to facilitate floor cleaning, marble tops, and quarter-turn taps to save on water consumption. They were also equipped with emergency lighting and fire-extinguishing sprinklers. Their arrangement on each floor allowed for the installation of directors' services or a small service office next to them, with direct access from the offices.

***Elevators.*** All the elevators in the tower and adjoining buildings are of the highest quality. The elevator installation in the tower was one of the most complex issues for the project, as it comprises two groups of four elevators with double cabins. In an elevator with a double cabin, passengers going to odd-numbered floors enter the lower cabin and passengers for even-numbered floors enter the elevator's upper cabins via an escalator. This type of elevator optimizes the installation to the maximum, reducing the number of elevator shafts (in this case from twelve to eight) with a consequent increase in floor area for offices on each floor, since the elevator must be planned to cope with the morning and afternoon rush hours in the building.

This type of cabin has a unique tradition in Barcelona since elevators with double cabins were used extensively in the early decades of this century, the upper cabin being for the building's residents and the lower one for services and goods. Modernist buildings can still be seen in Barcelona with this type of elevator in glass and wood.

The elevator speeds of 4 and 6 m/sec never exceed an acceleration of 1.1 m/sec to guarantee a good level of comfort. The tower also has a freight elevator with a capacity of 2000 kg (440 lb) and 2.5-m/sec (8.2-ft/sec) speed with a high headroom, and two elevators connecting the ground floor and the basements. As the freight elevator runs from the first basement straight up to the plant floor on the roof, it greatly facilitates maintenance of the building's machinery.

The tower's elevators are fully equipped and include a speech digitizer, emergency lighting, firefighter's priority service, connection to the generating set, capacity for 13 persons, and integrated computer control of breakdowns and movement recording. Elevators for tall buildings are discussed in further detail in Chapter 2.

The cabins are decorated in white marble, wood, and stainless steel, matching the rest of the building and providing durability, quality, and a good level of lighting; they have good access and exits with 1.10-m (3.5-ft)-wide doors, allowing use by the handicapped. Design issues related to the disabled are further developed in *Building Design for Handicapped and Aged Persons* (CTBUH, Committee 56, 1992).

***Fire Control System.*** In conjunction with the traditional fire control systems such as hoses and extinguishers, the tower also has an integrated detection and automatic extinction system with sprinklers on all floors, using 300 $m^3$ (10,600 $ft^3$) of water stored on the roof. The generators installed guarantee emergency lighting, power the pumps, and move the elevators to the ground floor. Three elevators are reserved for the exclusive use of firefighters.

The tower has four emergency staircases on each floor with different alternative escape routes. As these stairs are independent and on the building's exterior, they provide a great degree of security. Care has also been taken with the details of fire control, such as the use of fireproofing in the structure with 2 hr of protection and of breastwork between floors on the facade to prevent propagation and to split up the building. The accesses have been designed to allow fire engines to come as close to the building as possible. There is also a network of fire hydrants on the site. The detection centers are located on the ground floor.

A more developed discussion on fire prevention systems may be found in Chapter 9. Also, the interested reader may refer to *Fire Safety in Tall Buildings* (CTBUH, Committee 8A, 1992) for further details on this topic.

***Surfaces and Floorings.*** Both the building's surfaces and its floorings have been carried out with a small range of materials to give a high degree of uniformity and to facilitate construction. The quality of these materials can be seen in the pieces' sizes and scantlings and also in the quality of the workmanship, which in many cases was complex because of large sizes. Following is a brief description of the materials used for surfaces and floorings.

1. A very Mediterranean and luminous white marble from Greece has been used throughout the building. Only in very heavily used areas have black and gray granites been used. This marble is exceptionally hard and, unlike granite, can be used as flooring. It is polished to provide a good finish over time.
2. The wood chosen also had to be luminous and with a clear local reference. Thus sycamore was chosen, a tree that is popular in Barcelona. This choice has given the interior a great uniformity and quality, which is unusual in this type of building. As it is a delicate material, which is used to decorate intensively used vestibules, constructive solutions had to be sought to allow the replacement of the wooden panels without using either screws or nails. As the result of careful design of the walls and ceilings of the main vestibules using the wood, an elegant form has been achieved with the wood pieces and their grains.
3. Stainless steel, used on the building's exterior and interior, is also the material used for joints and pediments in moldings, accessories, metal fittings, and such. The quality pediments in the interiors testify to the durability and ease of maintenance of all the pieces used.

***Security Systems.*** The security systems in tall buildings play an important part in the services provided (Sheehan, 1971). For this building project, access to parking is controlled by motorized barriers for cars, activated by ticket machines, and infrared barriers which detect any other object or person entering. Simultaneously, a security guard at every door can watch the area using TV cameras. The entire parking area is equipped with closed-circuit TV, intercoms, and readers for the guards' patrols.

Access by people to the tower is controlled by turnstiles with card readers to provide information about visitors and the amount of time they spend in the building. Everyone working in the building has personalized cards. Activities are monitored continuously by closed-circuit TV and guards. Panic buttons are also provided in the reception area.

All the interior public areas, landings, and elevator, floor, and basement vestibules are equipped with movement detectors which send signals to the data center. All the building's doors are part of an intricate locking plan with master keys for all locks. Within this plan, some access doors leading to the building's installations and internal services are restricted to the holders of special customized cards, providing information on who opened these doors and when.

All these systems are connected to the data center, where there is always at least one guard, day and night, to undertake whatever actions are necessary. It is here that alarms sound for any infrared barrier, movement detector, emergency and service door magnetic contact, panic button, or patrol reader. From here the electric locks are opened or communications made with any part of the building—ramps, vestibule, heliport, exits, basements, exteriors, and so on. It also houses the closed-circuit TV display center.

However, all these physical and human security measures present no obstacle to the visitor or tenant. Rather, they facilitate the localization of persons within the building using a computerized directory and alert the receptionists to any problems.

Discussions of the integration and automation of building systems in regard to security systems in general terms may be found in Chapter 8 as well as in Chapters PC-6 and PC-11 (CTBUH, Group PC, 1981).

### 3 Landscape Design

The landscape design embodying the characteristics of Japanese gardens adds to the quality of the urban character of the entire Olympic village and intensifies it by using very few materials, such as stone, wood, palm trees, and bamboos, creating a surprising air of elegance and individuality in the progression to the building. The landscaped sloping roof with red bougainvilleas and *zicas* which interpose a surprising base of color between the ground and the tower seen from the Olympic port is also noteworthy.

A stainless-steel sculpture of organic forms, 26 m (85 ft) high, by sculptor Andrea Alfaro decorates the exterior and acts as a reference and reinforcement to the geometrical volumes of the different buildings.

### 4 Maintenance

The building's design takes into account the low maintenance requirements of its parts and especially of its installations, the facade, and the internal common areas (IREM, 1981). Every attempt has been made to simplify the operational control systems in order to carry out the correct cleaning and replacement of all the elements. The same applies to the accessibility and provision of spaces such as workshops, stores, telephones (space for Ibercom), mail room, conference room, cafeteria, and pneumatic waste disposal system connected to the Olympic village network. Some pertinent discussions on building maintenance can be found in Chapters 8 and 9 as well as in Chapter PC-11 (CTBUH, Group PC, 1981).

## 1.7 ARCHITECTURAL DESIGN ISSUES: ITALY

Italy has a rich legacy of historic architecture and urban planning, ranging from the *cardo* and *decumanus* of the Roman street grid to the topological irregularity of the medieval town, to the regularization of streets and buildings introduced with renaissance and baroque planning concepts. Italy's lush and rugged topography varies from the mountainous terrain of the north to the rolling hills of the south. Its strategic geographic relationship to both the Adriatic Sea and the Mediterranean Sea provided ready access to the Middle East, Asia Minor, North Africa, the British Isles, and the European Continent as well as creating strong maritime economies affecting trade, political expansionism, and the rise of a wealthy merchant class.

The issues of urban design and building in Italy are inextricably intertwined, where the internal relationships of the existing context are acted upon by the builtforms introduced by the architect. In the Italian city, the urban condition always retains a position of paramount importance to the architect, whether the design is a single building or the redevelopment of an entire zone of the city. For a general overview of architecture and the urban context the reader is referred to Chapter 3.

In *The Tradition of Modern Architecture In Italy,* Ernesto Rogers writes that there are common characteristics which link all Italian architecture together (Smith, 1955):

> One of the characteristics of Italian architecture is the balance in the constant relationship between the single parts and the whole. This represents a desire for synthesis which seeks decorative values not so much through arabesques or ornamentation as in the interplay of volumes and surfaces, by emphasizing structure and constructive [sic] feeling (even when it does not correspond to actual structure).

Rogers views modern architecture in Italy as a return to tradition: "the tradition of the spirit against the false tradition of dogma." Therefore there is an intrinsic and historical relationship between the architecture of the past and contemporary architecture of Italy today. "The difference lies only in the fact that the same principles are going deeper, according to the same unchanging method" (Smith, 1955).

## 1 *Italian Tall Buildings: Growth and Change*

Historically speaking, Italy is a land of towers and tall buildings (Smith, 1955):

> In Italy this realization of the three-dimensional aspect of town planning and the importance of what might be called the Vertical Accent flourishes as in no other country. Several different means of expression are employed....They are: (1) obelisks and commemorative columns, (2) outdoor sculpture and fountains, (3) campaniles and even baptisteries.

Some Italian cities, such as Bologna, Pavia, and San Gimignano, are so rich in towers that often architectural historians cannot resist the temptation of comparing the new American city centers, engaged in a continuous skyward growth, to the historic centers of Italy. As a matter of fact, the most famous Italian towers or bell towers were even emulated in their shape, or at least in their crowning. They were generally built to a height of 100 m (328 ft) or even higher between the early 1200s and the end of the 1300s. For example, the campanile of the Piazza San Marco in Venice is often emulated by architects due to its stereometric massing and distinctive pyramidal spire. As Broadbent (1990) points out, it acts as a "hinge" linking the piazza to the piazzetta and "acts as a focal point which unifies, most successfully, the irregular plan of the Piazza, the Piazzetta, and the disparate forms of the buildings which surround them."

Italy is a land of historic towns, and it is coincidental that the most interesting tall buildings of the early twentieth century, whether built or only proposed, state the need to create free areas in the ground plane by radical interventions on the cities, which otherwise are unable to accommodate such concentrations of matter, spaces, and integrated functions. This actually poses the question quite relevant from a cultural point of view, whether to demolish urban fabric in historic centers, and to what extent, or to retain them as they are. Since demolishing would not be justified either by economics or by aesthetic and architectural points of view, tall buildings have not been allowed in nearly any Italian city for a number of years.

Mumford's (1952) commentary on the destructive potential of the tall building to the urban context was noted earlier. The problem of introducing new architecture into existing cities is compounded by the need of the city to continue to grow on the one hand, and its desire to preserve artifacts of its past on the other. This dilemma has been addressed by many architects and theorists, including Rossi (1982) and Krier (1979). Rossi, for example, recognizes the requirement to preserve what he describes as an "urban artifact—a building, a street, a square." These artifacts are the actual elements from which the city is made and can be studied scientifically through type. Yet, for the city to grow, it must by necessity assimilate the old into the new. Rossi views these urban artifacts as mnemonic devices which contribute to the collective memory of the city, but when they are isolated, as in the case when all the physical context around them has been destroyed, they lose their original meanings and are reduced to set pieces. Le Corbusier, with his plan for Villa Radiuse, proposed the destruction of large areas of central Paris (Barnett, 1986; Broadbent, 1990). Krier (1979) and other neotraditional planners find this practice abhorrent, and instead seek to unify new architectural interventions with existing architectural forms and typologies.

If no tall buildings were actually built in Italy, many were designed. A large number of Italian architects involved themselves in finding new solutions for tall buildings, proposing them both in Italy and abroad. One must also recall the avant-garde movement of futurism and its promotion of the *tabula rasa* by Antonio Sant' Elia and Mario Chiattone, who take a more resolute stand (Pagani et al., 1955):

> In 1914 the futurist architect Antonio Sant' Elia published a manifesto stating the future program of the new architecture. His work as an architect is confined to a series of sketches which demonstrate the idea of a rebirth as opposed to eclecticism and eternal traditionalism. Despite his formal and esthetically immature mistakes, his revolutionary challenge to the poverty of the plagiaristic style,...Sant' Elia is the forerunner of a new trend which even today...is still alive.

Their projects support the thesis of wiping out the city of the past and to build the city of the future with an overall rethinking of the urban context. Architecture is a middle ground to create complex urban communication grids and to replace the old traditional and classic perspective vistas with new tridimensional impacts on sight. Their towers are not, as are the tall buildings designed by Picasso, simple schemes or building types. They are the result of research which is free from urban planning decisions or restrictions. These architects understood that history, tradition, and economic reasons alone would not allow tall buildings to be constructed. Furthermore, they undertook a new and experimental way of designing, which gave place to architectural objects using the various languages proposed by the architectural culture at the time.

Sant' Elia's drawings of megalithic towers and multilevel circulation systems for imaginary projects like *La Città Nuova* in 1914 anticipated the forms of the skyscrapers of the 1930s and integrated buildings with high-speed urban transportation systems (Banham, 1960). His incorporation of monumentally scaled forms, geometric massing, and abstraction continues to be emulated by architects seeking radical, nontraditional, and distinctly futuristic imagery. The futurists' denial of tradition and historic forms is exemplified by their eradication of the existing city and its historic artifacts and substituting the *tabula rasa*—the "erased plane."

Research follows different directions. Suggestions that support this change in direction may be found coming from the German rationalism in Piero Portaluppi, as well as classical reminiscences in the first drawings by Fiorini. Mario Ridolfi's drawings can be referred to as the French rationalist avant-garde, while celebrating and monumental elements are emphasized in Palenti's buildings and entries in the international competition for the Columbus lighthouse in the Dominican Republic.

The triangulated projecting balconies of Ridolfi's six-story Casa a Roma, for example, shows an endeavor to express the spatial values of the interior, where strictly right-angle partitioning was abandoned in favor of a more organic design (Pagani et al., 1955). The reinforced concrete structure with paneled infillings is divided into two houses facing different directions and separated by a small central area of a different color.

Opportunities arose in the thirties for tall buildings in the field of public works. This was a result of the fascist government's intentions to change and mark the architectural quality of large and medium city centers. This caused drastic cultural interventions on the urban fabric of a number of historic cities. A qualification of the new design of these areas is now possible through vertical architecture, which means concentration. Exhibitions and competition entries were the opportunities for architects to propose tall buildings and architectural languages to the rulers. On one hand, neofuturist and rationalist architects suggested modern visions expressing the revolutionary spirit that fascism had claimed; on the other hand, a number of academicians, raised in stoicism, offered self-celebrating and monumentalist buildings as an opportunity to express the authority and prestige of the new government (Smith, 1955):

> Fascism, having been founded in 1923, when Mussolini marched on Rome, sought to arouse Italy to the largely untapped glories of the twentieth century in all fields of progress. This attitude towards architecture was...in direct contrast to that of Hitler who, coming to power eleven years later, after modern architecture in Germany had already flowered, felt that he had to stamp out the "radical" and "bolshevik" tendencies of the previous regime.

The race for height started in Brescia in 1932 with a building that featured clear monumentalism and the Italian tradition. It was designed by Massimo Piacentini, the leading interpreter of the fascist ideology. Other buildings followed within a framework of city center plans.

Tall buildings in Torino and Milano suggested a more modern expression and in spite of some compromise in style, demonstrated the functional lesson developed by European architects from the American way of building. Later on, the monumental outlook prevailed in two buildings in Genova, whose designers tried to weigh down the modern structural mass of their buildings by decorative use of the envelope materials.

Functional opportunities for tall buildings occurred when no restrictions were placed due to the need for redesigning the neighboring urban fabric. The reader should reference the hotels by Bonade Bottino at Sestriere, and by Nardi Greco at Chiavari, Salice d'Ulzio, and Massa. The small number of projects actually implemented during a time span does not truly indicate how many good tall buildings were proposed during the same time period. Some examples of these proposals include the Restaurant Tower and the Genoa building.

After World War II Europe had to face the reconstruction of its destroyed cities. The major problem was housing rather than an increase of the density in downtown service sectors (Krier, 1979). The International Building Exhibition (IBA) in 1987, for example, was a competition which divided the city of Berlin into several "demonstration" districts for which architectural reconstruction projects were proposed for areas that had been decimated during the world wars (Kleihus and Klotz, 1986). Vertical construction had not been the major choice for low-income housing, which had sprawled into the outskirts of Italian cities. Also the public support was not strong enough to push vertical buildings to the height of housing as, for example, in Hong Kong at that same time. Only in the midfifties, when American cities already looked like tall building forests, were some isolated proposals accepted in Italy. The most famous examples of these building types are the Torre Velasca (Fig. 1.25) and the Pirelli building (Fig. 1.26), both built in Milano. The first one, built in the 1950s, immediately aroused an international debate. This multiuse tower (office, residential, commercial, and parking) in the center of Milano is a strong interpretation of the city spirit, its walls and colors being a delicate balance between integration and contrast with the city. The second one, built also during the 1950s, in front of the railway station, has a slender mass perfectly expressing both structural principles and architectural intentions.

Following these remarkable examples, up to now isolated ones in the city skyline, there was a pause until the seventies for new tall buildings to rise in the immediate outskirts, where new urban highway networks suggested the implementation of new business centers. Such urban areas containing housing and office sectors can be found in Milano, Genova, Bologna, and Naples, the last showing the largest and most interesting grouping of tall buildings in Italy, now at an advanced stage of implementation.

## 2 Single-Building Projects and the Urban Scale

Modern buildings in Italy are becoming increasingly larger and more complex in their needs and design requirements. Moreover, the continuously growing size of urban developments and the functional design approach, generally prevailing together with the

**Fig. 1.25 Torre Velasca, Milano, Italy.** (*Courtesy: Pica Ciamarra Associati.*)

principle of problem separation, drive the architect's interest toward problems arising only inside a single building. Due to its cultural roots, this approach is coherently present at both the urban and the building scales throughout the major metropolitan cities of Italy.

Following World War II there was an immediate need in Italy for housing. This problem was compounded by the fact that, between the world wars, there was no building activity in parallel to the continual increase in population (Pagani et al., 1955). For instance, in 1931 about 6% of the dwellings were empty, indicating both the seriousness and the insufficiency of the housing problem, and that a considerable fraction of the population lived in overcrowding. Therefore providing comfortable, low-cost, multifamily housing that conformed to local contexts and urban usage patterns became one of the most important concerns of the postwar architects in Italy and throughout Europe (Smith, 1955; Pagani et al., 1955).

Two high-rise housing blocks, called Casa a Milano, were planned according to the shape of the site, the need to achieve a high construction index, and the desire to conserve the green area provided by the courtyard garden (Pagani et al., 1955). Reinforced

**Fig. 1.26 Pirelli building, Milano, Italy.** (*Courtesy: Pica Ciamarra Associati.*)

concrete pillars arranged according to the cross-beam system define the structure of the two blocks. The two buildings are different from each other based on their locations. The lower block is noteworthy for the surface unity of its street facade, which harmonizes with the continuity of the adjacent building. The courtyard face is more elaborate and geometrical. The eleven-story block facing the garden has a clearly defined three-dimensionally expressed structural grid, providing open-air balconies for each unit (Pagani et al., 1955).

The totally glazed tall building, with its pure and essential builtform, may be taken as a symbol of the design approach following functional, formal, and content autonomy. These buildings do not merely house offices. Many different space usages and related activities are included such as retail stores, parking, bars, and restaurants, with the intent to ensure the maximum functional intensity and autonomy. The aim is to bring the city within the building rather than the building within the city. The glazed surfaces enveloping tall buildings reflect the skyline and the images of surrounding buildings by conforming with (but at the same time rejecting) the surrounding reality: the denial of the context may be the same as in stone constructions. Although the context is not always formed by such emblematic buildings, often the urban landscape is no more than a sum of isolated entities: the city denies its very essence, its primary interest for the relationships between its parts, both for space and for usage relationships.

Barnett writes that the idea of the megacity—the city as a building—was proposed as early as the mid-1950s. Its origins can be traced back to the royal palace of Roman antiquity, which was always a self-contained community within a city and in preindustrial times assumed the dimensions of the city itself. "Recent visions of the city as a gigantic structure were almost always tied to a future in which the imperfections of modern cities would be swept away by the force of new technology" (Barnett, 1986). On a more modest scale, multifunctional tall buildings, such as the DG Bank in Frankfurt (see Chapter 7), appropriate the "city within a city" concept.

The Olivetti factory at Ivera is still regarded as one of Italy's finest modern buildings and set an international standard for the design of industrial architecture. The Olivetti buildings are concrete frame structures with glass curtain walls, having double rows of iron frames oriented to the northwest, and concrete brise-soleils on the southwest facades protecting the building from excess heat and glare (Smith, 1955; Pagani et al., 1955). The buildings are comprised of an open-plan office block attached to a workshop volume amply illuminated with daylight.

When designing a single building the needs of technology and function are complex and difficult to fit, and an interest likely exists to reduce constraints and requirements marking the quality of the building itself. This is the case of the relationship between structure and environmental or electrical systems; the ratio between usable and service space; and the insulation and thermal inertia of the enveloping surface to be transformed in a system of heat buffer spaces.

On a larger scale, in major urban developments, the image of the building must be seen from a distance. It must be the clear expression of architecture and structure blending together, complying with certain guiding principles (Ali, 1990). In a shorter distance it must be rich in signs, facade details, chiaroscuro, materials, and coherent with the pedestrian impact. So the builtform as a whole, and especially the middle and the top of the building, are mainly related to the sky and to the speed and the distance of the citywide vision, whereas the bottom of the building is mainly related to the grafting onto the ground. Here materials and technology may be traditional, colors may be rich and varied, forms may be broken and articulated.

The six-story Casa a Roma has two apartments on each floor. Constructed of reinforced concrete with brick ceilings, the exterior finishes of the building are in limestone and grounded marble plaster (Pagani et al., 1955). The plastic movement of the walls, the projecting balconies, the set back upper story, and the uneven and broken surface of the roof break up the cubic massing of the building.

On a citywide scale, where the builtform of the building is no longer the main interest, environmental problems arise with reference to the relation to the site and the surrounding city, the relation to morphology, history, tradition, and the climate (Krier, 1979; Rossi, 1982). At the same time the transformation of the idea of function, as it has been conceived by modern architects, drives architects to acknowledge that the reasons and needs of a single building only seem to be autonomous. Architects' attention must clearly move toward the relationships between the different parts, the intermixing of different activities, the integration at different levels, toward simultaneous and complex evaluations.

The urban distribution of the four-building group of the Complesso Immobiliare in Milano had to follow the lines of a detailed plan for the reconstruction of an urban zone almost completely destroyed by bombardments during the war. The highest building, of 49 m (160 ft), is situated on a north-south axis and placed sideways to the new road on the town plan, which passes underneath the center of the block. The other three buildings flank the new road. The structure of the buildings is of reinforced concrete with foundations of reversed beams, and two naves of pillars with continuous beams the same thickness as the floors (Pagani et al., 1955). The general architectural concept of the buildings arose from the necessity of flexible functional planning. Flexibility was enhanced by the slab ceilings, which do not require visible beams, thus allowing the internal divisions to be changed. External verandas or loggias are formed by the projections of the ceiling slabs beyond the exterior facades. The architectural expression of the buildings is based on a free architectonic composition of elements liberated from the strict adherence to the cartesian axes where forms do not lose their plastic values in the accommodation of internal necessities (Pagani et al., 1955).

To express the needs and design criteria of single-building projects in terms of space and form, it is necessary to change the land use and urban design code of reference. Any code is the latest official formulation of a previous need or of a praxis: after a time span it likely becomes obsolete, being the response to a need of a past time. In Italy the transformation of land use and design codes is strongly supported by the rapid changes currently going through the system. Architects and designers call for a more precise definition of the project and of the roles of designers and for greater efficiency. The consequence of this process is likely to be a better quality of tall buildings. But quality must not be limited merely to the building process; it must change the urban planning process and the image of the city fundamentally. Urban design and context are also discussed in Chapter 3.

## 3 Recent Tall Buildings in Naples

The new business center of Naples, the new downtown of the city, is located at the confluence of the major road and highway networks crossing the region and in front of the main railway station. It is a site on the edge of the city, facing the much degraded nineteenth-century district of the town. The scheme relates these two areas and gives them new qualities, ignores the context, and provides definite stereometric shapes.

The shape and size of each block suggest sheer height alone, meaning that high-rise buildings of a traditional style should be utilized, rather than modern tall buildings expressing in their builtform both structural and functional complexity and high-tech architecture, with a special regard to the impact on the environment.

Only the Law Courts buildings (Fig. 1.27), designed before the approval of the business center, escape these restrictions. The lower part of the complex is a large bulk over which three high-rise towers reach different heights. Their double-shell curved shapes

(Fig. 1.28) express the structural principles and improve daylighting. In the middle floors of the base a public plaza provides a link of transition between outside and inside, with thousands of people passing through on their way to offices and other functional spaces.

For all other buildings, including those discussed here, the shapes and dimensions of the floor plans were preestablished and very restrictive; the entrance towers are 14 m (46 ft) wide and 110 m (361 ft) high, whereas the other two buildings have rectangular floors about 20 by 50 m (66 by 164 ft), which are two glazed prisms about 90 m (295 ft) high. Three points must be stated before going into further discussion:

1. Except for the design process adopted for the new downtown of Naples, any urban scheme should be able to interact with the components of the design process described in the next section, particularly those which are supposed to define the builtform and its impact on the surrounding city and environment. A priori definition of the floor, volumes, and shells sometimes ignores many elements and opportunities strongly arising at a second stage. This calls for defining more suitable urban design rules, particularly in those cases where long time periods elapse between the two phases.
2. The Italian land use code still provides building limitations in terms of volume related to lot area, whereas in many other European countries urban land use plans limit the net usable floor space in relation to the lot area.
3. Public buildings in Italy too often require long time spans to be completed; bureaucracy takes too long for approval, construction also takes too long, and the design time wrongly becomes the shortest part of the implementation process.

The two entrance buildings, which will house the ENEL (National Electricity Company), involve a very advanced and fast-track technology. The other project that is

**Fig. 1.27 Law Courts buildings (distant view), Naples, Italy.** (*Courtesy: Pica Ciamarra Associati.*)

discussed in the following is built completely by a private developer. Only a few of the buildings projected in the new area are now being built, in spite of the importance of the new downtown core of Naples for the economic development of the city. The remainder of the building projects are currently working through the bureaucratic and technological paths.

## 4 Design Approach and Constraints: The New Business Center of Naples

***Entrance Towers to the New Downtown.*** The two buildings (Fig. 1.29) are two prisms whose volume is predetermined by the new downtown scheme, where they are meant as the side posts of the entrance gate to the Green Axis, the main pedestrian mall crossing the length of the entire new downtown.

Above the pedestrian mall the building has a floor of about 700 $m^2$ (7535 $ft^2$), rising to about 110 m (361 ft). The parking and common space are located on the two levels below. A steel roof beam of trapezoidal section is supported by two tall concrete piers. These two reinforced concrete "nuclei" have triangular shapes that include stairways, mechanical and electrical main ducts, and outside elevators. The building's 30 floors hang from this beam, leaving the underlying three-story-high atrium free from vertical structures and clearly expressing the function of each structural element. Suspended steel columns support steel-plate floors. A second system of steel trusses borders the floor where system plants are located and supports the meeting rooms hanging over the atrium.

The lift-slab construction process, never before used in Italy, provides floors assembled at the ground level, two to four at one time, and then raised to the roof beam or to the previous suspended floor set. The technical approach lies in the possibility of completing the building from the top down. The reader should refer to Chapters 2 and 4 for more information on the relationship between structural technology and architecture.

**Fig. 1.28 Law Courts buildings (detail), Naples, Italy.** (*Courtesy: Pica Ciamarra Associati.*)

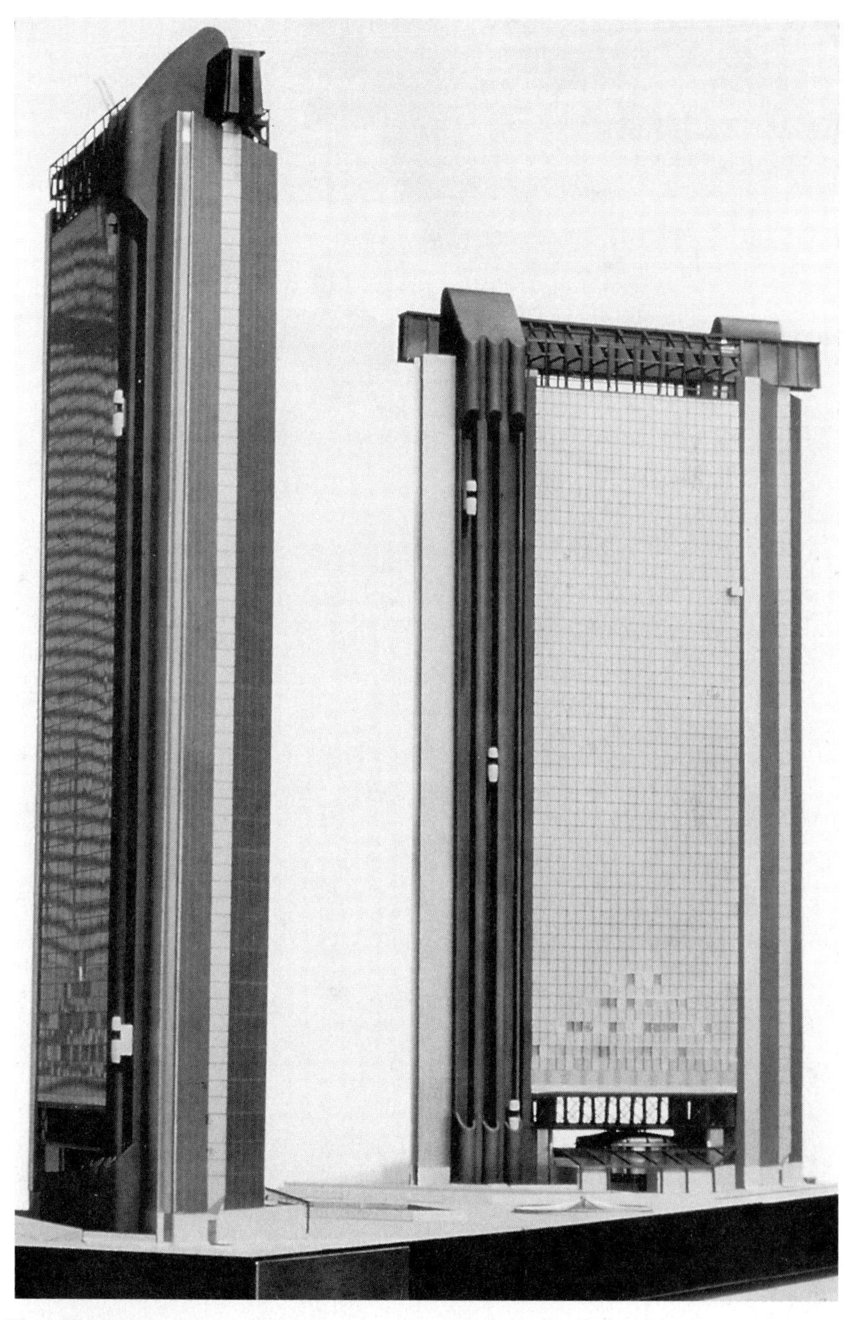

**Fig. 1.29 Entrance towers to new downtown (proposed), Naples, Italy.** (*Courtesy: Pica Ciamarra Associati.*)

The architectural image of the towers is based on the relationship between different parts of the builtform, each expressing its role and function. The glazed curtain wall is confined to the suspended part of each tower and takes the role of a bidimensional element set within a builtform which is no longer unitary but is fragmented and dissolved. The solid concrete triangular "nuclei," decked in mat steel plate, are marked by a thin vertical glass cut, offering a flow of light to the entrance space of all floors. The continuous enveloping surface of the nuclei wrinkles to make room for outside elevators.

***Twin Towers on the Green Axis (Fig. 1.30).*** The towers are located at the east end of the axis going through the scheme of the new business center of Naples. The new downtown is a large double deck about 7 m (23 ft) high, covering and unifying traffic routes and two parking levels. Above the second deck only pedestrian traffic is allowed along

**Fig. 1.30 Twin towers on Green Axis (general view), Naples, Italy.** (*Courtesy: Pica Ciamarra Associati.*)

the main central axis about 60 m (197 ft) wide (the Green Axis), where buildings rise to different heights. The shells provided for the two towers are completely glazed prisms based on a floor of 18-m (59-ft) width, 50-m (164-ft) length, and 70-m (230-ft) height.

While respecting these limits, the architectural aim is to deny the flat repetition of volumes simplistically defined as reflecting entities, so alien to the culture and tradition of the Mediterranean cities. Conversely, a strong chiaroscuro and extrovert builtform had to be sought, an architectural expression rich in its relationships to the sky and corroded in its contact to the ground. In order to increase the ratio of usable space to service space, the floor surface was fixed at 660 $m^2$ (7104 $ft^2$), the maximum allowed for two stairways of 1.2-m (47.3-in.) width. Moreover, following the opportunities offered by the district land use code, outdoor elevators are placed on the opposing fronts of the towers whereas the environmental system central plants are located in a large empty space, in the middle of each tower.

The steel structure is not visible due to fire regulations, but is clearly expressed within the architectural aims, the bottom of the buildings being a traditional steel structure and the top suspended from a large steel roof truss. The result is a building rising higher than it was originally designed and having roof space available for use. In the middle of the towers, the empty space at the 66-m (216-ft) level underlines the architectural difference between the top and the bottom of the buildings. Facade technologies mark this concept using different solutions for both the top and the bottom of the towers and for their short and long elevations. The long elevations are different from one another due to the presence of outdoor elevators; the empty space in the middle of the towers is the stronger element of their architectural image. The top (Fig. 1.31) is completely glazed whereas the bottom enveloping stone surface has the only glass elements necessary to provide interior daylight. The short elevations show a continuous major

**Fig. 1.31 Twin towers on Green Axis (detail of facade), Naples, Italy.** (*Courtesy: Pica Ciamarra Associati.*)

orthogonal grid mixing transparent and opaque surfaces and a subgrid of continuous glazing. Here the middle empty space does not appear, but the opening at each corner gives a feeling of its presence. Once again, the atrium is a unifying link between the three first floors above the pedestrian mall and the ones below, being the major common space for the auditorium and the cafeteria.

As the new business center introduces a complete separation between the traffic road network and the pedestrian level, the entrance hall of the buildings must provide spatial unity for the two systems of movement. This need results in a sort of exploded space, four or five levels high, completely different from the compressed 3-m (10-ft)-high space between floors, into which all other major functions in the buildings are brought.

The two towers differ in their connection to the ground. The building at the eastern end of the pedestrian mall plunges two floors deeper than the other building. The western tower, housing activities and departments of the National Council for Research (CNR), is marked by the bulging volume of the conference hall, placed at the underground level but rising from the basement and grafting onto the ground the main volume and its geometry.

Within the site of the two towers the paved surfaces of the pedestrian mall are limited to those really needed by pedestrian access and movement. Water basins and fountains are connected to the environmental systems of the buildings and to fire reservoirs, whereas trees frame the Green Axis. The towers are spectacular to each other; elevations seen in a distance are marked by the middle empty space and its inherent plastic form. A spatial grid, including the stairways to the roof terraces, connects the buildings to the sky. The facing elevations, usually seen from a more limited distance and in a sharper perspective, are both marked by the five deep grooves in which the four outdoor elevators run. Nearby are the continuous grids within which there are two more elevators, doubling as emergency elevators.

Once again, land use codes, limiting the built volume to the surface of the available site, limit floor-to-floor heights and the outer wall and floor thicknesses. Insulation and environmental problems are the reasons for the technology of the enveloping surface. This surface includes built-in cupboards as heat buffer spaces on the entire periphery of the buildings and interstitial spaces for air-conditioning system ducts and machines.

## 1.8 ARCHITECTURAL DESIGN ISSUES: JAPAN

### 1 Rise of Tall Buildings in the Asian Pacific Region

Architectural and economic development throughout Southeast Asia and the Far East has burgeoned in the recent past at an unprecedented rate. The intense tall building programs in Japan, Malaysia, Singapore, Thailand, Indonesia, and particularly Hong Kong resulted from the export boom since the 1950s. Hong Kong, for example, is one of the world's most important financial centers. The Pacific Rim countries of Asia and the Far East, despite political and cultural differences, have at least two things in common. By and large, their major cities, strategically located along natural coastlines and harbors, are physically restricted by the sea, the vertical walls of volcanic mountains, and rugged terrain. Second, most of these countries and territories support large populations on a relatively small land base. The resulting urban densities have driven land values upward, and consequently the cost of occupancy has spiraled upward (Bonavia, 1986). The economic prosperity of these regions in recent times has made the realization of tall buildings in urban centers possible.

The noteworthy tall building projects that have been completed in the past decade in Asia include the high-tech Hong Kong and Shanghai Bank (1986), the 65,030-$m^2$ (700,000-$ft^2$) multiuse Concourse in Singapore (commissioned in 1979), and the megastructure new Tokyo City Hall complex (1991). Other recent tall buildings in these regions include the Petronas Towers in Kuala Lumpur, Malaysia (now under construction), which will exceed the height of the Sears Tower when completed, the Central Plaza in Hong Kong, the Jin Mao building in Shanghai, China (now under construction), the Ruogyong Hotel in Pyongyang, North Korea, and the Metropolis International Building in Bangkok, Thailand, to list just a few tall buildings that are currently under construction or have just been built.

Associated with the megastructuralists of the 1960s who sought a repetition of a single function and module on a gargantuan scale (Jencks, 1988), Kenzo Tange has lamented the Japanese system of small-scale land ownership where "again and again the area of the property is divided into smaller and smaller pieces, but still people like to hold onto their small lot. This results in 'pencil-type' buildings which can be found all over Japan" (Jencks, 1988). Tange's dream of monumental expressionism—"to express a social necessity with structural purity and an heroic diagram"—was finally realized with his competition-winning entry for the new Tokyo City Hall complex.

Although Tange's Tokyo City Hall succeeds as a monumental and symbolic expression as well as creating a public urban space encouraging social interaction, Jencks (1988) is generally critical of large megastructure projects. Their enormous scale and the interconnected massing of multifunctional volumes are often at odds with the variety of scales and discreet functional massing of buildings in existing urban contexts. According to Jencks (1988):

> The connection between monumental building, large numbers and the ideology of Late Modernism is not fortuitous: it's an inheritance of the Modern belief in servicing the "greater number" (as it was known when the ideology was more democratic and socialist). But two contradictions have forced a change in doctrine and led the style towards its Expressionistic phase: the fact that now corporations and Arabian kings—not an egalitarian state—demand such intense concentrations, coupled with the change in opinion.... These contradictions force the Late-Modernist to reach for ever greater expressive powers, more complicated and impressive articulation. He must forestall criticism with a creative outburst of the "Unquestionable Sublime."

## 2 The New Tokyo City Hall Complex

The issue of remodeling the existing Tokyo City Hall was first discussed in the late 1970s. An appointed council at that time concluded that it would be best to demolish the existing city hall in the Marunouchi district and build a new one, but the plan seemed quite difficult to implement.

In 1979 it was decided to build a new city hall on three sites owned by the government in the Shinjuku district of Tokyo. As the administrative districts in Tokyo extend in the east-west direction, the plan to transfer the administration buildings to the centrally located Shinjuku district seemed preferable to the old Marunouchi site.

The metropolitan government has long planned to create cultural facilities near the new city hall, but the sites were too small to accommodate both civic and cultural facilities. The government, therefore, decided to build a cultural center on the old city hall site in Marunouchi after it relocated to Shinjuku.

The new city hall complex was expected to symbolize the autonomy and culture of Tokyo, and to represent Japan, since Tokyo is the nation's capital. In addition to providing administrative offices and the assembly hall, the complex was required to accommodate international conferences, emergency and disaster relief services, and a public plaza for the meeting of citizens. Nine architects were invited to participate in the design competition beginning in November 1985, which was won by Kenzo Tange Associates in April 1986.

The site occupies three out of nine blocks in the new center in Shinjuku, and in the middle of this center the architects designed a public plaza to symbolize the open relationship existing between the citizens and their city government. The assembly hall was built around the plaza, as it requires less volume and hence less height than the administrative towers. City Hall Tower I is in front of the plaza and City Hall Tower II is adjacent to Tower I.

City Hall Tower I required a large volume, but the height is dictated not to exceed 250 m (820 ft), in accordance with the regulations of the Shinjuku-ku Sub-Center Council. If a building was constructed having a width of 110 m (361 ft) and a height of 240 m (787 ft), the effect would be overwhelming. Therefore, taking into account that there are several buildings of about 150-m (492-ft) height in the surrounding area, it was decided to branch Tower I into two towers above that level so as to avoid an oppressive mass. Chapter 5 discusses the aesthetic issues related to tall buildings and their tripartite division into "base, stem, and crown."

The complex has a total floor space of 425,000 m$^2$ (4,574,666 ft$^2$) and is divided into two administrative towers, Tower I having 48 stories, 243 m (797 ft), and a floor space of 195,000 m$^2$ (2,098,964 ft$^2$), and Tower II having 33 stories, 163 m (535 ft), a floor space of 140,000 m$^2$ (1,506,948 ft$^2$), and a seven-story assembly building of 45,000 m$^2$ (484,376 ft$^2$).

City Hall Tower I containing the governor's office divides into two towers at the 33rd floor to decrease the impact of this massive volume. In City Hall Tower II, which houses public corporations and other departments, the architects wanted to achieve a sense of integrity and continuity with Tower I as well as creating a unique skyline. The upper part of the building steps, in the shape of three towers, sequentially downward toward City Hall Tower I. Across from City Hall Tower I is the assembly building and the citizen's plaza, which form the central axis of Shinjuku. The semicircular plaza and its colonnade are an integral part of the complex, whereas at the same time the surrounding high rises create a sense of urbanity in the plaza, which is unique not only for Tokyo but for most major cities. The three buildings are linked by two passageways extending from the third-floor level of the City Hall Tower I so as to integrate the circulation and define the outer edge of the plaza together with the colonnade. In Chapter 3, urban and contextual issues of tall buildings and their impact on scale in general, the skyline, and urban space are discussed at length.

Access to City Hall was designed at two levels. Cars were given access from the upper level at the central entrance. An underground passageway leading from the west exit of Shinjuku Station is connected to the lower level of City Hall accruing the safety of pedestrians. The complex will be approachable from a new subway station, which is scheduled for completion in 1995.

As the functions of Tokyo as a city have become increasingly complicated and sophisticated, the administration of the metropolitan government is required to enhance its efficiency. As the use of computers and office automation has become mandatory, the work environment and facilities are also required to function intelligently. A superstructure was employed in order to create a flexible office space, integrating diverse functions into one cohesive working environment. The City Hall complex is designed using a 6.4-m (21-ft) grid, with 6.4- by 6.4-m (21- by 21-ft) cores acting as large posts and providing 19.2-m (63-ft) spans for column-free office space.

***Elevation Design.*** The process of designing the elevation began with the collection of photographs of townhouses from the Edo period as well as their elevations, ceiling plans, floor plans, and various other drawings. The architects wanted to incorporate into the elevation historical elements of Japanese traditional design. While buildings in Europe basically consist of walls and holes in the walls that become windows, Japanese architecture is essentially comprised of vertical posts and horizontal beams, and occasionally lintels. Even with such fundamental elements limited in number, the patterns could vary with the inclusion of other elements as, for example, vertical grid screens on the windows facing the road for the assurance of privacy. It was decided to adopt such elements, that is, to start from the basics of posts, beams, and lintels, but to finish the remaining portions freely. Montages and studies were meticulously made.

During the initial design stages it was suggested that the design looked like a microchip. It was also noticed that the design had a certain resemblance with the pattern of electronic circuits. The facade was gradually formed from that time on to be less reminiscent of the Edo period and more audacious in expression, like the pattern of integrated circuits. For general information on facades and their expression refer to Chapter 7.

The image of the microchip pattern persisted and reappeared in the design on the huge ceiling of the second floor of Tower I. This pattern was designed to act as a guiding system for the circulation of the users of City Hall. This pattern is repeated in different scales in many places, including the stainless-steel doors of the elevators. This detailing contributes to the overall unity of the building. An attempt was made to make the interior as flexible as possible because the office layout is expected to be rearranged and the office automation equipment replaced periodically. For this purpose, the cores were designed as 6.4- by 6.4-m (21- by 21-ft) posts to house staircases, emergency elevators, ducts, and shafts and created a span of 19.2 m (63 ft) for office space between the cores. This space is unusually large as well as flexible.

***Other Design Features.*** The complex is equipped with the most advanced intelligent systems in Japan. All the floors of the office spaces are made with free-access floors. A 2.7-m (8.8-ft) floor-to-ceiling height is employed in a standard story floor height of 4 m (13 ft). Refer to Chapter 8 for more information on "intelligent" building systems.

The central plaza can accommodate as many as 6000 people as it extends underneath the bridges, and it is always open to the citizens of Tokyo. In the center is a stage large enough to support a full orchestra. A colonnade is built in front of the plaza and a sidewalk cafe and coffee shop look out onto it. Few plazas in the world are surrounded by so many tall buildings, and in this sense, the urban character of the plaza is unique.

During design it was anticipated that the durable life of the buildings would be 100 to 200 years, especially the basic frames and outer walls, and therefore stone was used to finish the walls. However, it is expected that the interior facilities will need to be replaced in a cycle of 5 to 20 years. The metropolitan government celebrated the completion of the new Tokyo City Hall complex on March 9, 1991.

For detailed information on building economics and life-cycle costs the reader is directed to Chapter PC-5 (CTBUH, Group PC, 1981).

## 1.9 DESIGN CONCEPTS AND CHALLENGES: SEVEN CASE STUDIES

In the following, seven additional case studies are presented to illustrate the concepts and challenges in regard to tall building design.

### 1 AmSouth/Harbert Plaza Mixed-Use Center

***Design.*** The design of this high-rise office and mixed-use complex located in Birmingham, Alabama, United States, a joint venture of AmSouth Bank and Harbert International, Inc., was shaped by a number of urban design, aesthetic, and economic considerations directly related to its location in the center of this midsized southern city (Fig. 1.32). Two major concerns were the relationship of the 31-story office tower to the city's skyline and the relationship of the complex's low-rise elements to the existing streetscape.

The 31-floor office tower was designed as a square campanile with a stair-stepped profile, allowing from within expansive views in many directions. The structure's multiple faceting also provides an increased number of corner offices, enhancing the interior space appeal. A relatively light-colored exterior further distinguishes the building on the Birmingham skyline. The material is a golden Brazilian granite called carioca.

The tower's visibility on the city's skyline creates an architectural reference point for city residents and visitors. The client preferred that the tower not be intrusive or foreign in its form or details. The designers responded by relating the shape of the tower to the city's many existing Art Deco structures, and by borrowing details from the city's rich architectural heritage. For example, the spheres that mark the four corner parapets around the tower's pyramidal copper roof were inspired by stone spheres that crown several notable downtown buildings. At night, triangular glass peaks on the four sides of the roof parapets are equally visible on the skyline.

The tower's distinctive gold and dark green granite cladding refers to the stone of two neighboring structures, a neo-Gothic church and a landmark Art Deco building. The terracing of the tower's upper levels and the verticality of its mullioned windows reflect the forms of nearby buildings. To relate this 78,965-m$^2$ (850,000-ft$^2$) complex to the intimate scale of the surrounding streetscape, designers encircled the building's perimeter with a two-story granite-clad pedestrian arcade. Its striated stone facing reflects the weathered sandstone facade of a neighboring cathedral. The pedestrian arcade leads into 4180 m$^2$ (45,000 ft$^2$) of retail stores, specialty shops, and restaurants located around an inner courtyard in the complex's first two levels. The retail area is connected by a landscaped pedestrian atrium to another office building that occupies the other half of the one-block site. The retail area provides welcome new shopping and dining options for downtown workers. Overall, the office and retail complex reaffirms the client's commitment to and belief in the economic and cultural viability of the city's central business district. Its design evokes the romance of the city's past, expresses continuity with the present, and symbolizes optimism for the city's future. Parking for 400 cars is provided on three levels below grade.

***Construction, Materials, and Mechanical Issues.*** The 31-story office tower has a reinforced concrete structural frame and is clad in golden carioca and dark green ubatuba Brazilian granite. The lobby features five different Italian marbles. Tinted reflective glass is used in its punched and mullioned windows. The pyramidal roof is clad in copper, and the 3.66-m (12-ft) spheres that top the corner roof parapets are made of fiberglass.

An "intelligent" building, its direct digital control system monitors all mechanical equipment and provides automatic energy conservation features. (See Chapter 8 for a detailed case study of another intelligent building.) The building's central mechanical plant includes two 750-ton centrifugal chillers and a heat exchanger. Each floor also has its own air-handling unit and electric heating system, with direct digital controls on the fan power boxes to provide individual zone comfort. Special mechanical features include a building pressurization unit which pumps additional air into the lobby area in the winter to offset the air pressure differential and eliminate stack effects. The building also contains an advanced smoke evacuation system. In case of a fire, this system automatically pressurizes the floors immediately above and below the floor where the fire is located, preventing smoke from migrating to the stairwells for safe evacuation of building occupants. Instead of having two air shafts, one for ventilation and one for smoke evacuation, the building has one air shaft with an automatic damper system that is reversed for automatic smoke evacuation from the affected floor.

**Fig. 1.32 AmSouth/Harbert Plaza mixed-use center, Birmingham, Alabama.** (*Courtesy: George Cott.*)

## 2 801 Grand Avenue

***Description.*** The first requirement was to develop a commercial real estate investment to meet a projected need for future offices in the city. The second was to provide an attractive, high-quality, functional office for new tenants as well as client expansion. This premier tall building was to stand out in the city's skyline and symbolize the client's growth and commitment to the community, while at the same time complementing the urban fabric of the central business district in Des Moines, Iowa, United States (Fig. 1.33). The solution was a 45-story, 85,700-$m^2$ (922,500-$ft^2$) tall building standing more than 192 m (630 ft) high. To meet the commercial investment requirement, the lower three levels of the building were designed to accommodate retail businesses, with the third level providing access to the city's skywalk system. A private restaurant and club were designed for the top two levels of the building.

This 45-story structure reaches toward the sky as the tallest building in the Des Moines skyline. To soften the distinctive structure's effect against the city's skyline and provide a "signature" for the tall building, a series of terraces and setbacks were incorporated into the design. This allowed the building to appear gradually smaller while ascending into the skyline, and at the same time provided multiple corner offices on each floor, a plus for new building tenants (Fig. 1.33).

To create urban unity, a light, golden-colored granite, similar to that of existing surrounding buildings, was chosen as the building's primary exterior material. A glass curtain wall complements the granite and highlights each side of the building, providing a triangular bay window effect. To symbolize the client's growth, a stunning eight-pointed copper-clad roof points toward the sky and creates a unique tall building identity. At street level, granite columns and paving, expansive windows, and landscaped plazas create pedestrian areas that reinforce and enhance the city streetscape.

Throughout the high rise, electrified floor decks make the building one of the new generation of "smart" facilities. This feature allows tenants the flexibility of customizing their own computer data and voice communications networks, an attraction to new building tenants.

***Technical Information.*** The upper or tower portion of the building's energy-efficient exterior is comprised of a factory-fabricated, panelized, aluminum, glass, and granite curtain-wall system. The lower portion of the building's exterior is comprised of granite-clad precast concrete with an aluminum and glass storefront window system.

The foundation structural system includes drilled concrete caissons extending 18 m (60 ft) to bedrock. The building's superstructure is comprised of a slip-formed concrete core and a structural steel frame with a cellular floor deck for tenant electrical, data, and communication systems distribution.

Three banks of passenger elevators meet the building's transportation needs. A state-of-the-art computer system monitors and controls elevator usage, thus allowing reprogramming for increased efficiency. The entire elevator system is interfaced with the building's security system, thus maximizing the overall building security as well as individual tenant security.

The building's office floors are served by a variable air volume system extending from a central mechanical level. The upper two club floors have an independent mechanical and electrical system. The mechanical, electrical, and life safety systems are continuously monitored and controlled by an integrated state-of-the-art building automation system. (For more information on building automation systems, see Chapter 8.) The electrical distribution system is designed so that individual tenant spaces are metered separately. The building has a full sprinkler system.

**Fig. 1.33 801 Grand Avenue, Des Moines, Iowa.** (*Courtesy: Balthazar Korab.*)

The building's design meets all federal and state guidelines for handicapped accessibility and all codes for energy efficiency.

### 3 Oryx Energy Company Corporate Headquarters

***Description.*** Oryx Energy Company in Dallas, Texas, became the largest independent oil producer in the United States when the firm was recently spun off from Sun Oil. This transition to complete autonomy was the first step toward a new corporate culture, image, and logo for Oryx.

Previously housed at several different locations, one of the company's main goals was the consolidation of all departments within the same building. Because public identity was very important to the "new" Oryx, this same building was to have a high visibility and a progressive image, one that would reflect the independent yet unified company that it had become (Fig. 1.34).

In a time when the market was saturated with vacant office space, this new 26-story, 51,095-$m^2$ (550,000-$ft^2$) office tower was a rarity. The high visibility is achieved by the dominance of the Oryx tower at the intersection of LBJ and Dallas Parkways. The tall building's focal point is its octagonal faceted tower rising four floors above the rest of the building. The double cross-vaulted roof, an adaptation of the other vaulted roofs of the Galleria, helps achieve the illusion of a freestanding tower. The articulation of the typical floor plan reflects the exterior forms and is designed to create additional corner offices on each floor.

Satisfying immediate needs but also looking ahead, the building initially meets the requirements of a single corporation, but can easily convert to multitenant use at any time in the future.

As outlined in the master plan, the Oryx tower has evolved as a unique architectural statement by expanding the vocabulary of the first two towers. From the first office tower with its simple form and gray glass to the second tower's more complex forms and patterns and reflective glass, the evolution has unfolded. The Oryx tower with its multifaceted form, combination of nonreflective and reflective glass, and rose aggregate creates a rich, colorful facade appropriate for its position of visual dominance.

The building's main entry on the auto plaza is flanked by the glazed escalator link and by the expanded parking structure entrance. The lobby escalator links office workers of the tower to the middle retail level of the Galleria by creating an internal pedestrian walkway.

An extensive landscaped park along Noel Road connects Oryx with office towers I and II to create the visual continuity found throughout the Galleria complex (Fig. 1.35). Located amidst this green parkway is a day care center for which Oryx was a catalyst.

The addition of the Oryx tower allows the Dallas Galleria to mature and adapt to new opportunities and challenges, and to fulfill the high expectations of the Dallas community. The Oryx tower was completed in 1990.

***Interiors.*** The challenge was to create an environment for a corporation that was starting a new chapter. Oryx Energy Company was newly formed, independent from its parent corporation. It created the opportunity to form a new image for the major independent oil and gas exploration and production company.

Comprised of 46,450 $m^2$ (500,000 $ft^2$) on 26 floors in a new suburban high rise, this project was especially challenging due to the client's need for a high-profile image while restricted by the tight budget of \$20.00/$ft^2$. The budget included the criteria of 90% private office environment, a prestigious executive floor, and an employee credit union.

The primary focus was to develop a design that expressed the new image of this forward-thinking corporation. This was achieved architecturally through the use of abstract forms representative of the nature of their business. A feature wall located in all typical elevator lobbies relates the message of horizontal drilling technology of which Oryx Energy Company is the world leader. A horizontally benched granite wall visually expresses the layers of the earth. A horizontal stainless-steel element spanning the length

**Fig. 1.34 Oryx Energy Company corporate headquarters, Dallas, Texas.** (*Courtesy: Nick Merrick.*)

of the stone wall cuts into the earth, abstracting the drilling process. It intersects a vertical glass element that symbolizes an oil fissure. These lobbies are further sculpted with angled walls and ceiling representative of seismic movement. Their forms are strong and pristine, enhancing the continuity to the overall design. Security doors were a necessity in all elevator lobbies but remain transparent by the use of clear glass against the strength of the forms that surround them. The end of each lobby is punctuated by an accent color that articulates the terminus of the lobby and serves as a departmental identifier. Wood floors provide a strong base for the stone wall and establish a tactile warmth that continues throughout the finish palette.

Warm, midtone, neutral hues constitute the naturalistic color scheme, established in the main building lobby with the use of polished granite and makore wood. On the high-rise floors, textural warmth and rich contrasting color accents create a light and bright atmosphere, defining the humanistic, people-oriented attitude of this progressive company. Bold black furniture adds a formal attire to appropriate spaces.

On the executive floor, formality and prestige are the vocabulary. The benched stone wall and wood flooring elements are recalled here, then wrapped with a rich envelope of makore wood. The three-dimensional architectural character is designed with a sophistication of detail and line to receive stately furnishings and express the power center of the organization.

The design image created for Oryx is intentionally a complete departure from its former parent company. While modest in expenditure, the design exhibits strength and clarity through the composition of strong forms. This design is a catalyst supporting the new corporate mission and establishing a new corporate home.

**Fig. 1.35 Oryx Energy Company corporate headquarters, Dallas, Texas.** (*Courtesy: Nick Merrick.*)

### 4 BP America, Inc., Corporate Headquarters

***Description.*** An eight-story landscaped entry atrium provides a personable scale for this 45-story headquarters building, which faces historic Public Square in downtown Cleveland, Ohio, United States. Originally designed for Standard Oil of Ohio (Sohio), the building became the U.S. headquarters for BP America, Inc., after the company's acquisition by British Petroleum. The headquarters building, located on 1.01 hectares (2.5 acres) has played a significant role in the revitalization of the city's central business district, setting a precedent for other corporate headquarters and mixed-use facilities to be located downtown.

The design concept called for a building with two distinct zones: a low-rise entry atrium and a high-rise corporate office tower. The atrium includes an indoor esplanade and formal gardens with fountains, bordered by amenities such as retail shops, food service facilities, and an employee fitness center. It was designed to be in scale with the Public Square and nearby low-rise buildings, and serves as the tower's "front door." The 140,093-m$^2$ (1,508,000-ft$^2$) tower (Fig. 1.36) is set back from the square and another local landmark, the 52-story Terminal Tower. A slight crease in the BP tower's facade visually reduces its overall mass, aligns the building with its slightly wedge-shaped site, and puts it on axis with an adjacent government mall. A north entrance links the tower to the mall, and another major entrance connects it to downtown department stores and a public transportation hub to the southwest. Vertical setbacks and notched corners provide the tower with added dimension, highlighted by exterior lighting at night, and provide many corner offices.

The steel-frame building is clad in contrasting shades of polished rose granite to create a sense of warmth that carries through the area's long, cold winters. Inside, in the atrium and lobby, flamed rose granite flooring pavers are combined with pink marble wall cladding. Brass and polished woods provide accents and detailing in the public spaces.

### 5 Metropolitan Square

***Description.*** Metropolitan Square is a 180.8-m (593-ft)-tall luxury office tower developed by Metropolitan Life Insurance Company in St. Louis, Missouri, United States (Fig. 1.37). It is the largest single speculative building ever developed by the company. This luxury office tower, which is topped by a two-story glass penthouse, rises from a seven-story granite base that conceals parking for approximately 1000 vehicles. Its pitched roofline is spotlighted at night to draw attention to downtown St. Louis.

The tower is designed with 18 bay windows per floor and just four interior columns to a typical office floor. The 15.2-m (50-ft)-high glass-walled entry lobby is framed by marble columns and provides public access to 1720 m$^2$ (18,500 ft$^2$) of ground-level retail and restaurant space. A covered arcade that encircles the building's base encourages pedestrian traffic.

A design concern was to create a welcoming entry that would reflect the height and mass of the building. A pedestrian arcade was constructed around the entire building and leads to the 12.2-m (40-ft)-high entry lobby and atrium. The building at ground level is very appealing to downtown pedestrians. The lobby murals add a humanistic and artistic quality that is rarely found in public spaces.

Because the building is the tallest in the area, a painted metal roof was created as a profile in the sky, visible for many miles, and when lighted at night, serves as a beacon to downtown. The night lighting also references "The Light That Never Fails," a beacon on the roof of the Met Life building in New York City, which has become the company's signature.

**Fig. 1.36 BP America, Inc., corporate headquarters, Cleveland, Ohio.** (*Courtesy: Hellmuth, Obata & Kassabaum, Inc.*)

To achieve a timeless and colorful quality that would complement the city's enduring architecture, granite was selected for the exterior of the building and marble for the interior.

The building's 12.2-m (40-ft)-high glass-walled entry lobby is framed by marble columns. Brass fixtures and various shades of Italian marble are used throughout the lobby area, which features a series of huge murals (Fig. 1.38). The lobby level also contains a retail and restaurant arcade. Just outside the main entry, along Broadway, an

**Fig. 1.37 Metropolitan Square, St. Louis, Missouri.** (*Courtesy: Robert Pettus.*)

**Fig. 1.38 Entrance lobby of Metropolitan Square, St. Louis, Missouri.** (*Courtesy: Robert Pettus.*)

open plaza provides seating areas and changing displays of annual flowers. A covered arcade completely encircles the building at street level, further encouraging pedestrian traffic.

Metropolitan Square's floor plans were designed for flexibility and a maximum of column-free space. The floors range in size from 112 to 37,160 $m^2$ (1200 to 400,000 $ft^2$), accommodating virtually any size tenant.

***Materials.*** The tower has structural steel framing and concrete slab floors to provide maximum usable space. The granite and marble used for the 42-story skyscraper reflect traditional St. Louis building materials and blend with the surrounding architecture. The exterior is clad in 60% gold/bronze granite and 40% reflective gold/bronze glass. The lobby front is composed of Pilkington glass. Inside, imported marble and granite are combined with brass accents in the main lobby and elevator lobbies.

## 6 Foley Square Federal Office Building

***Description.*** The development of this 34-story federal office building offers a unique opportunity to complement and reinforce the special character of the Foley Square district of New York City. Located directly north of City Hall, Foley Square has long served as the civic heart of the city, providing the center for federal, state, and municipal courts as well as administrative and judiciary services. The expansion of this vibrant governmental complex constitutes the first major change in the area since the early 1970s.

Foley Square will be constructed on Broadway between Duane and Reade streets (Fig. 1.39) and will house the Internal Revenue Service, the Environmental Protection Agency, the General Accounting Office, and other federal agencies that are currently

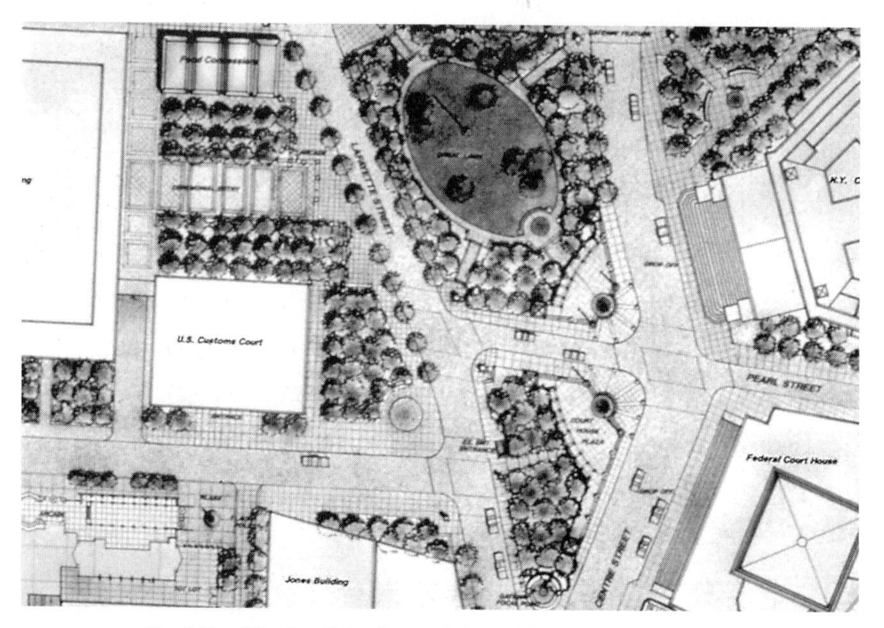


**Fig. 1.39 Site plan, Foley Square federal office building, New York.**

scattered in offices throughout the city. The 90,485-m$^2$ (974,000-ft$^2$) office tower will be crowned by a special conference center offering dramatic views of lower Manhattan. At ground level, 1300 m$^2$ (14,000 ft$^2$) of retail and restaurant space is organized along an enclosed urban arcade, which will provide a public link between Broadway and Foley Square.

The design integrates the classic, dignified tradition of American federal architecture with contemporary aesthetics for high-rise architecture in a dense, urban locale. The tower (Fig. 1.40)—buff-colored granite with decorative banding—will be capped by an illuminated barrel-vaulted roof. The design also incorporates a variety of grand public spaces to serve the building occupants and the citizens of New York. Enhancement of pedestrian, subway, and vehicular linkages in and around the site was also incorporated into the design.

### 7 Tokyo Telecom Center

***Description.*** Tokyo Telecom Center, a 92,900-m$^2$ (1,000,000-ft$^2$) telecommunications center in Tokyo, Japan, will be the focus of the 1995 Telecom Expo, and will consist of twin 12-story office towers over a four-story atrium at ground level and bridged together at the top by a four-story satellite and antenna platform from which a restaurant floor is suspended (Fig. 1.41).

The lower four floors of the building and the connecting atrium represent the public spaces and areas of service amenities, such as public television studios, exhibit spaces and galleries for telecom displays, and health club facilities. The atrium features a slanted circular glass roof visible from the exterior.

The building is intended to service the telecommunications industry and be a high-tech statement for Tokyo's leading edge in the field of telecommunications. The building is the centerpiece and premier building of Tokyo Teleport Town, a new telecom community located on landfill in Tokyo Bay.

Services provided by the architect include: landscape design; atrium design; and interior finishes for the atrium, the 20th-floor lounge, public corridors, and typical core space such as elevator cabs. The architect has also played a significant role in the schematic development of the building and is involved in issues pertaining to the design and detailing of the glass curtain wall and selection of materials.

## 1.10 FUTURE DIRECTIONS AND RESEARCH NEEDS

### 1 Future Directions in Design

A threatening deficiency in the area of architectural design of tall buildings is taking hold of our urban centers around the world. In the process of building our cities, architects and designers allow the limited resources to diminish unchecked. As a result, permanent adverse physical conditions are created. Tall buildings of the current generation are larger and stronger than any artificial structures in history. The tall buildings produced today are sophisticated machines capable of renewal through replacement, and therefore architects must regard them as permanent fixtures of our cultural condition and our urban fabric. Nevertheless, architects are approaching the end of an era in the architectural design of tall buildings, and have not quite yet established the metaphor for their next generation.

**Fig. 1.40 Section through building, Foley Square federal office building, New York.**

Some future directions can be speculated, however. Today's architectural attributes of tallness are likely to evolve into a magnified attribute of massiveness. Growth of city population due to increased birth rate and migration of people from rural areas, particularly in the developing world, will inevitably lead to this massiveness to accommodate more people. This will result in superlarge buildings over a complete city block or a number of city blocks, forming a continuous megastructure. Another possibility is the development of superthin buildings in crowded downtown cores of cities. Both these future possibilities will pose an enormous challenge to the architect's imagination and the engineer's ingenuity. New design criteria are expected to evolve with the growth and development of future tall building typologies.

Because of the increasing complexity of urban life and living conditions, architecture of future tall buildings can no longer limit itself to its microdesign involving its interior space, form, and decorative skin. It must expand to macrodesign involving inevitable issues related to the urban habitat in which the building is located.

## 2 Research Needs

The historical development of high-rise buildings has occurred incrementally and has been determined, by and large, by innovations in technology, fortuitous economic conditions, and changes in zoning within cities. Some architectural critics argue that many of the aesthetic and structural problems of the tall building, as defined by Sullivan, Root, and Burnham, have been solved through advances in technology and structure, as well as a more comprehensive understanding of the typology itself. Others have argued that many design and structural issues remain to be investigated and that high-rise build-

**Fig. 1.41 Model of Tokyo Telecom Center, Tokyo, Japan.** (*Courtesy: Hellmuth. Obata & Kassabaum, Inc.*)

ings will have to continue to be adapted to increasingly complex design, structural, and contextual criteria. Given these diametric and conflicting points of view, it seems evident that more investigation into these issues is required.

The roles of the architect, structural engineer, and urban planner must be redefined continuously in terms of both changes in their respective professions as well as changes in society. There is a need to increase collaboration among the professions, in addition to increasing their awareness of larger changes which affect social demographics, economic conditions, and physical contexts for which they are designing and building. Additional research is needed to help define the collaborative roles these professionals will be required to play in the planning, design, and construction of tall buildings, and how they can develop interdisciplinary approaches to resolving building planning, construction, and usage issues.

Once again, architects and planners are realizing that the historic patterns and fabrics of existing cities must be understood and preserved when introducing tall buildings as interventions into these contexts. The manner in which architects and planners can preserve existing buildings, neighborhoods, and patterns of use while introducing new buildings is an important and difficult issue. Many large cities, particularly in the United States and parts of Asia, have been overdeveloped with high-rise towers placed side by side, crowding one another for space, light, and air rights. Without proper planning and zoning, the tendency is to build increasingly taller and larger buildings, compromising all the buildings adjacent to them. These trends have had an adverse impact on the physical and social environment of the city. Densely packed large buildings loom over narrow streets blocking out sunlight and create unacceptable wind vortices at their bases. In older cities with few high-rise buildings, such as in Spain and Italy, strategies are being developed of how to introduce tall buildings into low- and mid-rise contexts. The historic patterns and textures of these old and picturesque cities must be preserved, yet new architecture which accommodates changing needs and growth must be introduced. These issues are complex and cannot be resolved simply by using historical planning strategies developed without high-rise buildings in mind. Innovative research will be required to address these issues in creative and effective ways.

The tall building is a relatively new building type. In many respects it is still being explored in terms of its urban and contextual impact, as well as its functional planning and organization. The supertall building and the megastructure continue to be explored by architects and planners. Large multifunctional buildings, as opposed to single-use buildings, may become more predominant as dictated by changes in context and economics. The impact that large-scale buildings will have on the urban context and their specialized planning requirements will present new challenges to designers and engineers.

At this time there are no specialized design criteria, developed expressly for tall building architecture, that are available to the tall building designer. Design methodology varies greatly among architects and engineers as well as among building types. Many architectural firms that specialize in tall building design have developed their own planning and design methodologies, but these design guidelines have never been documented in the form of a matrix or similar format.

Tall building design involves a unique collaboration between architect and engineer. Due to the scale and complexity of tall buildings, this collaboration often begins at the earliest stages of the design process and continues into construction. Architects and engineers must know what the design goals and objectives of the project are and steps necessary to achieve them as planning is initiated. A set of specialized criteria in the form of a design matrix would be extremely valuable to the designer so that all important criteria essential to the planning, design, and construction of tall buildings would be met. This matrix would perhaps be a two-dimensional chart with design criteria on one axis and tall building typologies along the other. Development of such a matrix will in effect systematize the design process and clearly articulate the relationship of the various factors that influence design.

Investigations related to building systems, components, and technologies will also be required. New developments in technology continuously impact building design and planning. Today architects and planners are required to respond to complex environmental and ecological concerns. New building materials and systems which help to conserve energy and increase the operating and maintenance efficiency of tall buildings are currently being explored.

Perhaps the greatest advances have been made in new technologies for the building's exterior skin. Developments in high-performance glazing systems which more effectively control heating and cooling loads continue to bring building operation costs down. However, additional research is necessary in this area. Intelligent building systems, incorporating sophisticated computer monitoring systems, are now being incorporated into tall buildings (refer to Chapter 8). These systems have an enormous impact on the planning and design of tall buildings and must be considered in the earliest programming stages of design. Proposals for buildings constructed from reused materials are also being introduced. These buildings suggest that architecture can contribute to the judicious use of materials without depleting valuable resources.

Finally, architects, planners, engineers, and developers must develop a broad, holistic understanding of the planning, design, and construction of tall buildings and their impact on a global scale. Tall buildings have become an intrinsic part of the city, and their full impact on the social, economic, and physical environment of the city is becoming increasingly apparent. Some architectural critics, like Lewis Mumford (1952), recognized the potentially negative impact that the tall building has on the city. It is evident, as in the case of design at any scale, that each individual part of the design contributes to the whole. If we view the city as an organism in which buildings contribute to its life and vitality, then architects and planners must find ways to introduce increasingly large and complex tall buildings into the fragile context of the city. More research is required to fully investigate and develop strategies to address these complex issues.

## 1.11 CONDENSED REFERENCES/ BIBLIOGRAPHY

The following is a condensed bibliography for this chapter. Not only does it include all articles referred to or cited in the text, but it also contains bibliography for further reading. The full citations will be found at the end of the volume. What is given here should be sufficient information to lead the reader to the correct article—the author, date, and title. In case of multiple authors, only the first named is listed.

Ali 1990, *Integration of Structural Form and Esthetics in Tall Building Design: The Future Challenge*

Architectural Record 1974, *The High Rise Office Buildings—The Public Spaces They Make*

Aregger 1975, *Highrise Building and Urban Design*

Aynsley 1973, *The Environment around Tall Buildings*

Banham 1960, *Theory and Design in the First Machine Age*

Barnett 1973, *The Future of Tall Buildings: Systems and Concepts*

Barnett 1986, *The Elusive City: Five Centuries of Design, Ambition and Miscalculation*

Beedle 1971, *What's a Tall Building?*

Boles 1985, *Cincinnati Centerpiece*
Bonavia 1986, *The Hong Kong Bank: Introduction*
Broadbent 1990, *Emerging Concepts in Urban Space Design*
Codella 1973, *Architecture of Tall Buildings*
Codella 1983, *Where Do We Go from Here Architecturally?*
Condit 1964, *The Chicago School of Architecture: A History of Commercial and Public Building in the Chicago Area*
CTBUH, Group CB, 1978, *Structural Design of Tall Concrete and Masonry Buildings*
CTBUH, Group PC, 1981, *Planning and Environmental Criteria*
CTBUH, Committee 8A, 1992, *Fire Safety in Tall Buildings*
CTBUH, Committee 56, 1992, *Building Design for Handicapped and Aged Persons*
Cullen 1961, *Townscapes*
Giedion 1967, *Space, Time, and Architecture: The Growth of a New Tradition*
Goldberger 1989, *The Skyscraper*
Gregory 1962, *Thoughts on the Architecture of High Buildings*
Guise 1985, *Design and Technology in Architecture*
Handlin 1985, *American Architecture*
Hart 1985, *Multi-Storey Buildings in Steel*
Hubka 1973, *Man as Criterion of Tall Buildings*
Huxtable 1984, *The Tall Building Artistically Reconsidered: The Search for a Skyscraper Style*
IREM 1981, *Managing the Office Building*
Jencks 1980, *Skyscrapers—Skyprickers—Skycities*
Jencks 1984, *The Language of Post-Modern Architecture*
Jencks 1988, *Architecture Today*
Kavanagh 1972, *Environmental Systems*
Kleihus 1986, *International Building Exhibition Berlin 1987*
Krier 1979, *Urban Space*
Lamela 1973, *Town Planning and Vertical Architecture, The Problems Raised Thereby and the Necessity of a New Outlook*
Lynch 1969, *The Image of the City*
Lynch 1981, *A Theory of Good City Form*
Lynch 1990, *City Sense and City Design: Writings and Projects of Kevin Lynch*
MacMillan 1974, *Amenity and Aesthetics of Tall Buildings*
Maki 1972, *The Menace of Tall Buildings*
Mumford 1952, *Roots of Contemporary American Architecture: A Series of Thirty-Seven Essays Dating from the Mid-Nineteenth Century to the Present*
Norberg-Schulz 1974, *Roots of Modern Architecture*
Pagani 1955, *Italy's Architecture Today*
Peets 1968, *On the Art of Designing Cities*
Pelli 1982, *Skyscrapers*
Polshek 1988, *Context and Responsibility*
Posokhin 1973, *Design and Trends in the Development of Tall Buildings in Moscow*
Powys 1971, *Economic Aspects of Planning and Design of Tall Buildings*
Prescott 1962, *Formal Values and High Buildings*
Robertson 1973, *Heights We Can Reach*
Rossi 1982, *Architecture of the City*
Rowe 1978, *Collage City*

Schmertz 1975, *Office Building Design*

Scuri 1990, *Late-Twentieth Century Skyscrapers*

Sheehan 1971, *Security*

Siegel 1962, *Structure and Form in Modern Architecture*

Smith 1955, *Italy Builds*

Soleri 1973, *Tall Buildings and Gigantism*

Spreiregen 1965, *Urban Design: The Architecture of Towns and Cities*

Syrovy 1973, *Development, Architecture and Town-Planning Conditions of Design and Construction of Tall Buildings*

Zalcik 1974, *Town-Planning and Sociological Considerations of Tall Building Construction in Slovakia*

Zelanski 1987, *Shaping Space*

# 2

# Building Planning and Design Collaboration

Chapter 1 of this monograph provided a broad perspective of the architectural design process. This chapter presents a comprehensive overview of the planning of tall buildings, the relationships of structural, mechanical, and vertical transportation systems to their design, as well as technological approaches to the design process. It provides a brief account of the development in design and structural expression of tall buildings from the early 1960s to the present time, including a look to the future. It also describes in detail, through case studies and illustrations of studio research models, the ideas and approaches to the design process of tall buildings.

This chapter is organized in four sections: Building Planning and User Needs, Architectural-Structural Integration, Collaboration between Academia and Office Practice, and Future Research. Section 2.1 describes the influence of the plan and the integration of building systems, such as mechanical, electrical, and vertical transportation systems, on the architecture of tall buildings. While it illustrates the accepted practice for the sizes of lease spans, design of cores, and best location for mixed-use tenants, it also discusses novel ways of approaching these problems. Several structural systems are then discussed in Section 2.2 from a structural planning point of view, and their influence on the expression of architecture is presented. (See Chapter 4 for a more complete discussion on this topic.) The chapter continues with examples of research models developed in an academic setting by graduate students working in collaboration with a team of professional advisors in the architectural design studio in Section 2.3, and it concludes with topics for future research in Section 2.4.

It was through the influence of Mies van der Rohe that the concept of structural architecture began to be investigated at the Illinois Institute of Technology (IIT) in Chicago in the late 1940s and early 1950s. Guided initially by Myron Goldsmith and, beginning in 1961, by Goldsmith, David Sharpe, and Fazlur R. Khan, of Skidmore Owings & Merrill (SOM), students began exploring innovative ideas in structural architecture through graduate research projects. The applications of structural architecture concepts in steel and concrete were later realized in buildings by Khan and architect Bruce Graham of SOM in the John Hancock Center of 1968 (using X-braced steel structure), the Onterie Center of 1986 (using X-braced concrete structure), and several other buildings, including the Sears Tower (using bundled-tube steel structure).

## 2.1 BUILDING PLANNING AND USER NEEDS

### 1 User Needs and Building Types

Growth in population and economic resources along with limited building sites and infrastructure in centralized business districts have given rise to the construction of tall buildings in major urban centers. Clients and their representatives are desirous of solutions to their building problems that will permit them to function well in an atmosphere that enhances their working and living environment. In response to the client's needs, the architect seeks solutions that are aesthetically pleasing and functionally accommodating, and which result in spaces and plans that are flexible, uncluttered, and efficient. To accomplish these goals, the structural engineer seeks the most efficient and economical structural solution, and the mechanical, electrical, and environmental engineers, while concerned with the energy budget and conservation, create the desired microclimate and enthalpy for a pleasing and healthy working environment.

The issue of priorities of user needs has been discussed in detail in Chapter PC-6 of *Planning and Environmental Criteria for Tall Buildings* (CTBUH, Group PC, 1981). Human behavior and user needs are also discussed in Chapter 6 of this Monograph.

***Single-Use Development.*** Long before the Industrial Revolution, living quarters and working or trading spaces were combined in single buildings. However, beginning in the late nineteenth century, new land-use patterns emerged where places of employment and dwelling became separate and distinct, thereby giving rise to single-use developments in building design. Residential functions were excluded from the concentration of commercial and office functions. Although being located near the high-density core, they were not physically joined to it. This resulted in a unique building type for a specific use.

***Mixed-Use Development.*** To keep abreast with the demands of contemporary urban life created by the concentration of many functions in city cores, new ideas and solutions are being developed to provide a high quality of life commensurate with the high-density character resulting from this concentration. The multiuse high-rise building with many amenities is now appearing on the urban scene. In such cases, commercial, office, hotel, residential, recreational, and sometimes parking are included in one building, each function having its own entry and circulation. The first mixed-use building built in the United States is Broadway Plaza in Los Angeles. It accommodates a hotel, retail concourse, and parking in one self-contained megastructure (*Architectural Record,* 1974). In Chicago, several mixed-use developments have been built in the last 25 years, including the John Hancock Center, Water Tower Place, One Magnificent Mile, and 900 North Michigan. The DG Bank Building is a recently completed mixed-use development, consisting of office, retail, and residential space, in Frankfurt, Germany. A more thorough discussion of this project is included in Chapter 7.

Our consciousness of urban life in high-density areas has brought about concerns of lifestyle, security, and conservation of natural resources. These concerns have led to projects where one could take advantage of working and living within a pleasant, active, and secure environment. Multiuse projects have thus become attractive to developers from marketing and economic points of view.

### 2 Planning Considerations for Tall Buildings

An important phase of research in the design of tall buildings is the exploration of the optimum design of building components. Usually these components, such as floor construction, columns, bracing, skin, and mechanical systems, are examined individually rather than as they relate to the entire building system. There is a definite relationship among these components. A change in one component or building system will generally result in changes in many others. For instance, a variation in the floor depth will change building height, and therefore the overall structural, architectural, and mechanical costs of the building.

The vertical stacking of a series of repetitive components, and the need for vertical communication between these units and the ground, will largely dictate the planning procedures for high-rise structures (CTBUH, Group PC, 1981). The design of the core is dictated by vertical transportation of people and services, as well as the location of public spaces. The structural system, in turn, affects the relationship between the core and the repetitive vertical units, as well as the appearance and configuration of the building.

***Basic Planning Considerations.*** Most high-rise buildings can be defined as either slab or point (tower) buildings. The form of the building may be determined either by the functional layout of interior spaces, or by the a priori selection of an exterior influencing layout of interior areas. This is the case for both residential and commercial buildings.

According to Kotela (1972), the utility assets of apartments are largely determined by the layout of buildings and the rules by which they have been designed. Point buildings have the greatest advantage as they have comparatively small floor areas and the apartments surround a central community space. This structure allows the architect the opportunity to design airy, sunlit apartments.

Rubanenko (1973) characterizes point buildings:

> ...by the presence of a central core which bears at the same time a structural function. Corridors of flats radiate from the core. Their number is dictated by planning and economic considerations. On the basis of this lay-out scheme there might be buildings created of various shapes in plan: square, rectangular, triangular, trifoliate, cross-shaped, elliptical, circular, semi-circular, fan-shaped, et cetera.

Slab buildings, on the other hand, tend toward corridor layouts. Rubanenko (1973) classifies slab buildings as a variant of the point type of building, whereas Kotela (1972) finds the corridor solution the least desirable because of extended circulation routes, increased noise levels, and limited opportunities for cross ventilation and sunlight. Generally, slab-type buildings are best suited to midrise apartments and do not require elevators (although they may still be relatively tall buildings in a given context). Kotela finds the same shortcomings in tall nonresidential buildings such as hotels, office buildings, and multiuse structures in regard to circulation, ventilation, and natural light. However, he does point out that for such buildings a corridor layout is most efficient.

In office buildings in the United States and Canada, where the open office landscape provides for the "creation of free spaces where people's activities can be carried out in illuminated and conditioned environment," Rubanenko finds additional planning considerations important. In other countries, office work areas tend to be shallower in depth and, therefore, utilize more daylight and natural ventilation. In these cases, a structural scheme with corridors radiating from the vertical utility core is also widely used (Rubanenko, 1973).

Chapter PC-6 (CTBUH, Group PC, 1981) discusses several aspects of basic planning considerations, such as space allocation, optimum size of typical floors, critical planning dimensions, flexibility of space divisions, security, as well as unique planning conditions.

***Planning Module.*** The space one needs for living varies with the culture and with the economic class. In residential tall building design, the space allowed per person for normal living functions varies considerably among nationalities and cultures. In the United States, for example, the average living area per person is 24 to 28 $m^2$ (258 to 301 $ft^2$). In countries where housing is subsidized by the government, such as Germany, 108 $m^2$ (1114.2 $ft^2$) is allowed for rented apartments (2 to 5 persons) or 21.6 to 54.0 $m^2$ (222.9 to 557.3 $ft^2$) per person. The United Nations recommended minimum for Asiatic countries with low incomes is 6 $m^2$ (65 $ft^2$) per person, and in Hong Kong the public housing estates for squatters raised the area from 2.23 to 3.25 $m^2$ (24 to 35 $ft^2$) in 1970 (CTBUH, Group PC, 1981).

In the United States, with the "office landscape" concept, where an entire floor area is flexibly subdivided by free-standing and movable partitions, tall commercial buildings achieve a high degree of efficiency by means of a concentrated compact floor plan in which the traffic areas have been adjusted to the actual flow of traffic. According to Rubanenko (1973): "On an average there is 73% utilization area (net area) to every 27% traffic area....the highly efficient working area utilization factor of 10.23 $m^2$ (105.57 $ft^2$) per employee is to be ascribed to the unusual form of the floor plan."

Analysis of many buildings constructed since World War II indicates that a planning module of 1.5 by 1.5 m (5 by 5 ft) nominal, or 1.4 to 1.7 m (4.5 to 5.5 ft), is an appropriate module for commercial, office, and residential functions. A planning module allows for a reasonable variety of office or room widths at the building's perimeter, starting with a minimum of two module spaces of 3 m (10 ft) and ranging upward to four or five module spaces of 6 to 8 m (20 to 25 ft).

There are smaller modules such as 0.60 or 0.90 m (2 or 3 ft) that will approximate the same width for a minimum office or room and give a somewhat noticeable variety of space widths as they are combined to create larger perimeter exposure. However, the additional mullions used in the exterior walls for connecting partitions render these modules uneconomical.

The 1.5-m (5-ft) modular ceiling system can economically accommodate light fixtures, diffusers, sprinklers, and such. In many speculative commercial buildings, modular ceiling systems may become economically unfeasible due to the disparity of space use requirements. To arrive at an economical solution in such cases, an acceptable pattern of the ceiling grid, light fixtures, and diffusers can usually be reached during development of the tenant spaces.

***Lease Span.*** This is the distance from a fixed interior element, such as the building core, to the exterior window wall. The lease span differs in dimension depending upon the function of the space (commercial, office, hotel, and residential) and is a very important consideration for good interior planning. Determination of acceptable lease spans is governed by office layouts, hotel room standards, and residential code requirements for outside light and air. Usually the depth of the lease span should be between 10 and 14 m (33 and 46 ft) for office functions, except where very large single tenant groups are to be accommodated. Lease spans for hotels and residential units range from 6 to 9 m (20 to 30 ft).

Unlike space allocations which are primarily determined by the building's internal functional requirements, the optimum size of a typical floor can relate to external factors [such as floor area ratio (FAR) as dictated by zoning ordinances], to structural considerations for the overall building, or to planning considerations for the individual units per floor. As Kotela (1972) points out, there is an economic relationship:

> The floor area of a tall building...should not fall below a certain economical minimum. The situation of the tall building and its shape of plan must ensure not only proper light and

> insulation of rooms...as well as suitable strength and stability, but should also comply with the criteria of economic floor area to obtain proper ratio of transport surfaces (staircases, lifts, and corridors) to the usable floor surfaces on each story.

Kotela recommends that, based on research and experience, the optimum gross floor area for offices is 1000 $m^2$ (11,000 $ft^2$) and at least 20 rooms per story for hotels. Mankowski (1972) finds that there are existing minima far below this optimum; for example, tall office buildings whose floor area does not exceed 300 $m^2$ (3200 $ft^2$) and residential buildings with a floor area less than 200 $m^2$ (2200 $ft^2$).

There are no international standards that determine lease span depths. Some countries may have a requirement that all offices must have an outside exposure to light and air. In such cases, office function arrangements may affect the dimension of the lease span. Adherence to this requirement may also affect the aspect ratio of very tall buildings.

The megastructure of the future, the multiuse building, provides the ultimate flexibility of space divisions within a very large structural grid. According to Powys (1971), from the developer's point of view:

> ...flexible design of a floor of a building is related to the number of subdivisions into which the area can be divided for tenancy or other reasons. Corridors are wasteful, whether within the core or around it, and profitability can be greatly enhanced by careful core design so that maximum subdivision is possible but corridors are eliminated or reduced to a minimum.

Rubanenko (1973) points out that flexible floor layout allows quick alteration of the office structure and rearrangements:

> ...The use of various types of demountable partitions, lighting and air conditioning, sun protection and acoustics will give a possibility to rearrange any unit of the plan (multiple of the adopted planning modules) into separate spaces or part of a bigger hall which sometimes can cover almost the whole space of the floor, such as in the case of an office landscape.

In residential tall buildings, flexibility can serve two purposes. According to Beedle (1974), within individual apartments it can allow residents to create a sense of ownership and pride, and, according to MacMillan and Metzstein (1974), within a building it can allow a structure to adapt over a period of years to changing circumstances.

Tall, slender buildings have small floor plates and shallow lease spans. These buildings require extra effort in their structural design and marketing and leasing programs geared to small, prestigious tenants (for example, see Fig. 2.1).

***Floor-to-Floor Heights.*** The overall economics of a building are impacted by the floor-to-floor height. A small difference in floor-to-floor height, when multiplied by the number of floors and the area of the perimeter enclosure of the building, can have a great effect on the exterior as well as on the structure. This relatively small difference can change the area of the exterior enclosure and, therefore, impact the structure due to the weight of the exterior wall and wind loads. It also has an effect on the mechanical system due to heat losses and gains.

The floor-to-floor height of a building is a function of the required ceiling height, the depth of the structural floor system, and the depth of the space required for mechanical distribution.

Floor-to-floor height determines the overall height of the building. The height of the building determines structure, round-trip times of elevators, and quantities of all verti-

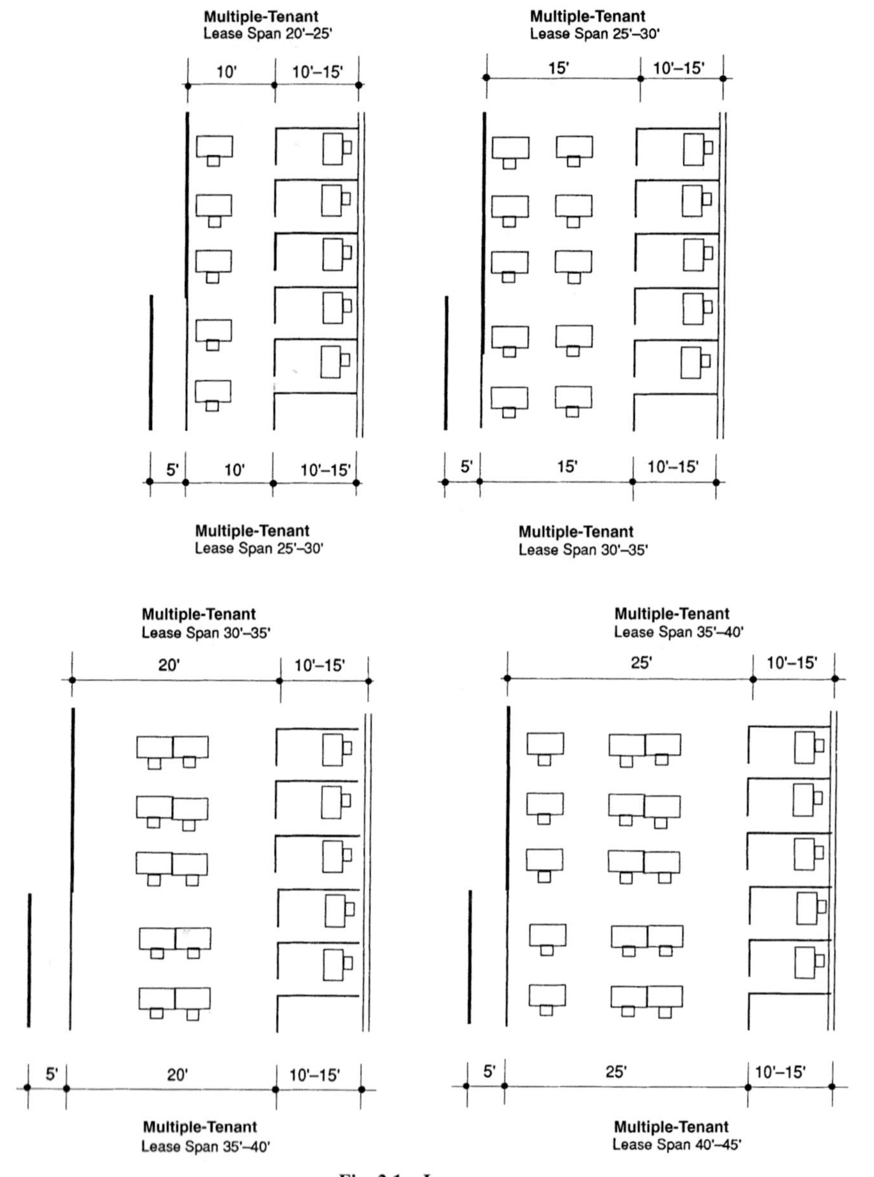


**Fig. 2.1 Lease spans.**

cal elements in the building such as the exterior wall, interior partitions, shaft enclosures, and HVAC and plumbing, electrical, and communication risers (Fig. 2.2).

***Ceiling Heights.*** Commercial functions require a variety of ceiling heights ranging between 2.7 and 3.7 m (9 and 12 ft). Ceiling heights in office buildings range from approximately 2.5 to 2.7 m (8.5 to 9 ft). Residential and hotel functions require ceiling heights of 2.4 to 2.7 m (8 to 9 ft).

***Depth of Structural Floor System.*** The depth of the structural floor system varies widely depending on floor load requirements, size of structural bay, and type of floor framing system. An allowance should be made for dead load deflection, and in the case of steel systems, an allowance should also be made for fire-proofing.

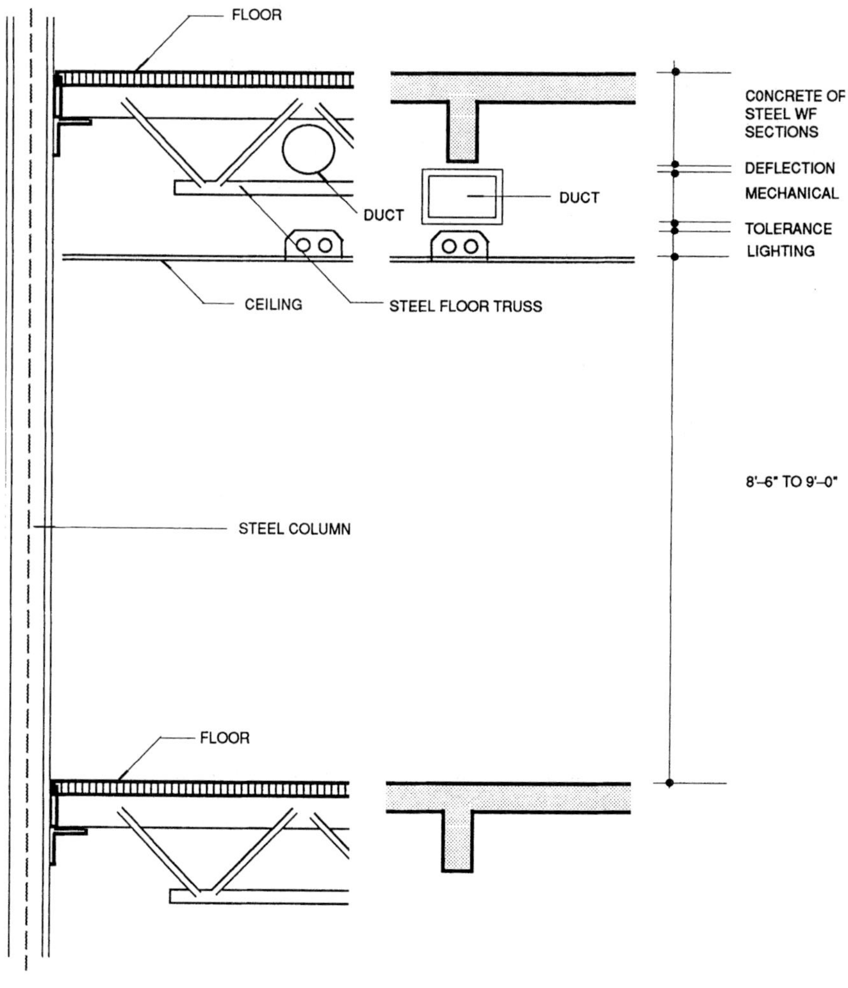


**Fig. 2.2 Floor-to-floor heights.**

In steel systems, increasing the structural depth will, up to a point, result in decreased weights of rolled sections, and with trusses there is the possibility of even more weight savings. Since they permit the passage of ducts, trusses provide structural depth without a proportional increase in floor-to-floor height.

When using a steel system, an allowance should be made for the slab thickness, and the girder depth may be equal to or greater than the beam depth. For further information on structural systems, refer to *Structural Design of Tall Steel Buildings* (CTBUH, Group SB, 1979), and *Structural Design of Tall Concrete and Masonry Buildings* (CTBUH, Group CB, 1978). Also see Chapters 10, 11, and 12.

As tall buildings tend more and more toward the megastructure concept of multiple use, the critical planning dimensions will probably come from the interaction of space needs and structural framing (Schueller, 1990). Khan (1974) describes what the future situation could be:

> Take a 30-story building; say there are floors every 3 m (12 ft) on centers and columns every 6 m (20 ft). But in a megastructure there might be columns every 30 m (100 ft) on center and a main support system every 30 m (100 ft) vertically. In between you'd fill in with building of smaller spans.
>
> The advantage here, the idea for the future, is that in one area, the spans could be 12 m (40 ft), in another area 6 m (20 ft), and in still another 18 m (60 ft), depending on the functions of the areas.
>
> The whole idea of a megastructure is to be able to divide a building into different physical functions. In the lower 20 stories, there could be parking, with spans of 18 m (60 ft). The next 40 stories could be for offices, which require a span of something like 12 m (40 ft). Above the offices could be apartment floors, which need to be only 6 m (12 ft) on centers.
>
> So what you do is make a megastructure defined by broad outlines—120, 130 or 150 stories—and then fill in spaces of 10, 20 or 30 stories of different types of structural systems, as needed to make the building economical, and to provide for various functions.

***Elevators.*** Vertical transportation in any tall building is totally dependent upon its elevator system. The selection of a system is a very critical issue in tall building design. The development of high-rise buildings has had a direct relationship with the development of elevator system design by the elevator industry. Designing an elevator system involves interpretation of program criteria, calculations of population, and areas to be served. This becomes even more complicated in mixed-use projects.

For preliminary planning, a rule of thumb for estimating the number of elevators needed is one elevator per 4645 $m^2$ (50,000 $ft^2$) of gross area. In calculating the actual number of elevators, consideration is given to the general population density of the building and the handling capacity of the system at peak periods, which in turn determines the waiting interval, elevator size, and speed.

Population density is the amount of net usable area dedicated to each occupant. Net usable area is the space devoted to occupancy, and it differs from net rentable area. Net usable area varies from elevator zone to elevator zone and from floor to floor, and should average from 80 to 85% over the entire building. Population density estimates usually range from approximately 13 to 15 $m^2$ (140 to 160 $ft^2$) per occupant.

Handling capacity is the percentage of the total building population that an elevator system can handle in one direction in a time period of 5 minutes. The desirable capacity is contingent upon several factors: building location, proximity to mass transit systems, and type of building tenants.

The interval is the average waiting time at the ground-floor lobby at the peak of upward traffic. The acceptable interval varies with the type and location of the building.

The interval is determined directly from round-trip time and is inversely proportional to the number of elevators in a group or bank.

When calculating the number of elevators needed, the population density of the building and the handling capacity of the system at peak periods must be taken into consideration. This in turn determines the interval, elevator size, and elevator speed.

As the floor plate increases in area, so does the height of the building, and the elevator count increases also. Capacity can be increased by reducing the number of stops for each elevator. Elevator systems may be designed so that some of the elevators are designated to serve the lower floors, others to serve the midrange, and still others to serve the upper floors. This zoning is largely determined by the usability of the space that becomes available on the upper floors after the low-rise elevator drops off. Separate lobbies for each zone are generally used. The high-rise zone is usually in the center so that the increased rental area that becomes available on the upper floors is within a reasonable distance from the perimeter to allow flexibility in multitenant development.

The sky-lobby concept utilizes high-speed express shuttle cars to transport passengers from ground level to a lobby higher up in the building for transfer to local elevator zones. This concept creates two or more buildings vertically connected, each having essentially its own independent local elevator system served from the ground level by express shuttle elevators. The sky-lobby approach reduces the area dedicated to elevator shafts and lobbies on the lower floors of the building. The sky lobby can also serve as the starting point of a different function in multiuse buildings (Fig. 2.3). Additional information on elevator systems is presented in Chapter SC-4 of *Tall Building Systems and Concepts* (CTBUH, Group SC, 1980). The psychological aspects of elevators are discussed in Chapter 6.

***Core Planning.*** A typical floor in a high-rise building consists of perimeter, interior, and core zones. The perimeter zone is defined as an area approximately three planning modules deep from the window wall, with access through the interior zone. The interior zone is the area between the perimeter and the public corridor. The core zone consists of those areas between elevator banks which become rentable on floors at which elevators do not stop.

Building cores can be arranged in several ways. Most typical are the central-core and split-core plans. A central-core plan works best in rectangular buildings when the depth of the building is limited by site or choice, whereas the split-core plan works best in buildings that are relatively square in plan.

Central-core plans provide circulation to stairs, toilets, and tenant spaces by way of a perimeter corridor that encircles the core. The resulting core-zone spaces are rather long and narrow, between 2.7 and 3.4 m (9 and 11 ft). These spaces have limited use since they are separated from the interior zone by public corridors. The split-core plan eliminates the peripheral hallways by combining them into one wider central corridor. This provides the most efficient use of core-zone spaces as they become adjacent to the interior zones and usable for reception, conference, and storage.

The major elements within the core are elevator shafts, mechanical shafts, stairs, and elevator lobbies. Electric, communication, and plumbing risers are accommodated by much less space. Core elements that pass through or serve every floor should be located so that they can rise continuously and thus avoid costly and space-consuming transfers (Fig. 2.4).

Stair entrances should be located as remotely from each other as possible. Mechanical fan rooms should be located where they can be easily changed in area or shape and where they are not surrounded by stairs, shafts, or telephone and electrical closets since these prevent or limit duct distribution from shafts or rooms.

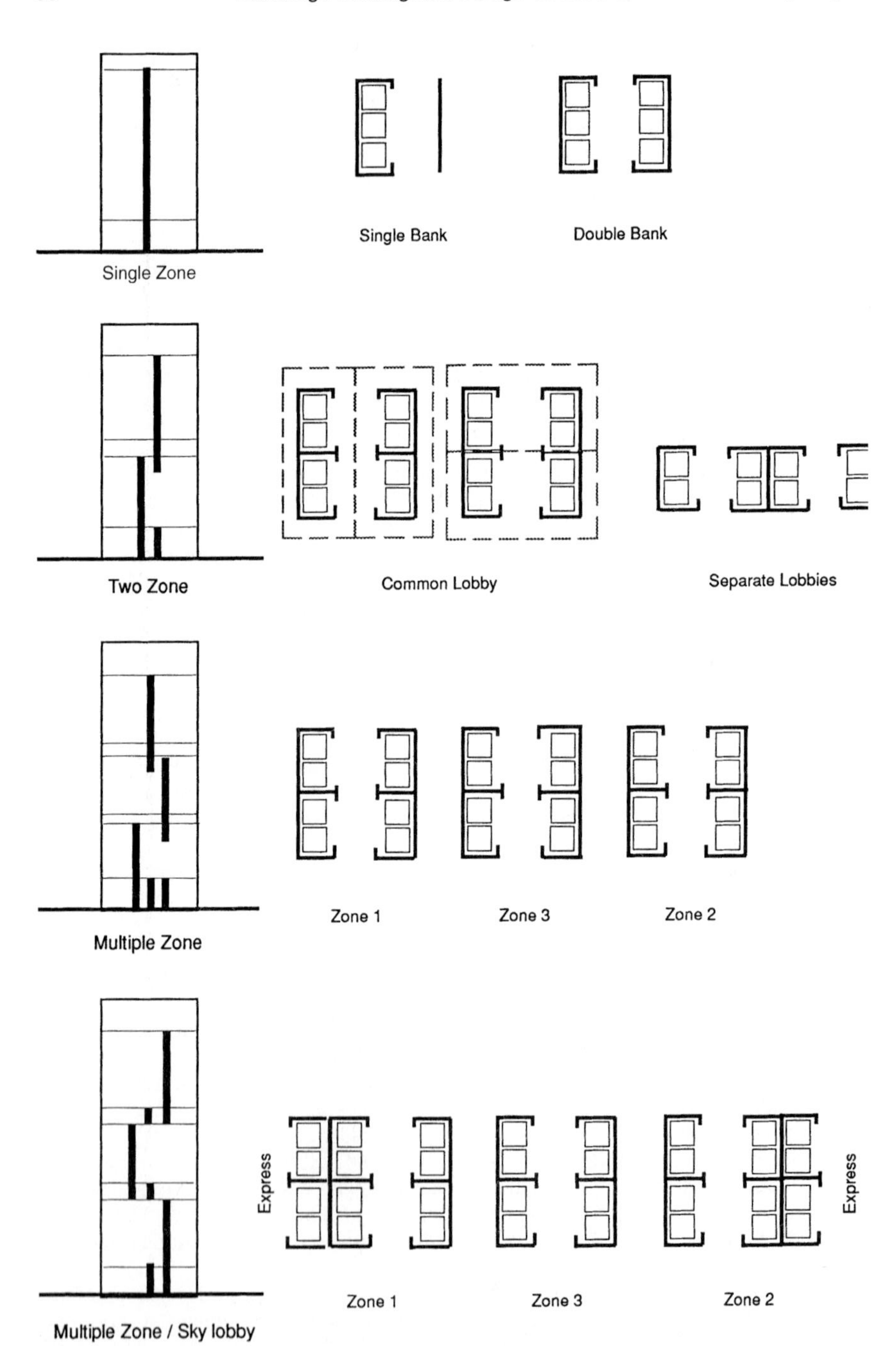


**Fig. 2.3 Elevators for tall buildings.**

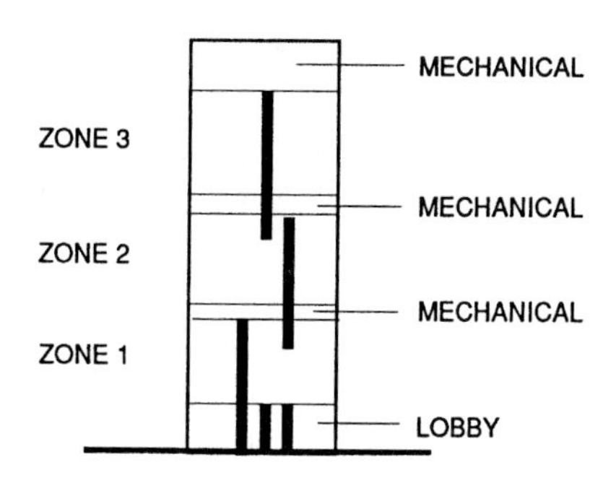


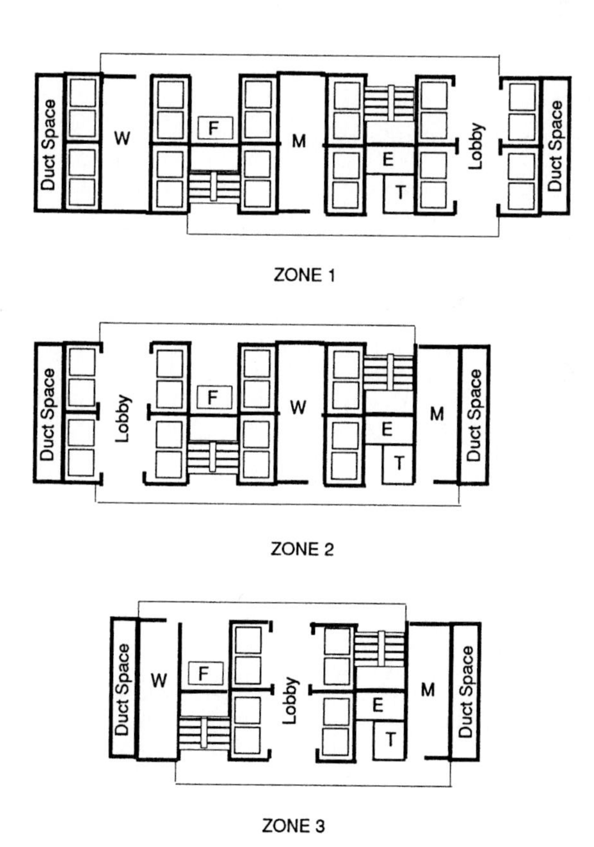


Fig. 2.4 Core planning.

***Parking.*** In many large projects being built today, the requirements of the owner and local zoning codes often necessitate the inclusion of parking facilities for use by tenants. This requirement differs for office, residential, or commercial functions. The inclusion of such a parking facility within the footprint of the tower has a very great impact upon the plan and the structure and should be avoided if at all possible. The presence of the tower columns and shear walls, elevator shafts, stairways, duct shafts, and other risers tend to make the parking inefficient. When parking within the footprint of the tower is unavoidable, a real effort should be made to find the structural bay that is efficient for both parking and functional space use, and to locate the core elements so as to minimize interference with car parking and circulation.

If the parking facility is not above grade with at least 50% of its exterior area open to the outdoors, it will be necessary to provide mechanical ventilation. Supply and exhaust shafts should be located on opposite sides of the facility to promote cross ventilation. Intake and exhaust locations at the exterior must be positioned high enough above grade to avoid nuisance to pedestrians and intake of exhaust fumes from the street. These openings will necessarily be placed 1 to 3 m (3 to 10 ft) above grade. Chapter 6 discusses the psychological aspects of parking structures in relation to the user, and additional information on parking for tall buildings is presented in Chapter PC-10 (CTBUH, Group PC, 1981).

## 3 Mechanical and Electrical Systems

The mechanical systems for buildings consist of heating, ventilating, and air-conditioning systems (HVAC), domestic water systems, waste water systems, and fire protection systems. For further information on environmental systems see Chapter 9 and Chapter SC-2 (CTBUH, Group SC, 1980).

HVAC systems are designed to ventilate spaces and to maintain within the spaces a certain range of temperature and humidity conditions. The systems consist of several basic components: refrigeration, heating, humidification, and air distribution.

The function of the refrigeration component of HVAC systems is to transfer heat from interior spaces to the exterior. The most economical and practical system for large building projects is the five-step heat transfer system (air to water to refrigerant to water to air). Almost without exception, refrigeration systems for large projects are central systems. Cooling towers are located where there is an abundance of fresh air. Usually this means that they will be located at the roof level or uppermost mechanical level of the building.

Heating systems differ from refrigeration systems in that they can either be a central type or an individual unit system. Individual unit systems consist of electric radiation at a building perimeter or electric coils in the air-mixing boxes. Central systems require a boiler, producing hot water or steam, fired by either gas, oil, or electricity. Since all boiler types, with exception of the electric one, require flues, it is usually economical to locate the heating plant at the top of the building.

Air-handling systems can be of either the central or the decentralized type. When using either type, an outside air source must be furnished. In the case where fan rooms are located on each floor, the outside air can be supplied from an interior shaft; however, it is preferable to locate the fan rooms against an outside wall. Fan rooms for central systems should be located through the height of the building so that the duct risers do not exceed 10 to 12 floors in height.

Regardless of whether the air-handling system is a central or a decentralized type for air distribution, a typical floor is divided into two basic areas: the exterior zone, which is a 4.6-m (15-ft)-wide area around the perimeter of the building, and the interior zone, which consists of all the floor area. Since the exterior zone is adjacent to the exterior wall, it is strongly affected by outdoor temperature and solar exposure. The interior zone is not subject to the widely fluctuating outdoor conditions. Further, there are different influences on the interior and exterior zones, and hence different methods are used to condition them. The types of air distribution systems currently used for the exterior zone include a fan-coil unit system, a single-duct system with terminal reheat, a dual-duct system, and a constant-volume system. Air distribution systems that are used for the interior zone include constant-volume systems, variable-volume systems, single-duct with terminal reheat systems, and dual-duct systems.

Plumbing systems for buildings consist of domestic water and waste water systems. The domestic water system consists of an incoming service, water meters, and a vertical and horizontal distribution system. Where tempered water is required, water heaters must be provided. In buildings located where the street pressure is inadequate to provide water service to the upper floors, water pumps must be provided. In very tall buildings or in buildings where there is an unreliable source of water, it may be necessary to provide domestic water tanks.

Waste and storm water systems collect water from the building plumbing fixtures and roof and plaza drains, respectively. They carry it to either combined or separate waste and sanitary sewer systems. These are gravity systems, except where the areas served are below the level of the main sewer. In this case, ejector pumps must be provided to lift storm water and sewage to the municipal sewage system.

The purpose of the fire protection system is to provide building users either an automatic or a manual alarm and fire fighting capability. Manual systems include pull box alarms, hand-held extinguishers, and water risers with both fire hose connections and fire hose cabinets for use by building personnel and the fire department. Standpipes and fire hose cabinets are always located within a fire stair enclosure. Automatic systems consist of smoke and fire detectors connected to an alarm system and automatic sprinkler systems. The fire protection system begins at the incoming service point of the domestic water system. Fire pumps are provided as part of the fire protection system where the height of the building cannot receive adequate water pressure. Siamese connections, located at grade level at various points around the building, are provided so that the fire department can use their pumps to deliver water into the system when necessary. See also Chapter 9 and Chapter CL-4 of *Tall Building Criteria and Loading* (CTBUH, Group CL, 1980).

Electrical systems for buildings consist of high- and low-voltage systems. The high-voltage systems provide power for equipment, lighting, and appliances for functional spaces in the building. The low-voltage systems include the telephone system, the public address system, and most of the control systems. Generally, an electric utility company provides high voltage (13,000 V) to a building site. The utility company provides a series of electrical transformers that transform this power to 480 V for vertical distribution in the building. The main distribution switch-gear room should be located adjacent to the transformer vault. It is from there that the power is distributed to the various load centers: mechanical equipment, elevator equipment room, lighting, and receptacles in the building. Vertical conduit or bus ducts distribute electric power to transformers and panels located in electrical closets on each floor for transformation and distribution for lighting and receptacles.

The telephone system is the primary low-voltage system. Telephone lines enter the building at the main telephone terminal room, and from there they are distributed to telephone closets located on each floor and finally to the equipment of the tenant.

***Determination of Structural Systems.*** There are several major advantages of high-rise mixed-use developments, such as conservation of material and energy, optimization of land use in urban areas, and the creation of a 24-hour active environment. They also serve as a catalyst for adjacent development as well as a major force in revitalizing urban areas that are going through the process of aging and becoming inefficient.

Parking, office, commercial, hotel, and residential functions all have their specific space requirements, which, over the years, have developed into specific identifiable building types that are efficient and economical for specific functions.

Constructing a single-function high-rise building without paying a large premium for height would require a building program of sufficient area and a structural system suitable to carry gravity and lateral loading. This is not very difficult if the function could be placed on a large, deep floor plate. In the case of residential structures, where the depth of the lease span is critical, it is more difficult to construct extremely tall buildings due to lateral loading. The aspect ratio is an important factor in the design of a tall building. A structure becomes efficient when the aspect ratio is equal to 1, that is, the plan is, for example, a perfect square, circle, or right hexagon. Similarly, the slenderness ratio is another important parameter for tall building design. For tall slender buildings with a slenderness ratio exceeding 8, an innovative and highly efficient structural system is warranted to provide the necessary stiffness and mass needed to minimize lateral movements and building vibrations.

To combine mixed functions into a single building, a structural system that can accommodate the space requirements for each function is necessary. The location of each function in the buildings is critical. Functions may be stacked above one another, as is typically done, or they can be placed side by side (see Section 2.3).

From a structural point of view, when considering stacked functions, smaller structural bays should be located in the lower portion of the building and larger structural bays in the upper portion so that intermediate columns can be dropped off along the height of the building. In practice, however, office, commercial, and parking functions, with their larger bays, are placed in the lower portion of the building. This is done primarily to promote easier access to the heavily used floors. Structurally this type of arrangement requires the use of large girders to transfer the column loads from the smaller structural bays of the upper floors to the larger structural bays of the lower floors. Consequently, the strategy to resolve the conflict between a desirable structure and its functional requirements is the key to high efficiency of the high-rise mixed-use building. When functions are located side by side, each can be placed in a tube of a size that satisfies functional requirements. An advantage of this system is that there are no vertical structural transitions, thereby eliminating the need for transfer mechanisms. The common structure between the adjacent tubes becomes an interior frame with the continuity of the exterior system allowing the spandrel to have the same depth and stiffness. In addition, each tube can be of any shape suitable for the architectural purpose.

To resolve the discrepancy between the structural requirements and functional needs, the architect and the structural engineer should evaluate the importance of each function in terms of its relative floor area and its contribution to the total cost of the project. It is then possible to find a structural system that can be used for the combined functions. In the next section, structural systems will be discussed individually.

## 2.2 ARCHITECTURAL-STRUCTURAL INTEGRATION

There is a historical relationship between the architecture and the structure of tall buildings. The nineteenth century represented one of the most technically inventive centuries. As a result, it was expected that innovations in tall building architecture would come from those who envisioned the varied use of new materials and means of construction. The development of cast iron as a building material and the invention of elevators paved the way for tall buildings that were unthinkable only a few years earlier.

Structural systems were developed to correspond to the various architectural trends described in Chapter 1. The tall buildings prior to the invention of fluorescent lights and air-conditioning were of limited width and could accommodate interior columns at a rather close spacing of 6 to 7.6 m (20 to 25 ft). Deep girders were rigidly moment-connected to the columns to provide lateral stiffness and stability (Taranath, 1988). The rigid frame action was further augmented by adding cross bracing for lateral resistance, particularly in the utility core areas. For very tall buildings this was inadequate, and as such, the buildings' structural systems were frequently overdesigned.

As a result of architectural response to the machine age, the international style became very popular, in which simplicity meant elegance derived from "pure" forms. Structures designed in this era incorporated three distinct features of the new style (Taranath, 1988):

> (1) a new vocabulary of forms borrowed largely from abstract art, consisting of planes, lines, and rectangles without ornaments or moldings; (2) the representation of interior space and exterior facade as a cohesive unit; and (3) the use of new structural materials such as steel and concrete.

This style also eliminated heavy partitions and cladding from the building. Further, column-free spaces in the interior became desirable for meeting leasing requirements. This greatly influenced the planning of buildings. It was also found that a natural structural response to the economic efficiency of tall buildings was to move the bracing away from the interior core to the building's perimeter. Such a notion eventually led to a new system called the tube system. This system was remarkably well suited to the international style. Building forms were prismatic and the architectural design tried to accommodate the concept of tube structures.

As has been noted in Chapter 1 and is discussed in Chapter 5, the current architectural movement known as post-modernism, however, tipped the balance in favor of arbitrary forms (Ali, 1990). Such new design concepts have resulted in innumerable possibilities of building form. The issue of architectural-structural integration has not, however, been lost. It is only going through a critical phase. Collaboration of architects and structural engineers is nevertheless an essential requirement for the architectural-structural integration of tall building design, regardless of the styles and fashions of various periods (see also Chapter 4).

### 1 Definition of Structural Systems

As far as the structural behavior is concerned, a tall building can be clearly defined as one that is influenced by wind loads. It is understood that a building, as any other object on the face of the earth, is subjected to gravitational forces. Their magnitude depends on the weight of the structure and any superimposed vertical loads. When the structure of a building is well proportioned, the gravity loads introduce compression forces on the main vertical members that carry the load to the ground, thereby creating a structure that is well balanced and stable.

For short and stocky buildings, such vertical forces are of high enough magnitude to dominate the design and strength of the structure, and the impact of wind loads can be neglected. However, when the building grows taller and narrower, the wind load will induce significant lateral loads, which will tend to overturn the building and will impact its behavior and design.

Supertall or very slender buildings are highly susceptible to the dynamic impact of wind loads, and as such special consideration is required in the development of their structural system. The efficiency of the lateral load-resisting system becomes of great importance in the selection of the total structure of a tall building. Such a system shall have to be designed in accordance with the architectural, mechanical, and programmatic requirements of the tall building. The supertall building was also discussed in Chapter 1.

To be accepted within the architectural building program and form, a structural system must be a synthesis of the design considerations combined with analysis and the actual performance of such a system (Ali, 1990). Structural systems for buildings, regardless of height, can be classified into two categories:

1. *Gravity system:* Consists of the floor framing required to carry all floor gravity loads and columns or walls that transfer that load to the foundation.
2. *Lateral system:* Consists of the vertical framing system or wall system that is required to resist lateral loads.

The floor framing system, to a great extent, is independent of the height and is governed by:

1. Column spacing for desired clear floor spans
2. Headroom requirement (mechanical sandwich)
3. Choice of structural material

The choice of the appropriate structural framing system of the building skeleton is dependent upon a combination of many factors, not all of which are necessarily structural. The most important factors are:

1. Design lateral loads (mainly wind and/or earthquake)
2. Strength design criteria
3. Serviceability criteria (drift ratio, perception of sway motion, cracking)
4. Height-to-width ratio (aspect ratio for the building and for the structural system)
5. Type of occupancy (single use versus multiuse)
6. Soil conditions and related foundation systems
7. Local fire rating considerations and other related code requirements
8. Choice of materials and availability and cost of such materials
9. Methods of construction and their economic impact on the choice of materials and time schedule
10. Cost of land and cost of financing (speculative development versus corporate ownership)
11. Building form (prismatic, tapered, stepped circular, open top)
12. Architectural design and its integration, or lack of it, with structure

In an earthquake zone, in addition to the foregoing, there are the following design constraints:

1. Compliance with the contemporary design philosophy:
   *a.* Resist minor earthquake without any damage
   *b.* Resist moderate earthquake without significant structural damage
   *c.* Resist major earthquake without collapse, and without loss of life and property
2. Maintenance considerations during postearthquake periods

It is apparent from the above-mentioned factors that there cannot be a standard structural system valid for all locations and that would fit all buildings. Every building is unique. And now, more than before, with the new trends in architecture, which emphasize greater excitement and freedom in form, a greater variety of structural systems is being developed (Khan, 1973b). The focus, however, from a structural point of view, is in the structural behavior and the performance of such buildings under the design loads specified (Taranath, 1988).

The objective from the structural point of view is always the same: to achieve efficiency, cost-effectiveness, simplicity of detailing, and simplicity of construction (Ali, 1990). That requires a clear understanding of the behavior of the structure and the most important parameters affecting it. Structural behavior still is and always will remain dependent on the laws of structural mechanics and on the engineering properties of the material, as well as on the structure's damping characteristics, the nature and magnitude of the applied loads, and the soil structure interaction (Youssef, 1991). However, the selection of the system is to a great extent dependent on the sensitivity of the engineer to the architecture of the building. See also Chapters 4 and 5 for further discussion.

The structural system should be able to:

1. Provide the strength to carry and resist all loads applied
2. Provide the lateral stiffness to control drift due to wind and earthquake loads
3. Provide the ductility and energy-absorbing or dissipating capability to withstand earthquake loads
4. Provide appropriate dynamic characteristics or supplement with a damping system to limit motion to acceptable levels
5. Provide the overall stability against overturning or against any load condition that renders the structure or some parts of it unstable
6. Satisfy all serviceability requirements dictated by the design criteria

The need to develop an efficient structural system is not just an academic exercise. It is by and large an economic issue as well as a safety issue. The safety issue is the most critical one in major or severe earthquake zones (Taranath, 1988).

If the buildings in earthquake zones are excluded, the other most important lateral load is wind load. The taller the building, the more wind load it must be able to sustain. A structural lateral system could be thought of as a free-standing cantilever, or a group of cantilevers, resisting the overturning moments and the shear forces at every floor level throughout the structure's height. Low-rise structures in general do not require any particularly sophisticated lateral system, except under special circumstances.

For medium-rise structures, the regular floor framing may provide an adequate lateral stiffness. Sometimes partial bracing, core bracing, or a combination of shear wall and frame systems might work as well. In fact, for most low-rise buildings and some medium-rise buildings the choice of one structural system over another would not generally have any significant economic bearing. It is primarily made for architectural and aesthetic reasons in conjunction with secondary considerations for the building cost.

For high-rise structures, a well-developed and efficient structural system is needed. An efficient system could be seen as the one that maximizes the moment of inertia and provides enough shear resistance (Fischer, 1962). Hence it is clear that when the structural system is spread out, one can achieve a higher efficiency, provided that the shear deformations are minimized.

At this point it is appropriate to introduce the notion of premium due to wind. The structural system must provide support for all gravity loads. With height, wind loads increase and the structural system must provide lateral stiffness to control drift and provide adequate strength. Hence there will always be a structural premium due to wind. The efficiency of the lateral system is judged mainly by the minimization of this premium.

For supertall buildings, very slender buildings, or tall buildings subjected to high turbulent winds, there is also the issue of perceptible motion due to wind (Taranath, 1988). The problem is usually due to the building oscillations perpendicular to the wind direction, caused by a phenomenon known as vortex shedding. Some solutions to this problem can be provided by tapering the structure or by disrupting the formation of vortices by stepping (the building mass) or lacing (the building structure) or by adding mechanical damping in the form of a tuned-mass damper or viscoelastic damping system.

Structural systems can be viewed by the choice of material:

1. *Masonry:* With special reinforcement, it is used for low-rise and medium-rise buildings, mainly housing. The increase in the thickness of the masonry wall with height and its inflexibility tend to interfere with the planning of the interior space and somewhat limit its use for tall buildings, although this problem is now being gradually overcome with the use of high-strength masonry and steel reinforcement as well as new design approaches (see Chapter 12).
2. *Steel:* Because of its high strength, ease of erection, and lightness it represents the state-of-the-art structural system for supertall buildings. Principal drawbacks are that fabrication and welding are expensive, required fire-proofing is costly, and standard shapes are limited in number (see Chapter 10).
3. *Reinforced concrete:* It is relatively cheap, rigid, moldable, and particularly suited to resist lateral loads. With higher strength, it would represent the future material for tall buildings if the construction technology can meet the challenge (see Chapter 11).
4. *Mixed or composite construction:* It combines the merits of both steel and concrete.
5. *Prestressed concrete:*
   *a. Precast:* Used mainly in low-rise and medium-rise buildings, in particular for parking structures.
   *b. Posttensioned:* Used mainly for floor framing as flat slab or joint system, mainly for apartments and offices.

The following points are also worthy of note:

1. Structural cost increases with height.
2. For low-rise buildings gravity systems generally govern the design.
3. For medium-rise buildings the lateral load-resisting system chosen often could be incorporated into the architectural expression. A well-planned building will yield a simple and economical structural system.
4. Every building is usually unique, and the need for novel structural systems is always there.
5. For tall buildings, structure-architecture interaction in the design development phase is very important. In expressing the structure, special care and thought must be given to realize a structurally sound system, that is, to develop an "honest" structure. For further information on structural systems, see Chapters 10 and 11 and Chapter SC-1 (CTBUH, Group SC, 1980). Also, see *Structural Systems for Tall Buildings* (CTBUH, Committee 3, 1994).

## 2 Structural Systems and Architectural Form

The issue of interaction among structure, aesthetics, and architecture is a complex one. When structure is expressed, it is expected to look elegant, appealing, and, above all, structurally correct. Harmony between architectural form and structure is key to the success of the expression (Ali, 1990). The structure has to blend in without overpowering the architectural form; ideally, the structure should visually clarify and enrich the form. Refer to Chapters 4 and 5 for further discussion of these issues.

Structure and aesthetics are also related through efficiency, lightness, elegance, and the concept of minimum weight or least material, among other factors. To a great extent, when the structure is well-proportioned, it will look right.

Expression of the structure does not follow any rigid rules, and, in many cases, different elements of the structure are shown in different ways. For example, one could choose to express all of the spandrel beams by one typical depth regardless of the actual required structural depth of these beams. For framed-tube structures, in transferring loads at plaza levels from the closely spaced columns to widely spread columns, it is possible to choose a deep transfer girder or to let the loads flow through an archlike action.

The corner column is an important element in tube structures, and special care is taken in the way it is expressed, particularly in its relationship with the other columns, spandrel beams, and diagonals. Diagonally braced structures face many problems that have to be worked out carefully. For instance, where the diagonal is terminated at the base, an appropriate pattern must be considered. The manner in which diagonals and other members cross each other and how the tie member is placed as a distinct element from the rest of the spandrel beams, must be given due consideration.

Reinforced concrete structures have their own character, which can be preserved and even strongly emphasized. In infilled-braced frames, the issue is how to handle the infill panel; the ties are but a few of the problems that must be worked out. In some cases the structure is not explicitly expressed, and the architectural form is such that it is necessary to suggest the structural system in a very subtle way. Two projects fall into this category, the Miglin-Beitler Tower in Chicago and a theoretical supertall building discussed later in this chapter (Lim, 1992).

## *3 Developments in Structural Systems*

In 1885 structural engineer William Le Baron Jenney designed and built the first modern high-rise building utilizing structural steel framing. The 10-story Home Insurance Tower in Chicago established the direction for the modern framing system used today. It followed the first Leiter Building of 1879 that started the beginning of skeletal forms with a glass and structural facade (see Chapter 5). The ability of structures to carry significant gravity loads and resist high wind loads subsequently made it possible to design and build taller buildings. In fact, the development of metal frame construction and modern elevator systems removed almost all limitations of height from tall building construction. For further discussion on this topic, see Chapters 4, 5, and 7, as well as Chapter PC-2 (CTBUH, Group PC, 1981).

The Woolworth Building (1913) in New York was the first to reach a height of approximately 60 stories, soaring to 242 m (792 ft). Designed in the Gothic style, the structure was a rigid frame combined with bracing. After a temporary slowdown during World War I, quite a number of tall buildings were built in New York, most notably the 77-story Chrysler Building, which reached a height of 319 m (1046 ft). This was followed in 1930 by the Empire State Building, the total height of which, not including the television antenna that was added later, is 381 m (1250 ft).

All of these building structures were basically moment-resisting frames in combination with bracing located around the core area and at other required locations. At that time this was the only structural scheme available, and the concept of a structural system was not yet recognized.

In the late 1950s it was realized that a structure placed at the perimeter of the building was an efficient means of resisting lateral loads. That realization was based on a simple mathematical model of the efficiency of a hollow tube. A framed tube was a logical and practical solution for building construction. The tube action could be created by closely spaced columns, typically in the range of 3 to 4.5 m (10 to 15 ft), in conjunction with deep spandrel beams, creating a hollow tube perforated for window openings. The system came to being in the late 1950s, and few engineers can claim that they applied it first. The first well-documented use of this system was by F. R. Khan in 1963 in the reinforced concrete 43-story DeWitt Chestnut Apartment Tower, located in Chicago. During that time this system was very popular and was used in many buildings in Chicago, New York, and other cities. The most notable buildings are the John Hancock Center and the Sears Tower in Chicago and the twin towers of the 110-story World Trade Center in New York. Another good example is the Amoco Building in Chicago. The tube system is still in frequent use today.

Between 1964 and 1974 many modern tall buildings were built utilizing novel structural systems for the first time. Interaction between academic research at institutions such as IIT and design practice at the office of SOM and others was very important in this regard. In the case of IIT, the interaction was implicit and based on the fact that most of the staff at IIT were practicing architects and engineers with SOM and other architectural firms in Chicago.

The John Hancock Center in Chicago is probably the most well-known diagonally braced steel-tube tower. The tapered architectural form indicates clearly the multiuse nature of the building and expresses the structure in a beautiful and elegant way. See Chapters 4, 5, and 7 for more information on the John Hancock Center. The use of diagonals and widely spaced columns produces a structure that is very efficient, since the diagonals resist wind shear with minimal deformation and also carry a major portion of the gravity loads. This structural system is extremely efficient, especially when it is well proportioned and utilized effectively for all combinations of loads. The efficiency of a structural system can be measured by the amount of structural steel required. In the case of the John Hancock Center, only 13.47 kg (29.7 lb) of steel per square foot of floor area was used. At that time such steel quantity was equivalent to that required for a 35-story traditional frame building. The optimization of material was clearly marked by the awareness of the necessity to make the structural system efficient (Khan, 1967, 1982).

Although the idea of a bundled tube was understood for some time, the Sears Tower is the most spectacular structure where the concept was applied. Sears stands alone for many reasons. First and foremost is the fact that it is the tallest building in the world at the time of this writing, and it posed many unique problems and concerns for the structural engineer. Second, Sears is a supertall building that brought important analysis and design issues to the forefront, issues such as acceptable drift criteria, the dynamic impact of wind on structure, and wind pressure on the window wall, particularly glass windows. Third, novel construction methods were developed to make it feasible.

The construction of the Sears Tower raised the interest in megastructures to a much higher level than before, because it became apparent that it was structurally and economically feasible to build supertall buildings. The relative cost of the structure to the overall cost of building became an important issue, and the search for more efficient structures continued. As a result, new architectural and structural concepts were developed. The concept of the superframe came into existence. When used for very tall buildings, a superframe resists all of the gravity loads and the wind loads at the same time. However, the gravity loads are not directly supported by the frame. They are instead transferred through floor trusses located at every 12 stories or so. This mechanism of transfer is not necessarily elegant or efficient.

Although structural steel was the main material for most tall buildings built in the 1960s and 1970s, reinforced concrete was making headway, especially for residential buildings (see Chapter 11). It satisfies the desire for a flat ceiling with minimum floor-to-floor height. For tall multiuse buildings consisting of commercial, office, and residential occupancies, reinforced concrete provides an efficient and monolithic structure. One of the major problems of reinforced concrete is the size of its members. A concrete column could be as much as 1.22 by 1.22 m (4 by 4 ft) square. That poses an architectural problem due to the large size of the column. With the advancement of concrete technology, particularly in the area of high-strength concrete, it is expected that smaller sections will be developed. High-strength concrete is discussed in detail in Chapter 11.

Application of reinforced concrete for high-rise buildings started with a shear wall and a flat-slab system. It is a logical system because of its simplicity, ease of construction, and efficiency up to 30 stories. As buildings start getting taller, a shear wall is coupled with a frame system (Taranath, 1988). The most effective frame is a framed-tube or a tube-in-tube system. The same structural systems that have been successfully utilized for steel structures are also used for reinforced concrete. The One Magnificent Mile building in Chicago adopted the concept of the bundled tube in reinforced concrete, utilizing a different column spacing at each function zone of the building.

The idea of using diagonal bracing in concrete structures was first examined seriously in 1968. The study showed that the diagonal members are efficient in the form of infilled panels. The structural system developed was a diagonally X-braced infilled panel tube structure. This system in concrete is much stiffer than its counterpart in steel because of the added stiffness of the concrete frame (Hodgkison, 1968). This idea was later applied to the Onterie Center in Chicago.

The 1980s witnessed changes in direction in the design of tall buildings. More emphasis was given to the facades and the looks of the building in conjunction with a deliberate attempt to steer away from rational forms that incorporate structure as a determinant of building aesthetics, or attempt to express the structure. Many design trends and fashions were introduced. The 1980s saw an increase in the construction of tall buildings in the United States. The structural engineer worked diligently in collaboration with the architect to develop new and efficient systems appropriate to the unique design criteria of each building.

It became difficult to utilize most of the structural systems that were developed in the 1960s and early 1970s in their purest forms. Many novel ideas that utilized the basic concepts, with some modifications, or combinations in hybrid forms were used. The result was that few exciting and efficient structural systems were developed.

Le Messurier developed a structure for the Bank of the Southwest Tower in Houston, Texas (designed by Helmut Jahn, but was never constructed). The proposed 82-story office tower had a structural system which included four mega vertical steel trusses. The chords of the trusses were the exterior columns and were of composite construction at the lower portion of the building. The system, in reality, consisted of plane trusses that intersected at the core of the building, forming a braced box. It was supplemented with interior vertical trusses to facilitate transfer of gravity loads. The system was very efficient and allowed for an open exterior without corner columns. However, the diagonal members of the trusses through the leasable spaces could be objectionable from a space-efficiency point of view.

A "space structure" was developed for the Bank of China building in Hong Kong (Tuchman, 1989). The system consists of a three-dimensional triangulated frame whose members resist both gravity and wind loads. Secondary vertical trusses were utilized in core areas to collect lateral loadings and inner-region gravity loads in between the main diagonals and then transfer them to the space frame at the main intersection joints (Tuchman, 1989).

During the last few years, due to the demand for small office floor plates, some tall and slender buildings are being designed. In Chicago, a new 606.5-m (1990-ft)-tall office tower is being seriously proposed by the Miglin-Beitler development group. If built, it will substantially exceed the height of the Sears Tower. The tower tapers as it rises to a vanishing point, and from a structural point of view it is very slender. Slenderness combined with height exposes buildings to the dynamic aspects of wind forces. A major task is to develop a structural system that will reduce the impact of these forces. An elegant and effective structure was developed consisting of a reinforced concrete wall coupled with exterior reinforced concrete fins to form a cruciform shape for the resistance of lateral loads (Thornton et al., 1991). The fins extend beyond the faces of the building at the lower portion of the tower, but they taper off with height.

## 4 Classification of Structural Systems

It is perhaps useful to identify some of the structural systems that have been referred to in this chapter. The structural systems will be classified according to their unique forms, and their efficiencies will be discussed. Three major types of systems prevail: structural steel systems, reinforced concrete systems, and composite systems. The word *system,* as referred to in this text, describes the overall lateral load-resisting system only and does not include the floor framing system. Although narrative accounts of these systems can be found amply in the literature (Taranath, 1988; Schueller, 1977, 1990), it is worthwhile to list and summarize them here.

***Structural Steel Systems.*** Figure 2.5 illustrates the range of applicability of a particular system (Khan, 1973a). From the late 1950s to the early 1970s most of the architecture of tall buildings, particularly in Chicago, was in concert with the modern movement. The simple prismatic and rectilinear form of the building made it possible for a single identifiable structural system to be developed to fit in. However, with the change of expression in architectural form, many innovative structural ideas were introduced to meet the new requirements (Iyengar, 1986).

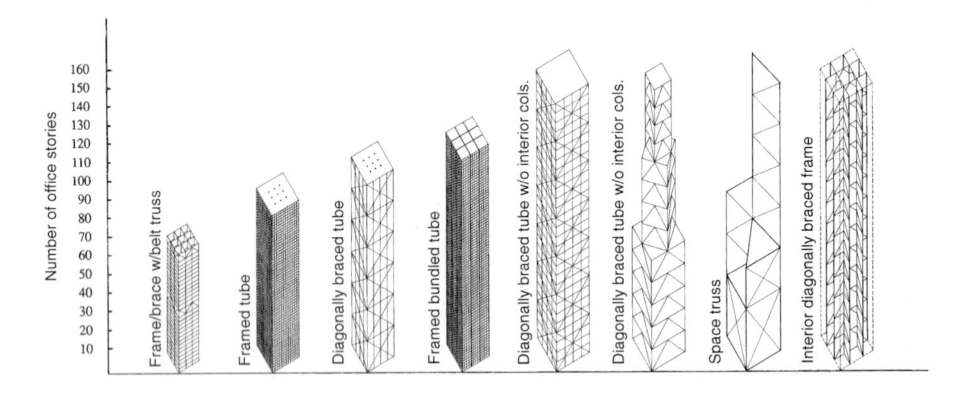


**Fig. 2.5 Heights of structural steel systems.** (*Courtesy: M. Elnimeiri.*)

*Rigid Frame:* Essentially a moment-connected beam-column system, the rigid frame system emerged in the late nineteenth century in Chicago, and its application quickly gained popularity worldwide. However, the system's efficiency, by virtue of its load-carrying mechanism, is limited to relatively midrise buildings.

*Frame-Shear Truss:* This system combines the benefits of the rigid frame and the vertical shear truss that is typically located around the elevator core. The lateral wind load is carried mostly by the frame in the upper portion of the building, and most of the wind shear is resisted by the shear truss in the lower portion of the building.

*Frame-Shear Truss (with Outriggered Belt Trusses):* The efficiency of the frame-shear truss system can be improved dramatically by the introduction of outriggered belt trusses at certain strategic locations along the height of the building. At these locations, a zone of almost zero rotations, under lateral loads, is generated, which in turn reduces the lateral sway significantly.

*Framed Tube:* As mentioned earlier, the idea of maximizing the overturning capability of the building by locating the structure around the perimeter was introduced in the 1950s. The framed tube became a powerful structural system because of its structural efficiency and least interference with the interior planning of the building spaces. It also contributes to a strong structural expression.

*Bundled Tube:* When the dimension of the building grows larger in both height and width, a single framed tube will not be an efficient structural system to use. The wider the structure is in plan, the less effective is the tube because of the increase in the shear lag, a phenomenon caused by excessive differential tube column shortening. A good example of this is the type of construction used in the World Trade Center in New York City. An attractive and efficient solution is to subdivide the single tube into smaller tubes. Such an arrangement reduces the shear lag effect considerably. It also allows the interior frame lines of the bundled tube to carry a significant portion of the gravity load in addition to being part of the tubular system, which will enhance the efficiency greatly. The system also provides flexibility to the architectural design of the building because any tube module can be dropped out whenever required by the planning of the interior spaces. The best example of a bundled-tube system is the Sears Tower in Chicago. For more information on the Sears Tower refer to Chapters 4 and 7.

*Diagonally Braced Tube:* The exterior columns are widely spaced and connected by diagonal members that intersect at the centerline of these columns and the spandrel beams. The system works together as a tube, and the diagonal members contribute to the resistance of both lateral and gravity loads. The system obtains its efficiency from the fact that it utilizes axial stiffness to transfer loads. The John Hancock Center in Chicago is an excellent example of this system. An extension of this system is the diagonally braced tube without interior columns, where the lateral load-resisting tube system also carries most or all of the gravity loads. This system is well suited for tall, slender buildings with small floor areas.

*Space Truss:* The primary load-carrying system consists of three-dimensional space trusses. The loads, both gravity and wind, flow through these trusses and find their way to the corner columns. The corner columns have to be of a substantial size, and in many cases a composite column is used. The 72-story Bank of China building in Hong Kong is an important example of such a system.

*Interior Diagonally Braced Trusses:* The system is more suitable for a square plan, although it could be used for other shapes. The system consists of four megatrusses, each extending the entire width of the building. The chords of these trusses comprise the exterior columns. The system is very efficient, again because all of the load is transferred through the exterior columns. These columns are usually of composite steel and concrete construction.

Some of the ongoing research work in academic institutions and architectural offices includes the study of the efficiency of different structural systems and how this research relates to actual existing buildings. Figure 2.6 gives some of the results of this work. The figure is generated from studying the systems while keeping the slenderness ratio constant. The sensitivity of varying the slenderness ratio was studied independently, and Fig. 2.7 shows a plot of the efficiency against the slenderness ratio. It is clear from the chart that the efficiency decreases significantly when the slenderness ratio increases above 10. This is the range where the impact of the dynamic loads is very important, and, accordingly, damping of the system becomes effective. For further discussion on steel structures see Chapter 10.

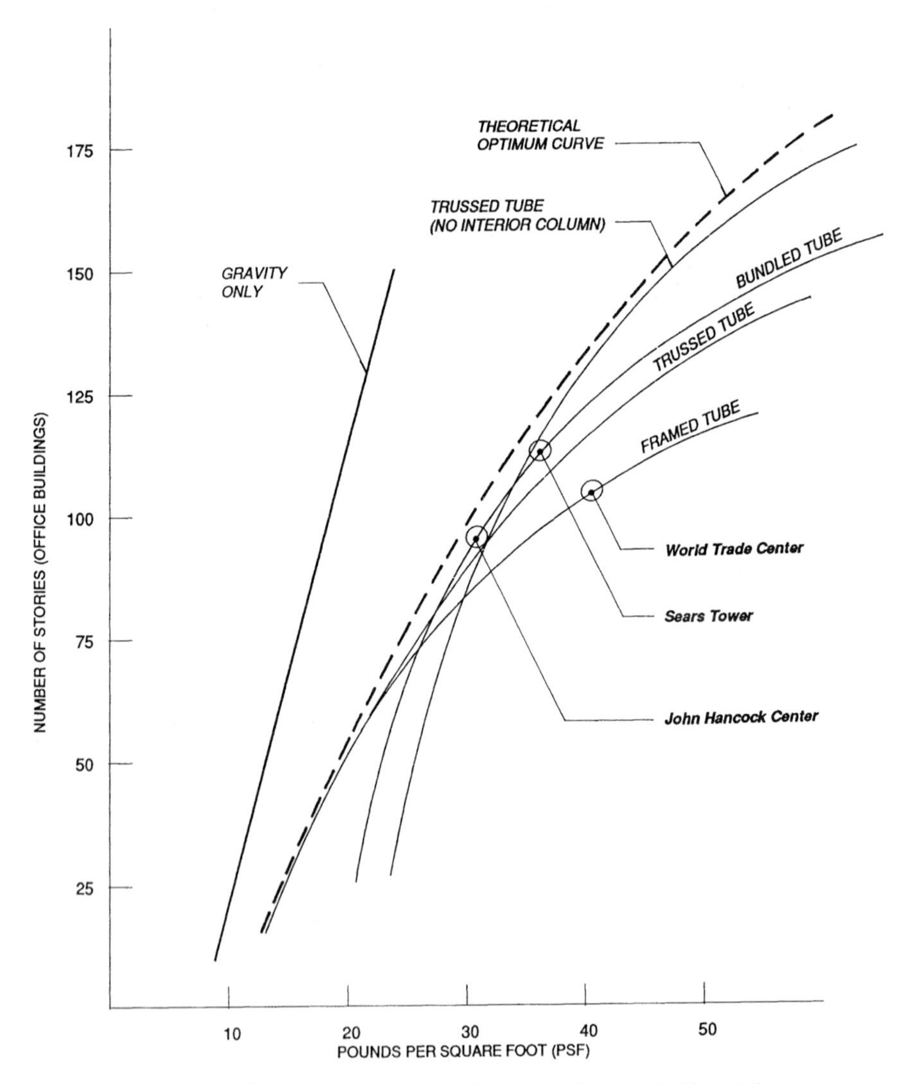


**Fig. 2.6 Efficiency of structural steel systems.** (*Courtesy: M. Elnimeiri.*)

***Reinforced Concrete Systems.*** Reinforced concrete, because of its strength capacity and its constraints in regard to construction techniques, has been limited to lesser heights in the past than those achieved with steel systems. However, with progress in the development of high-strength concrete as well as construction methods, coupled with the inherent advantage properties of its mass and rigidity, reinforced concrete could be used for more and more supertall buildings.

Figure 2.8 illustrates reinforced concrete structural systems (Khan, 1973a). The last two systems described on the chart are of special interest because they have the potential of reaching greater heights than shown. This could be achieved with the use of higher-strength concrete and steel reinforcement, and with advancements in construction technology. For further discussion on concrete structures see Chapter 11.

***Composite Systems.*** In addition to several other advantages, composite construction of structural steel and reinforced concrete, when it is used appropriately, can decrease the impact of dynamic loads. Accordingly, the damping effect of the entire system combines the benefits of the two materials, resulting in a superior system (Iyengar, 1977; Taranath, 1988).

Two popular systems have been well developed and have been in use for quite some time.

1. The reinforced concrete core shear wall enclosed within a structural steel building. The core wall provides the lateral support of the system both during construction and in the permanent condition. The reason for the success is due to the simplicity and the speed of construction of this system. Typically, the core wall is built well ahead of the steel floor framing in such a manner that there is no interference between the two trades during construction.

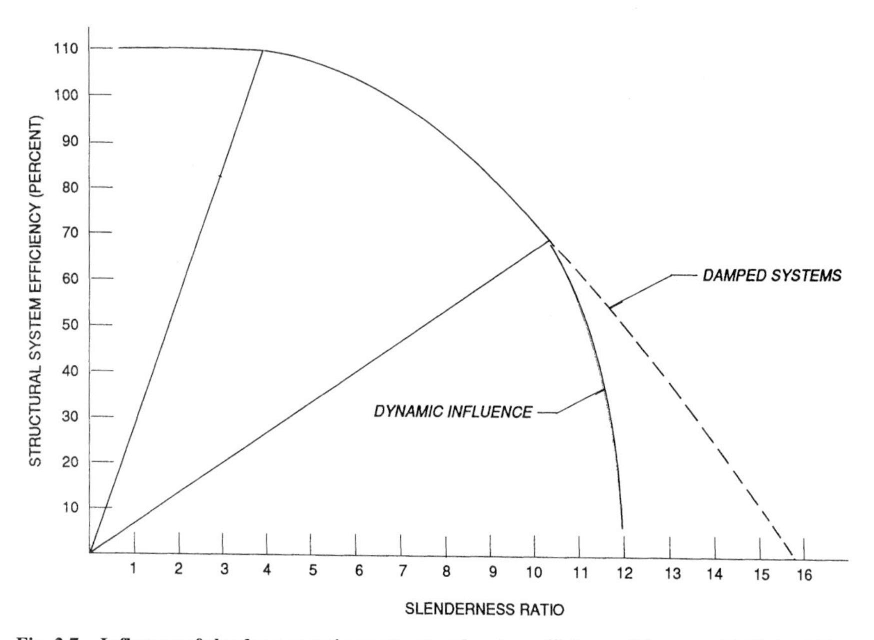


**Fig. 2.7 Influence of slenderness ratio on structural system efficiency.** (*Courtesy: M. Elnimeiri.*)

2. The composite framed-tube structure. In this system the interior framing is of structural steel and the exterior tube is of reinforced concrete. The reinforced concrete tube is very effective in providing stiff structures. The construction sequence is the reverse of that mentioned for system 1. In this case the steel framing is erected ahead of exterior walls, again to facilitate the separation of the two trades and to speed up the overall construction.

For supertall buildings an attractive composite system would be to utilize a reinforced concrete core wall coupled with exterior mega concrete columns or walls. The rest of the building structure would be of typical structural steel framing with composite metal deck slab floors. An example of this system is shown in Fig. 2.8 (mega columns/core wall). See Chapter 10 for additional discussion on composite systems.

With proper sequencing of different trades on the job site, the emergence of composite construction is likely to take the lead in the building of high rises. There are other structural systems that are derivatives of the principal systems listed here and can be found in the literature (Taranath, 1988; Khan, 1973a; Khan and Elnimeiri, 1983). See also Chapters 10 and 11 and Chapter SC-1 (CTBUH, Group SC, 1980) as well as *Structural Systems for Tall Buildings* (CTBUH, Committee 3, 1994).

## 2.3 COLLABORATION BETWEEN ACADEMIA AND OFFICE PRACTICE

Up to this point it has been implied that a repetitive module or column spacing is desirable for economy and efficiency in the design and planning of tall buildings. Architecturally, expressing the structural order and discipline on the facade is a tradi-

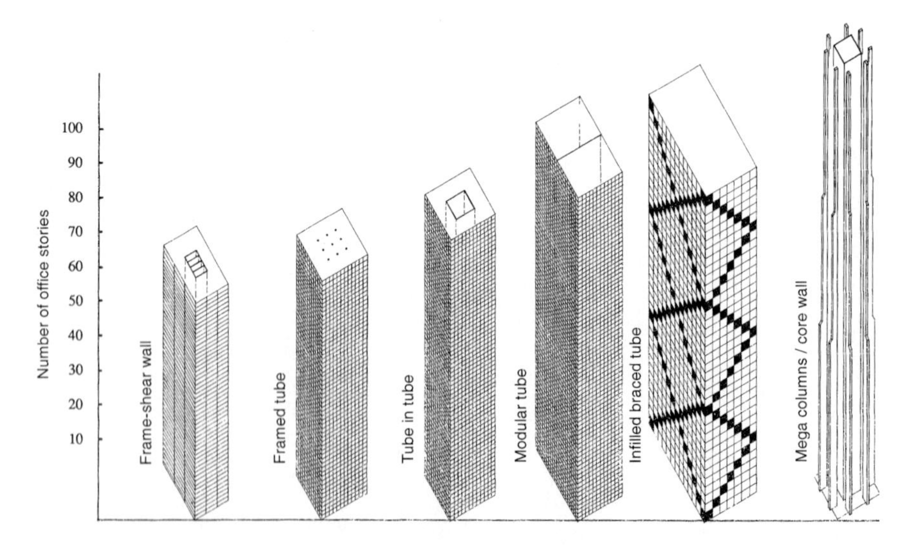


**Fig. 2.8 Heights of concrete structural systems.** (*Courtesy: M. Elnimeiri.*)

tional and rational method of unifying plan, facade, and structure. In recent years architects and their clients have searched for "new" and interesting designs for buildings, especially in the facade (see Chapter 7). Often faced with the choice of whether to express or to emphasize the structure, many current architects have used different curtain-wall patterns and compositions that work independently from the structural organization. Other architects have chosen to use structural expression in the facade design, an approach that emphasizes the inherent overall order in the building and imposes a larger order on the finely scaled curtain wall. Structural order in office buildings works in parallel with the customary 5-ft (1.5-m) planning module represented by the window divisions, and the vertical 3.7- to 4.3-m (12- to 14-ft) planning module representing the floor lines. In diagonally braced structures a different order may be superimposed to define the relationship between the braces and the columns, thereby adding scale and interest to the facade.

The necessity in tall buildings to start with a large base, usually at least 1/7 to 1/8 of the height, will dictate large floor plates in the lower stories, normally occupied by large corporate tenants. The need to incorporate smaller floor plates for smaller tenants (for example, the upper stories of the Sears Tower in Chicago) resulted in a unique structural organization. A nine-square grid employing three 23-m (75-ft) bays in each direction provides nine bundled-tube structures which are terminated at different elevations to create smaller floor plates. The John Hancock Center in Chicago also uses a unique structural organization to create smaller floor plates in the upper stories. A continuously sloped tube structure that is diagonally braced provides smaller residential floor plates at the top and larger office floor plates toward the bottom. One Magnificent Mile in Chicago is another built example of a bundled-tube structure in concrete. The tubes are irregular in plan, and when bundled, fill the irregular site and provide good views for the apartments in the upper stories of the mixed-use building.

The challenge to create different floor plate sizes in tall buildings has been studied extensively (Vierk, 1986). One research design project, addressing a structural system for a mixed-use office, apartment, and hotel building, starts from a base of 61 by 61 m (200 by 200 ft) and rotates to smaller squares as the building gets taller. The structural system of an exterior fully braced steel-tube structure, utilizing K bracing at the faces, transfers wind and gravity loads to the corner columns, and then along the exterior tube from top to bottom. Equally important is the opportunity to use structural organization to create very sculptural buildings. The simple transfer of structural loads through exterior bracing has opened up a multitude of geometric possibilities (Vierk, 1986).

A seemingly different direction is presented in another research design project (Lim, 1992). The building is characterized by a series of subtractive volumes 18.28 by 36.57 m (60 by 120 ft) and 4.57 by 4.57 m (15 by 15 ft) as the form ascends to a pure square, 36.57 by 36.57 m (120 by 120 ft), at the top. The architectural expression is similar to the buttressing on a cathedral, where a similar large base is reduced in volume as it reaches the top. Many model studies demonstrated that the implied structure of the building was clear, convincing, interesting, and beautiful. Studies where the diagonals were expressed on the exterior revealed that the clarity of the concept was weaker than when the forms alone expressed the structure.

In pursuing the notion of structural architecture for tall buildings, the visual factors are critical. The necessity to brace a structure and express the diagonals on the facade must be balanced with architectural and aesthetic considerations, including overall proportion, clarity of design intentions, and the relationship of the parts to the whole.

The concept of structural expression, that is, the aesthetic expression emanating from structure in tall building design, adds a new dimension for the architect and the structural engineer in that the structure must not only be economical and efficient but it must have an aesthetic quality as well (Billington, 1985; Ali, 1990). It must enhance and give expression to the form of the building. It follows that not every structure that meets the criteria of economy and efficiency will be satisfactory from an aesthetic standpoint (see also Chapter 4). The idea of structural architecture is not new and can be traced back to Greek architecture, Gothic architecture, and also in the works of a number of nineteenth-century and more contemporary engineers and architects.

In the study of architecture most schools have various degrees of interaction between academics and practitioners. Much of the instruction is given by practicing architects and sometimes by structural and mechanical engineers brought in as adjuncts. Today, in many architecture schools, design studio–based research has been incorporated into the academic programs at the graduate level, leading to new developments in the planning, design, structure, materials, and the environmental and psychological impact of tall buildings. This studio-based research explores all aspects of tall building design and provides an almost equal partnership of architects and engineers, acting as advisors, and students.

Design studio graduate research and its subsequent application to tall building architecture demonstrates that different structures are appropriate for buildings depending upon their heights. Research models have been developed primarily for tall building structures in steel and concrete. For example, it was through design studio research that the first X-braced tubes were developed in detail, first in steel (Sasaki, 1964) and later in concrete (Hodgkison, 1968). Some of these concepts were later realized in actual buildings in the John Hancock Center of 1968 and the Onterie Center of 1986, both located in Chicago.

The interaction between graduate student research and design professionals consisting of architects, structural engineers, mechanical engineers, urban planners, behaviorists and sociologists, interior designers, and developers collaborating in research teams provides a unique laboratory environment in which to develop hypothetical models that can be later applied to actual design projects involving tall buildings. Due to the diverse expertise of the professional advisors, students are immersed in a problem-solving environment not unlike that of office practice through which comprehensive prototypes can be developed over a period of a year or more.

In the following, eleven graduate research projects pertaining to tall buildings designed during the period between 1964 and 1992 and seven design office projects during the same period are presented. These projects, involving a wide variety of design concepts, help to illustrate how structural architecture could be reflected in design and planning, and how structural planning is a crucial aspect of overall tall building design. The seven design office projects involve buildings that were actually built or were waiting to be built at the time.

## 1 Graduate Research Thesis Projects at IIT

***Mikio Sasaki: A Tall Office Building (1964).*** The tower (Fig. 2.9) is a 213-m (700-ft)-high building comprised of 53 stories. It is square in plan, measuring 52 m (170 ft) on a side. It is based on a 1.5-m (4.67-ft) module in plan with a central core which contains elevators, stairs, toilets, and the vertical distribution of mechanical services.

This building is braced by three giant 18-story-high X braces on each face. These structural braces constitute a strong architectural feature and give an instant comprehensive reading of the overall organization. The angle of the bracing has been selected carefully to coincide with the orthogonal geometry of the columns and spandrels at every third floor. The typical story height is 4 m (12 ft). However, the lobby and the first story heights are 6 m (19 ft) and 4.5 m (15 ft), respectively, to suit special functions.

**Fig. 2.9 Sasaki project, photograph of model.** (*Courtesy: Hedrich-Blessing.*)

Variations in floor heights are expressed on the exterior. The building is proportionally calculated to create a harmonious relationship between height and width as well as the spacing of columns and floor heights in conjunction with the angle of bracing.

The building is of steel construction with six 8.5-m (28-ft) bays. There are four 8.5-m (28-ft) bays within the core, leaving a 14-m (45-ft) column-free lease span for office space. The exterior structure acts as a diagonally braced steel tube and is very efficient in resisting lateral loads. This project may have been one of the first proposed diagonally braced tube structures. The project included a structural plexiglass model, which was used to test the stiffness of the system. In addition, computer and long-hand calculations were made, which supported the findings of the model.

***Robin Hodgkison: An Ultra High-Rise Concrete Office Building (1968).*** This project (Fig. 2.10) had two primary purposes. The first was to try to find an alternate solution in concrete for the World Trade Center in New York, a similar building completed in steel. The second purpose was to offer an alternative to the proposed development of the Illinois Central Railroad site. Proposed was a large number of medium-sized high-rise buildings. The Hodgkison design proposed fewer buildings but of higher density, including three 116-story office buildings and four 100-story X-braced apartment buildings. A typical tower measures 66 m (216 ft) along each side of a square plan. It has 108 office floors, six mechanical floors, and an entrance floor. The building has a height-to-width ratio of 6.5:1.

The building elevations consist of a grid of columns at 3.0 m (10 ft) on centers and spandrels that are 0.9 m (3 ft) deep. For wind bracing, windows are infilled with concrete, producing a series of five Xs on each facade. The interaction of the orthogonal grid and the Xs adds enormously to the aesthetic impact and structural expression of the building.

The building's structure is an external tube of closely spaced columns stiffened by diagonals. A concrete service core helps stiffen the structure and carries the internal vertical loads. The lease spans are 14 m (46 ft) and are spanned by concrete beams perforated for lighting and for passing utilities.

***Wayne Petrie: Office Building (1981).*** This proposed 43-story office building (Fig. 2.11) would occupy a quarter-block next to the Xerox Building in Chicago. Two diverse goals were set for this project. The first was to build a building that solves the problem of the proximity to the property line of the adjacent 43-story Xerox Building and to relate to it architecturally and yet keep a separate identity. The second was an architectural-structural goal to design a steel-framed tube structure which increased the usual 4.5-m (15-ft) column spacing on the exterior to as much as 15.2 m (50 ft).

This building, like the Xerox Building, is 43 stories high. However, because its structure is of steel instead of concrete, with flat-plate construction, it is somewhat taller in actual dimensions.

One of the most striking aspects of this building is the floor plan, which has the shape of an elongated half-circle. This was done to provide more space between it and the Xerox Building. The building facade continues the uninterrupted bands of glass and spandrels of the Xerox Building in order to maintain visual continuity. Three-story commercial elements fill the gap between the two buildings.

The two buildings face a large plaza in downtown Chicago so that they can be readily seen from ground level as one composition. While they are far from identical, they have many similarities. They are similar in size and have horizontal bands of reflective glass and horizontal spandrel panels of aluminum. However, the Xerox Building uses white spandrel panels and this project uses a dark brown color. Both projects have arcaded ground floors of similar height and have prominent curves on the facade.

The structure, an external tube with widely spaced columns, has been proposed to eliminate the architectural disadvantage of the usual 4.5-m (15-ft) column spacing, which makes the outermost 0.9- to 1.2-m (3- to 4-ft) space of the building practically unusable. A 1.8-m (6-ft)-deep trussed spandrel is located partially above and below the floor line. An off-center core contains the building services and also the interior columns, leaving the office space column-free.

**Fig. 2.10 Hodgkison project, photograph of model.** (*Courtesy: Bill Engdahl, Hedrich-Blessing.*)

**Fig. 2.11 Petrie project, photograph of model.** (*Courtesy: Orlando Cabanban.*)

***Xu Ping: Two Multi-Use High-Rise Concrete Buildings (1991).*** The central theme behind this project (Fig. 2.12) was to complete a study that would make a substantial contribution to the bank of knowledge that could guide architects and structural engineers who are involved in the planning, design, and execution of multiuse high-rise buildings. The student chose to express her research findings in two separate buildings, both utilizing reinforced concrete as a structural material. However, the research results are not limited to concrete buildings—they can be applied to steel or composite structures as well.

Typically, functions are stacked one above the other in multiuse buildings. Office floors are located above commercial floors, and residential floors are placed at the top. This is generally done to resolve the problems associated with human traffic and its effect upon vertical transportation within the building.

In contrast with traditional practice, this project proposed to place a hotel and its functions, dwelling units and their functions, and commercial space and its function in the lower portion of the building and the majority of office functions in the upper portion of the building, totaling a gross area of 399,470 $m^2$ (4.3 million $ft^2$).

In a single tower building, when smaller column bays are placed at the top, a structural solution is required to transfer vertical loads from above onto the larger column bays below. Structural logic would reverse this practice. If the smaller column bays are placed in the lower portion of the building and the larger column bays in the bays in the upper portion, there would be no need for transfer girders or solutions that impose great demands on structural engineers or add to the structural cost.

One approach to provide additional resistance to lateral forces and deflections in high-rise buildings is to increase the material in the most effective resisting components. By placing office functions with larger bays above residential functions with smaller bays, more mass and stiffness are added to the lower part of the building, resulting in a more efficient structural system.

The architectural solution in this project places residential and hotel functions at the ends of the lower portion of the building, along with public, office, and commercial functions located between them. In doing so, the differences in floor heights must be resolved. Public and commercial floors are made twice the height of residential and hotel floors, and the height of three office floors equals that of every four residential and hotel floors.

The engineering solution places the closely spaced column bays at the lower levels of the building. Alternate columns in the lower levels are eliminated, while the remaining widely spaced column bays are continued through the upper levels of the buildings. The architectural and engineering solutions allow for the arrangement of the various programmed spaces to function appropriately and the structural system to function naturally and logically.

As stated before, both buildings in this study were developed using reinforced concrete as a structural material. The study also indicates that the combination of lightweight, high-strength, and regular-strength concrete can be used in a single structure to reduce the number of horizontal elements and the dimensions of vertical support members. The building, with an infilled panel truss-tube system, has closely spaced exterior columns at 3 m (10 ft) on centers and relatively shallow exterior spandrel beams. The other building, a formed-diagonal truss-tube system, uses higher concrete strengths in its diagonals and vertical support members. The exterior columns in this second system are placed at 12 m (40 ft) on centers and the spandrels are relatively deep. The infilled panel system results in a building facade that has a distinct character expressing a concrete structure. The formed-diagonal system presents a facade recognizable as being associated with steel construction.

After studying high-strength concrete compared to normal-strength concrete, in combination with different percentages of steel and various concrete strengths, it was

**Fig. 2.12 Xu Ping project, photograph of model.** (*Courtesy: Peter Beltemacchi, Jr.*)

found that it is more economical to use high-strength concrete and less steel to achieve a desired strength for a particular purpose. The higher the concrete strength, the higher is its cost. However, the lower the concrete strength, the more formwork and labor costs are required. Thus there is clearly a tradeoff.

***A. L. Menon: A Ninety Story Apartment Building Using an Optimized Concrete Structure (1966).*** This project (Fig. 2.13) attempts to push the slenderness of a concrete apartment building toward its extreme limit. Measuring 28.5 by 56 m (94 by 184 ft) in plan, it is 247.5 m (812 ft) high and has a slenderness ratio of 8:6. The exterior wall pattern is an expression of the building's structure. Columns are spaced at 2.0 m (6 ft) on centers and taper off toward the top, where the stresses are smaller due to reduced wind and vertical loads. Architecturally the building is a tall slender structural system proportioned according to its mass, which is also related to the structural grid.

The building has a central service core with a corridor for access to apartments on all four sides of the core. The service core provides tenant storage space and a laundry on each floor. There is a mix of apartments from efficiencies through three-bedroom units, with a predominance of smaller apartments in the lower floors of the building. Flexibility was built into the design so that apartment spaces could be combined to create units with more than three bedrooms.

The concrete structure is a framed tube with two full-width shear walls at the one-third and two-thirds points of the long sides. The whole assembly acts in unison to resist horizontal wind forces in an efficient manner. The floors comprise flat-plate construction in which the top surface receives a floor finish and the underside of the concrete is the finished ceiling.

***Alfonso Rodriquez: A Form Stiffened High-Rise Apartment Building (1970).*** Narrow linear apartment buildings are generally limited in height to an approximate slenderness ratio of 8:1. When this limit is exceeded substantially, some kind of modification of the plan from a single rectangle is required to obtain a stiffened building. This proposal (Fig. 2.14) is for a 60-story 165-m (544-ft)-high apartment building, 10 m (34 ft) wide, with a ratio of 16:1.

The building at the top floor is a rectangle 399 m (1310 ft) long and 10.3 m (34 ft) wide. The ground floor is serpentine in plan, and the form is created by connecting the two plan forms by straight lines. In order for the elevators to rise perpendicularly from the ground, the elevator cores are located at the points of contraflexure of the curve. In addition, apartment units are duplexes, halving the number of elevator stops and increasing the efficiency of the single-loaded corridor provided for the building.

The lateral load-resisting system is made up of concrete shear walls that run the width of the building, spaced at every second apartment. These shear walls work in combination with exterior moment-resisting frames spaced at 3.65 m (12 ft). This structure was found to be very efficient and stiff. The drift ratio due to a 10-year wind return is $\frac{1}{671}$. The calculated concrete volume used for the building is 0.00289 $m^3$ (1.1 $ft^3$) per $m^2$ ($ft^2$) of floor area.

***Kay Vierk: A 142 Story Steel K-Braced Multi-Use Building (1986).*** This project (Figs. 2.15 and 2.16) continues the evolution of the tall building type, reaching 532 m (1745 ft), and contains a mix of uses, including commercial, office, apartments, hotel, and parking. The building is a series of rotated squares, one inscribed in the other, creating setbacks and smaller floor plates to suit multiple uses as the building rises. Integrating these setbacks with diagonal steel bracing was used to manipulate the mass and form of the building.

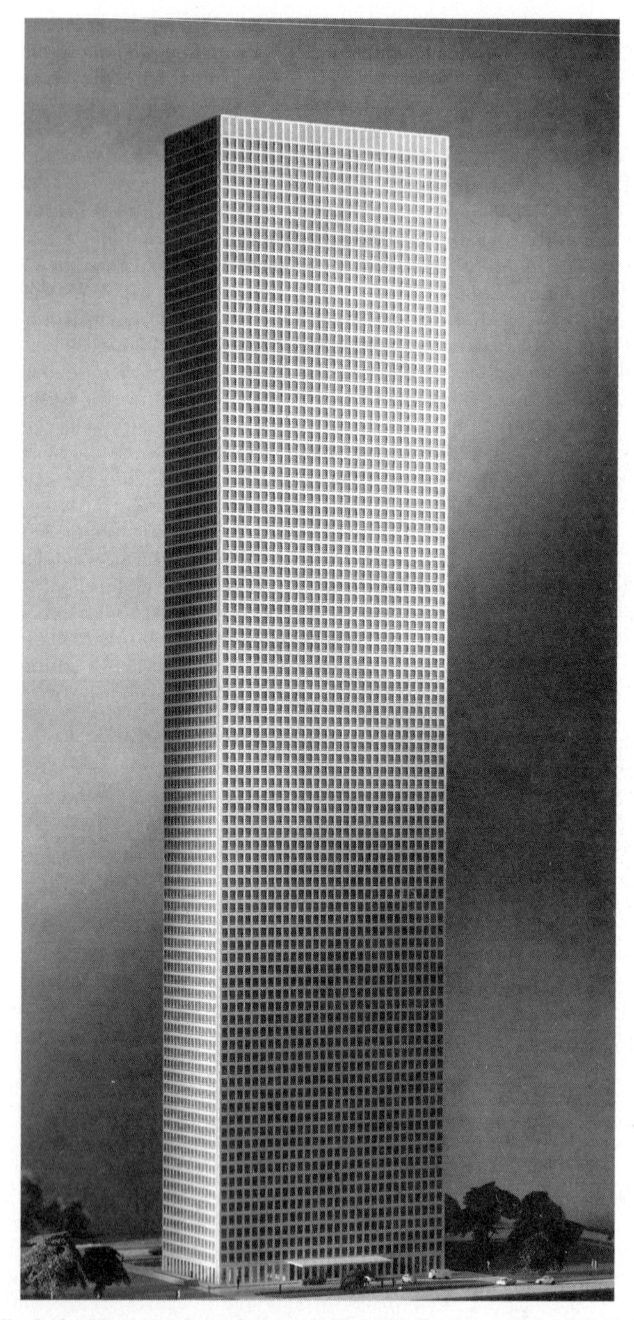

**Fig. 2.13 Menon project, photograph of model.** (*Courtesy: Richard Nickle.*)

The base of the building contains eight floors of commercial and retail space along with elevator lobbies. The elevator scheme of the building is a multiple-zone sky-lobby system. Two sky lobbies are used, one on the 41st and 42nd floors for offices and one on the 81st and 82nd floors for residences. Office space begins on the ninth floor and continues to the 78th floor. These floors range in area from 4096 to 2034 $m^2$ (44,100 to 21,900 $ft^2$) and are disposed in a variety of configurations.

The residential floors in varied configurations range from 1782 to 1278 $m^2$ (19,185 to 13,758 $ft^2$). The hotel runs from the 113th to the 136th floor, with an average area of 1024 $m^2$ (11,025 $ft^2$). The floors that support the building's mechanical system are located below the sky lobbies, combining the space requirements for elevator pits and overruns.

The intention governing the design of the structural system is to bring all or most of the loads to the perimeter of the building and then have those loads transfer smoothly through the exterior bracing to the outside corners and eventually to the foundation. In the diagonalized system, as discussed earlier, the same structural system resists both gravity and wind loads. For the rotated square, K bracing was found to be the most direct structural solution and the most flexible with respect to the massing of the building. Visually, the K bracing enhances the geometry of the forms and expresses the structural elegance.

The framing system for the lower half of the building is on a 15-m (50-ft)-square structural grid with the elevator core loads carried by interior columns directly to the foundation. The top half of the building uses a system of interior hanging trusses at every 10 floors to take all of the load to the perimeter. The trusses connect to the major horizontal tie beams of the exterior K bracing.

The structural efficiency resulted in a drift ratio of 1/550 of the building height and a weight of 151.3 kgf/$m^2$ (31 psf). With a slenderness ratio of 8.3:1, the net result is an efficient and slender structure, creating a building with an interesting shape and sculptural quality.

***Hiroshi Fujisawa: 1400 Foot High Steel Office Building (1987).*** This office building proposal (Fig. 2.17) has a cross-shaped plan with overall dimensions of 73 by 73 m

**Fig. 2.14 Rodriquez project, photograph of model.** (*Courtesy: Illinois Institute of Technology, Department of Architecture.*)

(240 by 240 ft). The building is 427 m (1400 ft) high and has 12 sides; four are 36.5 m (120 ft) wide and eight are 18.2 m (60 ft) wide. On the exterior, there is an orthogonal grid of columns and spandrels. The grid is overlaid by the diagonal bracing, dividing the 36.5-m (120-ft) sides into nine half-asymmetrical diamonds. The shorter sides are diagonalized also. The structure is covered with aluminum panels painted white, and the glazing is reflective glass.

**Fig. 2.15 Vierk project, photograph of model.** (*Courtesy: D. Chichester.*)

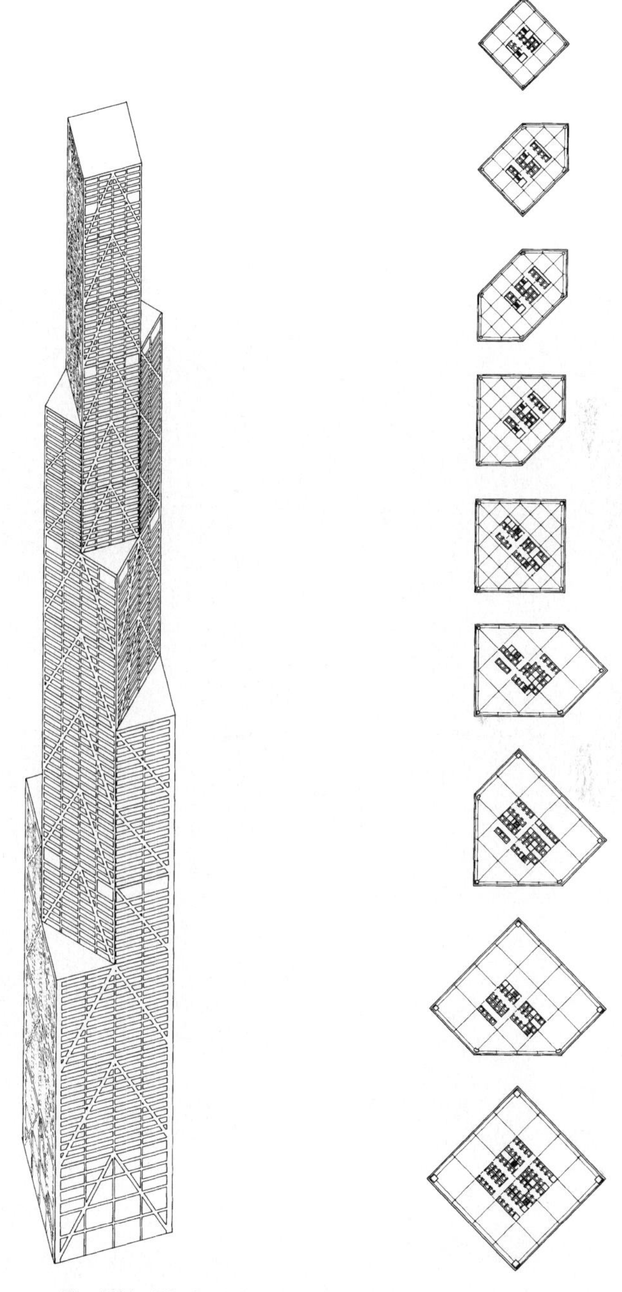

**Fig. 2.16 Vierk project, isometric view and typical plans.**

The proposed site is adjacent to the Chicago river on a square block. A lower-level street serves as access to the loading docks and to the parking areas. There is a commercial space on the first three floors, plus an extension to the main building on one side.

The steel structure is an exterior braced tube with the eight outside corner columns carrying most of the vertical loads on the faces and resisting all of the lateral loads. The loads on the interior columns are carried directly to the ground.

Aesthetically, the building has a very strong, rational, and positive image. The verticality is emphasized by the breakup of the mass into long slender segments. The aes-

**Fig. 2.17 Fujisawa project, photograph of model.** (*Courtesy: Orlando Cabanban.*)

thetic qualities are emphasized by the pattern of the structure and the reflection of the corner segments against each other.

In planning for its large floor plates, the cross shape with a central core has a significant advantage over a square plan of the same area, where the social and health problems of working long distances from the glass or even in windowless offices are a real concern. First, in a square plan as the floor area gets larger, the areas increase as the square of the dimensions whereas the perimeter increases directly with the dimensions. It can be seen that the larger the floor plan, the smaller is the available perimeter working area compared to the total usable area. The cross shape tends to alleviate this problem by providing 13% more perimeter than its square counterpart. Second, the cross shape divides the plan into four working areas, each with glass on three sides, providing more pleasant working areas.

The central split core is largely devoted to vertical circulation. Its configuration has several advantages in that floor circulation can be more efficient by circulating through the core instead of around it. Also, stairs, toilets, service areas, and other functions tend to open onto internal corridors rather than opening into office spaces. In addition, the division of the floor into two or more tenant spaces is facilitated by direct access from the elevator lobby to tenant offices rather than by circulating through a public corridor.

The elevator system is based on two sky lobbies at approximately one-third and two-thirds the height of the building. These are served by double-deck shuttle elevators. In addition, four banks of six elevators serve each third of the building.

***Joseph Renn: Wolf Point Center, A Multifunctional Tall Building (1990).*** This project (Fig. 2.18) was based on the idea that functions located above one another, while requiring different lease spans, can be incorporated into a single building, utilizing a structural system that accommodates each function without losing its integrity and visual impact.

The tower is 478 m (1568 ft) high with 123 floors total. The hexagonal plan yields 3028.5 $m^2$ (32,600 $ft^2$) per floor for 44 floors, 2025 $m^2$ (21,800 $ft^2$) average for 20 floors of hotel functions, and 1009 $m^2$ (10,860 $ft^2$) per floor for 52 floors of residential functions. Public spaces are integrated throughout the building. Pedestrian circulation begins at the plaza level. From there, three floors of public space, containing the various lobbies, retail, parking, and service areas, can be reached. Parking is below the plaza level adjacent to the building.

The exterior structural steel trussed frame allows for column-free interior lease span areas, as well as the expression of an aesthetically pleasing form. This system transfers all lateral loads, and a good portion of the gravity loads, to the six large exterior columns. Gravity columns, which primarily support the floors, are free to be placed where necessary within the building. The diagonal bracing incorporated with the hexagonal plan increases the economy and efficiency of the structural materials. For aesthetic reasons it was decided to sheath the building with bronze-tinted glass and dark-anodized aluminum, yielding a dark appearance against the building's surroundings.

***Sangchul Oh: Development of High-Rise Mixed-Use Building (1992).*** This project (Fig. 2.19) supports the notion that a mixed-use high-rise building can be an ideal solution to present a clear, compact, and efficient structure in high-density environments. The building utilizes high-strength concrete as the principal structural material. Since the structure is subjected to considerable movement due to wind forces, seismic forces, and thermal stresses, a structural system that can effectively resist such movements is viewed as the governing factor in this integration.

Stacking the structures for different functions requires complex structural transitions, often rendering the structure less efficient. Considering this fact, the balance

**Fig. 2.18 Renn project, photograph of model.** (*Courtesy: Orlando Cabanban.*)

between the structural efficiency and the functional performance of the buildings is the primary design criterion for the development of this project.

Two structural principles are incorporated in developing the system for this project. One is to use the most efficient system that can provide the ability to withstand lateral loads at extreme heights, and the other is to reduce the transfer of gravity loads inherent

**Fig. 2.19 Oh project, photograph of model.** (*Courtesy: Orlando Cabanban.*)

to most mixed-use structures. For the height of 312 m (1025 ft), the framed-tube system achieved by closely spaced exterior columns and interconnecting rigid spandrel beams is the most efficient system.

The base of the building is square in plan and rotates 45° at the 44th-floor level to take advantage of views for the residential units. The tube system of the lower half of the building is based on a 9.0- by 9.0-m (30- by 30-ft) structural grid with the exterior columns spaced at 3.0 m (10 ft) on centers on the perimeter of the base, measuring 46 m (150 ft), yielding a slenderness ratio of 6.8. The lateral loads are picked up by the exterior columns, whereas most of the gravity loads are carried by interior columns. The size of the interior columns remains the same throughout the office floors by varying the strength of concrete from 82,740 kPa (12,000 psi) at the lower levels to 41,370 kPa (6000 psi) at the 43rd-floor level. In the core area, the structural geometry is rotated 45° to achieve the continuity of the gravity columns on the upper portion of the building as well as to simplify the transfer mechanism of the gravity loads. Between the 15th- and 43rd-floor levels, Vierendeel action takes place in the midsection of each face of the building. From the ground level to the 13th-floor level, the midsection of the tube is opened to reduce the shear-lag effect. From the first lower level to the plaza level, where bending stress is greatest, the tube is complete to ensure adequate stiffness of the building.

For the upper portion of the building, which functions as residential space, tube action is achieved by extending the column system in the corner sections of the office floors and enclosing the diagonal side with a column system spaced at 3 m (10.5 ft) on centers. In addition, shear walls are utilized to bring the sway of the residential floors to an acceptable level. On the 43rd-floor level, where mechanical equipment is housed, Vierendeel trusses are used to transfer the gravity loads of the residential floors to the gravity columns within the square tube of the office floors.

Urban planning issues addressing the integration of space, circulation of vehicles and people, and amenities for a large population were resolved in a way that would enhance the viability of the project as well as the proposed site at Illinois Center.

***Jong Hwan Lim: Super-Thin and Super-Tall Office Building (1992).*** This office building proposal (Fig. 2.20) would be 491 m (1612 ft) high, comprising 124 stories. If built, it would exceed the height of the Sears Tower by 45 m (150 ft). The proposed site is a complete block adjacent to public transportation, convenient to expressways, and, more important from an architectural point of view, it is adjacent to parks to the north and northeast.

The tower is tall and slender. However, the proportions never exceed the slenderness ratio for an economical structure. At the top the plan is 38 by 38 m (126 by 126 ft) in measurements; it gradually increases in size toward the base by adding 9- by 18-m (30- by 60-ft) and 4.5- by 4.5-m (15- by 15-ft) areas in a staggered arrangement. This ordering results in a building with four facets in the upper stories and gradually increases to 36 facets in the lower stories. Architecturally the multiple facets enhance the verticality of the building, provide a greater number of corner offices, and give the building a strong architectural character. The building is clad in green-tinted glass, which completely covers the structure, yet the structure is implied by the manner in which the tower gets larger toward the base. The use of additive elements is analogous to the way a Gothic tower would be braced by buttressing, which gets larger toward its base. The building is carefully proportioned, resulting in a very elegant building form.

The plan of the building promotes efficiency and function with a central core and column-free office spaces. The top floors are 1477 m$^2$ (15,900 ft$^2$) in area. The middle floors average 1858 m$^2$ (20,000 ft$^2$), and the lower 36 floors, used for commercial ten-

**Fig. 2.20 Lim project, photograph of model.** (*Courtesy: Orlando Cabanban.*)

ants, increase to 2378 $m^2$ (25,600 $ft^2$). A garage for 600 automobiles occupies most of the six floors below grade. The total area above grade is 223,000 $m^2$ (2.4 million $ft^2$).

The structural system consists of a diagonally braced tube at the top. Toward the bottom of the building it changes to a mixed frame at the staggered corners and braced truss at the four faces, resembling four buttresses linked together through a so-called moment-connected shear-link system, which does not transfer gravity loads. The system proved to be very elegant and extremely efficient. The structure is not explicitly expressed, but the form, with its visible buttresses, clearly and subtly reflects the structure.

## 2 Design Office Projects

***John Hancock Center (1968).*** Rising to a height of about 344 m (1127 ft), the John Hancock Center (Fig. 2.21) occupies a dominant position on Chicago's skyline. Structurally, the exterior members of the steel frame represent a unique solution for a stiffened tube (Khan, 1967). The continuous tapered form of the building decreases the size of the floor plates to suit a mix of different functions. Lower floors requiring greater floor areas for efficiency are used for commercial and office space, with parking floors above them. Singles and efficiency apartments, which require a greater depth from the outside wall than two, three, and four bedrooms, are located above the parking levels. Finally, the top floors are used for large expensive apartments, requiring less space for ancillary rooms, but providing maximum measures of privacy, quietness, prestige, and view.

The elevator system is managed with a sky lobby for the apartment tower, thus allowing elevator banks to be stacked one above the other. Without this arrangement of elevators, the floor area required for continuous shafts would not have justified the multiuse scheme.

The structural frame is clad in aluminum with black anodic coating. The bronze-tinted glass is held in frames of bronze-colored anodized aluminum and is cleaned from a traveling scaffold running on rails attached to the columns. The combination of the diagonal pattern of the bracing and the rectilinear pattern of the spandrels and columns sets up a dynamic geometry that reinforces each other and lends appropriate scale and visual interest to the facade that might otherwise be overwhelming to nearby buildings.

The exterior members of the steel frame form a tube, where the necessary stiffness is provided by diagonal members and by structural floors which coincide with the intersection of the diagonals and the corner columns. In keeping with the functional organization of the building, the tubular body has its largest cross section where the stresses caused by wind forces are the greatest. The John Hancock Center is an excellent example of where the outer wall is designed as a structural element, as opposed to the traditional skeleton and skin system. The concept of a trussed tube, in this case, where all columns on each face of the building are equally loaded, avoids the usual premium for great height because the diagonals carry both vertical and wind loads. See Chapters 4, 5, and 7 for additional discussions on this building.

***Sears Tower (1971).*** This Chicago building (Figs. 2.22 to 2.24) is 443 m (1454 ft) high, has four below-grade levels, and is the tallest building in the world as of this writing. [Petronas Towers in Kuala Lumpur, Malaysia, under construction, is planned to reach a height of 450 m (1476 ft).] The Sears Tower plan is based on nine tube structures that are 23 by 23 m (75 by 75 ft) each in plan to form one large square 68.5 by 68.5 m (225 by 225 ft). Each tube is formed with columns placed 5.0 m (15 ft) on centers and

**Fig. 2.21 John Hancock Center, Chicago.** (*Courtesy: Ezra Stoller, ESTO.*)

uses a 1.5-m (5-ft) planning module. As the building ascends, tubes are terminated at different levels, producing the following floor areas:

| | |
|---|---|
| Floors 1–49 | 4893 $m^2$ (52,670 $ft^2$) |
| Floors 50–65 | 3848 $m^2$ (41,420 $ft^2$) |
| Floors 66–89 | 2802 $m^2$ (30,170 $ft^2$) |
| Floors 90–109 | 1141 $m^2$ (12,283 $ft^2$) |

**Fig. 2.22 Sears Tower, Chicago.** (*Courtesy: Ezra Stoller, ESTO.*)

With a total gross area of 413,848 m$^2$ (4,455,844 ft$^2$), the Sears Tower is one of the largest office buildings in the world. Sears did not start out to build the tallest building in the world. They required 3716-m$^2$ (40,000-ft$^2$) floor plates with a total building area of 185,800 m$^2$ (2 million ft$^2$) of usable space, plus space for future expansion. Original plans included two identical office towers of 50 stories each. Unfortunately this idea was abandoned when it was determined that tenants could not be found to occupy the empty large floor plates intended for Sears's expansion. A tall single building was an attractive idea when Sears considered the added revenue from observation decks and communication tower contracts. The final design convinced the client that an economical structural system could be found that provided half the building with large floor plates for its employees and half with smaller floor plates varying in size to lease to a wide mix of tenants. Today Sears is in the process of moving most of its staff to the suburbs and continues to generate revenue from a mix of tenants occupying all floors in the Sears Tower.

A strong attempt was made to express the structure and the plan on the exterior of the building. From a collective base of nine tubes, individual tubes end at various floors, with the top floors ending with two tubes with a total measurement of 23 by 46 m (75 by 150 ft) in plan. The moment-resisting frames are expressed on the entire facade as

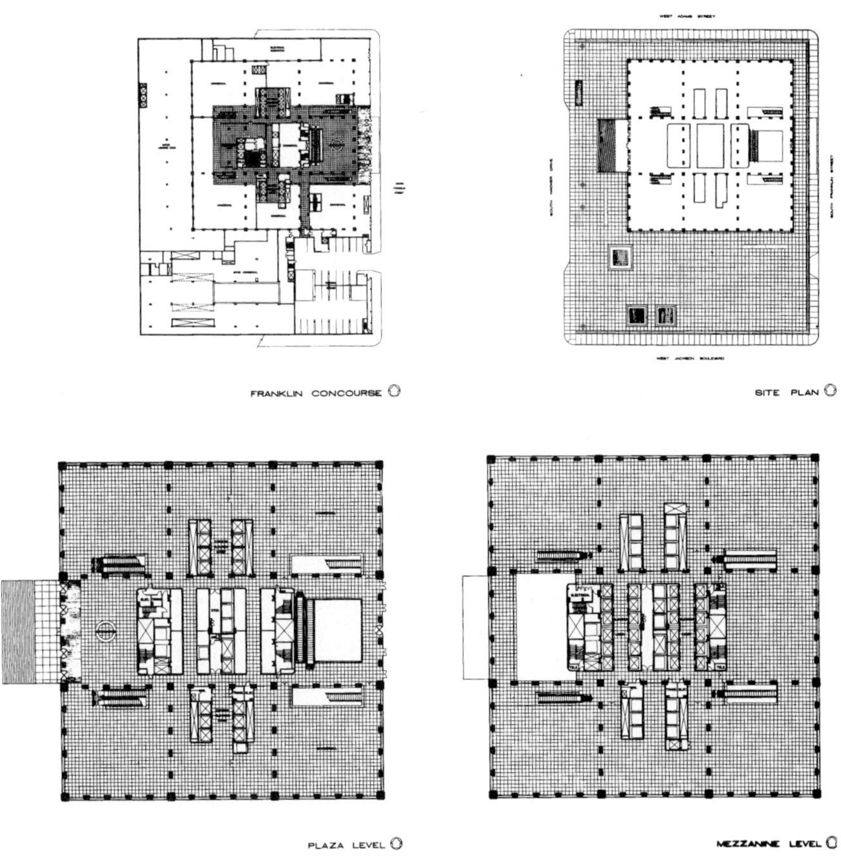


**Fig. 2.23 Sears Tower, typical floor plans.**

well as on each individual tube. Proportions were carefully studied, and as a result it was decided to emphasize the horizontal floor division rather than the vertical column division. The spandrel panel on the exteriors is painted black, and the glass is tinted gray, giving the building a monolithic, machinelike expression.

The structural system is a bundled frame tube with 4.6-m (15-ft) column spacing. The bundling of the tubes significantly enhances the structural stiffness because it reduces the shear lag (Schueller, 1990). For further information on the Sears Tower see Chapters 4 and 5.

***One Magnificent Mile (1983).*** The building (Fig. 2.25) is situated at the northern end of Michigan Avenue in Chicago. Designed as a multiuse building, the project contains commercial space, office space, and residential space in three towers rising 21, 49, and 57 stories high. The building floor plans and hexagonal shape were developed to take full advantage of the small angular lot and the logic and structural possibilities of a bundled-tube system (Iyengar, 1986).

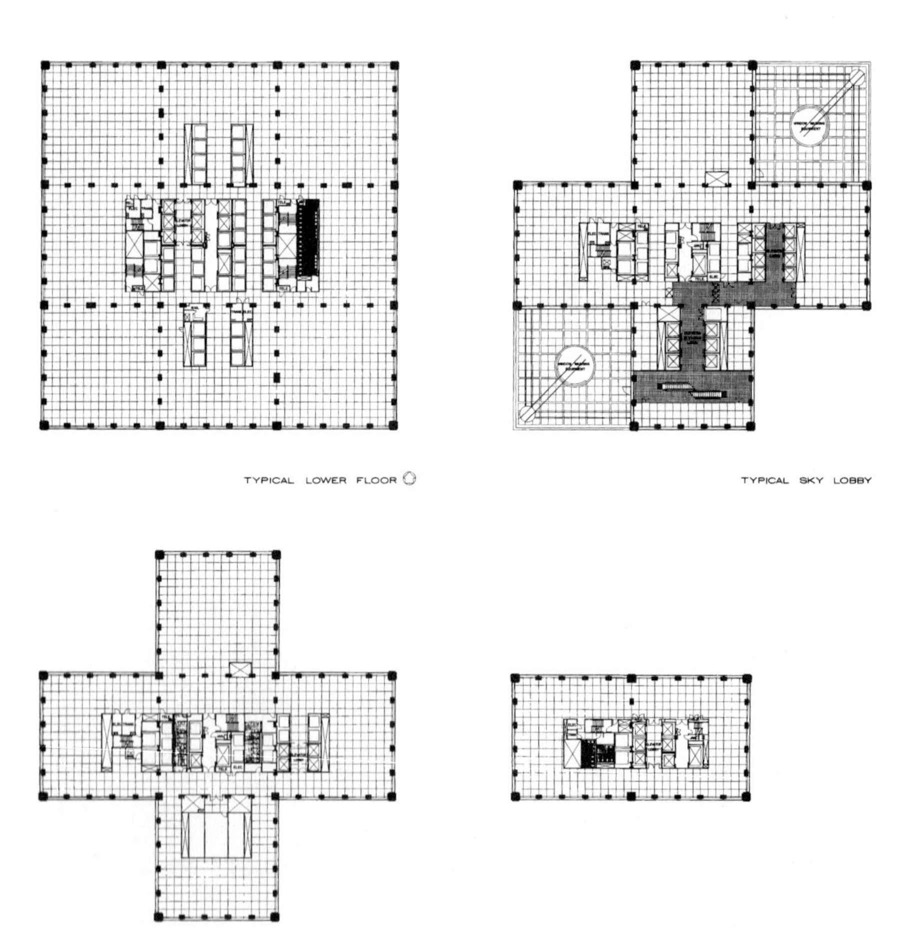


**Fig. 2.24 Sears Tower, typical floor plans.**

The building contains commercial space on the first three floors, office space on floors 4 to 19, and 181 luxury condominiums at the top of the building. The tube geometry and composition of three towers of varying heights resulted from extensive studies of desirable views for the condominiums, and the potential shadow cast over the nearby park and beach to the northeast. The 57-story tower element in the middle of the bundle is topped with a sloped, glazed roof with a view to the beach and lake. The faceted shapes allow multiple orientations to respond to views and to relate to significant adjacent buildings.

**Fig. 2.25 One Magnificent Mile, Chicago.** (*Courtesy: Gregory Murphey.*)

The top of the 21-story element bears a two-story mechanical floor that is carried across the other two towers, creating a clear division between office and residential floors. The exterior fenestration relates to the interior function and thus has a different treatment above and below this band at the mechanical floor. The building is clad in granite with clear glass at the commercial levels, gray reflective glass for the offices, and gray-tinted glass for the residences. Three hexagonal concrete tubes with punched window openings are joined together to resist wind loads as a bundled tube.

***Onterie Center (1983).*** The Onterie Center in Chicago (Figs. 2.26 and 2.27) is an example of a mixed-use high-rise building that uses a unique concrete panel system as part of its structure to increase its efficiency. With an aspect ratio of 8:1, the building rises to a height of 171 m (560 ft). The building is organized with retail and lobby space at grade and parking located above and below. Office floors are located above the parking levels, with some offices opening out to sloped atrium spaces. Apartments begin on the 21st floor and include efficiency and one- to three-bedroom units, with larger three-bedroom units located at the top of the tower. The exterior concrete structure is painted and infilled with gray-tinted double glazing, which gives the building a distinct character and at the same time expresses its concrete structure (Zils and Clark, 1986).

Onterie Center continues the evolution of the reinforced concrete structural system. The structural system consists of two exterior channels connected together with regular framing. The channels are constructed of diagonally braced, closely spaced frames. The diagonals, created by filling in the windows, serve a dual function. First, they increase the efficiency of the tube structure by diminishing the shear lag. Second, the diagonals reduce differential shortening of exterior columns by redistributing the gravity loads. This results in a stiffer and more efficient structure. The structural gravity system is very simple and requires only a few interior columns. There is no need for either shear wall or beam framing. This allows for more flexibility in space planning, which is an important feature for a mixed-use building. The issue of structural expression of this building is addressed in Chapter 4.

The development of this project illustrates the importance of interactions between the developer, the architect, and the engineer. Onterie was a new concept—trying to use the braced tube in concrete rather than in steel (as had been done in the John Hancock Center). But in any new concept there is a risk. Since it has not been tried before, will the contractor be able to do it? If so, will it cost more? It is the developer who takes the risk, so this means the developer, architect, and engineer have to work together in the early stages, which is what happened with the Onterie Center. They met frequently. Many schemes were reviewed. A contractor was brought in for his opinion. As a result of the team approach, the developer decided to take the risk, and Onterie came in under budget and ahead of schedule.

***Bank of China (1988).*** The Bank of China in Hong Kong (Figs. 2.28 and 2.29) is the tallest building in Asia at the present, reaching a height of 368 m (1209 ft). The building starts as a cube and diminishes in quarters until a single triangular prism remains. The tower is supported by an innovative structural system that also acts to strengthen the building against high-velocity winds (Tuchman, 1989; *Architectural Record,* 1985).

The building contains an impressive banking hall of 1560 m$^2$ (16,000 ft$^2$) and 180,000 m$^2$ (1.4 million ft$^2$) of office space, 40% used by the bank and the remainder for tenant leasing. The building had to be designed to meet many site challenges, including a steeply sloped, two-acre site with limited access due to a tangle of highways, ramps, and bridges on three sides of the site.

The tower's geometry emerges from a 52-m (170-ft) cube that is tapered to a point at the top. It is then sliced on its diagonal to create four triangular shafts, which are cut

**Fig. 2.26 Onterie Center, Chicago.** (*Courtesy: Gregory Murphey.*)

at different elevations. The north quadrant, facing the harbor, begins to taper at the 17th floor, creating an atrium lounge. The other quadrants are similarly sliced at the 38th, 51st, and 70th floors. The penthouse on the 70th floor uses a screen of aluminum rods on the sloped roof to filter out the sun, whereas great expanses of clear glass on the vertical walls provide panoramic views to the harbor and surrounding mountains.

The building incorporates a unique cooling system, based on evaporation. Because of the limited supply of fresh water in Hong Kong and the corrosive salinity of its sea, an evaporation system was selected over the standard water-cooling system. In addition, conventional roof-top cooling equipment was not compatible with the building's formal aesthetic. The building is serviced by 45 high-speed elevators arranged in banks of six. Express elevators carry passengers to a sky lobby on the 43rd floor.

The building is clad in silver-coated heat-reflective glass. Each pane is part of a rigorous 1.3-m (4.2-ft) modular system designed to unify the building and express its organizational logic. The building is 39 modules wide, its floor-to-floor height is three modules, and intermediate columns are six modules apart.

The structure of the tower is a square tube made up of eight vertical plane frames, four of which comprise diagonal cross bracing. All of the building's loads are collected and transferred through the frames to four massive reinforced concrete corner columns. A fifth column extends through the center of the tower to the 25th floor, where it transfers the accumulated loads diagonally out to the four corner columns. By sending gravity loads to the extreme corners of the building, resistance to high wind forces is increased and the building interior remains column-free. This building is also discussed in Chapters 4 and 5 from both a structural and an aesthetic viewpoint.

***New York Coliseum Proposal (1986).*** The New York Coliseum project (Fig. 2.30) was a submission in a competition for the New York Coliseum site at Columbus Circle. The entry proposed a rehabilitation of the existing 26-story office building and the

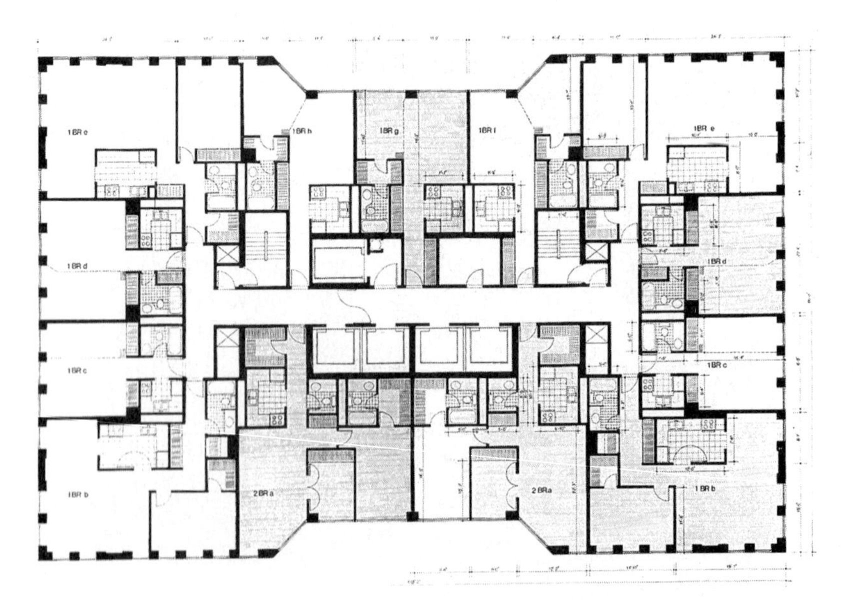


**Fig. 2.27 Onterie Center, typical apartment plan.**

design of a new 137-story, 467-m (1535-ft)-tall building. Designed as a mixed-use building, the proposal contained office, retail, hotel, and apartment space. The plan was octagonal in shape at its base and reduced through a series of setbacks to a hexagon at the top.

The structural concept for the new tower was based on the idea of a steel cage clad in the wall enclosure, reminiscent of the Eiffel Tower. The cage was a structural trussed-tube system, designed to carry both gravity and wind loads. In the upper portion of the

**Fig. 2.28 Bank of China, Hong Kong.** (*Courtesy: Pei, Cobb, Freed & Partners; photo by John Nye.*)

building all of the interior gravity loads were transferred to the outside tube through hanging tensile trusses. The system, apart from being well articulated and aesthetically exciting, proved to be extremely efficient.

***Miglin-Beitler Tower Proposal (1990).*** The Miglin-Beitler Tower (Figs. 2.31 and 2.32), if ever built, would become the world's tallest building, measuring 609 m (2000 ft) from the base to the tip of its spire. The proposed tower will be 125 stories tall, containing 127,273 m$^2$ (1.37 million ft$^2$) of office space, two floors of retail, 12 floors of parking, a sky lobby, and a two-story health club. Massive piers that brace the core are visibly expressed on the exterior as the building carefully sets back to enhance its sense of height and grace (Thornton, 1991).

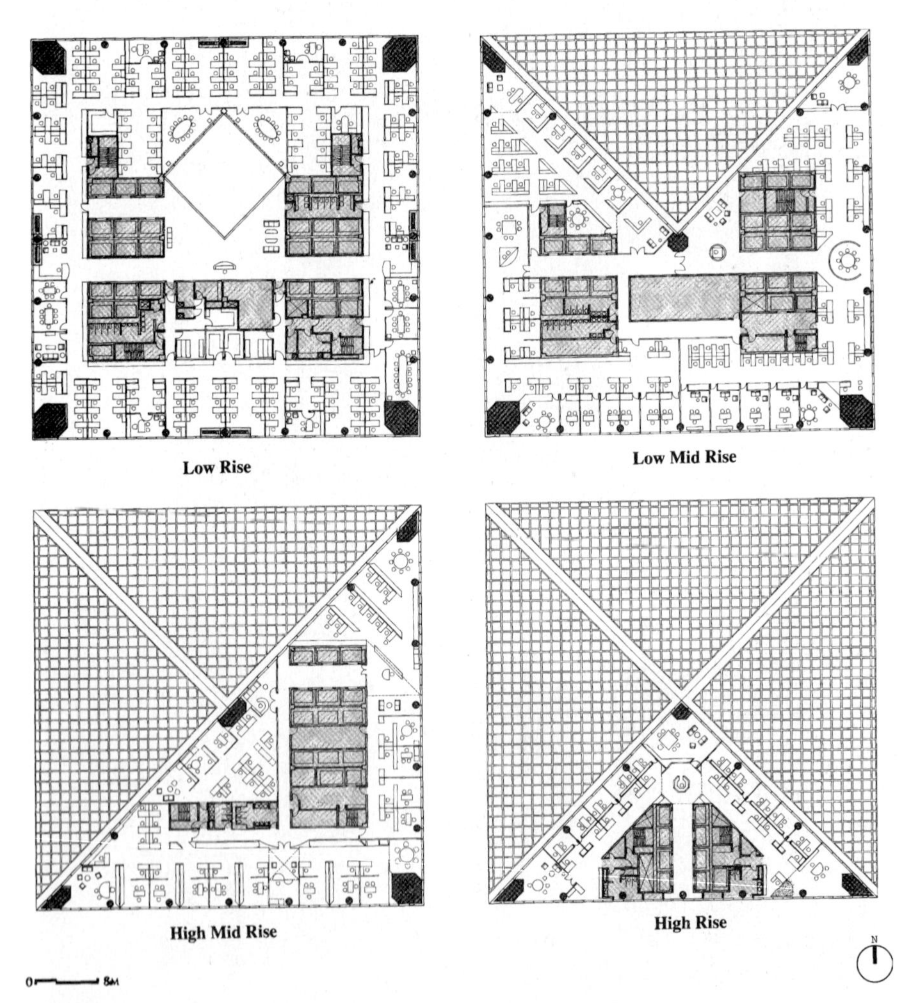


**Fig. 2.29 Bank of China, typical office plans.**

**Fig. 2.30 New York Coliseum proposal, photograph of model.** (*Courtesy: Hedrich-Blessing.*)

**Fig. 2.31 Miglin-Beitler Tower proposal, photograph of model.** (*Courtesy: Miglin-Beitler, Inc.*)

A tall building with small floor plates, a characteristic of the supertall and slender building type, has until recently been uneconomical. In comparison with an average high-rise building with typical floor plates of 2320 $m^2$ (25,000 $ft^2$), the Miglin-Beitler Tower has floor plates that range from 1670 $m^2$ (18,000 $ft^2$) to only 650 $m^2$ (7000 $ft^2$). This represents a new marketing strategy geared to smaller tenants who can afford higher rents and make use of smaller floor spaces.

In slender structures, reducing the amount of space taken up by the core and structure is critical. For this reason, in the Miglin-Beitler Tower the decision was made to locate the bank of elevators serving the sky lobby out of the core and to one side of the building. In addition, where the most efficient core-to-pier connection would have the piers aligned with the edge of the core, this arrangement would have conflicted with the standard 1.5-m (5-ft) module for efficient office layouts. The result has been to offset the piers slightly with the core.

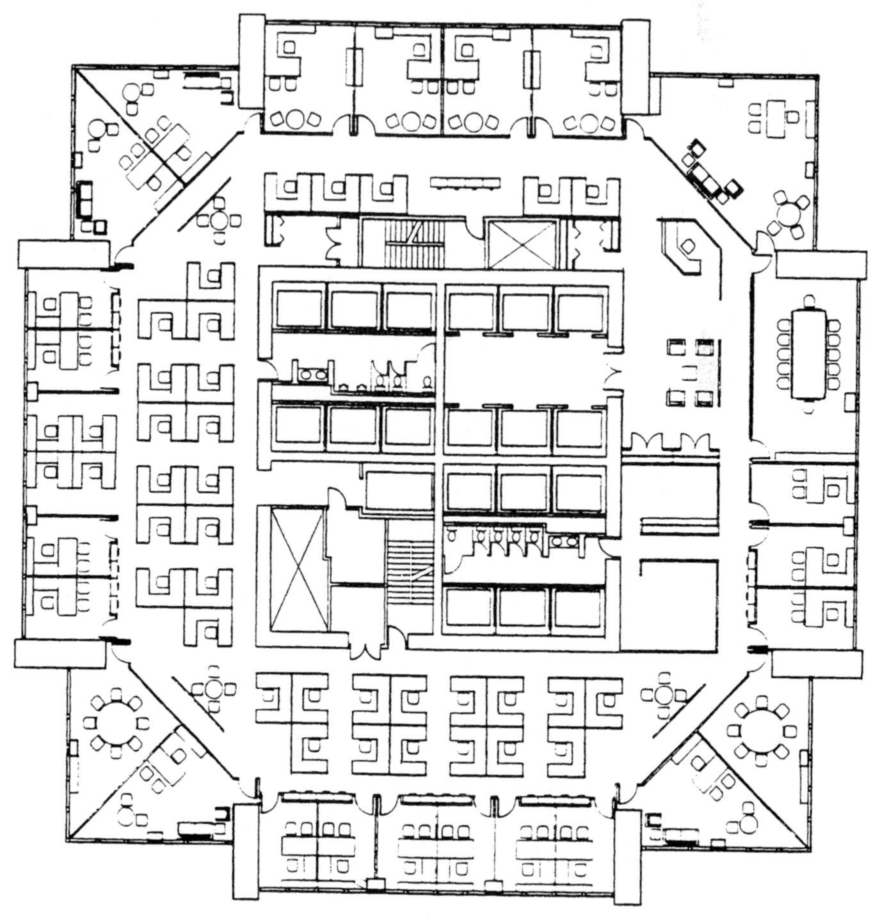

**Fig. 2.32 Miglin-Beitler Tower proposal, typical office plan.**

The building's structural support is a cruciform tube, which uses two massive composite steel and concrete piers at the ends of each side of the building and links them to a central core. The core is a 29-m (65.5-ft) square formed with concrete walls that vary from 0.5 to 1.0 m (1.5 to 3 ft) thick. Openings for access to elevators and services are punched out of the solid walls. The piers are constructed by pouring concrete around steel erection columns. The base of each pier is 2.0 by 10 m (6.5 by 33 ft), which is stepped back in series to a final size of 1.5 by 4.0 m (4.5 by 13 ft) near the top.

Concrete link beams which connect the piers to the core are cast integrally with each floor. Outrigger walls at the 16th, 56th, and 91st stories add rigidity to the composite structure by tying the piers and core together. Spandrels at each floor are backed by Vierendeel trusses that span 18 m (60 ft) between the piers. These trusses transfer the floor loads to the piers and add to the building's lateral load and torsional resistance. The integration of structural and architectural needs has resulted in a dramatic and structurally expressive building. For further details of this building, see Chapter 10.

## 2.4 FUTURE RESEARCH

There are a number of issues on planning and structural system development that need further study. Some of these are listed here.

1. Priorities between open space and ground-floor commercial space.
2. Future possibilities for mixed-use buildings as well as limitations and social implications of such buildings.
3. Optimizing the heights and floor areas of tall buildings.
4. Best location for elevator banks and service cores.
5. Suboptimization of structural and other systems to minimize the total cost of the building.
6. Innovative high-efficiency structural systems for supertall buildings. Such concepts may include the use of deep beams, deep truss, and Vierendeel systems spanning between columns and located in the exterior facades between the glazed areas for windows, cable and suspension systems, space frames, and composite systems.
7. Reducing building drift and minimizing building vibrations by the use of mechanical damping systems, as well as other mechanical systems that could be activated when the building undergoes large movements which could be sensed through the provision of intelligent system technology.
8. Improving structural-architectural integration in the design studios of architectural/engineering firms and of academic institutions. Modes of interaction between academia and practice.

## 2.5 CONDENSED REFERENCES/BIBLIOGRAPHY

The following is a condensed bibliography for this chapter. Not only does it include all articles referred to or cited in the text, it also contains bibliography for further reading. The full citations will be found at the end of the volume. What is given here should be sufficient information to lead the reader to the correct article—author, date, and title. In case of multiple authors, only the first named is listed.

Ali 1990, *Integration of Structural Form and Esthetics in Tall Building Design: The Future Challenge*

Architectural Book Publishing Co. 1974, *Architecture of Skidmore, Owings & Merrill—1963–1973*

Architectural Record 1974, *The High Rise Office Buildings—The Public Spaces They Make*

Architectural Record 1985, *The Bank of China*

Beedle 1974, *On Tall Buildings and the Esthetic Environment*

Billington 1985, *The Tower and the Bridge: The New Art of Structural Engineering*

Billington 1986, *Engineer as Artist—From Roebling to Khan*

Billington 1986, *Technique and Aesthetics in the Design of Tall Buildings*

Blake 1974, *The Rise of a Modern Architecture*

BOCA 1978, *The BOCA Basic Building Code*

Brotchie 1969, *A General Planning Model*

Brotchie 1971, *A Model for Integrated Building Design*

Brotchie 1974, *Systematic Design of Multistory Buildings*

BTM 1974, *Award Winning Building System*

Bush-Brown 1984, *Skidmore, Owings & Merrill Architecture and Urbanism—1973–1983*

CTBUH, Committee 3 1994, *Structural Systems for Tall Buildings*

CTBUH, Group CB 1978, *Structural Design of Tall Concrete and Masonry Buildings*

CTBUH, Group CL 1980, *Tall Building Criteria and Loading*

CTBUH, Group PC 1981, *Planning and Environmental Criteria for Tall Buildings*

CTBUH, Group SB 1979, *Structural Design of Tall Steel Buildings*

CTBUH, Group SC 1980, *Tall Building Systems and Concepts*

Fischer 1962, *Deflection of Tall Buildings*

Fischer 1972, *Optimizing the Structure of the Skyscraper*

Fujisawa 1987, *1400 Foot High Steel Office Building*

Gero 1975, *Interaction in the Planning of Buildings*

Goldsmith 1987, *Buildings and Concepts*

Hodgkison 1968, *An Ultra High-Rise Concrete Office Building*

Iyengar 1977, *Composite of Mixed Steel—Concrete Construction for Buildings*

Iyengar 1986, *Structural and Steel Systems*

Khan 1967, *The John Hancock Center*

Khan 1972, *The Future of Highrise Structures*

Khan 1973a, *Evolution of Structural Systems for High-Rise Buildings in Steel and Concrete*

Khan 1973b, *Newer Structural Systems and Their Effect on the Changing Scale of Cities*

Khan 1974, *A Crisis In Design—The New Role of the Structural Engineer*

Khan 1978, *The Role of Tall Buildings in Urban Space, 2001*

Khan 1982, *100-Story John Hancock Center in Chicago—A Case Study of the Design Process*

Khan 1983, *Structural Systems for Multi-Use High-Rise Buildings*

Kotela 1972, *Trends in the Development of Tall Buildings in Poland*

Lim 1992, *A Super Tall Thin Office Building*

MacMillan 1974, *Amenity and Aesthetics of Tall Buildings*

Mankowski 1972, *Discussion: The Problems of Tall Building Spatial Functions*

Menon 1966, *A Ninety Story Apartment Building Using an Optimized Concrete Structure*

Oh 1992, *Development of High-Rise Mixed-Use Building Utilizing High Strength Concrete*
Petrie 1981, *A High-Rise Office Building Using Long-Span Tube Structure*
Powys 1971, *Economic Aspects of Planning and Design of Tall Buildings*
Praege 1963, *Architecture of Skidmore, Owings and Merrill—1950–1962*
Renn 1990, *Wolf Point Center, A Multifunctional Tall Building*
Rodriquez 1970, *A Form Stiffened High-Rise Apartment Building*
Rubanenko 1973, *Design and Construction of Tall Buildings*
Ruchelman 1978, *Planning Urban Services in Support of High Rise Buildings, 2001*
Sasaki 1964, *A Tall Office Building*
Schueller 1977, *High-Rise Building Structures*
Schueller 1990, *The Vertical Building Structure*
Taranath 1988, *Structural Analysis and Design of Tall Buildings*
Thornton 1991, *Looking Down at the Sears Tower*
Tuchman 1989, *Leslie E. Robertson: ENR Man of the Year*
Vierk 1986, *Continuing the Evolution of the Skyscraper: A 142 Story Steel K-Braced Multi-Use Building*
Xu 1991, *Multi-Use High-Rise Concrete Building*
Youssef 1991, *Megastructure: A New Concept for Supertall Buildings*
Zils 1986, *The Concrete Diagonal*

# 3

# Urban Design and Development

The tall building can be described as a multistory building generally constructed using a structural frame, provided with high-speed elevators, and combining extraordinary height with ordinary room spaces such as could be found in low buildings. In aggregate, it is a physical, economic, and technological expression of the city's power base, representing its private and public investments.

Tall buildings are responses to unique development conditions found in the city. They are a natural response to scarcity of land, concentrated population growth, and high costs of land. Too often tall buildings have been designed without considering them as part of the larger context of the environment. To understand the complex phenomena of both technological and socioeconomic aspects of the built environment, organized efforts are needed to bring together and expand upon our current knowledge in the field of architecture. This chapter examines the impact of tall buildings on a city and the zoning considerations for tall buildings in an urban development scheme.

The chapter is organized in four parts. Section 3.1 serves to explain some of the factors leading to the skyscrapers' existence, their urban design implications, and some of the key factors influencing their form and their interiors. Section 3.2 outlines master planning principles from both a traditional and a modern perspective. Section 3.3 explores the emergence of tall buildings in Dhaka, Bangladesh, and its concurrent relationship with the prevailing zoning, or absence thereof, and the continued explosive urban growth. It provides as well a window of understanding of what is happening in many other similar third-world cities, such as Djakarta, Mexico City, Karachi, Bombay, and Cairo. Finally Section 3.4 presents case studies related to a two-semester urban and architectural studio project for the westward expansion of the city of Philadelphia, Pennsylvania, in the United States. These case studies represent the application of traditional planning principles to tall building types and the development of urban contexts. In a way, Dhaka and Philadelphia are two cities representing, respectively, an example from a developing nonindustrialized country and an example from a highly industrialized country. Master planning features of other cities of the world can be discussed by considering these two cities as reference points. Further discussions on urban design and planning are available in Chapters PC-6 and PC-8 of *Planning and Environmental Criteria for Tall Buildings* (CTBUH, Group PC, 1981).

## 3.1 URBAN DESIGN AND BUILTFORM IMPLICATIONS

It is commonly agreed that the tall building is a complex form. Essentially a commercial building type, programming input from the building economist and marketing input from the real estate specialist would likely have a significant effect on the tall building design, in addition to the efforts of the architect and the engineers. They determine the design in the context of the city block, the street, and the pedestrian, and with regard to its users and the interior spaces they occupy. Added to these design criteria are considerations that must be given to the skyscraper's impact on the city, its skyline, and its physical, social, and economic systems.

Whether by design or by default, the skyscraper or tall building has become an established aspect of our urban existence. It is a piece of international technology that almost every country's major city permits. It has come to represent the attainment of modernity since no other building type incorporates so many of the most up-to-date technology and commercial facets of the modern world. It has also been expressive of changing tastes and practices. It might be viewed as a factory of the post-Industrial Revolution era—an information factory that utilizes technologically advanced communication and information processing equipment.

Many city planning authorities employ the skyscraper's concentration of built space as an integral feature of their plans for density infrastructure utilization and land use. The size and density of a city becomes, for some observers, the basis for defining its urbanity; and a city is said to show its might by how many tall buildings it has and, more to the point, by how large these buildings can be made (Zalcik and Franco, 1974).

At the same time there is a growing concern that tall buildings are intrusive to the city's environment. The reason is that this building type has the ability to transform the city's built and natural environment significantly. One should therefore seek to understand what transformations these buildings might have on the life of the city and on its systems, and how they might be controlled.

### 1 The Skyscraper's Raison d'Etre

Broadly stated, the tall building type can be regarded as a consequence of land economy, finance, urban transportation, investment opportunities, and technological advancements in addition to other considerations.

Physically, tall buildings have a concentration of built space placed on a small site area (that is, many floors stacked over one another on a small ground floor). Functionally it enables usable floor spaces to be stacked high. Commercially it enables its owner to make more profit from the land, and to put more goods, more people, and more rents in one place. It can be defined as a "wealth-creating" mechanism operating within an urban economy. Its economic existence is a consequence of high land values, which are tied to urban accessibility, supporting infrastructure and land uses. This building type is the logical consequence of the optimization of high land costs with the balancing of building economics, of the locational preferences of its tenants, of the desire by its owners to have a highly visible image, and of the consequences of a number of technological advancements in architectural and engineering design (Giangreco, 1973; Robertson, 1973).

Governmental planning policy decisions carried out by economic and physical planners regulate the growth of the city and intensification of land use. Common assumptions made by planners and the public are that new intensive high-rise commercial developments will increase the city's employment and its productivity (or halt its decline), and will lead to the city's beneficial expansion. New technologies and materials have enabled architects to construct increasingly taller structures. The tall building is the culmination of a number of structural and technical innovations which have made its existence possible: the structural frame and wind bracing, new methods of making foundations, high-speed elevators, air-conditioning, flush toilets, large elements of glazing and window framing, advanced telecommunications and electronics, advanced indoor lighting, and improved mechanical ventilation and cleaning technologies. High land cost also drives its designers to devise more economical and better technological methods to build taller. However, as it becomes taller, the lateral stiffness in the structure must minimize lateral building movements due to wind forces and earthquakes. The taller it gets, the greater would be the natural forces of gravity, wind, and earthquakes. As a result, the structural action counteracting these forces becomes an important aspect of its design, which in turn is related directly to the building's economic viability.

## 2 Building Economics and Builtform

One of the key initial considerations influencing the viability of a tall building is the building economics. This determines the extent to which the owner can optimize profitability on the land value and the site's limited built footprint area. This demands the highest efficiency of floor areas with a minimum net-to-gross floor-area ratio and the maximum ratio of the building's gross area to the site area (plot ratio).

In order to achieve these economic objectives, minimum values are sought in the following areas for a cost-effective design:

1. External wall thickness
2. Column size
3. Depth of horizontal (floor) element
4. Vertical circulation and service-core area
5. Floor-to-floor height

The cost justifications for these are obvious. Minimum external wall thickness, reduced vertical member sizes (column sizes), and efficient core areas would increase the net usable floor areas per floor. The minimum horizontal member thickness and floor-to-floor heights would lower structural costs and the area of external cladding, and hence the construction costs. The architect should be aware that regardless of whether the building is low-rise or high-rise, when its economic considerations are given precedence over the aesthetic, human, or poetic aspects of architectural design, the building that results will likely end up as a bland, inarticulate, and plain box. Many of the commercially driven tall building designs are just that—devices for making money as "multistory real estate systems," creating rentable floors in the air. While we should acknowledge the economic benefits of these optimizing objectives, the architect should seek opportunities to emphasize the architectural considerations for a vital city life, while at the same time providing a habitable environment for the building users. Additional information on building economics can be found in Chapter PC-5 (CTBUH, Group PC, 1981).

## 3 Height Limitations and Urban Impact

Should tall building heights in urban contexts be regulated? Are there also limits to the extent of income derivable from a particular site?

Some observers speculate that the supertall building would likely exist more for prestige than for practical economic reasons (Gan, 1974). Theoretically, the 150- and 200-story skyscrapers are a palpable reality already made possible by engineering advancements, but their realization remains only as conceptual prototypes. Limitations to height seem to be more a factor of the building's services, life safety systems, and social acceptance rather than its structure. Proposals for side-mounted helipads, refuge-zone sky lobbies, spiraling monorail movement systems, exterior wall crawlers, and automatic window-cleaning and repair vehicles are some of the possible devices for the supertall building.

One aspect to consider is that the tall building today is more than just commercial offices consisting of the traditional vertical stacking of identical floors. As buildings grew in size, they began to include other space usages such as retail stores, restaurants, bars, apartment penthouses, hotels, and entrance lobbies, some of which are so large that they are gathering places in themselves. Consequently the essence of the tall building is becoming less about sheer height alone or a thin tower rising amidst smaller buildings, but more and more of a multifunction megastructure, a huge mass set beside other large masses. In some major downtown developments, the tall building complexes are concentrated on several blocks, incorporating shopping malls, residential apartment towers, and several hotels. Some megastructures provide so many different functions that the building becomes a virtual city within a city.

Economically speaking, there are optimum heights to the tall building that are compoundly related to each city's rising land values, interest and rental rates, and construction costs. These determine whether, beyond a certain height (for example, 60 stories), enough income could be generated to offset the high cost of land in that block. This optimum height assures that not too many floors would need to be stacked so that the number of elevators required to serve them would not cut into the efficiency of available rental space at the lower floors. At the same time there are optimum building heights beyond which there would be diminishing financial returns (for example, 130 stories), where the projected profits would eventually be nil.

City authorities' imposition of limitations to building height and mass were part of earlier zoning regulations. Planning controls to limit the height of tall buildings and their intensities were first introduced in the United States with the New York City Zoning Ordinance (1915) after the Equitable Building in New York City was built with an unrestricted plot ratio of 1:30. There are differing opinions on the limits of a city's growth, the intensity and concentration of built space in one locality, and its provisions for change. Some economists might contend that the city's expansion is a sign of a healthy economy and that the expanding city shall likely attract international investments. Another view is that the city is in constant flux and can never be in a static ideal state; it has to constantly adapt and respond to countless interest groups.

Generally, the limitations to tall building development in the city tend to be dependent upon its infrastructural systems' thresholds, changing external conditions to the city, and its internal market forces.

The proliferation of skyscrapers within a city presents a particular concern since the tall building, by virtue of its size and height (more than other building types), has a greater physical, social, and economic impact on the city besides, of course, having a highly visible public presence (regardless of any civic or architectural merit) (Contini, 1973). For further discussions concerning urban development see Chapters PC-6 and PC-8 (CTBUH, Group PC, 1981).

The impact of tall buildings will be examined here at the following levels:

1. On the city, its systems, and the skyline
2 On the city block, its pedestrians, and the streetscape

3. On the builtform and the structure
4. On the well-being of the building's users

## 4 The City, Its Systems, and the Skyline

All new buildings in a city have an impact on the city's form and its systems. However, the tall building, by its sheer intensity, demands a different set of considerations and attitudes toward the city.

***The Existing City.*** A tall building has to come to terms first with the city that is already there. For instance, it needs to resolve the issue of massing and how the scale of the new tower as a whole relates to the image of the city, to the city block, and to the neighboring structures. (Some of these may be only four, five, or ten stories above the ground.) Most importantly, it must resolve how it relates to the street's edge, the pedestrian scale that is around, the existing land use, and the character of the block where it is located. These considerations create a set of special conditions for the architect in such a way that propositions that work at the upper parts of the tall building may not be at all effective at its ground or lower levels.

The tall building type presents entirely new form, user, and technological problems that cannot be solved by exclusively delving into historical precedents. Although architects have turned to historical styles for tall building forms, true innovations in tall building design have been the result of collaboration between architects and structural engineers, and advances in materials and technology (see Chapter 4). Governmental planning restrictions often only create a uniform ensemble.

The impact of the tall building on the city and its systems is often circulation congestion. Tall buildings may increase or create congestion of their surrounding movement systems (private and commercial vehicles, public transport, pedestrians on the sidewalk) as well as place additional load on the city's infrastructure and utilities. Tall building development and the city's transportation requirements are interrelated. One entity cannot be considered apart from the other. The city's land-use plan and its transportation plan must be interlinked. Because tall buildings can create a dramatic impact on roads and other public infrastructure, changes must be properly anticipated and accommodated, or the result will be deterioration of the city's transportation systems and other services.

The tall building's development will also have an impact on the capacity of the city's public utilities and services, such as its telephone lines and exchanges, water supply and reticulation system, refuse collection and disposal, electrical supply and load shedding, sanitary system and discharge, and its postal services. Theoretically there are thresholds for each of these, beyond which these systems will not function effectively unless expanded.

Tall building developments can also have an impact on the existing city's urban fabric, its environmental ecology, and the historical heritage of buildings and spaces. Developments should ensure that they will not jeopardize local environmental quality, existing patterns of street life and subcultures, the existing townscape, and the landscape. If the traditional topology of the city consists largely of a mat of the low-rise shophouse type as the city's characteristic feature, then the new tall building's massing should blend with the scale of existing buildings and take it into consideration for conservation, preservation, and adaptive reuse.

The city is a resilient and complex organism capable of absorbing new built intensities inserted into it. More compelling is the view that the city provides a marketplace of ideas, a place in which the very idea of congestion is basic. Congestion becomes not a barrier to its correct functioning but an enhancement of it. A city of tall buildings need not be disturbing so long as congestion is kept within an acceptable range, and this could in fact be a catalyst to enliven the city for its inhabitants.

***Strategic Plan.*** Clearly, some form of governmental planning control of intensive developments is essential (Barnett, 1973). One mechanism to control tall building development, to direct intensification of development, and to encourage the city's overall urban design policies to be used by some city planners is a strategic plan (Zoll, 1974). The strategic plan defines planning policies that should govern the exercise of power. It defines the principles of environmental management, emphasizing the relationships between the city's urban activities and their integration with its support systems. More important, it specifies the practical actions needed to influence events in the city in the direction that city planners and the public want them to go. It fills the gaps that a two-dimensional land-use plan might have left out, by providing a set of ideas, strategies, and objectives for the future of the urban area.

The strategic plan affects the city's urban management, accessibility, diversity, and environment, and it ensures that the new developments will be compatible with the local area scale, character, and amenities.

The strategic plan can justify measures that limit the extent of development and ensure that new buildings (and their component parts) are broadly compatible with the surrounding land use and circulation system. Common planning tools are control of land use, outline of zoning plans, development controls, percentage of site coverage, height control, floor area ratios, building volume and permissible envelope, required linkages, controls of spaces between buildings, bonuses for particular contributions, and other devices such as the use of transferable floor-space ratios.

Zoning may also attempt to regulate the following concerns:

1. Loss of sunlight, light, and air by large, closely spaced towers placed right up against sidewalk property lines
2. Loss of historic buildings and districts with the rapid proliferation of new construction
3. Increased congestion on local streets and increased commuter problems
4. Tower massing
5. Strict sun and shadow control, in particular projected shadows over important public or private open space

## 5 The City Block, Its Pedestrians, and the Streetscape

The tall building has a significant impact on that city block in which it is located. Its sheer size and human population have obvious consequences on the massing of the city block, on the pedestrians at the street level, and on the streetscape itself.

We might refer to these as the tall building's contextual relationships, which should be the subject of an effective urban design plan for the locality. At this level tall building development can be controlled by planners through a local plan that would involve such urban design considerations as place making, the establishing of linkages in its movement systems (and accessibility), and formulating satisfactory figure-ground relationships in the new massing within the block, the street, and adjoining buildings.

***Public Spaces.*** The tall building's insertion into the city block must involve "place making" where the new development can provide a destination of spaces as "urban rooms" or as occasions for social interaction. By responding to urbanistic values, the architect might give the tall building some sense of civic use (if appropriate to the locality). Increases in urban population intensities would obviously increase the need for public spaces. This increase might be met by incorporating these spaces as noble civic spaces around or in the building itself, both at the ground level and at the upper parts of the building.

Particular attention might be given to the base and lower floors of the tall building as these areas serve the street, whereas its top generally serves the skyline. The tower's design must resolve how the tall building joins the ground to meet the street and its pedestrian life. In this regard, the base must respond primarily to the conflicting demands of its immediate surroundings at the street level and secondarily to the skyline of the city. Its base and lower floors might include retail spaces and public plazas, not merely as formal devices to set off the form of the tower, but to generate activity and "humanize" the scale at the street level of the building. These civic spaces or plazas, if they are included, could be enclosed, or they could even be squares or agoras. They may also be located above the street level, although such design devices can diminish the activity on the street, where the urban experience is more appropriately positioned. There can also be some sort of atrium, retail mall, or galleria, which itself may be multilevel. Opportunities for social interaction might be through the provision of cafes. (Refer to Chapter 5 for more information on the building base.)

There would be some who would argue that the indoor atrium is simply an indoor version of the plaza and does not show respect for the traditional urbanism of the city, and that the atrium represents an extinction of the street. For instance, the atrium might be regarded by some as being antiurban because it turns the street outside in and converts the street's edge and recesses into an enclosed space. The design challenge for the architect is how to move into an atrium off the street without a total denial of the street itself. For example, Lever House in New York City (1952) constitutes the first serious modernist assault upon the contextual continuity of the building on the streetscape, with each building organized into a series of solid blocks. Rejecting the alignment of the street, the vibrant bar of the principal body is set crosswise, whereas a subtle two-story slab continues the line of the surrounding buildings through which it drags the surrounding urban space into the void of the pillared story beneath (see Chapter 7).

***Figure-Ground Relationship.*** In designing the tall building, its figure-ground relationship with its context needs to be studied. This forces us to examine the building's configuration at the street elevation and the spaces between buildings. The tall building's base has to deal directly with the city block's urban fabric and be proportioned to the city's horizontal scale. Its design should pay close attention to the site coverage of the surrounding structures. If the new tall building is set on a plinth, then this can physically remove it from the street entirely, from the surrounding buildings in the same block, and from the ongoing life of the city around it. In many instances where the building stands back from all sides of its lot, it becomes disconnected from the street line and could become disjoined from its context. In terms of figure-ground, the ground space between tall buildings should not end up solely as vehicular driveways, which then could become a threat to urban linkages by resulting in disjointed urban forms and the creation of "island"-building situations.

***Street Life.*** Clearly, of paramount importance is the relationship between the pedestrian at the street level and the building. The tall building should form linkages with the surrounding buildings to reinforce the city's urban fabric and enhance street life at its base. If these are neglected, then the building stands alone, devoid of any connections to, or acknowledgments of, the city block of which it is a part. It then becomes like an island, such as one in which the tower is set back from the street by a plaza or park [for example, the Seagram Building in New York City (1957)]. The problem with such a site layout is that it rejects rather than reinforces the street life by having the activity in the tower's base set back too far from the sidewalk. The pedestrian is distanced and subsequently discouraged from window shopping, from gazing at the building's lobby (which may be filled with public art), or from visiting the small retail shops which could be located in the tower's base. Building setbacks from the street can separate the tower from the sidewalk, eliminating pedestrian plaza life itself, and discouraging communication and movement into and around the tower, particularly from the street's pedestrian traffic and its access points into the building.

The island building is better geared for the automobile than for the pedestrian. It is primarily an isolated building on an isolated plot. Many such buildings have destroyed the existing streetscapes of many American cities by eliminating older buildings that had once defined a hard edge on the street. In cases where the streetscape has been irreparably altered, ill-defined or unusable spaces are opened up on the street, resulting in the street losing its character and intimate scale. The new towers also have replaced those existing parts of the city where its public life had traditionally been active (for example, squares, markets, streets, and arcades). This situation was encouraged by bonus provisions of many zoning ordinances which gave additional building density in exchange for street-level plazas or open public spaces. In retrospect it is clear that the effect of such design elements has great potential for damage to the street and the pedestrian zones of the city.

Besides these design issues, architects also need to avoid designing "fortress" buildings such as those with lavish interiors, which are barricaded behind a screen of security, restricting public interaction. These buildings will only serve to create a defensive and divided city that will eventually embrace only those individuals who can afford the services offered within. On the contrary, buildings must provide the city with commercial and public spaces accessible to people at grade rather than alienating them. The interaction between the people within the base of the building and those on the street must be encouraged.

The common argument of many urban designers and planners is that the tall building's base should hold the street line to avoid the separation of the tower from its urban context, and that any setting back of the building from the street edge is disrespectful to the line of buildings that mark the street and are crucial definers of the city's urban fabric. Such rigid attitudes are too limiting. If the tall building's facade is placed right up to the street line and also incorporates transitional zones in the interior of its lobby, then the base of the skyscraper could relate its interior spaces to the life of the city outside.

In the case of a tower and building group set back from the edge of the street, a large covered lobby might be provided at the street's edge with a recessed veranda way to address the street. This can physically unify the buildings and towers rising above the street level and, thus, turn the whole complex into a unit. Tall buildings can also have public areas at higher levels in the building, which include green landscaping so that the park is now within the tower. Such devices should not take precedence over the street pedestrian zones. The tall building should always be seen in relation to its context, whether it is located as an infill building within the block or as a corner building defining intersecting streets. These issues need to be addressed so as not to diffuse or weaken the streetscape and the city grid or its existing street pattern (Cowan, 1974).

***Microclimatic Environment.*** Other urban design considerations include the impact of the tall building on the microclimatic environment of the city block and its surroundings, such as the shadows that the tall building might cast (say, over an important public or private park) during the major hours of solar access at the street. These shadows may significantly change the character of the area, affecting the microclimate as well as blocking out vistas. The tower may also create powerful wind downdrafts and uncomfortable eddies at its base, which would be objectionable to pedestrians. Wind tunnel tests performed during the preliminary design stages could contribute to incorporating design changes aimed at averting these ill effects. For more information on the effects of tall buildings on the microclimate refer to Chapter 1.

## 6 The Builtform and the Structure

***Marketing Considerations.*** From an economic viewpoint, the tall building's form might be regarded by the speculative real estate developer as simply the total mass permissible under the zoning laws which has a series of marketable floors. In commercial terms, the tall building becomes a series of identical vertically stacked floors, each representing a single prototype of valued leasable space. For many architects, complying with design objectives along such narrow commercial parameters means that they need only to respond to abstract user requirements (rather than, say, the customized requirements of the owner-occupied tall building). The architect finds himself torn between his aspirations for aesthetics and the demands of marketing. If the architect were to express the tall building's commercial values exclusively (for example, leasable floor areas and fashionable imagery), the building might only become identified with the expression of the prevailing real estate market. This approach may explain why so many commercially driven tall buildings convey such a bland neutrality or are decorated with "pastiche," which does not relate to either the building's form or its structural logic.

On the contrary, architects should strongly argue against this practice and support the idea that the new technology of the tall building should create a new architecture. Tall buildings demand not merely a different set of design attitudes toward the making of cities, but also a different aesthetic for individual tall buildings, which can derive from their context and climatic conditions. Instead of trying to solve new problems with old forms, architects should endeavor to develop new forms from the exigencies of site, context, and climate; and help define the appropriate technology to respond to these issues. The challenge for the architect then is to devise means for making a coherent design when the basic impulses behind the building (such as the need to relate to its context, the desire to relate to history, and the necessity to make use of contemporary high-rise technology) may not seem consistent with each other.

Marketing considerations may suggest that certain floor plans could become a serious rental liability. For instance, if a floor plan lacks corner windows, this would prevent any highly desirable corner offices. Floor plate planning should also allow for different tenancy situations (for example single, double, or multiple tenancies) and functions. Furthermore, real estate specialists might recommend added sales and marketing features into the building's design, such as large ground-floor atria, corner bay windows in the typical floors, sky terraces, user-friendly building features, and sky plazas.

***Skyline Identity.*** There exists a dialectical relationship between the tall building's base and its top which is similar to a discourse between two languages—individual at the top and more utilitarian at the base. Tall buildings already assume the role of high-level icons for the city. Tall buildings can create an epic-scale skyline, whereas at the street level they act on a human scale. At the tall building's skyline, its uppermost termination, the architect has greater opportunities for aesthetic expression. A skyscraper without a skyline identity might just be an unimpressive lump of mass in space. See Chapter 5 for more discussion on this subject.

## 7 Interior Life of Building Users

***Interior Life.*** A tall building's presence has consequences not only for the city and its block, but also internally for its inhabitants and their interior life. Ideally, the building design should create a high-quality interior life for its inhabitants. The creation of interior spaces should be more than just merely stacks of concrete trays as a workplace in the sky. Spaces for work in the tall building are generally remote from the ground level and from green open spaces. They should have some degree of humanity, some degree of interest, and some degree of scale. We can make social spaces and public spaces at intervals all the way up the building in the form of urban civic spaces in the building, such as sky courts and sky lobbies. This is an underexplored area in the design of tall buildings. In so doing, we may be able to change the quality of the spaces in the upper parts of the skyscrapers and in the city. The tall building design might create "events" for the building's inhabitants as they occupy, socially interact, and use the upper parts of the tall building.

***Penthouse Floors.*** The spaces at the top of the tall building are considered special spaces, and therefore these spaces and the interiors deserve greater design attention. These levels are obviously well suited to contain prestigious offices and luxurious restaurants or penthouse apartments. These penthouse floors can provide spectacular panoramic views of the city combined with the possible inclusion of the comfort of rooftop tropical gardens or communal spaces in the sky. The tall building's design thus provides opportunities to create a unique identity against the sky.

***Curtain Walls.*** The city and the tall building's interior spaces are separated from the outside environment by a wall at the edge of the building's floor plate that can become an extreme case of absolute permeability between the interior and the exterior. As buildings become taller, curtain walls which did not leak in buildings of fewer than 20 stories may have problems when installed at greater heights. The temperature differentials are higher, the winds are stronger, and the deflections are greater. To counter these problems, manufacturers frequently cooperate with architects and owners in constructing full-size mock-ups tested by simulating actual wind and rain conditions. Chapter 7 discusses the design and techniques of curtain-wall facades in more detail.

To create a softer edge to the tall building and to layer the relationship between the interior spaces and the exterior, the architect can add balconies and terraces. These transitional spaces reduce the sharp distinction between interior and exterior spaces as well as providing sunshading to the tower's sun-facing sides.

***Lighting Aspects.*** In the early tall building, natural lighting requirements would determine that around 8 m (27 ft) from the external window to the core was about the maximum preferred core-to-window distance. [For example, the distance from the window to the furthest desk is 7.5 m (25 ft) in the Lever House.] Often these early tall buildings have stepped profiles that step back when elevator banks are dropped as the buildings increase in height.

However, as interior artificial lighting and HVAC technology improved, they permitted wider and more efficient skyscrapers to be built, creating a larger amount of leasable space on very large floor plates. With the advent of these new technologies, office workers no longer need to be located directly adjacent to windows to receive light and air. This meant that uniformity in floor plate areas based on standard wall-to-core dimensions was no longer sacrosanct. Because of this and a growing market demand, very deep floor plans evolved. However, deep floor plans often require deep ceiling voids to house the mechanical and electrical (M&E) systems and equipment, thereby increasing floor-to-floor heights. Design then became a balance between acceptable dimensions for floor-to-floor heights in relation to structural beam depths, and the extent of ceiling voids for the M&E services. These ceiling voids (along with the structural ele-ments) can provide new opportunities for expression on the exterior of the building, such as expressed spandrels and variations in the structural and mullion grids.

## *3.2 CONCEPT AND CONTEXT IN URBAN DESIGN*

Urban design and development for a city in a developed country must be viewed from a different perspective than urban development in a developing country inasmuch as the issues and problems are of a different dimension. Section 3.4 describes a two-semester urban design exercise for graduate students in an American school of architecture, designing for an American city, under the guidance of one of the United States' preeminent city planners. During the academic year of 1991–1992 students at the University of Illinois at Urbana-Champaign worked on an urban and architectural design project for the westward expansion of Philadelphia with Edmund N. Bacon, author of *Design of Cities* (1976) and former executive director of the Philadelphia Planning Commission.

Although narrow in specific focus, this graduate studio project nonetheless has important international implications for urban design and for the design and location of tall buildings in urban settings. This case study is most relevant in the context of this Monograph because of the important role modern tall buildings play in the evolution of historic cities throughout the world.

The central concept is that individual buildings should be viewed more as component parts of a larger whole, the city, and not as isolated works of art. Urban design and architectural design of urban projects should be accomplished holistically and synergistically. Holistic design places emphasis on the whole and the interdependence of its parts. Synergistic design suggests that the whole is greater than the sum of its individual parts. To achieve these ends, architects, planners, and urban designers need to exercise their abilities to visualize cities as a totality, in three dimensions.

An important aspect of this project is the freedom that students in architecture have to explore design concepts beyond the constraints associated with professional practice. The students were encouraged to play the role of city builders instead of designers of individual buildings on single sites. Designing a city on a site-by-site basis with several independent agents often results in disjointed site design, a phenomenon associated with many urban development schemes. "Conflicts between parties are inevitable...(and) multi-developer design is fractionated" (Lynch and Hack, 1984). Students in the ideal setting of a university design studio are entirely free of these conflicts. Accordingly, they could be independent and unfettered in their individual pursuits of a holistic urban design solution for Philadelphia.

The University of Illinois students were encouraged to understand the existing form and structure of Philadelphia as a basis for their design concepts, much like a sensitive architect's design process for remodeling of a significant older house.

This process is valid for the design of tall buildings in cities throughout the world. Accordingly, the student projects discussed here provide internationally significant insight to architects, planners, and urban designers as they seek the proper role of tall buildings in urban design.

## 1 Tall Buildings in Utopia

Two internationally prominent architects of the early twentieth century developed utopian visions which incorporated tall buildings in quite different ways. Their approaches illustrate many of the well-meaning but flawed urban design ideals of the architects of the time.

Frank Lloyd Wright viewed the city as "obsolete." His utopian (or *Usonian*) replacement was Broadacre City, developed in the late 1920s. Wright was convinced that the automobile or private helicopters would allow people to live in decentralized communities of independent farmers and shopkeepers. Wright proposed a settlement of largely self-sufficient farms with community and cultural centers placed at strategic locations. Individual high-rise apartments were almost reluctantly proposed at crossroads in his open grid for "the city-dweller as yet unlearned where ground is concerned" (Fishman, 1977).

Le Corbusier celebrated and admired the city that Wright deplored. His plans for Ville Contemporaine and Ville Radieuse proposed large superblocks of apartment buildings separated by vast open space. "The Ville Radieuse took the open city concept of the Ville Contemporaine to its logical conclusion, and a typical section through the entire city showed all the structures raised clear of the ground, including the garages and access roads. By virtue of elevating everything on pilotis the ground surface would have become a continuous park in which the pedestrian would have been free to wander at will" (Frampton, 1985).

Cities did not disappear, as Wright predicted. Rather, they were "renewed" to provide new places for urban dwellers and tourists to embrace the cosmopolitan vitality of city life. Tall buildings did not become an architectural type associated with rural development. Wright reconsidered the role of tall buildings in urban design when he subsequently proposed the tallest of buildings, his 1956 project for a mile-high Illinois Tower for the City of Chicago.

Cities did not develop exactly as Le Corbusier imagined, as open green spaces with streets in the air. Le Corbusier's vision for the use of tall concrete slab buildings became widely accepted in the United States as the model for public housing . This is exemplified by Pruitt-Igoe Housing in St. Louis, Missouri, built in the mid-1950s but destroyed in 1972 (Jencks, 1984). The anonymity of building form and "streets-in-the-air" design contributed to a greater crime rate at Pruitt-Igoe than in other housing developments (Newman, 1973). From these and other examples it is clear that such idealism cannot solve the urban challenge. Cities are metamorphic—evolving over time. They grow, decay, and are redeveloped, hopefully retaining the best elements of their past.

## 2 Contemporary Tall Buildings in Their Urban Context

Individual buildings contribute to urban design by helping to define streets and public spaces, and by generating pedestrian movement to and from the street level. The design of individual buildings should provide a rhythmic stimulation as you move along the ground plane that Bacon calls "architexture" (Warfield, 1991). Tall buildings have the unique urban design opportunity of providing visual interest at the skyline. The configuration against the sky is the point where tall buildings "connect with infinity," a characteristic of towers Bacon calls "archetype" (Warfield, 1991).

An example of a tall building as archetype and with architexture is the Hong Kong and Shanghai Banking Corporation headquarters in Hong Kong (also known as HongkongBank). It is a structurally expressive building engaging to the eye whether viewed in elevation from the harbor against the background mountain, from the street as a profile against the sky, or from Victoria's Peak looking down on the city.

Pedestrians approaching the Hong Kong and Shanghai Banking Corporation headquarters are rewarded with interesting structural detailing which animates and "models" the form of the building. The building seems to float above the plaza, inviting pedestrians to its shade and shelter with a nearly seamless sequence from the plaza to its multilevel entrance lobby connected to a continuous interior volume. See Chapter 7 for another perspective on the Hong Kong and Shanghai Bank.

In the same city, the Bank of China building is an interesting archetypal profile as viewed from the same three perspectives as the HongkongBank. The Bank of China's design includes an arched entry sequence leading to a multistoried lobby, making a successful transition from the busy urban street to the peaceful and impressive interior of the bank.

By contrast, the John Hancock Building near Copley Square in Boston is a pure, continuous shaft of reflecting glass from street to sky, with little articulation of its profile, base, and plaza. Unlike the Bank of China, which joined a chorus of similarly scaled buildings in Hong Kong, the John Hancock Building's scale dominates the historic profile of the surrounding area. Chapter 7 discusses the contextual relationship of the facade of the John Hancock Building.

Norberg-Schulz (1979, 1980) described the effect of this building as destructive of the spirit of its urban context.

> During the last decade large parts of the urban tissue (of Boston) have been erased, and scattered "super-buildings" erected instead. The development culminated with the John Hancock Tower by I.M. Pei, which destroys the scale of a major urban focus, Copley Square. As a result, Boston today appears a hybrid city; the old remains, such as Beacon Hill, make the new buildings look inhuman and ridiculous, and the new structures have a crushing effect on the old environment, not only because of the scale, but because of their total lack of architectural character. Thus the place has lost its meaningful relationship to earth and sky.

The Sears Tower in Chicago has an archetypal profile and an innovative structural concept. This tall building has become an icon for the City of Chicago. Norberg-Schulz argues that the Sears Tower is well suited for the *genius loci* of Chicago with its architectural heritage formed by Jenney, Mies, and Skidmore, Owings & Merrill. Chicago is a city identified with innovative Chicago school architecture, such as the Rookery by Burnham and Root on LaSalle Street and the Carson Pirie Scott Department Store by Louis Sullivan on State Street, and more recently the First National Bank, the John Hancock Center, the Civic Center/Daley Plaza, the Federal Center, and the Sears Tower among others. All these modern tall buildings have individual, specially designed plazas and specially sited sculpture as part of the total project. Daley Plaza has the Picasso sculpture, Federal Center a Caldar stabile, and First National Bank a Chagall mosaic. Further discussions on these buildings are presented in Chapters 2, 4, 5, and 7.

Shanghai's urban skyline viewed from the Huangpu River is as recognizable as Chicago's skyline from Lake Michigan or Hong Kong's from Hong Kong Harbor. Shanghai's skyline ascends from the lower scale of lesser important buildings to the taller scale of the Peace Hotel and the Customs House where Nanjing Road intersects the "Bund," the embankment along the Huangpu (Warfield et al., 1989). While some new development threatens this historic skyline, one new tower plays a highly successful role in the urban context of Shanghai. Located several blocks west on Nanjing Road is the East China Power Administration building, a brick-clad structure with a profile that varies with each elevation. The tower culminates in a narrow shaft rising six stories above the main body of the building, containing electronic equipment expressive of technological power. At the base, the architects provided exciting entry spaces with bridges and other details appropriate for the pedestrian scale.

## 3.3 URBAN GROWTH AND CHANGE IN DEVELOPING COUNTRIES

The role of the tall building as a landmark as well as a consequence of population, economics, and land-use requirements in developing countries is discussed in this section. The city of Dhaka in Bangladesh illustrates modern and regional master planning strategies and high-rise development applied to a major city in a developing and nonindustrialized country.

### 1 Megacities

In developing countries throughout the world, major cities are experiencing phenomenal growth and change. Many of these cities are being transformed into "megacities" in the short span of 20 or 30 years. In some cases these are accompanied by growing commerce and wealth. However, the rapid growth in population and wealth (though in relatively few hands) is exerting tremendous pressure on land values and availability of buildable land, particularly in the inner city areas. Consequently we are seeing an emergence of tall buildings in which commerce and shelter are being provided in ever greater numbers. In the process they are changing the face and the urban character of these newly emerging megacities. In Asia, a number of such cases are appearing rapidly. One such city is Dhaka, the capital of Bangladesh. The emergence of tall buildings in Dhaka is a fairly recent phenomenon. With a few exceptions, tall buildings have been built in significant numbers only during the last 25 years.

A tremendous growth in population and density, zoning regulation changes, escalation in land values, and changes in business and market conditions have all played varying roles in the recent emergence of tall buildings in Dhaka. These conditions are similar to those in many other cities in the developing world (Liang, 1973). Zoning or its absence is largely responsible for the explosive urban tall building growth.

For the ensuing discussion on the emergence of tall buildings in Dhaka, it is relevant to review and discuss the issues that are leading to such development. It is clear now that major urban centers in many developing countries are growing rapidly and much faster than the rate of population growth of the country as a whole and faster than the overall rate of urban growth. Cities are currently absorbing two-thirds of the total population increase in the developing world. At this rate nearly two billion people are expected to populate the urban areas of the third world by the year 2000, representing a further 600 to 700 million urban dwellers over the next decade (Figs. 3.1 and 3.2). To drive the point further, half the urban population in developing countries are located in some 360 cities, each numbering over half a million inhabitants. At the beginning of the twenty-first century there will be over 520 cities of this size, many of which will have become major metropolitan areas (Fig. 3.3). Obviously the urban growth and expansion will not stop there. A point to remember is that world population is likely to turn from being predominantly rural to predominantly urban shortly after the turn of the century (Fig. 3.1). As a consequence, the growing megacities of developing countries will run into a myriad of problems, including huge urban sprawls and encroachment on sometimes very important agricultural lands, as in the case of Dhaka. At the same time, most of these megacities would have one or more commercial centers where we will see the growth and evolution of tall buildings. The reasons for their rise and rapid growth are varied, and some of them will be discussed in this section in the context of a review of the situation in Dhaka, which represents many of the urban problems of most of the major cities in developing countries such as Asia, Africa, and Latin America.

## 2 Emergence of Tall Buildings in Dhaka City

In matters of planning and urban design, few of the major Western and developed world cities, while having extremely large metropolitan areas and population, have had to deal with as complex social and economic issues as many of the major developing cities have faced. These issues are the result of tremendous and rapid changes in population in a relatively short period of time, increases in urban density, and, in some instances, land value escalation that accompanies those changes. The changes in the urban scene of the cities of developing countries have been too abrupt and rapid. Thus drastic population growth, expansion of the city, and related issues are additional factors involved in the emergence of tall buildings in many growing cities of developing countries, such as in the case of Dhaka.

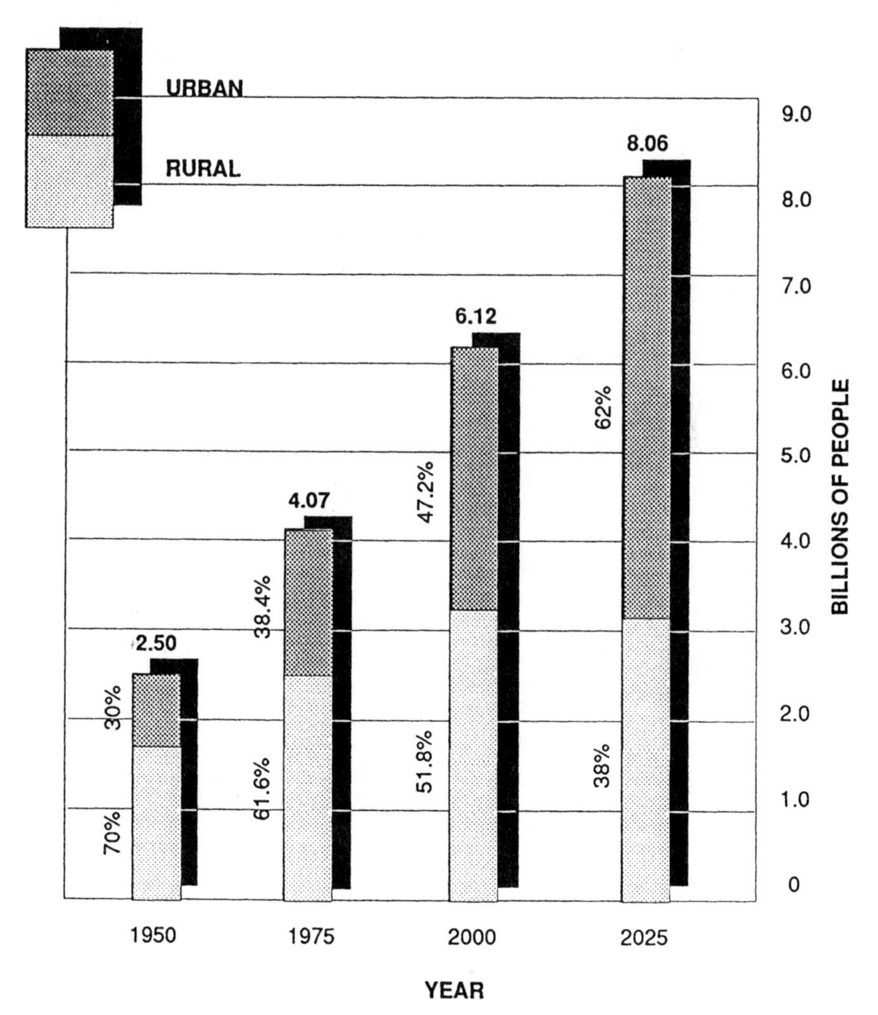


**Fig. 3.1 World population, rural versus urban split.** (*Courtesy: Tanwir Nawaz.*)

Since the creation of East Pakistan (now called Bangladesh) in 1947 and its establishment as the provincial capital, Dhaka was transformed from a district town to an administrative and commercial center. The growth of Dhaka from approximately 250,000 people in 1947 to more than 7 million in 1994 is a testimony to the tremendous growth and expansion of this capital of a very populous developing country. In fact, Dhaka is one of the fastest growing major urban centers in the world, with a population expecting to reach the 25 to 30 million range in the year 2025.

The emergence of tall buildings in Dhaka can be significantly attributed to the rapid growth of the city into a very large urban center with a high population density, escalating land values, and the concentration of wealth and decision making in a geographically concentrated area.

In the past, Dhaka has grown between 10% to current estimated levels of between 6.0 and 6.5%. While the rate of growth may have slowed from 10% in the 1960s and 6% in the 1970s, the actual population is higher than ever before. As a result, due to the tremendous pressure on the land and escalating land prices, scarcity of buildable land, market pressures, and other reasons, many planners and developers are looking at high-rise buildings for both residential and commercial solutions (Fig. 3.4).

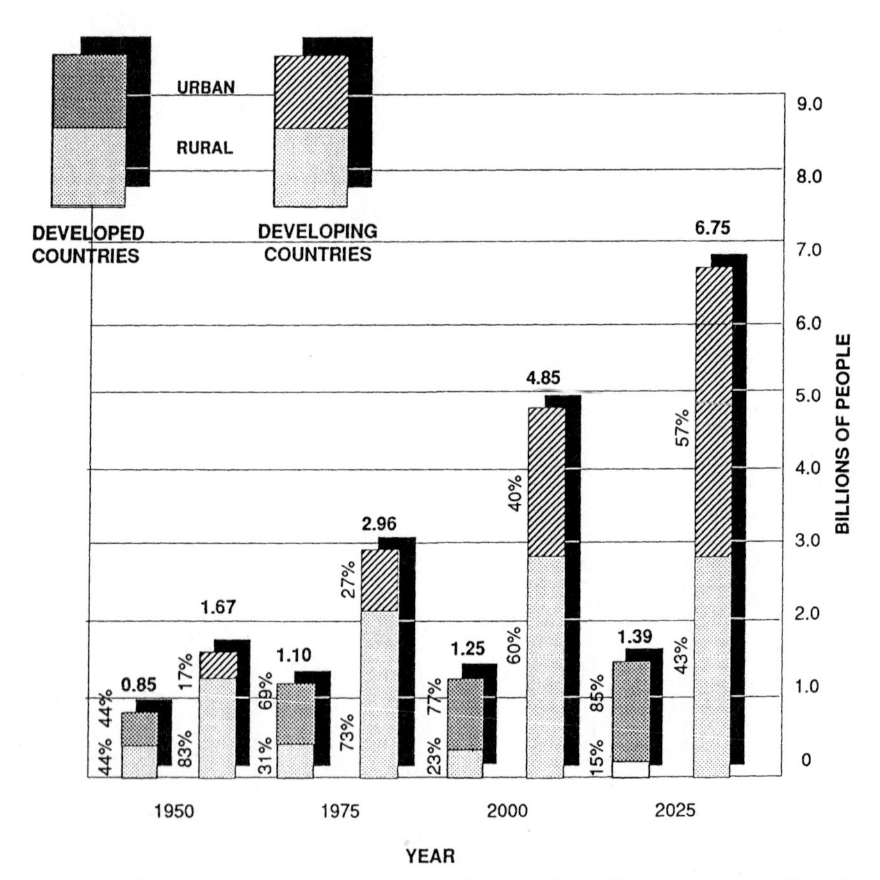


**Fig. 3.2 Developed versus developing countries, rural versus urban split.** (*Courtesy: Tanwir Nawaz.*)

A review of Table 3.1 indicates the relationship between the rapid population growth and consequent urban expansions for Dhaka. While the figures up to 1990 represent facts derived from various sources, the figures for the years 2000, 2010, and 2025 are based on flexible projections assuming a 4.5% annual growth rate. It becomes clear from these estimates that not only the physical expansion of Dhaka is continuing at an alarming rate, but also the density of population is increasing. This means that despite continued physical expansions, more and more people are being concentrated on the same piece of land. Herein lies part of the reason for high-rise developments and continued future growth of Dhaka.

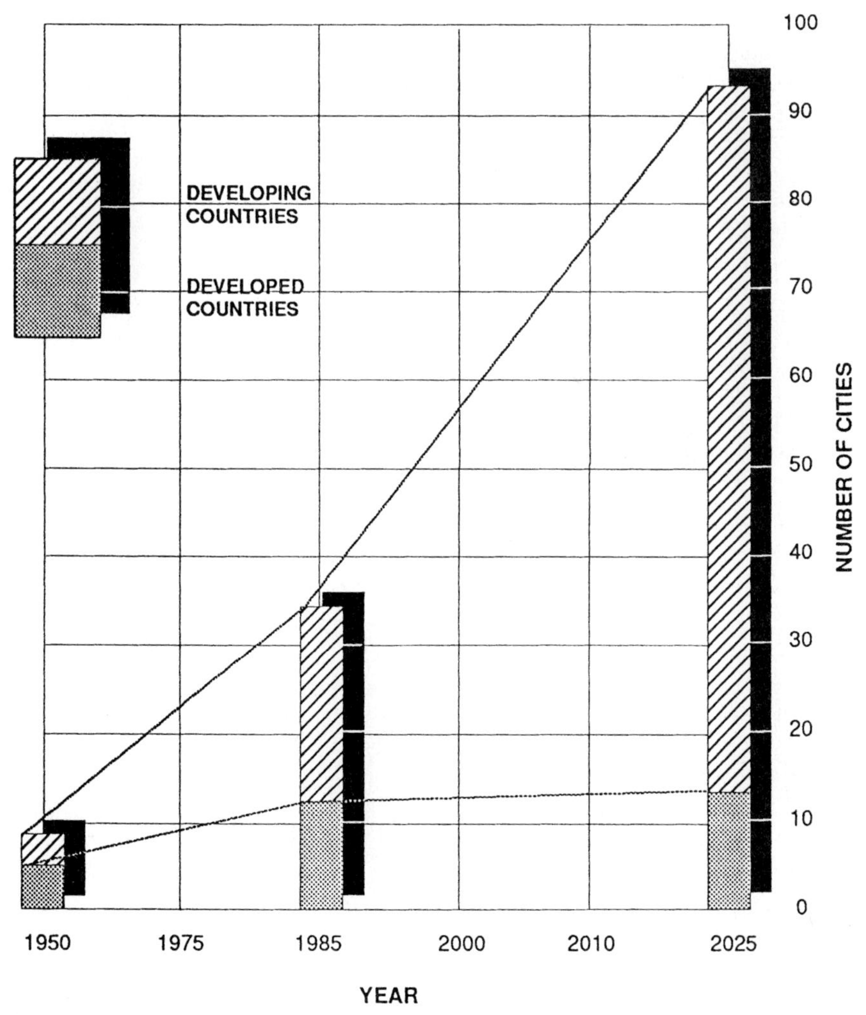


**Fig. 3.3 World's major cities of over 5 million people, developed versus developing areas.** (*Courtesy: Tanwir Nawaz.*)

### 3 Historical Context

With the formation of Pakistan in 1947 in the Indian subcontinent, Dhaka was transformed from a sleepy and quiet district town to the capital city of the eastern wing of the newly created state. Up to this point this was a major district town with a mercantile character on the one hand and a university town (since 1921) on the other. Physically there was a clearly defined older city, mostly to the south and west of the then existing railway line, whereas the university campus and the teacher and student residential areas as well as some older district administration and residential areas remained to the north of the railway line.

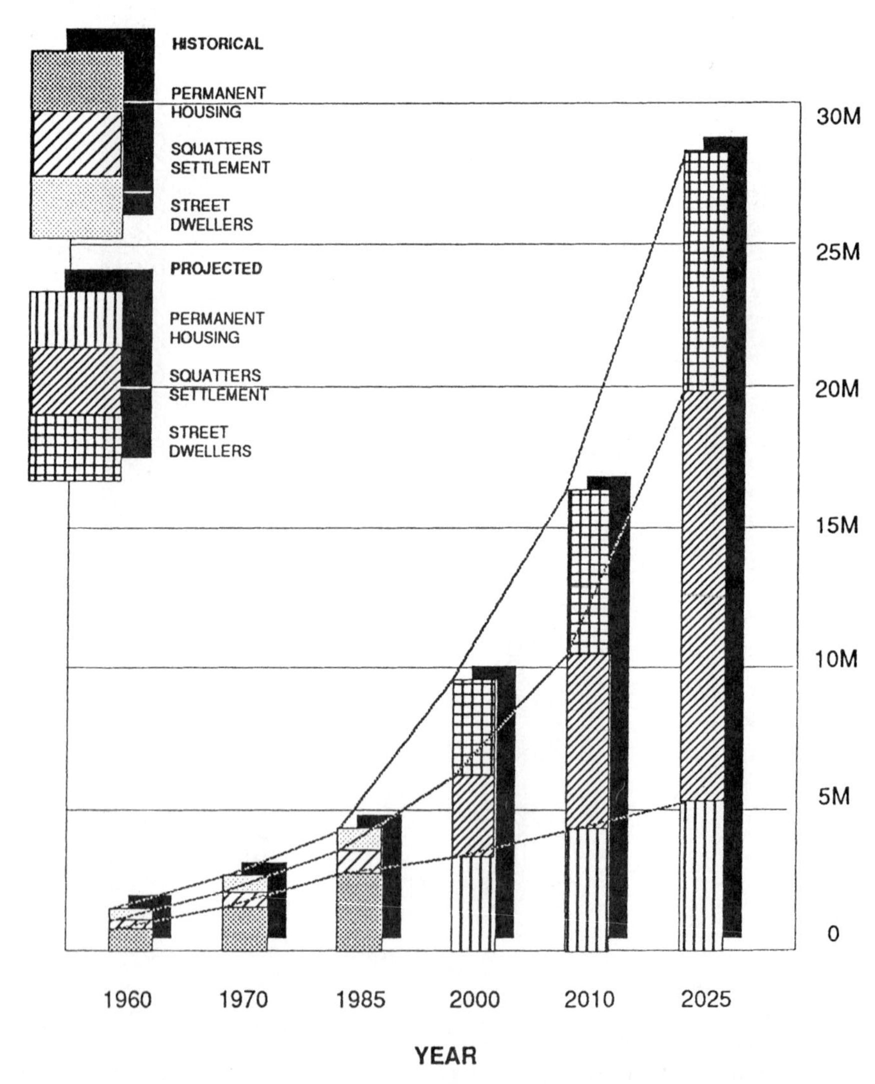


**Fig. 3.4 Dhaka, Bangladesh, projected population and impact of housing quality.** (*Courtesy: Tanwir Nawaz.*)

There were no tall or even midrise buildings at that time. The city had an area of approximately 72.5 $km^2$ (28 $mi^2$) and a population of around 200,000. The maximum height of any residential or commercial building was three stories. The city has undergone tremendous changes since 1947. Today it has a population of over 7 million (1994), an area of approximately 777 $km^2$ (300 $mi^2$), and many low-rise to midrise commercial and residential buildings. The city is expected to reach a population of 10 million, according to various projections, by the year 2000 and to have an area of over 906 $km^2$ (350 $mi^2$). Dhaka is one of those developing cities that is undergoing major and rapid changes in size and overall character primarily due to its tremendous urban migration and its role as the principal economic and technological center of the country. It can be seen from Fig. 3.2 that the rural sector in most developing countries, and particularly in densely populated underdeveloped countries such as Bangladesh, is an area of marginal growth in population and employment opportunities. Because of a slow pace of development and the lack of modern amenities and employment opportunities in rural areas and small towns, more and more people will crowd into this major city and change the urban character. Trade, commerce, and industry will also generally concentrate in this relatively small area in the future (Siddiqui et al., 1990).

In this context the questions arise: How does this change affect the city in the form of growth of tall buildings, and where is the current urban development headed? Historically, the first business commercial zone area to appear in Dhaka was the Motijheel commercial area (MCA) in the mid-1950s. As a follow-up, the first midrise building with elevators to appear in the city was the Adamjee Court building (seven stories) in the early 1960s. This was followed by other such buildings as the Habib Bank building, the State Bank of Pakistan building, and the Amin Court building. These were mostly midrise buildings (between six and ten floors) and were equipped with elevators for the convenience of the tenants and the users. These are some of the earliest buildings that heralded the beginning of a new era of high-rise buildings in Dhaka in the following decades.

## 4 Land Use and Zoning

To facilitate the growth and development of Dhaka in a planned way, an autonomous organization called the Dacca (Dhaka) Improvement Trust (DIT) was formed by an act of the provincial legislature in 1956. [This organization was renamed in 1987, and it functions today as RAJdhani Unayan Kartiphakkha (RAJUK), or Capitol Development Authority.] The responsibilities given to this body were to create a development plan for the city and to administer its growth through direct implementation of projects and

**Table 3.1 Dhaka: growth and urban expansion, past trends and future projections**
*(Courtesy: Tanwir Nawaz.)*

| Year | Population | Area, $mi^2$ | Density of population | Increase in density (over 1950), % | Urban land expansion (over 1950), % |
|---|---|---|---|---|---|
| 1950 | 250,000 | 30 | 8,333 | | |
| 1970 | 1700,000 | 90 | 18,888 | 226 | 300 |
| 1985 | 4000,000 | 200 | 20,000 | 240 | 666 |
| 1990 | 6000,000 | 250* | 24,000 | 288 | 833 |
| 2000 | 9500,000 | 350 | 27,142 | 325 | 1166 |
| 2010 | 14500,000 | 475 | 30,525 | 366 | 1583 |
| 2025 | 28000,000 | 750 | 37,333 | 448 | 2500 |

*320 $mi^2$, including Dhaka, Naryanganj, and Tongi municipality in 1991 according to RAJUK.

development control mechanisms. As a part of its mandate, the DIT commissioned a master plan for the city which was prepared in 1959. This was the first concrete and tangible land-use plan prepared for Dhaka which gave the direction for future growth. While it failed to anticipate the tremendous rate of growth for the city and thereby missed setting up an adequate development plan, the city in its present form is more or less an outgrowth of this plan prepared some 35 years ago. Through this plan, definitive land-use areas for the city were allocated for the first time. New land-use areas such as industrial areas (Tejgaon), single-family residential areas (Dhanmandi), and commercial areas (Motijheel) as well as other zoning areas were set up under this plan. The definitive allocation of MCA at the time, as an exclusive area for business and commerce, was the first step in the development of private commercial sector projects. While the provincial government offices remained mostly secluded in the enclosed Secretariat area, other private commercial buildings, including various semi-independent government organizations such as the Water and Power Development Authority (WAPDA), the East Pakistan Industrial Development Corporation (EPIDC) offices, and the head offices of the State Bank of Pakistan (later renamed Bangladesh Bank) were set up in this area. The private sector was allocated and encouraged to set up their corporate headquarters in this area. It can be observed that all the commercial banks, insurance companies, private industries, and national airlines gradually moved into this area and either built their own corporate head offices or rented major office spaces here. Some of the more influential embassies such as the U.S. Embassy also set up their offices in this area. Business, commerce, as well as the legal and engineering professions also began to move their offices to this area. Thus in the early years of 1960, a tremendous growth in construction and commercial office space took place.

Despite the foregoing developments, land was still plentiful at the time and land prices were still very low. Much of the land was allocated to private enterprises, banks, and the semigovernment/autonomous organizations for nominal sums of money by the DIT. Thus while a lot of floor space was being built, and land prices were quite accessible and affordable, there was virtually no pressure to build high-rise structures. However, for reasons of economic viability and corporate prestige, some initial steps toward high-rise construction occurred. Among the first buildings to use elevators and move beyond the traditional walk-up office structures with four to five stories were the Adamjee Court building and the EPIDC building. Both of these buildings were seven stories high and were unique for their installation of elevators. They were soon followed by other commercial office buildings such as the Habib Bank (renamed Agrani Bank) building, the WAPDA building, and the Amin Court building.

While the development of these buildings was taking place at a rapid pace, in retrospect most of these can be grouped under a broad heading of low-tall buildings. These seven- to ten-story office structures, clearly standing above the surrounding buildings, were the forerunners of the more advanced high rises of today.

The first group of tall buildings, around 20 stories high, did not appear until the mid-1960s. These were the Inter-Continental Hotel building at the junction of the Airport Road and the Minto Road, and the Eastern Federal Union Insurance building in Motijheel (Fig. 3.5). Other nonsignificant midrise buildings continued to sprout in the MCA before 1970 (and prior to the creation of Bangladesh).

There was a lull in major office building construction following the creation of Bangladesh. The 1970s did not see much in the way of major office building development, primarily due to its unstable political and economic climate. However, the city continued to grow very rapidly in population and expanded physically, mainly due to the migration of rural population, putting a tremendous pressure on the land and land values, which escalated in some instances in the prime areas more than tenfold in value. Furthermore, land for office construction in the zoned areas was becoming scarce.

The late 1970s and the 1980s saw some rapid changes. A number of things impacting development were beginning to happen. Finances were more readily available to corporations and business. Other areas, such as the Shiddheswari and Kawran Bazaar areas, were designated both as commercial areas and also for office construction. The high land values and transfer of new construction technologies to Bangladesh saw the emergence of a fresh crop of high-rise buildings, some more than 20 stories in height. Tall buildings have finally come of age in Bangladesh. The tallest building now is the Bangladesh Bank building, which is 30 stories tall. Other taller buildings are currently being planned in the 30- to 40-story range.

## 5 Multistory Apartment Buildings

The evolution of the commercial building type in Dhaka was gradual, first in stages of up to 10 stories, then up to 20 stories, and finally in the 1990s some 30-story structures. Residential high-rise construction did not go through such staged developments. Even before the 1959 master plan of Dhaka was prepared, multiple housing complexes in the areas of Azimpur and Motijheel were built in the early and mid-1950s. But these were all basically three-story walk-up apartment buildings. While later on in the 1970s and 1980s most of these have been extended to four and in some cases five floors, all of these early midrise buildings have still remained as walk-up buildings.

Elevator-equipped high-rise apartment buildings did not come about until the early 1980s. These are now mostly in the form of condominium buildings in the private sector, ranging anywhere from 7 to 20 stories. Because of the tremendous rise in land value and the scarcity of buildable residential land, even in the previously single-family areas such as Dhanmandi and Gulshan, these new types of high rises are being built, sometimes after demolishing existing single-family homes built several years ago. Thus the tremendously high-priced land and the scarcity of buildable vacant land are forcing the construction of more and more high-rise residential condominium buildings in several areas of Dhaka.

## 6 Land Values

While the population of Dhaka has increased nearly 30 times (or 3000%) from 250,000 in 1950 to more than 6 million in 1992, the land values in the central areas, which have become commercial areas and the site for high-rise developments such as the MCA and Airport Road near Kawran Bazaar, have increased 10,000 times (or 1,000,000%) or more. Even after accounting for inflation, the land values have increased more than 10,000%. An acre of unserviced land in the Motijheel area in 1950 could cost Tk.1500 (approximately $300 at the exchange rate of that time). The same parcel of land, now serviced and in the heavily built Motijheel area, costs more than Tk.40,000,000 ($1,000,000 at current exchange rates). It is reflective of not only market changes in terms of cost but also scarcity of available buildable land for business use. Despite the opening up of other areas for commercial uses, the upward spiral of land cost will continue due to the scarcity of urban land. This will increase the pressure to build more building space on the same area of available commercial land as well as appropriate more residential lands for high-rise construction. The economics of high land prices will be one of the driving factors for the continued development of high-rise structures. Thus the age of tall buildings in Dhaka has only begun and, given stable political and growing economic conditions, will continue. This is also true for many other major cities of developing countries around the world.

**Fig. 3.5 Eastern Federal Union Insurance building, DIT Avenue, seven-story base in foreground, 27-story tower in background, built late 1960s to mid-1970s.** (*Courtesy: Tanwir Nawaz.*)

## 7 Future Trends in Planning and Development

What is the direction of future tall buildings in terms of planning and development? It seems that due to the scarcity of buildable lots, the commercial business district previously located in the Motijheel area expanded beyond the area in recent years. The development of tall buildings which took place initially in the MCA, expanded beyond the previously designated zoned area and into the proximity of three principal roads: North-South Road, Bijoy Nagar Road, and Airport Road.

None of these subsequently developed areas was zoned as commercial in the first master plan prepared in 1959. However, pressure to develop beyond the forecast projections has forced encroachments, and today in the absence of a review of the master plan, these areas are allowed to develop and grow as office and commercial districts. It is likely that future high-rise office developments will continue to happen in these areas. These areas have not reached a saturation point as yet. In MCA, some early low-tall buildings will probably be torn down to make way for the increasingly higher buildings of tomorrow. At the same time, since the city is growing mostly to the north and north west, two things are likely to happen in the near future. The first will be the continuation of development of commercial districts to the north and consequent zoning amendments to allow it, and the second will be the development of a small-scale commercial area with some tall buildings in the industrial areas of the north west (in Tongi).

In the case of residential high rises, we are only beginning to see the growth and development of tall buildings. Unlike the commercial districts, the residential areas of new Dhaka (north of the old railway line) are large and underused. With changing economic conditions and a tremendous rise in residential land values, multistory high-rise developments are not only likely but seem to be the major logical direction. As stated before, this is already happening in the reuse of single-family one- and two-story buildings that are either being extended to four- and five-story walk-up apartments or torn down to make way for higher (and elevator-equipped) condominium apartments. Whereas the first type is usually being undertaken by owners themselves with borrowed funds and financing from private banks and financial institutions, the latter is being carried out by a limited number of development companies. As things get more complicated in the future, the development companies and possibly cooperative groups with public sector assistance will provide the bulk of future developments. In such a scenario, higher apartment types, both condominium and rental, will result. As to where the bulk of these developments will take place remains unclear at this point. A complementary role between public and private sectors is likely. The private sector, being profit oriented, is expected to provide high-rise developments only for a limited number of the economically fortunate local population. However, if the RAJUK (the government development and planning agency) were to change its obsolete policy of continuing to provide single-family development lots and concentrated on more innovative high-rise development, as has been done in such cities as Singapore, Toronto (St. Lawrence Development), and Hong Kong in cooperation with the private sector, then both midrise and high-rise residential development for more affluent portions of the population as well as large-scale housing in high-rise structures for more representative portions of the population might be feasible. There is a tremendous opportunity waiting. But at this juncture of events, the initiative has to be seized and exercised. Under this program, not only new areas could be targeted but also older areas, both in private hands and in the public sector, could be utilized. The new master plan now being contemplated for Dhaka must provide a framework for this program by providing a land-use basis and a planning framework under which development can take place.

## 8 Architectural Features

It is interesting to review the evolution of facade fenestration types of the tall buildings in Dhaka. Obviously, the only type of tall buildings to emerge initially were the commercial office buildings. The earliest examples, such as the seven-story Adamjee Court and the BIDC, formerly called the EPIDC, were examples of an early type of architecture of low-rise buildings carried into midrise buildings. These were massive, squat-looking buildings built of reinforced concrete floors, beams, and columns. The exterior walls were masonry construction plastered over with cement concrete and painted. These walls had limited openings for windows. Often the exterior wall had vertical or horizontal sun shades, somewhat in the manner of *bris soleils* by Le Corbusier (see Chapter 7).

The next group of buildings with respect to fenestration types were the Habib Bank (Agrani Bank) building and the State Bank of Pakistan (Bangladesh Bank) building. Although both of these buildings had a facade fenestration of horizontal and vertical metal mullions, only the State Bank of Pakistan building was the first glazed curtain-wall skin construction in Bangladesh.

The third group in the evolution of facade fenestration was the continuous horizontal window or glazing interrupted by only the structural columns (as in the Eastern Federal Building, Fig. 3.5) or as in the later instance of continuous horizontal perimeter glazing, uninterrupted by structural columns which remain on the inside, as in the case of the curtain-wall system. In these buildings the sill and the head of the windows continue as horizontal bands of poured-in-place concrete or masonry construction.

The fourth identifiable type is a pure curtain-wall system of enclosure, even though in this instance there are vertical concrete sections in the facade and horizontal sections at the top.

The last two categories seem to be more in vogue in Dhaka at the moment, even though an occasional exotic type can also be seen. However, the glazed curtain-wall system is likely to prevail for some time, until a new type or technology emerges internationally.

The residential high-rise buildings do not yet offer a distinctive architectural style that we can recognize as an established style. The evolutionary process is still very fluid, and architects and planners are still searching for an identity in this sector. This is because residential high-rise structures are very recent. Architects and developers have not yet found a specific style to emulate, and the profit-driven high-rise residential development industry is still in a relatively early stage of growth and maturity. However, once the private sector development industry stabilizes and the supply of high-rise buildings balances demand, the picture could change rather quickly. There are strong reasons to anticipate that if a stable economic climate continues, the demand for affluent housing will grow stronger, and in the absence of available single-family developable lands in the metropolitan area, continued growth of more and possibly refined taller residential structures will occur. As that happens, the emergence of more definitive architectural styles in high-rise residential buildings will also probably result.

## 9 Future Development Trends

Some inevitable facts remain. Despite continued outward growth, the density of population and the land values will continue to climb in the foreseeable future. At the same time, despite general poverty, capital accumulation will also continue in certain sectors of the free market situation. This will continue to create increasing demands for more commercial and residential floor space. Since land in the inner city is less available, the economic pressure to put more floor space in smaller parcels of land will mean that the construction of high-rise structures will continue. It is likely that in the next 5 to 20 years, two parallel kinds of development will take place, one for the residential sector and another for the commercial sector.

For the residential sector:

1. It will continue to grow exponentially to meet increasing housing needs.
2. Many existing houses and low-rise buildings will be demolished to build new midrise apartment buildings (10 to 20 stories).
3. There will be a few buildings of over 20 stories when circumstances permit.

In order to meet the housing needs of only the increase in population in Dhaka, it is estimated that about 60,000 housing units have to be built every year until the advent of the twenty-first century. Because of the low income level of the majority of the population, it is inconceivable that the problem of housing can be resolved by traditional building practices. If the current trend of increase in land prices continues, high-rise living is likely to become increasingly popular. A large number of residential construction projects that are under way or are on the drawing board only confirm this prediction.

For the commercial sector:

1. It will continue to expand beyond current and designated (zoned) areas into other areas.
2. There will be more consolidation and intensification of available land and buildings in the existing commercial areas.
3. There will be many more midrise buildings (buildings with heights of between 10 and 20 stories).
4. There will be many fewer in numbers, but a few significantly higher tall buildings in the urban landscape.

Those buildings in the range of between 30 and 40 stories, which have been technologically achieved elsewhere much earlier, will be more common in the context of Bangladesh. The reasons for the scarcity of higher buildings at present are fairly obvious. Despite continued growth in demands for office space, it is much more difficult and financially riskier to put together a package for one enormous building than for the construction of two or more midrise buildings (10 to 20 stories). Buildings with less than 10 stories in the high-priced commercial districts will become more and more inefficient. Thus the expectation is that we will not see many low-rise office commercial buildings, but a lot of new construction of midrise buildings, and few but significant and spectacular higher buildings. The scenario, however, may change with more pressure on land and with the availability of better construction technology, which may render taller buildings of even more than 40 stories more viable in the future.

In terms of urban planning, zoning, codes, and regulations, Dhaka City is still very much in its infancy. The old master plan is obsolete and has long outgrown its goals and objectives. The government has currently awarded the master planning to a Western firm under an aid package. Whereas a review of the old master plan is long overdue, there is concern about the new plan and its package. This is because traditionally master plans are basically land-use plans and not multidimensional in nature or scope.

Tall buildings will continue to emerge in the city out of their own dynamics unless more effective regulation is employed. As to where they will be or should be located and what role they will play in commerce and for shelter are issues that can and should be thoroughly examined. For now it can be said that this is an opportune time to address the issues so that architects, planners, and developers can concentrate on creating not only technologically and aesthetically beautiful tall buildings, but that these buildings also contextually fit well into the changing urban fabric of the city and enhance its livability.

## 3.4 CREATING URBAN SYNERGY: THE PHILADELPHIA PROJECT

As illustrated in the previous case study, incorporating high-rise buildings into nonindustrialized countries raises many issues of economics, scale, and context, which architects, developers, and planners must consider. Development of urban areas must be undertaken through a partnership of the public and the private sectors. The incorporation of the tall building into the city is a holistic, three-dimensional problem, which impacts future development, zoning and land use, pedestrians and the streetscape, building users, and building functions, as well as determining the impact the building will have on the life of the city and its image on the skyline. The following section presents a case study design project for a major city in an industrialized country (Warfield, 1991):

> As a designer you are capable of developing a concept which will result simply in a whole series of individual buildings with no particular reason or relation to each other or you can develop a concept which melds them together, creates synergy, and is exciting. If you worry about the details at the beginning, you just get fragments. You let the (plan) flow out of you and (it becomes) a totality.

With these words, Edmund N. Bacon, Plym Professor (1991–1992) at the School of Architecture, University of Illinois at Urbana-Champaign, challenged students to become "creator/city builders" and to develop "holistic, holographic" images of cities. Bacon collaborated with graduate students in studios conducted by James P. Warfield and colleagues to create a vision for Philadelphia for the next 40 years. Bacon's charge (Schmitt, 1991):

> Prepare drawings and models of the entire one mile extent of the Center City West bank of the Schuylkill, including the massing and design of the architecture that is in harmony with the magnificence of 30th St. Station, and the Post Office, yet which stretches into 21st Century concepts, showing the opportunity to build a splendid city like none that has ever been built before.

As the former executive director of the Philadelphia Planning Commission, Edmund Bacon was acutely interested in the westward development of Philadelphia where Market Street crosses the Schuylkill River between the recently restored 30th Street Station and the U.S. Post Office. The train station is not only a valued landmark from Philadelphia's distinguished past, it has a role to play in the city's future as a model city of what Bacon calls the "postpetroleum era." Accordingly, he considers it important to locate new development to be directly connected with 30th Street Station. He foresees the Metroliner, which now serves the station, as eventually being replaced by magnetic levitation trains which will operate at speeds competitive with airline travel. He proposed a graduate-level urban design study of this district, known as City Center West.

The 1682 Penn plan of Philadelphia created a strong and memorable image. Philadelphia is bracketed by two rivers, the Delaware on the east and the Schuylkill on the west. The central axis of Market Street connects the rivers from Penn's Landing on the east through the center of Philadelphia's city hall and extends westward between

30th Street Station and the Post Office. The Penn plan provided a "checkerboard" system of parks: Washington Square, Franklin Square, Rittenhouse Square, and Logan Circle, subsequently linked to the Museum of Art by Fairmount Park (Heckscher, 1977).

At the beginning of the project, studio critics introduced students to examples of the great planned cities of the world—Paris, Rome, Beijing, Washington, DC. In addition, a group of 40 of these students traveled to Philadelphia and studied the project site. While in the city, students were also able to observe Philadelphia's landmark buildings—historic Independence Hall; City Hall which served as central focus for Penn's axial plan; the major planning achievements under Bacon—Society Hill and Penn Center; and the most recent high-rise building complexes—Commerce Square, a project that works well within the Philadelphia tradition, and the twin towers of Liberty Place, designed in obvious contradiction to the *genius loci* of Philadelphia.

The original urban design concept of William Penn remains intact today, its clarity difficult to avoid or transform. The historic plan of Philadelphia is easy to diagram, so it was not difficult for students to understand its form, structure, and function, creating a sound basis upon which to develop new urban design concepts (Halpern, 1978).

### 1 Schuylkill River Edge Site

The site for student urban design projects is a one-mile-long segment along the Schuylkill River. This area once flourished as a zone of industrial vitality where river transportation and railroads joined. Today its factories are but a memory, and the river is quiet. This land is ready for transformation. The recently restored 30th Street Station at the center of the site, the second busiest station in the country, was the first sign of Philadelphia's commitment to this site. The north portion of the site offers the potential for development of the air rights over a complex network of railroad tracks. An elevated freight railroad bisects the site north to south. In the subsequent design phase, this became the west boundary of the detail study area. Major university campuses, Penn and Drexel, lie to the west.

This particular site had unusual potential for development of a new city center of extraordinary significance. On axis with the esteemed Penn plan, and on the bank of the Schuylkill River, this was the most prominent site in Philadelphia, a site for creative, forward-looking planning.

Elevated views east to central Philadelphia are spectacular, suggesting the possibility of towers with public rooftop restaurants or observation decks. Views west include the campuses of Penn and Drexel. The sweep of the Schuylkill River may be viewed to the north and south of the site. Across from the 30th Street Station, the east bank of the Schuylkill has potential for developing separate pedestrian promenades, equestrian trails, bicycle paths, and access to the water's edge. The west bank, under the 30th Street Station, has less potential for such development. Dense highway and railway infrastructures block visual and physical access to the water.

Busy activity at the street level is now quickly disbursed by taxis, subways, or private automobiles. Potential exists to provide street-level amenities, places to rest, eat, and view the endless motion of people on the way to more humane connections to the simultaneous urban movement systems. Additional potential exists to give this district a 24-hour per day population by programming housing as well as commercial and professional facilities. Accordingly, public spaces could be created to allow proper surveillance for safety and security (Newman, 1973) and to foster an active street life (Lynch, 1981).

The climate in Philadelphia is occasionally hostile. Protection is needed from the spring rains and winter snows.

## 2 Program

The entire Philadelphia project was accomplished in two semesters and was conducted in the graduate-level studios. During the Fall 1991 semester, several graduate studios participated, creating urban master plans linking the Schuylkill River edge site and West Philadelphia with the center city. William Penn's original plan of Philadelphia provided a historical reference from which to study and understand the organization and subsequent growth of the city. Developing the axis along Market Street and linking the Schuylkill River edge with the central city became a principal concern in each student master plan.

During the Spring 1992 term, a single graduate design studio was created to continue the development of the concepts proposed in some of the original student master plans. At this point it was necessary to increase the scale of the model and drawings so that each building and public space could be studied in detail. The students were asked to respect the organization and concepts presented in the original master plans, and especially to continue their investigation of the builtform of high-rise buildings to open, public spaces with respect to usage, public transportation, and pedestrian scale.

The Philadelphia project was guided by an implicit, open-ended project statement, written by Professor Ronald E. Schmitt, course coordinator. It read in part, "Derivation and definition of a program is part of the master planning process. While architectural design deals with how to build, master planning and urban design encompass the decisions of what, why, where as well as when to build" (Schmitt, 1991).

***Fall Semester, 1991.*** The project was divided into three phases, ranging in scope from general to specific, from comprehensive master planning to architectural design. In phase 1, students worked with small-scale drawings and study models 25.4 mm = 121.9 m (1 in. = 400 ft) to develop initial urban design concepts from the center of Philadelphia to the area near 30th Street Station called City Center West. In phase 2, the City Center West district was developed in more detail, using drawings and study models at the scale of 25.4 mm = 60.9 m (1 in. = 200 ft). Finally, in stage 3, students developed architectural drawings and models for proposed buildings in City Center West at a scale of 25.4 mm = 30.4 m (1 in. = 100 ft). This three-phase approach was conceived to teach students to think of buildings as details of a larger structure.

***Student Master Plans.*** Students in architecture like to identify quickly "the problem of the problem." It is their way of distilling a complex set of relationships to its essence before beginning. For some it was useful to view the redesign of Philadelphia as they would view remodeling a historically significant building. Then one may ask: What is the spirit of this place? What are the original design ideas and how may it be possible to perpetuate them? The problems of this problem then became intelligible. The main issues derived from the master planning phase, all dealing with existing conditions were:

1. To respect the Penn plan axis
2. To continue a system of open spaces
3. To honor existing landmark buildings, specifically 30th Street Station, City Hall, the Museum of Art, and the power plant
4. To improve Schuylkill River edges
5. To connect the west with east Philadelphia
6. To integrate anticipated new mass transportation systems, including the Metroliner

Most projects could be categorized by a strong link to one or more of these six issues, often in creative and highly unexpected ways. Three additional intentions emerged from student urban design proposals, all dealing with new concepts:

7. To create a new landmark symbol for Philadelphia
8. To create an exciting new skyline
9. To develop new "towns in town"

Responding to an implicit program, and without a uniform mandate, student projects nonetheless shared a number of important attributes. Almost all projects respected and extended the Penn plan axis using Market Street as a formal link of east and west Philadelphia. Most projects continued a system of open spaces relating to the rhythm, scale, and axial locations of the Penn plan's checkerboard of parks. Most projects proposed a plaza setting for the landmark 30th Street Station and Post Office. Many projects created a linear river-edge park on vacant property along the east bank of the Schuylkill River. Most projects proposed high-rise buildings creating exciting skylines, landmarks, and gateways from the west into the City Center West district. Many schemes offered loggias and sky walks, connecting their towers and urban spaces to 30th Street Station from which magnetic levitating trains were projected to arrive and depart. The following projects are representative of this effort.

Mira Metzinger's master plan (Fig. 3.6) emphasized the east-west Penn axis along Market Street with a continuous wall of tall buildings rising above the height of the surrounding urban fabric from center city, terminating west of 30th Street Station by a wall of corporate towers fronting a two-block plaza. A proposed north-south cross axis orig-

**Fig. 3.6 Master plan emphasizing east-west Penn axis.** (*Courtesy: Charles Mercer.*)

inates at the Museum of Art to the north and terminates at a proposed new opera house in the south. Two new urban parks on this cross axis extend Penn's checkerboard to the west side of the river. The two central corporate buildings face each other in apparent dialogue, forming a gateway in front of the multilevel plaza (Fig. 3.7). Adjacent residential towers play a subordinate role.

Hariri Yahya proposed four towers in a row connected by bridges to low-rise buildings defining an urban space (Fig. 3.8). John Lesak proposed a research park and a "bundle" of four crystalline office towers centered on Market Street (Fig. 3.9). Lesak's architectural model showed development of the transparent skin, open floor system, and sloped roofs (Fig. 3.10).

Lesak developed four towers as figural objects in the center of an urban space on the Market Street axis surrounded by low-rise buildings (Fig. 3.9). These towers have multilevel lobbies, elevators, and stairs as primary access to the subway below (Fig. 3.11). Secondary access is provided by grand ceremonial stairs to a formal recessed court and pool. The subway station mezzanine receives borrowed natural light from the tower lobbies above and directly from the loggia surrounding the recessed court. The subway station is designed to recall traditional railroad stations with trusses supporting a gable roof (Fig. 3.11).

Mark Freudenwald's master plan proposed a continuous urban wall of buildings ascending to the pair of gateway buildings centered on Market Street, linked by an upper-level bridge with excellent views east to City Hall (Fig. 3.12). The elevation of both towers together forms a keystone motif in recognition of the state's motto (Fig. 3.13).

**Fig. 3.7 Corporate buildings facing each other forming a gateway.** (*Courtesy: Charles Mercer.*)

**Fig. 3.8 Four towers in a row connected by bridges to low-rise buildings.** (*Courtesy: Charles Mercer.*)

**Fig. 3.9 Proposed research park and "bundle" of four office towers.** (*Courtesy: Charles Mercer.*)

Freudenwald developed twin towers with multilevel colonnades connecting building lobbies at street level to a lower plaza level and ultimately to the subway (Fig. 3.13). The plaza has three levels open to the sky. Surrounding the plaza are restaurants, a health club, and an assortment of shops. Subway entrances are within the building lobbies and from entry pavilions at the street level. Bridges above the street level connect all buildings surrounding the plaza with 30th Street Station.

***Spring Semester, 1992.*** This semester was divided into two phases of investigation and scales. Phase 1 required students to develop urban and architectural designs from the previous semester at 25.4 mm = 15.2 m (1 in. = 50 ft). In phase 2, students examined how buildings should meet the ground, interact with outdoor public spaces, and connect to adjacent buildings. Eight students studied the area surrounding 30th Street Station and the Post Office with connections to the subway station. One student studied areas along the east bank of the Schuylkill River edge near Market Street. Phase 2 required architectural development at the scale of 1.6 mm = 0.3 m (⅛ in. = 1 ft).

### 3 Subway and 30th Street Station

The first subway stop west of the Schuylkill River is under Market Street near 30th Street Station. It is a dark and threatening environment entered by descending narrow stairs with ample hiding places for muggers and purse snatchers. The subway's interior concrete end walls have been faced with a second wall of glass block backlighted with fluorescent tubes unsuccessfully imitating sunlight.

**Fig. 3.10 "Bundle" of four crystalline office towers showing details.** (*Courtesy: Charles Mercer.*)

Metzinger developed her tower's striking anthropometric profiles and graceful sculptural modeling fronting an urban plaza centered on Market Street (Fig. 3.14). Multilevel glazed lobbies provide daylight and access to the subway level. Metzinger's plazas descend to provide physical and visual access to the track level (Fig. 3.15). A two-story glass wall affords subway passengers views from the subway station to trees, fountains, steps, restaurants, and shops. A continuous sky walk connects all towers to the commuter level of 30th Street Station.

Yahya reduced the number of towers from four to two, relating placement of entries and other detailing to an urban plaza developed by his partner, Wan Zawber (Fig. 3.16). Yahya's continuous upper-level loggia connects their project to the commuter track level of 30th Street Station. Zawber's plaza descends gracefully to the track level of the subway enclosed behind a two-story glass wall affording waiting commuters and train passengers a view of water, paving, trees, and other vegetation (Fig. 3.16). At the top of this glass wall is a glass roof overhanging the sidewalk to give shelter to those waiting for a bus or cab. One may enter the subway from within the towers, from the plaza, or from glass pavilions connecting street level to track level. Sky walks connect all buildings in this complex with 30th Street Station.

Rob Neu's twin towers (Fig. 3.17) have gracefully curved shafts which rest on columned bases with multistory lobbies admitting borrowed light to the subway station under Market Street. Similar in planning concept to Lesak's plaza with four towers, Neu's paired towers are objects in a space surrounded by a low-rise frame of restaurants, shops, and offices with a colonnade at the plaza level and a loggia at the roof garden level.

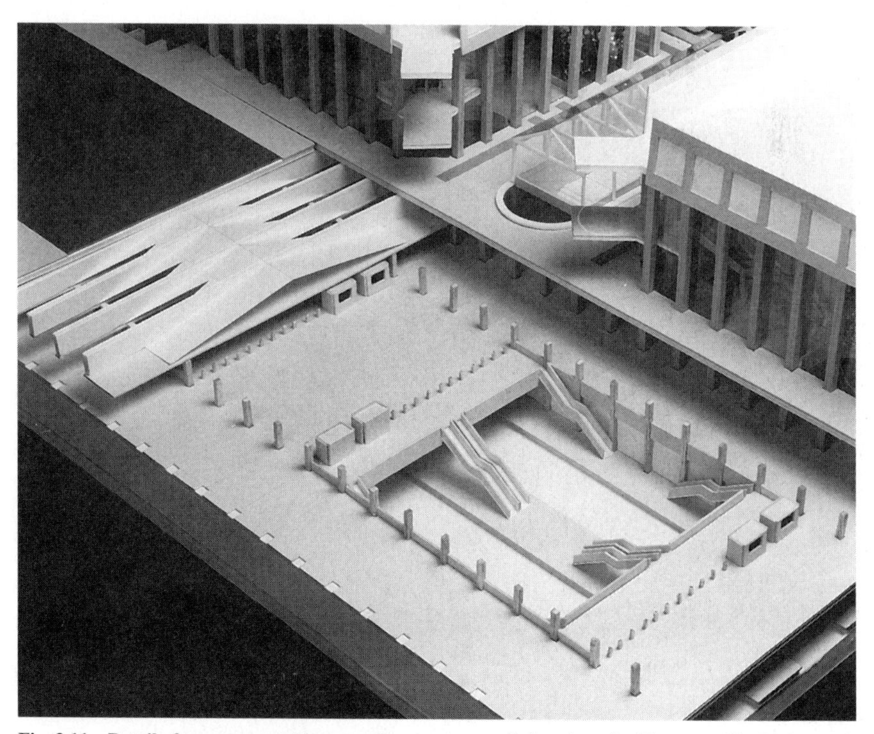

**Fig. 3.11 Detail of transparent skin, open floor system, and sloped roofs.** (*Courtesy: Charles Mercer.*)

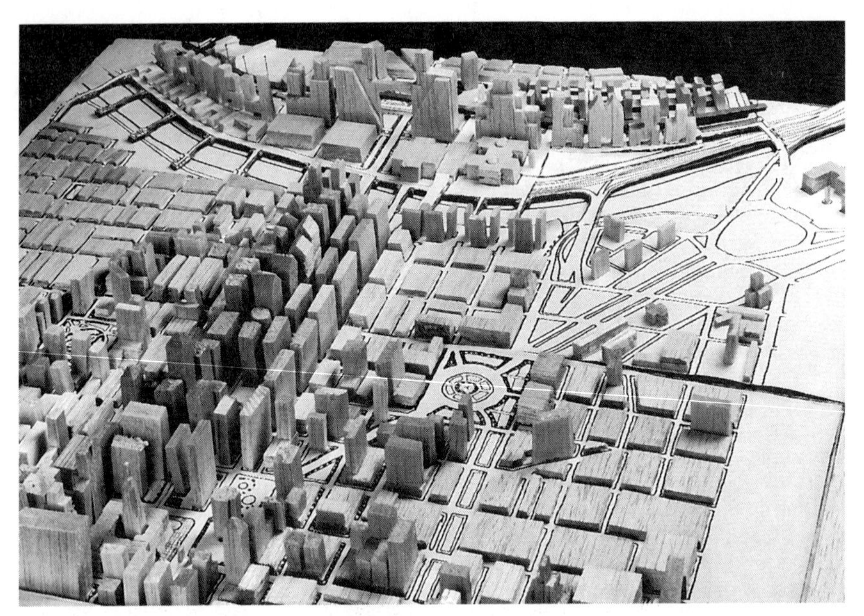

**Fig. 3.12 Continuous urban wall of buildings ascending to a pair of gateway buildings.** (*Courtesy: Charles Mercer.*)

**Fig. 3.13 Elevation of both towers forming a keystone motif.** (*Courtesy: Charles Mercer.*)

The team of Ken Allen and Michael Pipta developed a hotel with an underground connection from its lower lobby to the 30th Street Station (Fig. 3.18). Across the street from the hotel is a public plaza bisected by Market Street and the subway. Allen's and Pipta's plaza breaks down the public scale into more intimate human scale steps, seating platforms, and shade trees. Three edges of the plaza are facades for the subway entrance and surrounding lower-level shops and restaurants.

## 4 Schuylkill River Edge

Directly across the Schuylkill River from City Center West is its east bank, lined with visually unpleasant and psychologically threatening underdeveloped properties creating a physical barrier to the river. (For additional information on the psychological factors influencing tall building development refer to Chapter 6.) These sites have potential to be developed for public access to the river, to become visual and recreational amenities. One student, Peter Courlas, chose to work alone on this site because of its importance to the success of any proposal for City Center West. Courlas proposed a design center tower that is a "fraternal twin" of the Philadelphia Electric Company (PECO) building, creating a gateway on the east bank of the river at Market Street. In Figure 3.19 PECO is on the left and Courlas' design center is on the right. A street-level colonnade turns the corner and connects the design center to the adjacent merchandise mart. A continuous system of public steps and walks connects the colonnade to the Schuylkill River's

**Fig. 3.14 Towers with striking anthropometric profiles and graceful sculptural modeling fronting an urban plaza.** (*Courtesy: Charles Mercer.*)

edge. Parallel to the river are proposed pedestrian, bicycle, and equestrian paths, and a marina. A fast-food "boat-in" restaurant at the center of the river affords views of river bank activity as much as it is an object to be viewed from the water's edge.

Although differing in detail, all of the spring semester projects were successful translations of a master plan to architectural design at human scale. All projects provide urban stages for daily activities of people traveling, sitting and watching, reading, meditating, eating, walking, bicycling or horseback riding, and shopping, and settings for behaviors ranging from public to private.

Urban design at such enormous scale inevitably focuses upon issues relating to the role of the tall building in urban design: tall buildings as landmark objects, the aesthetic development of tall buildings appropriate to an existing urban character and relating to a city by historic reference (*genius loci*), archetypal tall buildings in the urban skyline, tall buildings as urban walls forming urban spaces and city edge, tall buildings as definers of urban corridors and symbolic city gateways, and finally appropriate human scale treatment of tall buildings at ground level and below.

## 5 International Implications from the Philadelphia Project

On one hand, the Philadelphia project is a record of student ideas created in an ideal academic setting. On the other hand, it documents the persistence of an individual architect/planner and his contention that cities are the result of individual persistence in pursuit of a strong design idea. Finally, and most significantly, it asserts the values derived from this enterprise for architects, planners, and city officials throughout the world in regard to the following notions.

**Fig. 3.15 Plazas descend to provide physical and visual access to track level.** (*Courtesy: Charles Mercer.*)

***"Programless" Program in Urban Design.*** Instead of focusing on a specific "problem" to be "solved," students began by viewing the city as an artifact, with three-dimensional form, and natural and artificial physical properties that convey a spirit of a particular place. Freudenwald's wall of towers, for example, defined an arc relating to the Schuylkill River with gateway towers responding to Penn's historic axis. Metzinger's urban form is based on Penn's concept of an axis and cross axis with a checkerboard system of parks.

What this suggests to architects, urban designers, and city officials throughout the world is this: From time to time there is value in examining a city holistically, without the programmatic, economic, or political constraints often associated with site-by-site urban development. This is an approach that says "function follows form." This allows successive generations to rediscover the spirit of their place, and the nature of urban form that defines it. This examination may be conducted by university students, as it was in this case, by consultant architects and urban designers, or by participants in international urban design competitions. Regardless of which course of action is preferred, the partici-pants must be accorded freedom to design holistically without significant constraints of competing local agendas.

***Strong Urban Design Idea.*** In an academic studio, individuals may follow their understanding of an existing form, and extend that concept in a holistic master plan and through designs of compatible individual buildings. In practice, individual architects and planners will seldom have this opportunity for total design control. What individu-

**Fig. 3.16 Two towers that relate placement of entries and other detailing to an urban plaza.** (*Courtesy: Charles Mercer.*)

als may have is an opportunity to project a strong urban design idea that may be understood, appreciated, and implemented by numerous architects and developers of individual sites within the city.

The student work validates the assertion that design ideas have the power to motivate others. Students were inspired, as was Bacon, by Penn's strong design ideas for a city with street grids and a formal axial organization. Students were further stimulated by Bacon's ideas of a multilayered city with archetypal towers connecting earth and sky, and architectural buildings for delightful pedestrian movement. Zawber's design is a strong example of the continuity of Bacon's design ideas. Her richly detailed and textured plazas complement Yahya's towers and loggias as they connect the subway to the sky on both sides of Market Street.

Cities seeking strong design ideas may specifically invite students or professionals to submit urban design concepts. Alternatively, through either a public agency or professional association, cities may choose to host open "ideas competitions" to encourage creative conceptual thinking for the entire city or for specific districts. These events can be of significant value for exciting public discourse and creating a shared vision of the future when design proposals are publicly exhibited and widely published.

***Concept of Context for Locating Tall Buildings.*** This project demonstrates the effectiveness of placing tall buildings according to a grand design. Metzinger, Yahya, Freudenwald, Lesak, Neu, and Courlas developed distinct and creative alternatives for how archetypal and architectural towers work in concert in an urban design. All are landmarks for a new urban district. All are urban gateways on the Market Street axis. Towers

**Fig. 3.17 Twin towers that have gracefully curved shafts resting on columned bases.** (*Courtesy: Charles Mercer.*)

designed by Metzinger, Yahya, and Freudenwald create an edge condition for a new urban space; towers by Lesak and Neu are formal objects in a new urban space. Courlas's tower complements its neighbor across Market Street, and completes a nearly symmetrical composition compatible with its urban context. Freudenwald's idea for a curved wall of buildings provides a framework for much individual architectural creativity within a system of urban harmony.

In spite of the architectural merit of individual students' tower designs, none of these tall buildings were intended or conceived as virtuoso masterpieces. The ultimate success of these tall buildings is measured by how well they contributed to an urban design idea. The international and universal implication of this study project is fundamental: tall buildings should be reserved for places of honor, discerningly placed where they have the maximum civic impact, placed only where they will serve as landmarks for, and components of, important urban districts, where they will meld together with other buildings in a holistic scheme, and where they will create urban synergy.

## *6 Tall Buildings and the City*

The tall building is without doubt a special architectural type that places heavy demands upon architects, engineers, urban designers, landscape architects, and many specialists in the design professions. So vast in scale, cost, and technical complexity, it is a building type which often overwhelms. Indeed, an expert who specializes in a particular aspect

**Fig. 3.18 Hotel with an underground connection from its lower lobby to the 30th Street Station.** (*Courtesy: Charles Mercer.*)

of tall building planning and design may only be responsible for a small part of the design, with little appreciation for how the whole project will come together. Students in an architectural design studio are not burdened by the constraints that daunt professionals. Accordingly, they believe that everything is possible and that they as individuals can be creators/city builders, viewing the design of cities as holistic, holographic images. Their comprehensive visions for the design of cities points the way for the design of future cities throughout the world. Their lesson: cities cannot be built on a site-by-site basis. Design for cities should be the result of strong design ideas steeped in a clear understanding of the physical, cultural, historical, and spiritual qualities of the city as a particular place in time and space. In Philadelphia this came about through the efforts of an individual, Edmund Bacon. "The success of the (comprehensive) plan (for Philadelphia) is due more to architect/planner Edmund Bacon's long commitment to Philadelphia and to his creative guidance of the planning process, than to the plan itself" (Halpern, 1978). There is a lesson to be learned from this project. Organization of tall buildings in a city must be done with appropriate programming and with a strong impulse for urban design that will lead to urban synergy. The message is universal in that the design ideas could be applied to other cities of the world.

## 3.5 FUTURE RESEARCH NEEDS

From the case study examples provided for Dhaka and West Philadelphia it is apparent that although general urban planning principles can be applied to each city, the needs

**Fig. 3.19 PECO (left) and proposed design center (right).** (*Courtesy: Charles Mercer.*)

and requirements of cities in nonindustrialized countries are often different than those of industrialized countries. Whereas the growth rates of cities in many of the industrialized countries (especially in the United States) have been stable over a long period of time, the rate of population growth of cities in the nonindustrialized countries continues exponentially. Compounding master planning problems are issues of zoning, demographic distribution, land use, and public and private sector involvement in district and regional planning.

Several issues relating to future research for industrialized and nonindustrialized countries should be considered. Urban design and planning principles derived from experiences and applications of cities in industrialized countries must be evaluated for their appropriateness for application to nonindustrialized cities in developing countries. Researchers are encouraged to investigate the roles of architects and urban planners in the master planning process of industrialized and nonindustrialized cities, and their interaction with public agencies and private developers to develop and implement holistic master plan concepts. Architects, planners, and academics should study the relationships that are developing or are likely to develop between private and public sectors in terms of addressing urban-scale design and planning issues.

Documentation of the contribution of architects in introducing high-rise buildings into the context of existing cities while maintaining or enhancing the vitality of the city is also encouraged. Research must be conducted on single-use versus multiuse high-rise buildings to be incorporated into urban design, on the impact of zoning on building use, on the viability of certain districts of the city in terms of street life, economic development, and sustainability, and on perceptions of comfort and safety. Studying the impact of the supertall building and megastructures on the context and land use of the city is related to the previously stated concern. Skyline studies would investigate the feasibility of adding new tall buildings and how they will impact the skyline. Researching existing infrastructures of cities in their ability to accommodate new buildings without creating social and environmental problems would enhance the sustainability of existing cities. Studying the relationship of tall buildings with the surface and underground transportation systems of a city would result in a better synthesis of transportation to builtform in urban environments.

## 3.6 CONDENSED REFERENCES/BIBLIOGRAPHY

The following is a condensed bibliography for this chapter. Not only does it include all articles referred to or cited in the text, but it also contains bibliography for further reading. The full citations will be found at the end of the volume. What is given here should be sufficient information to lead the reader to the correct article—author, date, and title. In case of multiple authors, only the first named is listed.

Alam 1991, *Dhaka: Planning Housing and the Environment. Shelter and the Living Environment*

Ali 1993, *Mass Rapid Transit System for Dhaka City*

Bacon 1976, *Design of Cities*

Bangladesh Bureau of Statistics 1986, *1986 Statistical Yearbook of Bangladesh*

Barnett 1973, *The Future of Tall Buildings: Systems and Concepts*

Barnett 1986, *The Elusive City: Five Centuries of Design, Ambition and Miscalculation*

Bartone 1991, *Environmental Challenge in Third World Cities*

Broadbent 1990, *Emerging Concepts in Urban Space Design*

Chowdhury 1985, *Land Use Planning in Bangladesh*

Cohen 1991, *Urban Policy and Economic Development—An Agenda for the 1990s*

Condit 1961, *American Building Art: The Twentieth Century*

Condit 1964, *The Chicago School of Architecture: A History of Commercial and Public Building in the Chicago Area, 1875–1925*

Contini 1973, *Tall Buildings: Urban Planning Factors*

Cowan 1974, *Tall Buildings for People—Aesthetics and Amenity*

CTBUH 1986a, *Advances in Tall Buildings*

CTBUH 1986b, *High-Rise Buildings: Recent Progress*

CTBUH 1988, *Second Century of the Skyscraper*

CTBUH Group PC 1981, *Planning and Environmental Criteria for Tall Buildings*

Cullen 1971, *The Concise Townscape*

Curtis 1986, *Le Corbusier: Ideas and Forms*

Development—Population Research and Review 1989, *Understanding the Dynamics of Population—Projections and Predictions*

Development—Environment and Development: The Crucial Decade 1992, *The Main Culprits—The Urban Explosion*

Durand 1836, *Recueil et Parallèle des Edifices de tout Genre, Anciens et Modernes*

Eckbo 1964, *Urban Landscape Design*

Fishman 1977, *Urban Utopias in the Twentieth Century*

Frampton 1985, *Modern Architecture: A Critical History*

Gan 1974, *Tall Buildings and Urban Planning*

Giangreco 1973, *General Report*

Giedion 1967, *Space, Time and Architecture: The Growth of a New Tradition*

Halpern 1978, *Downtown USA: Urban Design in Nine American Cities*

Halprin 1963, *Cities*

Heckscher 1977, *Open Spaces: The Life of American Cities*

Islam 1990, *Dhaka Metropolitan Fringe, Land and Housing Development*

Jencks 1980, *Skyscrapers—Skyprickers—Skycities*

Jencks 1984, *The Language of Post-Modern Architecture*

Khan 1978, *The Role of Tall Buildings in Urban Space, 2001*

Liang 1973, *Urban Land Use Analysis—A Case Study of Hong Kong*

Lynch 1981, *A Theory of Good City Form*

Lynch 1984, *Site Planning*

Maki 1972, *The Menace of Tall Buildings*

McHarg 1992, *Design with Nature*

Morris 1974, *History of Urban Form: Prehistory to the Renaissance*

Mumford 1938, *The Culture of Cities*

Mumford 1961, *The History of Cities*

Nawaz 1991, *Dhaka: 2000 A.D. and Beyond, an Urban Black Hole or a Livable World City*

Nawaz 1992, *A Proposal for Future Alternate Growth Options for Dhaka, Bangladesh*

Newman 1973, *Defensible Space: Crime Prevention through Urban Design*

Nicholas 1987, *Mega-Cities of the Future*

Norberg-Schulz 1979, 1980, *Genius Loci: Towards a Phenomenology of Architecture*

Planning Commission 1985, *The Third Five Year Plan*

Rasmussen 1962, *Experiencing Architecture*

Robertson 1973, *Heights We Can Reach*

Rossi 1982, *Architecture of the City*

Rowe 1978, *Collage City*

Schmitt 1991, *Schuylkill Riveredge Project, Philadelphia, Pennsylvania*

Scully 1988, *American Architecture and Urbanism*

Selby 1992, *Beyond Edmund Bacon's Philadelphia: A Case Study of Urban Metamorphoses*

Selby 1993, *Urban Synergy: Process, Projects, and Projections*

Siddiqui 1990, *Social Formation in Dhaka City*

Sitte 1889, *City Planning According to Artistic Principles*

Stackhouse 1992, *Foreign-Aid Army Regrouping in Face of Failure, the Globe and Mail*

Venturi 1977, *Complexity and Contradiction in Architecture*

Warfield 1989, *Shanghai: Urban Directions in New China*

Warfield 1991, *Edmund Bacon: 1991–92 Recipient of the Plym Distinguished Professorship in Architecture*

Warfield 1992, *Tomorrow's Skyline: A Search for the Role of Tall Buildings in Urban Design*

Zalcik 1974, *Town-Planning and Sociological Considerations of Tall Building Construction in Slovakia*

Zoll 1974, *The Utility of Very High Bulk*

# 4

# Structural Expression in Tall Buildings

This chapter seeks to describe the visual aspects of tall buildings from the perspective of structural engineering. Although there are other ways to view tall buildings, as presented in Chapters 2, 5 and 7, this chapter focuses only on a set of tall buildings which exhibit strong structural expression. The buildings described here are examples of structural art, a 200-year tradition which began to emerge after the Industrial Revolution, and which embodies three main ideals: efficiency, economy, and elegance. Within the disciplines of efficiency and economy, designers have realized the freedom of artistic expression open to them. Rather than adding only architectural details to their buildings, they saw the great potential for attractive structures through expressing and accenting the structural form. The goal of this chapter is to identify that expression and to call attention to the aesthetic qualities of the overall form and the structural details. In these works, this expression does not arise merely by satisfying functional requirements but also by fulfilling aesthetic desires of the individual designers through the collaboration of the structural engineer and the architect. Although all of the buildings described here are functional, efficient structures built economically, each of them could have been done differently, resulting in entirely different appearances without noticeable changes in cost or utility. However, due to the creativity and artistic sensitivity of the designers, these buildings are in a way truly works of structural art.

Section 4.1 provides a historical background of structural art as a component of architecture. Section 4.2 presents several examples of steel tall buildings and Section 4.3 covers a few examples of structural expression in concrete tall buildings. Section 4.4 discusses the need for collaboration between architects and structural engineers, describing structure as a new art form, and offering some ideas for current and future trends. Future research needs are discussed in Section 4.5.

## *4.1 HISTORICAL BACKGROUND*

The tradition of expressing structure in tall buildings has been a long one, beginning with the Gothic cathedrals of the middle ages and continuing with the theories of Viollet le Duc (1814–1879). In the twentieth century, engineers and architects have carried on this tradition with a variety of new forms of structural expression.

Particularly in the past 200 years, many of the buildings and bridges which clearly express their structure have come to be thought of as works of structural art (Billington, 1985). These structures began to emerge after the Industrial Revolution with the development of a new material, industrialized iron. With the later development of such materials as steel and reinforced concrete, this tradition of structural art was carried into the design of tall buildings. Although many tall buildings today have structural elements exposed on the facade, not all of these buildings are structural art. Works of structural art not only expose structure, but they express the behavior of the structural system so that a visual analysis of the building helps to explain its structural behavior. As will be described in this chapter, and as also noted in Chapters 2 and 5, the expression of structural behavior allows for many new forms.

## 1 Structural Art

Structural art, a relatively new name for a long existing tradition of expressing building and engineering structures, is grounded in the understanding of engineering principles—particularly structural theory and behavior—and embodies three basic ideals: efficiency, economy, and elegance.

There are many aspects in which structures need to be efficient. The first involves the use of natural resources. The structural materials must be used to their best advantage, and they must be formed to carry their loads as efficiently as possible. As engineers are being called upon to build larger structures—longer bridges, taller buildings, and longer spanning roofs—using less material more efficiently has become a priority. Medieval designers stretched stone into skeletal Gothic cathedrals, whereas engineers since the Industrial Revolution have been stretching iron, then steel and concrete into new, light forms that are not only structurally sound but show off their lightness.

The need for conserving natural resources is balanced by the need for conserving public resources. Whether the works are private or public, the demand for economical structures by the owners has always been paramount. Therefore, engineers have been required to design for economy coupled with usefulness, or more utility for less money. Great works of structural art came into being for the very reason that their designers learned to build efficiently. Through learning from previous works and working closely with politicians, business people, and clergy, these artists were able to use the economies of construction as a spur rather than an obstacle for creating structural art.

Designing with a minimum of materials and costs is not enough, however. Efficiency and economy alone have produced too many unattractive structures. These disciplines must be complemented by a sense of elegance. The opportunities and freedom for engineers to express the aesthetic criteria of a project and the ways in which they may do this without compromising structural performance and cost are numerous.

This freedom is important for civil engineers as they are indeed engineers for civilians. Civilization requires civic works for water, transportation, and shelter. The quality of life in a civilization depends on the quality of these works, the efficiency of their design, the economy of their construction, and the visual appeal of the final work. The best works therefore function reliably, provide the best possible cost value to the public, and are elegant additions to their environment. There are unfortunately many examples of faulty, excessively costly, and often ugly structures. However, if the general public as well as engineers and architects understand the extent and potential of what has already been built, they can encourage more strongly the creation of innovative, new works of structural art.

## 2 Engineering and Science

In looking at the history and tradition of structural art and how it has been carried on in the design of tall buildings, it is necessary to understand the difference between engineering and science and between structural engineering and architecture.

Engineering is often confused with science; however, there are distinct differences between the two. Engineering involves the making of things that did not previously exist. Science, on the other hand, is the discovery of laws that have long existed. Whereas engineering principles are grounded in scientific fact, such as material properties and gravity, there is no one answer to an engineering problem. In science there are optimum solutions to problems; however, in engineering there may be many roughly equal solutions to the same problem. An engineer is thus free to be creative and imaginative in finding solutions. Obviously, the result must obey the laws of nature, but the additional dimensions of economics and visual appeal lead to the numerous possibilities available to design engineers.

## 3 Structures and Architecture

Beginning with the Industrial Revolution, structural engineering has become a separate discipline from architecture. Such visible structures as the Eiffel Tower, the Brooklyn Bridge, and the King Dome result directly from technological ideas and from the experience and imagination of individual structural engineers. At times, the engineers have worked with architects just as with mechanical or electrical engineers, but in many cases the forms have come from structural engineering ideas.

Structural engineers are essential in collaborating with architects in the design of large-scale structures, whether the programmatic functions are simple or complex. Structural engineers are principally concerned with the structural form and components of a building, whereas architects are primarily concerned with spatial order and the harmonious integration of the building functions, form, mechanical systems, as well as the building structure itself. Although the roles of the structural engineer and the architect are autonomous, they are not mutually exclusive. It is through their collaboration that large-scale and complex buildings can be designed to the highest aesthetic and structural standards.

The prototypical engineering form—the public bridge—generally can be built without an architect. The prototypical architectural form—the private house—generally requires no engineer. We have seen that scientists and engineers develop their ideas in parallel and sometimes with much mutual discussion; and that engineers of structure must rely on engineers of machinery in order to get their works built. Similarly, structural engineers and architects learn from each other and sometimes collaborate fruitfully, especially when, as with tall buildings, large scale goes together with complex use. But the two types of designers act predominantly in different spheres.

Works of structural art have sprung from the imagination of engineers who have, for the most part, come from a new type of school—the polytechnical school, unheard of prior to the late eighteenth century. Engineers organized new professional societies and worked with new materials. Their schools developed curricula that decidedly cut whatever bond had previously existed between those who made architectural forms and those who began to make the new engineering forms synonymous with the modern world. For these forms the ideas inherited from the masonry world of antiquity no longer applied; new structural ideas were essential in order to build with the new materials. But as these new ideas broke so radically with conventional aesthetics and building practices, they were rejected by the cultural establishment. This is, of course, a classic problem in the history of art: new forms challenge established orders. In this case it was the classically derived architecture of the Beaux-Arts against the technologically inspired architecture of the structural arts. The skeletal metal construction of the nineteenth century was criticized by many architects, cultural leaders, and the popular press. New buildings and city bridges suffered from valiant attempts to cover up or contort their structure into some reflection of stone form. In the twentieth century, the use of reinforced concrete led to similar attempts. Although some people were able to see the potential for lightness and new forms, many architects tried gamely to make concrete look like stone or, later on, like the emerging abstractions of modern art. There was a deep sense that engineering alone was insufficient.

The conservative, plodding, hip-booted technicians might be, as the architect Le Corbusier said, "healthy and virile, active and useful, balanced and happy in their work, but only the architect, by his arrangement of forms, realizes an order which is a pure creation of his spirit...it is then that we experience the sense of beauty" (Le Corbusier, 1927). The belief that the happy engineer, like the noble savage, gives us useful things, but only the architect can make them into art is one that ignores the centrality of aesthetics to the structural engineering artist. True, the engineered structure is only one part of the design of such architectural works as a private house, a school, or a hospital; but in towers, bridges, free-spanning roofs, and many types of industrial buildings, aesthetic considerations derived from structural efficiency and expression provide important criteria for the engineer's design. The best engineering works are examples of structural art, and they have appeared with enough frequency to justify the identification of structural art as a mature tradition with a unique character and a 200-year tradition.

## 4 The Tradition of Structural Art

The tradition of structural art can be classified into two main periods. The first, from the late eighteenth century to the late nineteenth century, was a direct result of the Industrial Revolution. The second period began in the late nineteenth century and continues today. The fundamental difference between these two periods is in the materials that were involved. The first period was centered around structural applications of iron, the second arose through structural innovations in steel and structural concrete.

The first contributions to structural art were made through structural development of cast iron. The advantages of this new structural material became clear when the Severn Iron Bridge (1779) in England was the only bridge to survive the disastrous 1795 Severn River flood. A cast-iron arch bridge, its light open structure with its proven high strength spurred many engineers to study further this new material and the different forms it could take. The Scottish engineer Thomas Telford (1757–1834) was the first to make significant contributions to structural art through cast iron in his great works such as the Bonar Bridge (1810) and the Craigellachie Bridge (1814) in Scotland. He also led the way with wrought iron in his Menai Bridge (1826) in Wales, which at that time was the world's longest span. Telford (1811) clearly displayed his desire to "improve the principles of constructing iron bridges, also their external appearance [and] to save a very considerable portion of iron and consequently weight" (Billington, 1985). Thus Telford's central ideals—efficiency in materials, economy of construction, and attention to final appearance—continued to be those of structural artists ever since. Telford made great contributions not only to structural art through the construction of bridges, but also to structural theory. In his writings he made clear his understanding that laws of nature and needs of society give rise to form rather than preconceived aesthetic rules.

In all of his work, he was the first modern engineer to show that a concern for aesthetics does not compromise technical quality but, rather, can improve it (Telford, 1838).

Telford's ideals were carried on during the growth of the next major technological advance of the iron age—the railroad. The design of railroad bridges meant the consideration of heavier, dynamic loads. This new load requirement led to new forms. Two British engineers, Robert Stephenson (1781–1849) and Isambard Kingdom Brunel (1806–1859), recognized this new need and the possibilities it created for new forms.

Stephenson's greatest contribution, the Britannia Bridge, was a railroad bridge completed in 1850. It was more a technical success than an aesthetic one. Economy was less crucial to Stephenson and the result is monumental but not expressive of structure.

Brunel on the other hand had a stronger urge to express structure and focused on designing visually thin forms. His major contributions included the design of the Clifton Suspension Bridge completed after his death in 1864, the Saltash Bridge (1859), and Paddington Station in London (1854). Brunel's artistic style appears in these different types of construction through their light and open appearance. These permanent symbols of their age remain standing and embody Brunel's sound technical capabilities, attention to detail, and love of structural expression—qualities of a true structural artist. With the Paddington Station, the expressive power of metal for buildings began a tradition that eventually culminated in the late-twentieth-century tall buildings.

The tradition of structural art was carried on in France by Gustave Eiffel (1832–1923), who was the first to introduce structural expression in tall structures. His contributions mark a change in the direction structural art would take from then on. Three of Eiffel's most expressive works include two iron railroad bridges, the Rouzat Viaduct (1869) and the Garabit Viaduct (1884), as well as his famous Eiffel Tower, also of iron, for the 1889 Paris Exposition. All three of these tall structures clearly express their need to resist lateral wind forces. The tower supports for the Rouzat Viaduct splay out at their bases in anticipation of the form for the tower in Paris. The Garabit Viaduct is an arch bridge and the arch splays out laterally toward the base as well. These two works led then to Eiffel's design of the 300-m (970-ft) Paris tower, a strong and striking symbol of this new art form. Eiffel's use of structural expression for wind resistance was an influential precedent to twentieth-century structural art, particularly in the design of tall buildings.

The first American to attract attention as a structural artist was John Roebling (1806–1869), who used iron in a way quite opposite to Eiffel. While Eiffel supported structures with rigid arches, Roebling used iron in the form of catenary cables to hang structures in suspension. Roebling's most remarkable work (and his first in steel) was the Brooklyn Bridge (1883), completed by his son 14 years after his death. The main suspension cable with thinner-diameter vertical cables was typical of previous suspension bridges such as Telford's Menai Bridge. However, recognizing the need for resistance to the vertical motion due to lateral wind forces, Roebling added a secondary structural system of diagonal stays. The stays and cables not only display two different structural systems, they also provide a striking appearance from afar as well as intriguing views of the city from the bridge itself.

Another significant American bridge to influence the tradition of structural art is the Eads Bridge in St. Louis, Missouri (1874). This was the first metal arch bridge in the United States to be completed specifically as an object of civic art and as well, the first major structure to be built of steel. The Eads Bridge was a source of inspiration to both engineers and architects. Just as Telford was inspired by the Severn River flood to think about iron as a structural material, architect Louis Sullivan was inspired by the Eads Bridge to consider the expressive nature of steel. Sullivan's thinking was influenced by the Eads Bridge, which led to numerous late-nineteenth-century Chicago buildings referred to as the first Chicago school (Billington, 1985). The tall buildings of this time (10 to 16 stories) are not, however, considered truly developed works of structural art. In buildings like the Home Insurance Building and the First Leiter Building the designers, like Jenney, were just beginning to address the problem of structural expression. However, the first Chicago school was the foundation out of which the great tall buildings of the twentieth century rose (see Chapter 5). The sense of innovation in the first Chicago school in both engineering and architecture helped to inspire the later Chicago architects and engineers in their creation of works of structural art.

Also during the late nineteenth century the new material of reinforced concrete was introduced to the building profession. Robert Maillart (1872–1940), a Swiss engineer, realized probably better than anyone else the numerous possibilities of reinforced concrete. Maillart imagined new forms for three major building types—the flat-slab (or column-supported) floor, the beam-supported roof, and the thin-shell vault. In the first type, Maillart saw the need to support floors in a way that is "the most rational and the most beautiful" (Billington, 1985). By the most "rational" form Maillart meant the most efficient form which transfers the forces from slab to column by a hyperbolic profile. The structural difference between the hyperbola and, for example, the parabola is, however, insignificant to Maillart, and he selected the one that appealed to him most aesthetically, the one that had the smoothest transition. In the case of the Giesshuebel warehouse in Zurich (1910), Maillart did this by adding capitals to the columns to provide a smooth transfer from column to floor. The integration is both structurally efficient and visually pleasing. This idea of integration is carried on in Maillart's design for the Magazzini Generali warehouse shed in Chiasso (1924). The light roof is designed as a truss that deepens in the middle of the span where the forces are greatest. The members of this truss—the roof, the bottom chord, and the vertical connecting struts—are well integrated and have thin dimensions. The supporting columns as well connect smoothly by their lateral widening under the roof. Both the integration of columns with floors in Maillart's flat-slab structures and the control of forces with only vertical and horizontal members in his shed roof structure can be applied to many building types, especially the design of tall buildings.

Maillart's third building type, the thin-shell vault and his many bridges, are an important part of the tradition of structural art as well. The Cement Hall (1939) inspired many designers with its thinness and the elegance of its pure structural form. The Salginatobel Bridge (1930), in particular, exemplifies Maillart's talent as a structural artist using a form of his own invention (a three-hinged hollow-box arch). Parenthetically, it was the least expensive of all the proposals, and it is an outstanding addition to a beautiful setting. The possibilities of using concrete as a structural material were greatly increased with the invention of prestressed concrete by Eugene Freyssinet (1879–1962).

While Maillart and Freyssinet were designing in Switzerland and France, respectively, numerous other possibilities of new forms were being realized in Spain by three exceptional Spanish designers: Antonio Gaudi (1852–1926), Eduardo Torroja (1899–1961), and Felix Candela (b. 1910). Gaudi, a Catalan architect, reacted against the common practice of facade making of the nineteenth century and, in his mature works, searched for new forms, which would coincide with rational structures. His roof designs for part of the Eusebio Guell housing development (1900) are one result. These roofs are saddle-shaped (hyperbolic paraboloids), and with their double curvature they offer not only a stiffer structure than the more common domes of the time, but they also allow for elegant forms. The double curvature can be broken down into many straight lines, allowing for easy construction, especially in concrete. For Gaudi, form did not follow structure and construction. They were identical.

Gaudi's works provided a strong stimulus for the works of future Spanish designers. While Gaudi had a strong desire to search for new forms, Torroja, an engineering professor and designer, had a passion for thinness and for ribless thin shells made from reinforced concrete. His Zarzuela conoidal cantilever shells represent this tradition of thin vaulting.

Candela, trained as an architect but skilled in mathematical analysis, followed Gaudi in his search for new forms and Torroja in his belief in new materials. Like Maillart, Candela succeeded in creating a new style out of forms appropriate to concrete.

Candela carried on the Spanish tradition of expressing thinness and smooth surfaces. In particular this can be seen in the Xochimilco Restaurant roof near Mexico City (1958), an extremely thin roof made up of eight hyperbolic paraboloid vaults, which was entirely smooth (required no ribs). Much of this success was due to his experience as a builder rather than to his understanding of mathematical theories of structures. More important, however, Candela represents the advantage of collaboration between architects, engineers, and builders. Candela himself has acted as all three, and it is as a result of his understanding and sensitivity of each profession, as well as his rich imagination, that he became such an exceptional and influential structural artist.

One current designer of thin-shell concrete structures influenced by Candela is Heinz Isler (b. 1926). A brilliant Swiss engineer with a strong desire to "play" with concrete, Isler has designed numerous shells to cover tennis halls, swimming pools, and even gas stations. Examples of these can be seen in his roof for the Sicli Company building in Geneva (1969) and the indoor tennis center in Heimberg, Switzerland. The rigorous yet simple methods with which he designs and his continual consideration of the construction process led to works in concrete of exceptional thinness, yet requiring no waterproofing over the bare concrete.

Christian Menn (b. 1927) is a practicing Swiss bridge engineer who has designed the two longest spanning bridges in Switzerland, the Felsenau Bridge (1974) and the Ganter Bridge (1980). Both bridges are designed with remarkable aesthetic sensitivity and display their structural functions. The haunching of the box girders of the Felsenau Bridge reflects the flow of the forces into the pier supports. The splaying of the base of the extremely tall piers at the Ganter Bridge represents the need to resist lateral wind loads, reminiscent of Eiffel's railroad viaducts and Paris tower, yet Menn worked with an entirely different, more modern material—prestressed concrete. The design of such tall piers is similar to the design of tall buildings, structures that must resist great lateral forces as well.

A contemporary of these Swiss designers, Isler and Menn, as well as Candela, is Fazlur R. Khan (1930–1982). Khan was born in Bangladesh and practiced most of his life in the United States, specifically in Chicago. Influenced by Mies van der Rohe and working in collaboration with architect Bruce Graham and architect/engineer Myron Goldsmith, Khan helped define the second Chicago school, a tradition of innovative structural art in tall buildings (see also Chapters 2 and 5). Many of the works to be described in this chapter come from these designers and are in this tradition of structural art (Billington, 1986).

## *5 An Approach to Structural Architecture*

In the development of building forms, tall buildings are the most complex because of the interrelationship of a large number of interdisciplinary and often conflicting requirements of design. Tall buildings of modern times are becoming more and more slender, leading to the possibility of more sway in comparison with the massive high-rise buildings of the past. The taller the building becomes, the greater the impact of the natural forces of gravity, wind, and earthquake. Because of this, the structural action counteracting these forces becomes an important aspect of tall building design. The dynamic response of tall buildings to lateral loads can be controlled by improving the structural systems as well as by selecting efficient building configurations, although other nonconventional ways of controlling such response are also possible. The form of a tall building is thus very much related to efficient structural performance, which in turn dictates the economy of the building. The structural function of the building is in this case defined by the load-carrying capacity of the building and its resistance to sway, and it determines the composition of the building masses. (See Chapter 5 for further discussion on this topic.)

Although there are other ways to view tall buildings, we have chosen here to focus only on a set of works that exhibit strong structural expression. Our goal is to identify that expression and to call attention to its form and details. In these works this expression does not arise merely by satisfying functional requirements, but rather by fulfilling aesthetic desires of the designers. Everything described here could have been done differently without noticeable changes in cost or in utility. Moreover, nearly every building discussed here was designed with economy in mind, thus exhibiting one of the leading ideas for structural expression—that economy can be a spur to elegant design.

In the following buildings there exists a collaboration of talented design equals, rather than merely an architect-designer working with a consulting engineer. In the next two sections, case studies dealing with steel and reinforced concrete buildings are presented to demonstrate how structural expression is considered as a principal determinate of tall building design.

The examples of this chapter present different solutions to similar problems. The Marine Midland Bank, the Brunswick Building, and Two Shell Plaza all solve the problem of load transfer using different structural forms. The problem of wind loads is solved by the John Hancock Center, the Sears Tower, the Wang Building, and One Mercantile Place with different aesthetic results. Finally, the problem of spanning large open plazas is addressed at the Federal Reserve Bank and the Broadgate Building in two unique and striking solutions.

## 4.2 BUILDINGS IN STEEL

### 1 Inland Steel Building

One of the earliest examples of structural art in steel buildings is the Inland Steel Building (Fig. 4.1) in Chicago (Graham, 1986). The building's facade clearly expresses one of the main components of the building's structural system, the vertical load-carrying columns. (The Inland Steel Building is also discussed from the aspect of its facade in Chapter 7.) This exposure of the columns on the long side of the rectangular-shaped building was innovative, as glass curtain-wall buildings typical of Mies van der Rohe were much more common at that time. The difficult engineering problem of the connections between the vertical columns and the floor structure behind was solved by Fazlur R. Khan. Not only are the columns pulled out from the main building, but the entire building core is also pulled out from its usual position in the center of the building. The core is at the back of the building (opposite the main street facade). The core is clearly visible as it rises above the building and it is accented by vertical mullions on its walls. On the shorter sides of the building, there are no main structural columns

**Fig. 4.1 Inland Steel Building, Chicago, with its vertical load-carrying columns expressed on the facade.** (*Courtesy: D. P. Billington.*)

reaching the ground; however, the columns that are on the building face are also pulled out from the plane of the windows. All of these expressed columns provide different and interesting views from various angles.

The fully exposed columns let the main building, particularly the well-articulated horizontal spandrels covering the floor slabs, appear to be floating. The lack of a heavy corner column in combination with the slight cantilevering of the main building beyond the outermost column adds to this floating effect. At the street level, the structural columns and the setback of the building create a covered walkway. This walkway through the structure integrates the building with its surroundings while also providing a visual base for the building. This base is open unlike the many solid-based buildings of that time. Another unusual feature is the lack of decorative elements on the facade. Rather, the structural elements are used to provide visual interest. The window mullions are pulled out slightly from the window wall, and they pass over the spandrels, further accentuating the vertical, structural columns. The variation in the plane of the facade not only provides different views from different angles, it also allows for variations in shadows depending on the time of day. With all of the exposed structure, the building still maintains a simple look. The glass wall is clearly visible and gives the building a light appearance.

## 2 Business Men's Assurance Tower

The Business Men's Assurance Tower (Fig. 4.2) in Kansas City, Missouri, is another project in which structure is fully exposed. The building's rigid steel frame defines the facade of the building (Graham, 1986). The expression of this frame is marked as the main building is set back approximately 1.8 m (6 ft). The technical problem of temperature movements in fully exposed exterior frames demanded a new engineering approach (Khan and Nassetta, 1970).

The frame provides both a facade and shading for the office spaces behind. This shading accentuates the facade with the dark areas contrasting with the light steel frame. Within the facade, there is also the expression of structural and architectural components of the building. At the base, the columns are slightly taller, and the entire area is open with only the frame and a core from the building meeting the ground. This openness expresses the frame as the building's structural system while providing a clear entrance. The columns of the top story are slightly taller than those at the middle of the building. Not only does this complement the taller first story, it also signifies the difference in the space beyond. This top story is a mechanical story requiring more height than normal office space stories.

Although the steel required fire-proofing, it was covered as minimally as possible and therefore still expresses itself as a steel frame. Another feature adding to the appearance of this simple facade is the selection of an odd number of bays. This choice eliminates the necessity of a column in the center, which would interrupt the facade. Instead, the open bay across the center gives the facade a balanced appearance. The use of slender members adds to the simple elegance of the building.

## 3 First Wisconsin Bank Building

The First Wisconsin Bank (Fig. 4.3) in Milwaukee, Wisconsin, is another tall building with an exposed steel frame (Beedle and Iyengar, 1982). Here, however, there is an addition of exposed trusses wrapping around the entire building. These "belt" trusses

**Fig. 4.2 Business Men's Assurance Tower, Kansas City, Missouri, with its slender steel frame fully visible.** (*Courtesy: Ezra Stoller.*)

**Fig. 4.3 First Wisconsin Bank building, Milwaukee, Wisconsin. The "belt" trusses have both structural and visual significance.** (*Courtesy: D. P. Billington.*)

are at three locations—directly above the base of the tower, one-third of the way up, and at the top. Near the base, the truss acts as a transfer zone from the many closely spaced columns above to the fewer more widely spaced columns below. The use of this open truss in combination with the fewer columns at the lowest level of the tower lightens the base and provides a smooth transition from the lower level to the main building. This open base is also desirable for marking the transition between the tower and the lower building. The lower truss at the base of the building balances the uppermost truss which, as in the case of the Business Men's Assurance Tower, marks a mechanical level. The truss within the main shaft of the building marks a mechanical level as well. The location of this truss, below the center of the shaft, strengthens the vertical expression of the tower, rather than breaking it in half. All of the trusses are in the same plane as the rest of the steel frame; however, they are accented by a setback of the building envelope which occurs only at these levels. The shadows thus created make the trusses stand out visually.

The two upper trusses connect the core to the outer columns and force them to carry some of the wind load, hence reducing the lateral deflections and allowing the exterior facades to act slightly as a tube. Similar to the trusses, the well-articulated horizontal members wrap around the building. The building's relatively small plan-to-height ratio is its only vertical expression and is complemented by the horizontal dominance of these spandrels.

### 4 John Hancock Center

The most striking feature of the John Hancock Center (Fig. 4.4) in Chicago is the X bracing fully exposed on the facade. (See also Chapter 7 for further information on the Hancock center's facade.) The idea for this design was developed at the Illinois Institute of Technology (see Chapter 2).

Completed in 1968, the John Hancock Center is rectangular in plan, 80.8 by 50.3 m (265 by 165 ft), and tapers gradually to 50.3 by 30.5 m (165 by 100 ft) as it rises 344 m (1127 ft). The X bracing tapers with the tower and is exposed on all four sides. The main vertical columns are also fully expressed, as are the horizontal ties at the beginning and ending of each X.

The tower rests on a masonry base set slightly above the street level. The first story in steel is somewhat taller than the stories of the main shaft. The tower or shaft itself rises uninterrupted to the top. There are few variations as it rises. Aside from the continual tapering, the mechanical levels are expressed by steel panels in place of windows. The mullion pattern in the windows changes as well, depending on the use of the spaces behind. For instance, they are more closely spaced at the residential levels in the upper half of the tower. None of these variations, however, are in the actual structure, and therefore they do not interrupt the continuity of the shaft, which is dominated by the exposed structure. The top of the tower is another mechanical level made up of two taller stories. These stories are similar in size to the base and provide a complementing cap for the building. However, this is not always apparent to the viewer due to the size and location of the building.

All of the exposed details of this building are held together by the X bracing. While the Xs obstruct some windows, the exposed structure in general appears thin and light. As a fully integrated expression of form and structure, however, the braced-tube structure maintains a look of strength and permanence. This delicate balance of stability and lightness through the exposure of its structural system is what makes this tower so aesthetically unusual (Khan, 1967).

**Fig. 4.4 John Hancock Center, Chicago. The exposed structure appears thin and light while maintaining a look of strength and permanence.** (*Courtesy: J. Wayman Williams.*)

## 5 Sears Tower

The Sears Tower (Fig. 4.5) in Chicago consists of nine equally sized 22.85-m (75-ft) square tubes bundled together. The bundled tubes, cut off at different heights, result in a series of stepped volumes and changing views of the building from different angles. This expression of each tube's independence is contrasted by the mechanical stories, which are expressed at the same level on each tube, thus uniting the volumes. Steel panels at these levels wrap around the entire building, bundling the individual tubes together. The mechanical levels cap many of the tubes as they are cut off. Originally there was no significant base to this building. A barrel-vaulted entrance has since been added to the Sears Tower, but is not integrated with the overall structural and aesthetic intentions of the building.

As with the John Hancock Center, the Sears Tower tapers as it rises. While the tapering of the Sears Tower, through the ending of individual tubes, provides new images as one moves around the city, the tapering of the Hancock Center creates a smoother, monolithic building mass. The Sears Tower makes a lively expression from a distance, while close up it is impressive only in scale. By contrast, the Hancock Tower is far more expressive at close range because the X bracing is practically invisible from a distance.

One of the major differences between the Sears Tower and the John Hancock Center is the shape of the building in plan; the John Hancock Center is rectangular and the Sears Tower is square. In fact, the Sears Tower was originally meant to be rectangular as well. However, part of the plan, a hotel adjacent to the Tower, was abandoned, thus leaving a square plan.

The design of a rectangular plan was particularly important to Graham (the architect). Not only does the increase in wall area per interior volume allow more sunlight into interior spaces, but the rectangular shape also allows for more interesting views of the building. Many engineers try to avoid rectangular plans at first, as they argue that the square plan resists loads equally in all directions and is, therefore, more economical. Khan (the structural engineer), however, was able to appreciate the aesthetic advantages of a rectangular plan and believed the aesthetic gain was worth more than the possibility of a slight economical loss. (See Chapters 2 and 7 for more information on the Sears Tower.)

## 6 First Boston Company Building

Another steel tower with expressed vertical bracing is the First Boston Company building (Fig. 4.6) in Boston, Massachusetts. The building was completed in 1970. Whereas the Hancock building has X bracing throughout the length of the building, the Boston building has a K truss appearing as a series of upside-down Vs running up its height. Here the bracing is not visible. The vertical columns, particularly the heavier corner columns, dominate this 41-story square tower. Although the bracing is not exposed, it is clearly expressed by means of steel panels in the windows behind which the bracing passes. A thin-steel mullion runs along these panels, corresponding directly to the bracing behind. This expression of bracing is apparent up close, yet as one gets further from the building, the visibility of the bracing diminishes. In certain lighting, particularly strong sunlight, the bracing seems to disappear altogether. In comparison with other steel towers, such as the John Hancock Center in Chicago, the bracing becomes less visible as one gets further away from the building.

Despite the similarities of exposed structural bracing with the Hancock center, the First Boston building is quite different in character. The heavy corner columns give the building an overall appearance of one giant column rather than a tube structure. The cor-

**Fig. 4.5** **Sears Tower, Chicago. The bundled-tube structure provides different views from different angles.** (*Courtesy: J. Wayman Williams.*)

**Fig. 4.6 First Boston Company, Boston, Massachusetts. The building has expressed vertical bracing.** (*Courtesy: D. P. Billington.*)

ner columns interrupt the continuity in the expression of the floors wrapping around the building. However, they do act as a completion or frame for each facade. Each L-shaped column borders two adjoining facades. These columns also express their load-carrying function as they gradually increase in size toward the ground, where they must resist the most force. While the columns strictly define the four sides, the V-shaped bracing, beginning and ending on the same levels on all four sides, brings the facades together both visually and structurally. The emphasis, however, in this building is mostly on its vertical expression. The orientation of the bracing creates a continuity of upward pointing triangles.

Although the First Boston building does not express its structural system as fully as the Hancock center, it still maintains visual interest through its subtle structural expression. Unfortunately, the interesting bracing is somewhat masked by the closely spaced columns so popular in many skyscrapers. Expression of height is made to be the dominating characteristic rather than the pure structural expression of a braced tube.

## 7 One Mercantile Place

One Mercantile Place (Fig. 4.7), a high-rise office tower in St. Louis, Missouri, was completed in 1975 as the first building of a six-block complex. This 35-story tower recalls many elements of previous structurally expressive buildings, yet results in a new form. Here the corners of a basic rectangular plan have been cut off diagonally, giving the building an elongated octagonal shape, and allowing more sunlight to reach the center of the building. These cut corners are accentuated by a main structural element of the building—vertical K trusses. These trusses provide lateral stability for this tall building and emphasize the verticality of the thin tower.

The trusses provide the main feature of visual interest and are well integrated with the entire tower. The horizontal members of each truss tie into the floors at every third story. There is a balance and interdependence between the horizontal and vertical members. The horizontal elements dominate the main sides, yet are held together and only able to rise to a limited height because of the stabilizing vertical corner trusses, thereby accenting their importance. Another feature calling attention to the trusses is the set-back office windows beyond the trusses. The windows are like a curtain wall with no structural elements expressed, thus showing the structural independence of the trusses. The dark shadows created by the set-back windows also visually emphasize the trusses, and give them visual prominence.

The architects also created a spatial integration of the tower with the street level. The cut-off corners make room for more plaza space, allowing areas for small parks that will invite people to the building. This building was also originally designed to be integrated visually with surrounding buildings. The other buildings of this office complex have also been proposed to express the vertical trusses in cut-off corners.

This cutting of corners represents the idea of cutting something open to show the building's internal structure. This allows the lateral support system, the K trusses, to be brought out to the open. This building demonstrates that there is a potential for beauty in structure and it need not be hidden, but rather, simply expressed.

## 8 Federal Reserve Bank Building

The Federal Reserve Bank (Fig. 4.8) in Minneapolis, Minnesota, completed in 1974, expresses its structural system in the form of two steel catenaries, one on each main facade, suspended between two large piers. The glass curtain wall of this building is particularly impressive. The bottom of the curtain wall is flush with the edge of the catenary. Thin vertical and horizontal mullions mark the location of columns and floors of the main building. Above the catenary, the glass wall is set back about 1 m (3 ft) and only the vertical mullions appear, remaining in plane with the wall below. While these two sections are visually quite different, they are well integrated by means of the vertical mullions.

The building reaches the ground in only three places—the two main concrete piers from which the catenaries are suspended, and the elevator core which is attached to one of the main facades. The overall appearance of the structure is dominated by the main building, which is suspended between two piers. The function of the catenary supporting the entire steel building is apparent to the viewer as the base between the piers is left entirely open. The large open base of the building is in contrast to the solid horizontal top. Although originally not intended to crown the building (an arch-supported section was to be built on top), this solid panel does act as a cap to the building. Viewing the building straight on, one could imagine the towers leaning toward each other due to the weight of the building on the catenary. However, this beam holds them apart, acting as a compression tie.

**Fig. 4.7 One Mercantile Place, St. Louis, Missouri. The cut corners expose the main structural element of the building—vertical K trusses.** (*Courtesy: J. Wayman Williams.*)

The combination of these elements—the open base, the solid top, the delicate steel catenary, and the light glass and steel box suspended between solid concrete towers—all complement each other visually and express their structural importance. The result is a well-balanced building which recalls previous works of structural art. In particular, the suspended catenary reminds one of the Brooklyn Bridge with its delicate cables suspended between stone piers. The mirrored glass, reflecting the surrounding buildings, adds to the imagery of a bridge. The mullions appear as thin cables through which the city can be seen.

## 9 Broadgate Building

A more recent example of structural art in steel is a building of the Broadgate project (Fig. 4.9) in London, England. Completed in 1990, this building is similar to the Federal Reserve Bank in scale and appearance. However, in this case the designers reversed the function of a tension catenary by providing as the main structural support steel parabolic arches carrying loads in compression (Iyengar, 1989). On each of the main facades, a large steel arch is fully exposed. There are two additional arches on the interior. Rather than being integrated with the window wall as at the Federal Reserve Bank, the steel arch is pulled out from the main building. The floors of the building are also clearly articulated and are tied to the arch by steel beams. A tension tie runs between the two supported ends of the arch. Not only is this tie an essential structural element, preventing the arch from splaying outward, it also visually ties the building together. The V-shaped arch brace adds stability both structurally and visually and allows the arches to be slender.

**Fig. 4.8 Federal Reserve Bank building, Minneapolis, Minnesota. Suspended catenary spanning an open plaza.** (*Courtesy: D. P. Billington.*)

Since it is open underneath, the entire building seems to float behind the arch. The location of the stair towers on either side of the arch supports combined with the open base further expresses the structural importance of the arch. The open base also hints at the similarly open, column-free spaces in the building above. The strong arch, spanning an open plaza below, again recalls the idea of a bridge. In this case, the "bridge" is spanning a plaza above railroad tracks.

The connections between the horizontal tie and the base of the arch are also fully exposed, showing the function of the connection. On one side is a pin connection and on the other, a roller. Rather than being covered over, these connections are exposed to add visual interest to a structurally expressive facade. Most important, as with the Inland Steel Building and the John Hancock Center, the fully exposed structure gives the impression of stability without sacrificing the appearance of light weight and aesthetic elegance. (See Chapter 9 for an account of the fire-resistive design of this building.)

### 10 Bank of China

A major tall structure outside the United States is the Bank of China building (Fig. 4.10) in Hong Kong, in which, once again, economy and elegance go together. This 368.5-m (1209-ft)-high tower brings together the diagonal bracing idea of the John Hancock Center and the cut-off tube idea of the Sears Tower in an original way. The building is square in plan and is divided by its diagonals into four triangular tubes, which rise up individually to different heights (*Architectural Record,* 1985). The cut-off tube idea is handled differently than at the Sears Tower as the sections are cut on a slope rather than horizontally. This is consistent with the accent on the diagonals on the facade as well as with the theme of triangles throughout the entire building. The sloped cuts also con-

**Fig. 4.9 Broadgate Building, London, England. Its main structural supports are steel parabolic arches carrying loads in compression.** (*Courtesy: D. P. Billington.*)

**Fig. 4.10 Bank of China, Hong Kong. Composite construction led to the economical and structurally expressive building.**

tribute to the elegant tapering of the structure toward the sky. This tapering is in turn complemented by a visually strong base.

The diagonal bracing exposed on the exterior is framed by the four large corner columns. There are no visible horizontal ties above and below the Xs as at the Hancock center. These diagonals and corner columns not only make up the primary vertical and horizontal force-resisting system of the building, but also give the tower a striking sculptural appearance. They are accented from behind by the light grid of the exterior glass wall. Both the diagonal ties and the major vertical columns are made of structural steel box members filled with concrete. The composite construction resulted in this economical, structurally expressive building.

While I. M. Pei credits Leslie Robertson with the economy of the structure, both designers working together came up with this unique form, which provides both the structure and the aesthetic elegance of the tower (Tuchman, 1989). This tall building illustrates yet another of the many possibilities for new forms with structural expression, particularly through the close collaboration of engineers and architects. (See Chapter 5 for a discussion on the aesthetics of this building and Chapter 2 for further information on its structural system.)

### 11 Hotel Artes-Barcelona

The Hotel Artes-Barcelona complex (Fig. 4.11) at the Olympic village in Barcelona, Spain, is dominated by a 45-story hotel-apartment tower completed in 1992. The tower is square in plan and is set up on a base of columns. The tower's structural frame is pulled out from the main building and is fully exposed. The cross-bracing for lateral loads is concentrated at the corners. These strengthened corners are then connected at three separate levels to create a partial trussed-tube form. Although not as structurally efficient as a fully trussed tube, such as the John Hancock Center, the connection of corner trusses provides sufficient stability for this midheight building (Iyengar et al., 1993). The attention to detail gives this building its visual appeal. The full, clear expression of all members and joints allows for interesting shadows and ever-changing views, despite the regularity often associated with buildings which are square in plan. In fact, the frame itself is not a true square since the corners are slightly beveled. This adds an extra visual framing of each individual facade while smoothly connecting the sides. As with the First Wisconsin Bank, the connecting trusses across the top and bottom of each facade balance each other. While the positions of the trusses help to accentuate the height of a building, the higher positioned middle truss of the Hotel Artes-Barcelona tends to accent the top of the building whereas the lower middle truss of the Wisconsin Bank accentuates its base. The location of these trusses can be a function of the spaces beyond, derived from the stability criteria for lateral loads, or simply the preference of the designer.

## 4.3 CONCRETE BUILDINGS

### 1 Hartford Company Building

The 20-story Hartford Company building (Fig. 4.12) in Chicago, built in 1961, was one of the first examples of a large-scale concrete building with the structure fully exposed (Khan, 1972). The contrast to the skyscrapers sheathed in glass, where the glazing is either infilled within the building structure or suspended from the structure as in a curtain

**Fig. 4.11 Hotel Artes-Barcelona, Spain. Partially braced tube.** (*Courtesy: Skidmore, Owings & Merrill.*)

**Fig. 4.12 Hartford Company, Chicago. The building's facade is simply its exposed structural concrete frame.** (*Courtesy: D. P. Billington.*)

wall, the glass wall of the Hartford building is set back about 1.8 m (6 ft) from the reinforced concrete frame. The structure, therefore, becomes the facade. This frame clearly displays its function and the interrelationships between the columns and floor slabs. The slabs are haunched, increasing in depth as they meet the columns. This is desired to accommodate the increased load the slab brings to these columns. Not only is the increased depth structurally efficient, it also adds visual interest, providing smooth connections. (This approach is similar to Maillart's designs of column-supported flat slabs.) At the upper levels the columns do not need to carry as much load and can therefore decrease in size as they rise (Grube and Schulze, 1977).

Whereas each facade differs only in the number of bays, there are slight differences in story heights. The lowest and the uppermost levels are both taller. This height difference accentuates the building's vertical rise while also signifying the importance of the taller levels, namely, the entrance and a mechanical level. As with the Business Men's Assurance Tower, there are an odd number of bays on each facade. The use of a bay over the centerline of the building balances the facade rather than dividing it into two sections.

### 2 Brunswick Building

The 1966 Brunswick Building (Fig. 4.13) in Chicago is another example of an exposed reinforced concrete framed building (Grube and Schulze, 1977). This 38-story building is of a much larger scale than the Hartford Company building and consequently addresses different structural problems.

The uniformity of the facades shows the building to be a tube structure. In fact, it is a tube around a tube, the inner tube being a concrete core. The most striking feature, however, is the two-story solid transfer girder toward the base of the building. Structurally as well as visually, this girder transfers the loads from the many closely spaced columns above to the widely spaced columns below. This deep girder is partly integrated with the columns above as they continue down and flair out to the outside of the solid girder. There is little integration, however, with the columns below. The girder simply rests on top of them. Nonetheless, the girder and ground-level columns do provide a strong visual base and, from a distance, are balanced by the tall mechanical level at the top. The widely spaced lower columns also provide a large amount of open space to accommodate pedestrians and connect the building spatially with the street level.

As with the Hartford Company building, attention to structural detail adds greatly to the visual interest of this building. This is particularly true of the columns of the shaft. As they reach the base where they must carry the most load, they not only get wider within the plane of the facade but they also extend forward from this plane to meet the heavy transfer girder.

### 3 DeWitt Chestnut Building

In later buildings designers gave more attention to the visual integration of a transfer girder with the rest of the building. In the DeWitt Chestnut apartment building of 1963 (Fig. 4.14) in Chicago, the designers used a deep transfer girder within the tube structure of the building. The structural frame is exposed similar to the Brunswick Building; however, in this case, the tube system is more obvious since the many columns at the ground level are part of the building wall (Grube and Schulze, 1977). Thus the building appears as one continuous tube.

**Fig. 4.13 Brunswick Building, Chicago. The deep transfer girder carries the loads visually and structurally from the many small columns above to the fewer large columns below.** (*Courtesy: D. P. Billington.*)

**Fig. 4.14 DeWitt Chestnut building, Chicago. The light concrete frame forms are continuous tubes.** (*Courtesy: Hedrich-Blessing.*)

The vertical load path is visible within the facade since the columns become wider as they approach the base, where they must carry the greatest loads. The floor beams also increase in depth toward the bottom of the shaft.

This concrete frame has a light appearance, a result of the thinness of its members, which are only slightly pulled away from the windows. The effect is that of a frame sheathing a glass box. Rather than dominating the exterior, the frame acts with the window wall to provide a simple and structurally light appearance. Just as the Brunswick Building's base reflects the widely spaced columns of the Civic Center it faces, so too the lightly framed DeWitt tower echoes the similar lightweight structure of the Lake Shore apartments nearby.

### 4 *Two Shell Plaza*

The problem of transferring many column loads to one column below, which was previously taken care of with a heavy transfer girder, was skillfully handled at Two Shell Plaza (Fig. 4.15) in Houston, Texas (Khan, 1972). Completed in 1964, this 21-story building expresses its load transfer without the use of a heavy girder. Instead, the transfer is accommodated over the course of many stories within the plane of the structural grid. The loads are slowly brought toward the more widely spaced columns by means of widening the upper columns and deepening the beams toward the base. The effect is one of creating shallow arches that become thinner in section as they rise. One can easily see the loads following the more solid areas. The strong spandrel beams transfer the vertical loads through shear. This use of vertical and horizontal members to control forces recalls Maillart's Chiasso shed roof, in which the gravity and lateral loads were taken only by vertical and horizontal elements.

The base is clearly the main focus of visual interest in this building as there is no parallel expression at the top. The transition of the base to the rest of the building is particularly smooth as the loads are taken by many first-level columns, as opposed to only four columns on the long facade of the Brunswick Building. The use of additional smaller columns in the base helps integrate the larger columns in the shaft of the building above, yet does not take away from the openness below.

### 5 *Marine Midland Bank Building*

The Marine Midland Bank (Fig. 4.16) in Rochester, New York, completed in 1970, is another example of an exposed reinforced concrete frame building expressing the vertical load transfer from many columns above to fewer widely spaced columns below (Khan, 1972). Unlike Two Shell Plaza, this 20-story building's load transfer occurs both in and out of the plane of the facade, creating an undulating effect. The upper columns carrying loads to the main columns below grow wider and thicker as they approach the base. The result is a facade supported by treelike forms; the thick columns below branch and thin out as they rise. This play with the columns is visually balanced by the horizontal bands of uniform thickness, corresponding to the spandrel beams.

There is neither a main corner column at the base nor are there smaller ones along the edge of the facade above. The floor beams, when viewed straight on, appear to cantilever slightly toward the corners. From an angle, one sees the beams and the glass wall wrap around the corner. The strong horizontal bands near the base provide the shear transfer needed to channel the loads into the heavier columns and down to the isolated supports.

Although each facade is not visually completed by corner columns, there is a sense of vertical continuity. The large, widely spaced columns of the open plaza branch out into many columns at the tall second level. This branching continues up through the building and is capped by a tall mechanical level, composed entirely of thin, closely spaced columns on the facade. The articulation of these parts is subtle. It is the even integration of these elements which leads to the simplicity of the facade. The plaza level, slightly raised from the street level, appears as a platform upon which the building stands.

**Fig. 4.15 Two Shell Plaza, Houston, Texas. The expression of load transfer occurs over many stories in the plane of the structural grid.** (*Courtesy: J. Wayman Williams.*)

### 6 One Shell Plaza

One Shell Plaza (Fig. 4.17) in Houston, Texas, is another reinforced concrete framed-tube building. It is currently (1994) the tallest lightweight concrete structure and was the tallest concrete building at 218 m (714 ft) when it was completed in 1971 (Fintel, 1986). Like the Marine Midland Bank, the facade undulates. This undulation, however, occurs for the entire height of the building rather than at the base alone. With the Marine

**Fig. 4.16 Marine Midland Bank, Rochester, New York. The building has an undulating load transfer region toward its base.** (*Courtesy: J. Wayman Williams.*)

**Fig. 4.17 One Shell Plaza, Houston, Texas. The undulating facade expresses the increase in load on exterior columns opposite the corner of the interior core.** (*Courtesy: J. Wayman Williams.*)

Midland Bank, the undulation accommodates the increased vertical loads transferred to the few columns at the base. At One Shell Plaza the undulation serves to compensate for the increase in load at locations corresponding to the corners of the interior core. Where the floor slabs no longer are framed into the core, they must frame into a large beam, which in turn frames into the core on one side and the facade on the other. Therefore, on the facade a stronger column support is required, thus the increase in column size where the core ends.

Here designers had a choice. They could have provided the extra column strength by using additional reinforcing steel, by increasing the column depth perpendicular to the facade plane, by increasing the column width, or by combinations of these means. On the other hand, they could have extended the columns inside the building to achieve the appearance of a smooth facade. These choices, having negligible influence on the overall cost, are the essence of structural art, expressing the extra loads by pulling the column depths out from the facade, and have resulted in a structural design of originality and surprise.

This undulation for structural reasons is visually interesting as it creates shadows that are striking. Although the two places of undulation occur symmetrically on the facade, the shadows they create are usually asymmetrical. These long stripes of shadow accentuate the height of the building.

The tall windows toward the base integrate the main tower with the solid base through the similar proportions of these two stories as well as through the undulation. These stories are both of similar height (taller than the stories of the main tower), and the undulation in the solid base corresponds to the undulation of the columns above. The solid base seems naturally to visually and structurally support the undulating columns. As the column sizes are increased to accommodate larger loads, the base must also be increased as a support.

### *7 Wang Building*

A much different form of structural expression in concrete occurs in the Wang Building (Fig. 4.18) of New York City, completed in 1984. This narrow tower resembles the John Hancock Center in Chicago in its expression of X bracing to resist lateral loads (Grossman, 1985).

The Xs are created by filling in windows with reinforced concrete covered by dusty red stone, leaving the diagonal form visible. Not only does this form of bracing add visual interest, but it is also done within the framework of the building. Rather than adding new structural members, the stability of the tower is increased simply by filling in windows and adding diagonal reinforcing steel. The bracing is on all four sides, and the rectangular plan requires Xs on the long sides and "zigzag" forms on the short side. As with the John Hancock Center, this rectangular form creates different views from different angles and allows more sunlight to reach the center of the building.

The X bracing is the main feature of visual interest for this building. The structural facade is flush; however, the windows are set back. The dark shadows often created by this setback accent the X bracing. Depending on the angle of vision, the bracing may or may not appear as continuous lines. At times, the unfilled windows barely appear to meet at the corners.

### *8 Onterie Center*

Similar to the Wang Building in New York, the 1986 Onterie Center (Fig. 4.19) is also a concrete tower with X bracing visually expressed on the facade. Here, however, the

**Fig. 4.18** **Wang Building, New York City. The stiffness of the concrete was increased by filling in windows and adding diagonal reinforcing steel.** (*Courtesy: Skidmore, Owings & Merrill.*)

**Fig. 4.19 Onterie Building, Chicago. X bracing increases lateral stiffness.** (*Courtesy: Skidmore, Owings & Merrill.*)

bracing is thinner and, therefore, more delicate in appearance. The thinness results from the more closely spaced columns, forming smaller windows and thus creating thinner diagonals (Zils and Clark, 1986).

Rather than being one single shaft, the tower's center on its long side is set back and not braced diagonally. This plain setback accentuates the two braced shafts, which form channels in plan. Although the setback seems to divide the building, a solid top floor brings the sides together, capping the building and visually holding it together. The base of the tower splays out in front to provide more office space with much sunlight below, and the X bracing is carried through to the street level. (See Chapters 2 and 11 for more information on the Onterie Building.)

## *4.4 COLLABORATION AND STRUCTURAL EXPRESSION*

The designs of the last 40 years demonstrate the evolution of structural expression from the modernist gridded box to the technological advancements of the trussed-tube and bundled-tube structural systems. These new and innovative structural systems express the ideals of economy and elegance and, in turn, liberate the imaginations of the architect and the structural engineer.

### *1 Collaboration of Engineers and Architects*

The expression of structure requires close collaboration between the structural engineer and the architect at the outset. Collaborative designing is crucial because it is the structural system and structural detailing that stimulate the imagination to create new forms with aesthetic appeal. The choice of a structural system for a tall building must satisfy both the site conditions and the architectural program for the building. As with the John Hancock Center in Chicago, structural engineer Khan proposed the X-braced structure to withstand strong Chicago winds, whereas architect Graham pushed for a rectangular plan to allow more sunlight to reach the office and living spaces. Another example is the choice of an arch system for the Broadgate Building in London. The arches fulfill the site requirement of having the building span a plaza over railroad tracks, while at the same time providing column-free interior spaces for the occupants.

This collaboration of engineers and architects continues from the broad range of choosing a structural system down to the smaller structural details. The diagonal bracing and composite columns of the Bank of China came from the close working relationship of Pei and Robertson. Other examples of details from collaboration include the use of X or V bracing at the Hancock center, the Wang Building, and One Mercantile Place as well as the "tree" forms of the Marine Midland Bank and the structural undulation at One Shell Plaza. Even smaller-scale details expressing the building's structure add interest as well. This can be seen in the raised mullions of the Inland Steel Building, the flaring columns at the Brunswick Building, the mirrored glass at the Federal Reserve Bank, and the thin mullions along the K bracing at the First Boston building. All of these forms and details were chosen by both engineer and architect to express the structure of the building while balancing structural efficiency and economy with aesthetic appeal.

The structures discussed in this chapter are samples of many successful buildings that were possible as a result of the engineering imagination combined with the collaboration of sensitive architects. (See also Chapter 2 for such collaborative efforts, particularly as design collaboration between academia and office practice is concerned.)

### 2 Current Trends in Structural Expression

The focus of this chapter is an examination of completed works with strong structural expression. Buildings of steel from 1955 to 1992 and of concrete from 1961 to 1986 have been discussed. This period represents a high point for structural expression. Designers worked closely together to achieve these innovative elegant designs. In spite of the many new possibilities, there are few contemporary buildings that exhibit structural expression. One explanation for this deficiency lies in the belief of many designers and clients that structure should be hidden and building form be a matter of the visual envelope only. This view reflects the attitude taken by critics of Gothic architecture who thought structure was crude and barbarous—hence "Gothic." Therefore the skeleton should not be expressed. All buildings should not necessarily express structure, but creative designers ought to explore new forms collaboratively so that they can enliven cities with many more examples of structural expression, while simultaneously meeting architectural and functional requirements.

Current trends in high-rise architecture are characterized by the decorative and articulated sculptured forms (see Chapter 5). The goal seems to be exhibition of the architectural envelope rather than structural form. Because of the irregularities of shapes, structural logic can no longer be applied to discontinuities in the building skeleton. The structural engineer's role has generally been reduced to finding possible structural solutions behind the facade proposed by the architect, and to develop interconnections of structural elements to achieve overall structural continuity. This, however, seems to be a passing phase. Architects may again feel bored with such graphic and sculptural architecture revolving around the central theme of post-modernism and return to the natural concept of structural architecture with the attendant expression of the structural form. With future tall buildings, reaching great heights, efficiency, economy, and elegance will still be the guiding principles of design. The overall form is always set in some environment, and therefore its attraction and repulsion relate to form in context. Thus the essence of tall building design must proceed from materials to elements by thinness, from elements to the builtform by integration, and from the builtform to the surrounding environment by contrast. The goals of structural art, that is, expression of thinness, integration, and contrast, will continue to prevail in future tall building designs.

To encourage such designs, further research is needed to serve as a basis for publishing more detailed case studies of significant buildings that express structure. These studies should focus on visual analysis not only from an architectural viewpoint, but also from a structural point of view, as well as on the structural performance and the construction economy.

## 4.5 FUTURE RESEARCH

Developing a cooperative working relationship between the structural engineer and the architect is important in advancing the art of structural expressionism. Historically, advances in structural materials and forms have coincided with the development of new materials and technologies. Architects and engineers can anticipate future development of new composite or alloy materials that will increase the structural efficiency and cost-effectiveness of tall buildings. Research into materials and applications is necessary if advances are to occur.

Innovations in structural systems, always at the heart of structural expressionism, must also continue to be stressed. Again, many of these innovations in tall buildings have come about as the result of the close working relationship as equals, which has been developed between the structural engineer and the architect. Collaborative research in terms of both actual high-rise building projects and theoretical projects must be carried out to develop and test innovative structural systems and to incorporate them into tall building design.

Further research into the development of new technologies in high-rise building systems and design will also have an impact on the art of structural expressionism. New building systems and components will require new ways of thinking about the building structure and its implication on the aesthetics of the tall building. More emphasis placed on prefabrication and composite integration of building components and systems will promote greater efficiency in erection and installation, reduce construction time and costs as well as waste of materials, raise the standards of construction, and promote durability. Research into how these changes may impact tall building structures and how the architect and engineer can effectively express the relationships of these components with the building's structural system is needed.

Finally, it is necessary to educate architects and the public to better understand and appreciate the vital role of the structural engineer in the design and execution of high-rise building structures. Architects should be taught to appreciate the integral role that the structural engineer plays in the design process and to accept the structural engineer as an equal partner throughout all phases of building planning and design. The public must develop a broader understanding and appreciation of the contributions that structural engineers have made to advances in engineering and architecture and how these advances contribute aesthetically as well as functionally to the physical context.

## 4.6 CONDENSED REFERENCES/BIBLIOGRAPHY

The following is a condensed bibliography for this chapter. Not only does it include all articles referred to or cited in the text, it also contains bibliography for further reading. The full citations will be found at the end of the volume. What is given here should be sufficient information to lead the reader to the correct article—author, date, and title. In case of multiple authors, only the first named is listed.

Architectural Record 1985, *The Bank of China*

Beedle 1982, *Selected Works of Fazlur R. Khan: First Wisconsin Center*

Billington 1985, *The Tower and the Bridge: The New Art of Structural Engineering*

Billington 1986, *Engineer as Artist—From Roebling to Khan*

Engineering News Record 1971, *Fazlur Khan: Avant Garde Designer of High-Rises*

Fintel 1986, *New Forms in Concrete*

Fox 1971, *Bank Building, Suspended 30 ft in Air, Spans 273 ft*

Fox 1975, *Tower's Corner Bracing Cuts Steel Weight by 14%*

Goldsmith 1986, *Fazlur Khan's Contributions in Education*

Graham 1986, *Collaboration in Practice between Architect and Engineer*

Grossman 1985, *780 Third Avenue—First High Rise Diagonally Braced Concrete Structure*

Grube 1977, *100 Years of Architecture in Chicago*

Iyengar 1989, *Exposed Steel Frame—A Unique Solution for Broadgate, London*

Iyengar 1993, *Exposed Steel Frame Creates Architectural Excitement*

Khan 1967, *The John Hancock Center*

Khan 1970, *Temperature Effects on Tall Steel Framed Buildings: Part 3—Design Considerations*

Khan 1972, *A Philosophical Comparison between Maillart's Bridges and Some Recent Concrete Buildings*

Le Corbusier 1927, *Towards a New Architecture*

Telford 1811, *Fifth Report of the Commissioners for Roads and Bridges in the Highlands of Scotland*

Telford 1838, *Life of Thomas Telford, Civil Engineer*

Tuchman 1989, *Leslie E. Robertson: ENR Man of the Year*

Vilar 1970, *Advanced Steel Construction Implements Belluschi-Roth Design for a 41-Story Office Building*

Zils 1986, *The Concrete Diagonal*

# 5

# Aesthetics and Form

This chapter focuses on the aesthetics and forms of tall buildings, two qualities without precise definitions, qualities that can be judged but not measured. The aesthetic appeal of a building can be elusive to the designer but critical toward the outcome of the design. The tall building's response to the environment and its visual impact to the user affect both the acceptance and the success of the building. Section 5.1 explores the ideas of beauty in high-rise architecture, the principles of aesthetics, and the aesthetic impact of tall buildings on the urban environment. Section 5.2 concentrates on the development of the builtform in tall buildings and the historical evolution of building forms through changing styles and time periods. Section 5.3 uses the John Hancock Center in Chicago as a case study to discuss the characteristics of building form and the aesthetics from an architectural perspective followed by a structural view. Section 5.4 discusses Chicago's search for a skyscraper style. After becoming the birthplace of the tall building, Chicago struggled to find a definite architectural expression for the new commercial skyscrapers. Section 5.5 considers recent trends in aesthetics, the impact of high-tech architecture, and a future outlook on tall building design. (For additional information on aesthetics see Chapter PC-6 of *Planning and Environmental Criteria for Tall Buildings* (CTBUH, Group PC, 1981).

## *5.1 BUILDING AESTHETICS*

### *1 Beauty and the Building*

The aesthetic impact of major structures has always been a matter of primary concern to architects and the public. Aesthetics, in particular architectural, is a very imprecise field of study for a number of reasons. The subject of aesthetics encompasses a broad sense of beauty and harmony in a structure and the perception of the viewer. "A thing of beauty is a joy forever," yet "beauty is in the eye of the beholder." What delights one person may be just interesting to someone else, or may not delight another person at all. Thus beauty is subjective and emotional. As a result, the subject of beauty has been the cause of much controversy since the days of the earliest philosophers. Even today, although there have been attempts to formulate the so-called rules of aesthetics, there is no generally accepted theory of aesthetics. As Liebenberg (1983) states in connection with the aesthetics of bridges:

> The development of art and architecture has gone through many phases which also included various attempts at the formulation of principles and rules. These included the well-known rule of the golden ratio during the Grecian period, which was an age of formulation, and various subsequent attempts to produce geometrical formulations for beautiful forms and shapes. The eighteenth century philosopher, Dave Hume, related beauty to utility. However, Emmanuel Kant was probably the first to assess aesthetics to be equal to and independent of reason and ethics. Most artists and architects today appear to agree that there are no rules by which one can create or measure the quality of art in architecture. Even if the relevant values are used absolutely in the Neo-Platonic sense, they would remain elusive to analysis in terms of our highly complex processes of perception.
>
> ...It is not the intention to decry the work of those that have in the past and recently produced such design rules, as there is no doubt that they serve a useful purpose, especially the novice. It must be remembered, however, that these rules or laws have been deduced from past results and do not necessarily have a fundamental basis. They only work to the extent that they define some visual properties of bridges which are aesthetically satisfactory and have withstood the tests of time, not unlike classical art. Generalizations should consequently be treated with great care. Every design can best be considered to be unique and even where such rules are applied, an imaginative adjustment will invariably result in some improvement.

Torroja (1958) states:

> ...the designer must rely more on his instinct and artistic background than on hard and fast rules, for it is far more difficult to formulate rules in the field of art than in technology, especially if these rules are not mere nebulous philosophical considerations on art, lacking direct contact with the specific problem.

Historically, from the ruins of early civilizations it can be deduced that there was always a passion for beauty in art and architecture. The Greeks discussed the subject of aesthetics at the philosophical level. Plato emphasized "beauty" and "immortality" as the chief passions of humankind and thought that the logical backbone of science was geometry. The Islamic world subsequently responded to geometry even more wholeheartedly. Its study influenced both architecture and the decorative arts, and Ibn Khaldun recommended the study of geometry as good training in logical thought. Geometry is a fascinating branch of mathematics since it correlates the mathematical principles and the visible external world and implies a mystic and curious affinity between art and nature. The application of geometric principles is a basic feature of Islamic architecture. As we will see later, the essence of geometrical proportioning is inseparable from the architecture of tall buildings.

Aristotle raised the issue of beauty in the realm of scale. According to him (Ross, 1955):

> ...to be beautiful, a living creature and every whole made up of parts, must not only present a certain order in its arrangement of parts, but also be of certain definite magnitude. Beauty is a matter of size and order, and therefore impossible either (1) in a very minute creature, since our perception becomes indistinct as it approaches instantaneity; or (2) in a creature of vast size—one, say 1000 miles long—as in that case, instead of the object being seen all at once, the unity and wholeness of it is lost to the beholder. Just in the same way, then, as a beautiful whole made up of parts, or a beautiful living creature, must be of some size, but a size to be taken in by the eye, so a story or plot must be of some length, but of a length to be taken in by the memory.

As alluded to earlier, David Hume related beauty to utility in the eighteenth century, and the first thinker to evaluate beauty to be independent of reason and ethics was Kant. It is widely accepted that the enormous technological outburst of the nineteenth century, controlled by a materialistic philosophy of life, resulted in excessive utilitarianism and insensitivity toward the value and essence of beauty. As a consequence of this, structures and buildings were being built primarily in big cities without much consideration of aesthetics. As distinct from the past, in the present century the two world wars caused tremendous devastations of the built environment in many parts of the world and resulted in economic downturn. As a result, aesthetics was not a principal concern for the design of buildings. Following World War II, however, there was enormous scientific, technological, and economic progress. A new age of awareness crept in and the public became more conscious of the aesthetic appeal of the environment in conjunction with human well-being and comfort. It became more evident that our psychic health is very much dependent on the aesthetic quality of our environment. This fact will be apparent if we compare the structures in public places today with those of the past. Thus it becomes very important for architects, urban planners, and engineers to study the fundamentals of aesthetics. It goes without saying that it pays to spend extra money for improved aesthetics of buildings. It results in better working and living environments, thereby improving the social and health effects. After all, millions of dollars are spent on tall buildings throughout the world. If these buildings are not accepted by the public, whether for aesthetic or for functional reasons, what good do they serve? They simply become, then, the living monuments of senseless waste of natural resources.

## 2 Principles of Aesthetics

Although it is not possible to formulate laws or rules for aesthetics, it is, however, possible to follow certain guiding principles. We all basically agree that the geometric shape of the tall building in plan and in elevation, the rhythm of the stories and bays, the flamboyancy of the main entry and the main floor at the lobby level, the color scheme, the exterior facade treatment, the roof design, and several other features may be chosen freely by the designer without significantly changing the cost of the building. The designer can create a true work of art in accordance with his or her own mood, temper, and style and in the context of the surrounding cityscape. There are so many attributes characterizing the aesthetics of tall buildings that the designer has to be very skillful in combining these attributes to obtain a composition that could be described as "beautiful." The resulting combination that will deliver the most aesthetic tall building is almost impossible to determine and will remain elusive. There is no formula available as there is in mathematics—the closest resemblance between mathematics and aesthetics being through the media of logic and order.

Notwithstanding the foregoing, there are some principles of composition that could define beauty. Unity of purpose, harmony, rhythm, character, proportion, order, balance, grace and dignity, contrast in form and mass, superior color and texture of materials, and appearance of strength and stability without being excessively massive are some of these principles. Some other principles which are somewhat related to those stated are slenderness, scale, curves, smoothness of shape, and forms that seem to "flow." All of these define the aesthetic quality of a structure and must be so chosen that the ultimate product will convey a message of excellence, which should transfix the viewer with solemn admiration.

Generally speaking, we do not seem to see tall buildings with the same eye as a totality of visual sensations. The dimensions and characteristic features that we observe emanating from an object are being processed in our mind, and our brain is stimulated to infer a perception of the object. Thus the perception of an object is very much related to our level of appreciation, our power of observation, and our psychic understanding of it. Lee (1985) has considered the various psychological aspects and the different cases of visual illusion. He suggests that it is possible to make the assumption that design possesses a psychological implication which places an onus upon the designer to attempt to move the world toward better appreciation of structures and objects made by humans through our endeavors to produce an aesthetic impact. It is possible for us to adjust to a habitat in a modern world of technology. An ultratall structure may be frightening to us because of its extraordinary size, yet when we observe it from a long distance or from an airplane, it appears much smaller and may, in fact, be delightful to us. This shows that nature can respond to the variations of aesthetics, and humans can be flexible in adapting to these variations.

The aesthetic principles are also dependent on time, because they are linked to the prevailing style or fashion of an era or a particular setting. Although tall buildings emerged in the nineteenth century, they essentially belong to the twentieth century. Stern (1986) reports:

> The skyscraper tower, rising free of the city, is the single most potent and widely recognized architectural expression of American pride. More than any other building type, the skyscraper, with its romantic synthesis of utilitarian office space and aesthetically powerful, historically derived form, emblematizes the American genius of marrying materialism to idealism.
>
> ...it was not the technology or even mere height that made a tall building a proud tower; what was needed was permanent independence of form, a striking silhouette, and most of all an aspirational quality.

According to Louis Sullivan (1896), yet another characterization of tall buildings is that a skyscraper, regardless of its formal vocabulary, must be "every inch a proud and soaring thing."

The prevailing styles in the past, according to Cowan (1974), took on characteristics of the prevailing architectural style, which often reflected the society's cultural aspirations. But as we approach the present, we find that "architectural 'styles' have endured for shorter and shorter periods until today we merely have 'fashion,' which changes so fast that we have no time to adjust to it."

Following World War I there was disillusionment which led to visions of an ideal world free from traditional social and political injustice. For Mies van der Rohe, such a vision gave rise to a dream of crystalline office towers composed of "layer upon layer of universal space, gridded by steel, girdled by walls of glass" (Stern, 1986). Mies's Lake Shore Drive apartments in Chicago are an example of this dream compromised to meet the requirements of the American setting. He was compelled to adjust the aesthetics of the towers to prevailing American building techniques, and as a result the buildings have less glass than what he envisioned in his original plans.

Kohn (1990) describes Mies's modernist approach to design as follows:

> The Modernist approach to office towers produced buildings such as the Seagrams Building of 1958, designed by Mies van der Rohe in association with Philip Johnson. The Modernist building was stripped of its ornamentation and emphasized an aesthetic of function where "less is more." Although Modernist ideals were well suited to projects with a tight budget and minimal time for construction, it can be argued that the Modernist aesthetic failed to achieve one of its most important goals, namely, the creation of a socially egalitarian style.
>
> Whereas the intent was to strip buildings of their "gaudy" ornamentation to create a language of the international working class, the effect was to create a stylistic vocabulary in ser-

vice of technology and the machine, not the worker. For the Modernist, the concept of "form follows function" means that form follows the building's structural function, not its social function. No matter how beautiful the building is structurally, it is almost devoid of any social identity.... Although the plaza in front of the Seagrams Building is pleasant, there is a lack of interaction between the people in the base of the building and people on the street.

It is clear that aesthetics cannot be measured but can be judged. Such judgment is essentially based on the conscious or unconscious evaluation of the underlying principles. These principles could be sacrificed when other criteria of design dominate. We may, however, refer to a building as aesthetically precise, which means the building does not violate the basic principles of aesthetics and does not necessarily imply that there is no other aesthetically "correct" solution possible for the building. Tall buildings are not intended to be temporary structures, and, therefore, their appearance should represent permanence and goodness. Aesthetics is an expression of goodness and quality. Good architectural work should always be aesthetically correct. This correctness must stand out clearly from aesthetic error. The design team must show concern about aesthetics and a search for aesthetic excellence. To be an aesthetically excellent structure, a tall building must express, among other things, its materials, its structural system, and its function clearly via correct geometric treatment. (For further information on the psychological aspects of the aesthetics of tall buildings see Chapter 6.)

***Aesthetic Quality.*** Aesthetic judgment and taste are acquired skills, although in some people these may be inborn, analogous to the appreciation of music, drama, literature, painting, and sculpture. Some suggestions for seeking aesthetic quality for bridge structures were given by Grant (1990). They may be paraphrased (for buildings) as follows:

1. Seek harmonious geometric relationships in the overall structural arrangement. Study of geometry, classical architecture, fundamentals of music, and the basics of light and color will reveal the presence of harmonious and disharmonious relationships. Harmonious relationships may be sought in the structure geometries and geometric proportions of the various parts of the structure. A rich source for learning classical rules in modern design is the study of classical architecture.
2. Omit all superfluous and unnecessary items. Avoid merely decorative items. If an item is omitted and the design is still complete, then omit that item.
3. Avoid superficial sculpture. To attempt sculpting may pamper one's ego but the results of such efforts will be worthless even to its designer. If, in certain situations, sculptural elements are desired, those should be developed by an accepted and respected artist who exercises good common sense and restraint.
4. Flat, large, unexpressive, and overbearing surfaces, such as long and high walls, are unattractive, and should be broken up with clear geometric subdivisions or figures.
5. For concrete structures, many designers, in search for aesthetic expression, have treated the hard concrete externally to soften its looks.
6. The most honest forms for load-bearing elements are those related to the character of Greek, Doric, Tuscan, or Ionic columns and related framing arrangements.
7. If a choice is to be made between some expected aesthetic expression or a particular geometric configuration, and structural and functional capacity, the latter should be chosen for the design. Further, if meeting the structural design objectives should result in a structure that is questionable in appearance, then the structural design has not been successfully completed.

### 3 Visual Impact

The visual stimulation caused by a tall building has an important bearing on aesthetics. A building viewed from a distance versus close proximity will create different visual perceptions. The scale and profile of a tall building will also greatly affect one's sensibilities and the physiological process of viewing the building. Another factor that will result in a certain visual impact is the context as well as whether the building in question conforms or contrasts with the surrounding landscape in general and the aesthetic environment in particular.

Regardless of the vantage point, the view from the street level is of crucial importance. As Kohn (1990) states:

> ...no matter how high we are poised in our office in the sky, we must always return to the street. Although the invention of the airplane allows us to soar through the air, we must always return to the ground. Up until the twentieth century, the tower was the exceptional building type. Today it is the norm. Thus, we must ask what kind of impact the proliferation of tall office buildings has had on our urban centers at street level. How can we construct tall buildings ever higher without damaging the fragile lattice of pedestrian street life and infrastructure below?

In line with Sullivan's classic "top-middle-bottom" definition of a tall building, the base at the street level, the skin around its middle, and the top clearly constitute the aesthetic field of a tall building exterior. The controversy as to whether a tall building should or should not have clear bottom, middle, and top has been in place since the beginning of this century and will probably continue into the future. However, with more and more emphasis on these three segments of a building by contemporary architects, this distinction appears to be more and more vivid. In the urban context (*Architectural Record,* 1974),

> ...the top and the middle make a visible presence on the skyline of a city, and this presence can be a cause for general civic enthusiasm. It can also warm the hearts of executives, stimulate them to pride or envy, and justify the millions spent in creating a corporate architectural image. The bottom of a high-rise building is where all this aspiration hits the ground. It is visible only in the vicinity of its site. Here the building has an immediate power over those who pass by and those who enter; its effect is more personal. What will it be like to be next to? Or across the street from?

The "top-middle-bottom" or "crown-stem-base" design approach becomes more meaningful when a building reaches enormous heights in a dense urban setting such that one cannot view and comprehend the entire building as a single entity. The twin towers of the World Trade Center demonstrate this fact insofar as they "follow the modern custom of differentiating between the bottom, middle, and top only by changes on the surfaces of a continuous shaft" (*Architectural Record,* 1974). Additional commentary on the building segments viewed in elevation may be found in Chapter 1.

***The Base.*** As stated before, Kohn (1990) stresses the significance of the base (see Fig. 5.1) when he writes:

> Of paramount importance with this tall soaring structure is the relationship of the pedestrian at street level to the building (Fig. 5.1)....we have addressed this problem by reinvesting commercial architecture with finely crafted ornament which shows evidence of the

**Fig. 5.1 900 N. Michigan, Chicago.** (*Courtesy: Bryan Richter.*)

human hand at work and which allows the building to be perceived on a small, as well as a grand, scale as shown in the fine detailing of Heron Tower designed by Kohn Pederson Fox. Furthermore, buildings like Heron Tower contain floor plates which are small enough to form a slender tower as opposed to bulky monolith thereby enhancing the pedestrian experience of the building at street level.

Yet there is no amount of finely crafted material at the base of the building, contextual skin around its middle, or clear symbolism at its top which will observe the true concern of the tall building, namely, the concentration of large numbers of people in the city center than the infrastructure can handle and the subsequent proliferation of those who slip through the cracks in the system....As an architect, one is always taught the lesson of designing the part for the whole...we must try not only to design small details which relate to the whole of the building but also individual tall buildings which take into consideration all of the problems of the central business district (CBD), the homeless included. Fortress buildings with lavish interiors that are barricaded behind a screen of security with no public interaction will only serve to create a defensive and divided city that will eventually embrace the few and banish the many. Instead, we must design buildings which reinvigorate cities with spaces for people at grade. These spaces can include transparent lobbies with public art, cafes, restaurants, and retail, and/or plazas which are specifically designed with amenities for public use as illustrated by the public open space in front of the Washington Mutual Tower in Seattle.

***The Stem.*** Proper articulation of the middle depends on the architectural style and the structural form, a topic to be discussed later. This constitutes the principal element of a tall building in terms of the building volume and as such tends to dominate the overall visual impact. The structural framework becomes the basis for both the building form and the architectural expression. Unlike the base, where needs of people at the street level are to be satisfied, the stem is designed to meet the aspirations of the community and to act as a landmark.

An important aspect of the aesthetics of the stem is the exterior skin. An endless number of choices are available to the architect to meet the aesthetic requirements, these choices being in the realm of systems, materials, and subsystems. The materials used are generally metals, glass, stone, concrete, and brick. Each material will create a different visual expression. The exterior skin covers most of the area of the building envelope and has thus a substantive influence on the architectural expression of the building. There are a multitude of possibilities for stone, bricks, and concrete in terms of colors, textures, sizes, and shapes. Similarly, glass could be clear or tinted gray, blue, green, or bronze. It may be coated or uncoated and monolithic, insulating, or laminated, to mention just a few types. An example of a building where glass has been amply used is the Inland Steel Building in Chicago (Fig. 5.2). Various types of metallic cladding material in terms of paints, finishes, and coatings are available on the market.

In tall buildings, often curtain walls are anchored to floors, and as such, the dimension of the vertical unit is dictated by the story height. Thus there is a correlation between the curtain wall's subdivisions and the structural frame. In the horizontal direction, the individual elements should have a harmonious relationship with the column spacing. Some architects ignore such relationships of creating visual patterns in curtain walls and the structural form. Others exploit this relationship and attempt to develop the necessary harmony between curtain walls and the structural frame. As Siegel (1975) describes:

...The usual horizontal module lies somewhere between 1.25 and 2.00 meters. There are at present relatively few examples of the use of wider units. Hence, the outward pattern always suggests a narrow grid, although the function of the curtain wall is actually completely different. If curtain wall elements of uniform size are repeated indefinitely across the elevation, the result is a very indifferent pattern...bearing no apparent relation to the structure. Inside the building, dead spaces are left between columns and windows. The columns get in the light and fail to make a purposeful impression. The element of conscious design is lacking. As far as the corner detail is concerned, the same principles apply to the curtain walls as applied to the structural frame. The heavy corner column may either be exposed even more emphatically than before...or be completely glazed in. In the first case, it is the only column visible in elevation, thus making an acute contrast with the slender mullions of the curtain wall. In the second case, the dead space left between glazing and corner column is not very felicitous, when viewed from inside the building.

**Fig. 5.2 Inland Steel Building, Chicago.** (*Courtesy: Hedrich-Blessing.*)

On the question of "dressed up" facades, where no relationship with the structural form is maintained, Siegel (1975) puts it this way:

> The 'enlivening' of skeleton structures by means of 'fancy' details, costly materials, lighting effects, decorative grilles, and ornamental metalwork involves a facelifting operation with no bearing on the problem of structural form. Yet the buildings like those...always attract flocks of admirers. Their formal language is not derived from genuine and deep-rooted relationships. The field is dominated to an appalling degree by false Romanticism, Neoeclecticism (which uses engineering forms in the wrong places), a taste for primitive ornamentation and a respect for monumentality, associated with newly acquired wealth. But no grid and no skeleton frame, however heavily accented, can deceive the expert eye.

(For further discussions on facades, see Chapter 7.)

A prime example of the aesthetics of the skin of tall buildings is the facade of the Xerox Center located in downtown Chicago at the intersection of Dearborn and Monroe Streets. The rounded corner of the building, rather than the conventional 90° corner, and the building's setback radiate welcoming gestures to the public. At the grade level, the curve, instead of going outside the column line, passes inside the tangent to "form inflections on both Dearborn and Monroe" (*Progressive Architecture,* 1980a):

> ...it should be noted that the curves at the base of the building, intended to accentuate the diagonal pedestrian path across the corner, are heightened by diagonal patterns in the ceiling and the floor of the building itself...the ground floor curves are repeated—somewhat—at the roof level. In viewing the building at some little distance, it is possible to see top and bottom at once, completing the image. Even from above, the rooftop penthouse adds visual delight, not ugly machines.

On the topic of exterior skin, the same article continues:

> An integral part of the building's visual strength, the skin is elegant and refined,...the east, south, and west facades are 50 percent glass, the north face and the curve, 75 percent. This clearly is a way of limiting heat gain, and yet it allows the greater glass areas (floor to ceiling) facing the best views. The insulating double glazing is clear inside lights with silver reflective outside lights. Spandrels and mullions are insulated aluminum panels coated in an off-white fluoropolymer; although designer Helmut Jahn would have preferred a purer white, the slightly beige cast that proved technically feasible is quite possibly better. From a distance, it 'reads' as white, but up close it has a warmer feel than would have a bluer white.
>
> This skin shows a solid and thoughtful control, something so often missing in the indiscriminate use of one type of glass or another to clad whole buildings. The horizontality of the aluminum and glass ribbons carries the eye along the flat surfaces and around the celebrated—yet integrated—corner in a sweeping motion. The Dearborn and Monroe facades... blend together in one continuous flowing expression.

The article concludes:

> It is also a major contribution to the Chicago scene. The perception of the building is one of comfortable scale, sparkle, lightness, and unselfconscious urbanity. Because of the horizontal banding, the lobby and bank spaces have a much more people-like scale than most. The building is approachable and less overwhelming. Even on bleak, gray, windy Chicago days...Xerox will be...a pleasant landmark to pass. With its lights and reflective ribbons, it picks up and passes along even meager light and images to passersby. On sunny days, it glitters.

> It may seem a paradox that a slick package of this building can be thought of in terms of visual comfort. Yet one of the best descriptive terms to apply is "friendly," especially at grade level, where pedestrian perception counts. It is a friendly building, both at this level and in the larger context. Yet it is also crisp, clean, and powerful in its expression....That seems a most satisfying mix of both high style and familiarity....Historically a leader of architectural thought and accomplishment, Chicago once again shines at Xerox Center.

***The Crown.*** Building tops are getting more and more attention from architects. Decorative tops or crowns not only act as symbols, but they also announce the majestic nature of the building. *Progressive Architecture* (1990a) reports:

> When the Moderns built this country's skyscrapers, flat roofs and sheer sides abounded. These design decisions simplified construction, HVAC, lighting, and window washing systems. But in 1978, the completion of The Stubbins Associates' Citicorp Center and the appearance of Philip Johnson and John Burgee's AT&T design changed all that. Johnson's Post-Modern gesture started the trend toward building tops of every size and shape.
>
> Tall buildings with articulated tops are becoming common in most U.S. cities. San Francisco even has laws intended to create an interesting skyline, developed as part of a 1983 downtown plan....San Francisco was victimized by "benching of the skyline," a result of boxy International Style skyscrapers built to the maximum zoning height, and this is certainly one of the factors changing the skyscrapers' aesthetic to shaped tops nationwide.

Some examples are 900 N. Michigan Avenue, Chicago; Building B, World Financial Center, New York; 123 N. Wacker Drive, Chicago; Griffin Office Tower, Dallas, Texas (see also Section 5.5 and Fig. 5.19); 801 Grand Avenue Office Tower, Des Moines, Iowa (see also Section 5.5 and Fig. 5.18); and Allied Bank Tower, Dallas, Texas. Figure 5.3 compares the decorative crowns of two European tall buildings, the Campanile, Piazza San Marco, Venice, Italy, and the Messeturm, Frankfurt am Main, Germany.

The crown of the building permits a range of options for treatments to provide a highly visible skyline. In Hong Kong, according to Smith (1990):

> Tall buildings generally have a flat coping and roof-line. This reflects...the maximization of plot ratio and percentage site coverage within the most cost-effective framework in terms of maximizing usable floor area. Height limitations which are applicable over some parts of the urban area also clearly act to regulate the degree to which usable floor area might be sacrificed in order to achieve a more articulated or decorative roofscape. This does tend to occur rarely, however, in prestige commercial buildings for major users who wish to project a symbolic corporate image. Important corporate buildings might also sacrifice site coverage potential on the lower floors in order to build higher and achieve an identifiable tall silhouette like the Bank of China.

***Color in Architecture.*** The issue of architectural color, often relegated as peripheral to the spatial issues of architecture and the related issue of habitation, or the constructive or tectonic aspects of architecture, is directly related to our experience of the world, which incorporates a full spectrum of color as light, shadow, and surface (Brady and English, 1993). The selection of a color scheme for a building is a difficult issue from the point of view of aesthetics. A building's overall scale, the distances from which it is viewed, its materials (natural and synthetic), the context in which it is placed, and even the daily and seasonal effects of light and shadow make the issue of color in architecture very subtle and complex.

In general, color as a phenomenon has two primary impacts—physiological and psychological. The physiological properties of color in architecture are directly related to scale, materials, and light. Le Corbusier's famous dictum "architecture is form revealed by light" underscores light as a physical phenomenon that informs our perception of the qualities of a particular space. Le Corbusier was also acutely aware of the spatial qualities of form as well as the psychological and emotional impact of color and materials on architecture, and our perception of space. Brady and English (1993) write of perception and experience of color relative to the physical environment and context in which a building is placed:

> Our perception of a visual field is enhanced by the effective reading of this field in terms of some distinguishing element, a figure or frame of some sort. The human experience unfolds over time as an orchestrated composition of painterly and sculptural elements within the enlarged frame of an architectural experience—a frame informed by the interaction of color, light, and spatial position. Habits of perception that reduce the world to a sort of color shorthand (for example, the sky is blue, grass is green, shadows are black) are challenged by on-site observations of color and light effects. Building facades become walls of written color that attempt to tell their story of tectonic emphasis and intention.

Brady and English stress that a coherent relationship should exist between the design concept and the means of visual expression. Frank Lloyd Wright's architecture is often cited for its analogical references to nature and is achieved, in part, through the use of natural materials and earth colors. However, this explanation does not recognize the possibilities for composition and expression latent within these materials and colors; nor does it take into account that these same colors and materials could be treated in unnatural, inorganic, and unearthly ways. Wright did more than make a facile connection between color and material; he made a connection between form and idea in which color and material are integral aspects.

The integration of color, materials, and context into the conceptual design process and building form is evident in 333 Wacker Drive in Chicago (Fig. 5.4). The siting of

**Fig. 5.3 Campanile, Piazza San Marco, Venice, Italy, and Messeturm, Frankfurt am Main, Germany.** (*Courtesy: Paul Armstrong.*)

**Fig. 5.4 333 Wacker Drive, Chicago.** (*Courtesy: Paul Armstrong.*)

the building on Wacker Drive on the south side of the Chicago River as it bends gently to the southwest directly influenced the architects' intention to create a curved facade of reflective green glass facing the river. The green glass curtain wall reflects the river and the adjacent buildings, and its bottle-green color complements the metallic green cast of the Chicago River. The black granite base of the building contextually integrates 333 with the dark monochromatic Miesian architecture of the late Chicago school as well as the dark terra cotta–clad Art Deco facades of the Carbide Building on Michigan Avenue and the dark glazing of the IBM Building. (Refer to Chapter 7 for more information on 333 Wacker Drive.)

Although the psychological aspects of color in architecture have been researched by psychologists and behaviorists, more scientific research is required in this area. Often color is cited as an issue of individual preference or "taste" with little or no acknowledgment of the complex emotional and experiential factors which influence those choices. For example, if asked about their favorite color, children's frequent response is "red" or "orange." However, as Brady and English (1993) point out, saturated colors are seldom found in nature. Therefore one might assume that children's preference for color arises from their earliest childhood experiences in which colorful objects were recognizable because they stood out in contrast to their relatively neutral surroundings. Research has also demonstrated a connection between the use of color in spatial environments and the effects upon the psychological and emotional well-being of the user. The color green, for example, is sometimes selected for library reading rooms because of its calming or restful effect upon the reader. Bright, bold colors, on the other hand, might be used to enhance spaces that feature activity and movement. Color also influences our perception of scale. Light, unsaturated colors make even small rooms seem larger and more spacious, whereas dark colors can seem to make large rooms close in upon the user.

Obviously, an architect's objective should be to make our surroundings interesting and delightful. Brady and English write that architecture must bring into play the integral role of color in design and its effect on the conceptual origin of a work of architecture, its formal tectonics, material integrity, and the emotive impact of color on the building and its environment. Ultimately architecture must synthesize the forces that are engaged as the compositional, structural, and spatial aspects of architectural form are discovered simultaneously and rationally with the phenomenon of color. Although there is a subjective element in the choice of color, it must be directly related to the conceptual and tectonic criteria governing the design of a building. Some work has been done in the past with regard to the physiological properties and psychological aspects of color, but more research is required to fully understand how they impact our perception of the environment and their implications on design and building. Because of the large mass and scale of a tall building, architects must exercise earlier in selecting a color scheme that should augment the building's aesthetics.

## 5.2 DEVELOPMENT IN BUILDING FORMS AND AESTHETICS

The evolution of the aesthetic expression of the tall building is a consequence of advances in technology in the form of cast-iron construction and the steel frame, as well as a reflection of deep ideological differences among architects and their clients in their interpretation of the individual versus collective symbolism of the tall building typology. The mid to late 1800s was a time of great discovery and invention. Parallel developments and applications of cast-iron and steel-frame construction were occurring in

Europe and the United States. Technology was changing at a rapid pace, most notably in engineering. Although adaptation of the steel frame to the aesthetics of the tall building facade would not be fully realized until the advent of the first Chicago school, its principles were already being applied, first to the early cast-iron bridges in England and then to the great steel suspension bridges in the United States.

A truly American building type, the high-rise building evolved differently on the East Coast of the United States than it did farther west (see also Section 5.4). Although the skeletal steel frame became the impetus for building tall, its stylistic expression as a structural aesthetic primarily came from Chicago. New York and other East Coast architects looked to European and historical precedents—particularly to the Italian palazzo—for skyscraper aesthetics. Chicago architects, less constrained by public opinion and historical imagery, recognized that the new technology of the structural frame required a new architectural style altogether.

### 1 *Structure and Aesthetics*

A building form, even if it incorporates the most efficient structural system, is not socially acceptable unless it is simple, ordered, and pleasant. This precisely characterizes the aesthetic quality of the building. Aesthetics of tall buildings is not only ambiguous, it also depends on the viewer's level of perception in a given context at a given time. It also must be recognized that there are other determinants of building form, such as functional, economic, and cultural. Only the aesthetic and structural determinants are addressed in this chapter.

As discussed in Chapter 4, structure has become a new "art form" independent of architecture. The structural forms of the Eiffel Tower, the Brooklyn Bridge, the John Hancock Center, and the Sears Tower evolved directly from the technological concepts and from the creative abilities and imagination of individual structural engineers (Billington, 1985; Ali, 1986). Since architecture and structural engineering are so closely intertwined, an understanding of the structural action in a tall building is vital for the architect in order to achieve a successful building form. With the emergence of innovative technology in our society, the structure of a building has assumed a new meaning. For centuries, humans produced machines and structures without sometimes knowing why they worked. The mechanics of how a device works was understood later through mathematical analysis. For tall buildings, this notion is so true and with more and more understanding of the underlying concepts of structural action, new heights could now be achieved for structures. The axiom that "form follows function" is a realistic approach to efficient design and reflects the ingenuity of humankind. The history of that axiom is a significant aspect of creative design and was known to Aristotle. It was further carried into the arts and crafts movement in England and transposed to the Bauhaus school of architecture, where it eventually spread throughout Western Europe and to the United States. The impact of this school had a profound effect on the harmonious relationship of form, art, and architecture (Ali, 1986).

In building design practice, while much emphasis is generally placed on the architectural expression of building forms, little consideration is given to the structural expression of buildings. In nature, animal bone structures, the shell of a snail or the egg causing hollowness, the mountains, the trees, and the ice structures in permafrost regions, are all intrinsically related to the structural optimization principle, where the logical nature of stress flow and a balanced distribution of the material in the structure consistent with the stress flow is evident. In building structures, likewise, the structural form must recognize the principle of structural optimization, a balanced material and, therefore, stress distribution (that is, complying with a fully stressed structural design concept), and a unity of purpose with the architectural form and functional requirements of a building.

As we will see in the following, engineers have left an indelible mark on the creation and shaping of tall buildings. In a sense it was an engineer who had the primacy for conceiving the first skeletal "tall" building in the United States in the city of Chicago, which is the birthplace of high-rise buildings.

## 2 The First Chicago School

The unwitting founder and a leader of this school was William Le Baron Jenney. The first Leiter Building of 1879 marked the beginning of the skeletal form with a glass and structure facade. As Hart (1985) describes:

> The determination with which this structure was realized within the confines of its formal expression has hardly ever been surpassed since that time. In its sheer strength and power the building is reminiscent of ancient Roman architecture....The daring slenderness of these piers and the great width of the window openings...were novel features.

Jenney was a civil engineer by training and practiced architecture in Chicago. His engineering background influenced his design. His next major work was the Home Insurance Building, which incorporated a revolutionary structural system in that it heralded the coming of true skyscrapers and in which the structure is the form. The third of Jenney's buildings is the second Leiter Building of 1891, where the facade is so vividly structural that it suggests the possibility of future skyscrapers.

John Wellborn Root, another engineer practicing as an architect, carried Jenney's ideas even further. Both Jenney and Root were greatly influenced by the principles of the French architect Eugène-Emmanuel Viollet-le-Duc. Although Root had an engineering mind, he had a general concern for visual design.

John Root's Monadnock building (Fig. 5.5), a 16-story masonry structure in Chicago, is the "most original of Chicago's multistory buildings." Although it is not a steel-framed construction, it had the distinction of being the world's tallest masonry building at that time.

Another great leader of the first Chicago school, often referred to as the "prophet of the skyscraper," is Louis Sullivan, who had his architectural education in Paris. According to him: "The skyscraper must be tall, every inch of it tall. The force and power of altitude must be in it, the glory and pride of exaltation must be in it" (Sullivan, 1896). His strong feelings about tall buildings led him to design several great early skyscrapers. The first high-rise commercial building that brought fame to Sullivan is the Wainwright Building of 1891 in St. Louis. He tried to "solve the architectural problem in the sense of a self-contained visually well-balanced composition" (Hart, 1985). Billington (1985) compares Root and Sullivan in terms of structural logic in their respective designs. (For additional discussion on Root and Sullivan see Section 5.4.)

Some other buildings by Sullivan are the Guaranty Building in Buffalo, the Chicago Stock Exchange, and the Bayard Building in New York. The Guaranty Building and the Bayard Building demonstrate Sullivan's strong interest in ornamentation and decorative architecture.

Sullivan, however, reverted to a strong expression of the steel frame in his last famous building, the Carson Pirie Scott Department Store in Chicago (1901–1904). The ornamental features are limited to the two bottom stories and around delicately molded windows. Thus in this building the classical principle of architectural design was overridden. This trend continued with the subsequent buildings by others, although these "were more modest buildings, attracting less notice from the public and the critics, in which, with less architectural pretension and with simpler means, the designers managed to accomplish what Sullivan had, in the [Carson Pirie Scott Department Store], distilled as the quintessence of the Chicago School" (Hart, 1985).

While the first Chicago school was at its climax, Europe was also developing steel-framed multistory buildings. The Eiffel Tower in Paris (Fig. 5.6) is an example of an aesthetically delightful structural form, although some decorative features were added to the base form to render it architecturally attractive. This tower became a lasting symbol of Paris and of the technological supremacy of France at that time. It also inspired

**Fig. 5.5 Monadnock Building, Chicago.** (*Courtesy: Chicago Historical Society*.)

**Fig. 5.6 Eiffel Tower, Paris.**

artists and architects to "acquire a new sense of space and structure." The French architectural thinkers were already addressing the issues of formal logic and rationality of medieval structures. The notable architect Viollet-le-Duc attributed all developments of architectural forms to structural necessities and accomplishments. Hart (1985) reports:

> Aided by the movement led by Ruskin and Morris in Britain, the rational doctrines advocated by Viollet le Duc constituted an important precondition for that movement in international architecture which preceded the modern school of thought and which for the first time made a complete break with the historical outlook.
>
> "Art Nouveau," when it arrived, was hailed as something really novel. It provided an artistic impetus which was as intense as it was shortlived, a necessary transition stage. In order to topple the deep-rooted concepts regarding traditional architectural forms, the leading architects evidently had to start with ornamentation....By exploiting and emphasizing the slenderness and the malleability of the icon loadbearing component, the Art Nouveau architects reverted to a course of evaluation....Although they did not achieve the true synthesis of the contemporary form of expression with the current loadbearing system, in their best work they did achieve an astonishing harmony of structure and decor.

The application of framed construction for multistory buildings in Germany started after World War I. New concepts of form and shape evolved by Walter Gropius and Mies van der Rohe, celebrated leaders of the Bauhaus group, became prevalent. The tenets of Mies van der Rohe, who taught at Illinois Institute of Technology (IIT) since 1938, gave rise to the second Chicago school of architecture. Following Root and Sullivan, the form of tall buildings was once again in the hands of architects until the appearance of Fazlur R. Khan, a structural engineer, and others like Bruce Graham and Myron Goldsmith, who propagated the concept of structural architecture for tall buildings once again. (See Chapter 2 for further information.) In the interim, New York became the city where a great era of new skyscrapers ensued.

### *3 The American Skyscraper: 1920–1940*

The affluent 1920s resulted in an upsurge of building construction in the United States. Once the first Chicago school architects and engineers established the reality of building tall structures, the architects of the new generation began employing style and ornamentation from ancient monuments of Greece to the Italian Renaissance. European cities were not equipped to build tall, and so the tall building design and consideration remained fundamentally an American matter. The idea of transforming ancient ideals to twentieth-century skyscrapers culminated during the Chicago Tribune international competition for its new headquarters in 1922. The interest of non-American architects was demonstrated by the large number of entries out of a total of 260 from around the world. The winner, a Greek revival designed by Howells and Hood, confirmed the stylistic ideal and prolonged its use for the following decade. The competition also articulated the fact that the skyscraper had become a symbol of the corporation or an entrepreneur. The second-place entry by Eliel Saarinen was a very influential design embodying a central tower stepped back, which had a sculptural quality, as if molded rather than constructed. This design opened up a whole new spectrum of aesthetic possibilities. The art of composition became the real goal of skyscrapers at this time.

The privately financed, corporate-owned skyscrapers became for the city a symbol of pride. "For the first time in the city's history, business buildings defined the cutting edge of architecture, setting standards and precedents for other building types, rather than simply adapting stylistic directions as they had done previously" (Stern et al.,

1978). The rise of the large corporation replacing small business became the principal economic organizational unit. This was reflected in the urban structure and architecture of New York City. The Standard Oil Building of 1926 in New York was an example of an imaginative urban response in that its curved base reflected the slope of lower Broadway, responding to the conflicting demands of its immediate surroundings and the skyline of the city at large. The classicism of this building gradually evolved into a somewhat different style as a result of Saarinen's design entry coupled with the growing awareness of the modern international style. Two Manhattan buildings at that time were also very influential—the Radiator Building by Raymond Hood in 1924 (now called American Standard) and the Shelton Hotel by Arthur L. Harmon, also in 1924 (now called Halloran House). A number of other tall buildings were built in the 1920s. There was no attempt to improve upon the technology, but rather attention was focused on refinement of the architectural expression and aesthetics. Building tops took on rich and varied shapes. They were not only rounded and pyramidal in form but also involved other shapes. For example, the Paramount Building in New York was topped with a 6-m (19-ft)-wide glass globe.

The late 1920s and early 1930s witnessed a "drive for height" after the optimism and self-confidence of the 1920s. The design of the Larkin Tower, which was 152 m (500 ft) taller than the Woolworth Building, showed the majestic scale of development that was feasible. Larkin, an architect and engineer, designed this telescopic tower which was intended to hold 100 office floors. He developed a special structural system for the building. The building was never built because of economic considerations. The drive for height led to the Chanin Building in 1929. This building was shorter than the Woolworth Building, yet had more floors and prominence in midtown Manhattan. Its massing composition had a strong influence on subsequent skyscraper design.

An extreme example of stylistic experimentation was the Chrysler Building by William Van Alen in 1930. The building has been referred to as Art Deco and, with the Empire State Building, dominated New York City's skyline for about half a century.

The Empire State Building was an abundantly skillful masterpiece of massing and at 102 stories became the most influential and sustaining image of New York. Its tower rose above a five-story base, its mass broken by indentations running the full height, topped by a crown of setbacks culminating in a gently rounded tower. This building has drawn awe and admiration not only from the United States, but from around the world. Some other major buildings of the 1930s are the Daily News Building, the Waldorf-Astoria Hotel, the Rockefeller Center, and the RCA Building, in New York. With the Great Depression setting in and the ensuing World War II, there was a decline in tall building construction that continued until the 1950s. When building started again, the scene was Chicago once more.

## *4 The Second Chicago School*

The second Chicago school illustrates the changing nature of structural forms for tall buildings and in a way redefined the building expression by emphasizing "honest" structural expression. Immediately before World War II, when air-conditioning and fluorescent lights were not yet in place, the requirement of sunlight and air ventilation dictated the form of high-rise buildings. It was necessary to have office spaces around hallways to get maximum daylight and air. The width of the building would be limited to about 18 m (60 ft). This would also require closely spaced columns. The buildings would have heavy masonry cladding that would give the building stability and an expression of heaviness without showing the structure. Thus there was rarely any structural expression. This started changing with the arrival in Chicago of Mies van der Rohe.

Some of the spectacular tall buildings in the United States following World War II were designed by Chicago firms. Influenced by the structuralist architecture of Mies, Myron Goldsmith, a structural engineer-architect, and his associates were working on projects in the 1950s that expressed structural characteristics of the buildings. Fazlur R. Khan joined the team and was eventually invited by Goldsmith to teach at IIT (Goldsmith, 1986). Khan, strongly influenced by Mies's ideas, propagated these notions of designing buildings in a structural way even further.

Goldsmith (1986) explains structural architecture in an article presented at the Fazlur R. Khan memorial session as follows:

> The idea of architecture coming out of the structure is a very old one, and it has persisted throughout architectural history. Perhaps the grandest period of the structure being so intertwined with the architecture that they are inseparable, is the Gothic. Of course, there are other ideas about architecture, and there was an equally great period in the Renaissance where great artists did buildings. That is another way to approach building, but our approach, and we think it appropriate for today, is the structural one.

The advent of computers in conjunction with a boom in the construction industry following the Great Depression and World War II facilitated the development of new structural systems and forms. It was now possible to analyze and investigate different structural systems and concepts with the aid of the computer, which had never been possible before. This is, in fact, a primary reason why conventional rigid-frame systems were the prevalent structural systems for tall buildings until then. Khan developed and refined the revolutionary tubular building concept. Here the building skeleton comprises closely spaced perimeter columns that provide much greater lateral resistance than is obtained with conventional systems because of the three-dimensional response of the building to lateral loads. The tubular building is an efficient and economical system when compared with conventional systems. The framed-tube building offers the traditional architectural articulation of exterior window treatment on the building facade in conjunction with clear structural expression of the same facade. Thus it gave rise to a new definition of building form and aesthetics. Modified versions of the basic framed-tube form, such as bundled tube, braced tube, composite tube, and tube in tube, to name just a few, appeared on the building scene. These concepts have been employed for both steel and concrete buildings. Regarding the endless possibilities for future development in building forms, Ali (1990) states:

> With better numerical analysis and construction techniques...improved energy conservation methods and material properties, there are almost limitless possibilities of building forms now. The bundled tube or cellular system in particular lends itself amenable to various forms both in plan and elevation, because independent tubes or cells of different heights and configurations can be assembled together that will act in unison. This leads to the notion of "free-form massing" and a building form may be derived almost similar to a child's play with toy building blocks.

In tall buildings, structural expression arises from the design problems, the resolution of which warrants structural imagination and understanding in regard to the structural response to lateral loads, a progressive reduction of floor area at higher floor levels, the necessity to include lower-level plazas and atria, and the uneven gravity load distribution on columns. Khan realized that the structural forms were based on a logical evolutionary premise and that for tall buildings, where tremendous "forces must follow the structural form," logic and the scientific method should prevail within the broad framework of art and architecture (Billington, 1985). As Ali (1990) observed in regard to the developments in tube form by Khan, "One development quickly led to another with some modification as well as commonness. It was like a chain reaction because the flow of logic pushed the essence of building form from one version to the next." Some of Khan's buildings in various U.S. cities are the Brunswick Building, Chicago; One Shell Plaza, Houston; the DeWitt Chestnut building, Chicago; One Magnificent Mile, Chicago (Fig. 5.7); the John Hancock Center, Chicago (Fig. 5.8); the Sears Tower, Chicago; and One Shell Square, New Orleans.

The 100-story John Hancock Center has a tapered structural form that is a truncated pyramid form with a large base. Khan worked with Bruce Graham, the architect for the project. Since the largest shear and overturning moment for a building is at the base, this was a logical shape of the structure to provide maximum structural stability. This also responded well to the architectural requirements. In apartment planning, deep space from window wall to building core cannot be used effectively, since proximity to windows for natural light and viewing is important. Office spaces, however, can accept a much deeper

**Fig. 5.7 One Magnificent Mile, Chicago.** (*Courtesy: SOM, Chicago.*)

space from the window wall. A natural consequence of this was to place the apartment floors above the office floors. A logical solution was to build a tapered tube in which the largest feasible apartment floor was located at the 46th floor. At the top of the building are located a panoramic restaurant and a television station. The building is a town in itself and an exemplification of the megastructure concept (Schueller, 1990).

The Hancock center is analogous to a vertical cantilever projecting out from the ground and is a natural structural form responding to the same principles of load distri-

**Fig. 5.8 John Hancock Center, Chicago.** (*Courtesy: Hedrich-Blessing.*)

bution as the Eiffel Tower. The exterior frames are braced with giant diagonals to augment the in-plane stiffness of the perimeter walls such that they work together as an equivalent tube (Khan, 1967). By making a tube, Khan took advantage of structural changes introduced in tall building construction as a result of having larger column-free interiors. He was thus able to express visually the efficiency of the structure that was built economically in comparison with other tall buildings and that offered a new meaning by its highly visible and unique structural expression. Khan (1982a) observed in a paper on this building's design process:

> The process of design for major architectural projects often does not take advantage of team effort of all disciplines working together to create the most relevant engineering/architectural situation. *A priori* architectural facades unrelated to natural and efficient structural systems are not only a wastage of natural resources but will also have difficulty in standing the test of time. The author, in this particular case, has attempted to highlight the structural/architectural team interaction which has resulted in this significant architectural statement based on reason and the laws of nature in such a way that the resulting esthetics may have a transcendental value and quality far beyond arbitrary forms and expressions that reflect the fashion of the time.

The 110-story 443-m (1454-ft)-high Sears Tower is the tallest building in the world at the time of this writing. It is a cantilevered tube building in which nine square tubes are "bundled" together to create the large overall tube. From an architectural planning point of view, the nonprismatic building form is the direct consequence of two different space requirements. First Sears, Roebuck and Company required large office floor areas for their operations, and second, smaller floor areas were required for rental purposes. A logical solution was to use a modular approach consistent with the bundled-tube concept that provided the organization of the modular areas, which could be terminated at various levels to create floors of different shapes and sizes. Structurally, the Sears Tower is a brilliant creation, and its visual impact is greater from far away, "where the great expanse of glass and metal wall has little differentiation." Hal Iyengar (1986), who worked closely with Khan, reports:

> The bundled tube would distribute axial loads between tubular lines, thereby increasing efficiency....The modularity of the bundled tube concept was also derived to effect a vertical modulation of space and massing in a logical way. The cells can be terminated when they are not required functionally. This provided a basic planning vocabulary that can be used to compose shapes to fit different requirements both in terms of spaces and esthetics.

He further observes:

> The architecture resulting from the step-backs or tube terminations can be observed in Sears Tower....The consistency of the tubular frame proportions can be observed....The depths of members were kept constant with only the material thickness varying.

Concrete is a natural material that can be employed to accomplish a framed-tube behavior because it can be molded into the desired form of a punched tube. The Brunswick Building in Chicago, employing the tube-in-tube concept, and One Shell Plaza in Houston are two applications of tubular design in concrete. Later the braced-tube concept for reinforced concrete was employed for the 60-story Onterie Center in Chicago. It may be noted here that the idea of the braced tube for both steel and concrete was investigated by Khan and his partner Goldsmith at IIT (Goldsmith, 1986). Some constructions of high-rise buildings in concrete by Khan have been discussed by Fintel

(1986), who cites a few examples, such as the Two Shell Plaza building in Houston, Texas, where "the elevation received its strong visual expression from the 'tree-like' transfer of the gravity load," the Marine Midland building in Rochester, New York, where "a natural looking structural configuration has become the visual expression," and the Brunswick Building, in which "the structural frame became the primary visual expression...the flaring of the closely spaced peripheral columns as they descend to the transfer girder lends a structural truthfulness to the visual expression of the building," just to name a few.

A modification of the bundled-tube form of the Sears Tower is the "bundling" of tubes of different shapes. An approximately hexagonal configuration is used in the case of One Magnificent Mile in Chicago. Iyengar (1986) reports: "The obvious result is to mold the form of the building in the L-shaped site, and also to create a three-dimensional modulation. Different tubes were terminated vertically to meet the functional needs of the multiuse program. The faceted overall volume was instrumental in visually reducing the massiveness of the building for its prominent exposure to Lake Michigan."

## 5 Efficiency of Forms

The conventional approach to tall building design is to limit the forms of tall buildings to rectangular prisms. There is a definite need to develop a form that will lead to an efficient structural system, resulting in substantial savings in cost. Since tall buildings involve millions of dollars, a careful coordination of the structural elements and the shape of the building may offer considerable savings. Such large savings can be realized by optimizing the entire building structure rather than its components or parts. Of course, the structural optimization requirements alone must not dictate the shape of a tall building, for then all tall buildings would look alike. Although other determinants of tall building forms must be duly considered, it can hardly be overemphasized that the structural shape and arrangement play a pivotal role in determining the building plan.

The primary goal of accomplishing cost savings in tall building structures is realized by minimizing the lateral displacement of the building caused by lateral loads. In seismic zones, the shape and layout of buildings is particularly related to symmetry in plan and uniformity of lateral stiffness in elevation (Ali, 1987; Arnold and Reitherman, 1982). The style and aesthetics of buildings are integrally related to the horizontal and vertical configurations. For wind loads, the aerodynamic effects are also important design considerations. Reduction of lateral drift may be achieved by sloping the exterior columns as in the John Hancock Center in Chicago, or by tapering the exterior frame as in the First National Bank of Chicago. A cylindrical form offers a true tube geometry, providing three-dimensional structural action, and is highly efficient aerodynamically. The Marina City Towers in Chicago are examples of this form. A cylindrical building has also a smaller exposed projected surface or "sail" perpendicular to the wind direction and thus encounters lower magnitudes of wind forces compared to other shapes. Another example is the proposed circular tapered structure called the Millennium Tower, over 792 m (2600 ft) tall, for Tokyo (*Progressive Architecture,* 1990a).

Other shapes are triangles, crosses, ellipses, crescents, and the like. Triangular spaces are efficient because they represent a tripod-type configuration in plan. The Mile High Tower consisting of 528 floors proposed by Frank Lloyd Wright had a plan of a double triangle. Wright intuitively felt that the tripod was the most stable form since pressure on one side is immediately resisted by adjacent sides. Although single triangles seem outwardly to be efficient forms, they have some inherent weaknesses for framed buildings. For example, wind loads acting on one leeward side of the building, that is, the other two windward sides being directly exposed to the wind forces, will result in large uplift forces in the perimeter columns at and near the vertex of the triangle on the windward side. This vertex, being pointed and sharp like the front end of a ship's head, will however reduce the drag forces and thereby the wind pressure on the buildings. The U.S. Steel building in Pittsburgh is an example of a triangular form. A double triangle is, of course, a very efficient form, and hence Wright's intuition for the Mile High Tower form was correct. Buildings with curvature in plan configurations yield additional strength and stiffness to resist lateral loads. Some notable examples are the two crescent-shaped towers of Toronto City Hall in Toronto and Le France building in Paris.

A study by the structural engineers Skilling, Helle, Christiansen, and Robertson in the later 1960s reported the peak deflections under wind loading for six different shapes (*Progressive Architecture,* 1980b). As expected, the circle was the superior shape for wind. Although structural engineers know the optimum shape for structures, many of them consider this to be outside their direct role in that the architect and owner usually provide the particular shape to start with in order to meet the site and occupancy requirements. For large projects this may result in major structural changes and modification of the form at a belated stage of design, which is certainly not desirable.

Tall buildings in seismic zones, as stated before, need special attention because the form of these buildings has a lot of influence on the seismic response of the buildings. Most buildings are intricate in form and have stairways, elevator shafts, mechanical ducts, pipe runs, and skylights that will result in inevitable dispersion of the building mass, causing complex seismic effects. Although the structural form is a crucial element of the overall building form and aesthetics, architects must consider a number of other aspects, which include control of traffic, acoustics, space allocation, energy efficiency, and general financial viability. These additional factors may also influence the form of a building.

The 1991 *Uniform Building Code* (ICBO, 1991) provides a significant expression of the seismic design of buildings as an architectural design concern by introducing the definition of *regular* structure. The characteristics of irregular structures described in UBC Tables 23M and 23N are generally related to architectural planning of the building. Although irregular building forms may be acceptable in some situations, such guidance by code nevertheless encourages the architect to appreciate the importance and ramifications of different types of vertical and horizontal irregularity as they relate to the architectural design process.

Efficiency in form can be achieved in many creative ways. An example is the 77-story Bank of China building in Hong Kong. (For further discussion on the Bank of China building in Hong Kong see Chapters 2, 4, and 10.) *Architectural Record* (1985) describes the building as "a stunning exercise of architectural geometry." The building's architect, I. M. Pei, conceived it "as a cube, rising out of the ground, and divided diagonally into quadrants....As the structure moved upward, the mass is diminished one quadrant at a time until it is reduced to a single, triangular prism...." Pei intended the architectural partition to fall precisely on the geometry of the form: a crystalline Euclidean vision in reflective glass and aluminum. Maintaining the purity of that geometry was the challenge around which the very structural feasibility of the project turned. The article continues (*Architectural Record,* 1985):

> Because the tower's structural lines are pulled just inside the building envelope, and because the shape of the envelope changes dramatically, the position of the five major columns must joggle out-of-plumb as they make their way down to the ground while maintaining the structural steel in a true and plumb, repeating geometry. The result is eccentrically loaded columns...The engineers at Robertson, Fowler & Associates hit upon a system that accepts the radical eccentricities inherent in the architecture.
>
> The structural designer, Les Robertson, based the system on what he refers to as "a *truism* not obvious to most engineers." A single eccentricity in a column will cause bending;

> but two or more lines of eccentricity, joined by a uniform shear force mechanism, will counteract and therefore eliminate the bending....The five composite columns of the system support the braced frame of structural steel that spans them. The centroid, shape, and position of these columns change as they move down the building—the source of eccentricity. But because the concrete "glues" the steel to itself, bending is eliminated. The concrete, then, serves as the uniform shear force mechanism. Dynamic eccentricity is arrested with creative engineering logic.
>
> The result is a structural configuration equal in its own right to the eloquence of the architectural design...the system is outstanding for its economy of material. Compared to buildings of the same height and area, the Bank of China will use approximately 40 percent less structural steel.

There have been a number of recent attempts to build skyscrapers taller than the Sears Tower in which efficiency of form plays a crucial role. Some examples of such developments are the 168-story World Trade Center in Chicago (Iyengar, 1986; Tucker, 1985); the 150-story Trump Tower in New York (Robins, 1987); the 210-story Chicago World Trade Center (Tucker, 1985); and the 125-story Miglin-Beitler Tower in Chicago (Thornton et al., 1991); to name a few. (See Chapter 10 for a case study of the Miglin-Beitler Tower.) The 450-m (1476-ft)-high Petronas Towers, which is currently under construction, will be the tallest building in the world once completed in 1996 as scheduled.

## 5.3 AESTHETICS AND STRUCTURAL FORM: A CASE STUDY

In the design of tall buildings, the basic need for providing lateral resistance to wind and seismic loads prevails during the structural design process. This will influence many aspects of subsequent design development and decisions. It will also impact the architectural planning and the selection of material and structural system. When such considerations are dealt with during the conceptual stage of the design process, better decisions on building form and planning are expected, which will lead to optimal conditions.

Building form and aesthetics may be considered from an *architectural* viewpoint in terms of the following characteristics:

1. Plan view
2. Elevation
3. External appearance
4. Balance and simplicity
5. Proportion and scale
6. Relationship of spaces
7. Visual impact
8. Style
9. Ornamentation and decor

While there may be other determinants of the architectural expression of a tall building, these attributes provide a spectrum of reference points against which a building's aesthetics and form can be compared. In effect, these attributes can be treated as a checklist for a qualitative and frank evaluation of a building's aesthetics. The John

Hancock Center in Chicago is used as an example in following the sequence of characteristics listed.

1. *Plan view.* The plan view is a straightforward rectangle, although it constantly varies in dimension with height. The largest feasible apartment is on the 46th floor near the midheight of the building. The taper is extended until the program requirements are met. Since the structure is based on a tapered-tube concept, the interior has large open spaces.
2. *Elevation.* The building achieves its architectural expression blended with, and somewhat dominated by, its structural expression with a visual recognition of its strength and stability through interplay of horizontal, vertical, and diagonal members.
3. *External appearance.* It is a building where structural expression dominates and its vitality gives it a distinct architectural presence in its urban setting. It is a building that has a technological rather than architectural facade, and no emphasis is placed on a bright and jovial color scheme.
4. *Balance and simplicity.* The building mass is balanced and the architectural expression is simple without any intricacy.
5. *Proportion and scale.* Although the scale of the building seems amplified in relationship with the surrounding buildings in terms of height and width, the braces on the facades separate the building into sections which are within conceivable limits, and the visible floor structure contributes to the scale of the building by bringing it down to a tolerable human level. The building's overall proportions are interestingly balanced. The gentle slope adds to the beauty of the proportion in that this elegance would be lost if the building were a straight tall box.
6. *Relationship of spaces.* There exists a vertical linear relationship of spaces in the following upward order: commercial, offices, and finally residential apartments.
7. *Visual impact.* The diagonal braces on the facades help stabilize the tapered form. They tie the entire structure to the ground and thus represent a feeling of safety. The cross braces thus accentuate the form, and the juxtaposition of vertical and diagonal members in its fenestrations provides a special visual impact. The visual effect is pragmatic and technologically satisfying and does not represent an extravagant, fanciful sculpture or exuberant romance. It simply represents structural elegance whereas the structural members are organized in a skeletal form. The building also provides a contrast in its setting because of its massive size and the cross bracing on its facades.
8. *Style.* The building has a distinctive style because of the tapered shape and the cross braces that unquestionably express the structure. The cross bracing is not readily accepted by some of the users as it does not look logical from within and obstructs the views outside at some locations. This sacrifice was made for gaining the structuralist architectural expression of the building.
9. *Ornamentation and decor.* Pompous beauty is absent in the building's aesthetics. No attempt to tone down the structural expression with ornamentation and decor on the facades was made, except sheathing the bottom in travertine.

Building form and aesthetics may be viewed from a *structural* point of view in terms of the following characteristics:

1. Shape and size
2. Dimensions

3. Strength and stability
4. Stiffness
5. Efficiency and economy
6. Simplicity and clarity
7. Lightness and thinness

The principle of structural design for the John Hancock Center is based on evenly spaced exterior columns interconnected with diagonal members and horizontal spandrel beams in such a way that all the exterior columns act together not only in carrying the gravity loads but also in resisting the horizontal wind forces. Thus the structural system comprises diagonally braced exterior frames that act in unison as an equivalent tube. The special feature that makes the structural system unique is that a few diagonal members introduced in the plane of the perimeter columns create a rigid tube form. The structural attributes of the building can be discussed as follows.

1. *Shape and size.* The tapered form and cross-bracing basically provide the structural expression to the building. The structural loads are distributed in the skeleton because of its shape, which gives the structural expression to the building. Although the size of the building looks enormous, the ratio of height to footprint dimension (7 and 4.3 in two principal directions) is in the acceptable range. Although a slenderness ratio of 7 is on the high side, the rigid box system makes it tolerable.
2. *Dimensions.* The ground floor measures approximately 49 by 79 m (160 by 260 ft) and the clear span from central core is 18 m (60 ft). The building is tapered to a dimension of 30 by 49 m (100 by 160 ft) at the top, and the exterior clear span reduces to 9 m (30 ft). The huge dimensions affect the structure, and diagonals are added to strengthen the facade structure. From a planning point of view, the multiuse nature of the building makes the tapered form possible. Further, the interior spaces are still not unrealistically far from the windows.
3. *Strength and stability.* A major attribute of the John Hancock Center is its demonstration of natural strength and stability. As stated before, the tapered form and the diagonal braces express the inherent strength of the building clearly. There is a harmony between the structural strength characteristics and the building form. The exposed structural members reveal how the load is transferred through these elements. The overall form of the building reflects stability with structural redundancy created by the diagonals in addition to the ties at locations where the diagonals intersect, as well as the external columns. The structural members connect all the corners and sides of the building. The larger diagonals are at the bottom and become smaller as they rise.
4. *Stiffness.* The structural form has adequate resistance to excessive deformations under loads that may cause discomfort to the users and result in damage to the nonstructural components. The unique system of a few diagonals added in the plane of exterior columns creates a tube effect, thereby giving enormous rigidity to the structure.
5. *Efficiency and economy.* For the multiuse occupancy of the building, a conventional solution would be to place a narrow building for apartments on top of a broad one for offices at the bottom to create a wedding cake type arrangement. This type of structure, when compared to a tapered-tube building form, would become cost-ineffective since a tapered form with gentle slope allows a continuous optimum structure to be used that closely follows the flow of stresses for both

gravity and wind loads. The few diagonals added to the plane of the exterior columns increase the efficiency of the building, resulting in an economical low-premium structural system.

6. *Simplicity and clarity.* The diagonal braces form continuous lines from one face to the other, creating an X brace between a pair of tie beams and a tier of X braces on each face of the building. Visible steel elements clearly show the direction of forces and reveal the structural logic of the building form. The building has simple and clear connections. Columns, diagonals, and ties were fabricated to an I section composed of three plates welded together. This shape simplified the joint detailing. The overall form demonstrates a strict discipline of design.
7. *Lightness and thinness.* The building utilizes 145-kg/m$^2$ (29.7-lb/ft$^2$) steel, which shows that the structure is relatively light for a 100-story tower. The size of the members on the face of the building maintains a proportion to the shape of the building. The lightness and thinness of the members result in a special visual effect that may be termed "structural elegance" and do not visually disturb the mode of transfer of loads to the base. At some levels, the floor beams are omitted, which appears to disturb the rhythm of the structural members in the vertical direction.

## 5.4 SEARCH FOR A SKYSCRAPER STYLE

Chicago is generally regarded as the birthplace of the skyscraper. It was there in the 1880s that real estate speculators, prompted by the invisible hand of the marketplace, persuaded their architects to invent the tall office building. The necessary technology was already at hand, and architects quickly molded elevators and rolled steel beams and other industrial elements into high-rise buildings.

At the same time these architects were faced with a formal problem: what was the proper architectural expression for the new commercial skyscrapers? Previous tall buildings—pyramids, ziggurats, pagodas, cathedral spires—had all expressed spiritual meanings central to their civilizations. However, in the late nineteenth century it was thought that spiritual and cultural values were to be found in libraries, museums, theaters, universities, and churches. The skyscraper was seen as a purely economic artifact, lacking any spiritual content. How could an appropriate style be found for it?

The complete novelty of this unprecedented building type, which appeared so suddenly, and its visual prominence in the heart of the city seemed to present Chicago architects with a remarkable opportunity to develop a new style, appropriate not only to the skyscraper, but perhaps for other building types as well. This did not prove to be an easy task, however.

### 1 Stylistic Development and the Commercial Building

Advances in engineering and iron construction in the late 1700s in England (see Chapter 10) and the invention of the Bessemer process in 1855 made steel-framed buildings possible. The I beam, the basic structural element of modern steel work and the first strictly standardized structural component, is the direct descendant of the old iron rail, which became the hallmark of the emergent industrial era (Hart, 1985).

By the time William Le Baron Jenney designed the first Leiter Building in 1879 in Chicago, the use of cast-iron and steel-frame construction for bridges and multistory buildings had already been in progress in England and France. The first use of cast iron in industrial buildings was introduced by Boulton and Watt in the British cotton mills in 1801. Cast-iron columns and beams were substituted for traditional timber posts and beams in order to gain more floor space and increase the load-bearing capacity of the floors for heavy machinery. Wrought-iron beams were first introduced by William Fairbairn in 1845 in an English refinery. In that same year A. Zorès brought the rolled I section into residential construction in France.

However, the transformation of the external wall into a load-bearing metal frame, or steel skeleton, construction was long in coming and would have the most profound aesthetic effect on the architectural treatment of the facade (see Chapter 7). The first attempts were in the use of cast-iron columns to create wide-spanning arcades in the stark brick fronts of such British dockside warehouses as St. Katharine's dock in London (1828) and Albert dock in Liverpool (1845). Wrought-iron frameworks were used from the early nineteenth century to span the increasingly wide shop windows that were coming into vogue particularly in Paris. In the United States cast-iron facades were incorporated into the riverfront buildings of St. Louis. Here the traditional features that were used to enliven the facades of historic buildings were reduced to the barest minimum.

The Menier chocolate factory at Noisel-sur-Marne near Paris (1871–1872) was the first multistory building in continental Europe entirely designed as a steel-framed structure. The Menier factory anticipated modern steel structural framework in several respects, including the cantilevered corner and especially the diagonal network of structural bracing, which anticipated the wind bracing of modern steel construction used in tall buildings. The principle of the metal frame and its aesthetic and architectural expression was to make a more lasting impact only when high-rise commercial building became an urgent constructional problem demanding new structural solutions. This development began in Chicago following the Great Fire of 1871.

Although the architects of the first Chicago school realized the full potential of steel-frame construction and used it as an aesthetic determinant for the stylistic development of tall building architecture, New York gained a certain early lead in the construction of tall buildings (Hart, 1985). The first high-rise buildings in New York were nine- or ten-story commercial buildings of masonry construction erected in the mid-1870s (Hart, 1985). The seven-story Equitable Building (1868–1870) by Arthur Gilman and Edward H. Kendall was the first building to have a passenger elevator, and for a time it was the tallest building in New York (Gibbs, 1984).

The earliest high-rise buildings were based on historical building models and emulated the permanence and grandeur of the Greek temple or medieval cathedral—great works of art expressing tremendous power and highly developed craftsmanship (see Chapter 1). According to Hart (1985), in comparison with other applications of manual skill and industrial technology, building technology has always been conservative. The aesthetic expression of structural iron and steel was realized by engineers in structures such as the Eiffel Tower and the Brooklyn Bridge before its potential was investigated in architecture. The historical prejudice of the architects widened the gulf between architect and structural engineer. The building owners and the architects of the nineteenth century developed a sort of artificial tradition based on historical eclecticism, which hampered truly innovative architectural design. Finally the external pressures, such as economics, increased spans, and load-bearing characteristics, driving the designers of bridges and single-story buildings to increasingly daring achievements, were lacking in multistory buildings.

The urge for great height appeared first in office buildings and in not all categories of business architecture (Gibbs, 1984). More than any other business, the insurance industry led other branches of business not only in the drive for great height, but also in the pursuit of an appropriate imagery that would move beyond the tradition of the Italian palazzo (see Chapter 7). During the late 1860s and the early 1870s the insurance industry in the United States had its first period of spectacular growth. The meteoric rise of the life insurance industry in the business world was paralleled by the visual impact of the new insurance buildings in the larger cities of the East. The manner in which the palazzo tradition was altered after about 1870 in Chicago differed dramatically from the ways it was altered in the larger cities of the East. According to Gibbs, all of the aesthetic and architectural differences between the approaches of the East Coast and the mid-West to the transformation of the palazzo image relate, in one way or another, to contrasting methods of business practices in the two regions.

Traditionally the eastern business clients and their architects are judged retrograde because of their reliance on past styles and their failure to grasp the imperatives of steel-frame construction. Conversely, their counterparts in the West are judged progressive because of their break with past styles and their willingness to accept the dictates of unprecedented conditions and technical innovations. Thus the East Coast architects lacked the mid-West's architects' vision of how closely architecture might reflect its own time. Gibbs contends that the object is not to account for genuine differences between the architecture of the eastern and western regions of the United States by the lack or presence of a single value or concept, but to understand how different values or concepts directed the building of business architecture based on the contrasting nature of business in the East and the West. In New York, according to Gibbs, public opinion shaped business practice and architectural style, whereas Chicago was relatively unhampered by public opinion. New York business architecture conveyed the goals of particular, individualized companies. By contrast, Chicago school business architecture as a whole was brought about in part to recognize and symbolize in Chicago's speculative office blocks the collective character and goals of the American business community.

Broader, philosophical currents also seemed to have played a role in the contrasting values and styles of tall building architecture in the United States. In his influential *The Seven Lamps of Architecture,* John Ruskin extolled the virtues of architectural imitation of historical styles, particularly the Venetian Gothic architecture and its aesthetic symbolism, traditional and expressive use of materials, and its overall sense of craftsmanship (Gibbs, 1984). In his preface to the second edition Ruskin wrote, "Proud Admiration—the delight which most worldly people take in showy, large or complete buildings, for the sake [of] which such buildings center on themselves, as their predecessors, or admirers" (Gibbs, 1984). The pragmatism which shaped New York architecture can be contrasted with the transcendental ideas—introduced by Semper and Thoreau—which seemed to have shaped the thought and architectural expression of Chicago architects. In *Walden,* Henry David Thoreau frequently drew parallels between common natural phenomena and larger truths. In nature as well as in building, Gottfried Semper sought the relationship between how the part stands for the whole—where the part is enough to convey the idea. Undoubtedly these transcendental principles played a major role in guiding the precepts of Sullivan and Root's organic architecture.

## 2 John Root and Louis Sullivan

John Root and Daniel Burnham built the first Chicago skyscraper, the 10-story Montauk Building, in 1882 for the Boston investors Peter and Shepard Brooks. Root knew that the Montauk had an ungainly architectural expression, and he was determined to find a new form for the tall office building. He adopted Richardson's Romanesque for his next building for the Brooks brothers, the 11-story Rookery Building completed in 1886. But he was apparently not satisfied with such historical solutions, as he indicated in his essay on the skyscraper, *A Great Architectural Problem* (Root, 1890):

> These buildings, the result of commercial conditions without precedent, are new in every essential element. Looking at the problems presented by these buildings...we may certainly guess that all preexisting architectural forms are inadequate for their solution.

He goes on to reflect on the question of inventing a new style—perhaps an American style—for the skyscraper (Root, 1890):

> Architectural styles, national or new, were never discovered by human prospectors however eagerly they have searched. Styles are found truly at their appointed time, but solely by those who, with intelligence and soberness, are working out their ends by the best means at hand, and with scarce a thought of the coming new or national style.
>
> In [these buildings] there should be carried out the ideas of modern business life: simplicity, stability, breadth, dignity. To lavish upon them a profusion of delicate ornament is worse than useless, for this had better be preserved for the place and hour of contemplation and repose. Rather should they by their mass and proportion convey in some large elemental sense an idea of the great, stable, and conserving forces of modern civilization.
>
> ...Permeating ourselves with the full spirit of the age...we may give its architecture true art forms.
>
> To other and older types of architecture, these problems are related as the poetry of Darwin's evolution is to other poetry. They destroy indeed much of the most admirable and inspiring architectural forms, but they create new forms, adapted to the expression of new ideas and new aspects of life...
>
> Here the vagaries of fashion and temporary fancies should have no influence; here the arbitrary dicta of self-constituted architectural prophets should have no voice. Every one of these problems should be rationally worked out alone, and each should express the character and aims of the people about it.
>
> I do not believe it is possible to exaggerate the importance of the influence which may be exerted for good or evil by these distinctively modern buildings. Hedged about as they are by so many unavoidable conditions, they become either gross and self-asserting shams, untrue both in the material realization of their aims and in their art functions as expressions of the deeper spirit of the age, or they are sincere, noble and enduring monuments to the broad beneficent commerce of the age.

Clearly Root expected that the skyscraper would serve as a catalyst to create a new architectural style. It is difficult to understand what exactly he may have meant by his tantalizing phrase, "the poetry of Darwin's evolution," but he seems to be appealing to a kind of unadorned functionalism. About the time he wrote these lines he designed two buildings which were quite different from his previous work. These were the Monadnock Building and the Great Northern Hotel, both completed in 1891, which stood diagonally opposite each other at a Chicago Loop intersection. Both were characterized by lively undulating surfaces of plain masonry with very little ornament, his attempt to convey a sense of the "broad and beneficent commerce of the age." A similar functional mode of expression was seen in the work of some of Root's Chicago contemporaries, such as William Jenney's Manhattan Building and Holabird and Roche's Pontiac and Champlain (Fig. 5.9) buildings. This mode of expression would later be identified by critics as one of the hallmarks of the Chicago school of architecture.

This purely functional approach apparently did not satisfy Root either, for in his last design he returned to historical forms. This was the 21-story Masonic Temple (Fig. 5.10), completed after his death in 1891. It was then the tallest building in the world and a great source of civic pride. The lower portions were similar to the functional style of the Great Northern Hotel, but to accommodate the meeting rooms of the Masonic Lodge on the top floors, Root added a steeply sloping roof with dormer windows suggestive of a medieval guild hall. The other Chicago school architects too would abandon their functionalist vocabulary by the turn of the century.

Another major effort to find a new style worthy of the skyscraper was undertaken by Root's talented contemporary Louis Sullivan, who with his partner Dankmar Adler also built a number of early tall buildings in Chicago. Sullivan explained his approach in an essay entitled *The Tall Building Artistically Considered* (1896). He began with a challenge to his profession:

> The architects of this land and generation are now brought face to face with something new under the sun...[the] tall office building....It is the joint product of the speculator, the engineer and the builder.

**Fig. 5.9 Champlain Building, Chicago; Holabird & Roche, architects.** (*Courtesy: Chicago Historical Society.*)

Problem: How shall we impart to this sterile pile, this crude, harsh, and brutal agglomeration, this stark, staring exclamation of eternal strife, the graciousness of those higher forms of sensibility and culture that rest on the lower and fiercer passions? How shall we proclaim from the dizzy height of this strange, weird modern housetop the peaceful evangel of sentiment, of beauty, the cult of a higher life?

**Fig. 5.10 Masonic Temple, Chicago; John Root, architect.** (*Courtesy: Chicago Historical Society.*)

He then replies with his solution: "It is my belief that it is of the very essence of every problem that it contains and suggests its own solution. This I believe to be natural law" (Sullivan, 1896).

Following this formula, he proceeds to describe the functional parts of the office building: the mechanical basement space, the commercial first and second stories, the variable stack of office floors, and finally the attic for further mechanical equipment and other atypical spaces. After completing this functional catalog, he notes: "Only in rare instances does the plan of floor arrangement of the tall office building take on an aesthetic value. The...arrangements of light courts...and elevators...have to do strictly with the economics of the building" (Sullivan, 1896). Having summarily disposed of the plan, he then proceeds to the building street facades, which he feels are the only elements worthy of architectural consideration (Sullivan, 1896):

> The first...story we treat in a more or less liberal, expansive, and sumptuous way...The second story...in a similar way but usually with milder pretensions. Above this throughout the indefinite number of typical office tiers, we take our cue from the individual office cell, which requires a window with its separating pier, its sill and lintel...and make them look all alike because they are all alike. This brings us to the attic, which...show[s] by...its dominating weight and character...that the series of office tiers has come definitely to an end.

Having established the tripartite form based on building function, he goes on to note its classical roots, and finally reaches his general stylistic conclusion in a famous dictum (Sullivan, 1896):

> It is the pervading law of all things organic and inorganic, of all things physical and metaphysical, of all things human and superhuman, of all true manifestations of the head, of the heart, of the soul, that life is recognizable in its expression, that *form ever follows function*. This is the law.

Although these statements would seem to place Sullivan in the functionalist realm with Root, he produced buildings that were quite different from the Monadnock. Working within the classical tripartite convention he had adopted, he embellished the facades of his tall office buildings with marvelous ornament, using botanical forms arranged in subtle geometrical patterns. The Wainwright Building in St. Louis (1890) and the Guaranty Building (Fig. 5.11) in Buffalo (1895) are two refined and elegant examples of his highly personal approach. Sullivan was also the first architect to frankly emphasize the verticality of the skyscraper; as he wrote: "It must be tall, every inch of it tall. The force and power of altitude must be in it. It must be every inch a proud and soaring thing, rising from sheer exaltation that from bottom to top it is a unit without a single dissenting line" (Sullivan, 1896).

Although the delicately ornamented verticality of Sullivan's towers fulfilled his vision of the skyscraper as "proud, soaring and alone," and the functionalism of Root had its followers, neither of these approaches led to the definitive style in the sense Root hoped for. He had described his Darwinian expectations in his essay *The Value of Type in Art* (Root, 1883):

> Principles...obtained from Nature we will find to underlie all good art of every sort—painting, sculpture, music, architecture, or any other.
>
> That great natural principle which will be found most important is Adherence to Type. By Adherence to Type is meant not only natural insistence on the unity of each created object, but persistence in certain solutions of given problems....In general it will be found that

once a solution of a given problem in structure is obtained, there is a manifest disinclination to solve it in a new way, provided the conditions remain the same. The various solutions of the questions vary by almost imperceptible degrees, as in the course of evolution, the conditions vary.

A moment's consideration of the development of the foot or hand will show the full significance of this. And what is true of structure is also true of all decorative devices, which are only structural conditions somewhat removed....This consistency, this adherence to

**Fig. 5.11 Guaranty Building, Buffalo; Louis Sullivan, architect.** (*Courtesy: Buffalo & Erie County Historical Society.*)

> type is the reason why architectural *olla podrias* are bad, composed of bits from many periods and many styles. It is not so much that tradition and history are violated, but that type is violated. Greek and Gothic architecture are good because they are completed and perfect expressions of certain methods of structure and are therefore typical.

The models of Greek and Gothic architecture clearly set the standard for skyscraper style, but even Root and Sullivan were unable to adhere to a definite type, and their skyscrapers show a wide variation from what critics regarded as their best efforts.

### 3 Daniel Burnham and Classicism

The next phase of skyscraper style in Chicago was inaugurated by Daniel Burnham after his partner's death in 1891. Burnham was one of the principal promoters of the World's Colombian Exposition held in Chicago in 1893, and he was a major participant in the decision to choose Beaux-Arts classicism as the principal style for the fair buildings. The sparkling white facades of the classical Court of Honor, brilliantly illuminated at night by electric light, made a profound impression on the Chicago architectural community. Dankmar Adler remarked as the fair was closing (Adler, 1893):

> The immediate example of the Fair buildings will be a general and indiscriminate use of the classic in American architecture. Efforts will be made to force into the garb of the classic Renaissance structures of every kind and quality devoted to every conceivable purpose...in palace and cottage, in residence and outhouse, in skyscraping temple of Mammon on city streets, and in humble chapel and schoolhouse of the country roadside.

Although Adler's remarks proved to be an ironic overstatement, Burnham was indeed instrumental in bringing classical motifs to the "skyscraping temple of Mammon." In his famous *Plan for Chicago* of 1909 he envisioned a compact central core of commercial buildings for the city with a uniform cornice line of 18 stories, crowned by the towering dome of the Civic Center. He himself built a number of the massive corporate palazzos including the People's Gas building, the City National Bank, and the Chicago Title and Trust building. But this projection of the image of imperial Roman power in the skyscraper soon began to lose its appeal, even before Burnham's death in 1912. The workings of the marketplace were forcing office buildings ever higher, particularly in New York, which had replaced Chicago as the skyscraper capital in the late 1890s. New York architects now took the lead in adapting the upreaching verticality of Gothic forms to the ever taller skyscraper, with Le Brun's Metropolitan Life building of 1909, and culminating in Cass Gilbert's Woolworth Building of 1916, which was proclaimed a "cathedral of commerce."

### 4 Art Deco Style

After the end of World War I, Chicago abandoned Burnham's uniform cornice line and followed New York's Gothic example with Graham, Anderson, Probst & White's Wrigley Building of 1921, and Raymond Hood's Tribune Tower (Fig. 5.12) of 1925 (see Chapter 7). This Gothic period too would be short-lived. At the same time he was designing the Tribune Tower, Hood was also at work on the American Radiator building in New York. In it he would begin the development of a new kind of "secular Gothic" vocabulary for the tall building, still keeping its verticality, with buttresslike setbacks and crowning spires, which would become known as Art Deco. Hood seemed much less certain about the possibility of a definitive skyscraper style than Root or Sullivan or Burnham (Hood, 1925):

> My experience, which in reality consists of designing only two skyscrapers, the Tribune and the American Radiator, does not justify my expressing an opinion as to whether a building should be treated vertically, horizontally or in cubist fashion. On the contrary, it has convinced me that on these matters I should not have a definite opinion. To use these two buildings as examples, they are both in the "vertical" style or what is called "Gothic" sim-

**Fig. 5.12 Tribune Tower, Chicago; Raymond Hood, architect.** (*Courtesy: Chicago Historical Society.*)

ply because I happened to make them so. If at the time of designing I had been under the spell of Italian campaniles or Chinese pagodas, I suppose the resulting composition would have been horizontal...Nothing but harm could result if at this stage in our development the free exercise of study and imagination should stop, and the standardizing and formulation of our meager knowledge and experience should take its place. It might be proper to say something precise about the different styles, but I am as much up in the air about style as I am about everything else.

Hood's diffidence notwithstanding, Art Deco soon became popular with Chicago architects. Holabird and Root's Palmolive Building (Fig. 5.13) and Board of Trade, and Graham, Anderson, Probst & White's Civic Opera and Field buildings were but a few of the examples of the style that were completed before all skyscraper building stopped with the Great Depression of the 1930s.

## 5 Miesian Model of Aesthetic Expression

When tall buildings began to be built again in Chicago after World War II, Art Deco had faded from the scene, and there was another attempt to establish a definitive skyscraper style. The leader of this movement was the German architect Ludwig Mies van der Rohe, who had emigrated to Chicago in 1938 to head the Department of Architecture at Illinois Institute of Technology. Mies had expectations of a skyscraper style which were quite similar to those of Root. Where Root's vision came from the American commercial milieu, Mies's view arose from the intellectual ferment of utopian modern architecture that appeared in Europe after World War I. Mies had begun his career as a neoclassicist, but he announced his conversion to modernism with projects for two skyscrapers enclosed entirely in glass, published in 1922. He too saw the novelty of this building type as the basis for a new style (Mies van der Rohe, 1922):

Skyscrapers reveal their bold structural pattern during construction. Only then does the gigantic steel web seem impressive. When the outer walls are in place, the structural system, which is the basis of all artistic design, is hidden by a chaos of meaningless and trivial forms. When finished, these buildings are impressive only because of their size; yet they could surely be more than mere examples of our technical ability. Instead of trying to solve new problems with old forms, we should develop new forms from the very nature of new problems....

In his essay *Architecture and the Times* (1924), Mies expanded on his views of the basis for a new architecture, in relation to past styles:

Greek temples, Roman basilicas, and medieval cathedrals are significant to us as creations of a whole epoch, rather than as works of individual architects. Who asks for the names of these builders? Of what significance are the fortuitous personalities of their creators? Such buildings are impersonal by their very nature. They are pure expressions of their time. Their true meaning is that they are symbols of their epoch.

Architecture is the will of an epoch, translated into space. Until this simple truth is recognized, the new architecture will be uncertain and tentative. Until then it will remain a chaos of undirected forces. The question as to the nature of architecture is of decisive importance. It must be understood that all architecture is bound up with its own time, that it can only be manifested in living tasks, and in the medium of its epoch. In no age has it been otherwise.

It is hopeless to try and use the forms of the past in our architecture....It is a question of essentials. It is not possible to move forward and look backwards. He who lives in the past cannot advance.

The whole trend of our time is toward the secular.... Despite our greater understanding of life, we shall build no cathedrals.... Ours is not an age of pathos; we do not respect flights of the spirit as much as we value reason and realism.

The demand of our time for realism and functionalism must be met. Only then will our buildings express the potential greatness of our time, and only a fool can say that it has no greatness....

**Fig. 5.13 Palmolive Building, Chicago; Holabird & Root, architects.** (*Courtesy: Chicago Historical Society.*)

In 1946 Mies met the developer Herbert Greenwald, who commissioned him to design the first high-rise towers built in Chicago since the Great Depression. Mies was at last able to realize the all-glass skyscraper with the twin apartment towers at 860–880 North Lake Shore Drive (Fig. 5.14) completed in 1951. With these buildings, Mies established his stylistic formula for the high rise. Like Sullivan's Wainwright, it emphasized verticality and tripartite form; like Root's Monadnock, it emphasized functionalism and structure. It seemed as though at last a synthesis had been reached. Mies's archetypal tower was a prismatic shaft with a recessed loggia at the base, which expressed the structural columns of the frame; above the frame was infilled entirely with glass and superimposed over it was a vertical array of metal mullions; its termination

**Fig. 5.14 860–880 North Lake Shore Drive, Chicago; Mies van der Rohe.** (*Courtesy: Chicago Historical Society.*)

was a mechanical penthouse. Mies presented his ideas for the validity of this style, comparing it (as did Root) to the Greek and Gothic, in his essay *Architecture and Technology*. Where Root saw the possible origins of skyscraper style in "the broad beneficent commerce of the age," and Sullivan in "the peaceful evangel of sentiment, of beauty, the cult of a higher life," Mies described the source of his architecture simply as "technology," although he clearly means it in the larger context of our civilization (Mies van der Rohe, 1950):

> Technology is rooted in the past.
> It dominates the present and tends into the future,
> It is a real historical movement—
> one of the great movements which shape and
> represent their epoch.
> It can be compared only with the Classic
> discovery of man as a person,
> the Roman will to power,
> and the religious movement of the Middle Ages.
> Technology is far more than a method,
> it is a world in itself.
> As a method it is superior in almost every respect.
> But only where it is left to itself as in
> gigantic structures of engineering, there
> technology reveals its true nature...
> Whenever technology reaches its fulfillment,
> it transcends into architecture...
> Architecture is the real battleground of the spirit.
> Architecture wrote the history of the epochs
> and gave them their names.
> Architecture depends on its time.
> It is the crystallization of its inner structure,
> the slow unfolding of its form.
> That is the reason why technology and architecture
> are so closely related.
> Our real hope is that they grow together,
> that someday the one will be the expression of
> the other.
> Only then will we have an architecture worthy of its name:
> Architecture as a true symbol of our time.

The Miesian model of high-rise expression came to be widely accepted by the architectural community in Chicago (and elsewhere) during the 1950s and 1960s, perhaps to an even greater degree than Art Deco in the 1920s or Burnham's classicism in the first two decades of the century. Mies's almost mystical evocation of "technology" in this essay, as well as his later statements about "truth relations" (Mies van der Rohe, 1960), which he set forth as the basis of his architectural expression, are rich in possible meanings. But the predominant interpretation of Mies's work was that clarity of structure was its most significant element. In this period there were also many structural innovations that resulted in more efficient solutions for ever taller buildings. They would become the work of the second Chicago school of architecture. Many prominent examples of this style were produced, including the Daley Center, the John Hancock Center, and the Sears Tower. Although Mies often built groups of towers such as 860–880 North Lake Shore Drive and the Chicago Federal Center, where the spatial interplay between the buildings created remarkable effects, other Chicago architects tended to make individual towers, which emphasized structure over spatial qualities.

## 6 Robert Venturi

But like all previous notions in skyscraper style, the Miesian model of expression too proved ephemeral, and was not accepted as definitive. Miesian assertions of the cultural centrality of technology were replaced by a new diffidence, reminiscent of that of Raymond Hood some 40 years earlier. Perhaps the leading exponent of this reaction was the architect Robert Venturi in his book *Complexity and Contradiction in Architecture* of 1966. In it he announced a more modest approach for architecture. Although he does not specifically mention the tall building, he offers a critique of Mies's formula with its minimal glass and metal facades embodied in his famous aphorism "less is more" (Venturi, 1977):

> But architecture is necessarily complex and contradictory in its very inclusion of the traditional Vitruvian elements of commodity, firmness, and delight. And today, the wants of program, structure, mechanical equipment, and expression, even in simple buildings in simple contexts, are diverse and conflicting in ways previously unimaginable. The increasing dimension and scale of architecture in urban and regional planning add to the difficulties. I welcome the problems and exploit the uncertainties. By embracing contradiction as well as complexity, I am for vitality as well as validity.
>
> Orthodox Modern architects have tended to recognize complexity insufficiently or inconsistently. In their attempt to break with tradition and start all over again, they idealize the primitive and elementary at the expense of the diverse and sophisticated. As participants in a revolutionary movement, they proclaim the newness of modern, ignoring their complications. In their role as reformers, they puritanically advocated the separation and exclusion of elements, rather than the inclusion of various requirements and juxtapositions.
>
> The doctrine "less is more" bemoans complexity and justifies exclusion for expressive purposes....[The architect] can exclude important considerations only at the risk of separating architecture from the experience of life and needs of society. If some problems prove insoluble, he can express this; in an inclusive rather than exclusive kind of architecture there is room for the fragment, for contradiction, for improvisation, and for the tensions these produce. Mies's exquisite pavilions have had valuable implications for architecture, but their selectiveness of content and language is their limitation as well as their strength....Blatant simplification means bland architecture. Less is a bore.
>
> Although the means involved in the program of a rocket to get to the moon, for instance, are almost infinitely complex, the goal is simple and contains few contradictions; although the means involved in the program and structure of buildings are far simpler and less sophisticated technologically than any engineering project, the purpose is more complex and often inherently ambiguous.

Venturi's critique of modernist minimalism signaled the start of a search for new architectural styles. Although Venturi's book did not advocate the eclectic use of historical elements, one wing of the post-modernist movement did revive the use of both classical and Gothic forms in simplified abstracted versions. In the recent skyscraper building period in Chicago of 1982 to 1992, many building facades were derived from previous styles. 190 South LaSalle Street (Fig. 5.15) invoked John Root's Masonic Temple; 77 West Wacker Drive offered neoclassical pilasters and pediments; whereas the AT&T building returned to an Art Deco vocabulary.

## 7 The Market Economy

Today the prospect of a definitive skyscraper style seems as remote as ever. Perhaps Root and Sullivan, Burnham, and Mies were mistaken in their expectations. Their fundamental assumption seems to have been that the market economy would somehow play an analogous role to that of religion in the development of the Greek and Gothic styles they so much admired. During the 1950s and 1960s, the work of the Austrian economist and Nobel laureate Frederick Hayek helped to clarify our understanding of how the market economy works, and how it is something very different from all previous patrons of architecture. He suggests that the market order is created by the decentralized actions of many minds, using far more information than is available to any one mind, thus generating an order much more complex than any one mind could imagine. The market order cannot be said to have an overall purpose in the common sense of the word (Hayek, 1973):

> Most important…is the relation of a spontaneous order to the conception of purpose. Since such an order has not been created by an outside agency, the order as such also can have no purpose, although its existence may be very serviceable to the individuals which move within such order. But in a different sense it may well be said that the order rests on the purposive action of its elements, when "purpose" would, of course, mean nothing more than that their actions tend to secure the preservation or restoration of that order.

Thus the differing and often conflicting purposes of the market, one manifestation of which is the building of skyscrapers, evolve in a direction that none can foresee, because the knowledge dispersed throughout the system, and sustaining the order, is much greater than any individual can comprehend (Hayek, 1973):

> Certainly nobody has yet succeeded in deliberately arranging all the activities that go on in a complex society. If anyone did ever succeed in fully organizing such a society, it would no longer make use of many minds, but would be altogether dependent on one mind; it

**Fig. 5.15 190 South LaSalle Street, Chicago; John Burgee Architects w/Philip Johnson.** (*Courtesy: Hedrich-Blessing.*)

> would certainly not be very complex, but extremely primitive—and so would soon be the mind whose knowledge and will determine everything. The facts which could enter into the design of such an order could be only those which were known and digested by this mind, and as only he could decide on action and thus gain experience, there would be none of that interplay of many minds, in which alone mind can grow.

Hayek's model seems to offer an explanation as to why no definitive architectural style evolved for the skyscraper, only a rough consensus on functional elements such as structure, elevators, and mechanical systems. The very diversity and complexity of the marketplace seem to demand constant variation in skyscraper appearance.

In this regard, it may be useful to recall how the market economy, in its first major exercise of political power, created the architect that we know today. Indeed these events marked the end of Gothic architecture, the last example of a "complete" style that Root and Mies could offer. Up through Gothic times, the professionals who executed buildings were known simply as *master builders.* They were skilled in the craft and technology of construction, and received clear guidelines from the patrons they served, be it Pericles representing the Athenian state or Abbott Suger representing the Roman Catholic church. The spiritual values of these central institutions were the source of style.

The turning point in this relationship of master builder and client came in the fourteenth century, when the government of Florence was taken over by the merchants' and craftsmen's guilds. The "many minds" of Hayek's marketplace began their characteristic pursuit of change. In 1292 the guilds decided that the old cathedral of Florence, San Reparta, built in the seventh century, should be demolished, and a new and much larger cathedral should be built with secular funds, for the church said it could not afford it. The new building was begun in the Gothic style under a traditional master builder, Arnolfo di Cambio. The guilds began to change the building process, with the cathedral bell tower or campanile built during 1334–1359. Here they engaged the famous painter Giotto di Bondone, who knew nothing of construction, to provide a painting of the design of the tower, and the master builder Francesco Talenti was simply required to execute it (Prager and Scaglia, 1970).

In 1367 the guilds took the unprecedented decision of abandoning the Gothic style for the cathedral and adopting the Romanesque (Fig. 5.16), a decision that was ratified in a public referendum of 500 citizens of Florence. This was done over the vigorous objection of the then master builder Giovanni Ghini, who contended that the Gothic was the only proper style for church building. Ultimately, as a part of their program of novelty, the guilds demanded a classical dome over the crossing of the new cathedral's nave and transept of the same span as the Pantheon, which the master masons said would be very difficult to execute (Prager and Scaglia, 1970).

In 1418 the Opera, the guild committee overseeing the construction of the cathedral, announced a competition for designs for the dome to replace the 50-year-old temporary roof. This competition gave the opportunity to Filippo Brunelleschi, a member of the goldsmiths' guild, with relatively little experience in construction, to establish the professional identity of the designer as distinct from the client or builder. His design for the dome which won the competition not only met the structural requirements, but permitted very economical execution without form work, which the traditional Gothic masons had maintained was impossible. After this triumph, Brunelleschi pointedly refused eager invitations to join either the masons' or the carpenters' guilds, saying he was an "architect" (Prager and Scaglia, 1970). Now the word architect comes from the Greek word *architekton,* which in fact means "master builder," but the linguistic difference denoted by using the Greek root not only set the design professional apart from the builder, but also identified him with the classical world that was the source of the unfamiliar forms he was reintroducing to building. Thus was established the role of the architect as the bringer of novelty to built-form, stimulated by the "many minds'" interaction of the marketplace.

Although the guilds of Florence eventually lost power to the Medicis in the fifteenth century, the Renaissance church and state retained the classical forms the guilds had introduced to architecture, transmuting them into symbols of their power. But as the market economy began to gain wider influence in the eighteenth century with the Industrial Revolution, it began to affect architecture once again. In 1747 the *Ecole des Ponts et Chausees* was established in Paris as the first institution to train civil and structural engineers, who would further divide the building design process. During the nineteenth century, with the coming of the heroic age of Victorian engineering, the consensual functional aspects of buildings would be increasingly placed in the hands of engineers, whereas the "many minds" of the marketplace would impose ever-growing demands for novelty and change of building appearance on the architect. The "battle of the styles" of the Victorian era was the logical outcome of the increasing dominance of the market economy. The role of change in the market economy was vividly portrayed by Karl Marx in the *Communist Manifesto* (1848):

> Constant revolutionizing of production, uninterrupted disturbance of all social relations, everlasting uncertainty and agitation, distinguish the bourgeois epoch from all earlier times. All fixed, fast-frozen relationships, with their train of venerable ideas and opinions, are swept away, all new-formed ones become obsolete before they can ossify. All that is solid melts into air, all that is holy is profaned, and men are at last forced to face with sober senses the real conditions of their lives and their relations with their fellow men.

As Marshall Berman has pointed out, the whole notion of fashion, of planned obsolescence, of "all that is solid melts into air," is an integral part of the modern market economy (Berman, 1982).

**Fig. 5.16 Duomo Campanile and Duomo Dome, Florence, Italy.** (*Courtesy: Paul Armstrong.*)

When the phenomenon of the skyscraper suddenly appeared in Chicago, the architects seized on it with the hope it was a harbinger of a new "complete" style similar to the Greek and Gothic they so much admired. It seemed to be an opportunity to break the cycle of fashion. But the Greek city state and the Roman Catholic church were very different institutions from the commercial patrons of the skyscraper. Each embodied an intense spiritual identity, and the distinct architectural styles of temple and cathedral helped to present this identity. The "many minds" of the marketplace have no such well-defined spiritual identity, and demand no definitive skyscraper style.

The problem of skyscraper style is an architectural one, and demands an architectural solution or, more precisely, an architectonic solution in the sense of the response of the architekton, the ancient Egyptian, Greek, or Gothic master builder, to the spiritual forces of their times. Such a new "complete" style will appear only when a new spirit intervenes. Then skyscrapers will become blank tablets on which architects will inscribe their interpretation of the new spiritual paradigm.

## *5.5 RECENT TRENDS IN AESTHETICS AND FORMS*

With the continued boredom with modern architecture that emphasized glass and metal grids, some architects initiated a movement known as post-modernism in the early 1970s. This architectural trend generally encouraged somewhat arbitrary and dramatically articulated buildings. Post-modernism in architecture emerged as a specific reaction against the established form of high modernism in art in general and the international style of Le Corbusier, Frank Lloyd Wright, and Mies van der Rohe in particular. There are different sporadic forms of post-modernism, and the only coherent thing in this new architectural movement is its unity of purpose in terms of its stimulus to replace the traditional forms of modernism. The 1960s constitute the principal transitional period, a period when computer revolution and electronic information experienced their peak impulses in conjunction with a newly emerging social order. The advent of post-modernism in the 1970s dealt a severe blow to the structuralist architecture of tall buildings where structural logic was a basic ingredient of a tall building's aesthetics and form (Khan, 1982b; Ali, 1990).

The post-modern buildings have the characteristics of stepbacks, horizontal angles, slopes, curves, notches, and other novel geometric shapes, and therefore they bring about nontraditional articulations that do not match any particular structural planning concept from a logical standpoint. The new post-modern buildings usually exhibit themselves through a radical departure from conventional architecture by enacting varied forms, particularly varied rooflines that stand out in the flat-top cityscape. These rooflines are created by assigning domes, pyramids, unsymmetric sloped roofs, or a combination of these geometric shapes.

### *1 Departure from Modernism*

Departure from the typical box form was introduced in Lake Point Tower in Chicago as early as 1968, although the building could still be classified as a conventional tower (Goldberger, 1989). In this building a curved facade with dark color was used. A number of tall buildings that exhibited flamboyance, exuberance, and other theatrical tendencies were built in the 1970s. Some notable ones are the Waterside Plaza, New York

(1974), 9 West 57th Street, New York (1974), Citicorp Center, New York (1977), and Overseas Chinese Bank Center, Singapore (1976), to name only a few. Such deviations in building forms marked the beginning of more drastic changes in forms of the 1980s.

Goldberger (1989) notes:

> By 1980, one thing was clear: the box, the rationalist dream of the International Style, was making more and more architects uncomfortable. Not only was it no longer the clean and exhilarating structure that would serve as a clarion call to a new age, but it was not even able to hold out much promise of practicality. It was generally inefficient from the standpoint of energy, and it was not as marketable from the viewpoint of real estate operators either.

Goldberger further states:

> So just as it had been economics...that had ultimately won the battle of modernism after World War II, it is economics of a sort that turned builders away from the Miesian boxes with which they had been filling American cities since the mid-1950s....Even Skidmore, Owings & Merrill began to indicate doubts: in late 1980 the partners of the New York office of the firm that...has been identified with the postwar corporate Miesian style invited a group of younger, so-called post-modern architects to present their work and engage in a dialogue at a private symposium.

A few examples of tall buildings that departed from the austere glass-box style are the Texas Commerce Tower, Houston; 101 Park Avenue, New York; 333 Wacker Drive, Chicago; and the Trump Tower, New York. The Trump Tower is especially noteworthy since in this building "a series of setbacks creates a cascading effect" and it desperately attempts "to put a sense of distance between the notion of the skyscraper and the notion of the box—the building has been pushed and pulled in every which way to give it form" (Goldberger, 1989).

Current trends in building design show that the idea of structural expression has substantially lost its influence on the architectural community. Goldberger (1989) states in this regard: "...technology seems only slightly more able to provide an impetus to creativity. Most skyscrapers that have been a direct expression of technological prowess...have seemed exhibitionistic and strained. Only the sleek computer aesthetic appears able to convince as a technological expression today—and it is more a symbol of the romance of technology than it is a literal expression of the actual technology of building."

A good example of concealing the structure is the Citicorp Center, where the X braces on the building perimeter are hidden behind the metal skin. In this regard, Goldberger (1989) writes: "The visually pleasanter form that resulted was justification enough for this act in the late 1970s...structure is entirely concealed in the cool buildings of the computer aesthetic, which do not reveal floor divisions. They romanticize the idea of post-mechanical technology, which is an altogether different matter than displaying the mechanical technology by which they were built."

A number of tall buildings represent the new trends of post-modernism and other imageries that deviate from the modernist glass-box architecture. Some examples are the 42-story Metropolitan Square in St. Louis, Missouri (Fig. 5.17); the 44-story 801 Grand Avenue Office Tower in Des Moines, Iowa (Fig. 5.18); the 55-story Griffin Office Tower to be built in Dallas, Texas (Fig. 5.19); the 22-story Albuquerque Plaza in Albuquerque, New Mexico (Fig. 5.20); and the 31-story Am South/Harbert Plaza in Birmingham, Alabama (Fig. 5.21). These buildings use in a variety of ways energy-efficient tinted glass, sculptured tops with glass pyramids, and globes, golden or pink granite cladding, pitched copper roof pinnacles, and other similar features.

**Fig. 5.17 Metropolitan Square, St. Louis, Missouri; HOK, architects.** (*Courtesy: Hellmuth, Obata & Kassabaum.*)

**Fig. 5.18 801 Grand Avenue Office Tower, Des Moines, Iowa; HOK.** (*Courtesy: Hellmuth, Obata & Kassabaum.*)

**Fig. 5.19 Griffin Office Tower, Dallas, Texas; HOK.** (*Courtesy: Hellmuth, Obata & Kassabaum.*)

**Fig. 5.20 Albuquerque Plaza, Albuquerque, New Mexico; HOK.** (*Courtesy: Hellmuth, Obata & Kassabaum.*)

**Fig. 5.21 Am South/Harbert Plaza, Birmingham, Alabama; HOK.** (*Courtesy: Hellmuth, Obata & Kassabaum.*)

On the decline of structural logic, Khan (1982b) lamented by stating: "Today it seems the pendulum has swung back again towards architecture that is unrelated to technology and does not consciously represent the logic of structure. Nostalgia for the thirties and even earlier times has hit a large segment of the architectural profession; in many cases facade making has become the predominant occupation."

Khan was critical of both the engineer and the architect when he said:

> It is apparent that Post-Modernism in architecture is very much the result of the architect's lack of interest in the realities of materials and structural possibilities: the logic of structure has become irrelevant once again. This attitude in architecture suits many engineers because of their overspecialization in engineering schools which treats the solution of the problem as the ultimate goal, and not the critical development of the problem itself...any structure can be made to work with many engineers gladly willing to play with their computers and come up with the answers to hold up the building.

Khan continues:

> But logic and reasoning are strong elements of human existence, always important when man must transcend into the next level of refinement. There are already some signs of that happening in architecture. New structural systems and forms are beginning to appear once again and with them new architectural forms and aesthetics. The pendulum of structural logic in architecture continues to swing.

## 2 High-Tech Architecture

While the swinging of the pendulum continues, it is apparent that tall buildings have grown and developed to their present forms passing through the stages of technological beginnings and rationalistic modernism and have reached a comparable state in which exuberant trends and liberal tendencies have outstripped their design tradition and conception of forms. Such tendencies have other ramifications. For example, a popular, primarily British movement called high-tech manifests itself in the Hong Kong and Shanghai Bank in Hong Kong (see Chapter 6). On this issue, Goldsmith (1986) observed: "In this case there was a preconceived idea that was not structurally efficient, and that structure was used regardless. This building is but one example of several in which a very expressive structure is used for expressionism."

The Hong Kong and Shanghai Bank building, located in the heart of Hong Kong's central district, is organized as a series of suspended banking operation floors, each having an interrupted view of Hong Kong Harbor and Kowloon. The building is primarily recognized for its unique structure, both visually and functionally. The high-tech theme of the building is even carried over to the mechanical services. Individual prefabricated modules containing bathrooms and individual air-conditioning plants for every floorstack were installed alongside the structural towers. Another feature of the building is the use of extensive aluminum cladding to maintain the high-tech precision of the project (Doubliet and Fisher, 1986). This bank has been said to be the most expensive building ever built and represents a symbol of capitalist prosperity.

Proponents of high-tech buildings view such developments in a somewhat different context. Nicholson (1989) reports: "During a period when technology seems to be the driving force of change and commercial development, it is reasonable to expect that the buildings required to accommodate such activities should themselves reflect the high technology. After all, history is rich with examples of buildings expressing their function on their faces." Nicholson notes, however, that high-tech imagery used "in many cases has nothing to do with internal function of the building and everything to do with image and styling." He asserts, "Many of the best buildings solve the problems first and rely on solution, largely, to provide the imagery; indeed, their aesthetic goes far beyond simple styling." Critics of the Hong Kong and Shanghai Bank have even referred to it as a "monster." The main issue under scrutiny here is the appropriateness of the image that this bank attempts to exhibit.

## 3 Future Outlook

Khan correctly observed that new structural systems and forms are beginning to appear once again, although to a much lesser extent compared to the glory days of modernist rationalism. As Youssef (1991) states: "...the next generation of super tall buildings will be characterized by: lightness, resulting in reduced massing, slenderer structural forms, which results in reduced stiffness; and reduced natural damping." He discusses in this article the design process for a 150-story steel-framed, mixed-use building proposed for the mid-Wilshire district of Los Angeles. After evaluating a series of forms, a diamond-shaped (rhomboid) plan was developed. In Youssef's words, "The genesis of the final scheme came from the idea that such a large structure need not have a rectangular plan, but might have a non-rectilinear geometric form. A single triangle, although stable, was not as strong as a four-legged structure. Thus, a diamond form was created by abutting two triangles." A combination of vertical trusses at the corners and rigid-frame Vierendeel infill in between provides an integrated megastructure system, which is efficient and cost-effective.

For the proposed 125-story Miglin-Beitler Tower in Chicago, "a simple and elegant integration of building form and function has emerged from close cooperation of architectural, structural, and development team members. The resulting cruciform tube scheme offers structural efficiency, superior dynamic behavior...for this 125-story office building" (Thornton et al., 1991). This building has a slick tapered form appearing to penetrate the sky.

The 310-m (1018-ft)-high First Interstate World Center, completed in 1989, is the tallest building in Los Angeles, the tallest tower on the West Coast, and the tallest tower anywhere located in a seismic zone 4 (Gordon, 1990). The tower's location in a most severe seismic zone called for a special structural system to achieve ductility and stiffness and hence a unique building form.

The challenge for the future is to create tall buildings that have grace and dignity and are intended to serve people and to improve the urban fabric. As we approach the twenty-first century, it is likely that more and more innovative tall buildings will be built utilizing the latest technological tool that will facilitate the analysis and construction of structural systems of varied forms and configurations. Making future tall buildings efficient and aesthetic is the principal goal. Architects and engineers can rise to this challenge only if they collaborate and if they care about people, who are the ultimate critic of what they design and build.

Skyscraper watchers must observe the current trends in the design and construction of tall buildings to make realistic projection of what could be forthcoming. Goldberger (1989) perceives the current trends in tall building architecture in the following way:

> ...Architects now, even those who reject it, have tasted of modernism, and its effects are felt throughout everything now designed. This is a time much more like the Baroque—a time of intricate and ironic visual tricks, exuberance, and self-indulgence. It is a time of excess, but it is also a time of promise.

Scuri (1990) states: "The design of office buildings finds itself now in a state of transition, not to say ferment." Ali (1990) observes:

> Even post-modern architecture may not be entirely incompatible with structural discipline if the extent of extravagant post-modernistic expression can be toned down in a building. With the availability of computers it is now possible to employ multiple structural systems in the same building with some degree of Post-Modernistic flamboyancy at the roof, facade, and entrance. The tube system, with its structural logic, free-form architecture, and its decorative style may be gradually harmonized to evolve into a new mature style for tall building forms. This seems to be the greatest challenge for the immediate future.

This is not to say that a structural system following structural logic could be simply covered with a beautiful skin that would make the building look aesthetically pleasing. Architectural expression emanates from the overall building form and skin, whereas structural expression arises out of the *inherent* form of the structure, which should be based on structural logic and should be uninhibited by the architect's passion with the skin. Bridging the gap between these two expressions honestly without feigning or suppressing one at the expense of the other such that they ultimately blend into a single wholesome visual imagery, should be the quintessence of future tall building design.

## 5.6 CONDENSED REFERENCES/BIBLIOGRAPHY

The following is a condensed bibliography for this chapter. Not only does it include all articles referred to or cited in the text, it also contains bibliography for further reading. The full citations will be found at the end of the volume. What is given here should be sufficient information to lead the reader to the correct article—author, date, and title. In case of multiple authors, only the first named is listed.

Adler 1893, *The Architecture of John Wellborn Root*

Ali 1986, *The Forgotten Half of Design: Structural Art Form*

Ali 1987, *Influence of Architectural Configuration on Seismic Response of Buildings*

Ali 1990, *Integration of Structural Form and Esthetics in Tall Building Design: The Future Challenge*

Architectural Record 1974, *The High Rise Office Buildings—The Public Spaces They Make*

Architectural Record 1985, *The Bank of China*

Arnold 1982, *Building Configurations and Seismic Design*

Beedle 1992, *Tall Buildings around the World*

Berman 1982, *All That Is Solid Melts into Air: The Experience of Modernity*

Billington 1985, *The Tower and the Bridge: The New Art of Structural Engineering*

Brady 1993, *Color*

Cowan 1974, *Tall Buildings for People—Aesthetics and Amenity*

CTBUH, Group PC 1981, *Planning and Environmental Criteria for Tall Buildings*

Doubliet 1986, *The HongKongBank*

Fintel 1986, *New Forms in Concrete*

Gibbs 1984, *Business Architectural Imagery in America, 1870–1930*

Goldberger 1989, *The Skyscraper*

Goldsmith 1986, *Fazlur Khan's Contributions in Education*

Gordon 1990, *Design for the Big One*

Grant 1990, *Beauty and Bridges*
Hart 1985, *One Hundred Years of Steel Framed Structures 1872–1972*
Hayek 1973, *Law, Legislation and Liberty,* vol. I: *Rules and Order*
Hoffman 1967, *The Meanings of Architecture: Buildings and Writings of John Wellborn Root*
Hood 1925, *Exterior Architecture of Office Buildings*
Hurd 1990, *Concrete Giant Rises on Chicago Skyline*
ICBO 1991, *Uniform Building Code*
Iyengar 1986, *Structural and Steel Systems*
Johnson 1978, *Mies van der Rohe*
Khan 1967, *The John Hancock Center*
Khan 1982a, *100-Story John Hancock Center in Chicago—A Case Study of the Design Process*
Khan 1982b, *The Rise and Fall of Structural Logic in Architecture*
Kohn 1990, *The Tall Building*
Lee 1985, *The Fine Art of Structural Engineering*
Liebenberg 1983, *The Aesthetic Evaluation of Bridges*
Lim 1992, *The Architectural Characteristics of Some Tall Buildings since Sears Tower*
Marx 1848, *The Communist Manifesto*
Mies van der Rohe 1922, *Two Glass Skyscrapers*
Mies van der Rohe 1924, *Architecture and the Times*
Mies van der Rohe 1950, *Architecture and Technology*
Mies van der Rohe 1960, *Architecture and Civilization*
Nicholson 1989, *High Tech Buildings: Product from Process*
Prager 1970, *Brunelleschi: Studies of His Technology and Inventions*
Progressive Architecture 1980a, *Facade at Right Angles*
Progressive Architecture 1980b, *Thinking Tall*
Progressive Architecture 1990a, *Building Tops*
Progressive Architecture 1990b, *Foster's Japanese Needle*
Robins 1987, *Observations*
Root 1883, *The Value of Type in Art*
Root 1890, *A Great Architectural Problem*
Ross 1955, *Aristotle: Selections*
Schueller 1990, *The Vertical Building Structure*
Scuri 1990, *Late-Twentieth Century Skyscrapers*
Siegel 1975, *Structure and Form in Modern Architecture*
Smith 1990, *The Expressive Role of the Tall Building in Hong Kong*
Stern 1978, *New York 1930: Architecture and Urbanism between the Two Wars*
Stern 1986, *Pride of Place: Building the American Dream*
Sullivan 1896, *The Tall Office Building Artistically Considered*
Thornton 1991, *Looking Down at the Sears Tower*
Torroja 1958, *The Beauty of Structures*
Tucker 1985, *Super Skyscrapers: Aiming for 200 Stories*
Venturi 1977, *Complexity and Contradiction in Architecture*
Youssef 1991, *Megastructure: A New Concept for Supertall Buildings*

# 6

# Psychological Aspects

Picture a camera with a zoom lens focused on a cityscape. With the setting at 28 mm for a wide-angle view, we can see the entire city skyline, and as we zoom all the way to 135 mm for a telephoto shot, we can focus on one building in detail. If we could zoom in even further, we might see inside the windows of the building itself. *Each point of view presents a unique experience.* The intention of this chapter is to reflect upon the different perspectives from which people experience tall building types.

Section 6.1 provides a framework from which to understand the ordering of topics within the chapter. Section 6.2 discusses the symbolic qualities of tall buildings. A brief overview of tall building symbolism is provided, and the Empire State Building is illustrated as a specific case in point. Section 6.3 addresses how people perceive tall buildings and city skylines from a distance—whether from an airplane, a train, a car, or simply from a postcard, photograph, film, or other medium. Section 6.4 addresses the relationships among climatic and geological events, tall buildings, and human behavior. Section 6.5 begins to address how people perceive tall buildings from within—from the building entry to the lobby or atrium space. It raises a number of issues about how people find their way around in tall buildings and how barrier-free design issues relate to tall buildings. Section 6.6 covers passenger behavior in elevators. It describes the phenomena of waiting behavior, personal space and crowding, and civil inattention as they are seen in elevators. It also discusses fears and phobias about elevators. Finally, it wraps up with a discussion of what happens when elevators are temporarily out of service.

The remaining sections examine sociopsychological issues in a few specific types of tall buildings. Four types of buildings are highlighted here, two that have been the focus of much environment-behavior research, and two in which environment-behavior research is noticeably absent but sorely needed. Section 6.7 explores these issues in and around tall office buildings. It covers psychological effects of sunlight, user satisfaction, and pedestrian use of urban plazas near tall office buildings. Section 6.8 examines the same issues in high-rise housing. Section 6.9 addresses such issues in tall parking garages, a topic on which extremely little research has been done. Section 6.10 looks at social and psychological issues in high-rise shopping centers, another area where little research has been conducted. It addresses mall psychology, how different users experience shopping, and the vertical shopping mall. Finally, Section 6.11 provides a brief summary and suggests some directions for future research.

The meaning of tall buildings can be viewed from psychological, sociological, cultural, and geographic perspectives. This chapter focuses especially on the social and psychological aspects of tall building design, that is, the meaning of tall buildings to groups as well as to individuals.

## 6.1 NOTES ON METHODOLOGY

Despite the burgeoning literature on the relationships between environment and behavior, to date only one book has been published which focuses specifically on the psychological and behavioral aspects of tall buildings (Conway, 1977). This landmark work documents the proceedings from a symposium by the same name, sponsored by the Office of Research Programs of the American Institute of Architects (AIA) and the Joint Committee on Tall Buildings, forerunner to the Council. The symposium was held, appropriately, at Chicago's Sears Tower in July 1975. The book is divided into four parts: a commentary and overview, the tall building and its neighborhood, housing and the livability of tall buildings, and the response to emergency. Contributors include a multidisciplinary group of researchers from the United States and abroad. Interested readers are referred to this excellent source, which is far more comprehensive than what can be covered here. Nonetheless, since so many years have elapsed in the meantime, this chapter updates and builds upon the valuable information from this classic text.

A diverse set of issues is raised in the study of tall buildings. In this chapter, some issues are discussed in separate sections, whereas others are included as part of discussions of specific building types or spaces. These issues include perception, satisfaction, imageability, fears and phobias, way-finding behavior, personal space, and crowding.

Wherever possible, an attempt has been made to draw upon relevant sources of information. Note that the literature cited in this chapter usually comes from two types of sources, one scientific and the other nonscientific. The first source is research from the rapidly emerging field of environment and behavior. More often than not, the environment-behavior research cited here consists of empirical studies about the social and psychological aspects of specific physical environments. The second source is journalistic and anecdotal. Relevant anecdotes about people's experiences in tall buildings are drawn from newspaper and magazine articles. Although much research has already been done, many pressing questions remain.

## 6.2 THE SYMBOLISM OF TALL BUILDINGS

Think of Paris—and an image of the Eiffel Tower appears. Picture New York City—and the Empire State Building and the World Trade Center come to mind. Now envision Chicago—and the Sears Tower looms overhead. Tall buildings are the predominant part of today's urban landscape. Along with prominent natural features, such as mountains, volcanoes, lakes, and rivers, tall buildings can symbolize cities and metropolitan areas, encapsulating the spirit of a geographic region. Consequently, compared to any other component of the built environment, tall buildings take on a special significance.

Tall buildings acquire different levels of meaning for different audiences. People who work in them, live in them, visit them, walk or drive by them, or simply view them from afar relate to tall buildings in different ways. Some individuals may have seen tall buildings only in postcards, on television, or in the movies. For these people, the tall building could well represent the mysterious unknown.

### 1 Importance of Symbolism

The study of tall building symbolism is part of a larger area of study called environmental symbolism, which is a subfield of the larger field of environment and behavior. Environmental symbolism addresses the hidden meanings and associations that people attribute to places and spaces.

Tall buildings represent many things to many people. They can stand for the architect who designs them, the developer who builds them, the corporation who finances them, the district or neighborhood in which they are located, and, in some cases, the city or even the nation as a whole. At the national level, the tallest, newest buildings symbolize economic status, technological progress, modernity, and growth. At the individual level, for those who own, live in, or work in tall buildings, these edifices symbolize private wealth, social status, and business prestige. As developer Donald Trump admitted (*National Geographic,* 1989):

> Ego is a very important part of the building of skyscrapers....It's probably a combination of ego and desire for financial gain. I mean, once you have enough money so that you can eat and live, then ego enters into it. It's involved with the building not only of skyscrapers but all great buildings whether they are tall or not.

Tall buildings can epitomize people's pride in their city or, on the other hand, people's disdain of just another eyesore, a source of visual pollution (CTBUH, 1978). At one level one might believe that fulfilling the functional requirements is the paramount concern of tall building designers. But the hidden agenda of the architect, the developer, the client, and others may well be much deeper than that. This agenda may include creating a monument by leaving a clear personal signature on the landscape, seeking an architectural award, or competing with another nearby building for height and space.

Because tall buildings are among the most visible in the urban landscape, their symbolic value can be extremely powerful. While individual sentiments enter into the design of any structure, at the scale of the tall building, these sentiments are often magnified. Similarly, public reaction to high rises often falls clearly on either end of an emotional spectrum; people either love them or they hate them. Any analysis of tall buildings must take this symbolic quality into account.

## 2 Tall Building Symbolism and the City

For as long as they have existed, tall buildings have been viewed as monuments (see Chapter 1). The ancient temples of Athens and Rome, the pyramids of Egypt, and other well-known tall buildings throughout history were often deliberately placed at the crests of hills so that local citizens could perceive their prominence. More recently, Paris' Sacré Cœur cathedral and Moscow's distinctive wedding-cake-style public buildings of the Stalinist era were also perched high above the rest of the city so that the public could marvel at them from afar.

Changing skylines were the subject of artists' paintings beginning in the seventeenth century, with popular urban panoramas of London and other European cities. A modern perspective developed through the nineteenth century, with the appearance of the first skyline drawings and photographs, and the first use of the word "skyline" to signify urban views. The word apparently was used first in the *New York Journal* on May 3, 1896 (Taylor, 1992). Dramatic changes in cityscape photography took place during the 1920s and the 1930s, when Hollywood captured the spirit of this new art form in musicals and films about sophisticated city life (Taylor, 1992). A group of photographers loosely affiliated with Alfred Stieglitz and his journal, *Camera Work,* helped make the cityscape an important genre of the new field of photography. In addition to Stieglitz, other cityscape photographers included Edward Steichen, Alvin Coburn, and Paul Strand. These photographers captured the spirit of tall buildings both by day and by night, and they were often fascinated by construction sites and unfinished skyscrapers (Taylor, 1992). Their photographs represent how new urban residents experienced both the skyscraper and the skyline, often in a highly romanticized way.

In the United States, where the high rise was born, the symbolic quality of tall buildings has always been great. Whereas in the late nineteenth century, church steeples dominated the New York City skyline, in the early twentieth century, cathedrals of a different type began to appear. Some have referred to this new genre of high rise as "the cathedral of commerce" (Goldman, 1980). These cathedrals of commerce—newspaper offices, insurance buildings, and other commercial structures—were in striking contrast to the images of most European cities, where a cathedral, a castle, or some other public or government structure usually reigned over the city. Montgomery Schuyler, an early architectural critic, recounted the experience of Europeans arriving in New York by ship (Taylor, 1992):

> As his steamer picks her way through the traffic which crowds the river, past huge utilitarian structures that dwarf into obscurity the spires of churches and the low domes of public monuments, it seems to him that he has never before seen a waterfront that so impressively and exclusively "looked like business."

In New York City these new cathedrals began to appear in rapid succession, as each outdid the other: the Flatiron Building (20 stories) (see Fig. 7.5), the Times Building (22 stories), the City Investing Company Building (32 stories), the Singer Building (47 stories), and the Woolworth Building (60 stories), which remained the world's tallest building until 1929.

Behind each of these new cathedrals were the personalities of the men determined to see them built (Goldman, 1980):

> And the motives were not strictly financial. Frank Woolworth, for one, plunked down $13 million of his own cash for his tower, knowing perfectly well that his profits would be slim. "Beyond a doubt his ego was a thing of extra size," wrote his builder. "Whoever tried to find a reason for his tall building and did not take that fact into account would reach a false conclusion." Woolworth was a farmer's son from upstate New York, but for his office in "The Cathedral of Commerce" he replicated Napoleon's audience chamber at Compiègne, complete with embossed ceiling, marble wall panels, throne room chairs and bust of Napoleon posing as Caesar.

In 1929 the Bank of Manhattan (71 stories) overtook the record-breaking Woolworth Building, but it too was soon surpassed by the glittering Chrysler Building (77 stories). Again note the motives behind the mammoth scope of this world-famous skyscraper, as depicted in Walter Chrysler's comments (Goldman, 1980):

> I came to the conclusion that what my boys ought to have was something to be responsible for. They had grown up in New York and probably would want to live there. They wanted to work, and so the idea of putting up a building was born. Something that I had seen in Paris recurred to me. I said to the architects: "Make this building higher than the Eiffel Tower." That was the beginning of the 77 story Chrysler Building.

### *3 The Empire State Building: A Case Study in Tall Building Symbolism*

No doubt the winner of the skyscraper race was New York City's Empire State Building (Fig. 6.1), the brainchild of John Jacob Raskob, Vice President of General Motors and the Democratic National Chairman. In 1928 he ran New York politician Al Smith's unsuccessful presidential campaign against Herbert Hoover. After the race was over, Raskob hired Smith to be a front man for the building project. Raskob's original plan was to create a 30-story building, but once he saw the other skyscrapers breaking records (Goldman, 1980),

> ...the temptation was too much for Raskob. He and Smith might have lost their campaign for president but there was another race they could win. Besides, one acquaintance said, "It burned Raskob to think the French had built something higher than anything we had in this great country of ours."...John J. [Raskob] reached into a drawer and pulled out one of those big fat pencils schoolchildren liked to use. He held it up and he said to Bill Lamb [the architect]: "Bill, how high can you make it so that it won't fall down?"

**Fig. 6.1 For many years, the Empire State Building in New York City, completed in 1931, was the world-wide winner of the skyscraper race.** (*Courtesy: Empire State Building, Helmsley-Spear, Inc.*)

By the time it was completed in 1931, the 102-story Empire State Building, with its 61-m (200-ft) mast towering overhead, achieved a total height of 381 m (1250 ft), well over the 319-m (1046-ft) Chrysler Building (Goldman, 1980). But the building was not only symbolic of the personality behind it, it was also symbolic of a desire to create an atmosphere that would make people feel optimistic about the future. The upbeat aura of both the Empire State and the Chrysler buildings, as shown in their exuberant, theatrical Art Deco style, provided a striking contrast to the severely depressed national economy at that time.

Soon after it opened, the Empire State Building became a major tourist attraction and a popular spot for dignitaries visiting New York City. Among the most well-known visitors were the French Premier, the Japanese Prince, Albert Einstein, George Bernard Shaw, H. G. Wells, Winston Churchill, Pope Pius XII, Nikita Khrushchev, Fidel Castro, and Queen Elizabeth II. The world's tallest man, Robert Wadlow, a towering 2.62 m (8.6 ft), also made an appearance. In fact, the only man known to have refused Al Smith's invitation to visit the Empire State Building was Walter Chrysler. It also became a chic spot to shoot fashion ads, photograph beauty contestants, and hold wedding ceremonies (Goldman, 1980). The building became one of the city's most popular souvenir items, as millions of replicas in the form of jars with snow, candles, pencil sharpeners, powder boxes, lamps, and posters were sold.

Perhaps the person who best captured the symbolic feel of the Empire State Building was someone who could not see it at all: Helen Keller, leader of the blind. After visiting the building, she wrote: "Let cynics and super-sensitive souls say what they will about American materialism and machine civilization. Beneath the surface are poetry, mysticism and inspiration that the Empire Building somehow symbolizes. In that giant shaft I see a groping toward beauty and spiritual vision" (Goldman, 1980).

Whereas to many people, buildings like the Empire State symbolized a happy future, to others, such buildings provided an opportunity to reflect on their past, drown their sorrows, and simply end it all. Even before the building was completed, a discharged worker threw himself to his death down an open elevator shaft. Eighteen months after it opened, the next suicide occurred, forcing guards and ticket sellers to be on the lookout for despondent, solitary types. Before a guardrail was built, 16 people leaped over the top. During one three-week period in 1947, guards hauled five jumpers back from the edge (Goldman, 1980).

The case of the Empire State Building is telling, because it epitomizes the psychological, social, and cultural meanings that people impute to tall buildings. For the individual, the building can evoke emotions that reflect both ends of the pendulum—from the exuberance of seeing the building for the first time, when even adults succumb to the awe and curiosity of small children—to the despondency of contemplating suicide. For society, the building reflects a conscious departure from the depressed economic state during which it was built, signifying a visible, tangible transition to a more prosperous era. For many Americans and international visitors alike, the building symbolizes the prominence of the United States on the world scene and the nation's ability to rise above worldwide competition.

## 4 The Skyscraper and Its Worldwide Influence

Cities have engaged in competition for the world's tallest building, the greatest number of high rises, and other such marks of distinction. The battle lines between New York City and Chicago have been most clearly drawn, with both cities having the top contenders. New York's World Trade Center towers, completed in 1977 (Jencks, 1980) at a soaring height of 110 stories each (Fig. 6.2), stand in second place to Chicago's Sears

Tower, completed in 1974 and currently holding the world's record at 114 stories (Fig. 6.3). Nearby in Chicago, and in close competition, is the John Hancock Center, completed in 1970 at a height of 96 stories.

Today most of the world's largest cities are dominated by tall buildings. This is especially the case in the United States, Canada, parts of western Europe, the Middle East, Asia, Australia, and parts of South America. Europe's tallest skyscraper is the new

**Fig. 6.2 New York City's World Trade Center towers superseded the Empire State Building in 1977 as the tallest skyscrapers in New York.** (*Courtesy: New York Convention and Visitors Bureau.*)

259-m (850-ft), 70-story Messeturm in Frankfurt am Main, Germany, completed in 1991 (see Fig. 5.3). Tall buildings are less common in most nonindustrialized cities of developing countries. In developing nations, the symbolic importance of tall buildings is perhaps more encompassing than it is elsewhere. The Torre Latino Americano in Mexico City, for instance, completed in 1957 at a height of 181.3 m (595 ft), closely resembles the design of New York's Empire State Building and, like its predecessor, has come to symbolize the ascension and rapid growth of a huge geographic region, Mexico and Central America.

Strauss (1976) has described the symbolic nature of cities as a reflection of their geographic region. Strauss argues that due to a city's growth and development, symbols are constantly changing over time. Strauss' analysis is primarily at the macroscale.

At a smaller scale, Dovey (1991) sought to uncover the prevailing corporate values used to market commercial office space in tall corporate structures. He collected and analyzed leasing brochures for major high-rise office towers constructed during a late 1980s' building boom in Melbourne, Australia. In addition, he analyzed all large leasing advertisements appearing in the property section of a major daily newspaper over a period of 2 years. Because competition among these buildings was fierce, advertising was especially important. Dovey's perceptive analysis identified several themes, including identity, authenticity, power, contextualism, and timelessness.

Dovey (1991) concludes that, ironically, many of the traits that advertisers use to attract clients to office towers are at direct odds with these clients' strongly held values toward their city. For example, according to the ads, towers should be landmarks that dominate their surroundings, yet they must be "good neighbors" as well. In another

**Fig. 6.3 Chicago's Sears Tower, completed in 1974, is currently the world's tallest skyscraper.** (*Courtesy: James Steinkamp.*)

ironic twist, clients are often attracted to office towers located in traditional, small-scale neighborhoods, but once these high rises are built, they can actually destroy the flavor of these desirable neighborhoods. Can such blatant contradictions ever be resolved?

A number of scholars have been highly critical of the skyscraper and what it symbolizes. Architectural historian Dolores Hayden (1977), for example, criticizes the social and physical disruption caused by the intervention of the tall building into the context of the city. She notes that a careful history of the skyscraper (Hayden, 1977):

> ...reveals a century of struggles and protests against the tendency to building ever higher. The builders' fantasies alternate with grim reality...there is no escape from the contradictions of the capitalist city; as an instrument for enhancing land values and corporate eminence, the skyscraper consumes human lives, lays waste to human settlements, and ultimately overpowers the urban economic activities which provided its original justification.

Hayden (1977) also points out that histories of skyscrapers often omit the casual attitudes toward safety conditions for construction workers. Ironworkers endured the greatest risks. When the skyscraper race began, strong pressure by developers to simply get the building built often superseded safety concerns. As a result, many construction workers, affectionately known as "sky boys," fell to their deaths, all in the name of progress.

The hidden meanings and symbols behind tall buildings are extremely powerful. This phenomenon is not new; it can be traced back throughout history as citizens planned the location and design of tall buildings, as skylines began to take shape in cities across the world, and as cities began to compete with each other for the world's tallest skyscraper. As we will see later, tall buildings have often become the battlegrounds of different interest groups. As such, they serve not only as historical, social, psychological, and cultural symbols, but as political symbols as well. Few other building types can claim such significance.

## 6.3 VIEWS OF TALL BUILDINGS AND URBAN SKYLINES

Tall buildings can be viewed both from afar and from up close. Individuals' attitudes toward them differ with each of these perspectives. This section describes how people view skylines, or collections of tall buildings from afar, how skylines have become civic emblems, and how tall buildings form images and imprints in people's minds. It also discusses how people form attitudes about tall buildings and the role of the mass media in shaping these popular images. Finally, it concludes with a discussion about why some people actually fear tall buildings. For more information on the factors involved in urban design planning with regard to tall buildings see Chapter 3 of this Monograph.

### 1 The Skyline as a Civic Emblem

Attoe (1981) argues that skylines have become "the chief symbol of an urban collective...civic emblems." People view the skyline as a symbol of their city, their region, and, sometimes, their way of life. As Attoe points out, in some cities a single feature dominates the skyline and encapsulates the image of the city as a whole.

The Gateway Arch in St. Louis, Missouri, for instance, dominates the skyline from virtually every angle, an extremely powerful symbol of "the gateway to the west."

Traveling westward across the Mississippi River along U.S. Interstate 70, the driver is overwhelmed by the colossal arch in the distance that appears to touch the sky. From the air, as the airplanes jockey for position to make their final descent into Lambert Field, St. Louis' airport, passengers view the arch from many different angles. When seen head on, the arch frames the city skyline perfectly, much as it does in typical postcard shots. When observed from the side, all that can be seen of the arch is a shining spire. The vistas of the arch are striking and dramatic.

The Eiffel Tower in Paris evokes a comparable feeling, whether it be viewed from the air, from a car, on foot, or from a boat along the Seine River. So do Athens' Parthenon and the church at Lycabetos, the two tallest buildings perched high on mountaintops, and the most visible landmarks in a city otherwise lost in a sea of midrise buildings.

The skylines of many of the world's cities are especially stunning when approached from the air. (See Chapter 3 for skyline studies for Philadelphia, Pennsylvania.) First-time air travelers to Hong Kong, Singapore, Chicago, and New York City are often struck by the sheer number of skyscrapers in these cities, parading one after another in rapid succession.

Once in the city, the closest approximation to an aerial view is atop a high-rise observation deck. According to Attoe (1981), these observation decks celebrate humans' ascendance over the sky and the surrounding terrain. He suggests that "being up there distances us from conventional perceptions of the world, from recurrent patterns of thought, and from old habits." Such vista points allow us to experience cities in a new way, offering a unique perspective on the world below.

Similarly, private spaces atop high rises offer insiders a secret look at the unfolding panorama below. Conference rooms, private offices, and clubs are often located atop towers. They allow privileged executives to overlook the cities in which their enterprises dominate (Hayden, 1977).

In addition to being shaped by aerial views and vistas from observation decks or private lookout points, people's images of skylines are also shaped by postcards. In this regard, the lenses of photographers can provide a strong influence on how the public forms images of cities and their skylines. In Chicago, for example, postcards shot from the east, looking west toward the city from Lake Michigan, yield a completely different image from those shot from the west, looking east toward the lake. The first type of postcard highlights the colorful blue-green water, the city's sandy beaches, and its prominent highway, Lake Shore Drive. By contrast, the second type of card shows a massive collection of high rises, often without a view of the lake at all. Postcards shot from the south looking north toward Lake Shore Drive and the Magnificent Mile area form an entirely different vista from those looking south toward the Loop. The orientation of postcards also shapes people's images of the skyline. Vertical shots often create an entirely different view from horizontal shots.

Yet another perspective on city skylines is from highways and major streets. Appleyard et al. (1964) argue that highway views can provide meaningful information to drivers about how the city is organized, what it symbolizes, how people use it, and how it relates to them. This information varies depending upon whether or not those in the car are drivers or passengers, tourists or commuters. In each case, however, scenery makes both a dynamic and a static impression. Elements are perceived differently when they are being passed at high speed and when they are viewed as static pieces of architecture from a great distance. The authors suggest that urban designers take into account how their city is viewed from a distance and that they emphasize desired landmarks accordingly (Appleyard et al., 1964). Landmarks provide a means for drivers to orient themselves in the city, by serving as goals that are approached, attained, and finally passed. Ideally, if the goal is not always in view, it should reappear consistently; when it is not in view, passengers experience excitement, anticipation, and mystery along the trip (Appleyard et al., 1964). Tall buildings, tunnels, overhead signs, and other elements help shape the way in which people in cars experience the city. It is often easier to find one's way toward a particular tall building as a driver than as a pedestrian. On foot, the scale of tall buildings can be overwhelming, as one high rise blocks the view of the other.

Characteristics of skylines are crucial in helping orient newcomers to a city, just as centrally located clock towers on university campuses and observation towers at worlds' fair grounds help visitors to these places find their way. In some cities, tall natural features serve this function as well. For example, the view of Grouse Mountain from Vancouver, British Columbia, or Mount Tamalpais from San Francisco, California, helps visitors mark their place. In other cities, tall monuments provide welcome orientation points. Examples include Toronto's Canadian National (CN) Tower and the Space Needle in Seattle, Washington. They form one of the first and longest lasting impressions of a city. A number of researchers have studied how people form visual images or mental maps of cityscapes. The strength of the visual image is often referred to as its "imageability."

Some recent research by Thomas Bever (cited in Blakeslee, 1992), a psychology and linguistics professor and expert on how biological variables influence behavior, suggests some fascinating gender differences in how people navigate through their environment. Both human and animal experiments appear to indicate that females tend to rely mostly on specific landmarks when they move through space. By contrast, men tend to rely on a more primitive sense of motion, using vectors or paths they remember. Bever stressed that these differences are not absolute, and that many men adopt the female strategy and vice versa. Also, both men and women travel from one point to another just as efficiently, and neither men nor women get lost more often (Blakeslee, 1992). While Bever's research was conducted in a psychological laboratory—a 15.24- by 9.1-m (50- by 30-ft) figure-eight maze using hallways in the basement of the psychology building—it would be fascinating to test this theory in an actual urban landscape. Could it be that tall buildings are more useful landmarks for women than they are for men?

## 2 The Imageability of Tall Buildings

The imageability of tall buildings has long been a chief concern of designers, planners, developers, users, and others. Architects have often taken great pains to design tall buildings that stand apart from the crowd. To those who work or live in tall buildings, those who visit tall buildings occasionally, those who pass by them regularly, and those who simply see them from afar, the building's image is crucial.

At the urban design scale, imageability means the extent to which a particular tall building contributes to the overall impression of a neighborhood, district, or the city as a whole. The concept of imageability was introduced by Lynch (1960, 1984). By studying the visual images of 60 individuals from Boston, Massachusetts; Jersey City, New Jersey; and Los Angeles, California, Lynch was able to identify five key components that described how people viewed cities: landmarks, edges, paths, nodes, and districts. Tall buildings often serve as landmarks and orientation points to the rest of the city. The imageability of cities has sparked the interest not only of researchers in environment and behavior, but also of environmental designers, urban geographers, urban sociologists, journalists, and others.

Building upon the work by Lynch, Appleyard (1969, 1976) developed a model of landmark form and urban cognition in order to understand how people perceive buildings and cities. Appleyard tested his conceptual model with residents of Cuidad Guyana, Venezuela. His model was later retested by Evans et al. (1982) with 72 adults in Orange, California. The study by Evans et al. (1982) generally confirmed and refined Appleyard's model. The researchers found that the following building characteristics were significantly related to recall (Evans et al., 1982):

1. *Movement.* The amount of persons and other objects moving in and around the building
2. *Contour.* The clarity of building contour, ranging from blurred, partially obscured to free-standing
3. *Shape.* The complexity of shape, ranging from simple block shape to more complex multiple shapes
4. *Use intensity.* The extent of building use, that is, from limited use by a small segment of the population to daily use by large numbers of people
5. *Use singularity.* The uniqueness of building function, ranging from only one function to many buildings with shared functions
6. *Significance.* The extent of cultural, political, aesthetic, or historical importance of the building
7. *Quality.* The amount of physical maintenance, the upkeep of the structure

Although neither the studies by Appleyard (1969; 1976) nor those studies by Evans et al. (1982) deal exclusively with tall buildings, their findings offer insights about how all buildings, including tall buildings, are remembered. Based on this research, then, one might logically infer that buildings such as Hong Kong's Bank of China (Fig. 6.4) and the Hong Kong and Shanghai Bank (Fig. 6.5), along with the Bank of Asia in Bangkok, Thailand, San Francisco's Transamerica Pyramid, and New York's Citicorp Building are among today's most imageable, easily remembered buildings. For example, because of their sheer size, each of these buildings generates a great deal of pedestrian and vehicular movement, resulting in a high amount of use intensity.

Regarding Evans et al.'s (1982) concept of building significance, it is interesting to note that the developer of the Hong Kong and Shanghai Bank commissioned a Feng Shui analysis to be performed for the building during its construction. Feng Shui is an ancient Chinese practice that is used to determine the best location, orientation, and other qualities needed to satisfy the "spirits," and to make the building and its activities successful. The writings of architectural critic Charles Jencks (1980) are worth noting here, as he provides an interesting conceptual framework from which to view the concept of shape in studying tall buildings. He notes three shapes in particular: skyprickers or skyneedles, skyscrapers, and skycities. Skyprickers or skyneedles have centralized plans and masses and include buildings such as New York's Chrysler Building, the Empire State Building, or Chicago's John Hancock Center. Skyscrapers feature longitudinal, rectangular masses and plans. Examples include Boston's John Hancock Building and Minneapolis' IDS Building. Skycities are compound buildings consisting of a combination of mounted towers, a slab and tower, twin towers, or clustered towers and include buildings such as Chicago's Sears Tower and New York's Rockefeller Center.

One can speculate that people probably perceive skyprickers or skyneedles, skyscrapers, and skycities in different ways. Factors likely to influence these perceptions might include the urban context in which these buildings are located, whether they are one-of-a-kind buildings, whether or not others like them are nearby, and whether or not they feature any major public access, such as an observation deck, at the top.

**Fig. 6.4 Due to its height and unusual shape, the Bank of China is one of the most memorable skyscrapers in Hong Kong.** (*Courtesy: Grace Huang.*)

## 3 Public Images of Tall Buildings

Public images of high rises are highly influenced by the mass media. For instance, the 1974 film "The Towering Inferno," which portrayed people trapped in a burning high-rise office tower, evoked a strong feeling of fear and danger toward tall buildings. But film is only one of several media that affects public perceptions of high rises.

**Fig. 6.5 The Hong Kong and Shanghai Bank is another one of Hong Kong's most imageable tall buildings.** (*Courtesy: Grace Huang.*)

News reporters have captured terrifying scenes of fires in high-rise hotels. Two of the deadliest high-rise hotel fires in American history were captured in vivid detail by the news media. On November 21, 1980, a fire in the MGM Grand Hotel in Las Vegas, Nevada, killed 85 people and injured over 700 (Block, 1985). On New Year's Eve, December 31, 1986, a fire in the Dupont Plaza Hotel along the San Juan, Puerto Rico, beachfront killed 96 people and left over 100 injured. The latter fire was set by arsonists in a dispute between the local teamsters union and hotel security guards trying to break the union's strike (Hackett and Gonzalez, 1987). Among the high-impact news magazine headlines used to describe the San Juan tragedy were "San Juan's Towering Inferno" (Hackett and Gonzalez, 1987), "Death in a Towering Inferno" (Levin et al., 1987), and "Hotel Fire: Cage of Horror" (Hurt, 1988).

Public fear of high rises is also heightened after major natural disasters, especially earthquakes. The October 1989 earthquake in Northern California produced relatively little damage to San Francisco's new high-rise office towers built in compliance with the latest building codes. Nonetheless, some occupants of older housing in San Francisco's marina district suffered major injuries when several apartment buildings crumbled. After the October 1992 earthquake in Cairo, Egypt, the public was bombarded with explicit newspaper photographs and television images of residents crushed to death when their poorly constructed high-rise apartment buildings were leveled.

Although high rises, at their worst, have been associated with death and destruction, they have been glamorized by the media as well. The luxury of high-rise or penthouse apartments and the thrill of skyscraper offices with commanding city views are among the positive images promoted in countless films and television shows. Occupants of these spaces are often shown as having made it financially, and their homes and workplaces serve as highly visible status symbols. Filmmakers like Woody Allen have sometimes used striking images of New York skyscrapers in their opening scenes, revealing their personal identification with the New York skyline.

One of the few pieces of empirical research to address the public's general attitudes toward tall buildings was conducted by Haber (1977). Her study of 300 college students involved a 22-item survey where respondents were asked what they liked and disliked about tall buildings. Results indicate that the most disliked aspect of tall buildings was the elevator; 62% of the respondents said they hated waiting for elevators. Interestingly, the lower their income, the greater the percentage of respondents who detested elevators. Most likely this is due to the kinds of elevator service people of different incomes encounter. Low-income people may all too often be plagued by elevators that are temporarily out of service or broken. No doubt that low-income residents are also the most likely to fear criminal attacks in or near elevators.

Haber's findings suggest a relationship between individual perceptions of what constitutes a tall building and how well people like height and power, how much they fear fire, how concerned they are with spoiled skylines, and how safe they feel in tall buildings. Perhaps in response to their fears, the largest group of respondents (40%) reported that they wanted to live or work in the lower third of the tall building (Haber, 1977). One can speculate that from this location they can escape most quickly.

Whereas Haber examined the attitudes of students, a number of other researchers have examined how professional architects and lay persons perceive particular buildings. Not surprisingly, a number of studies have demonstrated that architects and nonarchitects perceive buildings differently. [See especially the work of Groat (1982, 1984) and Groat and Canter (1979).] Although these studies did not address tall buildings, their findings are intriguing and could well apply to how architects and nonarchitects might perceive post-modern high rises.

For instance, Devlin (1990) examined reactions to the State of Illinois Center and 333 Wacker Drive, two major high-rise buildings in the Chicago Loop. Research methods included a content analysis of architectural reviews in leading architectural magazines, as well as 10–15-minute unstructured interviews with 40 nonarchitects (20 users, that is, employees who worked in the office building, and 20 viewers, that is, persons who were within viewing distance of the building). She discovered that compared with nonarchitects, architects used more categories to describe these buildings. Most architects responded to *aesthetic* qualities, interpreting the buildings in terms of conceptual issues. Architects discussed issues of form, style, historic significance, design approach, and design quality. Some categories, such as historic significance, design quality, anthropomorphism, and form and function, were used exclusively by architects. By contrast, most nonarchitects responded to their *preferences* about the building, providing more affective responses about whether or not they liked it. Nonarchitects also gave more descriptive responses about the physical features of the building. Devlin suggests that architects must understand that they do not always look at a building in the same way as do nonarchitects (Devlin, 1990). Ideally, successful building designs can reflect perspectives of both the design professional and the lay person.

Another study by Stamps (1991) produced some fascinating results. Fifty-eight participants were shown 15 slides of high-rise buildings from a major American city. Five slides showed old brick buildings, five presented plain buildings, and five pictured complex modern buildings. The latter two were governed by two distinctly different generations of design codes for high-rise buildings. Respondents were asked to rank-order the slides based upon their personal preferences.

Stamps' (1991) results indicate that the complex modern style was most preferred, the older brick style was second, and the plain style was third. When the data were broken down into demographic subpopulations (by age, gender, income, level of education, and political preference), they revealed that each demographic group preferred the complex modern buildings. Interestingly, compared to political liberals, political conservatives preferred the plain buildings over the older buildings.

High-rise buildings are often the targets of heated controversy, evoking strong public reactions (Appleyard, 1979). Such controversies cannot help but influence the attitude of average person on the street toward tall buildings. One city where public antagonism toward high rises has reached an all-time high is San Francisco. A study by Kreimer (1973) provided a content analysis of 880 newspaper articles in the *San Francisco Chronicle* over a 1½-year period, focusing on the proposed U.S. Steel building. The building was to be a 167.6-m (550-ft)-tall high rise along San Francisco's waterfront. In 1971 the Board of Supervisors voted a 53.3-m (175-ft) maximum height limit on that portion of the waterfront and the project was defeated. Newspaper ads sponsored by civic opposition groups fueled the public fire with headlines like "Stop them from burying our city under a skyline of tombstones" (Appleyard with Fishman, 1977). More than 10,000 newspaper readers clipped coupons from that ad and sent them in to City Hall (Reinhardt, 1971).

Kreimer (1973) identified five themes in these articles that helped explain different viewpoints on the controversy:

1. *Political environment.* Individuals in favor of and opposed to the building, along with arbitrating bodies
2. *Strategic game.* Actions involving inputs, processing, and outputs
3. *Economic environment.* Including positive and negative consequences of the building for the city
4. *Physical environment.* Geographic aspects; valued elements such as views, height, access to water, open space, scale, color, variety; and positive and nega-

tive precedents from other cities, that is, negative examples from New York (fear of "manhattanization"), Waikiki Beach ("Chinese wall"), and such

5. *Axiological environment.* Human (quality of life, emotion, human rights, aesthetics), civic (history, tradition, future), and urban values (deterioration, conservation, uglification, beautification, naturalness, imageability, visibility)

In San Francisco height is not the only controversial issue when it comes to skyscrapers. For decades, whites and light pastel colors dominated San Francisco's skyline, and when new black, gray, or other dark-colored buildings were proposed, citizens sensed a hostile invasion of "out-of-town" values (Kreimer, 1973; Appleyard and Fishman, 1977). The dark brown Bank of America building (1965–1966), the city's tallest skyscraper at 52 stories high, generated much civic discontent as a result of not only its height but its color scheme. Both were deemed inappropriate within the surrounding cityscape. Building shape, too, has fired up local San Franciscans. The pyramidal Transamerica Building (1969–1971) was also strongly criticized for its highly unconventional shape (Appleyard and Fishman, 1977).

The high-rise controversy in San Francisco is chronicled in detail by Brugmann and Sletteland (1971). The book by these authors opens with a telling 1971 quote from San Francisco columnist, Herb Caen: "When you look at photos of the skyline of 1957 and compare it with today's it is hard to believe you are looking at the same city, which of course you aren't. The old city grew beautiful by accident, the new one is growing ugly by design."

Appleyard and Fishman (1977) identify different kinds of groups impacted by a tall office building: *users* (executives, office personnel, visitors, maintenance workers, merchants, management, and owners), *neighbors* (neighboring building owners, office workers, neighboring households, and street public), and the *public* (local citizens, their representatives, public agencies, interest groups, and tourists). They point out that, ironically, "most of the people impacted by high-rise buildings have no say in their construction." Furthermore, we know little about how people are affected by them. The public perceives a variety of visual and symbolic impacts of high-rise buildings, including scale (relative height, relative length, and bulk); color and tone; shape; plaza and street images; view blockage; symbolic disruption, prominence, and visibility; reducing the uniqueness of the city; and the high rise as a political symbol.

## 4 Fears and Phobias about Tall Buildings

Depending upon their background, psychological disposition, exposure, and a myriad of other factors, some individuals are predisposed positively or negatively toward tall buildings (Fig. 6.6). While most individuals adjust readily toward tall buildings, some people do not. Persons with certain kinds of phobias, especially *agoraphobia, acrophobia,* and *batophobia,* are often troubled by high rises.

Agoraphobia is an extremely complex type of fear, which can range from fear of open spaces to fear of going out one's front door. Persons suffering from agoraphobia are sometimes affected by visual features of the physical environment. "Usually the wider and higher the space walked in, the greater the fear" (Marks, 1969).

Acrophobia, fear of high places, is classified as a simple phobia, meaning that its symptoms are not usually associated with other psychiatric disorders. What specifically triggers acrophobia is exposure to heights. Other than that, these individuals are just like everyone else. Symptoms of acrophobia include high levels of fear and discomfort. In some cases, symptoms resemble those of a panic attack and include palpitations, sweating, dizziness, and difficulty breathing. Sometimes just the thought of being in an uncomfortable situation can even bring on these symptoms. Fear of heights is sometimes related to fear of airplanes and flying. It can also be related to other activities, such as bicycles, skiing, amusement rides, bridges, and freeways (Doctor and Kahn, 1989). Obviously, tall buildings can be among the worst enemies of acrophobics.

Fear of heights is also known as *altophobia, hypsophobia,* and *hypsiphobia.* Experts find that this type of phobia is common, especially in mild forms (Doctor and Kahn, 1989). Patients with acrophobia emphasize that their visual space is especially important (Doctor and Kahn, 1989):

> They will not be able to go down a flight of stairs if they can see the open stairwell. They will be frightened looking out of a high window that stretches from floor to ceiling, but not if the window's bottom is at or covered to waist level or higher. They have difficulty crossing bridges on foot because of the proximity of the edge but may be able to cross in a car.

**Fig. 6.6 Some individuals may be predisposed positively or negatively toward tall buildings. Here is an Indian student's drawing of his native Bombay, showing high rises overpowering their smaller neighbors.** (*Courtesy: Abhijeet Chavan.*)

Batophobia is the specific fear of high objects, such as tall buildings. Batophobia may be related to acrophobia—the fear of high places, or to *anablepophobia*—the fear of looking up at high places. Individuals who suffer from these types of fears may often avoid reaching for items from high shelves, as in a grocery store. They also fear climbing ladders (Doctor and Kahn, 1989).

Possible causes of such phobias remain a mystery. Among the explanations are *psychoanalytic theories, conditioned reflex theories,* and *biological theories.* In a nutshell, some psychoanalytic theories attribute the phobia to ego defense mechanisms, specifically displacement and avoidance, against incestuous oedipal genital drives, castration anxiety, or separation anxiety. Such theories are based on the original works of Sigmund Freud around the turn of the century and the early 1900s. Conditioned reflex theories were grounded in the work of psychologists John B. Watson and Rosalie Rayner in the 1920s. These theories contend that phobias are learned behaviors, and that they are formed by either classical or operant conditioning. Biological theories contend that an innate biological alarm is triggered by the phobia, and that physiological factors are largely responsible. For a more detailed discussion of the possible origins of phobic disorders, consult Nemiah (1985) and the *Diagnostic and Statistical Manual of Mental Disorders* (1987).

## 5 Terrorism in Tall Buildings

At 12:18 P.M. on Friday, February 26, 1993, a dramatic new issue abruptly made its way to center stage. New York City's World Trade Center, the second tallest building in the world, was struck by a powerful explosion. The bomb ripped a huge crater in the buildings' parking garage and blasted a massive hole in the wall above the PATH train station, collapsing the station's concrete ceiling and damaging the Vista Hotel. The colossal crater measured 61 by 30.5 m (200 by 100 ft) by five stories deep. Billowing streams of smoke were instantly streaming all the way up the two 110-story twin towers. Six people were killed and over 1000 injured, most from smoke inhalation as they frantically made their way down seemingly endless staircases. The blast knocked out the vast complex's emergency generator system, plunging occupants into total darkness—and terror.

Four hours after the World Trade Center blast, a bomb threat forced the evacuation of New York City's Empire State Building. No bomb was found, however. Several other bomb threats were reported throughout the city. Shortly afterward security forces in tall office buildings in other major U.S. cities were put on the alert.

The World Trade Center towers were closed for about a month while the investigation continued and the buildings underwent major structural repairs. More than 350 businesses, government agencies, and other organizations were affected in one of New York City's most costly financial disruptions. Damage estimates were as high as $1.1 billion (Prokesch and Meier, 1993).

The catastrophe prompted much discussion about the safety of tall buildings worldwide from terrorist attacks. The security of underground parking garages is a key issue. How can such public parking facilities be made terrorist-proof? Until such preventive measures are in place, should they continue to remain in operation—and possibly place the lives of all those living or working above them in jeopardy? The capacity of emergency generator systems to function in such a powerful blast is another focus of discussion. What good is an emergency system if it fails to operate during an emergency? Is there a way to design command and control centers to withstand such powerful explosions?

A myriad of issues previously unexplored were raised by this disaster. More likely than not, they will continue to plague all those involved in planning, designing, constructing, and operating tall buildings for many years to come. It is likely that a new wave of scholarly research on this topic will emerge.

Related to the physical dimensions of securing tall buildings from similar attacks are the powerful social and psychological issues sparked by this event. The World Trade Center catastrophe dominated front-page headlines and television screens for weeks.

Images of soot-stained office workers gasping for breath under oxygen masks, awesome photographs of the vast crater caused by the blast, and horrific stories of survivors remain firmly etched in the public memory. A class of 17 Brooklyn kindergartners and five adults, on an annual field trip, were trapped in an elevator for nearly 7 hours. Another group of 14 children and their teacher were trapped on the chilly roof for 3 hours, pelted by snow and wind. A pregnant woman was airlifted by helicopter from the roof. Two friends carried a woman in a wheelchair down 66 stories. Such powerful images exacerbate public fear of tall buildings and underscore the need for emergency preparedness. (For vivid accounts of the World Trade Center disaster, consult *Time,* 1993, and *Newsweek,* 1993.)

Tall buildings and skylines raise a host of psychological issues. How skylines are perceived, how they come to represent the city as a whole, from what viewpoints they are presented, how easy or difficult they are to remember, the extent to which one associates them with mass media images, whether positive or negative, and the fears or phobias that plague some individuals, all contribute to the ways in which people experience tall buildings and skylines. The World Trade Center bombing ushered in an urgent concern for public safety and protection from terrorism in tall buildings.

## 6.4 IMPACT OF CLIMATE AND GEOLOGICAL EVENTS ON HUMAN BEHAVIOR

A substantial amount of research has been conducted documenting the impacts of tall buildings on the microclimate of streets and sidewalks and, conversely, the impacts of climatic and geological events on tall buildings. However, relatively little research has examined how these events affect the building occupants themselves. This section provides a glimpse into these issues.

### 1 Climatic Impacts on Streets and Sidewalks

Tall buildings can have a dramatic effect on the microclimate of the streets and sidewalks below them. They have a direct impact on the amount of shadows, sun, and wind velocities that pedestrians and drivers experience. Consequently, they have the potential to either enhance or detract from the urban experience. Once a tall building is constructed, light breezes at street level may turn into dangerous gusts of wind that can become unbearable on a cold winter day. Wind velocities can increase three to four times their normal amount due to the canyonlike effect of tall buildings lining both sides of the street.

Warm weather breezes can be affected as well. In Durban, South Africa, for example, a wall of buildings along the coastline cut off breezes that used to cool and dehumidify that part of the city (CTBUH, 1978). Ideally, building designs would enhance wind on hot, humid days, while inhibiting wind on cold, bitter days; achieving this is not easy, however.

Peter Bosselman and his colleagues at the University of California's Environmental Simulation Laboratory in Berkeley have been studying the effects of tall buildings on parks, plazas, squares, and streetscapes in downtown San Francisco (Bosselman et al., 1983b). For instance, in one study using superimposed sun paths from a proposed high-rise building, the number of hours of sunlight lost on a nearby playground in Chinatown was calculated. The playground was divided into six play areas, each receiving an average of 5 hours of sunlight per day. If the high rise were actually built, the amount of sunlight would have been reduced to 1.25 hours per day in five of these play areas. If other buildings surrounding the playground were built to the maximum allowable height, the playground would not receive any sunlight at al. This study reveals the different sets of values that underlie the tall building process. It is doubtful that children playing outside were a prime consideration when the new building was proposed. But it is only when zoning regulations incorporate issues like these that all nearby users in small, public open spaces will be protected.

Bosselman et al.'s (1983b) research in both the retail district and downtown parks, plazas, and squares examined the effects of tall buildings on several variables, such as:

1. *Thermal comfort.* The product of temperature, humidity, direct sunlight, and wind (see Chapter 9 for a discussion of this topic from an environmental technology point of view).
2. *Sun access criteria.* Accounting for the daily and seasonal movement of the sun and ensuring that streets receive sunlight.
3. *Wind mitigation.* Ways to alleviate strong winds, especially at the base of very tall buildings. This is especially a problem if a high-rise building is out of scale with the rest of the street. Landscape screening can sometimes be effective in reducing the effects of strong winds.

Bosselman's research has been incorporated into San Francisco's revised urban design guidelines. Based upon the findings from his research, different kinds of guidelines were prescribed for wall and tower heights for streets oriented east-west and north-south; sun access criteria have been made mandatory in San Francisco's zoning laws.

Energy expert Knowles (1981) suggests that zoning laws be revised to incorporate solar envelopes. Designers must consider not only the size and shape of the building, but also the influence of the changing position of the sun during each day and across all four seasons. Cut-off times, that is, hours of the day during which shadows are cast on the streets and open spaces below, would thus be reduced. Selective shadowing, that is, by carefully positioning buildings such that lower buildings are on one side while taller buildings are on another, can help as well.

While solar issues are important in the design of any urban environment, they are even more pressing in cities where heat and cold are exceptionally intense. For example, in cities such as Athens, Greece; Cairo, Egypt; and Phoenix, Arizona, where the blazing summer sun makes walking on city sidewalks a strain, building for summer shade is a must. In others, such as Oslo, Norway; St. Petersburg, Russia; and Minneapolis, Minnesota, with unusually harsh winters, building for winter sun is needed.

Whether or not their cities are designed with solar considerations in mind, people engage in adaptive behavior. Observations of pedestrian behavior bear this out. During days when the weather is extreme, pedestrians usually travel on the more comfortable side of the street, even if it means going out of their way. On exceptionally hot days, most pedestrians will walk on the shady side of the street. On unusually bitter cold days, they will walk on the sunny side.

The climatic impacts of tall buildings are affected by the ways in which city streets are laid out (Knowles, 1981). Most urban areas in the United States are laid out on the traditional Jeffersonian grid, with streets running east-west and north-south. In

Jeffersonian grid cities during the winter, streets that run east-west are shadowed by tall buildings most of the day, making them dark and cold. By contrast, north-south streets are lit by the sun during midday. In summer, east-west streets are continuously lit by the sun, whereas north-south streets are lit for only part of the morning and afternoon. In this respect, tall buildings can exacerbate negative climatic conditions, making streets and sidewalks colder in the winter and warmer in the summer. High rises can create a hostile climatic environment for pedestrians.

Chicago, for example, is a city laid out in the traditional Jeffersonian grid pattern. In the Loop, the city's central downtown core, not only is there an abundance of tall buildings along both north-south and east-west streets, but some streets are even further shaded by the elevated train line. On chilly, gray winter days the Loop seems dreadfully cold, dark, and dank. By contrast, north of the Loop is Michigan Avenue, the city's major shopping thoroughfare. It is an extremely wide street running north-south, with at least six lanes of traffic plus left turn lanes. It also features wide sidewalks. As a result, even in winter, Michigan Avenue is overall a much lighter, brighter street than most of those in the nearby Chicago Loop.

Planners and developers who have the luxury of designing entirely new street systems which include tall buildings would do well to use the Spanish grid—the Jeffersonian grid rotated 45°. With this orientation, the impacts of tall buildings on sunlight are much more moderate. A combination of heat and light in the winter, and some shadows on most streets in the summer render sidewalks much more comfortable (Knowles, 1981).

In some cities, a new breed of skyscrapers is addressing the problem of pedestrian sun access. Tall, thin buildings generally block less light and air than their bulkier counterparts. In New York City, for example, 712 Fifth Avenue is the first tower built in that part of Manhattan in accordance with new zoning restrictions dealing with this issue (Russell, 1990).

Tall buildings can have an enormous impact on streets and sidewalks, largely due to the ways they affect sun and wind. As such, they strongly influence pedestrian comfort, both psychologically and physiologically.

## 2 Impacts of Climatic and Geological Events on Occupants

Climatic and geological events affect tall buildings and their occupants. Several architects have made significant advances in designing tall buildings that are especially responsive to harsh climates, thus increasing the comfort of people inside.

One of these architects is Ken Yeang, who helped pioneer an architectural style of tropical urban regionalism for tall buildings in his native Kuala Lumpur, Malaysia (Powell, 1989; Sharp, 1989). Yeang's innovative style is sometimes referred to as the "tropical skyscraper." Yeang's high-rise buildings have been especially designed to enhance human comfort in hot humid climates. Among his key high-rise design principles are to provide (Yeang, 1990):

1. Natural ventilation and sunlight to service cores
2. Building orientation so that windows face north and south whenever possible
3. Deep recesses in the external walls of the hottest sides of the building, thus allowing for balconies or "sky courts" on the upper floors
4. Atria, air spaces and wind scoops
5. Operable windows in the form of full-wall openings on both windward and leeward sides whenever possible

6. Open-air ground floor to act as a transition between the hot outdoors and the cooler indoors
7. Vertical landscaping to the face of the tall building or integrated with its balconies and inner sky courts to help minimize heat reflection and glare

One of his projects, the Weld Atrium in Kuala Lumpur, features a roof with a complex web of devices to regulate climate, even though the interior of the atrium is air-conditioned. The pergola at the Weld Atrium filters out the harsh effects of the midday sun, thereby reducing solar heat gain and lessening the load on the air-conditioning plant. Unlike the atria developed by John Portman at the Hyatt Regency hotels, Yeang's atrium allows visual contact with the sky, establishing a closer relationship with the outdoor climate (Powell, 1989). Yeang's style represents a unique blend of Western and Eastern values.

In addition to hot, humid weather, two other climatic and geological events, although rare, appear to have some of the greatest impacts on people who live and work in tall buildings: strong winds and earthquakes. Even though contemporary buildings are designed to accommodate these natural phenomena, nonetheless, in some cases, occupants can still sense building motion and sway (Conway, 1977).

A large body of technical research addresses these issues, and they will not be discussed here. Instead, our focus will be on one of the few studies that examines the effects of earthquake motion on building occupants. This section is based largely upon the fascinating study by Arnold (1986), which examines how people experience earthquakes in tall buildings.

A key point to understanding human behavior during earthquakes is that even during a moderate quake, most tall building occupants simply do not understand what is happening. If they live in a geological zone where earthquakes are common, they will probably recognize the symptoms, but they will not know what exact sensations to expect. The confusion that people experience during earthquakes can often exacerbate the psychological impacts of shaking.

Furthermore, as Arnold stresses, every earthquake is different. Different quakes and different ground conditions produce different kinds of ground motions. The distance from the epicenter will also impact the severity with which the quake will be felt. Different buildings respond in various ways to certain ground motions, which typically are amplified as they move through a building. Individuals on different floors of the same building will experience a quake quite differently. Thus it is truly impossible to prepare individuals for the exact sensations they will experience in any given quake. One of the most useful analogies to help people prepare for a moderate earthquake is that of riding a train. In a train, people are used to motions, vibrations, and shaking, but in a building they are not.

Arnold examined how workers experienced a 1979 earthquake at the Imperial County Services building, a six-story reinforced concrete building severely damaged in the 1979 Imperial Valley, California, earthquake. At the time of the quake, over 100 people were in the building. Building occupants were asked to estimate how long the shaking lasted. "Responses ranged from one second to 30 minutes, and a dozen people had no idea or said 'forever,' or 'long enough.' However, the three most frequent responses were 30 seconds (15%), 1 minute (14%), and 20 seconds (8%)."

The actual duration of strong shaking, as recorded from seismological data instruments, was only about 4.5 seconds at ground level outside the building. But inside the building, the instruments recorded shaking for a considerably longer period. The length of shaking varied, depending upon the location of the instruments within the building. In most instances, the higher up in the building, the longer the period of shaking. The shaking from north to south was also different from the shaking from east to west.

The study compared the actual vertical location of various respondents at the time of the quake with their responses about the perceived length of shaking. The following conclusion was reached (Arnold, 1986):

> No collaborating pattern was found to exist. For example, the majority of second-floor occupants (32%) said the strong shaking was in the four-minutes or longer time period, whereas on the sixth floor the majority of occupants (39%) felt that strong shaking lasted in the 11–30 second range. Clearly a wide range of subjectivity exists; what is important is that human perception differs markedly from the physical "facts."

The relationships among tall buildings, climate, geology, and human behavior are highly complex. Much more research is needed in this area. The study by Arnold must be replicated in many other tall buildings. We need more information about how people in different parts of the same tall building experience earthquakes.

In addition to the need for quantitative information about how long people think earthquakes last, we also need more qualitative information about what is running through their minds during earthquakes and what they do about it. In a strong quake, does everyone duck below a desk or run to a doorway? What happens when a group of people charge for the same doorway at the same time? How many injuries can be avoided by running away from tall bookshelves, file cabinets, and otherwise lethal office furniture? How do people express their fears? Do they scream, cry, laugh, or panic? What kinds of domino effects might such behaviors have on others? In light of several devastating earthquakes in recent years, and undoubtedly more to come, it is especially important to understand better the relationships among building performance, ground motion, and human perception and behavior.

## 6.5 BUILDING ENTRIES AND LOBBIES

### 1 The Stylistic Centerpieces

The entry to a tall building—the approach from the outside, the architectural detailing around the immediate exterior and interior of the opening, and the initial approach to the lobby inside—is often laden with symbolic meaning. It is here that designers and developers often leave their most visible personal stamps or trademarks. Even at such a large scale, the building entry assumes the character of a front door to a home, the first clue as to what to expect inside.

As people approach a building, they ask the following types of questions: "What is it?" "What benefits does it offer me?" "How do I get in?" "What is inside?" "How will I be received?" (Deasy, 1985). The entry design should reflect a careful consideration of these issues.

Especially during the recent post-modern architectural era, the entry has acquired even greater symbolic and psychological importance. During the 1980s and 1990s, the entry and lobby have often become the stylistic centerpieces of high-rise buildings. Note the following sharp criticism of the AT&T Building in New York City, completed in 1985 (Kozloff, 1990): "This building's real character is announced in the Herculean stone arches of its lobby entrance, which belittle and intimidate visitors. It is as if Albert Speer were bearing down upon you by way of Cheops, and no frippery on top can alleviate the authoritarianism of the design."

This section addresses how people experience their initial contact with tall buildings, that is, how users perceive entries, building lobbies, and atrium spaces. It also discusses how people find their way around in tall buildings. Additional views on aesthetics and user perceptions of tall buildings can be found in Chapter 5.

## 2 Users' Perceptions of Building Entries

Few studies have examined the psychological dimensions of building entries, and to date any research addressing this issue specifically for tall buildings is not widely known. However, findings from an intriguing body of research (Bain, 1989) about the meaning of entry sequences in low-rise buildings may well relate to how people experience entries in tall buildings. Bain's research examined a conceptual model of a successful entry and the degree to which six variables contributed to it. Forty-seven participants were shown videotapes of 12 building entry sequences from buildings in Columbus, Indiana, designed by well-known architects. The six variables tested in Bain's conceptual model included

1. *A sense of place.* The distinction between inside and outside, and a successful transition from the path (the arrival) to the goal (the inside)
2. *Legibility.* Providing orientation to the building as well as to the surrounding area, using cues to help orient users
3. *Sequential art.* Providing an opportunity for a building to be experienced over time
4. *Mystery.* Requiring users to predict what is not yet in view, creating a sense of anticipation which can be, at best, excitement, or at worst, disorientation
5. *Sense of dignity.* A fit between the person and the environment, for example, when an entry accommodates the needs of disabled people in a manner similar to how able-bodied users are accommodated
6. *Functionality.* Ease of entry by locating doors conveniently, allowing them to be opened easily, and so on

Using multiple-regression analysis, Bain's research findings indicate that sense of place (a combination of sense of dignity, sequential art, and the original sense of place concept) and legibility are the chief predictors of a successful entry. Of these two variables, sense of place is more important. Results also indicate that legibility contributes strongly to a user's sense of place. Mystery and legibility are significantly correlated with each other; however, mystery alone does not contribute to a successful entry. Bain concludes (1989): "Rather than having a concern for the comfort and emotional well-being of those who use a space, it seems that making an architectural 'statement' is often more important. The user, as a result, is possibly being left lost, disoriented, and without a strong association with the environment that he or she must use."

Bain's research raises some interesting issues about the ways in which architects and pedestrians perceive entryways. A related issue is how pedestrians actually use entryways. In this regard, an issue that architects must consider carefully in the design of high-rise entryways is pedestrian traffic movement. In many tall buildings, traffic builds up at entrances, causing an overflow onto adjacent sidewalks as well as into the lobby space inside. Separating entrances from exits and providing adequate transitional space immediately before and after the entryway can help alleviate this problem.

This is related to how pedestrians use revolving doorways. People who use these every day tend to race through them very quickly. As a result, the revolving door spins around at high speed. This fast pace can be intimidating to first-time users, persons with children, people with physical disabilities, and the elderly. Those who move more slowly and with greater trepidation can have some close calls with revolving doors.

People are often bumping into each other or into the doors themselves. People with children often need to grab them by the hand and quickly escape the moving door, lest they be trapped inside. For many types of users, revolving doorways can be quite dangerous.

## 3 Users' Perceptions of Building Lobbies

Like entries, building lobbies also represent the designer's signature in a highly visible way. As a result, they too assume great psychological as well as symbolic importance. Lobbies are places where people arrange to meet each other. In a large lobby of a tall building, people are likely to identify highly visible landmarks so that other people can find them easily. For example, "I'll meet you next to the fountain" or "under the chandelier" or "in the middle of the atrium."

Such meetings usually involve waiting, since rarely do parties arrive at the same time. People need a place to sit while they wait. Unfortunately the design of far too many building lobbies neglects this need altogether. Consequently, people who are waiting for each other are forced to lean up against walls, pace back and forth, and stand around aimlessly. Ironically, due to managements' fears that vagrants, the homeless, and other so-called undesirables will occupy precious lobby space, and thus ruin the image of the building and its clientele, the needs of regular building users are often ignored.

One of the few studies to address psychological aspects of lobby spaces examined users' perceptions of the design of a large, unusually attractive greenhouselike lobby in a 48-story cylindrical San Francisco office tower, 101 California Street, built in the early 1980s (Anthony, 1985a; Fig. 6.7). Interviews with building users and observations

**Fig. 6.7 101 California St., San Francisco, features an extremely attractive lobby, but few places for people to sit.** (*Courtesy: Leigh McMillen.*)

of pedestrian behavior revealed some pressing problems. Little seating, if any, had been provided. Most people stayed in the atrium lobby space for under 2 minutes. Many searched for a seat and, not finding one, leaned against the glass walls for support or sat awkwardly on planters. The lobby's lack of seating and its "cold" quality were among its most disliked features. However, two-thirds of the users surveyed liked the lobby, mainly due to its light and airy quality and its unusual shape. The results of this survey indicate a greater desire to make an architectural statement than to accommodate the people who actually use the building (Anthony, 1985a).

Ideally, lobbies in high-rise buildings help accommodate people's transition from outside to inside. Not only is it important to provide an adequate number of seats, but it is also important to provide a variety of seating arrangements. Flexible seats for people sitting alone, small groups of seats for small parties of two or three people, and larger seating arrangements for groups of four or more help users feel they have a choice (Deasy, 1985). Movable seating provides users with the greatest number of options.

In addition to providing both high-quantity and high-quality seating, successful lobbies also feature aids to help guide people from the entry into the rest of the building. These aids include a reception desk located immediately adjacent to the stream of traffic; an information center near the entrance that includes a clearly marked "you are here" map, as well as adequate space for people to stand around and study the map; effective signs that are easily read and that are repeated in strategic locations; and information about the building and its history (Deasy, 1985).

The entry can also be designed to convey more subtle messages. Physical features such as the height of doorways and ceilings, the quality of building materials, and the quality of maintenance, as well as psychological features such as the accessibility or aloofness of building staff all send messages of their own. In their own way, such messages convey to visitors just how welcome they really are.

### *4 Users' Perceptions of Atrium Spaces*

The energy crisis of the 1970s propelled an active movement among architects to create energy-efficient buildings. A chief concern among these designers was to allow as much natural light as possible into their buildings so as to provide opportunities for passive solar heating. With heating and cooling bills continually on the rise, designing for energy efficiency is just as important today. One way to accomplish passive solar heating is through the design of atrium spaces.

In addition to providing increased opportunities for natural light, atrium spaces also allow for an abundance of plant materials typically found only outdoors. At their best, atria resemble lush, exotic gardens, complete with palms and other tropical plants, colorful flowers, cascading fountains, and the like. Especially in cold climate zones, where in winter trees and shrubs turn a dreary brown or gray, the bright, cheerful colors often found in indoor atria are particularly welcome.

The atrium also provides opportunities for building owners to personalize the building. In many buildings, the atrium decor changes with the season. At holiday time especially, the atrium is usually a place for lavish holiday displays, complete with richly decorated Christmas trees, reindeers, Santas, and the like. Just like the entry to a tall building resembles the front door to a house, so does the atrium resemble the living room at a much larger scale. It presents an image to the public that encapsulates the spirit of the building.

Some of the first atria in tall buildings in the United States appeared in hotels. The innovative design of the Hyatt Regency Hotel at Peachtree Center in Atlanta, Georgia, completed in 1967, featured a 23-story-high atrium lobby flanked by glass elevators. This atrium soon became a trademark repeated in luxury hotels all across the nation. Portman's 1974 variation of this design at the Hyatt Regency Hotel in San Francisco featured a 17-story-high atrium lobby angled steeply to increase the dramatic effect. This design captured national attention (Fig. 6.8). Soon after it opened, the lobby became one of the city's most popular hangouts. The atrium features live music, happy hours, and tea dancing, which often draw large crowds. At this hotel and others like it, riding the glass elevators to the revolving restaurant on the top floor has become a favorite tourist attraction.

**Fig. 6.8 When it was built in 1974, this 17-story atrium lobby, complete with glass elevators, at San Francisco's Hyatt Regency Hotel captured national attention.** (*Courtesy: Hyatt Regency, San Francisco.*)

Bednar (1986) has written extensively about atria. As he points out, riding in an observation car (glass elevator) allows people to experience atria in a dynamic, exciting way. The way in which the atrium is experienced changes each second as the position of the viewer changes. He writes that ascending "gives the impression of being catapulted through the vast interior space," whereas descending is "like landing in a parachute."

According to Bednar, the location of elevators in atrium spaces is extremely important. Elevators can be located inside the atrium, along its perimeter, in an adjacent space, or divided across any of these locations. Clustering the elevators in one core reduces user confusion and sense of disorientation; however, the bulky machinery may distract from the overall positive experience. Separating elevators into two or more clusters increases the efficiency of the transportation service. However, users may become confused about which elevator bank they should use. Ideally, some cars should be located directly inside the atrium whereas others may be located on the perimeter or adjacent to it. Precautions are needed so that when elevators land on the ground, no one is underneath them or in the shaft. Large planters surrounding the base are effective ways to keep children from wandering into danger (Bednar, 1986).

Atria have become relatively common in high-rise hotels in major American cities. They are now becoming increasingly popular in high-rise hotels in suburbs as well. They serve as the centerpiece of the hotel, often housing a greenhouse-style restaurant, lobby seating, entertainment, and recreation.

Many atria in large suburban hotels feature swimming pools, Jacuzzis, and exercise facilities. As a result, guests who wish to use these facilities must walk through the atrium in their swimming or exercise attire. While some hotels are designed with this in mind, placing the pool and exercise area near the room elevators, others are not. In these cases, self-conscious guests may be embarrassed to be seen walking around such a public area in their bathing suits. Some may simply choose not to use the facilities, whereas others may attempt to use them at off-hours when they are least likely to be seen by others.

This matter aside, a chief advantage of the hotel atrium is that it creates a circulation system to hotel rooms that provides an open, panoramic view across the lobby, rather than the typical long walk through a narrow corridor. In many cases, hotel room windows overlook the atrium, enabling the guests inside their rooms to people-watch. More often than not, the crowd below has no idea that they are being observed.

In some cities the atrium has become a bargaining chip for developers of large office buildings. By providing an atrium for public use, developers are sometimes allowed to construct a building taller than would be permissible otherwise. An interesting study, using interviews with pedestrians at nine locations in downtown San Francisco, examined the extent to which the public actually uses atrium spaces, who uses them, and whether or not they justify the breaks given to the developer (Kaplan et al., 1985). Results indicate that indoor arcades, atria, and gallerias fulfill many of the same functions as outdoor parks and plazas, and in some cases they fulfill these functions even better than do their outdoor counterparts. Researchers provided a list of design guidelines that can help ensure a high amount of pedestrian use. Guidelines address accessibility, height and complexity of space, seating, landscaping and water features, and commercial services and food. Here are some of the guidelines from the study by Kaplan et al. (1985).

1. *Accessibility.* Openness, visibility, and public uses should be incorporated into every atrium design intended to serve as a public gathering space. Such spaces need to be located near the street and use highly transparent glass and adequate signage. The point is to allow pedestrians outside to easily see what is going on inside. Shortcuts also should be incorporated into the design, as about 15% of the users only pass through.

2. *Height and complexity of space.* Build space more than three stories high, and with at least two levels of activity. The most successful atrium spaces are linked to a series of other public spaces, both indoor and outdoor, to form a continuous pedestrian environment.
3. *Seating.* Increase seating because there is rarely enough. About one in five people interviewed stated that they did not use a given site simply because there was no place for them to sit.
4. *Landscaping and water features.* Both should be used generously. People often cite greenery as a reason for liking atrium spaces.
5. *Commercial services and food.* Emphasize these within the atrium or nearby. Because many users like to combine eating, shopping, and running errands during their lunch breaks, locating eateries near services such as dry cleaning and shoe repair is a good idea.

Nonetheless, for all its aesthetic appeal, the atrium has sometimes created serious problems in building function, which have had a dramatic negative impact upon users. One of the most well-documented examples is Chicago's State of Illinois building, completed in 1985. Paul Gapp, the late architecture critic for *The Chicago Tribune,* nicknamed the building "Thompson's folly" after the name of Illinois Governor James Thompson, who shepherded the project through from start to finish. The transparent glass throughout the building skin was intended to symbolize the "openness and accessibility" of the government (Benson, 1987; Murphy, 1985; Saliga, 1990).

While the design of the State of Illinois building certainly was well intended, the mammoth 17-story atrium resulted in colossal problems with noise and climate control (Figs. 6.9 and 6.10). The open office plan found on virtually every floor, where offices open up to the central atrium below, produces a volume of noise so high that many workers find it difficult to concentrate on the tasks at hand. Concerning problems with climate control, at one point during the summer soon after the building opened, the interior became so hot—up to 43.3°C (110°F) or more—that some workers even brought their patio umbrellas from home in order to keep their work spaces comfortable. Conversely, during the winter, temperatures would sometimes drop below 15.6°C (60°F). Workers were finding it difficult to work effectively while they were so chilly. The State of Illinois filed a $20 million lawsuit against the architects and their associates; engineering repairs have since been made (Branch, 1987; Martin, 1987; Saliga, 1990).

## 5 Way-Finding in Tall Buildings

A number of environment-behavior researchers have investigated how people find their way in and around buildings. Some of the frequent questions that users are likely to ask include: Where is the elevator? Where is the building directory? Where are the restrooms? Where are the fire escapes?

Way-finding is important in every building, but it is especially so in tall, large buildings where the volume of users is high. Many people are intimidated by crowds rushing to and fro, and those who are already lost may feel even worse being surrounded by so many people who appear to know exactly where they are going. Way-finding is key in certain high-rise building types as well. In hospitals, for example, finding the emergency entrance can sometimes be a matter of life and death. Hospital visitors, too, must be able to find their way easily to patients' rooms. With few exceptions, such as the birth of a new baby, most persons consider a stay in the hospital to be a negative event.

**Fig. 6.9** **The 17-story atrium at the State of Illinois building in Chicago produced some colossal problems with noise and climate control soon after the building opened. Climate problems have since been resolved.** (*Courtesy: James Steinkamp.*)

Visitors who are already upset at having a loved one or a friend in the hospital can feel even worse if they cannot even find their way easily to the right hospital room.

High-rise buildings that serve tourists as well as regular workers must also be easy to negotiate. In tourist attractions, such as Chicago's Sears Tower or John Hancock Center, special elevators take tourists nonstop from ground-level lobbies to the observation decks at the top. The mental maps that visitors form of such buildings tend to leave a conspicuous gap of all the floors in between. Once at the observation decks, buildings like these tend to feature a great deal of information to orient visitors. Photographs are often used to identify views and landmarks in various directions. City maps are sometimes on display as well.

Although his intention was not specifically to highlight way-finding in high-rise buildings, environment-behavior researcher Gerald Weisman (1979) has identified four classes of variables that influence people's ability to find their way through a building. Nonetheless these variables can also apply to high-rise buildings:

1. *Signs and numbers.* Most common aids to orientation and way-finding
2. *Perceptual access.* The ability to see through or out of a building in order to find a destination
3. *Architectural differentiation.* Distinctiveness of the regions within a building
4. *Plan configuration.* Aspects of the building form that influence how people visualize the overall layout of a building

**Fig. 6.10 The State of Illinois building in Chicago's downtown loop is a hub of activity. Its transparency symbolizes the openness and accessibility of state government.** (*Courtesy: James Steinkamp.*)

In a later study Weisman (1981) discovered that the most important factor contributing to users' sense of disorientation was an illegible building plan. He argues that the "ability to effectively find one's way into, through, and out of a building is clearly a prerequisite for the satisfaction of other higher goals" (Weisman, 1981). He cites previous research findings demonstrating that when people encounter illegible, confusing buildings, they often become angry and hostile.

Passini (1984) also has investigated the phenomenon of way-finding, claiming that some buildings are actually labyrinths in disguise. He argues that people find their way as they obtain and process cues from their environment. These cues are obtained from signs, directories, watching other people, as well as their knowledge of similar types of spaces. Two distinct way-finding styles predominate; while some people seek spatial information, others seek linear information. Passini also discusses three kinds of signs: directional, identification, and reassurance. In order to make each type of sign effective, Passini argues, several issues need to be considered:

1. *Sign type and placement.* People perceive and process information more easily if signs are consistent in location and in overall form and design.
2. *Sign identity.* When people are familiar with the overall form of the sign, they are more likely to process similar signs.
3. *Quantity of information.* Information on a sign must be limited, and the information contained should be structured.
4. *Structure of information.* Information should be structured visually so that the message can be picked up in glances.
5. *The receiver.* Information should be geared to a certain population or audience, be it clients, workers, or others.
6. *Content.* Written information must be clear to all users. Any symbols should be understandable and consistent.
7. *Anticipation.* The sign system should continue to the destination, with reassurance signs along the way to refresh memories.

While information booths are important, designers must bear in mind that many individuals are hesitant or shy to ask for directions. Ideally, a building should be designed clearly enough so that generally users need not rely on information booths.

## 6 Accessibility in Tall Buildings

The goal of barrier-free design for the handicapped and the aged is to provide an environment that supports the independent functioning of individuals so that they can get to, and participate without assistance in, everyday activities such as acquisition of goods and services, community living, employment, and leisure. From a user standpoint, the goal of barrier-free design is easy access. An accessible environment is one that a disabled person can approach, enter, and use. [For a more complete discussion on accessibility and barrier-free design, refer to *Building Design for the Handicapped and Aged Persons* (CTBUH, Committee 56, 1992).]

***The Handicapped and the Disabled.*** Not everyone who is disabled is handicapped. A disability becomes a handicap only when the environment impedes the individual's ability to perform a specific task at a specific time and place (Raschko, 1982). The American National Standards Institute (ANSI), a nongovernmental organization that distributes recommended design standards, helps distinguish between the handicapped and the disabled and links the concept of handicapped to a person's interaction with the environment, defining a handicapped person as one "with significant limitations using specific parts of the environment" (ANSI, 1980). The Uniform Federal Accessibility Standards more specifically define the physically handicapped as: "...anyone who has a physical impairment, including impaired sensory, manual or speaking abilities, which result in a functional limitation in access to and use of a building or facility" (UFAS, 1984).

In contrast, ANSI defines a disabled person as one who "suffers from a limitation or loss of use of a physical, mental or sensory body part or function" (ANSI, 1980). The concept of disabled is thus intrinsic to a person's physical and mental capabilities and is dependent on one's ability to function within the environment. As such, it reflects a measure of a person's competence.

In the following, barrier-free design related to a few countries in different parts of the world is discussed briefly.

***United States.*** Although in the United States a national effort to create a barrier-free environment dates back to the early 1960s, it is only within the last few years that government sanctions have supported design standards for accessibility. Most of the private sector accessibility standards adopted throughout the United States have been developed by ANSI. Its first set of standards, ANSI A117.1, were adopted in 1961 and formed the technical basis for the first accessibility standards adopted by the federal government and most state governments. Federal design standards have been promulgated as the Uniform Federal Accessibility Standards (UFAS, 1984) and, as in other countries, principally apply to federally funded or constructed structures. To maintain uniformity between federal requirements and those commonly applied by state and local governments, UFAS follow ANSI A117.1 closely in substance and format. Together, these standards provide the basis for public and private efforts to create a barrier-free environment.

The American population is estimated to include 43 million people who are disabled (Kridler and Stewart, 1992). The disabilities range from wheelchair confinement to the use of eyeglasses. Furthermore, as the population continues to age, accessibility becomes a growing issue not only for the disabled, but the elderly and infirm as well. The Americans with Disabilities Act (ADA), signed into U.S. law in July 1990, is a landmark civil rights legislation. The ADA makes it illegal to discriminate in employment or in access to goods or services against any individual who is mentally or physically disabled.

***Japan.*** Barrier-free design of buildings is relatively recent in Japan, as specific design guidelines for central and local government buildings have only been developed over the last 15 years (CTBUH, Committee 56, 1992). Although slopes and elevators to cope with gaps and height differences for the disabled have become popular in public buildings, barrier-free design is not mandatory for privately owned buildings, and the concept is not universally understood by architects and designers. The true meaning of the elimination of a "gap" between two levels by a slope, or ramp, is not widely acknowledged. Sometimes called approach steps, wheelchair users often encounter several grade changes before they can reach the slope. Once large-scale buildings, such as high-rise buildings with elevators, have been reached, they are generally accessible and usable by the disabled, except in case of emergency. In low-rise and midrise buildings, the situation is worse. Step differences normally exist, and elevators are sometimes designed to stop at access halls between two floors where toilets are located. No emergency elevators are provided, and no area for temporary refuge is ready for use. The development of a general concept of life safety for the disabled and the aged is needed (CTBUH, Committee 56, 1992).

***Canada.*** Canada has made considerable progress in barrier-free design on several fronts since 1960. Collaboration in research and development standards between U.S. and Canadian organizations has resulted in parallel activities and the adoption of comparable accessibility standards in Canada and the United States. The National Building Code (NBC) of Canada has established barrier-free standards for buildings since the mid-1960s, and the first code, called Supplement #7 to the NBC, was largely based on ANSI of 1961. The NBC acts as a model code, which must be adopted by some local authority having jurisdiction until it becomes law.

Most building codes in Canada do not require residential facilities to be completely accessible. Typically, access to public entrances, shared laundry facilities, and suites (but not the suite itself) are required (CTBUH, Committee 56, 1992).

***The Netherlands.*** In the Netherlands there is no official description or definition of disabled people. One of the reasons for this is that the government wants to consider these people as individuals and not as members of a group, especially a group that may be judged negatively in social interaction. Another reason is that it is difficult to give a uniform description of a very heterogeneous group. For practical and political purposes, the following categories of physical disabilities are used by the World Health Organization (1980):

> *Impairment* refers to the loss or abnormality of the structure or function of the body.
>
> *Disability* refers to the limitations or lack of ability in performing activities that are considered normal for a particular individual. It is normally a consequence of impairment.
>
> *Handicap* refers to the impact of impairment and/or disability on the individual and on the wider community involved. When such impact implies problems in relationships, limitations on life opportunities, with consequent disruption of social integration into ordinary community life, then an impaired or disabled person is a handicapped one.

Interaction in public life has not been easy for the disabled in the Netherlands. The physically disabled person finds that it is practically impossible to function in public buildings because of many obstacles and inconveniences. While there is no legislation in the Netherlands that regulates the layout of public buildings for the disabled, unlike regulations in the United States and Canada, a number of guidelines have been enacted defining distances to entrances, accessibility of public buildings and housing, and usable space for the disabled and the handicapped. These guidelines have been set down in a substantial reference book called *Geboden Toegang*. Now in its seventh addition, its purpose is as follows (National Organization of the Handicapped, 1983):

> An indication of the possibilities to create a society which is accessible to everyone.
>
> Developing the necessary criteria which relate to the adaptation of both existing facilities and those that have yet to be built. These criteria should be included integrally in the completion of building instructions, standard specifications, builders' estimates, conditions for subsidies, and so on.
>
> Giving information to all those connected with the realization of these goals.

***Singapore.*** Gradual changes in attitudes with regard to the disabled have created an underlying philosophy adopted by Singapore that the disabled should be treated as equal members of society. As such, they ought to be accorded equal opportunities for developing and expressing their physical, mental, social, and economic capabilities to the fullest (CTBUH, Committee 56, 1992). In the past the community's response to the disabled population was either to institutionalize or to put the onus for care on the disabled person's family. However, a wider sense of responsibility is becoming more discernible.

New guidelines pertaining to the built environment have been established, which define and promote awareness of disabled people, physical impairments, and accessibility. The *Design Guidelines on Accessibility for the Disabled in Buildings* (Ministry of National Development, 1985) states that:

> The disabled are those people who, as a consequence of physical disability or impairment, may be restricted or inconvenienced in their use of buildings because of:
>
> 1. The presence of physical barriers, such as steps or doors which are too narrow for wheelchairs;
> 2. The lack of suitable facilities such as ramps, elevators, staircases, and handrails; and
> 3. The absence of suitable facilities such as toilets, telephones, and suitable furniture.

In relation to the physical environment, disabled people are classified into four groups. *Ambulatory disabled* are those who are able, either with or without personal assistance, to walk on level ground and negotiate suitable graded steps provided that convenient handrails are available. *Wheelchair-bound* are those people unable to walk, either with or without assistance, and who, except when using mechanized transport, depend on a wheelchair for mobility. *Sensory disabled* people are those who, as a consequence of partially or totally impaired sight or hearing, may be restricted or inconvenienced in their use of buildings because of lack of suitable facilities. *Temporarily disabled* people include those who are sick or victims of accident. Expectant mothers can be included in this category (Ministry of National Development, 1985).

Accessibility and mobility are two inseparable aspects relating to the design for the disabled. The Ministry of National Development administers the implementation of policies relating to the built environment in the public domain mainly through the Urban Redevelopment Authority (URA) and the Public Works Department (PWD). These agencies have implemented many design changes, providing greater accessibility for the disabled in many public buildings. Some noteworthy examples include the Hill Street Centre, South Bridge Centre, and Changi International Airport. The provision of curb cuts and on-grade entry to public buildings, however, has been hampered by Singapore's heavy tropical rainfall, which makes deep storm drains and high curbs or coamings necessary to prevent water from entering buildings.

Since 1988, Singapore has taken major steps to facilitate access to all buildings, including tall buildings, by disabled people. The Building Control Act of 1989 now states in Division 2 of its regulations that:

> Where a proposed building is one to which disabled persons have or may be reasonably expected to have access, that building shall be designed to the satisfaction of the Building Authority in such a manner as will facilitate access to and use of that building and its facilities by disabled persons.

The regulation then defines types of buildings and the specified areas within a building where accessibility is required (generally "areas intended for public access") and, more particularly, the requirement that the design be in accordance with the *Code on Barrier-Free Accessibility in Buildings* (Public Works Department, 1990).

Although barrier-free design is not yet an issue in many of the developing countries, the worldwide concern for the disabled is likely to gradually influence the building design process in these countries in the future.

***Research Needs.*** In many cases there is still a lack of understanding of the environmental problems faced by disabled and handicapped people and the consequences of these problems on their daily lives. Consideration of disabled and handicapped people as a part of the normal building population has not as yet been internalized by the design professional in the United States and abroad. New human rights legislation, such as the American Disabilities Act in the United States, should help encourage barrier-free design throughout the world.

More research is required to ensure that the funds being spent on retrofitting existing buildings and on solutions proposed for new buildings are reasonable. Care must be taken in transposing the research results from one country to another since different kinds of wheelchairs, rehabilitation techniques, cultures, and climates should be considered.

Building entries, lobbies, and atria present a variety of psychological issues. Whereas it is important for users to be able to locate the entry, it is also important for designers to create an entry that is welcoming, informative, and not intimidating. Building lobbies must be designed with visual appeal in mind, but they must also reflect the waiting behavior and small-scale gathering that occurs there by providing adequate and varied seating arrangements. Although atrium spaces, such as entries and lobbies, serve as signature spaces, they must also be designed with practicality in mind. Finally, finding one's way and orienting oneself in tall buildings are complex tasks. Architects need to provide barrier-free, user-friendly tall buildings to regular patrons as well as first-time visitors.

## 6.6 PASSENGER BEHAVIOR IN ELEVATORS

Although their purpose is strictly utilitarian, to transport people vertically through a building, elevators raise a number of social and psychological issues. Waiting behavior, personal space and crowding behavior, civil inattention, fear of being trapped, fear of crime, and the inconvenience of elevators out of service are among the issues covered here. Undoubtedly, many other issues can be raised as well. (For a discussion on elevators from the perspective of building planning, see Chapter 2.)

### 1 Waiting for Elevators

Waiting for elevators is a frequent source of tension. The stress people experience while waiting for an elevator to arrive, particularly when they are rushed or late for appointments, is often highly visible. People exhibit such nonverbal behavior as pacing, fidgeting with their hands, anxiously checking their watches, or pressing the elevator button repeatedly. In a large elevator lobby, people can often be seen nervously scanning the floor indicator above each elevator so that they can dart into the first available car. Even in the most civilized crowds it is not unusual to see people pushing and shoving to squeeze their way into a crowded elevator, lest they have to wait for the next one to arrive. The elevator is often the place where human behavior can be seen at its worst.

While waiting for elevators can be a source of tension in many different building types, the problem is especially acute in residential apartment buildings. Residents must depend upon elevators every day to get into and out of their homes. If something goes wrong, they are highly vulnerable. Among those most affected are elderly and disabled users.

A survey of 100 residents' perceptions of an 11-story housing complex for disabled and elderly people revealed that residents were dissatisfied with elevator service. The building housed only two small elevators for 250 residents. Residents complained that elevators were too slow, too crowded, and they often broke down. Perhaps unfairly, they also criticized their disabled neighbors who, they believed, simply took too long to enter and exit the elevator. Some residents went so far as to request separate elevators for disabled residents (Haber, 1986).

It is thus not surprising that elevators are sometimes contentious in the design and building process. The interests of various groups involved in the building can often be directly at odds. For the owner, the key factor is likely to be the overall cost of the elevator system. Often the more rapid the system, the higher the cost. For architects, a key issue is the ability of the elevator system to coordinate both the function and the aesthetics of the building. For users, the issues are even more complex. The speed of elevator service can play a major role in influencing tenant satisfaction in a high-rise building. Slow elevators also can affect the rentability of a building (Frederking and Penz, 1973). In most cases the users simply have no choice in the matter.

To better address user needs such as these in elevators, an innovative elevator design was created in the Worldwide Plaza on the west side of New York City. Kevin Huntington, an elevator consultant, used computer models of likely passenger behavior to help plan out the elevator system. The elevator system can be changed to accommodate different times of day, different types of tenants, services, and so on. In Huntington's words: "I think everyone would agree that the ultimate elevator system would be if each tenant had his own private elevator...[but] there wouldn't be much building left, just a few windows and a bunch of elevators" (Sabbagh, 1989).

## 2 Personal Space and Crowding in Elevators

Crowding occurs when people are unable to regulate their desired amount of privacy; this is exactly what occurs in elevators (Altman, 1975). The sensation of being crowded is exaggerated by time. The longer people are forced to ride together under crowded conditions, the more crowded they are likely to feel. An elevator loaded to capacity will seem much more crowded when it stops at virtually every floor than when it rides an express route nonstop.

According to anthropologist Edward Hall's (1966) classification of interpersonal space zones, crowded elevator passengers temporarily are forced from the comfortable zone of personal distance [between 457 mm and 1.2 m (18 in. and 4 ft) apart] into the intimate zone [between 0 and 457 mm (0 and 18 in.) apart]. This zone is typically reserved only for those with whom one desires to have close contact, such as a close friend, spouse, or other intimate companion. At this distance, one is acutely aware of the presence of other people.

Two related studies examined the effects of personal space violations on elevator behavior (Knight et al., 1979). In the first study, the time it took passengers to exit the elevator and to cover certain distances after exiting was recorded. Results indicated that when the number of riders increased, it took less time for the first person to exit the elevator, and more time for the last person to exit. It also took people longer to walk a short distance away from the elevator when more riders exited the car with them. In the second study, only an experimenter and a rider shared the elevator. When the distance between the two people decreased, both the speed with which riders exited the elevator and their walking speed a short distance away from the elevator increased (Knight et al., 1979).

### 3 Civil Inattention in Elevators

When strangers are forced into such close quarters, only a few inches apart, rubbing shoulders, pressing against each other, and sometimes even smelling each other, their natural inclination is to retreat psychologically. Sociologist Erving Goffman (1963) was among the first to coin the term "civil inattention" to describe this phenomenon. Civil inattention is a behavioral ritual that occurs when two or more people are mutually present but not involved in any form of interaction.

In a series of three related studies, a team of psychologists (Zuckerman et al., 1983) used confederates to observe the behavior of 504 elevator riders in a downtown department store, an office building, and a large apartment building serving a university and its medical school. Researchers recorded the number of glances received from each rider. Only looks oriented toward the experimenters' faces or upper bodies were recorded, while looks toward the experimenters' feet did not count. The researchers also recorded latency, that is, the length of time before the first glance, the length of the rider's first look, the location and length of the ride, and the sex and estimated age of the rider. Zuckerman et al.'s (1983) results indicated that at the beginning of an elevator ride, about half the people exchanged glances with the confederate but then averted their gaze for the length of the ride. Riders glanced more at an experimenter who entered the elevator than at an experimenter who waited for them.

The same researchers also examined how riders perceived both the elevator ride and the confederate passenger when the confederate glanced, ignored, or stared at them. Results indicated that riders rated violations of civil inattention less favorably than behavior that was consistent with civil inattention. They speculate that the size of the group, the cultural backgrounds of the riders, and other factors may well influence the extent to which people practice and expect civil inattention. According to the researchers, "the need arises to acknowledge the presence of others without turning them into targets of attention" (Zuckerman et al., 1983).

### 4 Fear of Being Trapped

Aside from civil inattention, personal space, and crowding issues, another source of people's anxiety in elevators is the remote but ever-present possibility of being trapped by a power failure, unable to escape. Anyone who has ever experienced this sensation is likely to remember it in great detail long afterward since at the time this can be absolutely terrifying.

The February 26, 1993, World Trade Center bombing brought this fear to the public's attention. So did one of the most famous power failures of all, the blackout at 5:24 P.M. on November 9, 1965, which knocked out power in 80,000 square miles, parts of eight northeastern American states and most of Canada's Ontario province, plunging 30 million people into complete darkness. Here is *Newsweek*'s account of that evening's events in New York City (*Newsweek,* 1965):

> Lights blinked and dimmed and went out. Skyscrapers towered black against a cold November sky, mere artifacts lit only by the moon. Elevators hung immobile in their shafts. Subways ground dead in their tunnels. Streetcars froze in their tracks. Street lights and traffic signals went out—and with them the best-laid plans of the traffic engineers. Airports shut down.

Nearly 800,000 New York passengers were trapped in more than 200 elevator shafts and 630 subway and commuter trains throughout the city. Some were loaded to capacity with standing room only. Those who had room on the floor often found it to be filthy. Even at the start of this misadventure, hundreds of passengers were tired, hungry, thirsty, and in desperate need of a rest room. Imagine how they must have felt several hours later, still there. While some elevator prisoners feared suffocation, others worried about Communist sabotage.

Miraculously, only two New York citizens died. One fell down a stairway and struck his head. Another died of a heart attack after climbing 10 flights of stairs. Incredibly, criminal activity was lower than usual; only one-fourth as many arrests were made as on a normal night. Instead, strangers turned to each other for comfort.

In at least three skyscrapers rescue workers had to break through walls to access elevator cars. At New York's Empire State Building, 13 elevators carrying 96 passengers were stalled from the 34th to the 75th floor. Passengers formed a Blackout Club in one and played word games in another. Passengers from six cars were freed using ladders, and those in the remaining seven elevators were reached by breaking down the walls. After being trapped at approximately 5:15 P.M., the last passengers were finally rescued at 11:45 P.M. (For vivid accounts of the November 9, 1965, power failure see also Bigart, 1965; Phillips, 1965; and White, 1965.)

Fear of being trapped in elevators is often a symptom of claustrophobia. (For an excellent discussion of claustrophobia, consult the chapter on this subject in Rachman, 1990.) People who are afraid of being confined in elevators are often also afraid of being trapped in tunnels, cellars, subway trains, and other enclosed spaces. Unlike many other types of phobias, the fear subsides as soon as the person leaves the uncomfortable space. About a third of claustrophobics acquired their fear during childhood. Most report that they became fearful after an unpleasant experience in an enclosed space.

People who experience claustrophobia in elevators are likely to engage in extensive avoidance behavior. They may be much more likely to walk up 10 or more flights of steps rather than use the elevator. Claustrophobics are frightened not of elevators themselves, but of what might happen to them while they are in them. Specific fears include entrapment, caused by a restriction of movement, and suffocation. Panic episodes may result. Heart rates tend to elevate, chest pains may develop. Individuals may also suffer shortness of breath and dizziness during these panic attacks. Researchers who have studied claustrophobia have uncovered some fascinating gender differences. It appears that this affliction is more likely to strike women than men (Rachman, 1990).

## 5 *Fear of Crime in Elevators*

In many tall buildings, people fear not only being trapped in the elevator due to a power failure, but, perhaps even more likely, being trapped against their will by a criminal. In the United States, elevator muggings have become quite common, and in some cases people have been raped and murdered in elevators.

Elevators can be the scenes for vandalism as well. Engineers have recently designed innovative elevator buttons that have no moving parts, are not sensitive to heat, can withstand shock and abuse, and are vandal proof (*Design News,* 1986). Graffiti in elevators is common as well. Teens are among the chief "graffiti artists," stopping the elevator in place to do their artwork, and exiting without anyone knowing who did it.

Elevators in housing projects, parking garages, subway stations, and elsewhere are sometimes used as public restrooms. This disturbing behavior can be seen in public parking garages even in the most luxurious neighborhoods. Some individuals dread traveling in parking garage elevators for this reason. Unfortunately the same problem can occur in stairways, so people are forced to choose between the lesser of two evils.

### 6 The Inconvenience of Elevators out of Service

Anyone who has ever lived or worked in a tall building where elevators have gone out of service knows that it is a highly unpleasant experience. Walking up long flights of steps, especially while carrying heavy items or while wearing heavy winter clothing, is simply no fun. For some, such as many elderly and physically challenged individuals, the task is simply impossible.

Construction work and power failures are not the only causes of out-of-service elevators, however. Elevator workers have been known to strike, causing colossal impacts on the people who use them regularly. For instance, on September 24, 1945, elevator operators at New York's Empire State Building went on strike, along with those from 2000 other skyscrapers. City workers had to improvise (Goldman, 1980):

> Some executives tried to set up shop in the lobby, giving dictation and sending secretaries to phone booths to place calls. Up on the 85th floor (of the Empire State), the NBC television engineers were prepared: they'd provided themselves with cots and a refrigerator full of food. Below, on the 68th floor, one tenant was expecting an important phone call, so he waited in his office for it to come. He was there three days later when it did. Food was a problem for anyone who did make the hike to work. One luncheonette operator lugged 150 sandwiches and several jugs of coffee all the way up to the 31st floor. The famished stock brokers up there rewarded him with a $75 tip.

Although some valuable research has been done, several social and psychological questions about elevators remain unanswered. For instance, what are the psychological effects of placing mirrors in elevator walls, ceilings, or floors? Does music reduce tension in elevators? Are there different psychological effects of "elevator music," show tunes, pop, or other kinds of music? Do glass elevators feel safer or more dangerous than traditional elevators? Are people more willing or less willing to wait for glass elevators? Architects and managers must be aware of the significance of elevators and take these issues into account when they create and manage tall buildings.

## 6.7 PSYCHOLOGICAL ISSUES IN AND AROUND TALL OFFICE BUILDINGS

The following sections focus on four specific types of tall buildings: high-rise offices, housing, the parking garage, and the urban shopping center. They are highlighted here because each of these building types presents its own array of psychological issues. Whereas high-rise housing and offices have been the subject of widespread environment-behavior research, high-rise parking garages and urban shopping centers have, to date, been somewhat overlooked.

The literature on the psychological aspects of work environments is burgeoning. While not all the office research has focused specifically on tall buildings, much of what has been found applies to them. Among the issues covered here are the psychological effects of sunlight in offices, user satisfaction in offices, and pedestrian use of urban plazas near tall office buildings.

### 1 Effects of Sunlight

Just as sunlight is an important influence on the quality of the pedestrian experience on streets and sidewalks below tall buildings, sunlight is also a strong influence on the quality of workers' experiences within high-rise office buildings. Access to natural light can influence how workers feel about their jobs. Offices that are bright and airy and those that are dark and dank can produce different moods.

Some offices provide a visual connection to the outdoors, enabling workers to view the changing weather. Others are completely cut off from the outside world, illuminated only by artificial means. These spaces look the same no matter what the weather. Even violent weather systems, such as hailstorms and blizzards, can go unnoticed. The extent to which workers have a visual connection to the climate outside their buildings may well influence how they feel about coming to work every day.

Some interesting research has been conducted to determine the psychological effects of varying amounts of sunlight on office workers. (For an excellent overview of research on this issue, consult Sundstrom, 1987.) Some studies have shown that office workers tend to overestimate the contribution that windows provide toward their lighting, and that they gain satisfaction from views that windows offer (Sundstrom and Sundstrom, 1986).

In one recent study, Boubekri et al. (1991) examined the reactions of 40 workers who performed a proofreading task, relaxed for 2 minutes, and then answered a questionnaire about their current mood and state of satisfaction. The amount of sunlight in the room was controlled by a moving screen, and four window sizes were tested. These windows occupied 10, 20, 40, and 60% of the wall. The time of day was kept constant.

The results of this study identified an optimal amount of sunlight that produced relaxation and the best performance, as well as limits of high and low sunlight that produced unsatisfactory moods, even distress. People were most relaxed when 15 to 25% of their floor area was lit by the sun. However, positive effects also occurred when 10 to 45% of the floor area was lit. Although this study has some methodological shortcomings, its results are indeed provocative. When replicated, this body of research could lead to some useful design guidelines.

Nonetheless, many tall office buildings have already been designed to maximize workers' exposure to the sun. The new superthin skyscrapers provide a greater percentage of occupants with sun, bringing more offices to the perimeter (Russell, 1990). Although it is not a superthin skyscraper, a similar motive influenced the design of 101 California, a high-rise office building in San Francisco. This building features a series of angled glass windows such that most offices have opportunities for lots of sun as well as a corner view. Rather than fighting over which executive gets to have the corner office, almost everyone along the building perimeter is granted this privilege.

When considering access to sunlight in tall office buildings, it is important to note that such access can also reflect the relative social status of employees. Those highest in rank are more likely to have offices with the most sun and the best views, whereas subordinates are most likely to be denied these perks. In many offices these status demarcations often reflect gender differences as well. Upper-level management and executives, more often than not, predominantly men, are awarded the greatest amounts of sunlight. By contrast, clerical staff—secretaries, word processors, and receptionists—usually dominated by women, are relegated to interior offices with artificial lighting.

## 2 User Satisfaction

A wealth of information is available about how people respond to office environments. Among the most comprehensive sources on this subject are those by Sundstrom and Sundstrom (1986), Sundstrom (1987), Becker (1990), Brill et al. (1984), and Wineman (1986).

The study by Brill et al. (1984) spanned 4 years and examined 18 environmental variables in offices, such as the amount of floor area per employee, layout, enclosure, temperature and air quality, lighting, windows, status, and privacy. A 500-item questionnaire was sent to over 5000 workers at over 70 office sites. Survey respondents were questioned 2 to 4 months before moving to new and upgraded offices, and they were followed up about 8 months to a year later.

Several variables in this research (Brill et al., 1984) were found to affect job satisfaction, environmental satisfaction, and ease of communication on the job. The study identifies the financial impacts of each of these variables, concluding that through improved office design, companies can increase their annual productivity by up to 17% of the average annual salary for clerical workers, and about 15% of the average annual salary for professional/technical employees and managers.

A number of studies have examined workers' perceptions of office landscape designs, which incorporate movable screens, furniture clusters, and planters instead of the more conventional internal walls, partitions, and grid systems of furniture. In fact, the trend toward office landscape design has become widespread over the past decades. Nonetheless, while they were intended to promote greater interpersonal communication and smoother work flow, such offices tend to generate major problems of their own. Among the more serious complaints are loss of visual and conversational privacy, high amount of visual and sound distractions, frequent interruptions by other employees, and problems with ambient conditions. For a more comprehensive discussion of these issues, consult Hedge (1986).

Another issue in the workplace that has been the focus of extensive environment-behavior research is environmental stress. Findings suggest that a number of stressful environmental factors may affect physiological responses, work performance, social behavior, satisfaction, and health. Sources of environmental stress include ergonomics, the arrangement of work surface, chair, and storage space; office automation, the integration of personal computers and other machines into the workplace; heating, air-conditioning, and ventilation; artificial lighting; natural lighting and view; visual and acoustical privacy; crowding; and physical safety (Wineman, 1982).

### *3 Pedestrian Use of Urban Plazas*

A number of researchers have analyzed how people respond to the open spaces near tall office buildings (Anthony, 1985a; Conway, 1977; Cooper Marcus and Francis, 1990; Mozingo 1989; Whyte, 1980, 1988). A few of these studies have resulted in design guidelines (Cooper Marcus and Francis, 1990; Whyte, 1980, 1988).

Whyte's work involved a comparative analysis of several urban open spaces across the United States, spanning almost a decade. By using a set of detailed behavioral observations recorded through time-lapse photography, Whyte (1980, 1988) and his research team distilled the key ingredients of successful, heavily used, urban plazas. These include a large amount of sitting space and a choice of seating areas, preferably with movable chairs; adequate sun; provision of trees; protection from wind; access to water; presence of food through cafes, kiosks, and street vendors; triangulation, or the provision of an external stimulus such as a piece of sculpture, a fountain, or entertainment; and a close relationship to and high visibility from the street. Whyte's research resulted in significant amendments to New York City's zoning codes, and has had a great impact in altering the streetscape of downtown Manhattan.

Project for Public Spaces (PPS), Inc., a not-for-profit organization and an outgrowth of Whyte's Street Life Project, has conducted numerous public space evaluations in cities and towns throughout the United States. Many of these studies have produced design and management changes based on user-need research. Among the major outdoor spaces that PPS examined are the Exxon minipark and Rockefeller Center, both in New York City.

PPS has also worked for private developers and has collaborated with building architects, helping design and retrofit public spaces in and around mixed-use and office building projects. By analyzing existing circulation patterns, PPS determines optimal locations for entrances and predicts how people will circulate and orient themselves throughout the site. The organization earned the well-deserved nickname "the space doctors."

Access to sunlight, user satisfaction, and pedestrian use of nearby urban plazas are three of the most important issues surrounding tall office building design. Many more issues can be identified. For instance, what are the relationships between workers' relative height in the building and worker satisfaction? Are those who work at the top of the building necessarily the most satisfied? To what extent does the architectural style of the tall office building reflect the corporate identity of the occupants? Whose identity is being reflected, and whose is not? Furthermore, a need exists to test some of the findings from environment-behavior office research, specifically in tall office buildings. Comparisons among workers in high-rise, midrise, and low-rise office buildings could be made.

## 6.8 HIGH-RISE HOUSING

Many scholars have addressed the psychological impacts of living in high-rise housing. A sizable body of literature on this topic is available. (See, for example, Ginsberg and Churchman, 1985; Vergara, 1989; Normoyle and Foley, 1988.)

Housing research has demonstrated that the experience of living in a high rise differs greatly for various individuals. While a high-rise building may be ideal for an elderly person, it may be less desirable for a family with small children who need a place to play outdoors. Income is also a factor. This section examines residents' perceptions of and satisfaction with high-rise housing. Along these lines, the users most often studied include the elderly, low-income residents, and students. Each will be discussed here briefly.

### 1 Residents' Perceptions of High-Rise Housing

A wide variety of factors influence how residents perceive high-rise housing. One of the most important is prior experience. For instance, someone who moves from a single-family, detached house into a high-rise apartment will likely have a different reaction from someone who has always lived in high rises and simply moves from one apartment to another.

Another important factor influencing people's perceptions of high-rise living is their cultural background. In many parts of the world, living in a high rise is the norm rather than the exception. For instance, in most Asian cities, townhouses or single-family homes are generally expensive. Furthermore, soaring land costs often make anything but high-rise housing prohibitive. The same is true in many European cities. Nonetheless, even those accustomed to high-rise living may differ greatly in how they perceive and adjust to their housing environment.

An excellent study by Williamson (1981) examined how 430 Germans living in high rises in and around Cologne and Dusseldorf adjusted to their living environment. Using structured and open-ended questions in an interview format, the study addressed how independent variables such as age, sex, marital status, and social class affected residents' responses to the physical design, social networks, and adjustment of children in the high rise. Results indicated that men were more negative than were women; however, women were more concerned with specific problems such as security. Compared to men, women were more conscious of the poor design of entryways, hallways, and stairs. Single people were aware of views offered from the windows of the high rise as well as the anonymity they experienced, whereas married people were conscious of balconies and their neighbors' backgrounds, and chose to live on lower floors. Older respondents complained more about management, children, the cleanliness of neighbors, and other aspects (Williamson, 1981).

Lower-income children tended to play on the immediate floor or outside the apartment, whereas children of upper-income families tended to play inside the apartment unit itself. Lower-middle income respondents, single people, and childless couples had negative opinions about children in high rises (Williamson, 1981).

Another study examined the perceptions of 100 residents of a Beijing, China, apartment building (*Beijing Review,* 1985). Survey results indicated that almost all (95%) residents never visited each other; almost three-quarters (72%) of residents did not know the names of their neighbors, and over two-thirds (68%) had no idea where their neighbors worked. One older respondent commented on his life in the apartment: "I feel as though I am locked in a huge box. I wish I could return to my quadrangle where I had friends and fun" (*Beijing Review,* 1985). The author expressed concern about the rising tide of isolation being experienced in high-rise apartment buildings.

For a detailed analysis of how low-income, elderly people experience high-rise housing, consult Nahemow et al. (1977). This study involved a follow-up study of 662 original tenants of U.S. Department of Housing and Urban Development (HUD)–sponsored housing. One of the most common complaints of residents was poor elevator service due to breakdowns and electrical storms. High-rise housing, especially for older residents, can benefit greatly from including a small-scale grocery or convenience store. Since many elderly residents no longer drive or are dependent on friends and relatives to take them grocery shopping, providing groceries within the building can help increase their sense of independence (Haber, 1986).

Another interesting study (Husaini et al., 1991) examined the social and psychological well-being of black elderly living in high rises for the elderly in Nashville, Tennessee. They sampled 600 residents. Individuals living in senior high-rise apartments were compared with those living in community housing. The focus of the study was on demographic variables, various aspects of social support, chronic and acute stressors, and depression. Results from the Husaini et al. (1991) research are somewhat disturbing. They indicate that compared to their counterparts in community housing, black elderly living in high-rise apartments were more frail and had lower levels of psychological well-being. They had less social support and were in poorer health. On average, the high-rise group reported seeing a doctor at a significantly higher rate. Those in the high rises were under greater stress and experienced higher levels of depression. They experienced higher rates of schizophrenia, simple phobia, and at least one mental disorder. It may be that the relocation factor, or the negative effects of moving, are at least partially responsible for these high-rise residents being more vulnerable to illnesses and mental disorders.

By contrast, an interesting essay by Fuerst (1985) refutes the negative findings of many studies of high-rise housing. Citing a variety of studies showing strong resident satisfaction with high rises, he argues that in many cases, "tenants can and do live happily in high-rises with diverse ethnic groups, with different income groups, and with children" (Fuerst, 1985).

### 2 Residential Satisfaction in High-Rise Housing

The study of residential satisfaction is one component of the study of residents' perceptions of high-rise housing. Some researchers choose to focus primarily on residential satisfaction, whereas others include it as an issue related to residents' perceptions.

Williamson's (1981) study of 530 high-rise apartment residents in Germany found that overall residential satisfaction was strongly related to the physical attributes of the building, especially spaciousness, room arrangement, and quality of construction. Satisfaction is also related to social interaction among neighbors or with those in other parts of the building, whether or not the building is perceived as impersonal, and the presence of a specific social network, as well as the perceived adjustment of the children.

Francescato et al. (1979) conducted an award-winning study examining low-income residents' satisfaction in 37 publicly assisted housing projects in 10 states in the United States in both central-city and rural locations. Many of these projects were high rises. Over 1900 residents responded to the questionnaire. Three major factors explained a high proportion (74%) of the total variance in overall satisfaction: satisfaction with other residents, pleasant appearance, and economic value. In addition, both physical and nonphysical qualities of the housing development influenced residents' satisfaction.

Sociologist Oscar Newman (1972) examined the amount of crime occurring in high-rise versus midrise low-income housing. His findings demonstrated that crimes at the high-rise projects were significantly higher than elsewhere. One of his key arguments is that providing residents with a strong sense of territoriality can help reduce crime. Territoriality can be achieved by designing limited entrances to the building for a small number of families; by subdividing massive, anonymous, public open spaces in the middle of housing projects into smaller, more identifiable pieces of land; and by encouraging a strong sense of surveillance from each apartment unit to the open space below.

Newman's findings have since been contested. For example, Normoyle and Foley's (1988) study of 945 elderly tenants in public housing found that contrary to what one would expect, fear was actually lower among high-rise dwellers than among those in low-rise walk-ups and row houses. However, this finding applies only to those elderly residents scattered throughout high-rise apartments also housing younger tenants. For elderly residents segregated from younger tenants, this was not the case. The researchers suggest that crime-related judgments of elderly high-rise residents may become exaggerated or distorted, simply because these residents are housed apart from younger families. They suggest that building height may have an indirect influence on the anxieties and perceptions of the elderly. Based upon their findings, the researchers present various refinements to Newman's (1972) theory of defensible space.

Cooper Marcus and Hogue (1977) provide a detailed list of design guidelines for high-rise family housing. One of the chief psychological and design issues they address is the need for a sensitive transition from public and semipublic places into the private living space. The authors point out that spaces such as the lobby and the elevator can have a great impact on tenants, sometimes making the difference between a positive and a negative living experience. They suggest that elevators contain high-quality hardware that can hold up under abuse by children, and that the alarm and floor buttons can be reached by children aged 6 and older. They also recommend a separate service elevator for moving furniture. As far as the elevator waiting areas, the authors also suggest adding shelf units for residents to rest packages, and providing information boards to help relieve boredom.

Undoubtedly the most famous example of residential dissatisfaction in high-rise housing is the Pruitt-Igoe housing project in St. Louis, Missouri. Completed in 1954 as part of a mammoth effort by the St. Louis Housing Authority, three of its high-rise buildings were purposely demolished in 1972. The demolition was in response to strong resident dissatisfaction and high levels of criminal activity. The coup de grace was a massive 9-month rent strike in 1969, which depleted the Housing Authority's funds to such a great extent that the U.S. Department of Housing and Urban Development (HUD) was forced to consider closing the project.

The demolition of Pruitt-Igoe has been ingrained in the minds of a generation of architects and planners, and the failure of the project has usually been blamed on its insensitive, brutal architecture. Nonetheless, Bristol (1991) argues that Pruitt-Igoe does not represent an architectural failure, but rather failure due to flawed policies, local economic crises, class oppression, and racism. She debunks the Pruitt-Igoe myth by arguing that it "not only inflates the power of the architect to effect social change, but it masks the extent to which the profession is implicated, inextricably, in structures and practices that it is powerless to change" (Bristol, 1991). Bristol stresses that the project's occupancy rate peaked in 1957 at 91%, after which it began a steady decline. One of the chief problems, she emphasizes, was that with fewer residents, the local housing authority had fewer funds with which to maintain the buildings and to complete basic repairs. The project was soon inhabited by primarily female African-American heads of households, a population which, lacking adequate protection, soon became the victim of rampant vandalism and violent crime.

### 3 Students' Perceptions of High-Rise Dormitories

A number of researchers have examined students' perceptions of high-rise housing. A classic study by Van der Ryn and Silverstein (1967) examined residents' perceptions of high-rise dormitories at the University of California, Berkeley. The researchers discovered a mismatch between many of the architects' inherent assumptions about student behavior and the ways in which students actually used their dormitory rooms and common spaces. They also discovered that many of the common spaces immediately outside the high-rise dormitories were rarely used.

Another study of the International House at the University of California, Berkeley, an eight-story structure housing approximately 600 residents, half American and half international students, focused on what helped residents perceive this high-rise building as home (Anthony, 1981). A number of significant correlates were identified, including: feeling at home, high residential satisfaction, meeting expectations, high opinion of the administration, not feeling crowded or restricted, having favorite possessions, owned for long periods of time, in one's dormitory room, high assessment of programs, relatively long residence, high amount of territorial behavior, strong territorial attitudes, high degree of risk-taking behavior (leaving belongings unattended in semipublic areas of the building), feeling able to modify one's room, feeling an adequate level of privacy, and satisfaction with the level of social interaction. In general, the institution's social qualities contributed to some residents' sense of home, whereas its physical qualities prevented others from viewing it this way. When asked what could make the building feel more homelike, International House residents described mostly physical changes, especially redesigning the excessively long corridors and adding small lounges throughout the residence areas.

A series of dormitories built on the campus of Indiana State University were based on user-needs research through the cooperation of a behavioral scientist and an architect (Wener, 1988). The study examined both high-rise and low-rise dormitories.

Psychologist Robert Sommer examined patterns of crime and vandalism in university residence halls at the University of California, Davis (Sommer, 1987). An earlier study of his (Sommer, 1968) had shown that residents of these high-rise dormitories tended to describe the buildings negatively, finding them impersonal, sterile, and institutional, comparing them with hotels, prisons, hospitals, and even ant farms.

In the later study Sommer (1987) compared the experiences of two sets of residence halls: cluster and high-rise. The cluster halls, eight two- and three-story frame units, each accommodating about 60 students, were built during 1965–1966 at a total cost of about $3067 per occupant. By contrast, the high-rise halls, four five-story towers, each housing 200 students, were built between 1961 and 1963 at a cost of approximately $4584 per occupant. Data were obtained from campus police and campus housing office records to examine crime and vandalism. Results indicate that even when adjusted for population size, the number of incidents and civil offenses in high-rise halls was double the number in cluster dorms. Findings revealed one crime for every 25 occupants in cluster hall dorms, compared to one crime for every five occupants in high-rise halls. In addition, damage and vandalism costs in the high-rise buildings, when compared with those in cluster halls, averaged 25% higher (Sommer, 1987).

Keegan (1991) paints a grim picture of high-rise dormitories. He examines the current status of a few of the hundreds of huge tower dormitories built across the United States in the mid-1960s and early 1970s. He argues:

> You may recall that high-rises were considered the most sophisticated approach to metropolitan planning in those days. But the unintended result was a radical experiment in adolescent socialization that over the past quarter-century has proven itself to be a colossal, shameful mistake....These massive complexes not only mock the basic concept of an academic environment and stunt the intellectual and psychological development of their residents, but also place students in physical danger to themselves and one another.

To date the world's largest college dormitory is the Watterson Towers, located at Illinois State University at Normal, Illinois (Fig. 6.11). A 1991 article in the *Chicago Tribune* described the building (Smith, 1991):

> ...a dormitory that, if it did not exist in reality, would surely have been invented by horror novelist Stephen King....Completed in 1968 at a cost of $21.6 million in the midst of a boom that tripled the student population in one decade, this twin-towered monster was Illinois State's high-rise solution to a chronically cramped campus.

Watterson Towers houses 2400 students in two 91.44-m (300-ft) 28-story towers, providing 55,647 m$^2$ (599,000 ft$^2$) of dorm and cafeteria space on a ground base of an acre and a quarter. Smith (1991) chronicles the hazards of move-in day, when students, parents, and friends descend upon the building en masse, and horrendous waits for loading and unloading elevators occur. Until recently, only four elevators served all 2400 residents, and school officials reported that it was not unusual for large numbers of students to spend all their lunch break waiting for an elevator to take them to the cafeteria. Four additional elevators were added in 1990. Exacerbating the problem further, elevators stop only at every third floor, making move-ins and move-outs a nightmare.

A review of some of the archives from Illinois State University at Normal, focusing on newsworthy events about Watterson Towers (1988–1992), is revealing. Among the incidents reported to the press were the following.

An 18-year-old freshman was shot in the middle of the afternoon on the 19th floor of Watterson Towers. Because of the large size of the dormitory, many students were probably not aware of the shooting until hours later. It turns out that a student and his friends were simply playing with a gun when it accidentally discharged.

A fire broke out on the sixth floor. Sixteen students and two firefighters were treated at hospitals for smoke inhalation. Two brothers, residents of the dorm, were later charged with setting the fire that caused nearly $500,000 in damage to Watterson Towers. The incident was apparently a prank that got out of control. The students burned several couches and chairs that had been placed in front of an elevator on a walkway between the building's two residential towers.

A 19-year-old woman survived a fall from her 18th-floor dormitory window at Watterson Towers. The student was standing on a heating unit in front of her window, apparently lost her balance, fell through a screen, and plunged 54.8 m (180 ft). Since that accident, warning stickers telling students not to stand on their window heating units have been placed on all Watterson Tower dormitory windows.

**Fig. 6.11 Watterson Towers at Illinois State University, Normal, is the world's largest dormitory. Nicknamed a "twin-towered monster," it houses 2400 students in two 28-story towers.** (*Courtesy: Illinois State University, Normal.*)

Dormitories and other forms of high-rise housing present a myriad of social and psychological issues. Among the most pressing social problems are vandalism, criminal behavior, and fear of crime. Although a great deal of research has already been done, much more than space permits here, many questions remain. For instance, how do individuals' residential histories affect their perceptions of high-rise housing? Are students who have lived in high-rise dormitories likely to adapt more easily, or with greater difficulty, to high-rise living when they are older? What may be the effects of the new waves of immigrants—Russian Jews moving to Israel, former East Germans moving to West Germany—on the social patterns found in high-rise housing?

## 6.9 TALL PARKING GARAGES

The parking garage has become an increasingly common tall building type laden with psychological issues. Although millions of people use parking garages each day, little research has been conducted on this building type. Until recently the design of the parking garage has not received much attention in the architectural community. It is also considered out of the realm of most landscape architects, and it has become a type of no-man's-land. A few isolated articles have been published in English-speaking planning and design journals, but these are still relatively rare. Garages are rarely assigned as design projects in architectural schools. Few architects earn their reputation by designing parking garages—but this need not be the case. Additional information on parking garages can be found in Chapter PC-3 of *Planning and Environmental Criteria for Tall Buildings* (CTBUH, Group PC, 1981).

### 1 Users' Perceptions of Parking Facilities

A body of related research has focused on perceptions of parking lots (Edmonson, 1991). Some of the findings from this research may well apply to how people perceive tall parking garages as well. This series of studies examined how users, developers, and landscape architects perceived large parking lots. Results indicated that developers and landscape architects tend to place minimal importance on the design of the parking lot. By contrast, users view parking lots as important spaces that inform them about the values behind the nearby buildings, and the institutions they represent. Users believe not only that parking facilities reflect the prestige of the facilities they serve, but also that they reflect the owners' concern for the well-being of their customers. Users generally find parking lots visually unattractive.

According to Edmonson (1991), the following features influence user satisfaction with parking lots: the extent to which the lot is maintained, the extent to which one feels safe, the ease of entering and exiting the lot, the extent to which adequate space is available for maneuvering the car, and the presence of nature.

Among the key psychological issues surrounding tall parking garages are legibility, way-finding, and safety. When they leave their vehicles, people need to be able to find the exits easily. When they return to their vehicles, they need to be able to locate them without much confusion. Given that many garages are multilevel but underground, the usual ability to identify landmarks in the surrounding environment is altogether absent. Hence graphic design and layout are key.

Safety both from the danger of automobile traffic and from criminal activity is extremely important. Garages are often the settings for vandalism, muggings, thefts, rape, and murder. Due to the fact that they are often poorly lit and that the outside world cannot see inside, they are often among the most dangerous buildings in today's cities. The number of possible hiding places for criminals, the overall visibility within the garage, and the distance to one's destination are all factors that influence the extent to which people experience fear in parking garages (Edmonson, 1991).

Age is often a factor in the perception of safety in parking garages. Older individuals sometimes fear garages more than do their younger counterparts. Gender is also an issue. Parking garages are usually more dangerous places for women than they are for men from the viewpoint of personal safety against assaults. In many locations, women traveling alone must exercise extreme caution when using a parking garage at night. Some garages offer escort services to help women feel more comfortable, but far better is the garage so well designed that escort services are unnecessary.

Another psychological issue in many parking garages is the use of assigned spaces. Some organizations purchase parking spaces, assuring either an unassigned spot in a series of reserved spaces or a spot specifically reserved for particular employees. In such cases the parking spot becomes a tangible symbol of one's status within the organization. A spot close to an elevator or to an exit is likely to be assigned to an employee with high status. The parking spot becomes a type of public territory to which owners become highly attached.

Garage users are exposed to traffic, noise, and the smell of gasoline fumes. In this respect, garages are among the least hospitable environments in our urban areas. The most successful garage designs, from a psychological perspective, are those that minimize the distance from the vehicle to the exit so that the amount of time spent in them, especially as a pedestrian, is reduced.

## 2 User-Friendly Parking Garages

Parking garages have long been aesthetically displeasing and perceived by pedestrians and users as unsafe. In recent years, however, an increasing number of jurisdictions in the United States are beginning to take a more active stance in regulating parking garage design. The results are likely to have positive psychological effects on both garage users and pedestrians alike. Among the suggested design amenities are unifying the facade with the structure that the garage is intended to serve, avoiding blank walls, adding more landscaping, and in some cases including rooftop planting so that the view from nearby high rises will be improved. Some of the best examples of new, improved parking garage design can be seen in Orlando, Florida; Irvine, California; Kansas City, Missouri; and Chicago, Illinois (Zotti, 1987).

Although Chicago has few formal rules governing parking garage design, the efforts of public officials working with some enlightened developers have resulted in some whimsical structures that can serve as models elsewhere. At 203 North La Salle in Chicago, a garage sits below a 15-story office building. The garage cannot be easily distinguished from the rest of the building since it is hidden by a two-story retail arcade. Once inside, finding your car is easy due to an innovative musical floor-reminder system. Each floor is identified by the name of a city associated with a catchy song, such as Frank Sinatra singing "New York, New York." The patented system has been incorporated into several other parking decks not only in Chicago, but on the West Coast as well.

At another garage in Chicago's theater district, the 12-story, 950-space Theatre District Self-Park, the musical reminders are famous showtunes and Broadway musicals. Entrances are highlighted with theater marquees. Myron Warshauer, president of Standard Parking Corporation, invented this creative system with the intention of making his garages "people friendly" (Zotti, 1987).

Also in Chicago's Loop is another unusual garage whose facade recalls the grille of a vintage car (Fig. 6.12). Garages like these can be fun, exciting places, but at present, they are rare.

Parking garages that appear to be the least offensive spaces to pedestrians are often those that are relatively indistinguishable from the surrounding building facades. Once in a while one may come across a high-rise garage that includes retail at street level and

**Fig. 6.12 This unusual Chicago parking garage, designed by architect Stanley Tigerman, recalls the grille of a vintage car.** (*Courtesy: Barbara Karant.*)

only a minimal number of entry and exit driveways. While these retail units reduce the number of parking spaces available at street level, they serve to create a more lively, interactive streetscape for pedestrians. Most people find window shopping a more pleasant, entertaining experience than simply looking at a row of parked cars or, even worse, a blank wall.

Designers, developers, and planners must begin to pay much more attention to the design and management of high-rise parking garages. Parking and parking garages have long been issues of great importance to the public. Safety, legibility, way-finding, and territoriality are among the psychological issues that help explain how individuals perceive tall parking garages. Some possible research questions include: In what ways can tall parking garages be designed to make hunting for a parking space easier and the different levels more identifiable? How can safety in tall parking garages be enhanced? How can the environmental quality in parking garages be improved?

## *6.10 HIGH-RISE SHOPPING CENTERS*

In many cities across the United States the high-rise shopping center is a burgeoning building type. In other cities, the in-city shopping center has become a series of tall, free-standing structures that incorporate shops, restaurants, cinemas, and other commercial uses. One of the most innovative such schemes to be built in recent years is Horton Plaza in San Diego, California. Opened in 1985, it is one of the United States' first in-city post-modern shopping centers. Another highly visible urban shopping center is the Beverly Center in West Hollywood, California.

This section addresses mall psychology at the urban scale, how different kinds of users—men and women, teens and elderly, for example—experience shopping and the vertical shopping mall.

### *1 Mall Psychology at the Urban Scale*

Tall urban shopping centers share many traits of the ubiquitous suburban shopping mall. The contemporary urban department store incorporates some of the more successful elements of suburban shopping malls, for example, the interior "streetscape" concept with specialty and gourmet items featured on either side. The ability to attract shoppers not only to the shopping center's major anchor stores but also to smaller boutique shops is a serious challenge to shopping center designers and managers alike. A fine line often exists between creating an environment that is stimulating but not overwhelming. Kowinski (1985) addresses the psychological experience of the shopping center, both in suburbia and in the city. (For another analysis of shopping centers, consult Jacobs, 1984.)

Kowinski uses the term "the zombie effect" to describe a floating, timeless sensation that can suddenly be interrupted by the shocking realization that "you've been here all day" (Kowinski, 1985). The shopping center can have a hypnotic effect such that time simply passes by without notice. Kowinski's numerous interviews with mall employees showed that many suffer from these symptoms, complaining about the uniform sound, the lighting, and being cut off from all natural elements. While large atrium gathering spaces for shoppers are often sun-drenched on a bright day, shopping center employees complain that they have no idea what time of day it is or what the weather is like outside. Similarly, from the outside, except on the ground level, pedestrians have little clue as to what is going on inside the building. Usually they are faced with a massive series of blank walls. The image sent to outsiders and insiders can be sharply marked.

Woodward (1978) identified several psychological issues of concern to shopping center designers. These issues can just as readily apply to designers of high-rise shopping centers. They are: duration of shopping time, traffic control, building scale and materials, sound and light factors, psychological barriers, movement motivators, and entertainment attractions.

The duration of shopping time is one of the most crucial. The longer people stay at a shopping center, the more money they are likely to spend. The integration of atrium food courts, cinemas, automated teller machines, and other services into the center have proven effective ways to induce shoppers to spend more time in the center. In the most successful centers, the business day is expanded into the late night hours.

Traffic control in and out of the center is another key issue. The designer's intention often is to design a variety of entries so that business is distributed evenly, rather than concentrated only in a few key spots. Reducing the scale of a large space through landscaping, street furniture, and other elements helps shoppers feel comfortable rather than overwhelmed by a series of massive, overscaled spaces.

For instance, the food court at the top level of the retail component of Chicago Place (Figs. 6.13 and 6.14) features a 19.8-m (65-ft)-tall atrium whose scale is softened for pedestrians by lush tropical landscaping, water features, mood lighting, and a generous supply of small cafe-style tables and chairs. Nearby, Chicago's Water Tower Place employs a series of cascading fountains near the main entrance to mask irritating noise from traffic and crowds.

Doorways are among the greatest psychological barriers to shoppers. In most urban shopping centers, storefronts are entered through wide gates that are left open during business hours. The divisions between the main circulation areas and individual shops are often intentionally unclear, enabling shoppers to flow freely from one space to another. Another psychological barrier is the storefront vacancy. At most shopping centers, decorative coverings are used on vacant shops to maintain a festive atmosphere.

In order to motivate people to move throughout the center, designers often pay great attention to escalators and elevators. As in hotel atria, the glass elevator has become increasingly common in urban shopping centers, and it is highly visible. It provides shoppers with a panoramic view of the center and encourages them to stop along the way. The layout of shops in the center also is intended to encourage movement throughout the center. Stores selling similar kinds of merchandise, such as shoes, are usually scattered throughout the center to encourage comparison shopping.

## 2 Users' Experiences of Shopping

Urban shopping centers serve many psychological functions for different users. Gender and age appear to influence the ways in which people perceive and behave in shopping centers. For women who remain home to take care of children, the shopping center may be one of the few public places to which they can escape, particularly in inclement weather.

Many large department stores cater primarily to women customers, with the vast majority of space devoted to women's clothing. Perfumes, cosmetics, hosiery, scarves, hats, and other accessories usually dominate the first floor, creating an enticing shopping environment for women. Usually more small boutique shops at urban shopping centers are aimed at women rather than at men. Hence the physical environment of the urban shopping center both reflects and promotes gender-related shopping behavior. (For more information on shopping and gender, refer to Greene, 1983.)

For teenagers, the shopping center has become a hangout, a social gathering space, a contemporary town square (Anthony, 1985b). For the elderly, the shopping center is often a place for recreation. Mall walking has become common throughout the United States as a major form of exercise for the elderly. The indoor urban shopping center has also become a place for elderly people to sit and relax. Brown et al. (1986) conducted a study of a large multipurpose center in Montreal, Canada, featuring a four-story atrium lined with stores and linking three office towers and a hotel. Researchers were surprised to find a disproportionate representation of elderly people in the center. They also discovered that compared to other user groups, the elderly sat the longest. Average sitting

**Fig. 6.13 Nighttime streetscape of Chicago Place.** (*Courtesy: Nick Merrick.*)

time for the elderly was 13.1 minutes, compared to 11.2 minutes for adults, 6.8 minutes for teens, and 10.2 minutes for children (Brown et al., 1986).

Thus for many individuals, the shopping center has become a kind of "third place"—a respite from the treadmill between home and school, or home and work, that provides a sense of belonging and an opportunity to gain a different perspective on oneself (Jacobs, 1984; Oldenburg and Brisset, 1980). For people of all ages, the shopping center is a locus for people watching, entertainment, and leisure.

**Fig. 6.14 Food court at top level of retail component of Chicago Place.** (*Courtesy: Nick Merrick.*)

### 3 *The Vertical Shopping Mall*

The American shopping center, with its grand staircases, escalators, majestic fountains, skylights, gardens, and atria, presents the illusion of affluence and grandeur. The meticulous degree of maintenance, such that glittery surfaces, mirrors, and window displays look brand new all year long, lends a theatrical aura as well. Some researchers have aptly referred to the shopping center as "the signature building of our age, a central hub of community life" (Weisman, 1992).

For example, along Chicago's Michigan Avenue, the city's most famous shopping street, commonly known as the Magnificent Mile, sit three vertical shopping malls: Water Tower Place, 900 North Michigan Avenue, and Chicago Place. Water Tower Place, a 62-story tower, was completed in 1976. Its first eight stories are devoted to retail—at the time of this writing, 132 shops, 13 restaurants, and seven movie theaters. Sitting atop this are four stories of offices, and above these are a 400-room hotel and 260 condominium units.

Only a few blocks away is 900 North Michigan Avenue, completed in 1989. This 265.5-m (871-ft) structure includes a mix of retail, hotel, office space, and residential uses.

The most recent addition is Chicago Place, located at 700 North Michigan Avenue and completed in 1991. This 185.3-m (608-ft) high rise features an eight-story atrium mall base topped by a 51-story residential condominium tower.

Even in recessionary times, all three of these Chicago urban shopping centers are usually teeming with activity. They have become hot spots for tourists as well as Chicagoans. At holiday time, they are almost bursting at the seams.

It may be that today's "cathedral of commerce" is no longer the skyscraper but the urban shopping center. Similar trends in shopping centers can be found in Canada, some European countries, Hong Kong, Singapore, and a few other Asian capitals. Although essays on shopping centers abound, much more research is needed about how people perceive and use them. Among the key questions are: Just how public are high-rise shopping centers? Do certain individuals feel more or less welcome? Are these centers in effect screening out "undesirables?"

## 6.11 FUTURE OUTLOOK

Additional research is needed to better understand the psychological and behavioral aspects of tall buildings. Based on the material presented here, some intriguing questions for further research are raised. Which symbolic aspects of city form can enhance legibility, making it easier for visitors and residents of the city to find their way around? How can we help prepare occupants of tall buildings for the psychological and physiological effects of building sway specifically under earthquake loading? What else can we learn from how tall building occupants experience earthquakes? What sizes and shapes of windows lead to positive and negative psychological impacts on workers? How much light is optimal in a high-rise work environment? What is the role of the view out the window on worker productivity? How can barrier-free design be improved? In what ways can we improve the design of high-rise housing to make units more satisfying to families with young children? In what innovative ways can we create parking garages, housing, office towers, shopping centers, and other high-rise structures that better meet the needs of the people that use them?

These issues are complex and varied. It is paramount, however, that all those concerned with the design of tall buildings become aware of the vast array of psychological meanings that these spaces convey for different kinds of people.

The collaborative efforts of today's architects, engineers, developers, clients, planners, urban designers, politicians, and local citizens will produce the next generation of high rises. Their awareness of the relationships among tall buildings, human perception, and behavior can help the creation of future generations of tall buildings that will better respond to users' needs.

After all, tall buildings are not only here for the users of today. Many of these structures will outlast those who designed them and those who currently inhabit them. As such, tall buildings represent the most tangible, lasting legacies of our era. Critic Max Kozloff (1990) put it well: "For each new generation, skyscrapers reveal to the citizenry a message about the general political culture.... As they demonstrate it with a more theatrical, abrupt, destabilizing, and overbearing presence than any other urban building type, they confront us with the most visible remnants of our successive political cultures."

## 6.12 CONDENSED REFERENCES/BIBLIOGRAPHY

The following is a condensed bibliography for this chapter. Not only does it include all articles referred to or cited in the text, it also contains bibliography for further reading. The full citations will be found at the end of the volume. What is given here should be sufficient information to lead the reader to the correct article—author, date, and title. In case of multiple authors, only the first named is listed.


Altman 1975, *The Environment and Social Behavior*

ANSI A117.1 1980, *American National Standard Specifications for Making Buildings and Facilities Accessible to and Usable by Physically Handicapped People*

Anthony 1981, *International House: Home away from Home?*

Anthony 1985a, *Public Perceptions of Recent Projects: A Berkeley Class Conducts Evaluations of Five Buildings and Spaces*

Anthony 1985b, *The Shopping Mall: A Teenage Hangout*

Appleyard 1964, *The View from the Road*

Appleyard 1969, *Why Buildings Are Known*

Appleyard 1976, *Planning a Pluralist City*

Appleyard 1977, *High-Rise Buildings versus San Francisco: Measuring Visual and Symbolic Impacts*

Appleyard 1979, *The Environment as a Social Symbol*

Archival File of Press Releases 1988–92, *Watterson Towers*

Arnold 1986, *A Guide to Earthquake Motion and Its Effect on Buildings*

Attoe 1981, *Skylines: Understanding and Molding Urban Silhouettes*

Bain 1989, *An Approach to Buildings: The Entry Sequence*

Becker 1990, *The Total Workplace: Facilities Management and the Elastic Organization*

Bednar 1986, *The New Atrium*

Beijing Review 1985, *High-Rises Spell Isolation for Residents*

Benson 1987, *The State of Space*

Bigart 1965, *A Night of Confusion, Frustration and Adventure*

Blakeslee 1992, *Navigating Life's Maze: Styles Split the Sexes*

Block 1985, *Casualty Reports*

Bosselman 1983a, *Shadowboxing—Keeping the Sunlight on Chinatown's Kids*

Bosselman 1983b, *Sun and Light for Downtown San Francisco*

Boubekri 1991, *Impact of Window Size and Sunlight Penetration on Office Worker's Mood and Satisfaction*

Branch 1987, *State of Illinois Center Update*

Brill 1984, *Using Office Design to Increase Productivity*

Bristol 1991, *The Pruitt-Igoe Myth*

Brown 1986, *The Community Role of Public Indoor Space*

Brugmann 1971, *The Ultimate Highrise: San Francisco's Mad Rush toward the Sky*

Conway 1977, *Human Response to Tall Buildings*

Cooper Marcus 1977, *Design Guidelines for High-Rise Family Housing*

Cooper Marcus 1990, *People Places: Design Guidelines for Urban Open Space*

CTBUH 1978, *Philosophy of Tall Buildings*

CTBUH, Committee 56, 1992, *Building Design for the Handicapped and Aged Persons*

CTBUH, Group PC, 1981, *Planning and Environmental Criteria for Tall Buildings*

Deasy 1985, *Designing Places for People*

*Design News* 1986, *High-Tech Elevator Buttons Deter Vandals*

Devlin 1990, *An Examination of Architectural Interpretation: Architects versus Non-Architects*

*Diagnostic and Statistical Manual of Mental Disorders,* 1987

Doctor 1989, *The Encyclopedia of Phobias, Fears and Anxieties*

Dovey 1991, *Tall Storeys: Corporate Towers and Symbolic Capital*

Edmonson 1991, *Eyesores or Assets? How Landscape Architects, Developers and Users Perceive Large Parking Lots*

Evans 1982, *Cognitive Maps and Urban Form*

Francescato 1979, *Residents' Satisfaction in HUD-Assisted Housing: Design and Management Factors*

Frederking 1973, *An Integrated Methodology for Office Building Elevator Design*

Fuerst 1985, *High-Rise Living: What Tenants Say*

Ginsberg 1985, *The Pattern and Meaning of Neighbor Relations in High-Rise Housing in Israel*

Goffman 1963, *Behavior in Public Places*

Goldman 1980, *The Empire State Building Book*

Greene 1983, *But Shopping Is for Women*

Groat 1979, *Does Post-Modernism Communicate?*

Groat 1982, *Meaning in Post-Modern Architecture: An Examination Using the Multiple Sorting Task*

Groat 1984, *Contextual Compatibility in Architecture: An Investigation of Non-Designers' Conceptualizations*

Haber 1977, *The Impact of Tall Buildings on Users and Neighbors*

Haber 1986, *Response of the Elderly and Handicapped to Living in a Barrier-Free Tall Building*

Hackett 1987, *San Juan's Towering Inferno: A Suspicious Hotel Fire Leaves Close to 100 Dead*

Hall 1966, *The Hidden Dimension*

Hayden 1977, *Skyscraper Seduction, Skyscraper Rape*

Hedge 1986, *Open versus Enclosed Workspaces: The Impact of Design on Employee Reactions to Their Offices*

Hurt 1988, *Hotel Fire: Cage of Horror*

Husaini 1991, *Social and Psychological Well-Being of Black Elderly Living in High-Rises for the Elderly*

Jacobs 1984, *The Mall: An Attempted Escape from Everyday Life*

Jencks 1980, *Skyscrapers—Skyprickers—Skycities*

Kaplan 1985, *Celebrating Urban Gathering Places*

Keegan 1991, *Blowing the Zoo to Kingdom Come*

Knight 1979, *Crowding in Elevators: Duration of Post-Exposure Effects*

Knowles 1981, *Sun, Rhythm, Form*

Kowinski 1985, *The Malling of America: An Inside Look at the Great Consumer Paradise*

Kozloff 1990, *Skyscrapers, the Late Imperial Mob*

Kreimer 1973, *Building the Imagery of San Francisco: An Analysis of Controversy over High-Rise Development 1970–1971*

Kridler 1992, *Access for the Disabled: Part 1*

Levin 1987, *Death in a Towering Inferno*

Lynch 1960, *The Image of the City*

Lynch 1984, *Good City Form*

Marks 1969, *Fears and Phobias*

Martin 1987, *Center Throws Chicago for a Loop*

Ministry of National Development 1985, *Design Guidelines on Accessibility for the Disabled in Buildings*

Mozingo 1989, *Women and Downtown Open Spaces*

Murphy 1985, *2000 and Beyond: The New State of Illinois Center: Infamous, a Noble Effort, or Both?*

Nahemow 1977, *Elderly People in Tall Buildings: A Nationwide Study*

*National Geographic* 1989, *Skyscrapers*

National Organization of the Handicapped 1983, *Requirements for Access: Manual for Building Entrancible and Usable for Handicapped People*

Nemiah 1985, *Phobic Disorders (Phobic Neuroses)*

Newman 1972, *Defensible Space: Crime Prevention through Urban Design*

*Newsweek* 1965, *The Longest Night*

*Newsweek* 1993, *Terror Hits Home: The Hunt for the Bombers*

*The New York Times* 1993, *Blast Hits Trade Center, Bomb Suspected; 5 Killed, Thousands Flee Smoke in Towers*

Normoyle 1988, *The Defensible Space Model of Fear and Elderly Public Housing Residents*

Oldenburg 1980, *The Essential Hangout*

Passini 1984, *Wayfinding In Architecture*

Phillips 1965, *Blackout Vignettes Are Everywhere You Look*

Powell 1989, *Ken Yeang: Rethinking the Environmental Filter*

Prokesch 1993, *Inching Back to Life, Trade Center Tallies the Cost: Businesses Concerned with How Far Insurance Will Cover Them*

Public Works Department 1990, *Code on Barrier-Free Accessibility in Buildings*

Rachman 1990, *Fear and Courage*

Raschko 1982, *Housing Interiors for the Disabled and Elderly*

Reinhardt 1971, *On the Waterfront: The Great Wall of Magnin*

Russell 1990, *Designing the Super-Thin New Buildings*

Sabbagh 1989, *Skyscraper*

Saliga 1990, *The Sky's the Limit: A Century of Chicago Skyscrapers*

Sharp 1989, *Tropical Heights*

Sletteland 1971, *A City of Skyscrapers*

Smith 1991, *Faulty Towers: Students at Illinois State U. Learn to Live with High Anxiety in the World's Largest College Dorm*

Sommer 1986, *Student Reactions to Four Types of Residence Halls*

Sommer 1987, *Crime and Vandalism in University Residence Halls: A Confirmation of Defensible Space Theory*

Stamps 1991, *Public Preferences for High Rise Buildings: Stylistic and Demographic Effects*

Strauss 1976, *Images of the American City*

Sundstrom 1986, *Work Places: The Psychology of the Physical Environment in Offices and Factories*

Sundstrom 1987, *Work Environments: Offices and Factories*

Taylor 1992, *In Pursuit of Gotham: Culture and Commerce in New York*

*Time* 1965, *The Northeast: The Disaster That Wasn't*

*Time* 1993, *Tower Terror*

UFAS 1984, *Uniform Federal Accessibility Standards*

Van der Ryn 1967, *Dorms at Berkeley: An Environmental Analysis*

Vergara 1989, *Hell in a Very Tall Place*

Weisman 1979, *Way-Finding in the Built Environment: A Study in Architectural Legibility*

Weisman 1981, *Evaluating Architectural Legibility: Way-Finding in the Built Environment*

Weisman 1992, *Discrimination by Design: A Feminist Critique of the Man-Made Environment*

Wener 1988, *Doing It Right: Examples of Successful Application of Environment-Behavior Research*

White 1965, *What Went Wrong? Something Called 345 kV*

Whyte 1980, *The Social Life of Small Urban Spaces*

Whyte 1988, *City: Rediscovering the Center*

Williamson 1981, *Adjustment to the Highrise: Variables in a German Sample*

Wineman 1982, *The Office Environment as a Source of Stress*

Wineman 1986, *Behavioral Issues in Office Design*

Woodward 1978, *Modern Shopping Center Design: Psychology Made Concrete*

World Health Organization 1980, *Classifications of Impairments, Disabilities and Handicaps*

Yeang 1990, *High-Rise Design for Hot Humid Places*

Zotti 1987, *Stacking the Decks for Better Design*

Zuckerman 1983, *Civil Inattention Exists—In Elevators*

# 7

# Facades: Principles and Paradigms

This chapter examines the integral roles that architectural style and composition play as determinants of the image, materials, and techniques of tall building facades and their relationship to the aesthetics, form, and context of the building. The chapter investigates the nature of the exterior wall and cladding from an architectural and theoretical point of view, relating the influences of technology and aesthetic intentions as codeterminants of architectural facades, and addresses the issues of style, imagery, and symbolism of tall building facades. The topic of the facade has been extensively treated from a technological point of view in Chapter SC-5 of *Tall Building Systems and Concepts* (CTBUH, Group SC, 1980) and in *Cladding* (CTBUH, Committee 12A, 1992).

Section 7.1 provides an overview of style and imagery as determinants of facade expression. Sections 7.2 and 7.3 discuss the composition and development of tall building facades from a theoretical perspective, and the evolution of steel-frame technology and its impact on tall building facades. In Section 7.4 commercial and corporate tall buildings are used to illustrate how facades and forms are used for architectural expression. The impact of modernism and the international style on the aesthetics of facades is presented in Section 7.5 by focusing primarily on the contributions of Mies van der Rohe and Le Corbusier. Section 7.6 looks at the facades of commercial buildings and the return to historical reference in tall building facades. Sections 7.7 and 7.8 focus on recent trends in facade design and on the interrelationships among aesthetics, cladding materials, and technology in tall building facades. Section 7.9 illustrates the role that new technologies are playing in the development of tall building facades and the redefinition of tall building architecture. New developments in tall building facades are discussed in Section 7.10 with examples selected from tall buildings throughout the world. In Section 7.11 potential areas of future research are discussed. Factors in analyzing the quality of construction of tall building facades are presented in Section 7.12.

## 7.1 STYLE AND IMAGERY OF FACADES

The emphasis given to the technical and engineering aspects of high-rise building design often overshadows the formal impact of tall building facades on the architectural vocabulary (that is, style) of the building and its relationship with the ideals and attitudes from which that vocabulary stems. It is important to recognize that tall building design is a holistic process, and it is impossible to consider formal and aesthetic principles in isolation without reference to technology and construction. The relationship of tall building facades to the urban streetscape involves a consideration of social values, followed by special reference to questions of scale and traditional design principles versus contemporary expressions of architectural design objectives.

Facades of tall buildings have a significant influence on the imageability and symbolism of tall buildings. Kenneth Turney Gibbs (1984), writing about the tall office building as corporate image and one of the most visible symbols of American capitalism in the late nineteenth and early twentieth centuries, states that it has frequently been used to illustrate both the ills and the virtues of American society. Expressly due to the ease with which the typology lends itself to nonarchitectural treatment, even had it not originated in America (see Chapter 5), the tall building would have still been considered a visible embodiment of the American economic system. As early as the late nineteenth century, the image of the tall building as an American corporate icon was apparent, as indicated by Oscar Lovell Triggs, Professor of English at the University of Chicago (Gibbs, 1984):

> The commercial temple, largely the product of the American mind, is the exact equivalent of the modern business ideal. The daring, strength, Titanic energy, intelligence and majesty evidenced in many of the modern business temples indicate precisely one, and perhaps the dominant feature of American character.

Not all critics, however, agreed with Triggs's admiration of the tall building as a positive symbol of American capitalism. Their writings convey a somewhat pessimistic and moralistic attitude and regard the tall building as a negative symbol of corporate greed and avarice. One critic considered skyscrapers "monuments to a dangerous prosperity and to a pride which, like Beauvais, mistook great height for civic grandeur" (Gibbs, 1984). Lewis Mumford understood the tectonic value of the steel frame, yet was critical of the social value of the tall office building (Gibbs, 1984). Antagonism toward American business caused him to view the buildings of business with suspicion. In his novel *The Cliff-Dwellers,* Henry B. Fuller found the buildings of Chicago to be tangible proof of the excesses of American political and economic systems. The people who built them possessed "an extreme, even undue liberty to indulge in whatever...idiosyncrasies [were] suggested by greed, pride, carelessness," whereas the buildings themselves expressed "democracy absolute [sic] manifested in brick, stone, timber. The sociological interest is great; its artistic value is nil" (Gibbs, 1984).

## 1 Development of Tall Buildings

Although it was influenced by European models and theory, the skyscraper or tall building type developed almost completely in two American cities—Chicago and New York (Pelli, 1982). Today the skyscraper remains a preeminently American building type.

The evolution of high-rise buildings, as mentioned previously, was made possible by the development of structural steel construction and by the invention of the elevator in 1853. Whereas masonry high-rise buildings, such as the Monadnock Building (1891; Fig. 7.1) in Chicago, could be built to 16 stories, the steel frame made tall buildings structurally stronger, more flexible, and more economical. Therefore the once load-bearing requirements of the structural masonry wall could now be reduced to a thin veneer of masonry, stone, or even glass. The evolution of the tall building, therefore, is the result of the integration of technological and engineering innovations with architectural expression. The impact of this development was holistic, affecting not only the form of the tall building, but also the construction, materials, and stylistic imagery of the facade.

Cesar Pelli (1982), who is known for his minimalist design entailing slick detailing and exquisite use of glass as a facade material, and Ada Louis Huxtable (1984) cite four periods in the development of the skyscraper (see also Chapter 1). The first skyscraper age began with the invention of the elevator in 1853 and lasted until 1908–1909. During this period, architects in Chicago attempted to adapt existing building types, most

**Fig. 7.1 Monadnock Building, Chicago.**

notably the palazzo, to new heights allowed by elevators. The second skyscraper age or eclectic period occurred in New York and started with the Singer Building (1908) and the Metropolitan Life Insurance Company tower (1909). These buildings were characterized by the exuberant possibilities inherent in the skyscraper style and exploited the potential for vertical expression related to the facades, the articulated base, expressive tops, and ornamentation (see Chapter 5). This phase lasted until the mid-1930s and was brought to a close by the Great Depression. The third skyscraper age started after World War II and lasted until the mid-1970s. It saw the emergence of high-rise buildings as the dominant variation of the skyscraper style. International style modernism, with its emphasis on abstract, stereometric massing and sleek glass curtain-wall construction derived its stylistic and theoretical inspiration from European modernists like Walter Gropius and Mies van der Rohe. The fourth skyscraper age—post-modernism, late-modernism, and high-tech—began in the late 1970s and continues today. The post-modern style, considered retrograde by some architects, places primary emphasis on the scale, context, and stylistic characteristics of the facade. Post-modernism represents an attempt to realign modern technology and construction methods with historical architectural styles, scale, proportion, and urban design principles. Late-modern architecture is an adaptation and continuation of modernist principles to complex issues of theory and context. High-tech architecture, on the other hand, exploits technology to its fullest by articulating the structural and mechanical systems of the building as an expression of the facade, function, and building form. (For a detailed discussion of these historical periods, refer to Chapter 1.)

## 2 The Machine Aesthetic

After World War II, architecture was at a crossroads: the historicists' picturesque revivalism, the eclectic application of neo-Gothic, classical, and Moorish styles, was beginning to be supplanted by the streamlined influence of modernist architecture. According to Goldberger (1981), the late 1940s and early 1950s were a time for pragmatism. The economics of large-scale construction had also changed. There were no longer as many artisans available to execute the elaborate handwork that ornamented buildings of the art moderne and art deco styles. Adapting more and more to standardization, buildings were becoming more machine-made, and this had a profound impact on the image, construction, and technology of the tall building facade. Simple, abstract, and austere surfaces free of ornamentation were embraced by architects as criteria for a new aesthetic appropriate to the late twentieth century.

***The First Machine Age.*** The first machine age, described by Reyner Banham in 1960, evolved its architectural theory and practice during the first quarter of the twentieth century and continued until about the mid-1960s. It followed preliminary developments in the preceding century (Banham, 1960):

> When Marinetti—the first Futurist in goggles and flapping scarf—and his wrecked car were lifted from their suburban ditch, stripped and cleaned by the effluents of industry they were illumined by an understanding of a crucial change in the relationship of men—especially thinking men—and their machines; both were now stripped for action—the man freed of encumbering ideologies, the car stripped of the dead-weight of mud-guards and bodywork.

According to Banham, the machinery of the preceding Victorian industrial age of "cast iron, soot, and rust had been ponderous, simple-minded, tended by a mass-proletariat in parts of the world that were remote from centers of enlightenment and culture." In contrast, the machines of the first machine age of the early twentieth century were light, subtle, clean, and could be handled by educated people in the new electric suburbs. One of the symbols for this new and dynamic age was the tall building. The pristine steel and glass curtain-walled machine aesthetic of high-rise construction was perfectly suited to modernist ideology.

***The Second Machine Age.*** Jencks (1988) writes that the second machine age (beginning in the early 1980s and continuing through the present) is different from the first, the work of the pioneers of modernism, in several important aspects. Flexibility, change, and movement, an important component of the work of Archigram in England, the Dutch structuralists, and the Japanese metabolists, contrast significantly with futurism's application of movement to static monuments. Standardized production, embraced by the early modernists as a panacea to supplant historical styles and eclecticism, is eschewed by the second machine aesthetic for variation. Since computerized assembly lines can vary components with ease, architecture can be adapted to individual needs like any other commodity. Furthermore, architects using a standardized "kit of parts" might easily reconfigure the component elements to suit building types, functions, contexts, and user preferences.

## 7.2 FRAME, WALL, AND CLADDING

Even if all materials are of equal value to the artist, they are not equally suited to all his purposes. The requisite durability and the necessary construction requirements often demand materials that are not in harmony with the true purpose of the building. The architect's general task is to provide a warm and livable space. In the Introduction to *Comparative Building Theory,* Gottfried Semper (1850) identifies four basic elements of primitive building—the hearth, mound, roof, and enclosure (Vuyosevich, 1991). The roof and the supporting column are treated as a unit, leaving enclosure as a separate non-load-bearing element, serving to demarcate space. According to Semper, these first walls were of wickerwork, mats, and hurdles, as distinct from the material of the roof and column (wood) or mound (clay bricks or stone). *Tektonik,* the art of cutting and joining timber, was developed in the making of the roof with its columnar supports; *stereometry,* the art of masonry, was learned in securing firm foundations and building the terrace or the mound.

Adolf Loos draws upon Semper's description of the role of the carpet as an element of enclosure in Assyrian-Chaldean architecture (Loos, 1928):

> The primary material establishing the norm for vertical enclosure was not the stone wall but a material that, though less durable, for a long time influenced the development of architecture as strongly as stone, metal, and timber. I mean the hurdle, the mat, and the carpet. Using wickerwork for setting apart one's property and floor mats for protection against the heat and the cold far preceded making even the roughest masonry. Wickerwork was the original motif of the wall. It retained this primary significance...when the light hurdles and mattings were later transformed into brick or stone walls. The essence of the wall was wickerwork.
>
> Hanging carpets remained the true walls; they were the visible boundaries of the room. The often solid walls behind them were necessary for reasons that had nothing to do with the creation of space; they were needed for protection, for supporting a load, for their permanence, et cetera. Wherever the need for these secondary functions did not arise, carpets remained the only means for separating space. Even where solid walls became necessary,

> they were only the visible structure hidden behind the true representatives of the wall, the colorful carpets that the walls served to hold and support. It was therefore the covering of the wall that was primarily and essentially of architectural significance; the wall itself was secondary.

***Enclosure and Structure.*** Semper's decision to focus on the enclosure system itself as the primary and essential architectural element, as opposed to the structural support of the wall, is significant. To Semper, the cladding of the wall not only was a response to the functional requirements of climate, but most importantly, defined the architectural parameters of the space and embodied an aesthetic intention as well (Vuyosevich, 1991). This direct linkage of the functional, spatial, environmental, and aesthetic characteristics of the most utilitarian and tectonic architectural conditions characterized the discourse that led to the modern movement in architecture.

Adolf Loos extended Semper's illustration of primordial architecture from the cladding of the wall to the frame supporting the cladding (Loos, 1928):

> The architect's general task is to provide a warm and livable space. Carpets are warm and livable. He decides for this reason to spread out one carpet on the floor and to hang up four to form the four walls. But you cannot build a house out of carpets. Both the carpet on the floor and the tapestry on the wall require a structural frame to hold them in the correct place. To invent this frame is the architect's second task.

The shift in emphasis from the purely functional and aesthetic properties of the wall to its structural integrity is significant. Loos believed that this was the correct and logical path to be followed in architecture. It was in this sequence, Loos argues, that humans learned how to build. In the beginning was cladding. Humans sought shelter from inclement weather and protection while they slept. They sought to cover themselves. Originally made of animal skins or textile products, the covering is the oldest architectural detail. Walls were added to support the covering and provide protection on the sides. According to Loos, this was the genesis of architecture.

However, Loos recognized that the problem of the wall and the cladding is neither exclusively an issue of protection nor one of tectonics, but is transformed by the architect as artist into the psychological effect of the space upon the user (see Chapter 6). Loos recognized that some architects chose to focus simply on the section of the wall itself, not the spaces defined by the walls. Visualizing the effect that he or she wants to communicate, the architect translates the imagined space conjured in his mind's eye into the architecture and the space it defines: "fear and horror if it is a dungeon, reverence if a church, respect for the power of the state if a government palace, piety if a tomb, homeyness if a residence, gaiety if a tavern" (Loos, 1928). These effects are produced by both the material and the form of the space.

## 7.3 COMPOSITION AND AESTHETICS OF FACADES

Norberg-Schulz (1979) emphasizes the wall as defining the boundaries that define the spaces where life takes place. Furthermore, he distinguishes among the characteristics of the wall in terms of settlement, urban spaces, public building, and the house. The city wall mainly appears as a silhouette, the urban wall as a varied repetition of a "theme," the public wall as a conspicuous order, and the private wall as a relatively informal reflection of a particular "here." In conjunction with the wall, the floor and ceiling (roof) play a role in the definition of the builtform. In vernacular architecture, the roof usually recalls the forms of the landscape, whereas the public roof may act as a symbolic landmark, for instance, in the shape of a dome (Norberg-Schulz, 1979).

The wall or facade is composed of elements and consists of stories that are more or less distinguished from each other. This distinction is addressed by means of subordinate elements such as columns, architraves, arches, windows, bases, and cornices. Together these elements constitute a built figure. Norberg-Schulz states that the figure is characterized by being a form that gathers earth and sky. The three-story wall of the Renaissance palazzo, for example, defines the distinction of the floors as expressed through the materials and the elements of the facade.

According to Norberg-Schulz, the composition of elements usually depends on a comprehensive *vision,* that is, on an imagined figure which determines the solution. It also follows certain general principles which, at their turn, depend on the structure of spatiality (Norberg-Schulz, 1979):

> First, any composition must take into consideration the difference between horizontality and verticality, and conceive the figure in terms of rhythm and tension, and proportions. Second, the composition should be hierarchical, as any situation consists of superior and subordinate elements. A main entrance is more important than a window within a row, and a columnar order more important than the background on which it appears. If elements are given equal importance, as in cases of historicist buildings at the end of the nineteenth century, the composition becomes confusing. Third, the composition must possess structural identity....Thus it has to be conceived as being massive or skeletal, or both. It is important to note that a built form always has a structural "basis," since the structure contends with the forces of gravity and wind.

### *1 Facades as Expression of Function*

Krier (1983) writes that the facade is still the most essential architectural element capable of communicating the function and the significance of a building. The facade not only reflects the organization of rooms behind, it also describes the cultural situation at the time the building was built. It reveals the criteria of order and ordering, and gives an account of the possibilities and ingenuity of ornamentation and decoration. A facade also tells us about the inhabitants of a building, gives them a collective identity as a community, and ultimately is the representation of the latter in public.

The root of the word *facade* comes from the Latin word *facies,* which is synonymous with *face* and *appearance.* Krier points out that the facade, or face of a building is generally assumed to be the front of the building facing the street. In contrast, the back is assigned to semipublic or private exterior spaces. Both these phenomena of front and back relate—in general terms—on the one hand to public responsibility and on the other to the private self-representation of the inhabitants. Compared with the more representative character of the street facade, the back of the building is more open and communicates with courtyard, garden, and landscape. In the case of stereometric tall buildings, however, each elevation may be considered a facade.

Krier would agree with Norberg-Schulz that the composition of the facade, taking into account the functional requirements (windows, door openings, sun protection, roof area), essentially has to do with the creation of a harmonious entity by means of good proportions, vertical and horizontal structuring, materials, color, and decorative elements. Vitruvius believed that developing metrical relations would give an ideal order and structure to the facade, and also to the floor plans and rooms. This was thought by many architects to be the way of achieving absolute beauty. During the Renaissance, attempts were made to identify numerical systems and rules of proportions. The notion of the universe as a harmonious and mathematical creation was promoted by the ideas of Neoplatonism derived from the writings of Plato and embraced by the Renaissance artists (see Chapter 5).

Harmonious beauty as an end in itself is not the only requisite of building elevation. Experiential factors must also be considered, such as the oblique view given from the bottom of the building, together with the transient effects of light and shade, preventing us from perceiving such truly calculated proportions exactly. The goal of a well-balanced composition, therefore, is to develop a "natural" sense of pleasant, harmonious proportions through an understanding and application of mathematical proportioning systems adjusted to the experiential requirements of the facade.

Structuring the facade must create a distinction between horizontal and vertical elements. Generally the proportions of the elements should correspond to those of the whole to achieve a visually pleasing relationship. Accordingly in low broad buildings, windows, bays, and the like, broad proportions should predominate, whereas in high-rise buildings slender elements better integrate micro- and macroscale elements where large-scale elements are found in small-scale elements and vice versa (Krier, 1983). This principle is also found in nature and produces stronger unity and harmony throughout the composition.

For example, the constructional conditions and ordering principles of a facade can be made visible by channeling the bearing forces into piers. This is not to emphasize the importance of construction over any other aspect of the design process, but to reveal the nature of construction and craftsmanship and to reflect them directly in the composition.

Horizontal layering of the facade differentiates functional areas. Krier advances the classical tradition of horizontally dividing the facade into ground (base), middle (*piano nobile*), and attic stories. (Sullivan's tripartite division of base, stem, and crown is described in Chapter 5.) Krier advances Norberg-Schulz's discussion of the articulated wall by describing the facade as a "built border" acting in a similar way to the portal. The German word for wall is *Wand,* which is related to the words *wenden* (to turn) and *Wandlung* (change) (Krier, 1983). The wall is therefore the place where exterior turns into interior and vice versa. This transitional zone functions as an area of exchange and becomes more lively if the surface has a certain plasticity and if movement is evident. The three-dimensional development of the plane of the wall is enhanced by way of wall projections, ledges, and pilasters creating a relief, whereby light and shadow, foreground and background become perceptible.

The facade as a whole is composed of separate elements, each having an expressive capability of its own. The composition of the facade, however, consists of structuring and ordering these elements. Base, window, roof, and such are also distinguished by form, color, and material. A well-ordered composition, therefore, requires that each element remain recognizable while contributing to the language of the whole.

## 2 Facades as Expression of Structure

The design of the tall building facade presents a host of unique challenges to the architect. Arguments for and against structural expression have been raised by many architects and architectural critics (see Chapter 4). The Miesian aesthetic of the expressed frame enveloped with a glass curtain wall has virtually become the a priori condition of the modern skyscraper. Post-modern architects have challenged the rational and reductive strategies of the early modernists and their protégés by rejecting altogether the notion of structural expression by emphasizing the wall as an opaque skin (typically clad with masonry veneer or some other proprietary material), which conceals or even intentionally undermines the supporting structural frame of the building.

Siegel (1975) points out that if modern architects are searching for a physically and formally perfect entity, they should not treat the structure as a secondary condition. Siegel's argument is that the modern skeletal structure is the result of the rational use of steel and concrete in building. The reduction of all load-bearing members to minimum sizes and a clear division between structural and nonstructural elements have characterized most modern high-rise buildings.

The facades of skeleton structures reveal two opposing tendencies. Either the structural skeleton may be visible from the exterior of the building; or it may remain concealed behind a curtain wall. From the point of view of structural expressionism, the immediate intelligibility of the design favors a visually received understanding of the structure and, therefore, the exposed skeleton is of greater interest. The curtain wall, in contrast, clothes the skeleton from the outside, obscuring or even totally concealing the structural principle. The structure thus runs the risk of being reduced to a factor which, if not totally redundant, is at least unimportant in terms of architectural design. The issues of building structure and architectural expression were discussed in Chapters 2 and 4.

## *7.4 THE ARCHITECTURAL EXPRESSION OF FACADES*

The first tall buildings were the immediate result of a demand for more accommodations and of high land values, combined with the invention of the elevator by Elisha Graves Otis. Gregory (1962) cites three common reasons, excluding aesthetics, for high-rise building: (1) shortage of land, resulting in inflated land values, as in the case of cities such as Hong Kong; (2) a need for a concentration of certain functions in close association; and (3) prestige. The Industrial Revolution induced a gradual change in building methods, but the demand for a building type produced an architectural answer within the terms already existing, rather than from those of a new material or technique.

In the late 1880s and early 1890s, the problem that Sullivan and the other architects of the Chicago school were facing was the problem of the expression of the facade of the steel-frame office building. Sullivan himself grasped at once that the new form of engineering was revolutionary, demanding an equally revolutionary architectural mode. To Sullivan masonry construction, insofar as tall buildings were concerned, had become a thing of the past. The old idea of superimposition had to give way to the imperatives of vertical continuity in the expression of the facade of a tall office building.

The greatest impact on tall building design, however, was the first American zoning ordinance passed by the New York City authority in 1916 (Prescott, 1962). Due to the extraordinary size of tall buildings of the first decade of the twentieth century and the overbuilding of sites, whereby they represented a "solid" cube of building, a significant step was taken in the control of urban environment, and an interesting interrelationship between legal codes and architectural forms was developed.

According to Prescott (1962), an architectural aesthetic emerged from the building code, which to the American writers of the period produced a real "native" architecture. A continued interest in verticality and silhouette led to a gradual subordination of detail to the main effect of a stepped form. Prescott argues that published analysis of these buildings inordinately places emphasis upon structural expression and gives little credence to the fact that their design was principally dictated by formal criteria evolved from the zoning law.

## 1 Sullivan and the Articulated Frame

Louis Sullivan addressed the theoretical problem of the architectural expression of steel-framed tall buildings in *The Tall Building Artistically Considered* in 1896 and became the first architect to give architectural expression to the steel-frame tall building facade (Charenbhak, 1981). Sullivan developed and perfected his system in two great buildings—the Wainwright Building (1890–1891) in St. Louis and the Guaranty Building (1894–1895) in Buffalo (Goldberger, 1983):

> Sullivan, whose greatest achievements in skyscraper design, ironically, to take place outside of Chicago, created what could presumably be called his masterpiece in 1895, with the Guaranty (now Prudential) Building in Buffalo. The forms are similar to those that burst through with such freshness in the Wainwright; here, however, the building has been stretched to 13 stories so it appears, at last, a true tower and not a cube. The treatment of the top is more graceful, too, with Sullivan's lush ornament seeming to grow out of the arches that top each vertical row of windows and then swirl about the round windows of the uppermost story before disappearing into a cove-like cornice. More remarkable still is Sullivan's base, which here, thanks to an innovative design for the shopfronts involving windows bent back at the top, seems to lift the entire structure into the air. The Guaranty thus seems at once to represent the triumph of soaring, cohesive Chicago skyscraper design and to prefigure the skyscrapers of the International Style, which are raised off the ground on pilotis.

Sullivan's well-known admiration for the Monadnock Building reflected an attitude which had by 1890 rejected the classical concept, at least as far as detailed modeling of the facade was concerned. The concept of the finite composition—the monolith—however, was to remain. The Wainwright and the Guaranty, both framed buildings, accepted the repetitive treatment in their facades of a main body of similar office floors. The ground floor and the mezzanine were articulated to form a base to the whole, and the composition was terminated by combining the expression of the top floor with an enormous projecting cornice and ornamentation unique to Sullivan (Goldberger, 1981).

The evolution of the expression of the steel frame in the building facade was incremental. Facades that were true frames did not appear all at once, but were derived in part from the prefabricated cast-iron fronts that were widely used in the late nineteenth century (Goldberger, 1981). The Rookery (1886) and the Monadnock (1891) buildings are clearly transitional buildings, which point toward the development of the expression of the frame in the facade. Whereas the Monadnock is treated entirely as a load-bearing, structural masonry building, Root lightens the Romanesque facade of the Rookery (Fig. 7.2) by opening it with large areas of glazing and carrying the vertical expression of the structural piers from the columns at the base all the way to the cornice. Only on the interior of the building (Fig. 7.3) is one made explicitly aware of the lightweight cast-iron frame contrasted with the masonry and terra-cotta cladding of the facade.

Sullivan's best known work, the Carson Pirie Scott department store (Fig. 7.4) in Chicago (1889–1904), was anticipated in the differentiation of mullion widths from structural supports in the Condict Building in New York (1897–1899), followed by the Gage Building in Chicago (1898–1899). In the Carson Pirie Scott department store the horizontal, rather than vertical, expression of the upper part is determined largely by the underlying grid of structural steel. The lush cast-iron ornamentation of the ground- and first-floor shop fronts contrasts with the straightforward expression of the frame crisply clad in terra cotta.

The ultimate achievement of the Loop architects—Jenney, Holabird and Roche, Burnham and Root, Adler and Sullivan, and others—was that they created a direct and

**Fig. 7.2 Rookery, Chicago.** (*Courtesy: Paul Armstrong.*)

**Fig. 7.3 Interior of Rookery.** (*Courtesy: Paul Armstrong.*)

**Fig. 7.4 Carson Pirie Scott department store, Chicago.** (*Courtesy: Paul Armstrong.*)

fundamental expression of the commercial tall building facade (Prescott, 1962). The rational, scientific, and pragmatic qualities of the tall building were particularly intriguing to Sullivan; Jenney, an engineer; Root, the business manager partner of Burnham; and Holabird and Roche, whose commercially minded firm continued into the next era of skyscraper building in New York. The architectural expression of the structural frame has been addressed in Chapters 2 and 4. The historical development and the material properties of steel are discussed in Chapter 10.

***Carson Pirie Scott and Company Store (1899, 1903–1904).*** In *The Tall Building Artistically Considered,* Sullivan gave the first true analysis of the tall building type as a functional organism (Norberg-Schulz, 1980). He describes the technical purpose of the basement, the public character of the ground floor, the mezzanine connected with the ground floor, the indefinite number of stories of offices above, and finally the attic, which completes the "circulatory system" of the building.

Originally designed by Sullivan in 1899 as the Schlesinger and Mayer store, the Carson Pirie Scott and Company store occupies a prominent site at the corner of State and Madison streets in Chicago. The easternmost section of three bays on Madison Street was built first, and the main section, extending around the corner and with seven bays on State Street, several years later. The steel frame is expressed on the facade by the wide windows and narrow piers. The simplicity and straightforward expression of steel-frame construction in the uppermost stories of the building are contrasted by the rich ornamentation and detailing of the first and second floors, illustrating Sullivan's idea that the display windows were like pictures and deserved rich frames.

Here the famous "Chicago window," a central fixed glass pane flanked by two operable double-hung units, emphasizes the expressed horizontality of the elevations (Schulze and Harrington, 1993). The horizontal reading of the facades is further enhanced by the continuous horizontal bands at each floor level that interrupt the slender vertical piers. The carefully proportioned window openings, the firm emphasis in the moldings around them, and the accent given by the line of delicate ornamentation on the horizontal wall sections create a perfect synthesis of design articulation and structural clarity. The oxidized bronze effect that Sullivan intended to achieve in the ornamentation results from applying a coat of brownish-red paint and then applying a thin layer of deep-green paint, allowing the undercoat of red to show through.

Sullivan expresses the corner of the building and its main entrance with a large cylindrical volume. This form provides a strong, vertical emphasis at the corner while continuing the visual effect of carrying the horizontally expressed floors around the corner of the building. A projecting cornice at the top of the original building has been replaced by a bald parapet. Unlike the soaring verticality achieved in the facades of Sullivan's Wainwright and Guaranty buildings with the uninterrupted extension of the vertical piers, the Carson Pirie Scott store suppresses vertical continuity to express the horizontal extension of the large sales area, and "the building thus comes to participate actively in the longitudinal movement of the busy streets" (Norberg-Schulz, 1980).

## 2 The "Skyscraper Style"

The exterior massing of the tall building retained a much stronger traditional overlay in New York and at the East Coast of the United States. The logic of cast-iron expression pioneered by Bogardus (see this chapter and Chapter 10) was ignored and emulated the logical excellence of the commercial street architecture which had been produced from 1850 onward, as well as the starting point in expression arrived at by the Carson Pirie Scott building. At the turn of the century the academic tradition of the Ecole des Beaux-

Arts remained strong in America and the classical design the dominant "style" for tall building elevations (Prescott, 1962). In contrast to strictly classical buildings, Gothic revivalism found popularity in such influential buildings as Cass Gilbert's Woolworth Building (1913). Whereas the application of classical principles led to sculptural monoliths, such as Burnham's Flatiron Building (1903; Fig. 7.5), the Gothic buildings seem to have had the greatest influence on both the expression of structure and the facades of modern architectural design. For example, Cass Gilbert in his South Brooklyn U.S. Army Supply building shows an almost Miesian truncation of the vertical expression of structure and mullions on a construction devoid of "prestige value" (Prescott, 1962).

Whereas office building imagery in Chicago was shaped by investment unencumbered by the need for public relations, New York approved a more self-conscious corporate image comprised of conservative styling and innovative planning. In both cities, the roster of major business buildings manifested the success and prosperity possible under America's freewheeling version of capitalism. Beginning in the late 1880s, however, controversy over the nature and role of business caused office building imagery in East and West to move closer together and in a new direction. The new emphasis was on business and the architecture of business in the service of the public, that is, on business architecture designed for the pleasure and edification of the observer. The promotion of a more benevolent and philanthropic image for business and the tall building resulted in part from efforts to counter widespread criticism of business practices (Gibbs, 1984).

Perhaps the Chrysler Building (1930) or the Empire State Building (1931) are the most forceful and unambiguous expressions of the "skyscraper style." Both are point towers with the vertical dimension completely dominant and with an almost identical image from all directions. In both buildings, the vertical upward thrust culminates in "a great crown that resolves and continues this thrust up into the sky itself" (Pelli, 1982).

## *7.5 MODERNISM AND THE INTERNATIONAL STYLE*

Huxtable (1984) distinguishes between two aesthetics in modern tall building design. "Modern," according to Huxtable, "was radical, reductive, and reformist; 'modernistic' was richly decorative and attached to conservative and hedonistic values." The international style architecture promoted by the European school of Gropius, Mies, and Le Corbusier characterized modern architecture as austere, abstract, élite, and avant-garde. "Modernistic" architecture, in contrast, was neither pure nor revolutionary; it fused the ornamental and the exotic into what was, essentially, the last decorative style—art moderne or art deco. Derived from the luxurious, exotic combination of new and old materials and the traditional fine craftsmanship that characterized the products of the 1925 Paris Exposition des Arts Decoratives et Industrielles, art moderne was despised by the avant-garde as fussy, reactionary, and decadent.

### *1 Structural Symbolism*

According to Huxtable (1984), the minimalism of the modernist aesthetic lends itself to two extremes—either it represents a subtle, aesthetic beauty or it is the result of the cheapest corner cutting. Huxtable points to Mies van der Rohe's pioneer tall building projects in post–World War I Berlin as the epitome of the modernist ideals of critical quality of detail, material, and execution. The prismatic glass tower scheme of 1919 for the Friedrichstrasse is as responsive to its site as to an ideal vision, and the curved glass

**Fig. 7.5 Flatiron Building, New York.** (*Courtesy: Arthur Kaha.*)

tower project of 1920–1921 is a consummate aesthetic statement of intent and ideal. Through these projects Mies articulated the quintessential twentieth-century tension between art and technology, between idea and reality. Structure in its most pure and perfect form was enclosed in a sheer curtain of shaped glass. The concept of the aesthetic qualities of material and building structure was inseparable from the technological innovations that made it a creative possibility, elevating structure and its expression to a higher, symbolic level on a plane with the aspirations of Gothic architecture.

***Promontory Apartments.*** The Promontory apartments (1946–1949) on Chicago's Southside was Mies's first realized high-rise structure. Economic considerations compelled him to substitute reinforced concrete for steel in what was originally designed as a steel-frame building enclosed within a glass curtain wall. The exterior of the building, conceived as an exposed skeleton with operating windows and brick spandrels, is noteworthy for its simplicity and restraint. In characteristically restrained expression, Mies subtly reduces the depth of the columns, which step back at three points on the facade as the load diminishes. According to Spaeth (1985), the Promontory apartments echo the structural clarity of Sullivan's Carson Pirie Scott store and the subtle refinement of Gothic architecture.

***860 Lake Shore Drive Apartments (1951).*** Mies's towers at 860 Lake Shore Drive in Chicago (Fig. 7.6) were designed as residential apartments, but they became the international prototype for thousands of high-rise office buildings. They were the first apartment buildings anywhere in the United States to be sheathed entirely in glass (Goldberger, 1981). Derived from theory, the architectural expression of the Lake Shore towers is preeminently intellectual and ideological. The glass and steel structure expressed in the facade represents an ideal proportional order governed by the constraints of material and technology. To Mies, the expression of a regular spatial grid takes precedence over the display of the skeleton (Pelli, 1982).

In his use of structural steel in subsequent Chicago buildings, Mies was faced with the problem of fire-proofing the elements of the structure and at the same time expressing the nature of the material from which the structure was formed. He developed what Spaeth describes as "an organic approach to detailing" by developing his vocabulary from standard rolled steel sections. I beams, angles, and plates were welded together to form the exterior skin, the framework to which windows were attached and the formwork for the concrete fire-proofing. As a result, not only do the details reveal the architect's conceptual and structural intention, but they clearly express the technological means by which they were achieved (Spaeth, 1985).

Quoting from Mies's own writings, Banham (1961) stresses the architect's ideological expression of technology and structure and their influence upon building form:

> Skyscrapers reveal their bold structural pattern during construction. Only then does the gigantic steel web seem impressive. When the outer walls are put in place, the structural system which is the basis of all artistic design, is hidden in a chaos of meaningless and trivial forms....Instead of trying to solve new problems with old forms, we should develop the new forms from the very nature of the new problems.
>
> We can now see the new structural principles most clearly when we use glass in place of the outer walls, which is feasible today since in a skeletal building these outer walls do not actually carry weight. The use of glass imposes new solutions.

***Toronto-Dominion Centre (1964–1969).*** The last great work in which he took an active part is a structural and architectural development of ideas which had matured over four decades. The two towers of the Toronto-Dominion Centre comprise approximately

**Fig. 7.6 860 Lake Shore Drive, Chicago.** (*Courtesy: Paul Armstrong.*)

287,990 $m^2$ (3.1 million $ft^2$) of gross office space. The Toronto Dominion Bank pavilion, a low-scale building within the complex, recalls the clear-span, steel-frame vocabulary of Berlin's New National Gallery. Mies was convinced that the banking function would require a more flexible type of space than could be provided by incorporating it in an office building. The site is developed with the placement of the towers and the bank pavilion creating a number of individually identifiable public plazas at grade. A shopping center concourse is situated below ground due to Toronto's cold climate.

The Toronto complex embodies several basic Miesian concepts (Schmertz, 1975) related to the expression of the facades in all of his work: (1) expression of building structure and components to enhance a sense of scale; (2) harmoniously relating inside and outside spaces to human scale by creating a scale range through the careful proportioning of the structural grid and window mullion spacing; and (3) use of fine materials and careful detailing to increase the immediate visual and tactile experiences of the pedestrian and building user.

The 56-story Toronto Dominion Bank tower is three 12-m (40-ft) bays wide by eight 9-m (30-ft) bays long; the 46-story Royal Trust tower is three 12-m (40-ft) bays wide by seven 9-m (30-ft) bays long. Typically, a bay consists of 12-m (40-ft)-long girders at 9 m (30 ft) on centers (spanning the transverse direction) and 9-m (30-ft)-long intermediate beams at 3 m (10 ft) on-centers (spanning the longitudinal direction). The steel frame of the towers utilizes a braced core system in both the transverse and the longitudinal directions.

The installation of the first 43 stories of steel skin for the 56-story tower was by the "stick" method, that is, fascia plates were attached to edge beams followed by welding to them individual two-story-high mullions. Because this system proved to have a number of disadvantages, a "panel" method of installation was substituted from the 44th floor up as well as for the subsequent 46-story tower.

The curtain-wall system for each of the towers is based on 1.5-m (5-ft) modules and consists of an extruded aluminum glazing frame with bronze-gray tinted heat-absorbing and glare-reducing glass, and built-up steel mullion units. Expressed nonstructural steel I beams are mounted on each mullion and at the corners of the towers and the pavilion. A similar detail is used on the Seagram Building in New York (1955–1958).

***IBM Building (1967–1970).*** Mies's last, and perhaps best, office building constructed in the United States was the IBM Building in Chicago (Fig. 7.7). The IBM Building occupies 50% of a 0.6-hectare (1½-acre) site on the north side of the Chicago River. Directly to the west are the 60-story twin Marina City towers and immediately to the east is the Sun-Times and Daily News building, a low, horizontal seven-story aluminum and glass building defining the river's edge. Easements and site restrictions dictated a simple rectilinear 38- by 84-m (125- by 275-ft) volume in plan. The remainder of the site is dedicated to a plaza (*Architectural Forum,* 1968).

Like all of Mies's high-rise buildings, the 51-story IBM Building is raised on steel columns framing a glass-enclosed lobby. The space of the lobby is linked both visually and physically to the space of the plaza and the city beyond. The structural steel grid sets up 9-m by 12-m (30- by 40-ft) unobstructed office bays divided into five modules. Mechanical floors establish horizontal data at the 18th floor and the topmost floors of the building. The mechanical floors divide the building horizontally into ⅓ and ⅔ proportions. The large-scale column grid creates wide-bay elevations which are further broken in scale by smaller aluminum mullions. Spandrel panels coinciding with the horizontal beams and floor levels create a distinctive pattern of dark horizontal bands (*Progressive Architecture,* 1968). The subtle interplay between horizontal and vertical elements enhances the verticality of the IBM Building while relating contextually to the horizontal emphasis of the loop architecture.

**Fig. 7.7 IBM Building, Chicago.** (*Courtesy: Paul Armstrong.*)

Mies's concern for the contextual environment is also reflected in the solution for IBM's curtain wall. Whereas his other buildings, such as the Seagram Building in New York, stressed the structural and ornamental properties of the I-beam and curtain-wall construction, IBM's curtain wall combines three elements used before only singly or in pairs—double glazing, a complete thermal break between the interior and exterior of the building, and a system to equalize pressure within the skin to prevent air leakage or infiltration normally caused by the pressure differential between a building's interior and exterior (Spaeth, 1985). The sophisticated mechanical systems, designed to control the environment for housing computers, reclaim heat dissipated by people, lights, and machines. A plastic thermal barrier separating the curtain wall from the frame reduces heat transfer from the interior to the exterior of the building. The net result was that all these elements worked together to produce a more efficient curtain wall with less heat loss and heat gain in comparison with other buildings which had considerably less glass area.

Another characteristically Miesian detail, the articulated corner, is featured on the IBM Building. To express the corner and visually continue the facade around the corner of the building, he set two continuous vertical I beams oriented at 90° angles to one another at each corner (Fig. 7.8). These I beams are identical to the vertical pattern of I beams attached to the elevations, corresponding to the structural columns and the vertical window mullions. Unlike later curtain-wall buildings emphasizing a taut, monolithic skin with little or no surface articulation defining glass and mullion, the IBM Building's window frames are slightly raised with the spandrel panels recessed, thereby differentiating glass, spandrel, and mullion.

In this project and the 860 Lake Shore Drive apartments, the wall is rendered after Semper's description of a woven fabric—a subtle integration of structure with fenestration that displays the same capacity as load-bearing masonry for limiting any extension of space. This fusion of the structural frame and glass infill results in the loss of the identity of individual elements. The columns and mullion dimensions determine window widths and regulate cadences of contracting and expanding intervals on each facade (Frampton, 1992).

***Technology and Space.*** According to Spaeth (1985), architecture for Mies was a function of industrialization and new materials, of structure and space. It had to be understood on its own terms, free from political associations or implications, as Mies confirmed in 1924:

> Architecture is the will of an epoch translated into space. Until this simple truth is clearly recognized, the new architecture will be uncertain and tentative....The question as to the nature of architecture is of decisive importance. It must be understood that all architecture is bound up with its own time, that it can only be manifested in living tasks and in the medium of its epoch.

Mies's architectural goal was to achieve a building expression which directly reflected the pure use of materials using an absolute economy of means. To achieve this end, he could not rely on ornamentation or superfluous detail. Rather, he was limited to the absolute rational and minimal use of materials in the creation of purely egalitarian, uninterrupted space. Economy was primarily reflected in the use of materials, however, rather than in the actual construction or detailing of the building. For Mies, "God is in the details," and his architecture could endure no compromises or shortcuts.

The fact that he could never replicate the conceptual purity of the early prototypes in his later built projects is not essential in evaluating their ideological significance. "A

**Fig. 7.8 Corner detail of IBM Building.** (*Courtesy: Paul Armstrong.*)

narrow functionalism and literal structural expression were not his intent," writes Huxtable (1984), "what Mies sought, was a structural symbolism as a high art form for a technological age."

### 2 Le Corbusier and the "Free Facade"

The corollary to Le Corbusier's "free plan" in the vertical plane was the "free facade," which was achieved through the separation of the load-bearing columns from the exterior walls enclosing the interior spaces (Frampton, 1992). Without interference of the structural elements in the facade, Le Corbusier was free to develop the facade in an autonomous manner from the building plan and section. This approach is in contrast with the functionalist or structuralist approach to composing facades espoused by Gropius in the case of the former, and by Mies van der Rohe in the case of the latter.

In the text accompanying the publication *L'Esprit Nouveau* (1921), Julien Caron wrote (Frampton, 1992):

> Le Corbusier had to resolve a delicate problem which was contingent upon making a pure work of architecture, as postulated by a design in which the masses were of a primary geometry, the square and the circle. Such speculation in building...has rarely been attempted except during the Renaissance.

To maintain proportional control over the facade in the manner of classical architecture, Le Corbusier employed "regulating lines." The proportioning device employed by Le Corbusier was in accordance with the golden section and, later, evolved into his system of the "Modulor"—a proportioning system based on anthropomorphic measurements and geometric progressions derived from the golden section and articulated in the "red" and the "blue series."

Using the concept of the "free facade," Le Corbusier was free to express the facade as a screen of pure form and geometry. Furthermore, he was able to apply innovative technologies to the facade to regulate direct sunlight and shade interior spaces within the building, such as the brise soleil and light shelves, which give the facade a layered, three-dimensional appearance.

***Cité de Refuge (1929–1933).*** The Salvation Army hospice, the Cité de Refuge, is larger in scale than the Swiss dormitory and pavilion and addresses a different set of architectural issues. Le Corbusier's first major public commission in France, its basic features were worked out in April 1930 on the architect's return from a trip to Moscow (Benton, 1987). The monumental building was a modern version of Beaux Arts axial planning with a series of elements disposed hierarchically in an architectural promenade (Jencks, 1973). The main dormitory block was conceived as a reinforced concrete seven-story volume with 11 levels. The stepped-back facade on the rue du Chevaleret and the gentle incline of the huge glass dormitory wall toward the top was in response to limitations imposed by the planning commission.

The final scheme presented a five-story wall of glass oriented just a little to the west of due south. This window wall was sealed hermetically, apart from the doors to the balconies and some windows on the staircase landings. The glass dormitory slab contained two technical innovations: the "neutralizing wall" and "exact respiration" (Jencks, 1973). Ideally, these inventions should have provided the inhabitants with pure, clean air at a constant temperature of 18°C (64.4°F). Le Corbusier intended to install a system of total air-conditioning, but was compelled by budget constraints to accept hot-air central heating and mechanical ventilation without the refrigeration plant he knew was necessary, which resulted in uncomfortable environmental conditions in the summer.

The *neutralizing wall* was to be a form of double glazing with circulating hot and cold air between, whereas *exact respiration* was to be a form of air-conditioning system, found in large buildings today, which supplies and extracts humidified air from a central system. Since the double glazing was not installed, the dormitories overheated during the summer and openings and brises soleil were later provided, thus destroying the effect of the clear curtain wall.

***Unité d'Habitation (1947–1953).*** The Unité d'Habitation (Fig. 7.9), or Marseillian block, represents the culmination of Le Corbusier's research into housing and communal living. For Le Corbusier it synthesizes the relationship between the individual and the collective, which he admired in the monasteries of Ema and Mount Athos (Jencks, 1973). Essentially a small village of 1600 people, the Unité provides total individual privacy as well as meaningful collective activities, varying from a gymnasium to a shopping center on the seventh and eighth floors.

The sheer physical presence of its shiplike monolithic massing of raw concrete, the *béton brut,* is overwhelming. Sculpturally bold and aggressive, it continues to dominate the larger buildings that surround it. Exposed concrete on the exterior expresses the least incidents of the shuttering, the joints of the planks, the fibers and knots of the wood. A straightforward functional simplicity is exaggerated in its plastic effect through the power of proportion.

The whole building is constructed from 15 basic dimensions derived from Le Corbusier's own Modulor proportioning system. Each element is related to the other in simple, harmonic proportions which, according to Jencks (1973), "give a semantic strength quite apart from their numerical ratios." Months of work went into facade studies using the Modulor to help regulate relationships between large and small elements.

**Fig. 7.9 Unité d'Habitation, Marseilles, France.** (*Courtesy: Paul Armstrong.*)

Dominating the facades is the brise soleil, the depths of which are calculated to exclude hot summer sun and glare but to let in winter rays (Curtis, 1986). The resulting composition is a subtle blend of opacity and transparency, massiveness and hovering planes.

In both the Cité de Refuge and the Unité, Le Corbusier abandons the idealized, flat, planar facades of his architecture of the 1920s in favor of more sculptural and expressive elevations. Alternating light and dark rectangles are inverted or doubled over the surface of the building. Jencks (1973) describes the Unité as the most "musical" of Le Corbusier's buildings, presumably due to its carefully articulated proportions, which are translated onto the facades like a musical score, and perhaps the one of which he was most proud.

***Mies, Le Corbusier, and Modernism.*** In retrospect, the contributions of both Le Corbusier and Mies van der Rohe to modern architecture cannot be disputed. Both architects contributed in a totally unique and unparalleled manner in defining the principles of modernism and in producing the great archetypical buildings of modern architecture. Perhaps Mies best illustrates the pure expression of Gothic structural expression, particularly as applied to the tall building, and an unparalleled obsession with construction and detailing. Le Corbusier, on the other hand, contributes more to the symbolic and cultural meaning of architectural form and ideas. These differences are best exemplified in an examination of the building section. Whereas Mies sought minimalism and reductiveness, Le Corbusier sought richness and articulation. Inevitably, Mies sought to express the building in the most elegant and reductive of terms; Le Corbusier sought to charge the building with the most profound and symbolic cultural meanings.

## 7.6 THE COMMERCIAL FACADE: ECLECTICISM TO MODERNISM

In the time period between the Civil War and the Great Depression, from about 1870 to 1930, American business moved from an era of relative freedom to one of close scrutiny and demand for governmental restriction. These 60 years in the development of the tall building constitute, in one sense, a single line of evolution, characterized by feelings of optimism and exuberance not found before or after (Gibbs, 1984). This is the period that witnessed the extension of the power of American business and the arrival of the skyscraper as perhaps the best known architectural symbol of America.

The identification of classical architecture with civic and institutional virtues had become an American cliché by the late nineteenth century. It is not surprising, therefore, to find this association passing from classically styled buildings serving truly civic functions to classically styled buildings serving business functions. Writing for *Inland Architect* in 1891, Henry Van Brunt gave the lobby of the Rookery special praise, noting that "there is nothing bolder, more original, or more inspiring in modern civic architecture either here or elsewhere" (Charenbhak, 1981).

### 1 Temples of Commerce

Architects, always searching for new ways to express both the image of the tall building and the client, sought out historical examples to apply to skyscraper forms and facades. Several successful skyscrapers of the 1920s combined the functions and images of businesses and institutions. The implied purpose was to soften the corporate, profit-driven image of the commercial tall building by evoking civic and even religious architectural images to symbolize big business as a philanthropic, as opposed to purely mercantile, enterprise. Included among these buildings was the Chicago Temple Building, combining the uses of a church and a speculative office block. Emphasizing functional contrast, a severely plain skyscraper was topped with an elaborate Gothic cresting of pinnacles and spires, the tallest of which nearly equals the height of the office block beneath (Gibbs, 1984).

However, not everyone was pleased with the tendency toward eclecticism and the laissez-faire appropriation of historical styles that had overtaken the American architectural scene during the early part of the twentieth century. Among the most vocal critics was Louis Sullivan, who declared in 1896: "The tall office building is the new, the unexpected, the eloquent peroration of the most bald, most sinister, most forbidden conditions" (Gibbs, 1984).

***Woolworth Building (1913).*** According to Gibbs, the gala public opening of the Woolworth Building and subsequent publications by the Woolworth Company indicate how firmly established the progressive era's version of the philanthropic image for skyscrapers had become by 1913 (Gibbs, 1984). Located in New York, the Woolworth Building surpassed the Metropolitan building in both height and bulk. The Gothic styling of the building prompted Dr. S. Parkes to coin the phrase "cathedral of commerce." Montgomery Schuyler was commissioned to write a 50-page brochure of the Woolworth Building and cites the civic values resulting from the tower, which he viewed as a "purely decorative object." Schuyler's admiration for the Woolworth Building and its tower stemmed from his impression that "the client...must have sacrificed some space and added much enrichment of this crowning feature....The tower commemorates his sense of civic obligation" (Mumford, 1952). Therefore, by reason of its spire, the Woolworth Building was less commercial.

The expressed verticality of the building's piers, designed as imitations of Gothic structural elements, lighten the otherwise massive volume of the building and tower (Goldberger, 1981). Other Gothic details, such as terra-cotta ornamentation, gargoyles, and buttresses, complete the effect. And here, perhaps for the first time, the slender tower becomes a natural outgrowth of the base rising from the center of the facade.

***Wrigley Building (1921–1924).*** The Wrigley Building (Fig. 7.10) is the earliest of the celebrated skyscraper group at Michigan and the Chicago River. Inspired by the Municipal Building, it is sheathed in white terra cotta with the two volumes of the building establishing a strong cornice 18 stories above the street. The tower, which is structurally expressed throughout the entire elevation, was modeled on the Giralda of the Cathedral of Seville. A small, narrow plaza with plantings and a fountain is formed by the main building and its annex and is shielded from the street by a thin screen (Schulze and Harrington, 1993). The proportions are awkward—Goldberger (1981) describes the tower as "gangling"—but the building reflected the new direction of historical eclecticism where Chicago was heading.

Never restored, the building's facade has received a continuous program of maintenance, including cleaning, pointing, repair, and, when necessary, the replacement of terra cotta. The Wrigley Building's eclectic details and volumetric massing, while historically derived, do not diminish its landmark status. It functions both as good architecture and as an integral component of the urban context.

***The Tribune Tower Competition.*** The competition in 1922 for the Chicago Tribune tower was, in a sense, the last hurrah of the philanthropic image for business architec-

**Fig. 7.10 Wrigley Building, Chicago.** (*Courtesy: Paul Armstrong.*)

ture. Although civic values were foremost in the minds of the directors of the competition, the interests of the sponsor and the newspaper industry also played a part. For the owners of the Tribune, a superlative office building was "the ultimate in civic expression," but they also "aimed to provide for the world's greatest newspaper a worthy structure, a home that would be an inspiration to its own workers as well as a model for generations of newspaper publishers" (Gibbs, 1984). The Chicago Tribune tower competition asserted the Gothic style, spurned as decadent by the Chicago school architects only a few years earlier, as the accepted mode for tall building design in the United States in the early 1920s. The art of composition seemed to be the real goal, and assembling historical pieces into a coherent, well-proportioned entity was what tall building design had become (Goldberger, 1981).

Ada Louise Huxtable views the Chicago Tribune tower competition as a watershed in American skyscraper form. To Huxtable (1984), "it crystallized a unique moment in architecture when the long classical tradition was poised on the edge of the unknown abyss of modernism." The entries, a mix of the adventurous and the retardataire, were exceptionally good and reflected a complete understanding of the massing, scale, and detail of the tall building. The debate between those who championed the winning Gothic revival tower by Hood and Howells versus the romantic Finnish modernism of Eliel Saarinen's second-prize design or the radical modernism of Walter Gropius's Bauhaus submission seems inconsequential in retrospect (Huxtable, 1984).

Gibbs (1984) writes that the public spiritedness of the Tribune was an anachronism, more a part of the progressive era than of the 1920s. The vision of a corporate architecture designed with the aesthetic needs of urban inhabitants in mind and intended for consumption by the public rather than by businesspeople themselves was to disappear in the clamor of the competition. It is appropriate that the neo-Gothic winning entry of Hood and Howells characterized by its "Gothic verticality terminated with a crown of flying buttresses and a 'lantern' in neo-stylistic detail" (Prescott, 1962), which Gibbs cites as the most effective example of philanthropic imagery in the competition, should have less influence on office building design during the remainder of the decade than those entries that indicated a new direction.

## 2 Modernism and the Commercial Facade

Fueled by the exhibition and book of the same title by Henry-Russell Hitchcock and Philip Johnson, international style modernism continued to spread through the United States for about four decades. However, as Goldberger (1981) points out, the impetus for its success lies less in its aesthetic doctrine than the economy of the postwar period as glass curtain walls became cheaper than masonry and as the decline in craftsmanship made it expedient to simplify building construction and ornamentation. "By 1950 modern architecture had become the American corporate style, its clean, sparse lines ideally suited to the cool and anonymous world of American business in the postwar years."

***Lever House (1952).*** One of the most important modern buildings of the postwar era is the Lever House in New York City. Its slab massing is articulated by the same curtain wall and windows on the narrow and the long sides of the building, indicating flexible, undifferentiated space. The Lever House is an excellent pragmatic American response to a building that sits on *pilotis,* regaining the ground floor for public use (Pelli, 1982). The curtain-wall enclosure is extremely light—all glass with minimal mullions in order to accent reflections. Because a true transparent glass wall was impractical, the spandrel portions of the wall were clad with opaque glass, achieving the desired perceptual effect without sacrificing pragmatics. The building does not bring its visual weight to the ground. Instead it signals (but does not express) its independent structural system and lightweight enclosure.

No doubt Lever's modernist design, human sense of scale, and its bold break with the street edge continue to rank it among the most important international style buildings. However, in retrospect, many critics now view Lever's idiosyncrasies with more skepticism (Goldberger, 1981):

> Lever's abstract beauty remains powerful, more than a quarter century after its completion, and its genuine modesty of scale brings to the streetscape a sense of humanism that has been desperately lacking in many more recent glass towers. Still, the building seems flawed by today's standards—the break with the street wall of Park Avenue, so liberating in the 1950s, now seems needless and not a little narcissistic....And the premise of "structural honesty" on which the building was said to be based is, of course, an exaggeration. The double-slab form is a pure composition, as much as the crown of the Chrysler Building; and the use of spandrel glass—the glass that covers the structure between the floors, making the entire outside look as if it were made of glass—is not structural honesty at all, but merely a modernist brand of ornament.

***Inland Steel Building (1957).*** Less ambitious than Lever House, but less compromised by the passage of time, the Inland Steel building (see Fig. 5.2) was the first tall building to be constructed in the Loop since the Great Depression (Goldberger, 1981). Its sleek, stainless-steel and glass curtain walls and immaculate detailing rank it among the finest examples of modernist architectural theory (Schulze and Harrington, 1993).

Service and vertical circulation are placed in the opaque, nearly square tower on the eastern edge of the site. The 25-story service tower is sheathed in 51-mm (2-in.)-thick stainless steel. Open column-free space is permitted on the office floors in the stainless steel and blue-green glass rectangular tower on the west side. Supporting columns, placed on the outside of the curtain wall and thin window mullions, give a strong vertical emphasis (*Architectural Forum,* 1958).

Located at the corner of Dearborn and Monroe streets, the 19-story office tower's major facade provides a visual backdrop to the east side of the First National Bank Plaza. Seven slender stainless-steel columns on each side of the office volume support 18-m (60-ft)-long plate girders, which span the entire width of the building, providing 948 $m^2$ (10,200 $ft^2$) of clear space per floor. The spandrel panels are 16-gauge stainless steel backed by 51-mm (2-in.)-thick lightweight concrete. The columns, spaced at 8 m (26 ft) on centers, are wrapped with 18-gauge stainless-steel covers. Thin vertical mullions set the modular dimension and are channeled to guide the cleaning scaffold (*Architectural Forum,* 1958).

The Inland Steel building synthesizes the heritage of Chicago school architecture with pure, modernist expression. Its volumetric simplicity, functional clarity, design efficiency, and superb detailing have enabled it to endure.

***IDS Center (1972).*** The IDS Center in Minneapolis represents a model for the humanistic skyscraper (Goldberger, 1981). Originally a simple office tower, the program was expanded to include a hotel and a retail complex. A 51-story glass tower, a 19-story hotel, and a lower retail wing are all arranged around a central glass-covered court. Sheathed in a light-reflecting pale-blue glass, the tower's shape is roughly that of a flattened octagon, with its far diagonal sides cut into facets, creating a zigzag effect. The faceting breaks up the large mass of the tower, reducing its visual impact as well as bringing variety to the form while at the same time adding 32 corner offices per floor (Fig. 7.11). The transparent roof sections of the court are arranged like stacked cubes resembling "a great crystal tent pitched in the middle of the city" (Goldberger, 1981).

**Fig. 7.11 Corner detail of IDS Center, Minneapolis.** (*Courtesy: Paul Armstrong.*)

Although the IDS Center bears some affinity to the more famous Seagram Building, the minimalistic approach and emphasis on sculptural form is clearly a transition to later projects, such as Pennzoil Place in Philadelphia and the post-modern AT&T building in New York. Unlike the insistence on absolute structural and aesthetic integrity in the facades of the Seagram Building, the facades of the IDS Center, with their dark-colored transparent and opaque spandrel glazing, express the sculptural massing of the building. The underlying steel skeleton is sublimated by the taut curtain-walled skin and the articulated stepping at the corners.

***John Hancock Tower, Boston (1972–1975).*** The John Hancock Tower at Copley Square, Boston, signaled the departure from the whafflelike grids of reinforced concrete facades to a return to the glass curtain wall. Unfortunately the building is perhaps most infamous for reasons other than its aesthetics when, in 1972, its windows began to fall out. Its fortunes reversed from disaster to aesthetic triumph when the building finally opened to the public in 1975. In the final analysis, the Hancock tower emphasizes pure sculptural form and massing over historical cliché or contextual reference. Using reflective glass, an abstraction in glass was created which, by its shape, took into account the lines and forces of the surrounding buildings and open spaces (Goldberger, 1983). The slender parallelogram shape of the Hancock tower has a short base to carry across neighboring cornices. A vertical V-shaped reveal runs along the entire 60-story height of the facade to provide some sense of texture.

## 3 Structural Expression versus Aesthetics

The debate between structural expressionism and aesthetics persists among architects and engineers in the design of high-rise building facades. Whereas Pelli (1982) argues that Mies van der Rohe viewed the structural expression of materials as a means to resolve an aesthetic goal, some architects and engineers contend that he sought pure structural expression as an architectural goal. It is telling that in the Seagram Building in New York (1958), Mies compromises the structural purity of the steel frame by bolting bronze I beams over the actual steel skeleton. Obviously, one of the reasons he resorted to this solution was the problem of fire-proofing the steel frame, which obscured the skeletal and pure material expression of the facade. Although Mies's solution is an ingenious, pragmatic response, which enhances the aesthetic qualities of the Seagram Building, it nevertheless contributes to a mixed reading of the intended purity of the architectural concept.

Other architects have consciously sought to distinguish their work from the international style "boxes," which characterized much of the architecture during the post–World War II era. The sculpted towers of this new generation of architects exhibit their preoccupation with the effects of manipulating form rather than a radical departure from the abstraction of modernism. Perhaps the most significant technological contribution, other than structural innovation, is the effort on the part of a large number of architects to vary the skin, or sheathing, of the building. These new enclosure systems are stretched as thinly as possible across the frame, with virtually no texture and no depth to them. Unlike the masonry towers that preceded them, the windows have no recesses, giving the buildings a machinelike presence. (The subject of structural expressionism has been presented in Chapters 2 and 4; for more insight into aesthetics, the reader should reference Chapter 5.)

***John Hancock Center (1969) and Sears Tower (1974).*** According to Goldberger (1981), Skidmore, Owings & Merrill's major accomplishments in Chicago are the city's two tallest buildings, the 95-story John Hancock Center (1969; see Fig. 5.8) and the

110-story Sears Tower (1974; see Fig. 6.3). The John Hancock tower is a tall, tapered shaft with structurally expressed X braces cutting across the exterior facade. Goldberger refers to the Hancock as "a looming giant, a great cowboy stalking the town" (Goldberger, 1981) and criticizes its poor contextual response to Michigan Avenue. (See also Chapters 2, 4, and 5 for additional information on these buildings.)

The 443-m (1454-ft)-tall Sears Tower fares no better in Goldberger's view at the ground level. It is the innovative bundled-tube structural system developed by Fazlur R. Khan which Goldberger finds most exciting and laudatory. The tubes stop at different heights, giving the building a varied, stepped-down profile. In making a direct allusion to the elaborate tops of high-rise buildings of the 1920s and 1930s, Khan reinforces the direct and logical expression of the building's structure.

While Pelli (1982) acknowledges that both buildings derive their design and expression from their structural systems, he believes that they are based on a misinterpretation of Mies. According to Pelli, Mies was not concerned with real but with the ideal structure of the archetypal post and beam as the giver of form and beauty.

## 7.7 POST-MODERNISM AND CLASSICAL ECLECTICISM

According to Charles Jencks (1988):

> Post-Modern is a portmanteau concept covering several approaches to architecture which have evolved from Modernism. As this hybrid term suggests, its architects are still influenced by Modernism....and yet they have added other languages to it. A Post-Modern building is doubly coded—part Modern and part something else: vernacular, revivalist, local, commercial, metaphorical, or contextual. In several important instances it is also doubly coded in the sense that it seeks to speak on two levels at once: to a concerned minority of architects, an elite who recognize the subtle distinctions of a fast-changing language, and to the inhabitants, users, or passersby, who want only to understand and enjoy it.

Jencks further distinguishes between post-modernism's inclusion of past historical styles which root post-modern buildings in time and place versus late-modernism, which disdains all historical imagery. Post-modern architecture is eclectic in its expression and employs ornament, symbolism, humor, and urban context as architectural devices. In contrast, late-modern architecture derives its principles almost exclusively from modernism and focuses on the abstract qualities of space, geometry, and light. Style is disregarded for design methodology, such as technical problem solving through teamwork, or abstract propositions of geometry and organization. (For more information, refer to Chapter 1.)

### 1 Post-Modern Facades

The post-modern buildings of the late 1970s through the 1980s signified a return to historical precedents as inspiration for the design, scale, and materials of tall building facades. The abstract, monolithic international style glass box gave way to traditional issues of urban and human scale, reaffirming the pedestrian scale of the streetscape. Furthermore, as the abstract imagery of modernist-inspired high-rise architecture began to lose its appeal, clients once again became receptive to alternative architectural styles (Goldberger, 1981).

Buildings like the AT&T building (1982) in New York and the Humana headquarters building (1982–1984) in Louisville, Kentucky, dramatically altered the public's impression of what high-rise architecture ought to look like. Both buildings restored the tripartite relationship of base, shaft, and crown. Both buildings evoke historical imagery combined with monumental scale, which represents a conscious realignment of modern architecture with historical architectural principles governing the aesthetics of the building and its facades.

The AT&T building and the Humana building raised the ire of the architectural critics and the public alike. Huxtable (1984) in particular reacted negatively to the overall lack of unity expressed in the facade of the AT&T building and its heroic scale:

> It becomes a point of wonder that such respected architectural devices as arcades and occuli of the AT&T Building, with the enormously exaggerated scale of the building's traditionally composed base, shaft, and top, can result in anything so flaccid and so unexceptional as this assemblage of painstakingly and expensively executed details—tons of granite and pratfall playfulness notwithstanding.

As Goldberger (1981) states, the revolution that modernism represented had been won (at least insofar as corporate architecture was concerned), and the public more or less had become reticent about the aesthetics of Miesian rationalism. But post-modernism and its historical pastiche took the public by surprise.

> In the Humana Building, Graves takes the issue of historical eclecticism a step further. Its loggia, balcony, pilasters, and earth-tone polychromy are clear references to classically inspired buildings. A monumentally-scaled six-story base mediates between the pedestrian scale of the street and the height of the tower. The granite-clad shaft completely encloses and conceals the building's steel skeleton and is punctuated by small, square openings. A projecting balcony is supported by a structural steel truss which is intended to relate contextually to Louisville's steel bridges. Finally, the building is capped by a multi-story loggia which acts as a symbolic beacon on the city's skyline.

## 2 Classical Eclecticism

The appropriation of classical proportions, motifs, and elements in recent tall building designs has impacted the manner in which architects think about and respond to tall building facades as they relate to the scale and the stylistic imagery of the building as well as the context of the urban streetscape. Historically there seem to have been two parallel conceptual tracks dealing with the expression of the tall building. The first, arguably the most powerful, is the concept that the architectural aesthetic of the building reflects its structural realities. This is the problem that Sullivan sought to resolve in the Wainwright and Guaranty buildings and that Mies continued to deal with when he arrived in Chicago. The second concept is primarily an issue of aesthetics, which links the design of tall buildings to stylistic precedents, such as Gothic and classical architectural styles. Throughout the history of tall building design, these concepts have converged and diverged. Eclecticism, which was predominant during the 1910s and 1920s, culminated in the art moderne or art deco skyscrapers of the 1930s. Structural determinism, on the other hand, has always been a basic tenet of high-rise design (Huxtable, 1984). Huxtable views classical eclecticism as the most radical of the current new directions in high-rise architecture because it represents a departure from the imperative of structure as a chief generator of architectural style.

***Norwest Center (1989).*** The Norwest Center in Minneapolis is intended to conform to the grace and technology-oriented character of Minneapolis architecture. Located near the IDS Center, the facades of the Norwest Center are reminiscent of the iconic power of modernist high-rise buildings of the late 1920s and early 1930s, specifically Raymond Hood's RCA Building in New York (Wolner, 1989; Boles, 1989). The vertical organization of the elevations are based on a 762-mm (2-ft 6-in.) window module characteristic of older modern buildings, while also relating to other facades of Minneapolis buildings. The use of Minnesota stone, similar to Kasota stone, imparts a golden hue, for which Minneapolis's buildings are famous. Glazing is set back a full 152 mm (6 in.) from the face of the building within a white granite frame. Continuous vertical stone-clad pilasters between the windows visually carry the upward momentum of the facades to the building's stepped crown.

The vertical pilasters do not express the building's actual structural system, since the Norwest Center's structure is based on only four supercolumns. Although Pelli believes that architectural expression must be consistent with technology, he does not believe that structural expressionism is essential in modern architecture. Think of Mies, Pelli advises: "There is no need to express every beam" (Boles, 1989). Since the skin is not structural, in Pelli's view, it need not be expressed as such. The curtain-wall facade is a veneer that has been stretched over the frame of the building. The recessed windows and spandrel panels, along with the stepped massing of the building, provide vertical articulation and sculptural depth.

***Leo Burnett Building (1990).*** Located on Wacker Drive in Chicago, the Leo Burnett building (Fig. 7.12) is a 50-story office tower developed by the John Buck Company and named for its principal tenant, the Leo Burnett advertising firm. The Leo Burnett building evokes a classical column motif with a carefully defined tripartite organization of base, shaft, and crown. Conceived as a free-standing monumental column, its facade is a checkerboard pattern of glass windows and granite panels. A giant order of abstracted columns forms the base of the building, whereas similar bands of columns occur at the building's single setback and at its crown. Each window is identical and is deeply recessed behind the granite skin with a surround of stainless steel. A half-round column, boldly striped in black and silver, bisects each window.

The inspiration for the Leo Burnett building comes from the CBS building in New York. The weighty dark-grey granite and tinted glass facades of the CBS building earned it the pseudonym of "black box" (Keegan, 1990). Vincent Scully stated that buildings of that type (Keegan, 1990):

> ...give the impression of having been designed in the form of models for dramatic unveiling at board meetings and of having never been detailed beyond that point, so that, whatever their actual size, their scale reveals no connection with human use and remains that of small objects grown frightfully large. The effect is of arbitrary mass and order and individual mass disorientation.

***Morton International Building (1991).*** Among the more successful and structurally innovative buildings to grace Chicago's skyline in recent years, the Morton International building (Fig. 7.13) had to overcome some major structural and aesthetic challenges. The structure had to deftly bridge several railroad tracks running through its site along the Chicago River. The sheer physical problems of putting a building at all on this site were daunting. The most dramatic problem was solved by suspending the entire southwest corner of the building's smaller volume from a large truss visible on the exterior of the building.

**Fig. 7.12 Leo Burnett building, Chicago.** (*Courtesy: Paul Armstrong.*)

**Fig. 7.13 Morton International building, Chicago.** (*Courtesy: Paul Armstrong.*)

The second challenge was aesthetic: how to recover something of the material substantially found in many tall buildings before 1930, when the architect's role has increasingly come to be the application of thin skin over layers of essentially uniform space. Emulating the PSFS (Philadelphia Saving Fund Society) building (1932) in Philadelphia with its tower and pedestal massing, a complex set of modifications were used to unite the two pieces, including "two different tints of glass, six different colors of granite, and a whole arsenal of different column and spandrel panel arrangements to express on the curtain wall the structural and programmatic continuities and discontinuities" (Bruegmann, 1991).

The symmetry of the exposed steel truss is reflected in the symmetrical treatment of the north and south facades. Bruegmann wonders "whether some kind of asymmetrical structure with more clearly expressed tension members might have better reflected the asymmetrical suspension condition rather than the implied bridging across the top of the [lower] structure. It might have given the building a lighter, more dynamic feel." Still, the vertical pilasters and horizontal spandrels visible on the facades clearly express the rational structural and functional organization of the building. (See Chapter 10 for an alternate account of this building.)

***R. R. Donnelley Building (1993).*** The R. R. Donnelley Building (Fig. 7.14) in Chicago is modeled after the tripartite organization of a classical column (Berger, 1993). However, its classically inspired literalism goes a step further by emulating a Greek temple vertically "stretched" into a tall building.

The building's exterior glass curtain wall and white granite cladding combines abstracted classical details, such as the pedimented windows in the granite-clad base and the cross-gabled, pedimented summit. However, the shaft remains an international style glass box minimally articulated with vertically stacked pilasters and horizontal granite spandrels. The vertical elements reflect the varied spacing of the interior bays, and the continuous horizontal courses indicate special 5.5-m (18-ft)-high floors. However, with no differentiation in depth separating horizontal from vertical, the pilasters and spandrels read more like ribbons on a gift box and flatten the overall dimensional reading of the shaft.

## 7.8 MATERIALS AND THEIR IMPACT ON EXPRESSION

Although much has been written about the influence of structural and technological innovations on the morphology of tall buildings, aesthetic and architectural criteria of high-rise facades are primarily relegated to discussions of style and syntactic convention. Huxtable (1984) writes that aesthetic interpretations of the design of tall buildings have always absorbed architects and critics. Sullivan recognized the aesthetic implications of the tall building facade early on in *The Tall Building Artistically Considered.* He rejected the purely mercantile and philistine image of the massing and elevations of the high-rise office building as defined within the parameters of a market-driven economy. To Sullivan, as well as critics like Huxtable, "the skyscraper is essentially an art form and architecture, above all, is an expressive art" (Huxtable, 1984).

The intrinsic relationship between materials, structural form, and aesthetic expression makes it difficult to discuss one aspect of tall building architecture without including the other categories. Because of its technological constraints, the tall building is perhaps the quintessential example of Sullivan's dictum of "form follows function."

**Fig. 7.14 R. R. Donnelley Building, Chicago.** (*Courtesy: Paul Armstrong.*)

Although the formal and morphological expressions of tall building facades are in many instances determined by or manifestations of the building's structural elements, not all architects feel compelled to observe structural expression as the primary means of developing elevations. The aesthetic intentions of the architect are sometimes placed above the tectonic requirements of the engineer. Therefore the aesthetics of the tall building may be modeled after historical precedents, as in the Humana building, or it may represent a more innovative departure, as in 333 Wacker Drive (1980–1983; see Chapter 5) in Chicago, a sculptural monolith comprised of a gently curving glass volume emerging from a stone veneer clad base.

Fundamentally the structural and technological innovations which defined the memorable skyscrapers of the 1920s were essentially the same as those incorporated into the international style buildings introduced into American cities in the post–World War II era. The Miesian aesthetic of the glass curtain wall suspended on a steel frame employs similar structural technology as buildings that are clad with more traditional materials.

Innovations in the manufacture and use of new and existing materials have dramatically changed the manner in which architects design buildings. Improvements in stone-cutting technology, for example, have enabled architects to specify stone veneers which are thinner and lighter in weight. These veneers can be cut to extraordinary tolerances, which makes fabrication and detailing much easier and more precise. New composite materials manufactured from polymer plastics or combining materials in innovative ways, such as rigid foam insulation bonded to metal panels or coated with cementitious materials, have greatly expanded the material and aesthetic repertoire that architects now have at their disposal.

### 1 Structural and Nonstructural Facades

Prior to the use of steel-frame construction pioneered by Bogardus in cast-iron prefabrication and applied by Jenney in the Home Insurance building (see Chapter 10), tall building technology was restricted to the use of load-bearing masonry. Structural masonry and stone required extremely thick walls in high-rise applications. The limitations of immensely thickened walls in tall building design are quite evident in early high-rise buildings like the Monadnock Building, whose walls are up to 18 m (6 ft) thick at the base. The extreme loading conditions of the walls require more material than a steel-frame or even concrete-frame building of comparable height. Furthermore, the ultimate height of the building is limited due to the weight of the masonry itself, thereby significantly limiting the vertical expression.

The introduction of the steel frame and the reinforced concrete frame as load-bearing systems freed the exterior load conditions of the building and enabled architects to develop lighter, more flexible enclosure systems, which include thin veneers and curtain walls. Free of their load-bearing masonry walls, buildings could rise higher and, significantly, be built with greater economy.

Although steel and concrete had been used historically for structural applications, it was not until 1900 that steel frameworks and reinforced concrete had become well engineered materials (see Chapters 10 and 11). During the 1920s and 1930s, when buildings such as the Chrysler and the Empire State were constructed, skeleton steel construction with steel and concrete was well established. Steel-frame construction naturally lent itself to the rational expression of vertical columns and horizontal beams. The development of the modern glass and steel curtain wall has enabled architects to sheath buildings in ever thinner, more transparent materials. However, in spite of technological developments of nonstructural wall systems, composite wall construction systems integrating structure, insulation, and exterior and interior finishes have made the structural wall economically viable once again. A list of factors influencing the quality of facades is presented in Section 7.12.

## 2 Cladding Materials

The materials most commonly used in cladding high-rise buildings are metals, glass, stone, concrete, and brick (Smith and Slack, 1990). Aluminum, the most common metal used in cladding, may be used in extruded shapes, flat plates, bent or formed sheets, or laminated to core materials. Stainless steel may be used as flat or formed sheets, either as elements in themselves, or as cladding over other materials. Mild steel may be flat, brakeshape, or roll-formed. Any of these materials may be employed as complete wall cladding components, as framing for other materials, as infill panels, or for all of these. Each may be applied with a wide variety of integral finishes, paints, and coatings.

Traditionally, glass has been used as an infill panel framed by other materials such as aluminum or steel. Recently structural silicone glazing technology has made possible the "all glass" facade, without any other framing visible from the exterior. Glass is available clear or tinted grey, bronze, green, or blue. It may be coated or uncoated with uniform colored, metallic, or patterned coatings on interior and exterior surfaces. It may be monolithic, insulating, or laminated, using annealed, heat-strengthened, or tempered lights.

Stone is available in granite, marble, limestone, sandstone, slate, and other types of natural and synthetic materials. Recent advances in cutting technology, discussed later in this section, allow stone to be cut to increasingly thinner tolerances, producing stone panels that are lighter in weight and, consequently, easier to install.

If properly specified and detailed, fine and coarse aggregate concrete may be used as a cladding material. The ability to mold and cast concrete makes it an ideal choice for decorative and ornamental detailing on facades, or employed as cast-in-place facades or as precast panels. Concrete may be used alone, or it may serve as a substrate for thin stone veneers.

Masonry continues to be a popular cladding material for many facade applications (see Chapter 12). Nonstructural applications of standard-type brick veneers can be combined with custom-made extruded brick forms to provide texture, ornamentation, and special details to facades. Glazed and unglazed terra cotta has been used widely in many high-rise buildings. It was a popular material choice during the 1920s and 1930s for reasons of aesthetics, durability, light weight, ease of installation, and ability to be molded into an infinite variety of patterns, textures, and colors. Although not widely used today, terra cotta is used primarily in applications where restoration or additions to existing buildings require it.

## 3 Curtain-Wall Systems

The advent of lighter-weight materials, new technologies, and an ever-increasing emphasis on keeping high-rise building costs at a minimum has led to the development of the exterior curtain wall as one of the most efficient and affordable cladding solutions during the postwar era. Aesthetically the curtain wall proved to be extremely adaptable to modernist ideology of an emphasis on pure geometric forms and prismatic massing. Curtain walls are relatively light and can be stretched taut and virtually flat, emphasizing a building's inherent geometric and sculptural massing, or it can be layered three-dimensionally, incorporating a wide variety of materials, patterns, and textures. For detailed discussions on curtain-wall systems from an engineering perspective, refer to Chapter SC-5 of *Tall Building Systems and Concepts* (CTBUH, Group SC, 1980) and to *Cladding* (CTBUH, Committee 12A, 1992).

The structural core system used extensively in tall building design is highly efficient from a structural viewpoint, although it poses new problems for the architect in developing and articulating a cladding system for the exterior of the building, which has a sense of vertical and horizontal scale and does not compromise the sculptural aesthetic of the building's form and massing. Furthermore, whereas the projection of floors reduces the areas of paneling and gives a strong horizontal emphasis to the design, it conflicts with the natural vertical development that the tall building suggests. Gregory (1962) suggests that the reduction of panel areas to core-supported high-rise buildings is an avenue for further investigation.

Krier (1983) criticizes the use of curtain-wall systems on small-scale building projects, particularly in residential building types. "The often-used framed facade made of light material and glazing," he writes, "is too standard in type and too abstract in character for housing developments. It does not allow for aesthetic differentiation and is too vulnerable and too transparent." However, for high-rise construction curtain-wall systems have proved to be indispensable for reasons of economy, weight, and efficiency. Another criticism of Krier's is that curtain walling is not energy-efficient in that it increases heat loads on buildings in the summer and induces heat loss during winter months. Gregory (1962) points out that in curtain-wall systems, with the multiplication of flexible joints required, many cracks and openings occur where cool air escapes; metal windows do not fit tightly enough to prevent air from escaping between sashes and frames. In tall buildings there must be an optimum for this type of cladding when used as a homogeneous skin. Where this type of cladding is desirable, the areas could be reduced by means of intermediate vertical and horizontal disconnecting of materials subject to minimal thermal movement.

Water and moisture penetration is another issue in curtain-wall systems, particularly in climate zones subject to tropical storms such as cyclone, typhoon, and hurricane belts. Standard architectural construction is designed for rains falling within a 45° arc from the vertical. In tropical storm conditions, rain driven by wind speeds of up to 193 km/h (120 mph) strikes buildings within an arc of nearly 180°, often horizontal and often reflected upward. Under these conditions the standard design of metal window sections which give protection under normal circumstances fails to prevent water penetration; doubling the window catches is not satisfactory. Solutions that line metal window sections with a resilient plastic material which can be squeezed into a seal when windows are closed or the use of various proprietary rubber gasket materials for fixed glazing have been employed (Gregory, 1962).

The evolution of curtain-wall cladding allowed architects to stretch the rigid rules of modernist principles and explore the sculptural possibilities of tall building form. Structure became analogous to sculpture, transforming the whole building into a sculptural object. Or the building might be conceived as a provocative, abstract play of light, planes, and reflections, as a glass box gave way to a mirror-glass building. The IDS tower in Minneapolis (1972) and the Pennzoil building in Houston (1976) pointed the way to structures that fused outer form and inner space in a most suggestive way (Huxtable, 1984). The same year that the Pennzoil building was completed saw completion of the United Nations Plaza Hotel in New York. The unbroken sleekness of the glass-enclosed skin manipulates the visual effects of scale and light with mirrored surfaces and unexpected angles. The curved tower for the Oversea-China Banking Corporation in Singapore, also dating from 1976, presents a monolithic building image (Huxtable, 1984).

#### 4 Thin Stone Veneer Cladding

Due to the increased use of thin stone veneers as exterior cladding on high-rise buildings, it seems appropriate to focus on recent aesthetic and technical innovations. The historical image of the permanence of stone and its ability to withstand the ravages of climate and time has been supplanted by the realities of construction technology and economy. According to Harriman (1991), granite, marble, sandstone, limestone, travertine, and slate are increasingly being applied to curtain walls as stone-cutting technology has advanced to produce panels less than 51 mm (2 in.) thick, pushing stone veneers closer to their calculated stress limits. Dimensions approaching 19 mm (¾ in.) thick are now possible, allowing a safety margin for construction tolerances and variations in the veneer as slim as a panel's profile.

***Amoco Building (1973–1974).*** The most recent and visible example of the susceptibility of stone veneers to the effects of weather and climate is evident in the refacing of Chicago's 80-story Amoco Building (Fig. 7.15). Originally faced with 32-mm (1.25-in.)-thick Carrara marble cladding in 1973–1974, the originally straight, 1.3- by 1.1-m (50- by 45-in.) stone panels have bowed into warped dish shapes. Uneven expansion of the exposed surface in relation to its more stable rear face results in a type of marble distress called "hysteresis." All 43,000 panels have been replaced at an estimated cost of $60 to $80 million with 51-mm (2-in.)-thick Mt. Airy granite panels, chosen for strength, durability, and to maintain the building's distinctive white appearance and vertical fluting.

The ability to achieve increasingly slender dimensions has created a need to evaluate the properties and connection methods of stone veneers. The original anchoring system used in the Amoco Building, a clip angle set into panel kerfs, has also been improved. Continuous extruded, stainless-steel shelf angles for supporting the top and bottom of the new panels were used to replace the existing 76- to 102-mm (3- to 4-in.) bent-steel angles that were intermittently placed at the base of the original marble panels. A pressure-relieving joint under each shelf was formed by a compressible foam pad which accommodates vertical movement by allowing each stone panel to expand, thereby preventing stress cracks and spalling.

## 7.9 THE INFLUENCE OF TECHNOLOGY ON FACADE DESIGN

Earlier in this chapter the importance was pointed out of recognizing that, particularly in tall building design, it is impossible to consider formal and aesthetic principles in isolation without reference to technology and construction. In the Preface to *Office Building Design,* Schmertz (1975) reflects a shift in emphasis from purely aesthetic criteria to functional and technological criteria as primary determinants of tall building imagery and form:

> The best of the new office buildings are more practical and convenient to work in than ever before because of sophisticated interior space programming and planning techniques. They also are more comfortable because of improvements in their heating, ventilating, and air-conditioning systems and their control of heat loss and gain through better types of thermal insulation, and solar heat-reducing glass....Recent structural solutions for the problems of wind loads in high-rise towers herald a new visual aesthetic...economical structural systems to resist wind loads have come a long way from the simple rigid-frames devised...for

**Fig. 7.15 Amoco Building, Chicago.** (*Courtesy: Paul Armstrong.*)

buildings of 20 stories or less. The bundled-tube and truss-tube systems for steel structures up to 140 stories, and the framed-tube, tube-in-tube and modular-tube systems for similarly tall concrete structures, are exciting forms, which have begun to dramatically alter city skylines across the nation.

## 1 Technology and Innovation in Facade Design

Aside from the impact of the building zoning ordinance on the form and massing of the tall building, the greatest impacts on the aesthetics and detailing, most evident on the expression of the facades of tall buildings, have been advances in building structure and materials, and the incorporation of new technologies. The rational and scientific attitude introduced by the Chicago architects to the problem of high-rise design in the late 1800s was adopted by the modern architects of the early twentieth century and immediately applied to the problem of building aesthetics and form making.

In the late nineteenth century the negative impact of technology on society led to a romantic backlash. Artists and intellectuals viewed technology with disdain and derided industrialization and machine production as antithetical to humans' harmony with nature. The social philosopher John Oglivy writes that technology has come "to play the same role in nature, namely that of a hostile environment out of which man must carve enclaves to make his home" (Doubilet and Fisher, 1986). Architects have responded to technology in essentially two ways. Some have looked to natural means, such as passive heating and cooling of buildings, for solving technological problems. Others, modernists and post-modernists alike, have retained technology's nature-taming role, but have minimized its expression and form-giving potential.

Oglivy suggests a third alternative through which he advocates "demystifying and decentralizing technology to return man's tools to the men and women who now see only...products. Technology can oppress as well as liberate us. Denying the form-giving power of technology, looking at it as just a tool rather than," as Oglivy states, "an autonomous agent [with] a life of its own," plays to its oppressive side (Doubilet and Fisher, 1986).

## 2 Structural Innovations and Facades

The influence of the steel frame on the aesthetics of the facades of high-rise buildings was profound. Although the earliest and most direct expression of the frame and infill character of the tall building is to be found in Chicago buildings of Le Baron Jenney and the facade of the Carson Pirie Scott department store, its technological roots are to be found in the development of cast-iron construction introduced in the mid-nineteenth century. Prescott (1962) writes that the expression of the commercial buildings of the nineteenth century was, from the beginning, conditioned by the demand for ever-increasing areas of glass, in turn caused by the more and more intensive use of office space within.

The oriel window, developed in 1864 in Liverpool, was a technical response to brokers who required maximum natural light so that they could examine the quality of cotton samples in their offices. The direct expression of the iron structural frame of oriel chambers is of interest in the genesis of tall buildings. Probably for the first time a vocabulary is developed for building tall, and the simultaneous acceptance of large areas of glazing demonstrates a remarkable inversion of material expression in that their street elevations are in fact stone whereas they appear to be of metal construction.

In 1884 James Bogardus initiated the mass manufacture of cast-iron prefabricated facades in sectional elements (Prescott, 1962). A similar technology was employed in

New York buildings of the 1850s to 1880s, where classical details, attenuated and reduced to framework for all glass facades, were subsequently copied in stone, expressing more the ideals of the machine era than those of traditional stone. The "patent system" developed by Bogardus consisted of various basic structural elements assembled together to make up a whole building of cast iron with certain infill materials. Bogardus's innovative use of frame construction was a precursor to the technology of tall building construction. However prescient his vision might have been, Bogardus never recognized the vulnerability of cast-iron frame construction to fire and the lack of rigid jointing through bolted connections. It is interesting to note the formal decorative use of steel on the exterior of the Seagram Building (1958) in New York to achieve the intended aesthetic of expressed structure. (Refer to Chapter 10 for more information on cast-iron and steel.)

In a presentation to the Royal Institute of British Architects, Harris (1961) said: "that engineering is basically an art and that the science of this particular art is that needed to impose form on material." Harris defines the role of the engineer as producing new *structural* forms to be used in the evolutionary process of the architect, resulting in new *architectural* forms while providing greater scope in solving the functional and technological aspects of architecture as well as a possible new economy in which to work.

The conception of the structural engineer as a "pragmatic person concerned only with the sizes of beams and columns" (see Chapter 5) was supplanted as early as the Industrial Revolution, when structure became a new "art form" independent from architecture. A structural form is not "socially acceptable unless it is simple, ordered, and pleasant" (Ali, 1986). Since architecture and structural engineering are so closely intertwined in tall building design, it is essential for the architect to understand the physical properties of materials and their composite structural behavior in order to achieve a successful building form (see Chapters 10, 11, and 12).

The fundamental requirement in building structure is to resist both lateral and gravity loads. In the case of the tall building these functions achieve a greater significance as the building becomes taller. The X bracing on the exterior of the John Hancock Center, for example, is a direct expression of building structure on the exterior facades of the building. Structure, building aesthetics, and building form become one.

Fazlur R. Khan, whose innovative engineering solutions for Skidmore, Owings & Merrill (SOM) redefined tall building structural systems, envisioned the central problem of tall buildings to be an engineering problem. To Khan, the problems of engineering and architectural aesthetics of tall buildings are related issues which persist in architectural history. In his view, the Gothic constituted the "grandest period of the structure being so intertwined with the architecture that they were inseparable." And although he acknowledges that great buildings were also produced by Renaissance architects, he clearly sees structural expression as most appropriate to modern architectural form (Goldsmith, 1986).

Structural expressionism and building image have been a preoccupation of many contemporary architects. Their buildings represent an attempt to synthesize building technology into a direct expression of building form, creating powerful and, sometimes, original building designs. Directed by the structural innovations of Khan, the use of bundled-tube structural technology and exterior X bracing enabled SOM to surpass the limits of tall building height and produced the characteristic stepped volumetric massing of the Sears Tower and the powerful X-braced facades of the John Hancock Center in Chicago. Other architects, including Helmut Jahn and Norman Foster, have been working to define a new aesthetic based upon technological expression and innovation. These architects of the second machine age envision themselves as innovators, leading architecture and tall building design into the twenty-first century.

## 7.10 NEW DIRECTIONS

As architecture evolves toward the next century, the requirements for tall buildings to meet exceedingly difficult and complex technological, sociological, and environmental demands will increase proportionally. As was always the case in tall building design, technological innovation occurs hand in hand with structural innovation and is invariably reflected in the design of the facade and the overall aesthetics of the building. Greater emphasis on the physical and contextual environment and the efficient use of energy for heating and cooling tall buildings has led to the development of so-called smart or intelligent buildings, which employ sophisticated computerized systems that monitor the building's environmental functions and can even have an effect on the building's aesthetic appearance under different climatic and seasonal conditions. (For more information on this topic, refer to Section 8.4 for a case study of the Lucky Goldstar building in Seoul, South Korea.)

### 1 High-Tech Facades

Jencks (1988) writes that high-tech is less of a style than an approach to design which exploits and humanizes technology, function, and ideology. According to Jencks, the two most important high-tech buildings of this century, Lloyd's of London (1979–1980) and the Hong Kong and Shanghai Bank headquarters (1980–1986), are great works of architecture, but questionable buildings. They represent the most daring and honest expression of structure, virtuoso inventions, and fantastic spaces, but they are also unbelievably expensive. Jencks (1988) cites six basic rules of "high-tech architecture":

1. *Inside-out.* The services and structure of a building are almost always exposed on the exterior as a form of ornament or sculpture. (Refer to Chapters 8 and 9 on building system integration and to Chapter 4 on structural expression.)
2. *Celebration of process.* Emphasis is placed on construction logic, the "how, why, and what" of the building, and the expression of its joints, rivets, flanges, and ducts. Although the externalized expression of technology and building systems is exciting and visually stimulating, it is criticized by architects and critiques as nonhierarchical and confusing.
3. *Transparency, layering, and movement.* Extensive use of translucent and transparent glass, a layering of ducts, stairs, and structure, and the accentuation of moving escalators and elevators are dramatized in almost every case. (Chapter 6 discusses the psychological and behavioral aspects of pedestrian movement.)
4. *Bright, flat coloring.* Many high-tech buildings use bright, flat colors to distinguish different kinds of services and structure and to allow them to be understood and effectively used. (Refer to Chapter 5 for more information on the visceral properties of color.)
5. *A lightweight filigree of tensile members.* Lightweight cross bracing and structural cables become visual signs as well as ordering devices, creating visual associations with sailboats and ephemeral, informal living.
6. *Optimistic confidence in a scientific culture.* Underlying high-tech building is the futuristic promise of an unknown world waiting to be discovered.

***Hong Kong and Shanghai Bank Headquarters (1986).*** High-tech modernism has become a hallmark of leading architects all over the world, particularly in the United

Kingdom, which has a long tradition of engineering innovation. Norman Foster's Hong Kong and Shanghai Bank headquarters building is a *tour de force,* combining state-of-the-art architectural engineering and exacting detailing. Almost every element of this glass and aluminum-clad building, from its unusual pin-connected suspension structure to its precise, robot-made cladding, from its sun scoop to its underfloor air distribution, from its mechanical modules to its signage system, has been custom designed with intensive investigation into every aspect of each problem (Doubilet and Fisher, 1986).

Rising 180 m (590 ft) above grade, the 47-story building's most striking visual and functional feature is its structure. The floors are virtually column-free, allowing flexible office planning and clear views of the harbor from all points. Four mast towers rising on either side of the building are each composed of four tubular steel columns connected by haunched beams to act as Vierendeel trusses. At five intermediate levels, suspension trusses suspend the weight of the floors in the zone below. Each of the five structural zones demarcates a corresponding functional one. The double-height truss levels have specialized common functions serviced by specially programmed high-speed elevators.

Technical innovations in the manufacture of metal, ceramic, and thin stone veneer cladding and glass curtain-wall systems enable architects to create more exacting and precise elevational details. The precision of the complex aluminum cladding for the Hong Kong and Shanghai Bank was held to a tolerance of 0.40 mm ($\frac{1}{64}$ in.) (Doubilet and Fisher, 1986). The cladding is finished with a top-quality fluoropolymer coating, applied electrostatically to pretreated metal and cured at 232.2° C (450° F), for superior protection against fading, chalking, weathering, and chemical attack as well as humidity and salt spray.

The glazing system was designed to satisfy two criteria: (1) the client's desire to conserve energy and maintain views; and (2) the architect's desire to expose the building's structure and internal activity. The design for a typical window has a fixed, outer light of clear glass; an airspace containing conventionally hung, perforated blinds; and a slightly tinted inner light of glass. Aluminum sun screens, which protect the glazing from the direct summer sun on every floor, have enough strength to support maintenance personnel. They also enhance the horizontal reading of the floor levels on the exterior elevations, and their light, translucent structural characteristics filter sunlight and add a certain delicacy as a foil to the expressed, massive structural members.

***L'Institut du Monde Arabe (1988).*** "Smart" or "intelligent" building technology is only at its most nascent stage. However, as L'Institut du Monde Arabe (Arab World Institute; Fig. 7.16) in Paris demonstrates, sophisticated technology and integrated computerized systems will become standard in future high-rise buildings. The building's curved mirrorlike glass facade, which faces the Seine, is described as "hermetic and elusive" (Loriers, 1988). However, it is the innovative use of technology and exquisite detailing that distinguish the Arab World Institute from other curtain-wall high-rise buildings. The south-facing wall enclosing the southern galleries contains aluminum photovoltaic panels that form a wall of mechanical apertures which open and close (Fig. 7.17) like camera lenses in response to sunlight and evoke in high-tech terms the mandalalike patterns of Arab screens (Jencks, 1988).

## 2 Designing Facades for Context and Climate

Modern architecture dominated the profession from the mid-1920s to the late-1950s (Jencks, 1988). Since the breakup of CIAM (Congrès International d'Architecture Moderne) in 1959, the consensus among architects regarding a dominant architectural

**Fig. 7.16 L'Institut du Monde Arabe, Paris, France.** (*Courtesy: Paul Armstrong.*)

**Fig. 7.17 Detail of the south-facing wall of L'Institut du Monde Arabe.** (*Courtesy: Paul Armstrong.*)

style has eroded progressively, and many approaches have come and gone. The pluralistic tendencies currently dictating architectural design and responses to context and scale have created profound ideological rifts among architects and have fueled hostile debates. Although architects, critics, and the public are becoming more and more agitated and confused about this acrimony, pluralism has opened a public debate which has had a positive impact on the way that architects approach design and context. The post-modernists, for example, claim that their buildings are rooted in place and history, unlike the abstract buildings of their predecessors, and that they reincorporate the full repertoire of architectural expression, including ornamentation, symbolism, and urban context. Late-modern architects, by contrast, disdain all historical imagery and concentrate on space, geometry, and light as the perennial attributes of architecture. Generally refusing to discuss stylistic issues, they approach the building as a series of technical problems to be solved by teamwork, or abstract propositions of geometry and organization (Jencks, 1988).

In spite of these ideological differences, most architects tend to agree that architecture must respond in a meaningful way to its total physical context, including climate and locale. The ubiquitous international style high-rise glass box has given way to new and innovative uses of architectural forms to ameliorate differences in function, scale, and context. After many decades of indifference to place and site, architects once again are recognizing that tall buildings must accommodate multiple and diverse functions, new technology, and structural systems, and that they must be adapted to meet the exigencies of context and climate.

***Messeturm (1991).*** Square in plan, the tapered obelisk of the 70-story Messeturm (Fig. 7.18) in Frankfurt am Main, Germany, is reminiscent of the "point" skyscraper towers of the 1920s and the classical obelisks (Murphy, 1990). The tallest building in Europe at 259 m (850 ft), the stereometric treatment of its elevations and concrete frame provide similar views of the tower from all vantage points throughout the city. The pyramidal top of the building completes its classically inspired references and establishes a visual focal point on the skyline.

The tripartite scheme develops from the concept in plan of a circle circumscribed by a square. The geometry of the "squared circle" is developed in the base and carried through the entire shaft to the pyramidal cap of the building (see Chapter 5). As the building rises, the granite facade reads as a "screen wall" punched with window openings enclosing a glass cylinder, which is revealed at the corners and at the uppermost stories of the building.

The base of the tower is a monumental glass-enclosed atrium forming a transparent, segmented cylinder within a granite-clad box, and enfronts a large public plaza. Granite-clad elevator banks frame views through the lobby to the exterior. The granite panels facing the elevator and service core walls are mechanically attached with expressed stainless-steel hardware. Alternating bands of light and dark red granite and glass provide a sleek, polished exterior skin, blending technology with traditional imagery.

The 85,006-$m^2$ (915,000-$ft^2$) tower is one component of a larger complex, which includes the existing postwar Kongresshalle and the historic Festhalle built in 1909. In addition to the tower, the complex includes a new exhibition hall and an all-glass entrance pavilion (Murphy, 1990). These smaller-scale elements ameliorate between the vertical expression of the tower and the low-scale adjacent buildings, thereby integrating the entire complex with the surrounding urban scheme (Murphy, 1990).

***DG Bank (1987–1993).*** Developed in 1987 for an entry in a limited design competition, the DG Bank (Fig. 7.19) in Frankfurt am Main, Germany, establishes new directions for contextualism and sustainable architecture of high-rise buildings (see also Chapter 9).

**Fig. 7.18 Messeturm, Frankfurt am Main, Germany.** (*Courtesy: Paul Armstrong.*)

**Fig. 7.19 DG Bank, Frankfurt am Main, Germany.** (*Courtesy: Paul Armstrong.*)

The tower portions of the DG Bank building were designed to be more responsive to context and, simultaneously, more abstract. Traditional architectural references were limited to the street level and no classical detailing was brought to the 20th or 40th floor. The site afforded many contextual opportunities. The DG Bank complex aims to recapture advantages of the traditional city in three ways: (1) by reinforcing the traditional street wall and street rooms, (2) by using classical scale and rhythms to respond to the surroundings, and (3) by fragmenting the mass, as found in the traditional urban fabric (Boles, 1988). Sited adjacent to a new concentration of towers near Frankfurt's main railroad station, the DG Bank is fronted on the broad, busy Mainzer Landstrasse and backed up to a quiet and consistently scaled residential district. The tower's variously scaled volumes respond to the fixed cornice heights of the nearby apartment blocks and the traditional *Traufhöhe* [the eave height of a typical inner-city building, generally set at 22 m (72 ft)] (Mönninger, 1993). The quarter-round form of its 208-m (650-ft)-tall shaft and its prominent crown inflect the tower toward the old core of the city and the Main River to the southeast.

The asymmetrical, heterogeneous massing of the DG Bank results directly from its context and its mixed-use functions. The building was originally commissioned as a 65,032-$m^2$ (700,000-$ft^2$) mixed-use development, incorporating an office tower, a 300-room hotel, apartments, and a winter garden. Later the hotel was considered no longer economically viable, and its volume was dedicated to office space. A medium-rise penthouse to the north houses 15 apartments. The structurally expressed winter garden is surrounded by shops and restaurants, with tables spilling out into the public space.

With this skyscraper solution a new way was found of articulating the volumes of high rises. The attempt to avoid mirror-glass facades and to reemphasize volume and materiality is a departure from typical modern high-rise facade treatments. The facade treatment reinforces the discreet nature of the parts. Skeletal steel volumes are played off against monolithically clad volumes, further articulating the functions and scales of the component parts of the building. The lower buildings, together with the lower levels of the office tower, are clad in polychromatic granites and marbles, whereas the tower's curved shaft uses reflective glass and painted steel. Since buildings no longer depend on skeleton construction, but on the double-tube system of inner cores and outer punctured shells, the curtain wall gives only the illusion of being hung (Mönninger, 1993). The illusion of depth also was enhanced in some American buildings by deep window embrasures with inset mullions, where the structural order is delineated through the organization of volumes. In the design of the envelope, the architects use deep relief in the facades, even though it tends to squander developable interior space.

***United Gulf Bank (1987–1988).*** Contextualism and technology are merged in the United Gulf Bank in Manama, Bahrain. Without referring to the low-rise courtyard structures that compose Bahrain's traditional architecture, a forthrightly modern building has been created, directly influenced by the color, form, organization, structure, and detailing of vernacular buildings in the hot desert climate (Fisher, 1988). Because of the intense heat of Bahrain, the building provides a deeply shaded arcade, recessed balconies, and glass fins and light shelves to reduce heat gain and increase daylight in the offices. The precast concrete curved wall of the principal facade contains deeply recessed punched windows. Extending beyond the roofline, it also hides rooftop mechanical equipment.

The concrete facades are punched with deeply recessed regularly spaced window openings. Scale is imparted on the otherwise monolithic facades by scoring the concrete with thin vertical and horizontal lines and by imprinting a rhythmic pattern of incised reveals and squares around the windows. The overall effect conveys an image of massive monolithic walls that are lightened and enlivened by the window openings and the textures added to the wall.

### 3 Toward a New "Skyscraper Style"

The search for an appropriate architectural style for tall buildings which has preoccupied architects since the late nineteenth century is likely to continue into the next century. Goldberger (1983) writes that the technical and aesthetic problems of the early skyscrapers have been solved—architects have the requisite knowledge of how to make buildings that express great height. Therefore architects have turned toward the integration of more conventional architectural concerns with skyscraper design, and toward the elaboration of a skyscraper form that follows from it.

The conflict between function and expression is, in Huxtable's view, a legitimate, historic condition of architecture, and the manner in which the necessary equilibrium is achieved is intrinsic to the quality of the result. The pluralism displayed in current tall building design which characterizes the multiplicity of routes to a more expressive architecture is valid to a point. Rather than a diversity of styles, Huxtable (1984) sees only a diversity of means to the same end:

> No matter how dissimilar the source or the inspiration, all of the design elements employed are removed from their original context and purpose to exist on their own as discreet, independent forms, or as pure stylistic devices....And it is this overriding fact that unites and characterizes as a single aesthetic product that the work of those engaged in today's stridently competitive dialectics, no matter how much they may divide and protest.

## 7.11 FUTURE RESEARCH

The design of facades for tall buildings requires achieving a balance between the aesthetic and the theoretical concerns of the architect, the properties and limitations of materials, and the requirements of the building's form and structure. Integrating structural innovation and new materials and technologies with the design intentions of the architect will continue to be a pressing need as both architects and engineers endeavor to coordinate their efforts during the design process.

The physical context of the building will continue to have a major impact on the design of tall building facades both from the contextual and the climatic points of view. The aesthetic issues of scale, streetscape, and stylistic expression are paramount concerns of many architects and planners. Furthermore, the image projected by the building in terms of aesthetics and style will continue to be important to the client as well as the architect. Designing facades for energy efficiency and climate directly impacts the choice of cladding materials, percentage of transparent to opaque surfaces, and orientation of the building relative to sun, wind, and seasons.

As new materials and technologies are developed for the cladding of tall buildings, research must be conducted into the economic, climatic, and sound transmission properties as well as the life-cycle viability of these systems and components. This research is particularly important to determine the feasibility and physical behavior of new systems under vastly different climatic conditions, as well as their application to culturally and geographically diverse contexts. Understanding the effects of the macro- and microclimate, such as wind, water, temperature variations, and seasonal extremes, on cladding systems and components will contribute greatly to life-cycle durability, maintenance, and costs. Testing new and innovative materials and methods of cladding and reducing the costs of manufacturing and installation of these systems will increase the applicability of new technology to an increasingly diverse range of architectural projects and geographic regions.

Research into computerized technology and intelligent building systems and their incorporation into the facade will also be required. The incorporation of building systems technology (such as photovoltaic cells and active louvres) into building facades to monitor and adjust to thermal and climatic changes, both outside and inside the building envelope, and to incorporate higher standards of energy efficiency have become paramount in facade design. This research has become mandatory as a response to better utilization of energy and resources required by both industrialized and nonindustrialized countries.

Maintaining the health and productivity of the occupants of all buildings (and especially tall buildings) has become a critical issue in building design. The treatment of the building facade not only has a significant impact on the aesthetic form and the structural expression of the building, but can also contribute to the physical and psychological well-being of its occupants. The effectiveness of glazing or other facade treatments to control the undesirable effects of solar gain and glare through the incorporation of low-emissivity glass, thermal paned glass, and active and passive solar screening devices can make the interior environment more conducive to work and living. Also, continuous development of sandwich-panel construction using various types of veneers, ceramics, and metals as cladding materials and improvements in insulation materials can contribute to environmental comfort as well as life-cycle and maintenance costs.

The behavior of materials and their ease and economy of installation must continue to be investigated. Although prefabrication of exterior cladding systems has been used for some time, architects continue to have difficulty with field installation of factory-made panel systems to structures and components assembled in the field. New methods of anchoring panels that allow for field adjustments may have to be developed. Rain screen walls, in use in small-scale buildings, should also be investigated for tall building applications in addition to other methods of preventing moisture infiltration in cladding systems.

Further investigation into the interaction between the nonstructural architectural elements of the facade (such as mullions, glass, spandrel panels, and cladding materials and hardware) and the tall building's structure also needs to be undertaken. The relationship between the building's structure and the building skin is an area in which little research has been conducted. Architects, working in close association with the structural engineer, must be aware of the critical roles cladding and structure play not only in terms of aesthetics, but also in terms of their physical behavior and compatibility with one another.

The role that the facade might play in the incorporation of building systems, such as integrated mechanical air-handling systems and natural ventilation coupled with opaque, translucent, and transparent materials and sun-screening devices, needs to be studied to determine the long-term performance and effectiveness of these systems. Further study is also needed to test the application of specialized technologies to difficult climates and diverse locales.

## 7.12 APPENDIX: FACTORS INFLUENCING THE QUALITY OF FACADES

### 1 Facade Materials and Components

Normally, when the facades or exteriors of a building are being designed, the process is carried out without knowing exactly what they will turn out to be. It is normal to hear comments such as "it will be granite if there's enough money; if not, then concrete," or "it's reflective glass." Further details are rarely given, during this phase of the design, of the technical aspects of the choice, the emphasis being on the architectural or economic aspects.

The selection of the exterior material of a facade is determined by a few limitations arising from the quality of the material itself and all its internal components and from their composition and chemical properties as well as their availability on the market. Various aspects can be considered when discussing the quality of facade materials:

1. *Architectural characteristics.* The formal composition, the arrangement of volumes, functionality, and more, all considered as a work of art, as a functional object in a specific location within an existing environment. Generally this quality is controversial due to its subjectivity, and popular opinion rarely coincides with that of the architects.
2. *Design.* Often confused with architectural quality, it differs from it in that it is concerned with the degree of success and objectivity with which the design of the building's structural, functional (cleaning, replacement, maintenance) and operational (solar protection, internal lighting level, thermal and acoustic insulation, durability, internal adaptation to division) aspects as well as the constructional scheme have been resolved.
3. *Quality of materials.* The quality of the materials used, their durability, their technical suitability to the environment, thicknesses, and the sizing of these materials. It also includes the quality control guarantee of these materials.
4. *Construction.* A fundamental aspect of the final quality of the product in time, correct construction decisively influences the public opinion of all aspects of quality and of the professionalism of the persons and companies involved, not only of the facade but also of the construction and promotion of the entire building.

It should not be forgotten that a fault in one element, even if it is very small, in its quality, design, or construction may seriously affect the building's image and may sometimes be difficult to correct later. It is time that will determine whether a facade is of great quality, once ephemeral fashions have passed, if one refers to its aesthetic quality, or whether all its features, its elements, and its materials have endured without deterioration, confirming the wisdom of their selection. All the materials and constructional solutions are subject to a time limit, after which they require maintenance or even replacement. But the designer has to know all the variables in order for this to occur in a way that is controlled and known to everyone involved—the promoter, the proprietor, the public, and the insurance company.

## 2 Design Collaboration

The concept of the overall design of a facade (as well as being a part of the building) requires that it be designed with other types of considerations, such as constructional design, the building's internal functionality, the external functionality of the facade itself, and its influence on the workplace. All of these require the design team, led by the architect, to include representatives of the manufacturer of the raw materials, the industries that prepare and install them, structural and HVAC engineers, and, of course, the building's future users. This will bring about an integration of the different parts which comprise the facade, that is,

1. The elements that anchor the facade to the structure
2. Hidden parts
3. Glass and its supporting elements
4. Decorative and functional elements

## 3 Aspects of the Facade's Quality and Performance

A list of general attributes related to the performance and design of facades follows. The designer of the facade could treat these attributes as a checklist during the design process.

### *Materials and Their Preparation*

- Quality of base materials
- Chemical composition of chosen materials
- Mechanical properties
- Guarantee of supply and availability
- Quality control of raw materials
- Coefficient of thermal expansion
- Contract for materials, installation, and maintenance
- Compatibility with other materials in contact
- Recycling of materials
- Standardization and uniformity of materials
- Ease of execution
- Transportation from workshop to site

### *Exterior Agents*

- Thermal insulation of all components
- Acoustic insulation
- Permeability of air and water
- Removal of condensation
- Water vapor barriers
- Passive systems to adapt each of the facade's orientations
- Seismic effects
- Exterior physical damage
- Graffiti and vandalism protection on lower floors
- Solar protection

### *Site*

- Ease of assembly
- Coordination with lifting media
- Protection of exterior face during work
- Safety during assembly and on site
- Control of assembly
- Storage in workshop and on site
- Acceptance on site
- Preparation of as-built plans and maintenance program

***Miscellaneous Criteria***

- Fire insulation between frameworks
- Glass falling into street in case of fire
- Accessibility by fire fighters
- Natural ventilation and feasible areas
- Ease of cleaning and maintenance
- Movable elements for exterior cleaning
- Ease of replacement and substitution
- Fixed safety elements for exterior cleaning
- Interior-exterior view
- Possibility of installing interior and exterior solar protection
- Design allowing movement of structural frameworks
- Modulation of facade
- Durability
- Color, finish, and texture possibilities
- Correct joints (silicones with supporting weather stripping)
- Exterior flatness (views from short, medium, and long distance)
- View at night (exterior illumination)
- Interface with ground and roof
- Possibility of installing illuminated signs

## 7.13 Condensed References/Bibliography

The following is a condensed bibliography for this chapter. Not only does it include all articles referred to or cited in the text, it also contains bibliography for further reading. The full citations will be found at the end of the volume. What is given here should be sufficient information to lead the reader to the correct article—author, date, and title. In case of multiple authors, only the first named is listed.

Ali 1986, *The Forgotten Half of Design: Structural Art Form*

Anonymous 1930, *Skyscrapers: Prophecy in Steel*

Architectural Forum 1958, *Inland's Steel Showcase: The First Modern Office Buildings in the Chicago Loop*

Architectural Forum 1968, *Think Mies: IBM's New Chicago Headquarters*

Architectural Record 1958, *Inland Steel Building, Chicago*

Aregger 1975, *Highrise Building and Urban Design*

Banham 1960, *Theory and Design in the First Machine Age*

Barnett 1986, *The Elusive City: Five Centuries of Design, Ambition and Miscalculation*

Benton 1987, *Le Corbusier: Architect of the Century*

Berger 1993, *Last of the Towers*

Boles 1988, *Frankfurt Mixed-Use Complex*

Boles 1989, *Made in Minneapolis*

Broadbent 1990, *Emerging Concepts in Urban Space Design*
Bruegmann 1991, *Local Asymmetries*
Charenbhak 1981, *Chicago School Architects and Their Critics*
Clark 1930, *The Skyscraper: A Study in the Economic Height of Modern Office Buildings*
Condit 1964, *The Chicago School of Architecture: A History of Commercial and Public Building in the Chicago Area*
Crosbie 1992, *Steel Reflections*
CTBUH 1990, *Tall Buildings: 2000 and Beyond*
CTBUH, Committee 12A, 1992, *Cladding*
CTBUH, Group PC, 1981, *Planning and Environmental Criteria for Tall Buildings*
CTBUH, Group SC, 1980, *Tall Building Systems and Concepts*
Curtis 1986, *Le Corbusier: Ideas and Forms*
Dixon 1993, *American Salute to a European City*
Doubilet 1986, *The HongkongBank*
Edgell 1928, *The American Architecture of Today*
Fisher 1988, *A Bow to Bahrain*
Frampton 1992, *Modern Architecture: A Critical History*
Gibbs 1984, *Business Architectural Imagery in America, 1870–1930*
Giedion 1967, *Space, Time, and Architecture: The Growth of a New Tradition*
Goldberger 1981, *The Skyscraper*
Goldberger 1983, *On the Rise: Architecture and Design in a Postmodern Age*
Goldsmith 1986, *Fazlur Khan's Contributions in Education*
Gregory 1962, *Thoughts on the Architecture of High Buildings*
Harriman 1991, *Set in Stone: Technical Details Distinguish Well Designed Stone Cladding from Distressed Veneers*
Harriman 1992, *Full Metal Jacket*
Harris 1961, *Architectural Misconceptions of Engineering*
Hart 1985, *Multi-Storey Buildings in Steel*
Huxtable 1984, *The Tall Building Artistically Reconsidered: The Search for a Skyscraper Style*
Jencks 1973, *Le Corbusier and the Tragic View of Architecture*
Jencks 1988, *Architecture Today*
Keegan 1990, *Illusions in Chicago*
Keleher 1993, *Rain Screen Principles in Practice*
Kerr 1990, *The Rain Screen Wall*
Krier 1983, *Elements of Architecture*
Loos 1928, *The Principle of Cladding*
Loriers 1988, *Through the Looking Glass*
Mays 1989, *The Uses of Glass*
Mönninger 1993, *Crowned Skyscraper*
Mumford 1952, *Roots of Contemporary American Architecture: A Series of Thirty-Seven Essays Dating from the Mid-Nineteenth Century to the Present*
Murphy 1990, *To Be Continued*
Norberg-Schulz 1979, *Genius Loci: Towards a Phenomenology of Architecture*
Norberg-Schulz 1980, *Meaning in Western Architecture*
Pelli 1982, *Skyscrapers*
Prescott 1962, *Formal Values and High Buildings*
Progressive Architecture 1968, *Chicago to Have Another Mies Office Building by 1969*

Risselada 1989, *Raumplan versus Plan Libre*
Schmertz 1975, *Office Building Design*
Schulze 1993, *Chicago's Famous Buildings*
Scully 1988, *American Architecture and Urbanism*
Siegel 1975, *Structure and Form in Modern Architecture*
Smith 1990, *Technics: Curtain Walls—Options and Issues*
Spaeth 1985, *Mies van der Rohe*
Vuyosevich 1991, *Semper and Two American Glass Houses*
Wolner 1990, *Urban Icon*

# 8

# Building Systems Automation and Integration

This chapter examines the issue of building systems automation and integration primarily from the building systems technology perspective. The cost of building service systems in a modern high-rise building can be over one-third of the building's total first cost, and over two-thirds of its 25-year total life-cycle cost. Most liability issues associated with building systems can be attributed to failures in the communication, coordination, and integration of building systems information between architects and engineers. Building systems coordination and integration are the responsibilities of the architectural technical coordinators and project managers. Automating and integrating building systems into the complete design process, and tracking them through the manufacturing, construction, and operating phases, can minimize costly errors and problems associated with high rises and other large and complex types of buildings.

Sections 8.1 and 8.2 address the organizational alignment of the building industry today, which evolved before computers played any role in the industry. Is this still the most efficient structure from a computer perspective? Also examined in these sections is the issue of directly exchanging computer-generated information between systems across the industry. Section 8.3 discusses some of the issues related to the application of computer technology to the building industry, and how the various subsystems of a building can be integrated by taking advantage of this technology. It is expected that a future monograph volume providing an in-depth treatment of this important subject will be published by the Council on Tall Buildings and Urban Habitat.

Section 8.4 presents a detailed case study based on a conceptual design report for the Lucky-Goldstar International Business Center in Seoul, Korea. It describes a process of integrated design and summarizes the results of an 8-week conceptual design effort to define intelligent building systems, site planning strategies, and floor plan options for a high-rise office building. This case study might well be used as a model for conceptual design studies for other tall building projects where systems integration is given importance during the preliminary stages of design. Such early emphasis on systems integration in the design process could lead to efficient design solutions and avoid expensive design modifications at a later stage.

## 8.1 INTERDISCIPLINARY DESIGN

The need to consider the total picture of the tall building design process before developing a computer program for a segment of it has been recognized by computer professionals since the early years of software development. Limitations in computer hardware technology have, in the past, not only prevented the development of integrated systems across the design disciplines (architectural, structural, mechanical, and electrical), but the programs within each discipline also had to be developed in small modules. With today's computer hardware, software, networking, and telecommunication technology it is possible to develop a specialized program module that can operate as an integral component of a larger building development and operating system.

Building design and construction evolved before computers played a role in the industry. Since the end of the last century, the building process has required the services of management and planning staff, architects, structural engineers, and construction workers. The professional qualifications of this team depended on the type, size, quality, and complexity of the tall building. Mechanical and electrical engineering services and products began to have an increasingly greater impact on the building process through the last 100 years. Relative to the long history of building development, then, the computer is a very recent phenomenon. It began to have a significant impact on this industry only during the last decade, although computers themselves have been in use for over 25 years (Brown, 1982; Toong, 1982).

Centuries of building are responsible for the development of today's well-established technical and business organizational structure. Architectural-engineering (A-E) software development catered to this old structure, and tended to imitate the manual precomputer building process. This might have been the only approach possible, since it was not practical to impose on an industry organization that took centuries to evolve a redefined structure and process that, theoretically, would be more efficient from a computer perspective.

Until recently, therefore, the scope of software development within the building industry was restricted to the boundaries of the different segments of the industry, and thus development progressed within the segments independently of one another. The information generated by such fragmented software continues to be transmitted to other segments by conventional means—through drawings, schedules, and specifications on paper, and through telephone conversations, correspondence, and meetings. This can, to some extent, be explained by the lack of an independent institute or society defining the mechanisms and interfaces needed to exchange computer-generated information between the systems used by the different segments of the building industry.

Hence, although computer hardware technology has progressed rapidly in recent years and the relative cost of this technology has decreased during this same period of time, software technology in the building industry has not kept pace with this computer hardware revolution. With today's hardware it is possible not only to develop tall building design software that is integrated across the A-E design disciplines, but also to communicate such information directly to software systems in the other segments of the industry.

A fully automated and integrated tall building development and operating system is now a feasible objective. The obstacles are mainly administrative and political, not technical. Problems of communication and coordination that in the past have plagued the manual process of designing a building can be minimized in the future through automated systems that are integrated across the industry. The integrated systems approach will also increase the efficiency and reduce the cost of developing and operating buildings.

Interdisciplinary design integration methods and tools include integrated databases for building elements and client criteria; integrated graphic building model construction with interference checking; direct exchange of information between design program modules; and interdisciplinary information and project management systems.

Building services engineering (BSE) design integration across the industry involves integrating design programs with equipment selection programs and transferring electronic drawings, schedules, and specifications developed by design systems directly to systems in manufacturing, construction, and operation.

## 1 Historical Background

***Early Plans for Building Systems Integration.*** The need to develop an integrated computer design system for buildings has been recognized by computer professionals since the early 1960s, when the tall building industry first started to use computers commercially. During this period, several individuals and organizations proposed the development of integrated building systems with research papers and master plans. As always, dreams tend to precede reality: 20 to 30 years ago it was almost impossible to implement such grand ideas due to the limitations of computer hardware, software, and telecommunication technology.

In 1970 the symposium *Use of Computers for Environmental Engineering Related to Buildings* (Kasuda, 1970) was organized by the National Bureau of Standards (now the National Institute of Standards and Technology). In the opening keynote address and in the closing banquet address, several significant statements were made, which are still valid today, indicating the slow progress being made in this industry.

***An Architect's Assessment of the Industry in 1970.*** In his opening keynote address at the 1970 symposium (Kasuda, 1970), Bruce Graham of Skidmore, Owings & Merrill (SOM), Chicago, had this to say about conditions in the building industry at that time:

> The fracturized construction industry in America with its multiplicity of goals is only matched in disarray by the even more fracturized industry of construction in other parts of the world. Self interest is the motivating force in construction....I was told five years ago by a high government official that if architects do not respond to the crying needs of society, the government would step in and solve it. At that time it seemed a ludicrous statement. Today, government's failure to respond is even more obvious.
>
> Architects and planners of the last twenty years have been preoccupied with their profession. They are designers of objects. It appears that they have proven without a doubt their own irrelevancy.
>
> Academic institutions are no longer believable. Academic isolation has led to irrelevance. Yet we know, or have faith that solutions could be found and that these solutions will depend heavily on our technological baggage. This premise has been held for some time, but our credibility has lapsed. The philosophical evaluation of priorities has eluded our grasp.
>
> I believe that we are engaged in the process of developing a single civilization throughout the world.
>
> Computer technology, if it has any promise, is this: it can make available to an individual the knowledge of all; and the ability to make decisions at the most personal of levels under that larger umbrella of knowledge.

***An Engineer's Assessment of the Industry in 1970.*** In his closing banquet address at the same 1970 symposium, Sital Daryanani of Syska & Hennessy (S&H), New York, supported Graham's observations with the following remarks about automation and fragmentation in the building industry (Kasuda, 1970):

> If the building industry is compared with any other automated industry, such as the manufacture of automobiles or televisions, the problems will be evident. The construction industry has scarcely reached the level of mass production. Strangled by lack of organized

> research, outdated building codes and conservative trade unions, it has remained, in a large measure, a preindustrialized craft. The industry's rate of technological advance is far below that of other industries....
>
> This diversity of intellectual disciplines involved in the building design generally fragments the building team in several ways, the most important being the intellectual fragmentation. The architect does not appreciate the mechanical engineer's objectives; and so on. The design profession is like a universe with several planets, each merrily surviving in its own orbit. The most damaging consequence of this fragmentation is the problem of communication. As we know, communication is the binding force of any social unit, be it design or construction of buildings. Various disciplines cannot be integrated without communication and professional interaction.
>
> The lack of communication always results in suboptimization of design. Conceived without communication, the best structural design can force a suboptimal mechanical design. A solution based on initial cost may not be viable on the basis of the operating cost. The optimized building design is more a matter of communication than of linear programming. The last consequence of lack of communication is expensive duplication of effort. The same information is handled over and over again by different persons involved in the process of design.
>
> While the major problem in the building industry, in design and construction, is that of communication, the present computer applications are geared mainly to engineering problem-solving. In addition to problem-solving, what we need is a system which will consolidate all the information about the project in a central file of data where it will be available to all the members of the design team at the same time....Thus, engineers could develop integrated designs in terms of overall project goals instead of being limited to their own discipline.

***Intellectual Fragmentation by Profession.*** Building systems "disintegration" appears to begin at a college or university where students are channeled into specialized fields of study such as architecture, engineering (civil, structural, mechanical, electrical, or telecommunications), or construction management. The commercial world of building design, construction, operation, and product manufacturing continues this process of channeling professionals into specialized fields through separate companies, organizations, departments, institutes, societies, and journals for each field. Before the advent of computers, professional specialization was necessary in order to ensure product quality, although it added to the communication problem. Today the barriers separating the professions are blurring due to the expanding scope of computer applications for building development and operation. It is not necessary to know all of the internal details of a computer program in order to use the program. This will breed a new type of building professional in the future who can perform a wider range of professional services.

Compared to the conditions in 1970 described by Graham, are the research and development efforts of academic institutions any more relevant today? Limited budgets tend to force academic institutions to focus on very small pieces of this very large jigsaw puzzle. The fragmented and isolated (and also duplicated) development efforts of academia tend to result in products that are not of great value to the commercial world. The solution would therefore appear to be coordination and integration of research programs in academic institutions and frequent interaction with professionals in practice. The consumer now expects large and complete, but easy to use, computer software. Team effort, instead of several duplicated efforts, is probably the solution.

The organizational alignment of the building industry today is still based on the conditions that existed before computers played any role in the industry, when the emphasis was on specialized professions and trades. Until the June 1991 symposium *Building Systems Automation—Integration,* held at the University of Wisconsin-Madison (Dorgan, 1991), there were few instances in the past of architects and specialized professional engineers from academia and industry getting together on a regular and organized basis to discuss common issues. Professional isolation and intellectual fragmentation can no longer be justified due to the expanding scope of automated building systems.

***Building Industry versus Other Industries.*** At the 1991 symposium (Dorgan, 1991) the building industry was frequently compared to the manufacturing industry. The general view was that the building industry was about 20 years behind manufacturing in the use of automation and information technology. The cost and reliability of automobiles or televisions would be unacceptable to consumers if they were produced the way buildings are developed today. Yet, the same consumers are willing to accept the inefficiencies of manual integration and coordination of the fragmented process of developing a building. There appears to be no pressure or funding from building developers, owners, and their financial sources to enforce changes in the development of buildings in order to reach the efficiencies of manufacturing.

In his keynote address at the 1991 symposium (Dorgan, 1991), Dr. John Bollinger, Dean of the College of Engineering at the University of Wisconsin-Madison, described the automation plans at the U.S. Postal Service. He pointed out that the Postal Service had the advantage of being a single entity, where one person at the top of the organization's structure can enforce the development of integrated computer systems for automating the Postal Services. Buildings in some countries with advanced technologies are developed mainly by large conglomerate organizations that are responsible for almost all aspects of tall building development. Such organizations are well positioned to develop (and are developing) integrated computer systems for buildings. If the U.S. building industry continues to operate with its present fragmented organizational structure, then it must find a solution to integrating the computer systems serving each of the various fragments.

Bollinger also gave an example of a recent situation where the lowest bid for design services for a new University of Wisconsin building was based entirely on manual procedures. This clearly indicates that automation might have improved the quality of design documents, but not necessarily the overall efficiency of the design process. Interprofessional integration of automated design systems will have a significant impact on the overall design process and will begin to close the efficiency gap between construction and manufacturing industries.

There are advantages to maintaining a fragmented organizational structure, based on specialization, in the tall building industry. On one hand, it allows enterprising individuals to break into this industry since they are not competing with large conglomerates. On the other hand, there are also advantages in recommending integrated computer model structures for the systems serving each industry segment. It enables enterprising individuals to break into the building software development business since they could develop a module of the system and make it work with other modules of the complete building system.

At the 1992 symposium *Building Systems Automation—Integration* held in Dallas, Texas (Dorgan, 1992), current computer applications for the complete building lifecycle, including financing, design, manufacturing, construction, operation, and management, were discussed. It was also emphasized that architects, engineers, and managers from all sectors of the building industry should learn about parallel advances in computer systems development in one another's profession.

***Current Trends.*** Have conditions in the building industry changed since 1970? The intense activities and excitement of the late 1960s to produce integrated computer systems for buildings appear to have peaked with the 1970 symposium. The industry had to accept the realities imposed by computer hardware limitations for the next 15 years. The significant advances since the middle 1980s appear to be computer hardware and graphics software technology, neither of which are due to exclusive efforts made within the building industry. Because of professional and intellectual fragmentation, the building industry appears to be resigned to the fact that all computer-generated information, within each fracturized segment of the industry, must be converted to paper before it can be exchanged between the segments. The interaction of the information developed by each profession to achieve optimum solutions continues to take place orally at frequent meetings between representatives from the various professions on the project team.

Communication and coordination between the divided segments of the tall building project team are still some of the major causes of the problems that occur during building development today as it was in 1970. Building systems software that imitates manual procedures and that is confined to serve only a segment of the project team cannot resolve the communication problems between individuals and organizations that constitute the project team. Project design database integration as proposed by Daryanani (Kasuda, 1970) is one solution. Such databases are being developed and used today by multidisciplinary firms. This is particularly true in the case of graphic models of buildings, but not necessarily of building products (such as chillers, boilers, and lighting fixtures) and building elements (such as walls, roofs, doors, and windows). Building products and elements require standardized formats so that independent organizations can develop, maintain, and expand such reference databases for access and use by independently developed design systems serving each profession.

Database and systems integration must be extended beyond design to all phases of the tall building development and operating cycle. An integrated building system that can analyze the tall building as a whole, and that can react automatically, correctly, and appropriately to the frequent changes that occur during building development, can eliminate some of the problems of the past that were due to poor and ineffective communication.

The last decade produced computer imitations of a manual system. The next decade should produce a redefinition of the industry structure and a reprogramming of the tall building processes that will result in more efficient use of the computer.

There are two features inherent in ongoing computer systems development that will produce changes in the organizational alignment of the building industry, and also impact on the process of developing and operating a tall building. The first is the continuous expansion of the scope of existing software applications ("black boxes") and the continuous increase in the number of new applications to replace manual procedures that never before used the computer. The second is the cumulative effect of software development on the industry.

As more tasks that were previously done manually end up in these so-called black boxes (computer programs), and as the efficiency and quality of these black boxes improve, less knowledge and experience will be expected of the building professional. Just as it is not necessary to know the theory of internal combustion engines in order to drive an automobile, the future building design professional will not have to know all the internal details of computer programs in order to use the programs. Specialists will be responsible for researching, developing, producing, and servicing (maintaining) the specialized computer program modules and linking or integrating the modules to produce the complete building system.

During the centuries before computers, each individual entering the building profession had to learn a trade from the beginning and gain experience over a limited lifetime career. This knowledge and experience is now being entered into computers that are increasing in power and performance. This growth in computer-based knowledge and experience is cumulative and, therefore, unlimited.

Because of the black box and the cumulative knowledge and experience features of computer software development, the organizational alignment of the building industry, and the roles of building professionals in the industry, will change continuously in the future. Educational systems will also have to adapt to these changes. Ten years from now it might be possible for a single individual to design extensive components of the building instead of a specialized segment such as HVAC controls.

In the future the computer will be able to generate several alternative designs for a tall building project based on client criteria and constraints, as well as rules for designing different types of buildings. Computer-based mass-production assembly-line techniques used in manufacturing today cannot be ruled out for building design and construction in the future. The only profession in the building industry that might be immune to such drastic changes, however, might be architectural design disciplines, since these involve an inherent creative and artistic expression. For example, it is not possible to develop computer programs that can *create* music like that of Mozart. The architectural designer's role is therefore expected to gain even more importance in the future, although he or she is going to use computers for accomplishing various design objectives.

## 2 Data Exchange Standards for the Building Industry

A possible mechanism to facilitate the direct exchange of computer-generated information would be to define one or more generalized global building development and operation computer models that describe the interface formats required to exchange information between the distinctly separate systems within the model. In other words, standardized interface formats for the data are exchanged between program modules in the different segments of the industry. Besides defining such models, a standardized building description language (English) is required so that independently developed system modules can communicate with each other.

Before guidelines and standards for exchanging computer-generated information can be established, the segments and specialized organizations within the segments that currently exist in the building industry, between which the information is exchanged, must first be identified. Officially there are no clear or rigid lines of demarcation that separate the building industry into segments and the segments into distinct and specialized functions. Unofficially the segments can be categorized as finance, design manufacturing, construction, and operation, as shown in Figs. 8.1 and 8.2. The design segment can be broken down as architectural, structural, mechanical, electrical, cost estimating, and project management. The mechanical function can be further broken down into HVAC, plumbing, and fire protection, and the electrical function into power distribution, lighting, and telecommunication. Since the design segment determines, to a large extent, what happens in the other segments, the breakdown of these other segments somewhat parallels the design segment.

One could argue that data exchange standards might restrict market forces from determining the growth of software systems. This presently occurs in the industry through the creation of several de facto standards set by the more dominant software development forces. The publication of standards will ensure that the creative and innovative efforts of individuals and small organizations will also survive, and that they can continue to make significant research and development contributions to automated systems for the building industry. It allows the mixing and matching of several small component models offered by alternative software vendors, to obtain the complete computer model required for a building project. Large software corporations can continue to develop all or most of the modules of the complete building model, using the standardized interfaces between the modules. However, the client will have the option of replacing some of the modules required for the building project with modules by other software vendors. This corresponds to the flexible option, presently available to clients, of selecting alternative organizations for each specialized service to form the building project team.

Since there are no data exchange standards, computer imitations of manual tasks continue even with software currently being developed for the industry. An example is the conversion and exact reproduction of the *paper* product description and performance catalogs into *electronic* catalogs on CD-ROMs (Kennett, 1993) and other data storage devices. The voluminous paper catalogs of building products, particularly mechanical and electrical engineering products, are derived from a few test points using theory, equations, interpolations, and extrapolations. The efficiency of an automated building design system can be improved if the electronic equipment catalogs contain only the test

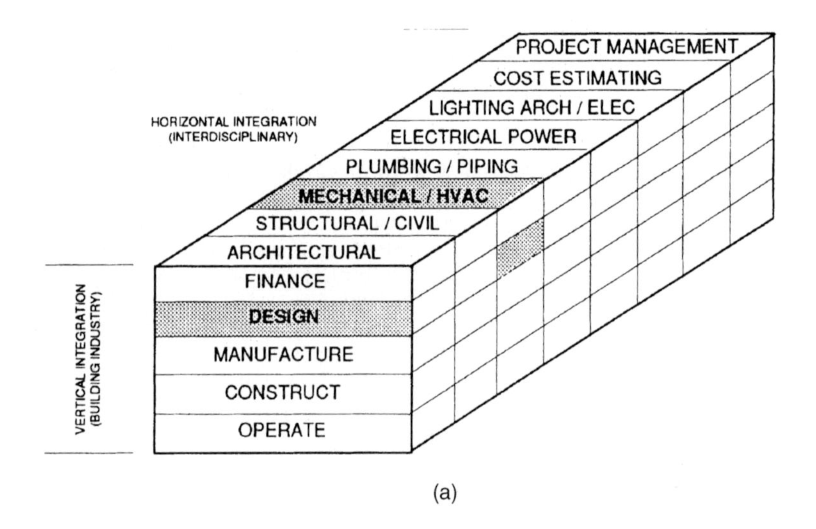


(a)

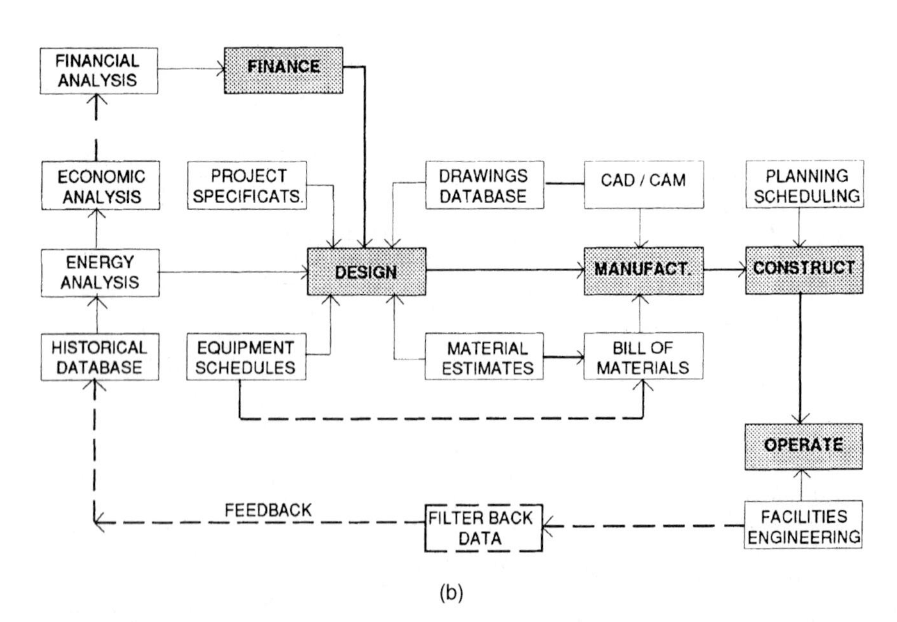


(b)

**Fig. 8.1 Building industry: (*a*) global and (*b*) system integration.** (*Courtesy: Skidmore, Owings & Merrill.*)

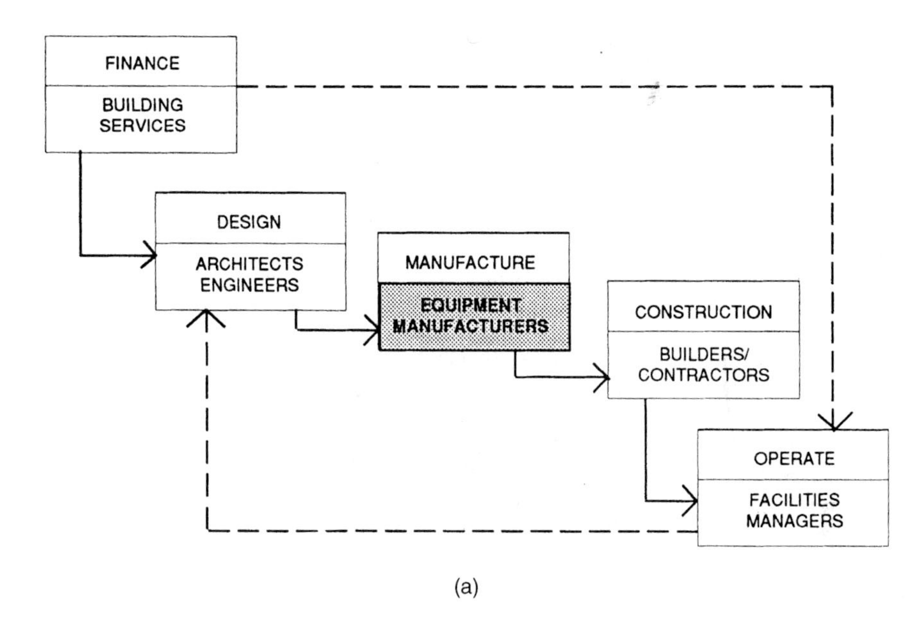


(a)

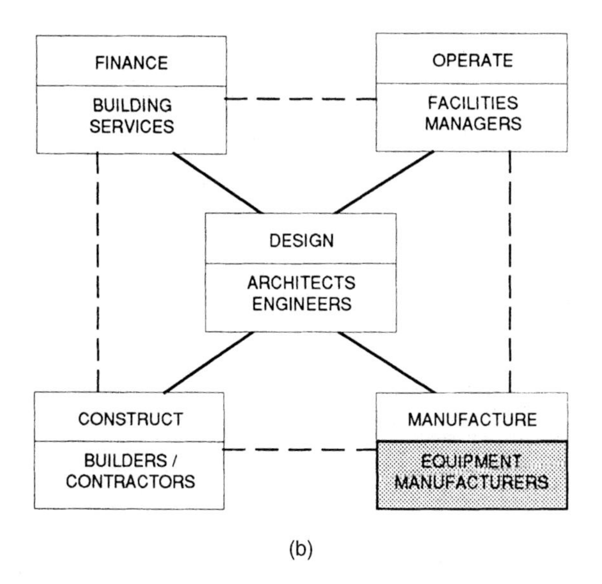


(b)

**Fig. 8.2 Building development team: (*a*) phases and (*b*) interaction.** (*Courtesy: Skidmore, Owings & Merrill.*)

points in a standardized format or if manufacturers provide their product selection programs with standardized input-output formats.

Note that standardization is required only where information must be exchanged directly between the various types of systems in the different segments of the building industry. In the manual process of developing and operating a building, this exchange of information now occurs through oral and written communication, meetings, and conferences.

Besides database and interface standardization, effective systems integration also requires a standardized building description language. The language syntax, verbs, adverbs, prepositions, and pronouns should be the same for all applications in the building industry within the English language world. Nouns and adjectives can be defined by the individual software developers of the various types of applications.

A common building description user language for all computer programs in the building industry will make it easier for independently developed programs to communicate with each other. It will allow universities and other research and development organizations to develop specialized building-related systems that can function within or are compatible with the larger commercially available building development and operating systems. Without this compatibility, the commercial use of these small specialized systems might never occur.

The alternative to a standardized global system for exchanging electronic information and a common computer user language in the building industry will be a few large "conglomerate"-type computer models for building projects, offered by a few large software organizations, where the software from one organization cannot communicate with the software by other competing organizations. Until recently this situation existed in one segment of the building industry, namely, building management systems (BMS). A client, such as the owner of a nationwide chain of supermarkets, could be locked into a single manufacturer of BMS hardware and software by the initial choice of a BMS. A BMS from one manufacturer could not communicate with that of another because they were based on different user languages. The introduction of standards for the BMS industry such as BACNET by the American Society of Heating, Refrigerating and Air Conditioning Engineers (ASHRAE) has begun to correct this situation.

## *8.2 ARCHITECTURAL-ENGINEERING DESIGN INTEGRATION*

### *1 The Original Source of Project Data*

The efficiency of the building development and operation cycle depends on the integrity and effectiveness of the information flowing between the client, design engineering, equipment manufacturing, contracting, and facilities management segments of the building industry. The design engineering segment is the originating source of the information developed for a building project, which is ultimately distributed to the other industry segments. A well-organized and easily accessible information system of reference and project databases for design will therefore increase the efficiency of the building cycle, and will result in greater financial and operational benefits to the client.

The need for interface standards and guidelines to exchange data between the building design disciplines (architectural, structural, civil, mechanical, and electrical) is not critical if a single organization with all of these disciplines is responsible for defining and developing the integrated design system. However, guidelines will be useful to firms developing systems for a single design discipline in order to make these systems compatible with the design systems developed by other firms in other disciplines.

### 2 Client and Systems Development Objectives

The objectives for developing A-E systems must satisfy the needs of A-E design firms in serving their clients, and it must also look beyond design so that information from design can be automatically transmitted to the manufacturing, construction, and operation segments of the building industry.

*Client Objectives.* Client-driven objectives include:

1. Design quality control
2. Project cost control
3. Design cost control
4. Compliance with codes and standards
5. On-time performance

A point to note about client objectives is that design quality can reduce the life-cycle cost of the building, with lower initial cost and efficient building operation. Reducing the design component of the total project cost could increase the building life-cycle cost. The A-E design system is therefore the most critical component of the total building life-cycle system.

*Systems Development Objectives.* Systems development objectives include:

1. Integration across the design disciplines of architecture and engineering (civil, structural, mechanical, and electrical)
2. Integration of the design systems with related systems in the other segments or phases in the industry, such as manufacturing, construction, and building operation
3. Information-based systems that are user-accessible and expandable
4. Intelligent systems where all system components can react and adjust to changes in any component (see Section 8.4)
5. Three-dimensional models of the building structure and operating equipment
6. Dynamic systems that can simulate and test the building structure and operating equipment before it is built
7. Systems that are easy to use

### 3 Project Design Integration Tools

Today's computer hardware, telecommunications, and software systems provide building design teams with new and powerful tools to develop integrated and, in the future, automatically checked design models and documents. Some of the software systems currently in use or being developed for design integration include:

1. Project management systems to plan, schedule, evaluate, monitor, and report on design development and costs
2. Information and file management systems to organize and coordinate computer databases for a project (The database structure is a multidimensional matrix of files organized by design disciplines, space organization, drawings, schedules, and systems.)

3. Reference databases of building elements and their properties that are integrated across the design disciplines
4. Design program modules that are integrated through standardized interfaces for exchanging information directly between modules
5. Machine-intelligible building description language for communication with computer systems and to create command files of instruction sets
6. Computer networking and telecommunication for the instantaneous transfer of data between members of the project team working on the building project

Building construction and operation problems and associated costs (which can be considerable) can be minimized or eliminated through project design integration.

## 4 *Interdisciplinary Design Databases*

Building elements such as walls, roofs, glass, windows, floors, ceilings, and partitions affect all design disciplines. Building products and equipment also affect all disciplines, although the primary responsibility for each product or equipment belongs to a particular discipline. For example, mechanical and electrical equipment affect the architectural and structural disciplines since they require space, and since the structure has to support their weight.

The information storage and retrieval capabilities of today's commercial work stations, and even personal computers, can accommodate most of the information used frequently by the building design profession. As the amount of information increases, the need for efficient organization of the information becomes more critical. Figure 8.3 shows an organizational breakdown of buildings by types, such as commercial, institutional, residential, mechanical, industrial, and recreational, with a further breakdown into building subtypes. Note that the information organization is first by building type and then by discipline and by project. Interdisciplinary design integration requires this organizational perspective.

Figure 8.3 also represents a typical management structure used by design firms. Project managers are responsible for design across the disciplines and the department heads are responsible for all projects within their discipline. This organizational structure existed before computers were used by the building industry. The limitations to its effective use were then the manual (sometimes unreliable) communication of information between the cells of the matrix structure.

Figures 8.4 and 8.5 demonstrate the creation of integrated interdisciplinary databases. Figure 8.4 shows a simplified record of wall information for the HVAC discipline. This record can be expanded to include information for the other A-E disciplines. Integration is achieved by using a common reference or identification code, for the same wall type, by all the disciplines. Figure 8.5 shows other building elements that affect all disciplines, such as roofs, floors, partitions, ceilings, glass, doors, and insulation.

In order to achieve interdisciplinary building systems integration, the emphasis must first be on objects, such as the building, its elements, and its equipment, and then by professional specialization and project phases.

Interdisciplinary design information exchange is achieved by integrating the text databases for building elements and equipment with the corresponding graphic elements representing the building's model. Figure 8.5 demonstrates the interdisciplinary exchange of information through the integration of text and graphic databases. In the system shown, the graphically based mechanical and electrical engineering design modules

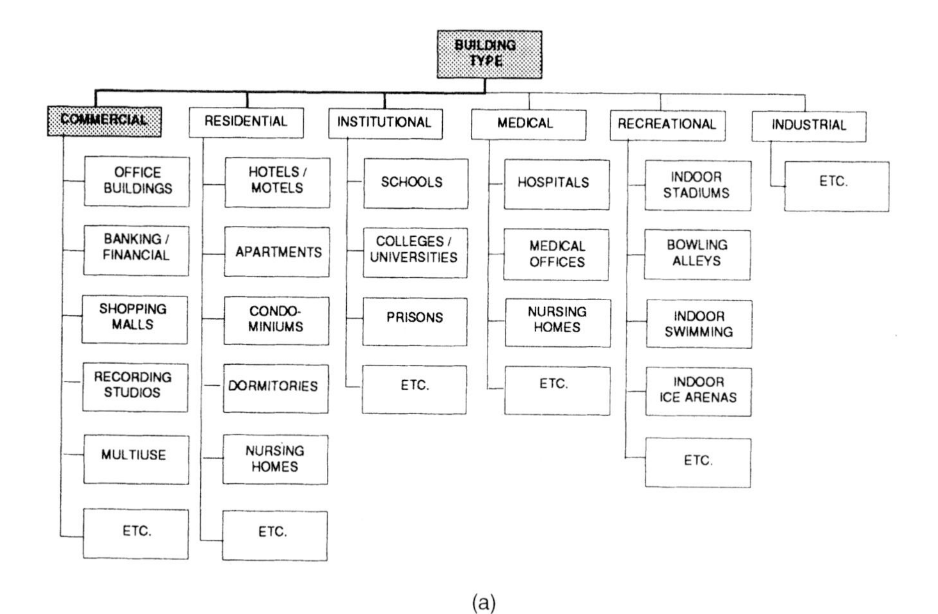


(a)

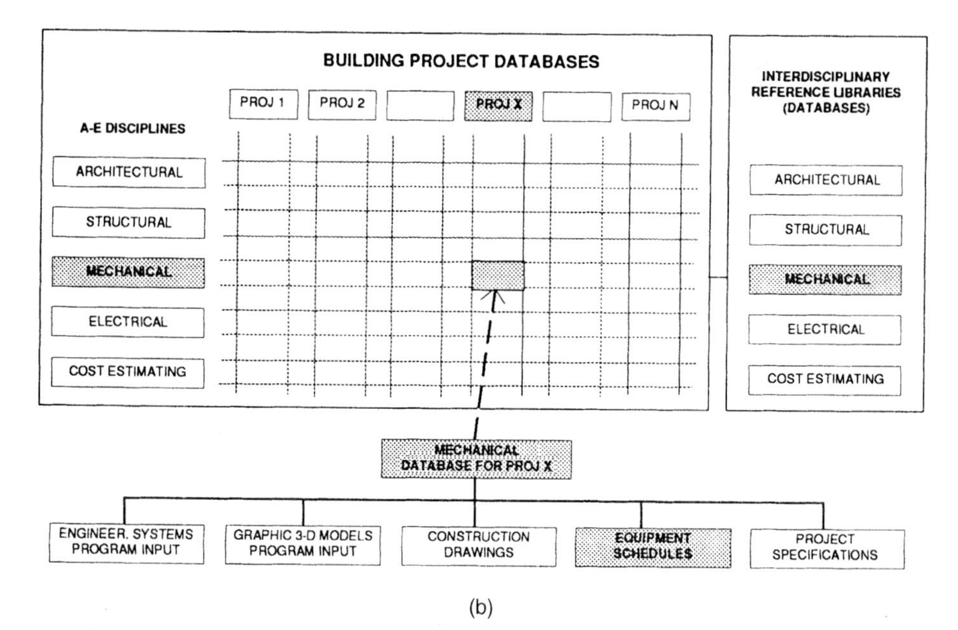


(b)

**Fig. 8.3 Design information database: (*a*) by building types and (*b*) interdisciplinary building project.** (*Courtesy: Skidmore, Owings & Merrill.*)

| WALL REFERENCE CODE | DESCRIPTION | ASHRAE GROUP | COMPOSITE THICKNESS (IN) | WEIGHT (LBS / FT) | HEAT CAPACITY (BTU / FT) | U-VALUE ((BTU / (HR F)) |
|---|---|---|---|---|---|---|
| WALL I | 4"" FACE BRICK + AIT SPACE + 4" FACE BRICK | C | 9.0 | 83.0 | 18.3 | 0.36 |
| WALL II | 4" FACE BRICK + 4" COMMON BRICK | D | 8.0 | 90.0 | 28.4 | 0.42 |
| WALL III | 4" FACE BRICK + 2" INSULATION + 4" COMMON BRICK | B | 10.0 | 88.0 | 18.5 | 0.11 |
| | | | | | | |

(a)

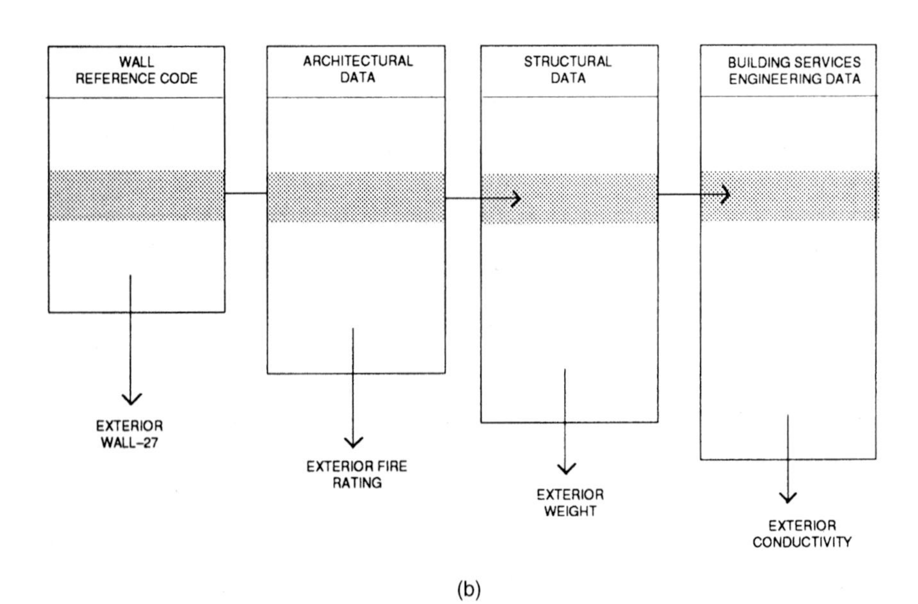


(b)

**Fig. 8.4 Wall library: (*a*) Intradisciplinary (BSE) and (*b*) interdisciplinary (A-E) databases.** *(Courtesy: Skidmore, Owings & Merrill.)*

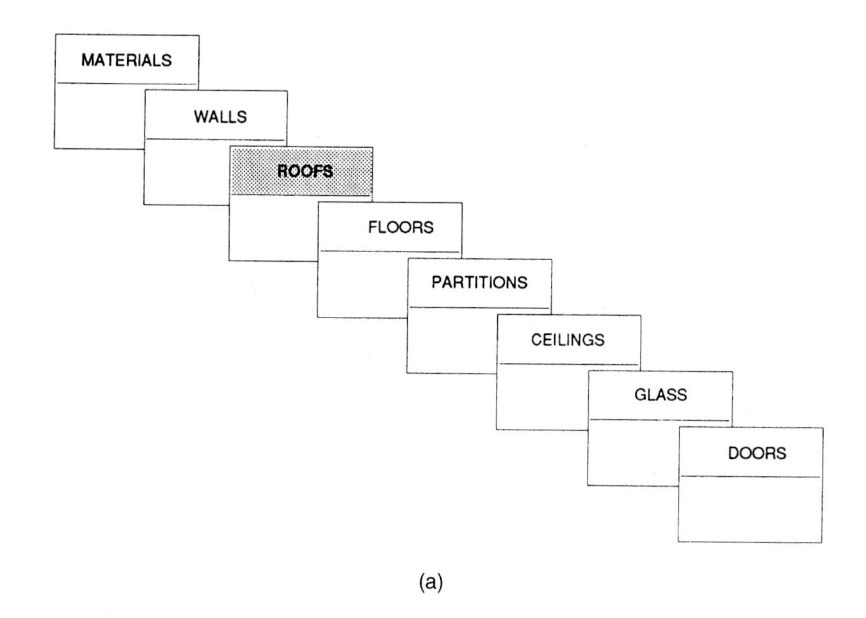


(a)

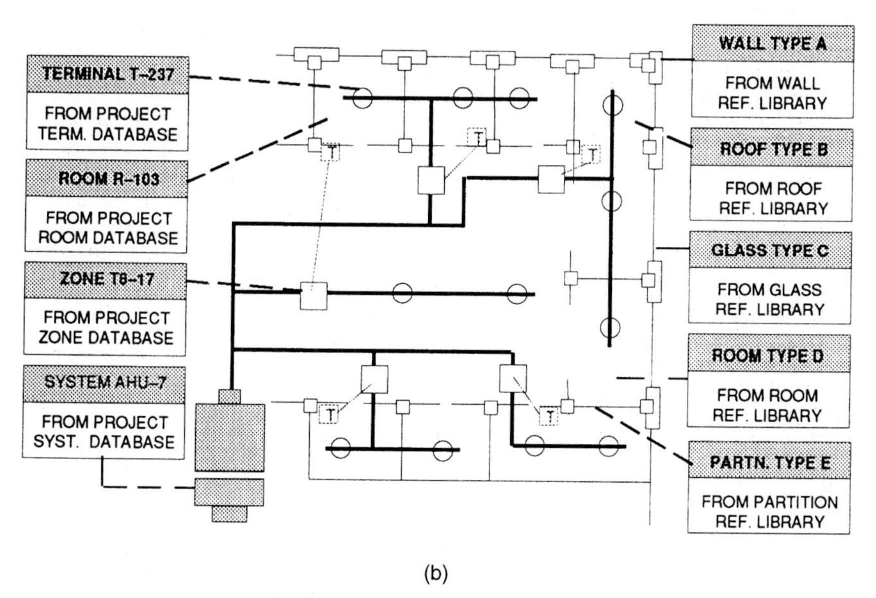


(b)

**Fig. 8.5 Interdisciplinary libraries: (*a*) building elements and (*b*) A-E graphic elements linked to databases.** (*Courtesy: Skidmore, Owings & Merrill.*)

can automatically retrieve the text information that they require for the analysis of a wall, for example, from the architectural and structural background drawings. This is possible since the wall database that is linked to the graphic wall element is integrated across the design disciplines with a common reference identification code.

Figure 8.5 also shows an air distribution system where the graphic HVAC elements such as ducts and diffusers are linked to their properties in text databases, and the information associated with HVAC systems organization such as terminals, rooms, zones, systems, and plants are linked to the graphic A-E space model.

## 5 Information Organization

The information required to design A-E systems can be organized into reference libraries and project libraries. Reference library information is for the use of all projects. Project library information is specific to the project, and it can be developed by customizing reference library information. Project information that can potentially be used by future similar projects should be copied back to the reference library. This information movement loop between reference and project libraries will produce growth within the library system.

The organization of information in the reference and project library system can be further organized as follows:

1. *Reference.* Historical information, engineering data, codes and standards, design rules and procedures, equipment performance data, graphic symbols of engineering elements, graphic database of standard details, master specifications, and cost data
2. *Project.* Application system input, engineering reports, construction drawings, equipment performance schedules, project specifications, and cost estimates

Figure 8.6 shows information organized into reference and project databases for building services engineering design. It also shows the flow of this information through an A-E file manager system to produce coordinated and integrated mechanical and electrical engineering drawings and schedules. Figure 8.7 shows an information retrieval system for graphic symbols and standard details. These libraries are used to build the graphic and engineering models of the building simultaneously.

## 6 The Use of Historical Information

Historical information must be segregated from the other types of information because it is a function of time and can, therefore, only be used for comparison and evaluation. It is not absolute knowledge such as the properties of steam or the dimensions of a terminal box. Although the validity of historical information applies to the time period when it was developed, the information can still play a very important role in preliminary design, preliminary cost estimating, and for checking and evaluating current design.

The objective of a historical database is to capture the knowledge and experience of experts (experienced engineers) and to save this information for future use. Historical information, therefore, represents experience. BSE design applications derived from historical information would be able to assist the less experienced engineers in decision making during the preliminary conceptual stages of a project when all the information required for absolute decisions is not available.

Figure 8.8 shows a method for identifying and classifying building projects. A historical database can become very large. A good system of organizing the database is

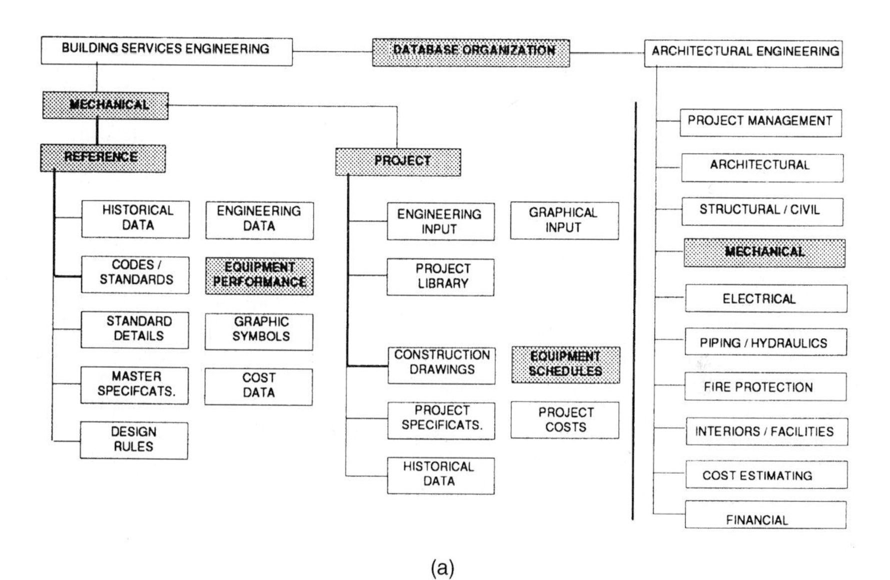


(a)

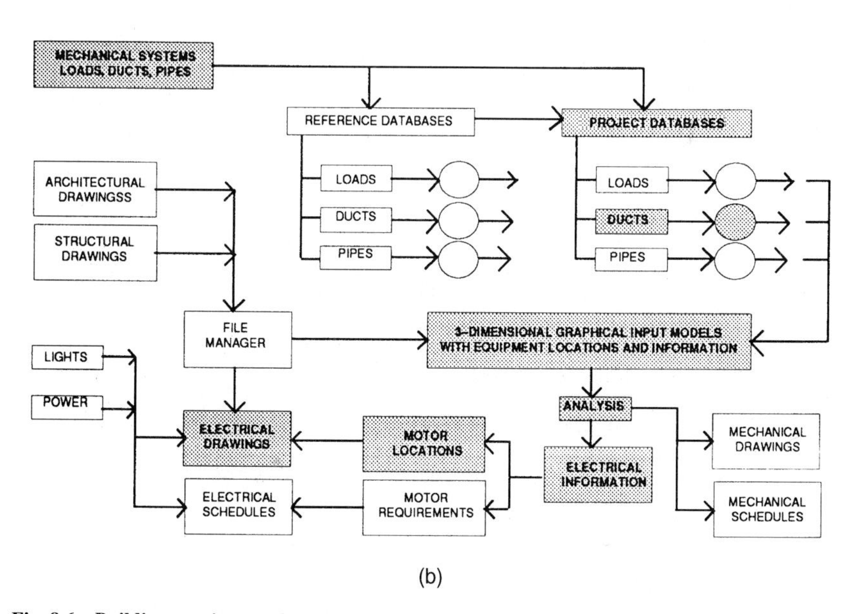


(b)

**Fig. 8.6 Building services engineering: (*a*) databases and (*b*) BSE design information transfer.** (*Courtesy: Skidmore, Owings & Merrill.*)

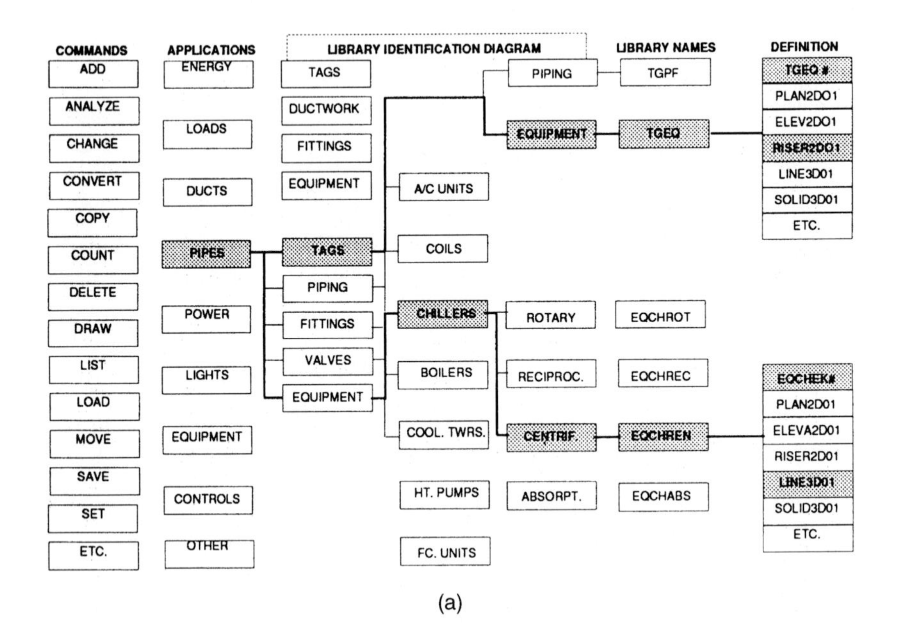


(a)

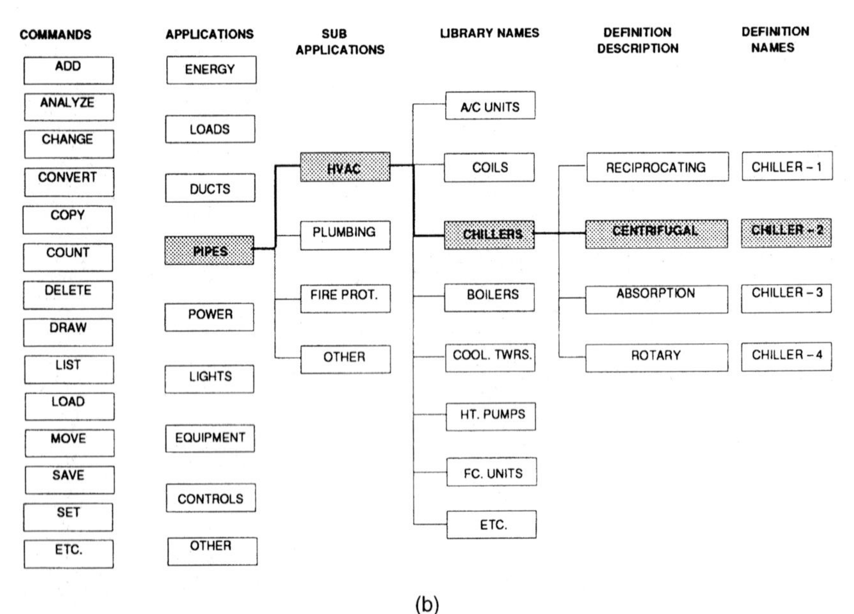


(b)

**Fig. 8.7 Reference library: (*a*) graphic symbols and (*b*) standard details.** (*Courtesy: Skidmore, Owings & Merrill.*)

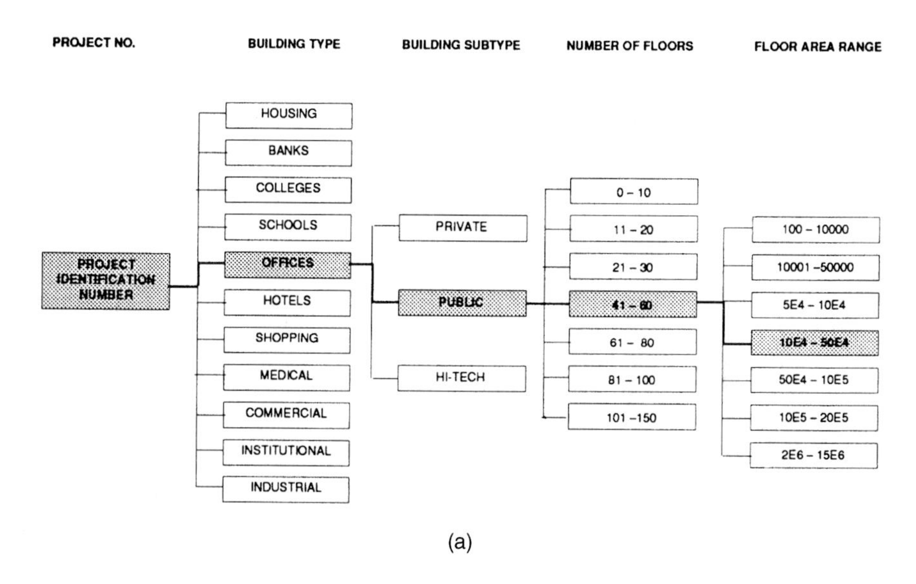


(a)

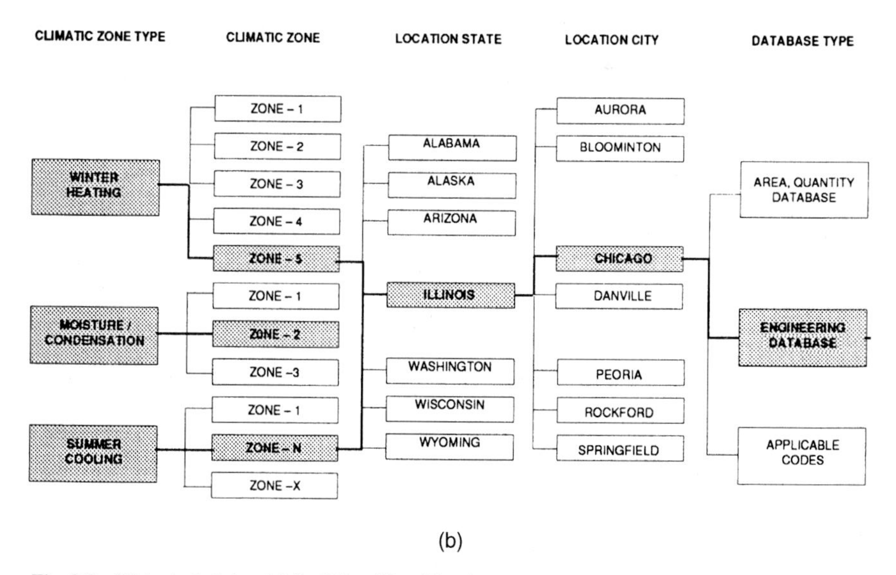


(b)

**Fig. 8.8 Historical data: (*a*) building identification and (*b*) regional classification.** (*Courtesy: Skidmore, Owings & Merrill.*)

critical for finding and retrieving the correct information quickly. The highlighted blocks in each illustration are examples of finding the correct information. The information extracted from history can be very precise if the search goes all the way down the tree structure shown in the example. The level of precision can be adjusted by stopping the search at some intermediate point of the tree structure. The same tree structure is valid for entering and extracting information.

Two distinct groups of information, physical (areas and quantities) and engineering, are examples of organizing information in a historical database. Three types of design applications that are based on historical information are equipment room layouts, core design systems, and cost estimating systems. All three examples are intended for preliminary discussion with the A-E team responsible for developing the project. A closed-loop system of historical information flow enables the historical database to grow with time.

The traditional BSE computer applications of energy, building, loads, lights, ducts, pipes, power, equipment, and controls enter information from current projects into the database and retrieve information from the database to compare and evaluate the current project. The next level of closed-loop information flow is from the database into preliminary design applications. After a few iterations though the loop, the preliminary applications will produce the more precise data required by the BSE design applications that turn out the final construction documents. The outermost loop interacts with building management systems. Interaction with the historical information should be through this "filtering" system.

## 7 A-E Program Modules in Perspective

The concept of integrating every program module serving the building industry with every related module in that industry is technically a feasible concept (Ali, 1988a). The main obstacle that prevents its realization is the lack of standards for exchanging information between modules. For any given type of program module, there are several competing software firms developing such a module. The goal of data exchange standards would therefore be to enable the creation of a project system by assembling modules from several such modules offered by alternative vendors.

Figure 8.9 schematically demonstrates the concept of mixing and matching program modules for each task to obtain a complete project system. This figure shows the location of an HVAC program module as part of all programs in the building industry, and illustrates the program module selection and assembling process to create a project system.

## 8 Automatic Communication

Timely and effective communication between the various professionals working within the several independent organizations that make up the building project team will increase the efficiency of the building process, prevent costly errors, and reduce the cost of the project. Before computers played a role in the industry, communication and efficiency depended on the experience and skills of the individuals that made up the project team. The responsibility for effective communication and process efficiency is now being passed on to the specialized program modules that make up the project system.

Establishing standardized interfaces between the program modules to facilitate the exchange of information between them is therefore not enough. Systems integration must be dynamic so that a specialized module will automatically reanalyze and readjust to the changes and new information that it receives and report its own reaction to the

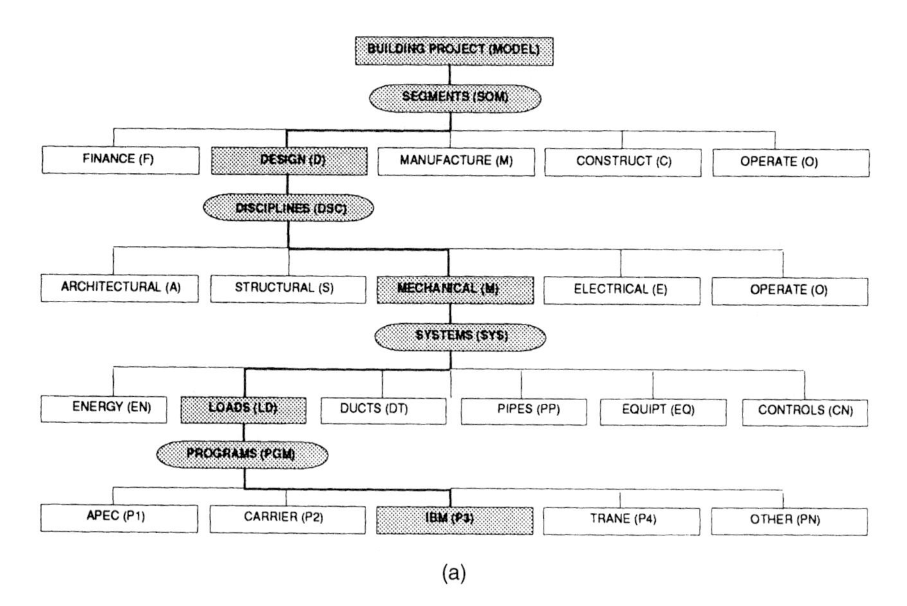


(a)

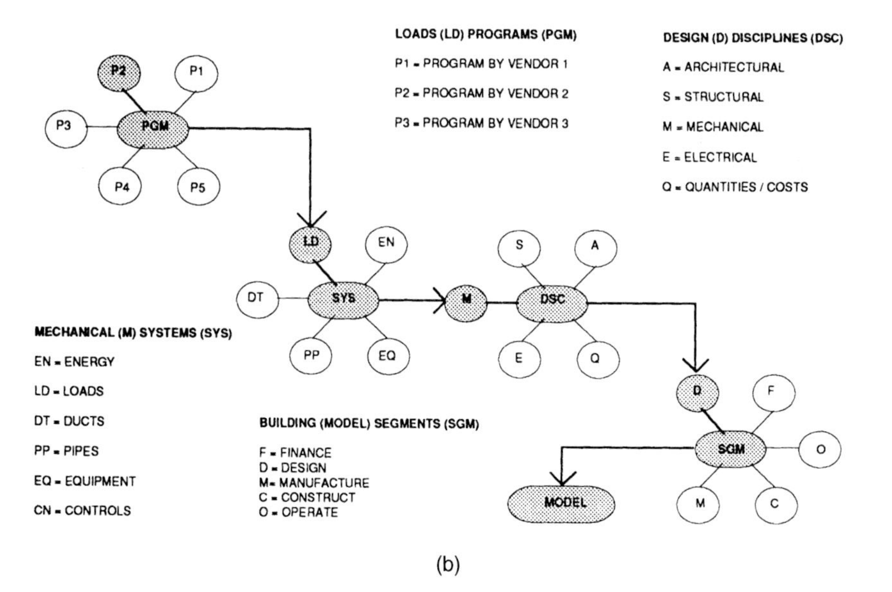


(b)

**Fig. 8.9 Integrated building system: (*a*) program location and (*b*) program selection.** (*Courtesy: Skidmore, Owings & Merrill.*)

other modules in the project system that are affected. In other words, all the modules that are assembled together to make up the project system must function like a single living organism.

Software systems that communicate with each other and react to changes that occur within the other components of the complete system already exist in the building management, automation, and controls segment of the building industry. The principle of the closed-loop control system is an example of reactions to changes based on an in-built mechanism of measurement, communication, and evaluation.

Figure 8.10 defines an intelligent system as one that can react and adjust to changes in order to maintain required objectives, using historical data and predetermined procedures, and shows how the closed-loop control system fits this definition.

## 9 Future Outlook

It is apparent from this analysis of the building industry that past trends are not necessarily the basis for determining the future of this industry. The rapid advances in computer hardware and telecommunications technology, and the black box and the cumulative knowledge and experience features of software development will force significant and continuous changes in the organizational alignment of the building industry and in the process of developing and operating a building. Computer hardware and software of the future will create, for example, the multidisciplinary building design professional. Specialists will be responsible for researching, developing, and maintaining the internal contents of the applications software modules (black boxes).

Automating and integrating the flow of information through the life cycle of large and complex buildings, such as today's high-rise office buildings, requires the coordinated effort of all the professions serving the industry. The underlying basis for developing efficiently integrated building systems from an object-oriented perspective is the science of interprofessional and interorganizational communication and project coordination. The middle- to senior-level technical staff of the various types of specialized firms working on large and complex buildings spend most of their time at project coordination meetings. This is where most of the information exchange and integration occur. Most errors, cost overruns, and liability claims can be attributed to failures in the communication and coordination process.

Interprofessional and interorganizational communication and project coordination is still not a recognized science that is taught at educational institutions today. At present this science can only be learned through experience on projects within the industry, in fragmented and specialized project environments. This science can be formalized, and the information communication processes can be fully automated in the future.

A holistic (macroanalysis) approach to the information flow processes throughout the complete building life cycle is therefore the basis for developing a fully integrated information model. The life-cycle model would be based on the sequential and interactive processes involved in developing and operating buildings today. However, the model must also look to the future, and must consider a reassessment and realignment of professional responsibilities and organizational structures in the building industry in order to make the optimum use of automation and telecommunication technologies.

Predefined information structures and flow models are necessary in order to allow computer systems to communicate with each other directly. Professionals from all segments of the building industry, including the architects, have to agree upon a few such information flow models (in practice the possibilities are unlimited) for different types of tall buildings, in different locations, and under different economic conditions. Data exchange standards and guidelines are necessary to make the communication system work. Using a well-defined information highway system (in spite of the fragmented organizational structure of the building industry today), computer-based mass-production assembly-line techniques used in manufacturing can be applied to tall building design and construction in the future, without affecting the aesthetic uniqueness of each building.

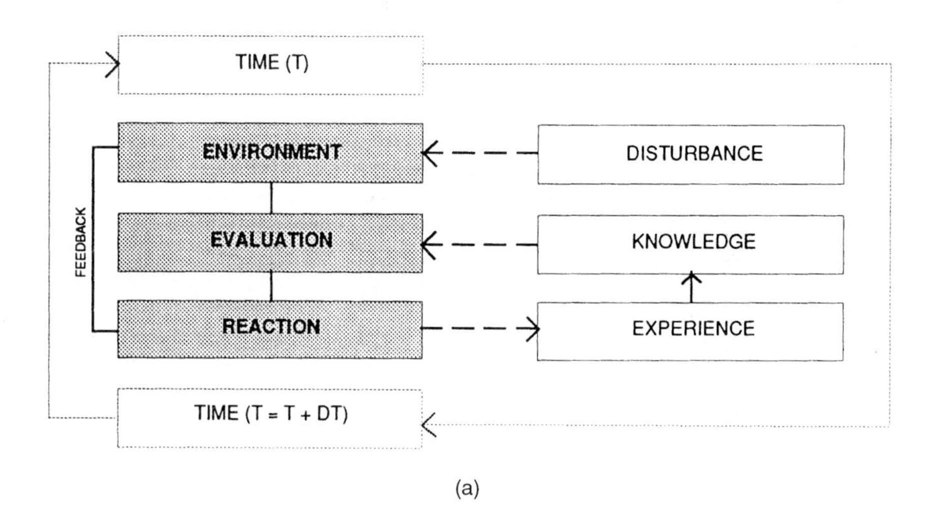


(a)

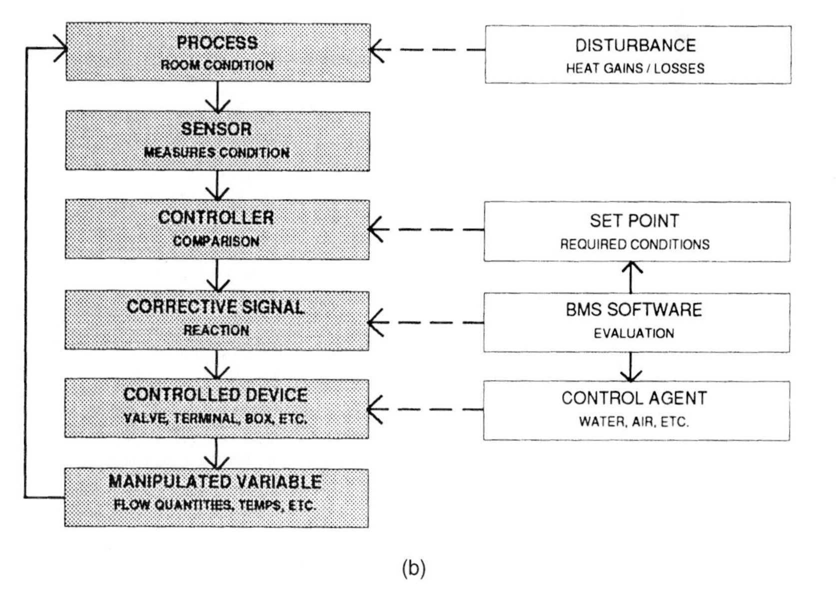


(b)

**Fig. 8.10 Intelligent system: (*a*) definition and (*b*) closed-loop controls.** (*Courtesy: Skidmore, Owings & Merrill.*)

## 8.3 APPLICATION OF COMPUTER TECHNOLOGY TO THE BUILDING INDUSTRY

Computer technologies are currently being used to modify and update the existing methods of designing, constructing, and maintaining buildings. With the advent of mainframe, minicomputer, and microcomputer technologies, and the attendant development of computer-aided design (CAD) and computer-aided manufacturing (CAM), the importance of computers and the computing process for building design and construction can hardly be overemphasized.

Computers are currently used in a limited way for design, construction, and management applications in the building process. A comprehensive approach to collecting all data about a building project still does not lead to a feasible or practical methodology to achieve a fully computerized and integrated building model. The use of advanced computing technologies in the building and construction industry is, however, becoming increasingly evident. It is now generally recognized that computers have the potential for improving the building process by making it feasible to develop complex building models that describe the project and preserve important information in useful databases. Efforts are under way by various organizations to automate not only the project-specific data but also general data about codes, standards, regulations, and building and engineering practices.

It is a well-known fact that computers have evolved very rapidly from the mainframe to the personal microcomputers over the past two decades. Unlike some other industries, the building industry has been slow to change and has shown little inclination to fund long-term research activities. To stimulate drastic changes, academic and industrial partnerships are required. The trend is toward increased understanding by both industry and academia concerning the importance of such collaboration (see Chapter 2).

### 1 Background

Computers have been used in the past for research and development in the building industry to respond to the need for optimizing building systems and products. Extensive research is, of course, not possible in the production-oriented environment of the A-E firms who often work on projects on fixed budgets (Ali, 1987). The importance of computers was recognized by large A-E firms involved in very large projects. However, mainframe computers were too expensive for smaller A-E or construction firms.

With the advent of minicomputers, the scenario changed. Today small to medium-size firms have personal microcomputers that are used to analyze and design building structures and components as well as mechanical and electrical systems. Some information on past trends in computing can be found in Abelson & Dorfman (1982), Brown (1982), and Toong (1982).

There are many barriers to academic computing in the area of building engineering, although some of these obstacles are gradually being overcome. Computers are currently being used in academia not only for research but also for teaching building design and construction courses (Ali, 1988b). The rapid influx of computers in academia is a natural outgrowth of the computerization efforts in the industry. It is now evident that computers will become more affordable, and there will be an increasing use of computer networks and telecommunications. Personal computers have become so affordable that even small to medium-sized firms can use them for structural design and drafting needs. These computers can now even be used for the design of high-rise buildings. Similarly, computers are being used extensively for the control of energy efficiency and illumination of buildings. The construction industry uses computers for construction management operations, scheduling, cost analysis, and specifications, to name just a few. Computer programs dealing with the critical-path method (CPM), which is a construction management tool for planning, scheduling, and controlling a construction project, are available to builders. The CPM network diagrams generated by the computer for the activities and their duration during the construction process assist a contractor to complete a project on schedule, thereby avoiding any unforeseen cash-flow problems.

Because of the complexity of present-day building projects and the need for articulation and coordination among the various phases of a project, the roles of the architect and the engineer can now be supplemented with the use of relatively more advanced tools of computing, such as CAD/CAM expert systems.

Some trends in the application of advanced computer technology seem distinctly clear:

1. Use of knowledge-based and expert techniques for design, belonging to a set of many specialties collectively called artificial intelligence
2. A systems approach to building design and construction
3. Realization of computer-aided buildings that are "smart"

## *2 Knowledge-Based Techniques*

Building component design can be made automatic by using knowledge-based techniques. In such techniques, a new approach is taken where a separation of the knowledge about the problem domain from the control logic of the application program takes place. The traditional computer programs written for component design have the advantages of speed of execution and reliability. However, they have some limitations in the areas of program maintenance, flexibility, and extendability. Inherent in the problem is the fact that the program is controlled exclusively by the owner of the high-level source code. Therefore the licensed user of the purchased computer program can neither correct known errors nor extend the applicability of the program. Another problem is related to the periodic revisions of engineering building codes, specifications, standards, and regulations. Amendments to the code may be difficult to incorporate since they influence the program logic and data. This is because the strategies, heuristics, and assumptions upon which the programs are based are not explicitly stated in the source code (Ali, 1993).

A knowledge-based system, on the other hand, comprises a knowledge base encompassing rules, facts, design heuristics, and procedural knowledge as well as an inference engine that contains reasoning or problem-solving strategies. By maintaining the standards and procedure-dependent knowledge in an explicit form easily understood and modified by the user (for example, the engineer), the primary limitations of the conventional approach are surmounted. A prime motivation for large A-E firms to turn to knowledge-based techniques has been to solve the maintenance and management problem while adding flexibility and modifiability. It is now generally recognized in the industry, academia, and other building research organizations that the knowledge-based approach is better than the traditional approach in that it places the technical knowledge back under the control of the engineering and architectural user. Some disadvantages of the knowledge-based approach are the initial difficulties associated with implementing a knowledge-based system, relatively slower execution times, and possible difficulties in maintaining a knowledge base comprised of occasionally conflicting or totally inert production rules.

A broader discipline of computer applications, called artificial intelligence (AI), is in the process of revolutionizing the practice of architecture, engineering, and construction. AI has the potential for developing systems that will perform a complete design project, including architecture, engineering, design, drafting, and construction management with little or no human involvement (Marksbury, 1984). AI encompasses knowledge-based systems, expert systems, language interpretation systems, as well as any other specialty that solves complex problems by analyzing a set of nonprocedural rules. AI-based systems have the capability of making judgments, drawing inferences, and developing solution choices and strategies that in a way imitate the human brain.

The AI-based approach also covers the expert systems that are capable of performing a single function on simple, segregated systems, for example, footings, columns, and walls. Each expert system can have access to its own specialized knowledge base oriented to its particular specification. With the introduction of expert systems, component design becomes automatic. A combination of knowledge-based techniques and expert systems gives rise to what is known as knowledge-based expert systems. The AI systems are justified by their competitive edge over traditional systems and the general productivity improvement (Fenves et al., 1984). A-E firms can carry out the entire design process in a very short time by utilizing such systems (Hayes-Roth et al., 1983).

## 3 A Systems Approach to Building Construction

Modern-day technological changes and legal responsibilities in building construction demand a more interactive and integrated system type solution to the problems of building design and construction. The military has developed many such approaches to perform complex activities that are part of their research programs (Napier and Golish, 1982; Ali and Anway, 1988). The most critical system considerations are the steps that must be followed to arrive at final strategies utilizing all the available data or information provided through research, field feedback, and experience. By taking such an approach to design and construction, complex problems related to buildings can be solved and future problems can be avoided through effective decision-making processes.

In a systems approach, all important phases of the design process and construction are integrated in a systematic procedure by considering all significant performance and functional characteristics of a building system. Volumes of information need to be produced, processed, and retrieved in such an approach. Computerization of information on building product description, specifications, details, design criteria, code regulations, construction documents, and the building model is an objective currently being pursued in the building industry. Consequently, research and education at the engineering and architectural schools are being impacted by this trend. It is recognized that such computerized information has some limitations within the scope of present-day technology, particularly when considering design amendments, structural modifications, changes in mechanical duct run, adjustment of concealed structural or piping elements and raceways, adjustment of energy, illumination, and acoustical requirements, to name only a few. More advanced systems, including rules about how one modification affects another, are required. Research on AI-based systems may provide tools to serve as mechanisms to incorporate such knowledge in building models with special programming languages.

A systems approach to design typically involves the following stages (Robbie, 1972; Ali and Anway, 1988):

1. Determine the requirements of the project.
2. Identify performance criteria and performance specifications.
3. Evaluate alternate solutions and prioritize the corresponding systems for further evaluation.

4. Based on an evaluation methodology, evaluate alternatives and select a solution based on cost, performance, functionalism, systems integration, and such.
5. Develop an implementation plan, a plan for designing the selected system following standards and codes of practice.
6. Develop completed drawings of the selected system.

Similarly, a systems approach to the construction phase can be developed. Eventually each stage of design and construction can be automated through the integration of AI and CAD/CAM. In a systems approach the AI system may process the requirements, develop a strategy, and select a solution. Further, it can produce detailed drawings, specifications, estimates, and material lists required for the construction phase.

## 4 High-Tech Buildings

The objective of computer-aided buildings is to realize "smart" or "high-technology" buildings, which are also sometimes referred to as "intelligent" buildings, where the building process is characterized as the integration of subsystems. As John and Shaw (1989) state:

> There are many definitions of an intelligent building but the one common factor in all of them is the assumption that the services will be monitored and controlled by a sophisticated computer-based management system. The goal of such a system is the optimization of control to balance the requirements for occupant comfort, security, safety, energy efficiency, and reliability.
>
> ...to develop a fully integrated management and control system worthy of the name "intelligent" it will be necessary to combine some of the more conventional techniques associated with existing controls and building management systems, with those based on artificial intelligence (AI) techniques.

A case study on the design of an intelligent tall building, the International Business Center in Seoul, Korea, which illustrates the technology and options currently available for the design of tall buildings, is presented in Section 8.4.

The intelligent building systems technology (IBST) that is generally incorporated in a tall building comprises, but is not limited to, the following:

1. Building management system
2. Communication systems
3. Office automation system
4. Central plant system
5. Air handling and distribution systems
6. Building skin system
7. Illumination systems

Conceptual design conducted at the early stages of the design development usually consists of defining the project goal and making decisions on the infrastructure, which includes the structural system, building facade, vertical transportation, ceilings and floors, and vertical utility distribution. Further, items related to building services will also have early impact on the conceptual design. The subsystems in a building include the space definition, structure, HVAC, lighting, plumbing, and acoustics. The coordinating information that interconnects these subsystems is presented in the form of drawings, specifications, and correspondence. The coordinating information also covers four principal phases:

1. Planning phase, the concept initiation stage
2. Design phase, the drawings preparation stage
3. Construction phase, the practical manifestation of the drawings stage
4. Postoccupancy management phase, the maintenance, repairs, and remodeling stage

The principal goal of the computer-aided building process is to determine how computer technology can provide new methods of efficiently handling the integration of subsystems and the coordinating information as well as the ways and means for handling larger volumes of such information. In this way, the utilization of computer technology offers a remarkable opportunity for expanding upon and improving the designed building, the management of its construction, and the maintenance of the built structure.

Examples of recent computer applications are geometric design and computer imaging, simulation, three-dimensional structural analysis with computer graphics, virtual reality, and project management. CAD programs have been used in the past for building design and are the most visible computer software in the building industry. There is a growing interest, however, in using CAM as a component of computer-aided engineering (CAE) in the building industry. In the building process CAM involves a spectrum of activities, such as the production of building components and fasteners, batch processing of steel structural members, laminated beams, and fabrication and field assembly processes.

With the transformation of the industrial process from traditional small-scale design-oriented production to mass-production technologies, CAD/CAM offers the potential for a continuous process of designing and manufacturing building components. There is an increased understanding among the clients of the building industry that, despite its many limitations, the concept of CAD/CAM/CAE does not terminate as soon as the building is completed, but may continue during the postoccupancy phase of the building. As a facility becomes the responsibility of a facility manager, a built environment suffers unless this manager has and can use the information generated during the planning, design, and construction phases. A common information link facilitates managing the building in terms of repairs and remodeling, maintenance, fire evacuation procedures, replacement of parts, and components (Ali, 1988a).

The concept of high-tech buildings has only partially been realized to date. But the notion is appealing to professionals, researchers, and educators, and is likely to have considerable impact on architectural engineering practice, education, and research in the near future.

## 5 Future Prospects

Computers have revolutionized the building process and will have substantial effects on the built environment of the future. It can be predicted that the tide is in favor of intelligent buildings that are able to sense changes in human needs, aspirations, and the environment and modify themselves to better serve humankind.

Building design and construction practices are bound to change with further advancement of computer technology. Further, new capabilities offered by computers will influence the roles of people involved in the building design and construction process.

It can be safely stated that building design and construction are going to be more and more automated to a limit not known at this point in time. Therefore, the activities of the

A-E profession need to be gradually automated too, lest it is overwhelmingly outstripped by the ever-expanding domain of computing technology. AI-based systems will eventually dominate the computing processes in the building industry. It is not hard to see that the three major trends discussed in this section, knowledge-based techniques, systems approach to building design and construction, and computer-aided buildings, are all closely linked to the notion of a comprehensive building process. Knowledge-based techniques and expert systems are "buzz" words today, and are expected to increasingly pervade the building industry in the future. Most A-E firms should start planning now for the inevitable integration of AI-based systems into their organizations. Some engineering schools in the United States have already started active research in this area. There is an impending need for a redirection of emphasis on these issues in the realm of A-E education and research. This will help today's engineers and architects to appreciate the benefits of an integrated building system and to cope with the challenges of tomorrow.

## *8.4 APPENDIX: LUCKY-GOLDSTAR INTERNATIONAL BUSINESS CENTER*

This appendix presents the Lucky-Goldstar International Business Center, a project that originated when the Lucky Development Company (LDC) of Seoul, Korea, engaged Bechtel Civil Inc., of San Francisco to provide a conceptual and schematic design for an intelligent office building in Seoul, Korea. The material is based on a conceptual design report prepared by Bechtel for the project (Bechtel Civil Inc., 1989).

A main purpose of the report was to describe a process and summarize the results of an 8-week conceptual design effort to define intelligent building systems, site planning strategies, and floor plan options for a 20-story office building. The building is called the Lucky-Goldstar International Business Center (IBC), and the tenants of the center are drawn from the component companies of the Lucky-Goldstar Company. Three architectural schemes are introduced, which explore these different concepts for the building. At the conclusion of the report, the final scheme selected by the Lucky-Goldstar Company is presented.

In Section 8.3 the concept of an intelligent building was introduced and distinguished by the integration of subsystems into the building process, with each system monitored and controlled by a computer-based management system. This integration of subsystems combines conventional and new technology, allowing the user to direct the comfort, security, and efficiency of the building systems (John and Shaw, 1989).

### *1 Project Description and Program Guidelines*

The project is located on Teheran Ro, a 50-m (164-ft)-wide boulevard about 10 km (6.2 mi) south and east of the city center near Sun-Neung Park. The surrounding area is presently being developed with new high-rise office and hotel projects that are either under construction or being contemplated.

The site is within walking distance of the new Korea World Trade Center and the 700-room Inter-Continental Hotel. There are many existing smaller hotels in the area. The Seoul sports complex and the Olympic park are located about 1 km (0.62 mi) to the east.

The site is a rectangle, approximately 115 by 47 m (377 by 154 ft), the long dimension fronting directly on Teheran Ro facing north. The ground slopes upward to the south with a change in elevation of approximately 5 to 6 m (16.4 to 19.7 ft). Utility lines are available beneath Teheran Ro, immediately in front of the north property line.

The site is located between the Sun-Neung and Sam Sung subway stations. Teheran Ro is a main route and fire bus stops are located within 250 m (820.2 ft) of the site. In addition, the site is easily accessible by automobile. Potential automobile access points are available from Teheran Ro and the 8-m (26.2-ft)-wide side street at the western edge of the site.

Guidelines were established for intelligent building systems to be accommodated by the IBC building. These guidelines, listed under the categories of building management system, communication system, and office automation system, are as follows:

*Building management system*

1. HVAC system monitoring, including thermal environment, air distribution, supply and return air temperatures, equipment maintenance, and data
2. Energy management, including night setback, floor-by-floor control, and peak-load limiting
3. Vertical transportation monitoring, including elevators, escalators, and automated mail handling
4. Life-safety systems, including fire extinguishing, emergency escape systems monitoring, smoke exhaust, and voice annunciator system
5. Security management systems, including closed-circuit television systems monitoring of all entrances, exits, emergency escape routes, loading docks, and parking areas, and monitoring of card access to designated building areas
6. Interlinking of control stations at strategic points with the building management office to provide satellite control and monitoring

*Communication system*

1. Telephone exchange system, including facsimile, video conferencing, paging, and multipurpose telephones
2. Multipurpose telephone capability for communication with building functions
3. Satellite transmission system
4. Local area and international networks for communication and automation
5. Leakage coaxial cable to supplement wireless communication

*Office automation system*

1. Microcomputer workstation capability throughout
2. Electronic data file and mainframe computer access
3. Word processing and laser printing

To provide a framework for decision making, the process of conceptual design begins with the definition of a project goal. The goal of this project is to provide a work environment that will not become obsolete. The next step is to define the two major types of project elements—infrastructure and services. The infrastructure consists of the building components and systems that are fixed in location, and which are fundamental to the consideration of project budget and development schedule. The services consist of those building components that are more flexible and can be added to or subtracted from the scheme with limited impact during design activities.

To assist in the process of system selection, the objectives of the system components are identified. These objectives are both specific to the functional requirements of the individual system and general to all of the building systems. Examples of specific objectives are solar heat-gain reduction for an automated window blind system and adaptability to zone control for a lighting system.

The general objectives would be defined as follows:

1. *Flexibility.* Ability to adapt to changes in location and level of service
2. *Cost.* Optimization of capital and life-cycle costs
3. *Compatibility.* Ability to integrate with other systems
4. *Expandability.* Growth characteristics of a particular system
5. *Maintenance.* Ability to minimize service and technical support requirements

A matrix is used to compare systems by rating the categories of specific and general objectives to illustrate strengths and weaknesses. The choice between systems is then illustrated.

The initial decisions required during the conceptual design period are those that will allow further study aimed at integration. These include structure, HVAC, horizontal distribution of power, lighting, air, and communication, and vertical transportation. The building structure is determined by the building floor plate, and building core locations will be unique to each scheme.

The selection of systems to be evaluated was based on the following:

1. Identification and comparison of the most advanced technology offered commercially for use in high-rise buildings (Table 8.1).
2. The technologies selected are professionally recognized, with proven "track records," to provide data for judgment and evaluation.
3. In specific cases (such as fiber-optic cabling), consideration was given to technology in development stages with a highly probable degree of marketability in the future.

The program guidelines received for the functions to be included in this project are listed in Table 8.2. To complement the project space program, the conceptual architectural design was organized such that building and site planning possibilities interact simultaneously. The formula for occupying the site, determining the allowable parking, and generating the building volume were prescribed by the Seoul Planning Board.

Three architectural schemes were generated from the site requirements for frontage, the potential impact of future development of adjacent property, the potential automobile access points, the possible floor plate geometries, and core space locations possible within those geometries.

The building volumes generated exceed the guideline criteria for minimum building volume, which are listed as follows:

1. 20 stories above grade
2. 1460 $m^2$ (15,715.3 $ft^2$) gross for each office floor
3. A total project floor area of 52,000 $m^2$ (559,723.8 $ft^2$)
4. Parking at 150 $m^2$ (1614.6 $ft^2$) of occupied building space per car with 30% additional spaces
5. Maximizing retail space at the first basement level

All building schemes are designed to provide an entrance directly from Teheran Ro that leads to an atrium space, which in turn collects building tenants from all other entries. This enhances building access control and directs visitors to central circulation points. All schemes use the sloping site to conceal below-grade parking and provide a retail level 2 m (6.6 ft) below the street level, easily accessible by escalators and stairs. Two building plan shapes are configured to reduce solar heat gain, and all building envelopes are designed to accommodate automated window washing equipment.

## 2 Site Planning and Architectural Design

Program guidelines for site and building floor plate possibilities were determined by the Seoul zoning ordinance. The three most important zoning determinants were:

1. Building slant line limits
2. Building street frontage requirements
3. Site coverage requirements

**Table 8.1 Worldwide comparison of intelligent buildings**

| Building system | Lloyds Bank, London | Hakozaki Bdg., Tokyo | Pacific Bell, California | Proposed IBC, Seoul |
|---|---|---|---|---|
| HVAC: | | | | |
| Floor-by-floor control | × | × | × | × |
| Digital control | × | × | × | × |
| VAV boxes | × | | × | × |
| Energy peak-load shifting | | × | × | × |
| Controlled by BMS | × | × | × | × |
| Lighting: | | | | |
| Noninterfering light distribution | × | × | × | × |
| Daylight reduction ability | × | × | × | × |
| Motion detection control | × | × | × | × |
| Monitored by BMS | × | × | × | × |
| Power and communications distribution: | | | | |
| Modular grid of key distribution points | × | × | × | × |
| Fiber-optic cable backbone | × | × | × | × |
| Access floor | × | × | × | × |
| Monitored by BMS | × | × | × | × |
| Security: | | | | |
| Monitored by BMS | × | × | × | × |
| Building management system: | | | | |
| Distributed controllers | × | × | | × |
| Flexible monitoring | × | × | × | × |
| Centralized control and readout | × | × | × | × |
| Exterior skin: | | | | |
| Thermal gain reduction glazing system | × | × | × | × |
| Controlled by BMS (e.g., blinds) | | | × | × |
| Vertical transportation: | | | | |
| Flexible control program | × | × | | × |
| Maintenance monitoring | × | × | × | × |
| Energy reduction | × | × | | × |
| Monitored by BMS | × | × | | × |

In compliance with the zoning ordinance, the maximum allowable ground-floor building area is a function of the total site area in relation to the area needed for required off-street parking, landscaping, and other open space. These site coverage requirements are summarized as follows:

1. The ratio between indoor and outdoor parking must be 75 to 25% if the site coverage of the building floor plate is less than 50%.
2. Occupied building space may be generated at the rate of 150 $m^2$ (1614.6 $ft^2$) per parking space.
3. A minimum of 15% of lot area must be allocated to landscaping.
4. A minimum of 10% of lot area must be allocated as open space.
5. The building must be located within the required setbacks of 3 m (10 ft) from Teheran Ro, 2 m (6.6 ft) from the east property line, and 1 m (3.3 ft) from the south property line.
6. The City Planning Board requested an additional 30% increase in total parking. Additional parking was to be contained below grade.

**Table 8.2 Building function program guidelines**

| Function | Net area | |
|---|---|---|
| **Basement** | | |
| Parking | | |
| Employee dining and kitchen | 1100 $m^2$ | (11,840 $ft^2$) |
| Loading dock | 100 $m^2$ | (1076.4 $ft^2$) |
| Building management center | 40 $m^2$ | (430.6 $ft^2$) |
| Fire control center | | |
| Commercial area | | |
| Snack corner | 100 $m^2$ | (1076.4 $ft^2$) |
| Sundry | 40 $m^2$ | (430.6 $ft^2$) |
| Bookstore | 40 $m^2$ | (430.6 $ft^2$) |
| Photo shop | 20 $m^2$ | (215.3 $ft^2$) |
| Barbershop | 60 $m^2$ | (645.8 $ft^2$) |
| VIP barbershop | 90 $m^2$ | (968.8 $ft^2$) |
| Other retail | 1000 $m^2$ | (10,764 $ft^2$) |
| **Low-rise** | | |
| Lobby and information desk | 580 $m^2$ | (6243.1 $ft^2$) |
| Bank | 1500 $m^2$ | (16,146 $ft^2$) |
| Information center | 1000 $m^2$ | (10,764 $ft^2$) |
| Business center | 2000 $m^2$ | (21,528 $ft^2$) |
| **High-rise** | | |
| General office | 11,200 $m^2$ | (120,556 $ft^2$) |
| Executive office | 500 $m^2$ | (5382 $ft^2$) |
| Library | 300 $m^2$ | (3229.2 $ft^2$) |
| Meeting and conference room | 1100 $m^2$ | (11,840 $ft^2$) |
| Information control center | 200 $m^2$ | (2152.8 $ft^2$) |

To comply with zoning regulations, the program guideline of 52,000 $m^2$ (559,724 $ft^2$) total building area with a 1460-$m^2$ (15,715-$ft^2$) floor plate area could only be realized by a high-rise building covering less than 40% of the site. Table 8.3 provides information of the site and building analysis of each architectural scheme.

City Planning Board requirements for automobile parking access are as follows:

1. Automobile access to surface parking is permitted from both Teheran Ro and the 8-m (26.2-ft)-wide side street on the west.
2. Access to below-grade parking is permitted only from the 8-m (26.2-ft)-wide side street on the west.

The site slopes upward toward the south from Teheran Ro. The change in elevation across the 47-m (154-ft) width is approximately 6 m (20 ft). The change in grade allows separate entries in each scheme for visitors and VIPs arriving by automobile and pedestrians arriving by transit system or walking from nearby hotels and the world trade center. Surface parking in each scheme is elevated from the street level and screened by landscaping, reducing its visibility to the street.

The various types of building uses and tenant access requirements divide the building into three distinct zones. The lower zone includes the high public access areas: entrance lobbies, bank, retail and dining space, parking, information center, and business center. These areas are interconnected by circulation elements (stairs, escalators, elevators) located in a common multilevel open lobby space.

The middle zone includes the general office space for Lucky-Goldstar tenants. The upper zone contains mechanical equipment, penthouse, heliport, communication equipment, and window washing system. The electronic video information system may be located in this zone.

The project includes the following basic planning considerations applied to the building layout (see Chapters 1 and 2 for a general discussion on building planning):

1. A minimum efficiency ratio of net usable space to gross floor area of 70%
2. A space planning module to coordinate the location of partitions, division of curtain wall, location of lighting systems, and location of structural columns
3. Core layout plans providing separate communication and electric power closets immediately accessible to core perimeter for inclusion in the access floor system for power and communication cable distribution

To meet energy-efficiency requirements, the maximum vision glass to area ratio for the building envelope was determined to be approximately 35%. In order to simplify and accommodate the requirements of an automatic window washing system, step-back floors, large reveals, and surface articulations had to be avoided. Stepback floors, large reveals, and surface articulations had to be avoided. A constant vertical-guide tract module was required. The walls were designed in a panel grid module system so that various finish materials might be utilized in the same design (either vision glass, spandrel glass, metal panels, or panelized stone).

### 3 Architectural Scheme Descriptions

During the process of the conceptual design report, three architectural schemes were selected with the intent of representing fundamentally different buildings, not just different architectural treatments. The three schemes attempt to take unique approaches to each of the following major design issues: any of the options in HVAC, plumbing, and fire protection systems are interchangeable in all schemes; any of the options in electrical and communications systems are interchangeable in all schemes.

The building layout is considered in terms of core locations, entrances and circulation, and building configuration. Exterior treatment is considered in terms of building skin and visual information systems. The site is organized by considering the building location and parking arrangement. The structural system relates to the architectural schemes as follows:

*Scheme A*. The unique geometry limits it to one structural option.

*Scheme B*. Most structural systems are adaptable.

*Scheme C*. Limited to two of the structural options.

The parking structure framing is the same for all schemes. For the basement parking and retail areas outside the footprint of the high-rise tower, a complete reinforced concrete framing system is recommended. Flat slabs with or without drop panels supported on concrete columns and load-bearing walls will provide an economical load-carrying system. This system also has minimal structural depth requirements, minimizing floor-to-floor height.

In the following, a scheme-by-scheme discussion is presented in regard to floor plan, facades, parking, and structural system (see Table 8.3).

**Table 8.3 Site and building analysis**

| | Scheme A | Scheme B | Scheme C |
|---|---|---|---|
| Site area | 5590 m$^2$ (60,170.3 ft$^2$) | 5590 m$^2$ (60,170.3 ft$^2$) | 5590 m$^2$ (60,170.3 ft$^2$) |
| Open space and landscaped area | 1332 m$^2$ (14,337.5 ft$^2$) | 1142 m$^2$ (12,292.4 ft$^2$) | 1267 m$^2$ (13,637.9 ft$^2$) |
| Available site area | 4258 m$^2$ (45,328.0 ft$^2$) | 4448 m$^2$ (47,877.9 ft$^2$) | 4323 m$^2$ (46,532.4 ft$^2$) |
| Percent site covered | 29.62% | 29.18% | 31.04% |
| Surface parking | 2602 m$^2$ (28,007.7 ft$^2$) | 2817 m$^2$ (30,322.0 ft$^2$) | 2588 m$^2$ (27,857.0 ft$^2$) |
| Number of cars | 69 | 69 | 69 |
| Interior parking | 13,017 m$^2$ (140,113.9 ft$^2$) | 14,112 m$^2$ (151,900.4 ft$^2$) | 15,527 m$^2$ (167,131 ft$^2$) |
| Number of cars | 298 | 308 | 297 |
| Total parking | 367 | 377 | 366 |
| Lobby | 3502 m$^2$ (37,695.2 ft$^2$) | 1631 m$^2$ (17,556.0 ft$^2$) | 1735 m$^2$ (18,675.4 ft$^2$) |
| Information center | 1656 m$^2$ (17,825.1 ft$^2$) | 1533 m$^2$ (16,501.1 ft$^2$) | 1530 m$^2$ (16,468.8 ft$^2$) |
| Basement 1 | 3797 m$^2$ (40,870.6 ft$^2$) | 3746 m$^2$ (40,321.6 ft$^2$) | 3990 m$^2$ (42,948.1 ft$^2$) |
| Basement 2 | 2976 m$^2$ (32,033.4 ft$^2$) | 2155 m$^2$ (23,196.2 ft$^2$) | 2655 m$^2$ (28,578.2 ft$^2$) |
| Total subgrade area | 6773 m$^2$ (72,904.0 ft$^2$) | 5901 m$^2$ (63,517.9 ft$^2$) | 6645 m$^2$ (71,526.2 ft$^2$) |
| Above-grade area | 34,612 m$^2$ (372,560.8 ft$^2$) | 32,958 m$^2$ (354,757.3 ft$^2$) | 34,628 m$^2$ (372,733 ft$^2$) |
| Total building area | 41,385 m$^2$ (445,464.8 ft$^2$) | 38,859 m$^2$ (418,275.2 ft$^2$) | 41,273 m$^2$ (444,259 ft$^2$) |
| Total building and parking | 54,402 m$^2$ (585,578.8 ft$^2$) | 52,971 m$^2$ (570,175.6 ft$^2$) | 56,800 m$^2$ (611,390 ft$^2$) |
| Number of floors | 20 + 2base | 20 + 2base | 20 + 2base |
| Typical floor area: | | | |
| Gross | 1702 m$^2$ (18,320.2 ft$^2$) | 1666 m$^2$ (17,932.7 ft$^2$) | 1751 m$^2$ (18,847.6 ft$^2$) |
| Net | 1272 m$^2$ (13,691.7 ft$^2$) | 1317 m$^2$ (14,176.1 ft$^2$) | 1312 m$^2$ (14,122.3 ft$^2$) |
| Efficiency | 75% | 79% | 75% |

***Scheme A.*** The arrangement of building and parking requires a narrow floor plan. The core is located along the south edge of the floor plate, reducing solar heat gain and maximizing glass on the north, east, and west facades. The core location allows only one lobby off the elevators on each floor. It does provide a large continuous floor area, suitable for the space needs of a single tenant.

The east and west facades of the building are angled toward the north. This reduces solar heat gain, and directs views from the office areas toward Teheran Ro and the north. In addition, the angled sides provide multiple reflections of the landscaping and street as the sun traverses the site. The building skin is a glass and glass spandrel curtain wall, with a coated metal frame system. The visual information system consists of electronic message boards located at the top of the building, along the two angled sides.

The planning objectives are to maximize the building visibility to Teheran Ro and conceal the surface parking as much as possible. To achieve this, the building is located centrally along the north setback line, fronting Teheran Ro. The surface parking is located along the south edge of the site. This places the parking behind the building mass and elevated from the street, minimizing visibility.

The structural selection of braced steel frames in the transverse direction and ductile steel frames in the longitudinal direction is recommended. Wind loads control the lateral force design. For structural connection compatibility the floor framing system of metal decking with concrete fill supported on steel beams is recommended.

***Scheme B.*** The floor plan configuration is a central core surrounded by office space. This provides the most efficient circulation on each floor, and the shortest building perimeter. It also allows each floor to have two principal lobbies. The fact that a central core results in windows on all sides increases the solar heat gain over the other schemes. A compact entrance lobby collects all visitors at a single point, enhancing security control.

The exterior is a curtain wall of horizontal tinted vision glass bands and coated metal spandrels. The penthouse is set back from the edge of the building so that the automatic window washing machine has access all around. The visual information system is located on the side of the penthouse.

The building is located on the west half of the site, with surface parking occupying the east. This arrangement provides the maximum space between the building and the adjacent property to the east. This ensures that the IBC building's identity will not be affected by what may be built on the adjacent property.

The structural selection for this scheme is moment resisting steel frames in both directions. Floor framing is metal decking with concrete fill supported on steel beams.

***Scheme C.*** The core locations are the most significant aspect of this scheme. They are located on the east and west ends of the building. The passenger elevators are located on the east, and toilets, service elevator, and the mail-handling system are on the west. Stairs are located on each side. Electrical closets, communication closets, and a fan room are located in the center of each floor. This arrangement allows efficient distribution of services from the center without the interruptions of elevator and stair shafts. The area on the upper floors, above the low-rise elevators, is used as an office area with the benefit of an exterior wall. The location of cores also reduces solar heat gain and glare from the east and west sun.

The exterior is clad in several systems. The cores are covered with a panelized stone system. The lower floors use horizontal tinted vision glass bands and stone spandrel panels. A curtain-wall system of tinted vision glass and glass spandrels in a coated metal frame is used for the upper floors. The visual information system displays are located halfway up the building at the top of the low-rise elevator core. This provides for better visibility from the street level.

The building location is similar to scheme B, occupying the west half of the site with surface parking on the east. However, in scheme C parking is also located along the drive connecting the parking to the 8-m (26-ft) road.

The structural selection of moment resisting steel frames in both directions and metal decking floor with concrete fill on steel beams is recommended. This combination of lateral load and floor systems best satisfies the architectural, structural, mechanical, and electrical parameters.

## 4 Structural Systems

The structural design of a high-rise office building requires an integrated systems approach. It is essential to consider both the specific objectives of the structural system and the related architectural, mechanical, electrical, and cost objectives. The resulting product is an integrated building, which serves the required function and is within the budget. The specific objectives of the structural system are as follows:

1. Integration of structure with vertical and horizontal service plenums to allow unobstructed passage of power and communication conduits as well as HVAC ducts from sources to workstations
2. Provision of maximum flexibility in workstation relocation
3. Selection of a structural system that will contribute to optimizing other building system costs
4. Optimization of structural member sizing to minimize frame costs within a selected system

Some of the related architectural, mechanical, and electrical objectives affecting the structural design are floor setbacks, space planning flexibility, floor-to-floor heights, fire ratings, large floor openings for elevators, escalators, mechanical plenums, and shafts, duct and pipe penetrations through girders, and power and communication cable passage requirements in riser ducts.

In the conceptual and schematic design phases the structural design is based on a combination of the Korean Building Code and the U.S. Uniform Building Code (UBC), 1988 edition (ICBO, 1988). Uniform live loads are based on the Korean Building Code; however, the required concentrated loads and live-load reduction are based on the UBC. Horizontal wind loads are based on the Korean Building Code; seismic loads and design requirements are based on seismic zone 2A of the UBC. Structural steel design is based on *Specifications for the Design, Fabrication and Erection of Structural Steel for Buildings* by the American Institute of Steel Construction (AISC, 1978), and reinforced concrete design is based on *Building Code Requirements for Reinforced Concrete* by the American Concrete Institute (ACI 318-83). Both are referenced by the UBC. Foundation design is based on the recommendations in the soil and geotechnical investigation report for the proposed building site.

The high-rise tower structural framing system consists of two elements, the floor slab and framing system and the lateral load (wind and earthquake) resistance system. The essence of the high-rise building structural design is the selection of the proper lateral wind and seismic load resistance system in conjunction with the options for floor slab and framing. Two options for floor slab and framing in high-rise tower structural systems are metal decking with concrete fill and concrete floor framing.

Metal decking topped with concrete fill, with or without sprayed-on fire-proofing, is used in conjunction with a structural-steel lateral load system. The floor slab can be made to act in a composite or noncomposite system with the floor beams and girders. This system has the advantage of being lightweight and imparting less gravity loads on the foundation. It also generates a smaller seismic lateral force on the structural system than the concrete framing. The only disadvantage of this floor system is higher cost when compared directly with concrete framing on a unit-floor-area basis for material and labor. This cost differential will be partly offset by savings due to a lower floor-to-floor height when using steel beams and girders.

The structural options of concrete floor framing systems are as follows:

1. Cast-in-place solid concrete slabs on concrete floor beams and girders
2. Precast and/or prestressed concrete planks with concrete topping supported on precast prestressed floor beams and girders
3. Precast prestressed T sections with concrete topping supported on precast prestressed girders

Although concrete floor framing would cost less on a unit-floor-area basis compared with metal decking, the overall building cost may be higher. Concrete floor framing requires deeper concrete floor beams and girders, resulting in larger floor-to-floor height. The heavier dead load of concrete generates higher foundation design loads and higher seismic lateral loads on the structural system. Other disadvantages of concrete floor framing include a lack of flexibility to accommodate large floor and girder penetrations. For these reasons concrete systems are used to a lesser extent in high-rise office construction.

The three basic lateral load-resisting systems are a total steel system, a total concrete system, and a composite steel and concrete system. [See Chapters 10 and 11 as well as *Structural Design of Tall Steel Buildings* (CTBUH, Group SB, 1979) and *Structural Design of Tall Concrete and Masonry Buildings* (CTBUH, Group CB, 1978) for additional information on steel and concrete structural systems in tall buildings.] Within the three basic systems, options may be created by the design of lateral bracing elements. Several major structural options are considered for high-rise structural design.

*Total steel framing*

1. Steel moment frame in both directions with metal decking floor
2. Steel-braced frames in both directions with metal decking floor
3. Steel perimeter tube frames with metal decking floor
4. Steel moment frames in one direction, and steel-braced frames in the orthogonal direction with metal decking floor

*Total concrete framing*

1. Cast-in-place concrete moment resisting frames in both directions with concrete floor framing
2. Cast-in-place concrete shear walls with concrete floor framing
3. Precast concrete shear walls with concrete floor framing
4. Cast-in-place concrete perimeter tube frames with concrete floor framing

*Composite steel and concrete framing*

1. Cast-in-place concrete perimeter frames with metal decking floor and perimeter steel erection columns
2. Cast-in-place concrete core walls with metal decking floor and steel columns

The selection of an overall structural system for a high-rise office building is primarily based on the interface with architectural layouts and configurations, setbacks, and floor-to-floor heights. Column size and bracing interference are minimized to obtain maximum usable space. Mechanical penetrations in girders are considered. See the flowchart in Fig. 8.11.

From the structural design standpoint, the choice of a lateral load resisting system can be made without completing the design of all of the possible options. A matrix of structural lateral load system comparative indices can be used as a guide. Table 8.4 compares the actual wind and seismic base shears and overturning moments for each architectural building scheme, and for each of the applicable structural options. The wind forces are influenced only by the external surface configuration of each building scheme, regardless of internal structural system. However, within each scheme the seismic base shear and overturning moments vary with each structural option because of the different building weight and the different vibration and energy absorption characteristics of each system.

After narrowing down the number of structural options, the final choice can be determined by consideration of the objectives of space planning, mechanical and electri-

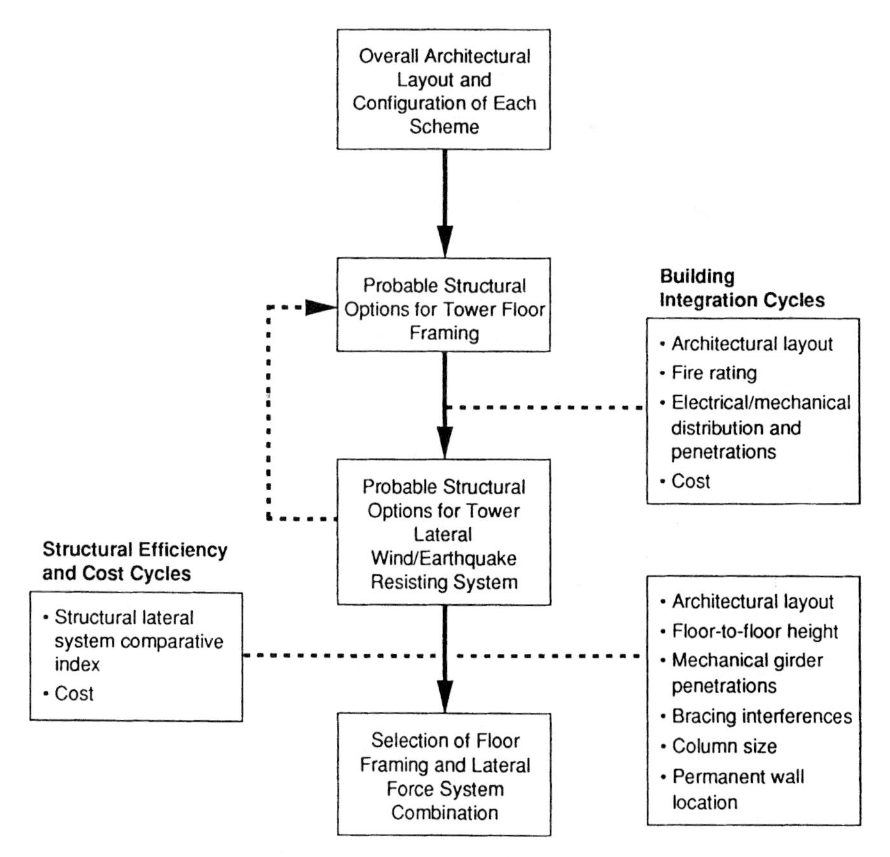


**Fig. 8.11 Flowchart for structural system selection of high-rise office building.**

cal requirements, overall relative building costs, material availability, and construction practices. A total steel-framing system was selected for the building.

The total concrete-framing options are not recommended because these systems require deeper girders than the steel frames and well-planned shear walls, which place strict limitations on doorways, passageways, and wall penetrations for architectural, mechanical, and electrical systems.

## 5 Building Management Systems

The BMS is a master control program for a series of modules that perform numerous building management functions. Each module is a network of linked microprocessors, the direct digital controllers (DDCs). This network of DDCs performs the task of controlling and monitoring such diverse tasks as security and lighting control. Several computer workstations control one or more modules. All of the modules (except fire management) are in turn controlled from a central host microcomputer located in the building manager's office. The following text contains discussions of each individual module.

The environmental management module provides automatic temperature and humidity control. It uses the latest in DDC technology for all mechanical equipment and systems being monitored and controlled. Monitoring information is dynamic (live) and is displayed graphically on a CRT screen. Graphics consist of floor plans showing the location of devices and systems, or schematics showing specific mechanical equipment. This provides the building engineer with not only a visual indication of the system, but also a verbal description of how it works. Mechanical systems use occupied and unoccupied modes of operation. Modes may be initiated automatically by time of day, by access control card holder activity, by wall switch, by phone, or manually by the engineer at the BMS central.

**Table 8.4 Comparative index of structural lateral system**

| | | Base shear | | | | | | | | | |
|---|---|---|---|---|---|---|---|---|---|---|---|
| | | | Seismic | | | | | | | | |
| Scheme | | Wind* | 1 | 2 | 3 | 4 | 5 | 6 | 7 | 8 | 9 |
| A | T | 4.27 | —† | — | — | 2.18 | — | — | — | — | — |
| | L | 1.17 | — | — | — | 1‡ | — | — | — | — | — |
| B | T | 2.67 | 0.95 | 2.06 | 0.99 | 2.06 | 1.20 | 1.68 | 5.51 | 5.79 | 1.78 |
| | L | 1.83 | 0.95 | 2.06 | 0.99 | 0.95 | 1.20 | 1.68 | 5.51 | 5.79 | 1.78 |
| C | T | 3.11 | 1.04 | — | — | 1.04 | — | 1.84 | — | — | — |
| L | | 1.45 | 1.04 | — | — | 2.27 | — | 1.84 | — | — | — |

*Regardless of building system.

†Dashes represent inappropriate options or impractical structural systems.

‡Base shear and overturning moment for option 4 of scheme A (with moment-resisting steel-frame system in E-W or longitudinal direction) are taken as unity for basis of comparison for the nine structural options considered for each scheme.

Energy management routines are easily optimized as a part of the BMS. Energy consumption in the building is metered and carefully monitored by the BMS. A building-wide demand-limiting program keeps the building kilowatt usage below an owner/utility definable limit, thereby limiting demand charges.

The fire management system is a versatile state-of-the-art audio evacuation system that gives building managers and fire fighters easy and accurate communication and control. The system provides:

1. A combination of multiple emergency communications systems in one fully integrated easy-to-operate fire command station
2. A stand-alone system that will continue to provide buildingwide emergency communication in the event the fire management central computer goes out of commission (see Table 8.5 for specific operation)
3. Reliable and easily understood "digitized" custom voice messages to increase the speed and safety of building evacuations
4. Capability of using multichannel audio technology
5. Capability of modular expansion for future expansion needs
6. Meets or exceeds all U.S. and local code requirements

A telephone communication system is provided which is capable of communication between all required locations and between these locations and the fire control center. In addition, all exterior phone jacks are weather-protected. An emergency telephone system provides two-way communication between the fire control center (FCC) and the stairways on every five floors. Visual indicators for the hearing-impaired are located throughout the building generally along the egress path. The zoning of the FCC's communications system is illustrated in Table 8.5.

| Overturning moment at base | | | | | | | | | | |
|---|---|---|---|---|---|---|---|---|---|---|
| | Seismic | | | | | | | | | |
| Wind* | 1 | 2 | 3 | 4 | 5 | 6 | 7 | 8 | 9 | Remarks |
| 3.22 | — | — | — | 2.18 | — | — | — | — | — | Steel-braced frames in transverse direction |
| 0.90 | — | — | — | 1‡ | — | — | — | — | — | |
| 1.91 | 0.95 | 2.06 | 0.99 | 2.06 | 1.20 | 1.68 | 5.51 | 5.79 | 1.78 | Steel moment frames in longitudinal direction |
| 1.31 | 0.95 | 2.06 | 0.99 | 0.95 | 1.20 | 1.68 | 5.51 | 5.79 | 1.78 | |
| 2.43 | 1.04 | — | — | 1.04 | — | 1.84 | — | — | — | |
| 1.13 | 1.04 | — | — | 2.27 | — | 1.84 | — | — | — | |

The security management system tightly monitors and controls access to the building. The access control system (ACS) uses state-of-the-art technologies applied in the building envelope security and internal occupancy control of restricted areas. Restricted areas consist of areas where inventory, databases, proprietary technology, sensitive documents, and executives reside. The system traces the movements of an individual card holder within the facility. The electronic ACS works in a stand-alone manner, yet is fully integrated with other BMSs. It will minimize the possibility of accidental alarm triggering. Reporting capabilities include reports on access, absenteeism, alarms, individual door activity, and database changes. The system also monitors activities within the facility utilizing door alarms and motion detectors to provide a complete report of what happened, when it happened, where it happened, and what to do about it.

A closed-circuit television system (CCTV) provides comprehensive visual inspection of the building entrances and internal critical areas. A CCTV switcher displays camera images on security-guard monitors in a timely pattern. Alarms activate video cameras and recorders at specified sensitive areas. All alarm information and system activity reports on graphic CRT display at the guard station.

As a part of the elevator and escalator operational status monitoring, after-hours card access to elevators is monitored as a part of the security management system, and operational status is reported through the elevator monitoring system. Access via elevators is restricted to floors designated by the card system.

The lighting control management system consists of a network of microprocessor-based lighting panels that use ambient light sensors to maintain the desired light level on a zone-by-zone basis. Levels are adjusted by dimming the light fixtures. A given zone of the building has a minimum of three different light levels (occupied, unoccupied, and janitorial). Levels are overridden automatically via a wall switch, a telephone, or in response to an event (for example, breach of security, fire alarm, energy management program, or use of an access control card in the elevator).

The maintenance management system contains data on each major piece of equipment, which is stored on hard disk. Data include manufacturers' operations and maintenance instructions, and a planned maintenance calendar.

**Table 8.5 Fire alarm operational matrix**

| Alarm | Ceiling smoke detector | Elevator-lobby smoke detector | Duct smoke detector | Sprinkler water flow switch | Sprinkler tamper switch | Manual pull station |
|---|---|---|---|---|---|---|
| Recall elevators | | X | | | | |
| Audible/visible alarm on fire floor | X | X | X | X | | X |
| Fire command center alarm annunciator | X | X | X | X | | X |
| Fire command center trouble annunciator | | | | | X | |
| Initiate smoke control sequence | X | X | X | X | | X |
| Unlock stairway doors* | X | X | X | X | | X |
| Release door holders | X | X | X | X | | X |
| Transmit off-site alarm | X | X | X | X | | X |
| Zoning by floor | X | X | | X | X | X |

*Electric locks shall be wired to fail safe (unlocked) upon loss of power.

## *6 HVAC and Plumbing Systems*

The objectives of the HVAC system in an intelligent office building are to provide the following:

1. A high level of occupant comfort, including temperature and humidity regulation, and the removal of the increased heat gains commonly associated with a high level of office automation
2. Energy-efficient design to maintain occupant comfort at the least possible cost
3. Flexibility in air delivery and monitoring to accommodate changes in the location and density of personnel and office equipment
4. Life-safety operation modes, integrated with the BMS, to assure the safety of occupants during fire and smoke emergencies

HVAC system technology used in high-rise intelligent office buildings was identified, analyzed, evaluated, and compared for application to the three building schemes (Bechtel Civil Inc., 1989).

The following are key components of an intelligent building HVAC system. They are important components to satisfy the objectives of occupant comfort, energy efficiency, and flexibility.

1. Multiple air-delivery terminal units for flexibility during regular-hours and after-hours occupancy
2. Variable air volume for energy efficiency
3. Heat recovery to reuse heat generated within building interior
4. Monitoring and control through a BMS to regulate air volume and temperature in conjunction with occupancy, equipment, and lighting loads, and to optimize energy utilization and minimize electric demand changes
5. Air or water economizing cycles to provide "free" cooling using increasing amounts of outside air

For details of these components see Bechtel Civil Inc. (1989).

It was decided that the HVAC design would comply with existing applicable Korean codes. Where no Korean code references exist, the appropriate U.S. national codes and standards would be applied. The HVAC options comparison study (Table 8.6) summarizes all major parameters—peak loads, annual energy consumption, investment costs, annual operating costs, life-cycle costs, and summary of system-related floor areas. This summary was done for the architectural scheme A. In Table 8.7 comparisons are made of the peak heating and cooling loads of the three architectural schemes using two outside ventilation air rates.

Heating and cooling loads, annual energy consumption, and demonstration economic analysis were performed using the Trane Company TRACE ULTRA computer program. The high-rise portion was analyzed applying the 10 HVAC options (Bechtel Civil Inc., 1989). The air-conditioned portions of the basements in each scheme were analyzed as being served by three air systems, the first serving kitchen and dining, the second serving the retail and other miscellaneous areas, and the third serving the computer room. The source or method of cooling the three systems was assumed to be the same as that used in analyzing the high-rise portion of the building. The results of the analysis were summarized, and they represent the entire building.

The objectives of the plumbing system are:

1. Provide potable cold and hot water services
2. Provide nonpotable water service
3. Provide adequate water storage to maintain a lower portion of the building plumbing system operational during limited-duration losses of off-site power or city water supply
4. Collect sanitary waste, retain and pretreat before discharge to the city municipal sewer
5. Collect and discharge storm drainage directly to the city municipal sewer
6. Provide a subsoil drainage system if required

City water supply connects to a primary reservoir common to plumbing and fire protection, located at the lowest basement level. Potable water is pumped through the water treatment unit, from the primary reservoir to an intermediate storage tank. Lower floors

**Table 8.6 Comparison study of HVAC systems**

| | 1 | 1A | 1B | 1C |
|---|---|---|---|---|
| System | Floor by floor with water economizer | Floor by floor with ice storage | Floor by floor with air economizer | Floor by floor with ice storage and low air temperature |
| Heating | LGN | LNG | LNG | LNG |
| Economizer | Water | Water | Air | Water |
| **Peak Loads:** | | | | |
| Cooling (R/T) | 1162 | 1162 | 1162 | 1162 |
| Heating (Mcal/hr) | 1318 | 1318 | 1318 | 1318 |
| **Annual energy:** | | | | |
| Electricity (kWhr/hr) (×103) | 3502 | 3475 | 2451 | 3146 |
| LNG ($m^3$) (×103) | 17.1 | 17.1 | 7.9 | 17.1 |
| Mcal/$m^2$ | 75.6 | 75.1 | 51.9 | 70.0 |
| Mcal/$m^2$ | 57.9 | 57.5 | 39.7 | 53.6 |
| Total investment cost ($) | 4,050,000 | 4,500,000 | 4,180,000 | 4,250,000 |
| **Annual operating costs ($):** | | | | |
| Electricity | 350,200 | 347,500 | 245,100 | 314,600 |
| Natural gas | 7500 | 7500 | 3500 | 7500 |
| Total energy | 357,700 | 355,000 | 248,600 | 322,100 |
| Maintenance | 32,000 | 37,000 | 37,000 | 37,000 |
| Total | 389,700 | 392,000 | 285,600 | 359,100 |
| **Life-cycle costs ($):** | | | | |
| Present value×106 | 11.006 | 11.649 | 10.896 | 11.133 |
| **System-related areas:** | | | | |
| Air shaft area ($m^2$) | 310 | 310 | 65 | 310 |
| Fan room areas ($m^2$) | 855 | 855 | 1070 | 790 |
| Ice storage areas ($m^2$) | — | 185 | — | 185 |
| Total ($m^2$) | 1165 | 1350 | 1135 | 1285 |

and the basement are gravity-fed from the intermediate storage tanks. The upper floors have water delivered by pumps, taking water from the intermediate storage tank. Nonpotable water has the same configuration and components except for omission of the water treatment unit. Sanitary waste and kitchen waste, after grease interception, runs through a septic tank sized to limit biological oxygen demand (BOD) at the tank outlet to not more than 100 ppm. A storm drainage system connected to roof drains and exterior ground-floor drains discharges directly into the city storm sewer. Collection sumps and sump pumps provide for basement drainage.

## *7 Electrical, Communications, Lighting, and Lightning Protection*

Electric power, communications, and other systems for intelligent office buildings must be designed to meet the present and future automation needs of every occupant. In conceptual design those concepts applicable to the high-rise intelligent office buildings are identified, and the owner is provided with preliminary technical and demonstration cost data to assist in the selection of systems.

| 2 | 3 | 3A | 4 | 4A | 4B |
|---|---|---|---|---|---|
| Floor-by-floor direct expansion units | Heat pumps on exterior central on interior | Heat pumps all zones and central vent air | Central with air economizer | Central with ice storage | Central with ice and low air temperature |
| LGN | LGN & electric | LGN & electric | LNG | LNG | LNG |
| None | Air | None | Air | Air | Air |
| | | | | | |
| 1162 | 1222 | 1189 | 1222 | 1222 | 1203 |
| 1318 | 1371 | 1371 | 1005 | 1005 | 1005 |
| | | | | | |
| 3446 | 3617 | 4258 | 2977 | 2739 | 2509 |
| 17.1 | 0.6 | 0.6 | 7.2 | 7.2 | 7.2 |
| 74.5 | 73.7 | 86.4 | 61.4 | 57.5 | 52.8 |
| 57.0 | 56.4 | 66.2 | 47.8 | 44.0 | 40.4 |
| 3,750,000 | 3,570,000 | 3,340,000 | 4,550,000 | 4,970,000 | 4,400,000 |
| | | | | | |
| 344,600 | 361,700 | 425,800 | 297,700 | 273,900 | 250,900 |
| 7500 | 300 | 300 | 3100 | 3100 | 3100 |
| 352,100 | 362,000 | 426,100 | 300,800 | 277,000 | 254,000 |
| 41,000 | 42,000 | 45,000 | 21,500 | 26,500 | 26,500 |
| 393,100 | 404,000 | 471,100 | 322,300 | 303,500 | 280,500 |
| | | | | | |
| 10.608 | 10.600 | 10.592 | 11.571 | 12.083 | 11.054 |
| 310 | 325 | 310 | 705 | 705 | 425 |
| | | | | | |
| 855 | 740 | 185 | 1485 | 1485 | 930 |
| — | — | — | — | 185 | 185 |
| 1165 | 1065 | 495 | 2190 | 2375 | 1540 |

The power distribution system's objectives include:

1. Sufficient load to drive a variety of electronic terminals at every workstation
2. "Clean power" to protect sensitive electronic components during operation
3. An efficient system of wire management to both conceal and provide ready access to wiring and cabling systems
4. A flexible system of wire and cable connections to provide for relocation of workstations and increases in service demands

The key features of the power supply system are as follows:

1. Modular transformers that supply initial power requirements and provide capability for additional incremental future connections
2. Power conditioning employed to provide a clean power supply
3. Electrical closets stacked to form an efficient distribution tree

The conceptual design of the communication system will define the communication service feeder system. Specific terminal devices, switching equipment, local area network (LAN) devices, terminal controllers, fiber-optic transducers, and other active electronic components that will be addressed as IBC requirements are further defined and equipment systems selected. Provision is made at the conceptual stage for 100% capacity increase in communication closets and employment of a modular floor distribution system.

The primary building vertical distribution cable, or backbone, is a fiber-optic cable. An option to this would be a coaxial cable backbone. The horizontal distribution of communications from the communications closet provides a standard service capability at each user workstation. The cabling will be compatible with present and future technologies and may be twisted pairs, coaxial cable, optical fiber, or a proprietary cable.

Special areas are provided with cabling to support services capabilities in addition to the standard capability described in the foregoing. Video-grade coaxial cable and loop-grade fiber-optic cable are installed between the major conference room and nearby communications closets.

Communications closets are aligned vertically with a minimum of two closets per floor, located so that no workstation is more than 100 m (382 ft) of cable from the near-

**Table 8.7 Comparison of architectural schemes**

| Architectural scheme | Total conditioned, $m^2$ | Peak loads | |
|---|---|---|---|
| | | Cooling, T/R | Heating, Mcal/hr |
| Ventilation air rate = 0.90 $m^3/hr \cdot m^2$: | | | |
| Scheme A | 42,300 | 1162 | 1318 |
| Scheme B | 42,360 | 1292 | 1506 |
| Scheme C | 43,617 | 1216 | 1270 |
| Ventilation air rate = 2.70 $m^3/hr \cdot m^2$: | | | |
| Scheme A | 42,300 | 1344 | 2392 |
| Scheme B | 42,360 | 1377 | 2397 |
| Scheme C | 43,617 | 1392 | 2354 |

Notes: (1) SC = 0.32 (glass shading coefficient). (2) Heat transmission coefficients ($kcal/m^2 \cdot hr \cdot °C$): Wall = 0.50; Roof = 0.35; Glass = 2.20.

est closet. The closet accommodates cable termination and patching for all connected and spare lateral and riser or tie cables, the anticipated LAN equipment, fiber-optic terminals, communications controllers, and other active devices. A raceway system of conduit or enclosed metal cable tray connects the central distribution room and the communications closets aligned vertically. Communications rooms and closets have overhead "Unistrut" cable racking along the ceiling. Fully enclosed metallic raceway or plenum-type cable is used in all air-handling plenum. Fiber-optic cables do not share a raceway with other media; they use separate raceways or inner ducts.

The telecommunications system must provide for all communication device cabling needs, including backbone system to provide for an expanding data network (Table 8.8), local area network systems (voice and data transmission) with high noise immunity, real-time process control, and video conferencing.

The interior and exterior lighting systems must provide for the program requirements and accommodate human comfort and security concerns. Interior office lighting (Table 8.9) is designed for visual comfort and veiling reflection control at video display terminals. High-color-rendering fluorescent or discharge-type lamps enhance the color treatment. Exterior lighting, an integral part of the security system, is coordinated with building architecture and landscaping. Discharge sources use high color rendering where appropriate.

**Table 8.8 Electrical-communications backbone systems**

| Objectives | Multiple twisted pair + overall shield | Coaxial cable + transceiver | Fiber-optics + transducer |
|---|---|---|---|
| Accepts broad-bandwidth high data transfer | Up to 6.3 Mb/sec | Up to 500 MHz | Up to 200 Mb/sec |
| Minimizes crosstalk | Moderate | Reduces if shielded | Eliminates, no electrical field |
| Security | Can be tapped | Can be tapped | Difficult to tap |
| Eliminates EMI, RFI, and EMP | Low | Moderate | Best |
| Supports ISDN | Yes | Yes | Yes |
| Supports LAN up to 10 Mb/sec | Yes | Yes | Yes |
| Supports LAN greater than 10 Mb/sec | No | Yes | Yes |
| Supports full-motion video | No | Yes | Yes, with special equipment |
| Flexibility | Easily relocatable | Fair, relocate transceiver | Low, relocate transceiver |
| Cost | Low | High | High |
| Compatibility | High | Expandable by bridging and patching | Expandable by bridging and patching |
| Expandability | Easily expandable | Can be patched to other wire systems easily | Requires transceiver for some equipment |
| Maintenance | Easily maintained | Minimal | Light sources must be kept clean |

The lightning protection system includes roof air terminals, connecting copper cables, copper down conductors, and ground electrodes, all in conformance with NFPA 78 (NFPA, 1978) and local codes. Roof antennas are bonded to roof system for protection. Secondary lightning and surge protection are generally required for voice and data wiring on upper floors of the building since induced lightning surges can occur on the wiring when lightning strikes the building-structure lightning protection system.

## 8 Building Transportation Systems

The objectives of elevators, escalators, and document-conveying systems in the intelligent building are as follows:

1. Energy conservation through computerized system monitoring
2. System flexibility and ability to absorb future changes
3. Improved service algorithms to reduce wait time
4. Management of maintenance priorities via automotive software-driven record keeping

**Table 8.9 Lighting systems**

| Objective | Overhead direct fixture only* | Overhead direct fixture + task lighting |
|---|---|---|
| **Functional lighting levels:** | | |
| Ambient lighting | Good | Good |
| Task lighting | Good | Good |
| Emergency lighting | Overhead provides | Overhead provides |
| Maintenance | Low | Low |
| **Energy management:** | | |
| Adaptability to zone control | Acceptable | Overhead acceptable |
| Adaptability to daylight control | Acceptable | Overhead acceptable |
| Energy consumption | Least | High |
| **Visual comfort:** | | |
| Direct glare | Minimal | Minimal |
| Reflected glare | Source visible in VDT | Source visible in VDT |
| Veiling reflections | Overhead requires diffusion | Overhead requires diffusion |
| **General objectives:** | | |
| Flexibility | Easily relocated | Fair |
| Annual cost of ownership† | $17.90/m$^2$ | $20.60/m$^2$ |
| Compatibility | Coordinate with HVAC | Coordinate with HVAC |
| Maintenance | Low | Above average |

*Ambient light level cut by 50%, supplemented by task lighting.
†Demonstration costs from IES model.

The size, quantity, and speed of elevators are determined primarily by the number of persons expected to occupy and use the building. For this project, a population density of approximately one person to 8.4 m$^2$ (90.4 ft$^2$) was used to estimate the total number of persons who will occupy the office tower. Large groupings of elevators, each serving all building floors, are inherently inefficient, requiring large numbers of passengers to be carried in each elevator trip. Therefore a low-rise/high-rise elevator configuration was selected. The elevator system is designed to handle arriving traffic equal to 10% of the building population during a 5-minute period. To avoid long queues during peak traffic, calculated performance shows that elevator intervals must be 30 seconds or less.

Two groups of high-speed passenger elevators furnish the primary vertical transportation to office tower occupants. The elevators in both groups are sized at 1600-kg (3527-lb) capacity. Advanced group control systems employing 32-bit microprocessor technology and AI are used to ensure maximum efficiency of elevator service. For further information on elevators as components of tall buildings see Chapter 2 of this Monograph and Chapter SC-4 of *Tall Building Systems and Concepts* (CTBUH, Group SC, 1980).

Escalator service is provided at levels basement 1 through level 2. Escalators must be able to move the maximum building population from the lobby level to level base-

| Floor-mounted indirect fixture only | Floor-mounted indirect fixture only* | Overhead indirect fixture + task lighting* | Overhead indirect fixture only |
|---|---|---|---|
| Good | Good | Excellent | Excellent |
| Fair | Good | Good | Fair |
| Needs separate system | Needs separate system | Overhead provides | Overhead provides |
| High | High | Above average | Above average |
| Difficult | Difficult | Difficult | Acceptable |
| Difficult | Difficult | Difficult | Acceptable |
| High | High | High | High |
| Minimal | Minimal | Minimal | Minimal |
| Minimal | Possible with task lighting | Possible with task lighting | Minimal |
| Minimal | Minimal | Minimal | Minimal |
| Fair | Poor | Poor | Poor |
| \$22.30/m$^2$ | \$25.00/m$^2$ | \$30.20/m$^2$ | \$27.60/m$^2$ |
| Independent of HVAC | Independent of HVAC | Coordinate with HVAC | Coordinate with HVAC |
| Above average | Above average | Above average | Above average |

ment 1, dining facilities, and shops. The elevator system will deposit 60 people per minute at the lobby level, with the escalator system capacity at 70 people per minute based on one person per escalator tread. For greatest energy efficiency, escalators switch on and off automatically.

The mail-handling system delivers and collects mail by a custom-designed and -manufactured automated system of conveyors and traveling bins.

A fully automatic exterior window washing machine is the base system for the project. The fully automatic machine employs a tracked car on the roof and specially designed tracks integral with the curtain wall for precise alignment of the automatic washing unit.

As an alternative, exterior window washing can be accomplished by a manned platform suspended by four wire ropes from a powered roof car running on tracks connected to the building structure at the roof. To ensure safe operation, the platform runs on tracks designed into the building curtain wall.

## *9 Building Skin Systems*

The objectives of the building skin in the intelligent building include more than just acting as an enclosure between inside and outside. The specific objectives are:

1. Providing energy efficiency, that is, the building skin should contribute to the reduction in energy consumption
2. Providing central daylight to reduce direct and reflected glare
3. Minimizing water penetration and condensation
4. Providing a choice of colors, textures, and finishes (See also Chapter 5 for information on aesthetics.)
5. Being compatible with automated window washing equipment
6. Accommodating building movement
7. Minimizing loading on structural frame
8. Minimizing maintenance requirements

The exterior wall system components are:

1. Tinted vision glass, providing a shading coefficient of not greater than 0.50
2. Aluminum spandrel panels with anodic or resinous coating
3. Stainless steel with polished or bright annealed finish
4. Polished stone panels with moisture-proof backing and weeping points
5. Regular-spaced vertical mullions, including accommodation for automated window washing equipment
6. Horizontal or vertical automated blind system, incorporated into the horizontal window head or vertical window jamb

8. Prefinished interior surfaces to eliminate requirement for secondary interior finish surfaces

The system recommended for the building skin is preglazed panels of 3 by 4 m (10 by 13 ft). Two such panels make up a 6-m (20-ft) horizontal module to accommodate the automatic window washing machine. The components of the panels are tinted vision glass with either glass or prefinished metal spandrels. Both the vertical and the horizontal panel edges provide for building and thermal movement. The panels are well insulated and the frame mounting is tightly gasketed for energy efficiency. The weight of this system is well below that of precast concrete or conventional stone veneer construction. This exterior wall system is adaptable to field assembly ("stick building") in lieu of panelization. (See Chapter 7 for more information on building skins.)

### 10 Fire Protection Systems

The primary objective of the fire protection system is to provide life and property protection in accordance with building codes and insurance requirements. Two basic standpipe and sprinkler options are detailed here. Both are complete systems, providing zoned control valves, tamper switches, water flow switches, hose cabinets, and fire department connections.

*Option 1.* Low- and high-level automatic fire pump assemblies supply separate fire standpipes and sprinkler risers. A high-level system will take suction from an intermediate storage tank.

*Option 2.* The combination standpipe and sprinkler riser is designed as a two-zone system with an intermediate water storage tank. The low-zone fire pump provides water under fire conditions to low-zone floors. This fire pump is also used to fill the storage tank. A separate fire department connection is provided for the low zone. The high zone is supplied by a fire pump connected to the water storage tank. A separate fire department connection is provided for the high zone.

On a failure of city water supply for fire fighting, a fire department–operated automatic valve directs tank water from the high zone to the low zone as necessary to fight the fire. All fire pumps are supplied with emergency power.

The following special systems are provided:

1. Fire extinguishers in mechanical rooms, truck dock, and each stair landing
2. Halon systems in computer rooms and building management system rooms
3. Mechanical foam and fuel-spill retention system for heliport
4. Automated fire alarm system using heat detectors in electrical equipment rooms, elevator machine rooms, and switchgear rooms

The fire protection system design complies with existing applicable Korean codes. Where no Korean code references exist, the appropriate U.S. national codes and standards are applied.

Sprinkler coverage is provided for all areas of the building, except the utility shaft or as prescribed by Korean codes. All unheated areas subject to freezing are provided with dry-type sprinkler systems. Additional fire hose outlet valves are also provided in all required emergency egress stairways on levels where these stairways open to the interior of the building. (For additional information on fire protection systems in tall buildings see Chapter 9.)

### 11 Concluding Remarks

After a comprehensive review of the conceptual design report, the Lucky-Goldstar Development Company selected scheme C for further development. Figures 8.12 to 8.14 illustrate the final design solution for the Lucky-Goldstar IBC building. The selected scheme's floor plan is organized with the core elements located on the east and west ends of the 36,000-m$^2$ (109,728-ft$^2$) building. This floor area represents a reduction with respect to the floor area of 52,000 m$^2$ (559,724 ft$^2$) proposed earlier. The actual building layout was modified from the original scheme C in terms of relationships of the stairs and elevator cores. The air-handling system is a floor-by-floor system with central chilled water; the central plant system uses heating and steam humidification, chilled water, and ice storage; the communications backbone system consists of fiber-optic cable and conversion components; the lighting system specifies direct overhead fixtures and task lighting with daylight controls; and the building skin system uses vision glass and glass spandrel with coated aluminum panels on the exterior surface of the cores.

## 8.5 CONDENSED REFERENCES/BIBLIOGRAPHY

The following is a condensed bibliography for this chapter. Not only does it include all articles referred to or cited in the text, it also contains bibliography for further reading. The full citations will be found at the end of the volume. What is given here should be sufficient information to lead the reader to the correct article—author, date, and title. In case of multiple authors, only the first named is listed.

Abelson 1982, *Computers and Electronics*

ACI 318-83, *Building Code Requirements for Reinforced Concrete*

AISC 1978, *Specifications for the Design, Fabrication, and Erection of Structural Steel for Buildings*

Ali 1987, *Academic vs. Industrial Research in Architectural Engineering*

Ali 1988, *Building Technology Forecast and Evaluation of Structural Systems*

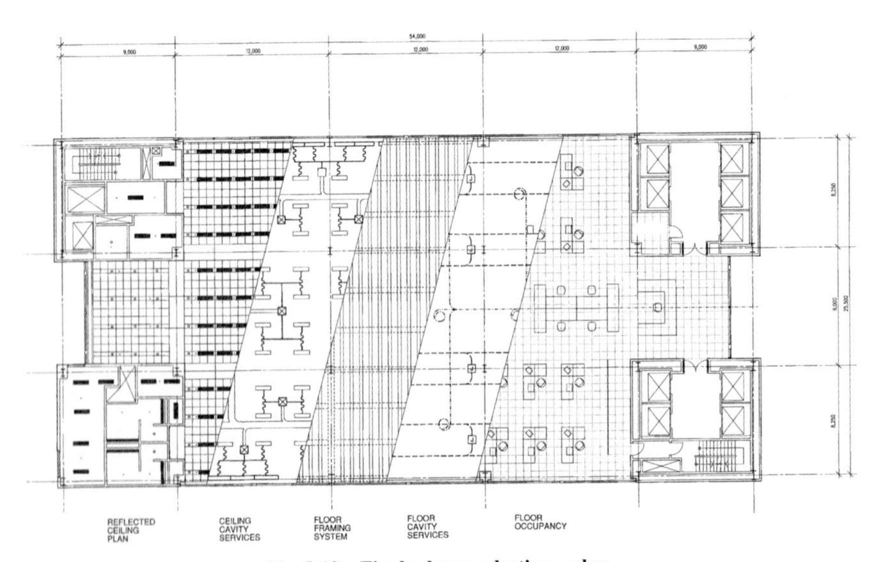


**Fig. 8.12 Final scheme selection—plan.**

**Fig. 8.13 Final scheme selection—exterior perspective.**

Ali 1988a, *Current Trends in the Use of Advanced Computer Technology for Building Design and Construction*

Ali 1988b, *Impact of Computers on Architectural Engineering Research and Education*

Ali 1993, *Developments in the Application of Computer Technology to Building Engineering*

Bechtel Civil Inc. 1989, *Lucky-Goldstar International Business Center: Conceptual Design Report*

Brown 1982, *Computers in Design and Construction: Everyone Reads the Same Music*

CTBUH, Group CB, 1978, *Structural Design of Tall Concrete and Masonry Buildings*

CTBUH, Group PC, 1981, *Planning and Environmental Criteria for Tall Buildings*

CTBUH, Group SB, 1979, *Structural Design of Tall Steel Buildings*

CTBUH, Group SC, 1980, *Tall Building Systems and Concepts*

Dorgan 1991, *Building Systems Automation—Integration*

Dorgan 1992, *Building Systems Automation—Integration*

Engineering News Record 1988, *SOM/IBM Field Queries on Joint CADD System*

Fenves 1984, *Expert Systems: C. E. Potential*

Hayes-Roth 1983, *Building Expert Systems*

ICBO 1988, *Uniform Building Code*

John 1989, *Making Intelligent Buildings a Reality*

Kasuda 1970, *Use of Computers for Environmental Engineering Related to Buildings*

Kennett 1993, *CD-ROM: The Future Is Now*

Marksbury 1984, *Artificial Intelligence: CAE's Outer Limits*

Napier 1982, *A Systems Approach to Military Construction*

National Bureau of Standards 1970, *Use of Computers for Environmental Engineering Related to Buildings*

**Fig. 8.14 Final scheme selection—interior perspective.**

NFPA 1978, *Life Safety Code*

Robbie 1972, *The Performance Concept in Building: The Working Application of the Systems Approach to Building*

Skidmore, Owings & Merrill 1988a, *Energy Application: Guide to the Energy Program*

Skidmore, Owings & Merrill 1988b, *HVAC Application: Loads Reference Guide*

Skidmore, Owings & Merrill 1988c, *HVAC Application: Ducts Reference Guide*

Skidmore, Owings & Merrill 1988d, *Lighting Application: Lights Reference Guide*

Skidmore, Owings & Merrill 1988e, *Power Application: Power Reference Guide*

Skidmore, Owings & Merrill 1989, *Piping Application: Pipes Reference Guide*

Thomas 1990, *Intelligent Computer Programs for Building Services Engineering Design*

Thomas 1991, *Mechanical Engineering Design Computer Programs for Buildings*

Toong 1982, *Personal Computers*

# 9

# Environmental Systems

The environmental technology employed in the design of tall buildings covers a diverse range of systems. This chapter focuses on four aspects of this technology—physiological comfort, environmental control, fire safety, and biomass energy—each addressing different issues related to the common backdrop of tall buildings. Section 9.1 discusses the physiological and physical and thermal comfort inside tall buildings, the components attributing to the comfort zone, and the criteria for determining thermal comfort. In conjunction with this topic of thermal comfort, Section 9.2 concentrates on the control of the thermal and chemical environment in tall buildings, particularly as it pertains to mechanical systems.

Preventive fire and smoke management and life-safety design of multiuse high-rise buildings today have become integral parts of environmental systems and architectural design. Environmental systems now have the dual responsibility of controlling space conditions under normal operations, and controlling escape routes, fire, and smoke under fire conditions. Air systems, which make up a significant component of environmental systems in modern high-rise buildings, will, under fire conditions, automatically convert to pressurization and exhaust systems to control the movement of fire and smoke, confine the fire and smoke, dilute the smoke to nontoxic levels, and maintain smoke-free escape routes. Thus the preventive fire protection design, requiring a multidisciplinary approach related to the overall building technology, is included in this chapter on environmental systems. Sections 9.3 through 9.6 focus on the fire preventive design of high-rise buildings by presenting three case studies which illustrate the new technology, materials, and equipment available in preventive design. Section 9.7 examines the use of biomass fuel and a small-scale gas turbine power plant, an ecological alternative to fossil fuels, to provide electricity and steam for an urban high-rise building, whereas Section 9.8 looks at the issue of sustainable architecture and its impact on tall building design.

For other related issues influencing the tall building environment (for example, lighting, acoustics, and sun and shadow) see Chapter PC-6 of *Planning and Environmental Criteria for Tall Buildings* (CTBUH, Group PC, 1981).

## 9.1 DESIGN FOR PHYSIOLOGICAL COMFORT

### 1 Environmental Control

Tall buildings generally accommodate a large population. It is then imperative that these buildings be designed for physiological comfort so that people can stay healthy and work in a comfortable environment. A limited control over the environment in a tall building can be achieved without full air-conditioning. It is relatively easy to heat buildings, although the amount of energy consumed, which can be very considerable in a cold climate, depends greatly on the sophistication of the methods employed and on the automatic temperature control. Heating the air lowers its relative humidity, and if the temperature increase is appreciable, the temperature is lowered below the level of comfort. This can, however, be remedied easily by allowing water in an open container to evaporate.

Cooling a building without refrigeration is possible. There are several age-old methods that were used before refrigeration was invented and are still used today. These are sunshading, natural ventilation (in hot and humid climates), and evaporative cooling (in hot and arid climates). There are, however, limits to what can be achieved by these traditional methods.

Most window air conditioners and automobile air conditioners consist merely of refrigeration units. Refrigeration by itself makes it possible to reduce the temperature by as much as desired, but it does not give any control over humidity.

To achieve full air-conditioning for thermal comfort, it is necessary to adjust both the temperature and the humidity. If the air needs to be cooled to achieve thermal comfort, it is generally also necessary to remove some of the water vapor, unless the climate is very dry.

The correct relative humidity for thermal comfort at the desired temperature is found, and from that the corresponding absolute humidity and the temperature at which this absolute humidity equals the dew point are determined. Cooling the air to that dew point temperature thus removes precisely the amount of water vapor required, no more and no less. When the air is then heated to the desired temperature, it has also the relative humidity required for thermal comfort.

This is an expensive process if the control is to be achieved accurately. Full air-conditioning is therefore not, at present, used for residences or small-scale buildings, but is limited to tall buildings and other large building complexes. A delicate balance among heating control, cooling, and relative humidity is the key to a comfortable, healthy, and productive environment in a tall building. The material presented in this section offers some guidelines to designers of environmental systems of tall buildings for providing maximum comfort to the occupants. (See Chapter 6 for a discussion of the psychological effects of thermal comfort on the occupants in tall buildings.)

## 2 The Physiological Basis of Human Comfort

The temperature of the central part of the human body is approximately constant at 37°C (98.6°F), although the temperature of the skin varies from place to place and from time to time. In conditions of thermal comfort, the skin temperature is approximately 33°C (91.4°F), but in cold weather the skin temperature of the hands and feet may be much lower.

Evidently, temperature is the main criterion for human comfort. It is difficult to be comfortable at a temperature higher than the central body temperature of 37°C (98.6°F), even if one wears very light clothing. There is greater tolerance for a lower temperature because it can be compensated to some extent by heavy clothing and physical activity.

Humidity is next in importance, particularly at high temperatures, because the heat loss from the skin and, to a lesser extent, by respiration greatly depends on it. Sweat evaporates quickly in hot and dry weather, and the latent heat of evaporation cools the skin. Sweat does not evaporate so easily in high humidity.

Air movement causes heat transfer by convection from the skin and the clothing. In hot and humid weather it provides great relief, increasing the evaporation of the sweat. However, when the air temperature rises above the body temperature of 37°C (98.6°F) in a hot and arid climate, air movement increases the body temperature, and a hot and dry wind therefore creates discomfort.

Radiation can also cause comfort or discomfort. Sunshine is pleasant on a cool day, but on a hot, clear day the temperature recorded by a thermometer exposed to the sun can be 20°C (68°F) above the shade temperature that is quoted in meteorological data. Radiation from an open fire or a radiator is pleasant, but they are only used when the temperature is low. In addition to this short-wave, high-temperature radiation, the walls, floors, and ceilings of rooms and the human body produce long-wave radiation at a much lower temperature.

Thus temperature, humidity, air movement, and radiation are the main external factors affecting thermal comfort.

### 3 The Psychrometric Chart

The temperature and humidity of the air and the heat contained in the moist air can be plotted on a psychrometric chart (from the Greek *psychros,* meaning *cold*). This greatly aids the discussion of human comfort criteria and the determination of the heating or cooling required to produce a comfortable environment. It has the dry-bulb temperature (DBT) in degrees Celsius as its horizontal coordinate and the absolute humidity (in grams of water vapor per kilogram of dry air, or g/kg) as its vertical coordinate.

Figure 9.1 shows a standard psychrometric chart. Its terminology and application are discussed in the following text. A number of curved lines are plotted on these coordinates, corresponding to the relative humidity (RH). When the relative humidity reaches 100%, water must be removed by condensation, and the 100% RH line therefore terminates the chart. On this chart the equal wet-bulb temperatures (WBT) form straight inclined lines.

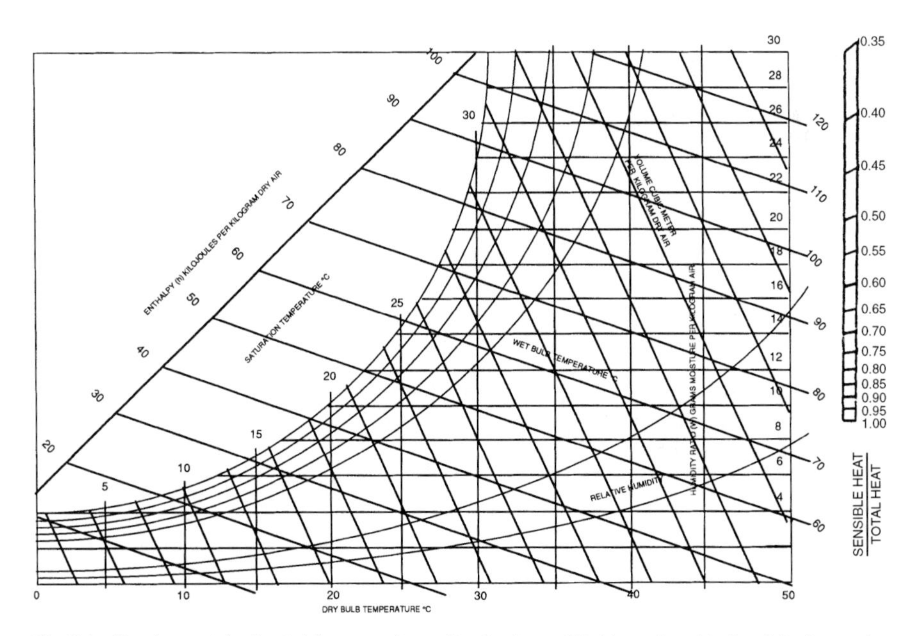


**Fig. 9.1 Psychrometric chart.** (*Courtesy: Australian Institute of Refrigeration, Air Conditioning, and Heating.*)

The density of the air changes with temperature and humidity. For air-conditioning calculations we need to know the volume of each kilogram of dry air to be handled, that is, the specific volume (SV) in cubic meters ($m^3/kg$). This is the reciprocal of the density of the air.

Another set of straight lines is formed by the enthalpy (from the Greek *thalpo,* meaning *to make hot*). The enthalpy is the heat content of the moist air, measured in kilojoules per unit mass of air (kJ/kg), and it is the sum of the sensible heat and the latent heat. The sensible heat is the amount of heat required or released during a change of temperature at a constant absolute humidity, such as the heat needed to raise the dry-bulb temperature from 20 to 25°C (68 to 77°F) without changing the mass of water contained in the air. The latent heat is the amount of heat required or released when the absolute humidity is changed at a constant temperature, because we must evaporate or condense some of the moisture in the air. The lines of enthalpy are also shown in Fig. 9.1.

In Fig. 9.1 the vertical lines of dry-bulb temperature, the horizontal lines of absolute humidity, the curved lines of relative humidity, and the inclined straight lines of wet-bulb temperature and of specific volume are combined to form the psychrometric chart.

It is now possible to plot on the psychrometric chart the range of temperatures and humidities considered comfortable for a particular occupancy, and the temperature and humidity of the outside air. From these the change in temperature and humidity required to produce a comfortable environment can be determined, as well as the enthalpy, that is, the energy required for this purpose (see Fig. 9.2).

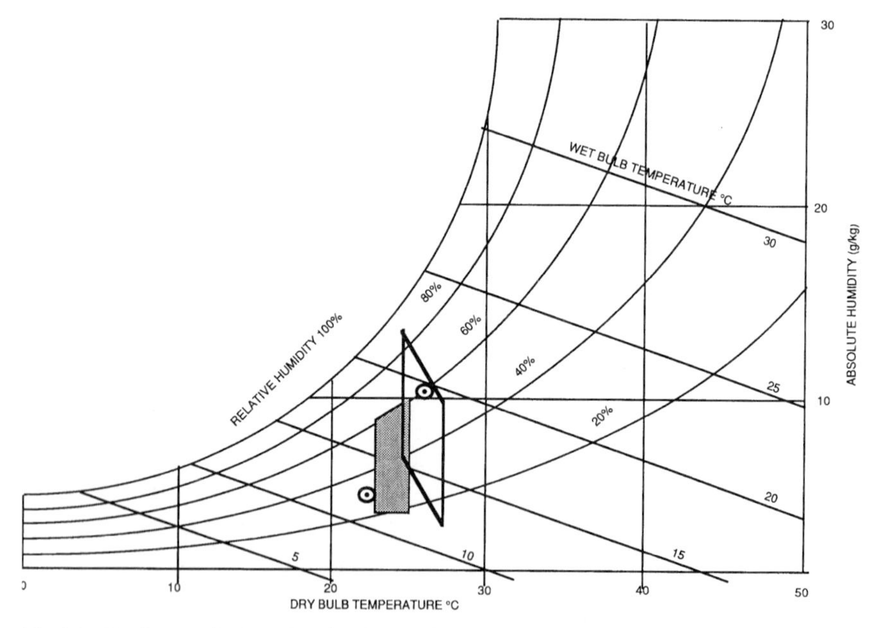


**Fig. 9.2 Psychrometric chart showing the comfort zone obtained from the experiments at Kansas State University (Rholes, 1973) with a solid-line parallelogram; the comfort zone recommended by ASHRAE Standard 55-74 (ASHRAE, 1974) with a hatched parallelogram; and the summer and winter design temperatures recommended by ASHRAE Standard 90-75 (ASHRAE, 1975) with circles.** (*Courtesy: Van Nostrand Reinhold, New York.*)

### 4 *Effect of Clothing on Human Comfort*

Thermal comfort in a tall building is greatly affected by clothing and by physical activity. Social and religious conventions impose limitations on the amount of clothing that can be shed in hot weather, and practical considerations impose limits on the protection that can be given to the face, the hands, and the feet in very cold weather. However, clothing can be of enormous help in creating comfortable conditions.

The unit for measuring the effect of clothing on thermal comfort is the *clo*. One clo is defined as the insulating value of a person's lightweight business suit (long trousers and coat) and cotton underwear. The relative insulating values of other forms of clothing are listed in the ASHRAE Handbook (1993). They were determined by thermal measurements on a copper manikin, appropriately clothed, made at Kansas State University in 1973. Some values for men's clothing are given in Table 9.1.

The type of activity is also important. A person wearing a swimsuit and lying on a beach may be very comfortable in weather that is most uncomfortable for a person wearing heavier clothes and doing heavy work. The metabolic rate of the body, that is, the rate at which it produces energy, increases with the body's activity. It can be measured in watts per square meter, or *mets,* where 1 met is the metabolic rate of a person sitting, inactive, in a room with still air; it is taken as 58 W/m$^2$. Table 9.2 shows metabolic rates for some activities. For additional information on the psychological aspects of thermal comfort, the reader should refer to Chapter 6.

The metabolic rate varies with the surface area of a person's skin. This is usually determined by an empirical formula derived by DuBois and DuBors (1915). A person whose mass is 75 kg (165.35 lb) and who is 1.8 m (5.91 ft) tall has a DuBois surface area of 1.85 m$^2$ (19.91 ft$^2$) (McIntyre, 1980), so that for this person 1 met corresponds to $58 \times 1.85 = 107$ watts.

### 5 *Criteria for Thermal Comfort*

Although recognized since the days of Vitruvius, the various components of thermal comfort had never been assembled into a unified criterion until Yaglou published the results of his experiments (Houghten and Yaglou, 1923) at the Pittsburgh research laboratory of the American Society of Heating and Ventilating Engineers (now renamed ASHRAE) in 1923. A number of people were placed in a room where the temperature,

**Table 9.1 Insulating effect of men's clothing**

| Type of clothing | clo |
|---|---|
| Brief, swimsuit | 0.05 |
| Shorts only | 0.1 |
| Shorts and short-sleeved shirt | 0.2 |
| Shorts and short-sleeved shirt, socks, and shoes | 0.3 |
| Long trousers, short-sleeved shirt, socks, and shoes | 0.5 |
| Long trousers, short-sleeved shirt, socks, shoes, and undershirt | 0.6 |
| Lightweight suit, cotton underwear | 1.0 |
| Three-piece winter suit, cotton underwear, and wool socks | 1.5 |
| Three-piece winter suit, cotton underwear, wool socks, and overcoat | 2.0–2.5 |
| Heavy clothing designed for outdoor use beyond the Arctic and Antarctic circles | 4.0 |

humidity, and air movement could be controlled and measured. They were engaged in light physical activity and wore light clothing. The clothing worn during these experiments was equal to the unit of 1 clo (Table 9.1). They were then asked to indicate the conditions of temperature, humidity, and air movement that gave the same sensation of warmth. Those combinations of temperature, humidity, and air movement were then designated as having the same effective temperature.

The experiments determining the effective temperature have been progressively refined. The current standard is based on experiments carried out under the auspices of ASHRAE at Kansas State University (Rohles Jr., 1973).

***Comfort Zone.*** Since the main variables are temperature and humidity, the comfort zone can be outlined on a psychrometric chart. The results of the 1973 experiments are shown by the solid-line parallelogram in Fig. 9.2. The recommendation of the ASHRAE standard published in the following year (ASHRAE, 1974), shown by the hatched parallelogram in Fig. 9.2, is slightly different.

In 1975, as an energy conservation measure, higher design temperatures were recommended for summer and lower ones for winter. These are shown in Fig. 9.2 by dots surrounded by circles.

Differences in comfort zones are due to variations in the amount of clothing normally worn, and this varies with time and place; but they also reflect different conceptions of comfort. A survey of people of both European and Asian descent conducted in Singapore (Ellis, 1953) indicated a preferred temperature of 27°C (80°F) at a relative humidity of 80% (a high percentage) and an air velocity of 0.4 m/sec (1.31 ft/sec).

While the ASHRAE (1992) recommendations provide the best data available for the design of tall buildings in North America, they need to be modified for people used to higher summer or lower winter temperatures; this can also conserve energy.

The concept of effective temperature and similar concepts elsewhere [equivalent temperature in Britain (Dufton, 1932) and resultant temperature in France (Missenard, 1948)] are based on experiments designed to determine the most comfortable conditions attainable, irrespective of the climatic conditions outside. However, if allowance is made for a frequent interchange between indoor and outdoor environment, greater comfort can be achieved by wearing lighter clothes in summer and much heavier clothes in

**Table 9.2 Approximate metabolic rate for various activities**

| Activity | Metabolic rate | |
|---|---|---|
| | W/m$^2$ | met |
| Sleeping | 41 | 0.7 |
| Resting, lying down | 47 | 0.8 |
| Sitting and not working | 58 | 1.0 |
| Standing | 70 | 1.2 |
| Typing | 80 | 1.4 |
| Light manual work | 120 | 2.0 |
| Cleaning a house | 180 | 3.0 |
| Heavy work | 270 | 4.5 |
| Walking on level ground at 3 km/hr | 120 | 2.0 |
| Walking on level ground at 6 km/hr | 200 | 3.5 |
| Walking up a 15° slope at 3 km/hr | 270 | 4.5 |

winter, and adjusting the indoor temperature accordingly. The energy conserved thereby is an additional inducement. The following additional points are worthy of consideration by designers of tall buildings.

In theory, air-conditioning aims at maintaining the perfect temperature and humidity. In practice, this is not always achieved when the weather is extreme, if the refrigeration plant is too small, and when the temperature rises during the morning and drops again in the late afternoon. This change in temperature should be planned deliberately, because people are used to a diurnal temperature movement and may miss it in air-conditioned buildings. Many complaints about "sick" buildings, which have no apparent justification, are probably due to this cause.

Allowing some diurnal change in tall building design would obviously save energy and reduce the prime cost of the plant. It might particularly please office workers and improve their productivity (see also Chapter 6). Some careful sociological research on this problem is appropriate to find out what the majority really wants. There may be a substantial difference between the responses of people doing interesting work and those doing boring routine jobs in the same tall office building.

Many people would like to be able to open windows partially on a nice day, when the weather is pleasant, that is, not too hot and not too cold. Although sealed windows save a lot of energy, windows that can be opened might be considered in some areas of air-conditioned buildings in such cities as Sydney or San Francisco, where the weather is rarely very hot or very cold, provided that the air-conditioning is turned off in those areas of the building. Rosiak (1972), Hubka (1973), and Jokl (1973) have addressed the issue of thermal comfort from different standpoints.

## 9.2 ENVIRONMENTAL CONTROL SYSTEMS

### 1 The Objective of Environmental Control

The basic objective of environmental control in a building, as was observed in Section 9.1, is to create an environment that is comfortable for the occupants. In most cases, however, the objective goes beyond meeting the physiological necessity of the occupants and includes meeting a greater expectation of rich environmental experience. Furthermore, in many buildings, including tall buildings, the objective is further expanded to include the creation of an environment that is most conducive to optimum productivity of the occupants for the intended function. Beyond this, there is even a greater objective of maximizing the benefit, in its broadest sense, to everyone involved, and to this end, environmental control is only a means. Then, to arrive at a successful building design, one should try to devise a design that can maximize the fulfillment of the global objective as well as every one of the subobjectives simultaneously.

### 2 The Domain of Environmental Control

While environment in general may include everything that surrounds us, its domain within the context of building environmental control includes the environment of energy as well as matters that affect our physiological well-being directly. These are thermal environment, luminous environment, sonic environment, electromagnetic environment, radiation environment, and chemical environment. Strictly speaking, the environment of gravity and motion should also be a part of it, particularly in a tall building, but this issue is usually dealt with as part of the structural problem. The elements of weather such as rain and snow also affect physiological well-being. However, in a well-designed building their effects are supposedly stopped completely at the building enclosure. These latter concerns, therefore, are considered as being part of the enclosure design rather than environmental control problems.

Thermal environment includes the ambient heat in the air and the radiant heat from any hot object, such as the sun, fireplace, or radiator. Luminous environment consists of the visible band of radiation from any radiant source, including the sun and artificial light sources. Sonic environment consists of the vibration of air in the audible frequency range. These are the three kinds of energy environments that have a direct and the most obvious impact on environmental comfort, and therefore have traditionally been the foci of attention in the design of environmental control systems.

Recently, however, we are becoming increasingly aware of the potential health risk of electromagnetic fields around us, such as around high-voltage electric transmission lines and videodisplay terminals of computers. This is the electromagnetic environment. Radiation environment is another area which did not receive much attention except in the 1950s and 1960s in fear of the life-threatening gamma radiation from fallouts in case of an atomic explosion. With the discovery of the carcinogenic effects of radon, a radioactive gas released from the earth, radiation concerns have been rekindled.

The importance of the chemical environment is unquestionable—it is the chemical constituent of the air we breathe. The points of concern include oxygen content, smoke, and toxic gases as well as odor, dust, and (although not a chemical) bacterial content. Moisture vapor in the air is, strictly speaking, a chemical. However, it affects the quality of the thermal environment much more seriously than the chemical environment. Therefore humidity control is always an integral part of thermal environment control. Furthermore, since most advanced thermal environment control involves "air conditioning" (that is, conditioning of the air), chemical and thermal environments are usually controlled simultaneously.

Although all these factors are important, this section will focus on the control of the thermal and the chemical environment, particularly through mechanical systems.

## 3 Subsystems of Tall Building Mechanical Systems

Building mechanical systems can range from a very simple heating system for a small residential building to a very complex fully automated all-air HVAC (heating, ventilating, and air-conditioning) system for large buildings. In a tall building it is desirable to isolate the indoor environment from the outdoor completely for the sake of better control. This, coupled with the fact that the tall building spaces are expected to be at the higher end of the spectrum of quality, necessitates the use of upscaled all-air HVAC systems.

To accomplish its purpose, an HVAC system consists of four subsystems:

1. Heating subsystem
2. Refrigeration subsystem
3. Air-conditioning and distribution subsystem
4. Regulatory subsystem

Obviously, the heating and refrigeration subsystems will add and remove heat to and from the system, thus controlling the quality of the thermal environment. The air-conditioning and distribution subsystem has two functions, conditioning of the air, both thermally and chemically, and distribution and circulation of the conditioned air throughout the building. Conditioning involves heating and cooling, humidification and dehumidification, and replenishment of oxygen and removal of carbon dioxide and other pollutants. All of these are done by an air-handling unit.

Finally, the system's operation must be regulated carefully to ensure optimum system performance toward impeccable control of the thermal and chemical environment. This requires, on the one hand, continuous monitoring of the environmental state, as well as various operational characteristics of the functional subsystems, and, on the other hand, activation of appropriate operational measures that will steer the entire system's performance toward the creation of the optimum environmental state. This is the role of the regulatory subsystem.

## 4 Heating Subsystem

A heating subsystem in a large building with an all-air HVAC system generally has three additional subsystems of its own—heating plant, air heating system, and perimeter heating system. However, under certain conditions, elimination of some of these in favor of some alternative means may become more desirable. For instance, consider a case where a relatively small amount of heating is necessary. Here, despite high energy costs of the electricity itself, the use of decentralized electric heating with heating strips wherever heat is needed could be far more attractive than the use of a conventional gas- or oil-fired centralized hot water heating system. The benefit is the savings in construction and maintenance costs that will result from the elimination of the central heating plant along with the smokestack and the entire hydronic system for perimeter heating. The savings could be enormous.

Tall buildings usually fit into this situation. Because of the preferred depth of 12 to 15 m (40 to 50 ft) for lease on both sides of the building core, as well as for structural reasons, tall buildings tend to be fairly deep, that is, about 34 to 40 m (110 to 130 ft) along the narrow dimension of the plan. This, coupled with their height, makes tall buildings "core-dominated" or "cooling-load driven," with relatively low heating requirements. Furthermore, every square meter literally counts in tall buildings, and the elimination of hydronic perimeter heating, particularly the baseboard type, will increase the amount of leasable floor areas substantially. This is not a small concern for both the developer and the user.

Nonetheless, the extra heat loss at the perimeter must be compensated for, particularly along the windows, where the occupants may experience local discomfort due to extra radiant heat loss directly from the body to the window, as well as draft. In modern buildings there are two generic approaches to heat loss. One is overhead radiant heating at the windows, most likely the electric type, to compensate for the extra radiant heat loss, and the other is blowing air directly against windows using the central air system, but with extra boost heat, as in the AT&T tower in Chicago (Fig. 9.3). This is discussed in more detail in Subsection 7. Of course there are instances where a hydronic perimeter heating system is warranted. This is the case with 900 North Michigan Avenue, Chicago (Figs. 9.4 and 9.5). In this case the low-rise portion of the building offers an economical location for the heating plant.

## 5 Refrigeration Subsystem

***Refrigeration Machine.*** The most widely used refrigeration system in tall buildings employs centrifugal refrigeration machines operating on a compressive refrigeration cycle (ASHRAE, 1989, 1992). Despite the obvious benefit of being very quiet, absorption machines are usually not applicable in tall buildings because their operation requires high-temperature heat, and tall buildings often do not have a heating plant sufficient to support this.

Centrifugal refrigeration machines consist of two parts—the evaporator, or water chiller as it is frequently called, and the condenser, which is typically water-cooled.

**Fig. 9.3 AT&T tower, Chicago. This 60-story tower of concrete tube and steel core structure has its refrigeration plant on the 16th floor and its cooling tower on the roof.** (*Courtesy: Hedrich-Blessing.*)

**Fig. 9.4 900 North Michigan Avenue, Chicago. This 66-story mixed-use building is also characterized by a mixed structure: steel structure to the 46th floor for long span required for retail and office functions; flat-plate concrete structure for hotel and condominium at the upper part of the tower to lower the structural height.** (*Courtesy: Kohn Pederson Fox Associates.*)

Chilled water generated in the evaporator, usually about 5.6°C (42°F), is distributed to various air-handling units in the building, thus cooling the air that passes through the air-handling units. The used chilled water, usually about 13.9°C (57°F), is returned to the chiller. The heat picked up from the air by the chilled water is transferred to the evaporating refrigerant in the evaporator, thus rechilling the used water. The gaseous refrigerant is compressed into the condenser by the centrifugal compressor, and the compressed hot refrigerant condenses in the condenser, giving away its heat to the surrounding condenser water. This warm condenser water, usually about 37.7°C (100°F), is pumped up to the cooling tower to be cooled down to around 29.4°C (85°F) by dissipating its heat to air by evaporation. This water is returned to the condenser to continue the refrigeration process (ASHRAE, 1992).

Depending on the size of the building, the cooling load in a tall building may range from a few to several thousand tons. Although the exact requirements depend on the

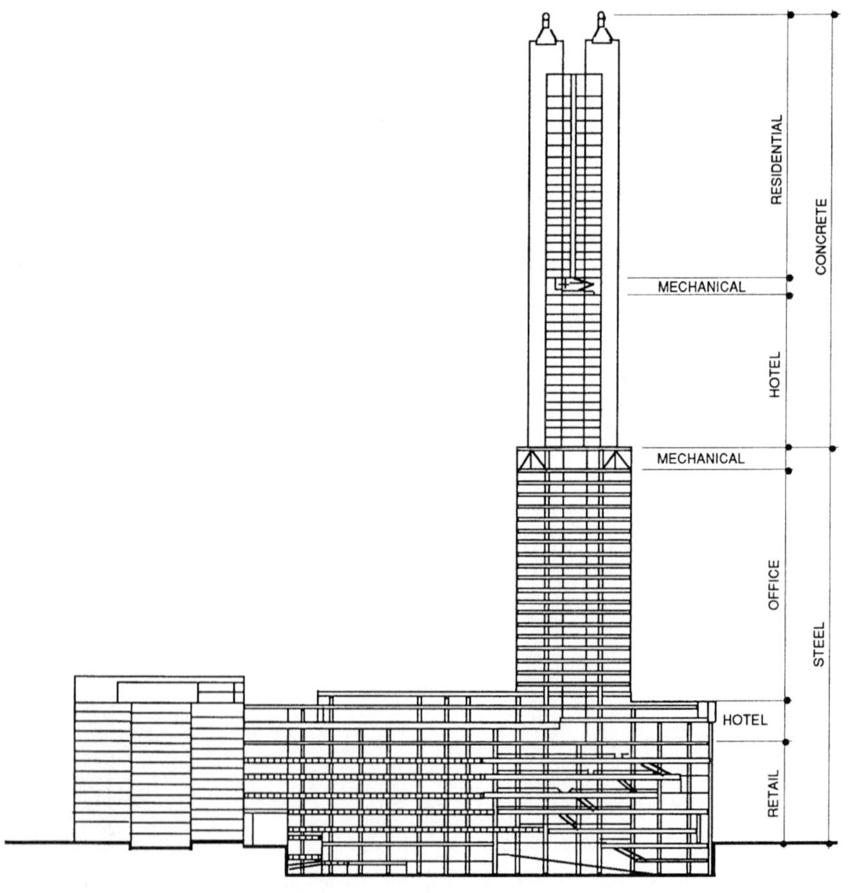


**Fig. 9.5 900 North Michigan. Refrigeration plant and cooling tower are located on the top floor and the roof of the low-rise in the rear (left in the drawing). Unlike many other tall buildings, it also has a heating plant conveniently located here.** (*Courtesy: Kohn Pederson Fox Associates.*)

specifics of the building, in the absence of any specifics, one may assume about 1 kW of cooling load for approximately 8 to 12 $m^2$ gross (86 to 129 $ft^2$ gross) of the building (or 1 ton for 300 to 400 $ft^2$ gross). Of course, one could use a single refrigeration machine for the entire load. However, for greater system reliability and operational flexibility it is much more desirable to use multiple units of smaller capacities. Usually the equipment is sized in such a way that the operational versatility under different cooling requirements over different seasons can be maximized while still maintaining relative simplicity of maintenance and stocking of spare parts. For example, for a relatively small building with a total cooling requirement of 7000 kW (approximately 2000 tons) one might use two machines of 3000 kW (800 tons) and 4000 kW (1200 tons) each so that three different operational levels of 3000 kW (800 tons), 4000 kW (1200 tons), and 7000 kW (2000 tons) of cooling can be possible. Oakbrook Terrace tower (Figs. 9.6 and 9.7) is an example. It has two refrigeration machines of 3300 kW (950 tons) and 4000 kW (1200 tons) each (Murphy/Jahn, 1985). On the other hand, for a building with a maximum of 10,000 kW (3000 tons) one might elect to use three identical machines of 3300 kW (1000 tons)

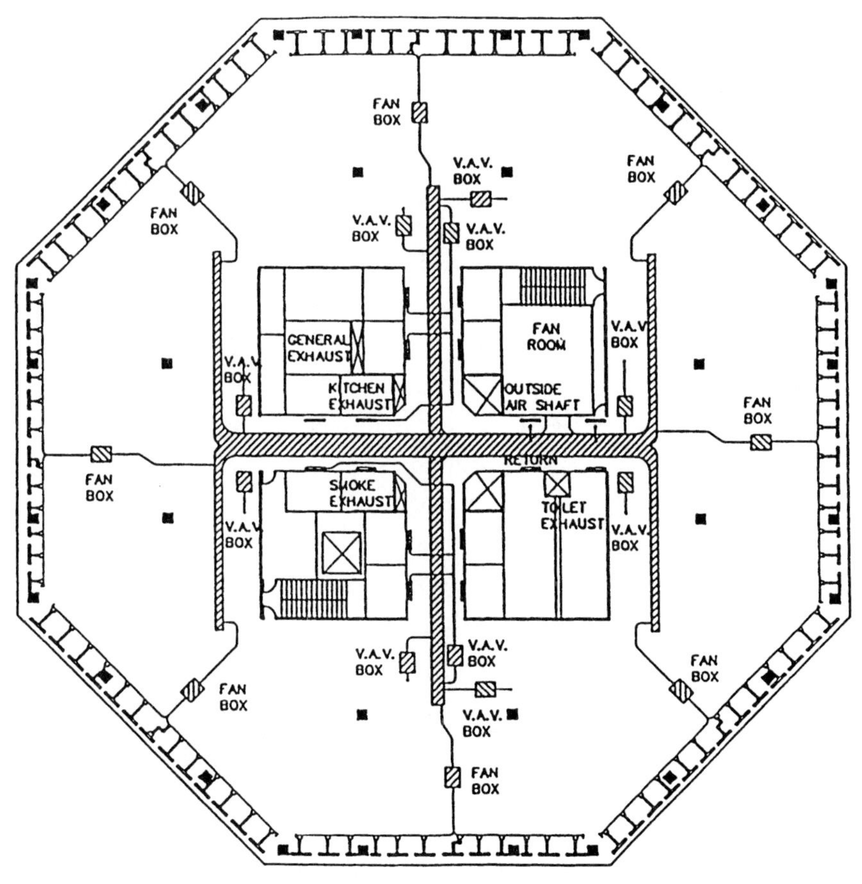


**Fig. 9.6 Oakbrook Terrace, Oakbrook, Illinois. Tree-type duct layout on a typical floor with single-point feed from the air-handling unit in the fan room.** (*Courtesy: Murphy/Jahn.*)

each, as is the case with the USG building in Chicago (Skidmore, Owings & Merrill, 1990). This will still give three levels of operational capacity while greatly simplifying the task of maintenance and the inventory of spare parts. Normally three to five pieces of equipment yield enough operational versatility with economy in terms of both operation and maintenance as well as the initial cost of purchase and installation.

***Refrigeration Plant: The Primary Mechanical Room.*** The equipment is installed in a large mechanical room along with a host of chilled water pumps, condenser water pumps, and other necessary equipment such as an air compressor and domestic water heater. (This room should really be called a "refrigeration plant" to distinguish it from the other kind of mechanical room, which houses air-handling units. However, this is not the case, and confusion continues.)

Obviously a refrigeration plant has an unusually heavy dead load, and the structural engineer must consult the equipment manufacturer for recommended design loads as well as the architect and the mechanical engineer for the specific equipment layout. Also the plant requires unusually high headroom because of the enormous amount of piping

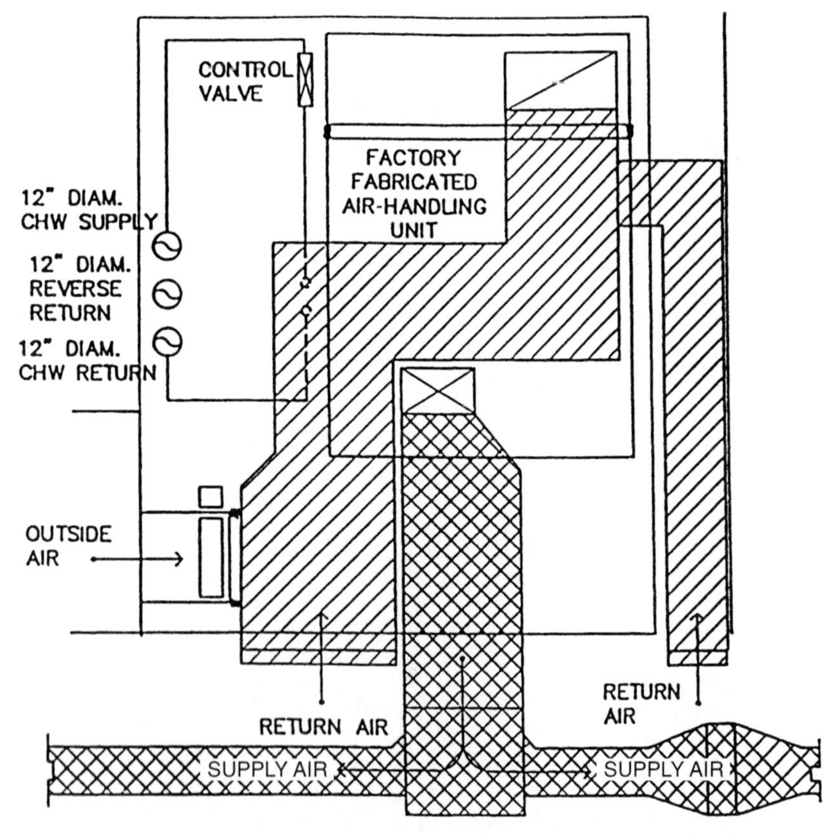


**Fig. 9.7 Oakbrook Terrace. Schematic diagram of airflow in fan room. Fresh air is supplied to the air-handling units on each floor through the common fresh air intake shaft in the core. Notice the chilled water pipe serving the air-handling units.** (*Courtesy: Murphy/Jahn.*)

work in addition to the height of the refrigeration machines themselves. For this purpose, usually at least 6.1 m (20 ft) of headroom is required. Thus unless the plant is in the basement, common practice is to design it as a two-story space. This gives the obvious benefit of being able to accommodate the mechanical space within the tower without disrupting the vertical rhythm of the building from the outside.

A basement is an obvious candidate for the plant from the structural point of view, and this is the case with the Lucky-Goldstar twin tower in Seoul (Hee Sung Industry, 1987; Fig. 9.8). However, there are two other points that must be considered. One is its relationship with air-handling units, the other is construction scheduling. Heavy mechanical equipment normally has a long lead time for delivery, and usually the floor slab over a mechanical room cannot be poured until the equipment is fully in place. This could mean a serious delay in the entire construction process. Unless the equipment is ordered by the owner long before the award of the construction contract, whoever the contractor may be, the basement may not be adequate from this point of view.

Furthermore, since every air-handling unit in the building, with the probable exception of small packaged air-conditioning units, needs to be served by the central refrigeration plant, the cost of chilled water piping as well as the operational efficiency of the entire system could be seriously affected by the location of the refrigeration plant. As will be seen later, there could be several major mechanical rooms distributed vertically in a tall building. One could be at the top of the building, another one just below the grade, and perhaps one to three more in midlevels at every 20 to 40 stories. It is clear that the basement may not necessarily be the optimum location for the refrigeration plant from this point of view. In fact, many tall buildings have their refrigeration plants somewhere near the bottom one-third of the building. The AT&T tower and 900 North Michigan Avenue are two examples. The 60-story AT&T tower has its refrigeration plant on the 16th floor (Skidmore, Owings & Merrill, 1986), whereas the 66-story 900 North Michigan has it on the 14th floor (Kohn Pedersen Fox Associates, 1985; Fig. 9.5). Of course, the refrigeration plant is also the best location for major air handlers.

***Cooling Tower.*** Undoubtedly the most obvious and convenient location for a cooling tower is the rooftop, and that is in fact where most of the cooling towers in tall buildings are found. In Fig. 9.3, the cooling towers are enclosed at the top of the building in an ornamental "cap." Louvered openings allow the air to freely flow through to enhance the efficient functioning of the cooling tower. Rooftop cooling towers, however, cause an extra burden on the construction costs because a pair of condenser water pipes, which are substantial both in size and in cost, must be installed from the roof to the refrigeration machines, wherever they are located. As mentioned earlier, this could be a distance of as much as two-thirds of the total building height or even the entire height.

Occasionally there could be a low-rise companion of the tower in a tall building complex. If that is the case, then the rooftop of the low-rise portion could be a convenient place for a cooling tower. This is the case with 900 North Michigan. As the building section view indicates, the mechanical floors are zoned in the tower to reduce duct and piping runs and to increase both the economy and the efficiency of the HVAC system. However, one should be cautious about possible condensation of moisture vapors from the cooling tower on the surfaces of the surrounding buildings, including the very high-rise tower that it is to serve. When the low-rise location is available and particularly attractive from the construction and operational points of view, then one should seriously explore ways of dealing with the moisture problem.

Similar to the case of refrigeration machines, system reliability and operational versatility are also very important for cooling towers. For this purpose, cooling towers are usually compartmentalized into several chambers. Furthermore, being core-dominated, tall buildings need year-round cooling. This means that some sections of the cooling

**Fig. 9.8 Lucky-Goldstar twin towers, Seoul, Korea. This 35-story tower has its refrigeration plant in the basement, cooling tower on the roof, and separate air-handling units on each floor.** (*Courtesy: Skidmore, Owings & Merrill.*)

tower need to be operating even in the middle of winter. For this purpose, that part of the cooling tower is "winterized," that is, protected from freezing by heating elements.

Finally, a cooling tower is not the only means of cooling the condenser water. In the case of the Hong Kong and Shanghai Bank, ocean water is used as the heat sink. Sea water is brought into the basement of the building through an intake tunnel of more than 305 m (1000 ft) in length at the rate of about 57 m$^3$ (15,000 gal) per minute, and it is this water that cools the condenser water (Walker, 1987). In the Christian Science Center in Boston the condenser water is cooled in a specially designed cooling pond, which occupies an entire city block. Although such a pond is visually appealing, the evaporation from the cooling pond impacts the microclimate of the surrounding areas significantly, and one should be very cautious about its use. Perhaps a more innovative solution to the problem could be actively harnessing the waste heat in the condenser water for other beneficial uses. This water is already about 38°C (100°F), and it is hot enough to preheat domestic hot water or even to heat small neighboring buildings in winter. If necessary, the waste heat in the condenser water can be pumped out, thus generating hot water of a higher temperature while cooling the condenser water on the other end. Developers, who usually own tall buildings, may not wish to do this because of the extra initial cost, but it is more than worthy of exploring.

***Provisions for Auxiliary Air-Conditioning Units.*** In designing a tall building, many different kinds of tenant requirements must be anticipated, not only initially but also over the life of the building. Appropriate responsive measures must be provided in the initial building system design to permit accommodating such requirements. One particular point of concern is added cooling requirements due to extensive computerization of offices.

Rather than trying to provide "enough" cooling through the central HVAC system without knowing exactly what the added cooling load might be, the usual strategy is to have the tenant handle the extra load directly with additional packaged air-conditioning units. These units come in two different types—one with air-cooled condensers and the other with water-cooled condensers. Since an air-cooled condenser requires direct access to outdoor air as the heat sink, this type is at best very difficult to use in a tall building. Appropriate ones are typically those with water-cooled condensers because these units can be cooled by the condenser water from the central cooling tower. For this, takeoffs are installed in the condenser water pipes as anticipated during the initial construction and capped off until needed, as in the Lucky-Goldstar twin tower and the AT&T tower. Of course, the primary condenser water pipe, as well as the cooling tower, must be designed with these extra loads in mind. Once the tower crane is taken down, adding an extra cooling tower on a tall building is difficult and expensive.

***Optional Decentralization of the Refrigeration System.*** Tall building development is primarily an economic enterprise. As far as the developer is concerned, the assessment of the pros and cons of various system organization options are essentially investment decisions. When a building is expected to be leased primarily by a few major tenants, centralized refrigeration systems become highly advantageous financially, managerially, and technically. This is the case with most tall buildings, such as the AT&T tower and the Morton International building in Chicago (Figs. 9.9 and 9.10).

However, if the building is designed for a host of small tenants, the case is different. The management, which most likely are the developers themselves or their agents, might wish to minimize their managerial and "custodial" responsibilities by providing a decentralized system so that individual tenants will be responsible for the operation and maintenance of their own systems. In this case there could be still other benefits: the rent appears lower on paper because the utility bill is on the tenant and also the capital outlay could be lowered, particularly if the tenants are expected to buy out the equipment.

**Fig. 9.9 Morton International building, Chicago, with famous cantilevered parking and office structure in the foreground (low rise). Its primary air-handling units are located on the 23d and 24th floors.** (*Courtesy: Perkins & Will.*)

Examples of this kind include the Oakbrook Terrace tower, Oakbrook, Illinois, and 123 North Wacker Drive, Chicago. In the former, prepackaged air-handling units are installed on each floor. Chilled water is supplied to each of these by the refrigeration plant in the basement through the chilled water riser in the fan room (Fig. 9.6). Because of the generous site, the cooling tower is installed on the ground in this case. The 123 North Wacker Drive building, on the other hand, employs packaged air-conditioning units with DX-coil and water-cooled condensers. These are served by the same cooling tower on the roof through the condenser water riser in the fan room.

When the mechanical rooms can be placed on the perimeter of the building, as in 123 North Wacker, one could consider using air-conditioning units with air-cooled condensers rather than water-cooled ones, thus making each unit stand completely alone without relying on the central cooling tower at all. By doing so, it could even be possible to eliminate the cooling tower altogether. However, an air-cooled condenser is much less efficient than its water-cooled counterpart, and it might require an excessively large tower for air circulation, so one should be cautious about its use.

### 6 Air-Conditioning and Distribution Subsystem

***Air Handlers.*** The heart of air-conditioning and distribution systems is the air handler. It is in this equipment that the air is conditioned, thermally and chemically, and it is also through this equipment that the air is propelled to travel to various parts of the building.

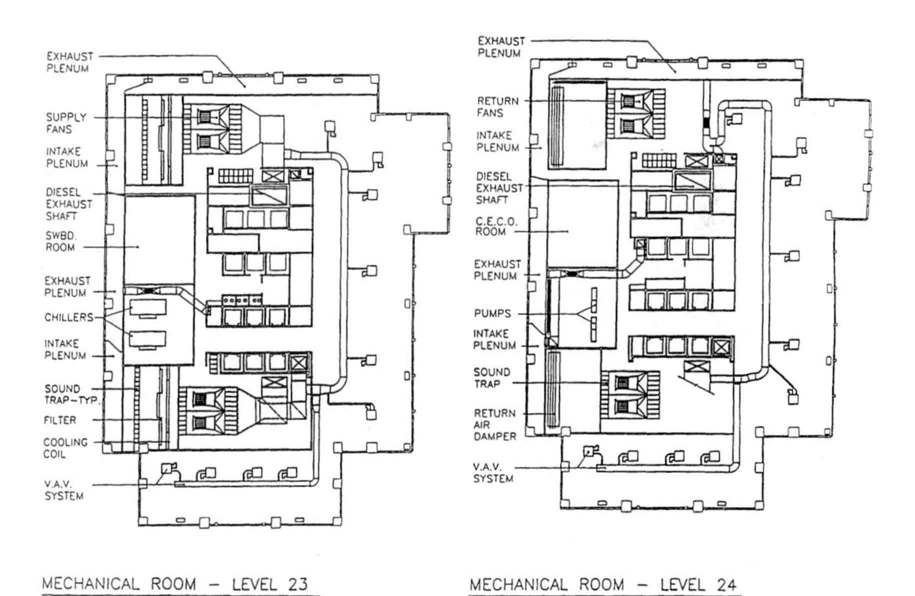


**Fig. 9.10 Morton International. Layout of equipment on the 23d and 24th floors. In reality, the 23d floor is the supply portion and the 24th floor the return portion of the huge air handlers. What appears to be two air handlers consisting of two supply fans and two return fans, each is actually working as two complementary parts of one. Both are supplying air into the same supply loop on typical floors.** (*Courtesy: Perkins & Will.*)

Generally speaking, there are three kinds of conditioning. The first is the replenishment of oxygen and the removal of carbon dioxide and any other pollutants through fresh air intake on the one hand, and exhaust on the other, as well as through active filtering of airborne particles. Another kind of conditioning is heating and cooling of the air. Since tall buildings have a much greater cooling than heating requirement, typically cooling alone is provided in the air handlers. As is discussed in Subsection 7, the heating of the air is accomplished in variable-air-volume (VAV) boxes when and where needed. The third kind of conditioning is humidification and dehumidification. Dehumidification is achieved in conjunction with the cooling process, and humidification is done by the spraying of steam.

Once conditioned, the air is blown into a carefully engineered supply air duct system by the supply fans. After conditioning various spaces, the used air is collected and drawn into the return air shaft by the return fans. After passing through the return fans, some of it is exhausted and the rest is fed into the air handler for recycling.

For tall buildings, large-capacity air handlers are typically field-assembled in a configuration and specification that is most suitable for the particular case. The fans could be either centrifugal or axial types. However, an axial fan offers greater versatility in configuring the air handlers, leading to greater constructibility. These are frequently used in pairs—one pair for the supply and the other for the return. Note the mirrored arrangement of these fans in each mechanical room shown in Fig. 9.10.

***Mechanical Room for Air Handlers.*** For operational efficiency and system reliability, as well as for construction economy, tall buildings are without exception divided into several HVAC service zones vertically, usually no more than 40 stories but no less than 20 stories each, so that each zone can be served autonomously by a set of air handlers (Fig. 9.11). As noted earlier, mechanical rooms require high vertical clearance not only for the air handlers themselves, but for huge air trunks, and are generally designed as double-story spaces. In tall buildings these are usually seen as slightly different bands compared to the rest of the building surface. This is due to the fresh air intake and exhaust louvers (Fig. 9.12). For more information on facades, refer to Chapter 7.

Often the decision on the location of mechanical rooms is strongly affected by nonmechanical considerations. This could include investment concerns, such as the number of floors between mechanical spaces that might be most attractive to the prospective tenant, programmatic concerns for the separation of occupancy, architectural concerns for the aesthetic implications, and structural concerns. Structural concerns include considerations for preferred location of a belt truss or outrigger for best performance with regard to lateral load. This also includes the consideration for vertical load transfer from the columns on one set of the grid to another, which often is necessary in tall buildings due to mixed occupancy types requiring different structural spans as well as the setback of the building profile. Mechanical space is a very convenient place to accommodate all of these. Structural systems are discussed in Chapters 2, 10, and 11, and in *Tall Building Systems and Concepts* (CTBUH, Group SC, 1980) and *Structured Systems for Tall Buildings* (CTBUH, Committee 3, 1994).

***Air Duct System.*** The air duct system consists largely of four parts—supply air trunk, supply air branches, return air plenum, and return air shaft. The conditioned air, after leaving the supply and fan, travels through the supply trunk and then is delivered into the supply branches. After servicing various spaces, the used air is returned to the air handler by the return air fans through the return air plenum, which in a tall building is usually the ceiling space, then travels upward or downward through the return air shaft, finally reaching the return air fan.

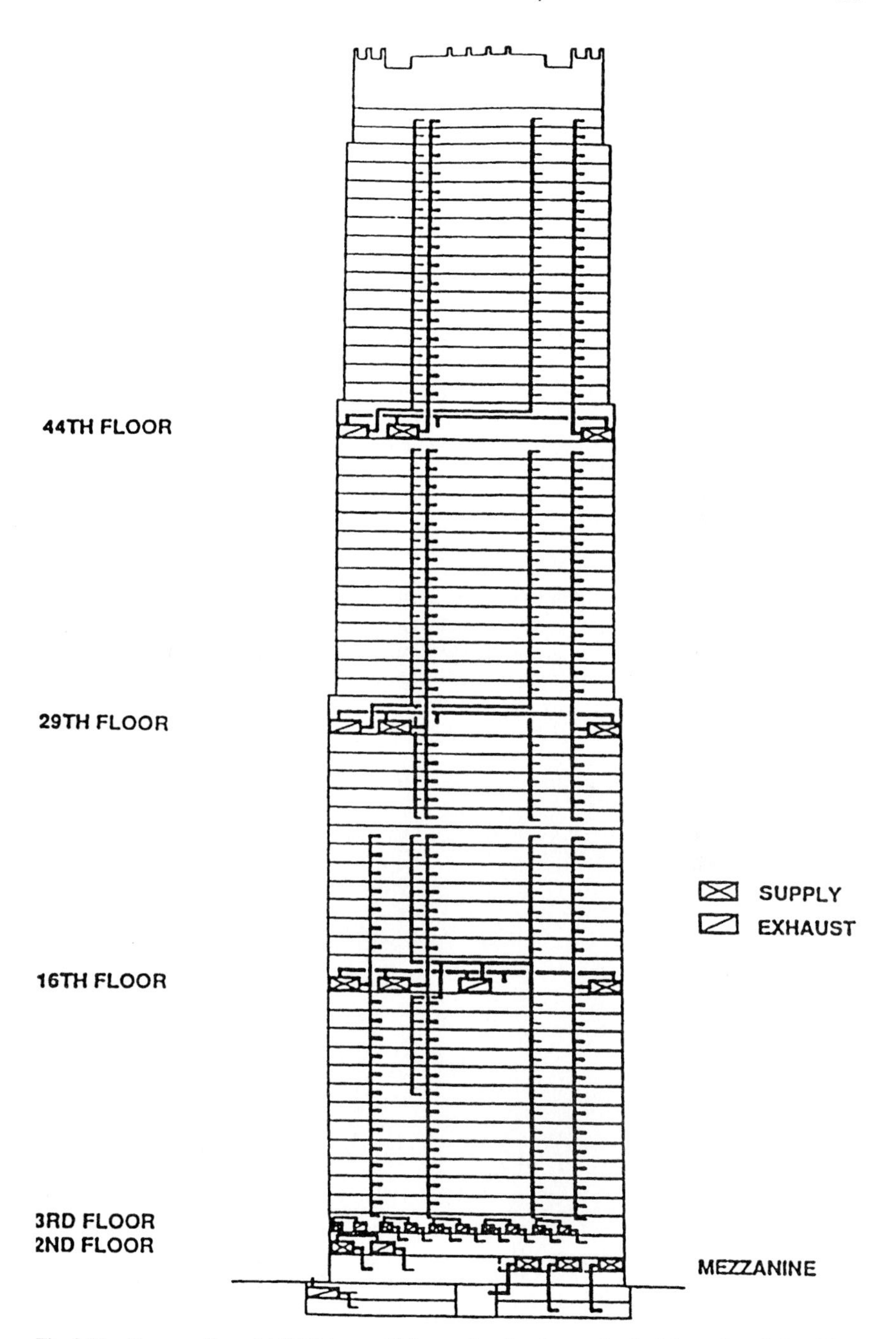


**Fig. 9.11** Cross section of AT&T tower, Chicago, showing the mechanical floors for air-handling units and the floors they are serving. (*Courtesy: Skidmore, Owings & Merrill.*)

The supply air system can be designed in a tree configuration (Fig. 9.6) or in a loop configuration (Fig. 9.13). The tree configuration has been used very widely. However, the downstream performance in this kind of system is heavily affected by the upstream activities, and therefore this calls for a high degree of engineering accuracy. For tall buildings this is not easy to achieve because it is often unclear at the time of engineering who the tenants will be and how they are going to use the space.

In a loop configuration the air in the loop is pressurized by the supply from usually two opposite ends of the loop. This creates relatively consistent pressure within the loop and increases the system performance considerably. For greater system reliability, this kind of system is often fed by two coordinated air handlers, as is the case with the AT&T and USG towers (Skidmore, Owings & Merrill, 1986, 1990).

The loop system gives the added benefit of being able to leave the final layout and engineering of the supply branches to the tenants. During the initial construction, only the loop and a few VAV boxes are installed. For the connection of future branches by the tenants, knockouts are provided on the face of the loop. This is beneficial not only to the tenant but also to the developer from capital outlay, construction efficiency, and management points of view.

## 7 Regulatory Subsystem

The air-conditioning load within a building varies, both spatially and temporally. The variation of particular interest is differential load variation over time across the space served by the same air handler. There are two generically different strategies for dealing with this kind of variation—by varying the temperature or by varying the amount of

**Fig. 9.12 World Trade Center, New York. Double-story mechanical spaces are usually nicely blended in the tower of tall buildings with a slight surface variation.** (*Courtesy: Hedrich-Blessing.*)

the air that is delivered to different spaces. From these two generic strategies came various systems alternatives—terminal reheat system, VAV system, double-duct system, dual-conduit system, and multizone system. For details refer to other literature, particularly ASHRAE handbooks (ASHRAE, 1992).

Most tall buildings nowadays employ a derivative of the VAV system, often referred to as *VAV systems with fan-powered box.* Traditional VAV systems respond to varying cooling demands by varying the amount of supply air by means of VAV dampers. However, problems occur when the difference in demand at two different areas under the same air handler is too drastic. In this case one area may become stagnant while the other area may become too cold. To alleviate this problem, new buildings are equipped with VAV boxes with a built-in fan. This can recirculate the room air locally through the box, as necessary to prevent stagnation of the air. By adding a heating coil in the box (typically an electric heating strip in a tall building), heating can be provided easily when necessary, thus eliminating the need for a separate perimeter heating system. VAV boxes serving the perimeter region of a building are typically of this kind.

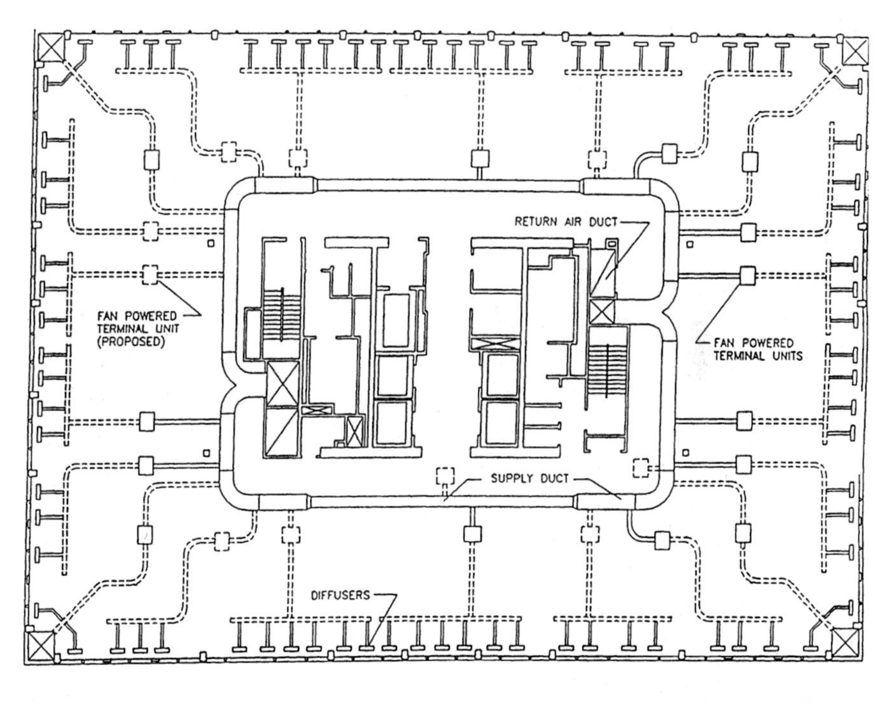

**Fig. 9.13 USG tower, Chicago. Typical loop-type air supply trunk is fed by two air-handling units for greater performance and reliability. Initial construction includes the loop and VAV boxes (squares) in solid lines. These limited VAV boxes tempered the air for construction. Those VAV boxes and supply branches in dotted lines are future provisions, which will be designed and constructed according to the specific needs and requirements of the tenants.** (*Courtesy: Skidmore, Owings & Merrill.*)

## 9.3 PREVENTIVE FIRE-SAFETY DESIGN

Another form of environmental control and technology is the integration of fire safety into the design of tall buildings. Sections 9.3 through 9.6, as an example, summarize the fire and life-safety reports prepared for the owners of three tall buildings. These sections focus on fire preventive design of high-rise buildings, with the emphasis on new technology, new materials, and new equipment. They cover fire protection systems (water storage and supply, fire pumps, fire department connections, sprinklers, standpipes, and extinguishers); smoke management systems (stair and vestibule pressurization, smoke containment, garage ventilation, and elevator shaft smoke relief); fire detection systems (manual fire alarms, smoke detectors, heat detectors, and water flow detection); fire alarm systems (fire alarm initiating devices, fire alarm alerting devices, emergency communications systems, emergency command centers, and emergency electric power systems); sequence of operation (manual operation, automatic alarm and transfer switch actuation, and smoke control operation); fire separation (fire-resistant materials and compartmentalization); and means of egress (escape routes and exits within a floor, final exits, means of escape for the disabled, and emergency escape lighting).

Three case studies are presented to illustrate fire preventive design of high-rise buildings internationally. These studies include buildings from Chicago, Illinois, United States; Seoul, Korea; and Barcelona, Spain. An additional case study involving a project in London, England, is presented to demonstrate the fire engineering concepts in structural design.

Some other topics on fire safety and protection, such as a unified approach to analyzing the fire resistance of structural systems, techniques for evaluating fire performance of structural materials, and critical aspects of human behavior in a fire, are included in *Fire Safety in Tall Buildings* (CTBUH, Committee 8A, 1992), where another international perspective is provided.

### 1 Objectives of Fire-Safety Design

The objectives of preventive fire-safety design include life safety, protecting valuable property, and ensuring the continuity of essential operations. These sections present important fire-safety features intended to provide reasonable protection of life and property when the building is fully operational. However, it should be recognized that fire protection is not an exact science. No fire suppression or detection system can eliminate entirely the possibility of fire, property damage, or loss of life.

The objectives of fire-safety design can be achieved through:

1. The use of automatic and manual fire extinguishing systems to contain and extinguish a fire in its early stages and in a reliable way
2. The provision of reliable mechanical smoke management systems to limit smoke spread
3. The proper layout and fire protection of escape routes and areas of refuge for all building occupants including the disabled
4. The use of construction and finish materials that provide the code-required and state-of-the-art fire resistance
5. The isolation by fire resistive separation and sprinkler density of higher-risk occupancies and areas

6. The use of automatic fire detection and alarm systems to detect emergency conditions, to alert the building occupants and the fire department, and to direct the evacuation of the occupants in an emergency

Tall buildings present special problems in planning for life safety from fire. The following three case studies describe fire preventive design of high-rise buildings. The issues of fire protection systems, smoke management systems, fire detection systems, fire alarm systems, sequence of operation, and more are briefly discussed for the three projects in generic terms. Reference is made to a building when a design component specifically relates to that particular building. The concepts presented here, being generic in nature, can also be applied to the fire preventive design of other tall buildings.

## 2 Project Descriptions

Fire protection methods for high-rise buildings are illustrated with three case studies. They include the AT&T corporate center in Chicago [60 stories, 265,000 m$^2$ (2,852,439 ft$^2$)]; the Lucky-Goldstar twin towers in Seoul [33 stories, 158,000 m$^2$ (1,700,699 ft$^2$)]; and the Hotel Artes-Barcelona in Barcelona [45 stories, 110,000 m$^2$ (1,184,031 ft$^2$)]. These buildings will be referred to herein as Chicago, Seoul, and Barcelona. In addition to these three buildings, reference will be made to the Broadgate Building in London, United Kingdom [11 stories, 55,000 m$^2$ (592,016 ft$^2$)] in Section 9.6.

***Chicago.*** The AT&T corporate center is located in the city's central business district bounded on the north and west by public streets and on the east and south by private interior property. The project consists of a 60-story composite concrete-and-steel-frame office tower and a 16-story steel-framed structure with exterior window wall of glass and granite. The project was completed in 1989.

***Seoul.*** The Lucky-Goldstar twin towers project consists of two 33-story towers united by a great outer court and two smaller inner courts forming one distinct project. A heliport is provided, as per local code, on top of the penthouse for emergency evacuation of the building's occupants. The project was completed in 1985. (This building is different from the Lucky-Goldstar International Business Center by Bechtel, covered in Chapter 8.)

***Barcelona.*** Hotel Artes-Barcelona consists of a 45-story tower containing approximately 500 hotel guest rooms in the lower 32 stories and approximately 30 duplex apartments in the upper stories. Multiple interconnected low-rise structures for commercial use are located on the remainder of the site. The tower was constructed using an exposed, painted steel frame 1.5 m (4.9 ft) to the exterior of a glass and metal mullion curtain-wall system. This project was completed in 1992 before the Summer Olympics.

## 3 Fire Protection Systems

***Water Supply Systems.*** Primary fire protection water service is provided from one (Seoul) or two (Chicago, Barcelona) points of connection to a potable city water source.

A water source backup system consisting of storage tanks is included in Barcelona. It consists of three storage tanks located at the base, midpoint, and top of the building to serve the low-rise buildings, and the low and high sprinkler zones of the tower. The tanks are sized to supply water flow for a period of 30 minutes with 140 m$^2$ (1507 ft$^2$) of sprinkler zone and to keep two hose reels active. Each tank has a capacity of 135 m$^3$ (4768 ft$^3$). A roof fire storage tank is detailed in Fig. 9.14.

***Fire Pumps.*** Electrically operated fire pumps supply water independently to each combined sprinkler and standpipe system. Jockey pumps are included with each system to maintain system pressure.

Fire and jockey pumps are connected to the emergency power supply system through automatic transfer switches. Starters are wye-delta closed-transition, reduced-voltage type. A typical fire pump room layout plan is provided in Fig. 9.15.

In addition to the electrically operated fire pumps, Barcelona is provided with standby diesel-powered fire pumps in parallel configurations.

***Fire Department Connections.*** Separate fire department connections are provided for each system. In Barcelona this includes the high and low wet pipe systems as well as the firefighter's dry pipe system. Fire department connections are provided on each building frontage, in accordance with the local authority having jurisdiction.

***Sprinkler System.*** The three buildings are fully sprinklered. The sprinkler systems are supplied from a fire protection system riser, with zoned control valves, check valves, tamper switches, and water flow switches. Dry pipe valves are provided in Barcelona. The systems are hydraulically designed in accordance with local codes for a fully sprinklered building.

A sprinkler loop is provided at each level with connections to both wet standpipes. The systems are arranged and calculated so that each connection will independently supply the full level or zone when required.

Tamper switches and water flow switches are alarmed remotely at the building's emergency command center, and are connected to the emergency standby power system. The sprinkler heads are standard 74°C (165°F) where maximum ceiling temperatures are 38°C (100°F), and 100°C (212°F) where maximum ceiling temperatures are

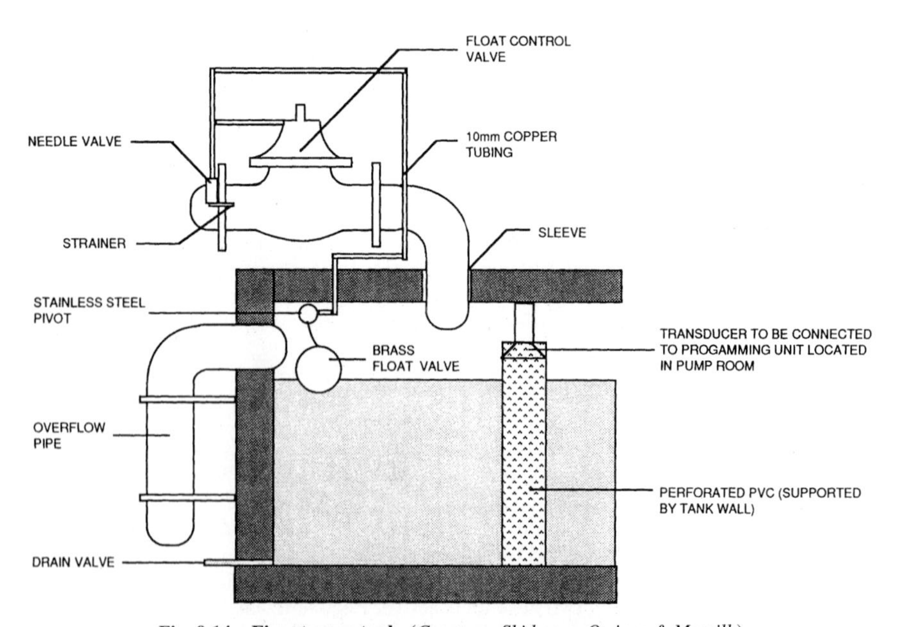


**Fig. 9.14 Fire storage tank.** (*Courtesy: Skidmore, Owings & Merrill.*)

66°C (150°F). Quick-response sprinkler heads are provided in all apartments and hotel guest rooms in Barcelona. Barcelona's building trash and linen chutes are provided with automatic sprinklers located at alternate floor levels with zone control valves, tamper switches, and water flow switches.

***Fire Standpipe System.*** A complete combined sprinkler and standpipe system is provided with fire department valves and other accessories located in each stairway on each level, in accordance with the referenced codes (Fig. 9.16).

All portions of the buildings are within 10-m (33-ft) hose spray range of a maximum 30-m (98-ft)-long hose located in the stairways and a 23-m (75-ft)-long hose located outside the stairways, as may be required by partitioning.

In Barcelona an additional 80-mm (3-in.) dry pipe riser is provided in each stairway with a firefighter's gate valve on each floor. Dry pipe risers extend from the ground-floor level to the roof.

***Fire Extinguishers.*** Wall-mounted 5-kg (11-lb) dry chemical fire extinguishers are provided at each stairway landing at the fire department valve and in fire department valve cabinets, the mechanical rooms, and the parking areas.

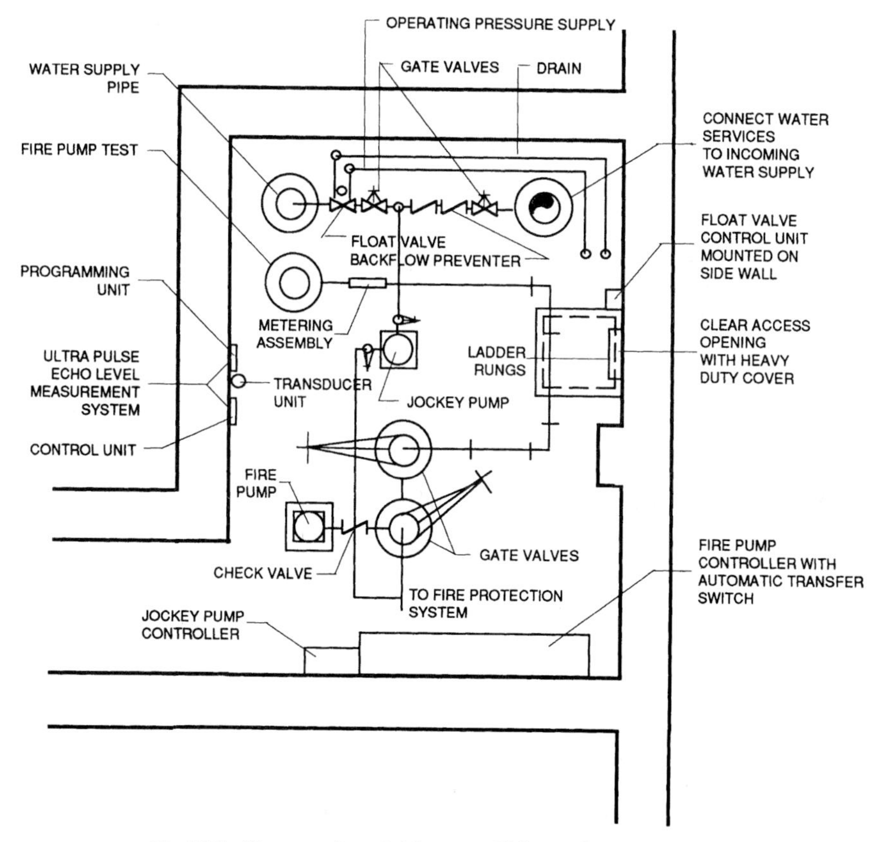


**Fig. 9.15 Fire room layout.** (*Courtesy: Skidmore, Owings & Merrill.*)

Wall-mounted 3-kg (6.6-lb) carbon dioxide fire extinguishers are provided in, or adjacent to, all transformer, switchgear, and other major electrical distribution equipment, and in all elevator machine rooms.

***Halon Fire Suppression Systems.*** Computer rooms are protected with Halon 1301 fire suppression systems. All systems are backed up by dry sprinkler systems using high-temperature heads to protect the structure.

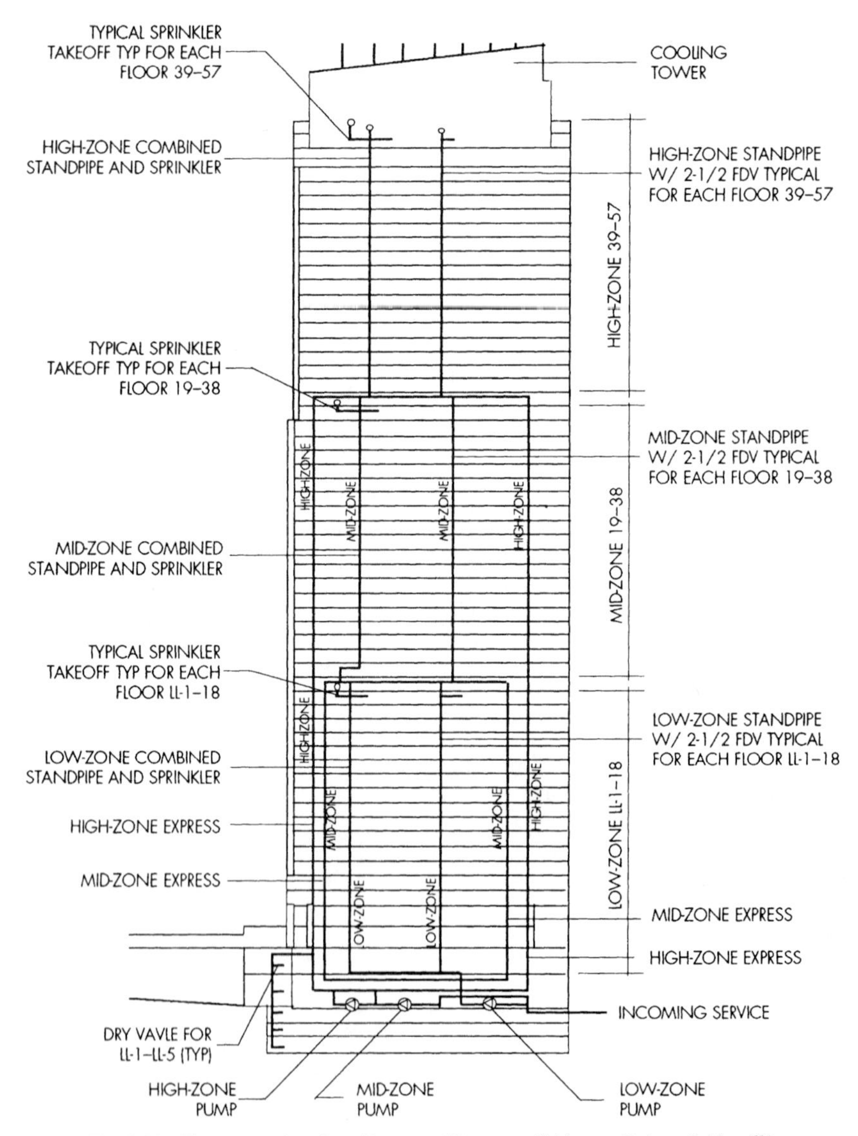


**Fig. 9.16 Fire protection riser diagram.** (*Courtesy: Skidmore, Owings & Merrill.*)

### 4 Smoke Management Systems

In a large-volume, multistory structure a mechanical smoke management system approach to smoke control is deemed superior to natural ventilation. The effective performance of smoke management systems is directly related to the parallel performance of the sprinkler systems serving the structure. Comprehensive smoke management can only be achieved in fully sprinklered buildings, where sprinkler response limits fire growth and smoke generation.

Smoke control for the three buildings studied is provided for every level of the office (Chicago, Seoul) and hotel and apartment (Barcelona) towers, retail areas, and lower-level parking. Under normal power supply conditions, all air-handling systems may be manually operated for smoke control via the BMS. In case of a fire or smoke condition during power failure, only those systems designated for smoke control for atria (Chicago) are connected to the emergency standby power system and will automatically operate to control smoke from the affected area.

***Stair and Vestibule Pressurization.*** Positive mechanical stairway and stairway vestibule pressurization minimizes the potential for smoke migration, thus facilitating occupant evacuation and fire department operations.

Dedicated, ducted fans (two fans are provided for each stair and vestibule, one is a standby) introduce 100% outside air to stairways at rates sufficient to maintain positive pressurization levels of 37 Pa (0.0054 psi) of water column within the stair enclosure with all doors closed. The fans are propeller exhaust type and are located at the top of the stairways. They are equipped with airtight motorized dampers and static pressure differential sensors.

All the emergency pressurization fans are connected to normal and emergency power and are activated by a signal from the emergency command center. If one fan fails to start, the standby fan is automatically activated. An airflow switch in the ductwork signals proof of flow to the emergency command center.

Supply air is introduced at every third story within each stairway. The top of each stair enclosure is equipped with a barometric damper to relieve overpressure by venting excess air to the atmosphere. Stairway vestibule ventilation is supplied at every floor with a minimum of 60 air changes per hour and exhausted at the rate necessary to maintain a pressure of 25 Pa (0.0036 psi) of water column relative to the stairway. Manual controls, which override the automatic operation of the pressurization systems, are provided for the fire department and building operations personnel at the emergency command center located at ground-floor level.

***Typical Office Towers (Chicago, Seoul).*** Typical levels are divided into horizontal one-level compartments. Dampers are provided to control air exhaust from each compartment under fire and smoke conditions.

Fire dampers are provided in air ducts penetrating the fire rated walls and vaults, as required. They are not required at typical floor shaft penetrations. Central supply and exhaust systems provided for typical floor-level air-conditioning, and their dampers are utilized under normal power conditions for smoke control in case of fire or smoke. VAV terminal units or fan-powered terminal units with normally open primary air motorized damper and normally closed induced return air damper are used in the air distribution systems.

Typical floor passenger elevator lobbies are pressurized by the following two systems. They are recommended, but not a code requirement. The systems are part of normal air-conditioning and ventilation.

1. Separate air supply outlets with motorized dampers from the typical floor-level supply system are provided to supply air to the elevator lobbies immediately adjacent to the fire levels, in case of fire or smoke condition.

2. Central supply systems, one for each elevator group (low-rise, midrise, and high-rise), with motorized dampers at the outlet of every level, are provided to supply air to the elevator lobbies of the fire floor in case of fire or smoke condition.

Typical floor service elevator (firefighter's elevator) vestibules are pressurized by a central supply system with motorized dampers at the outlet of each level. They supply air to the service elevator vestibule of the fire level in case of fire or smoke condition.

***Hotel and Apartment Areas (Barcelona).*** Pressurization of all corridors adjacent to apartments and hotel guest rooms, coupled with positive multilevel mechanical stair pressurization, creates and sustains pressure differentials which effectively limit smoke migration into the corridors and stairwells (Fig. 9.17).

Two-speed supply fans using 100% outside air for corridor make-up are utilized for smoke control. Air supply outlets introduce air to pressurize each corridor during a smoke or fire condition. Normal operation of fans is high speed during the day and low speed at night. Fans are automatically activated to high speed by a signal from the emergency command center. Manual controls which override the automatic operation of the fire floor smoke management system are provided for the firefighters and the building operations personnel at the emergency command center located at ground-floor level.

A corridor smoke evacuation system is provided at each level, consisting of exhaust fans, ductwork, exhaust grilles, motor-operated dampers (normally closed), wiring, and controls. Exhaust air quantity is 10 air changes per hour, or 30% greater than the corri dor make-up air, whichever is greater. The system has the capacity to exhaust three floors. Selection of the floor to be exhausted is initiated by corridor smoke detectors.

Manual controls operated by fire department and building operations personnel at the emergency command center on the ground-floor level are also provided. All fans for each system are connected to both normal and emergency power.

***Grade Levels.*** The grade level includes lobbies, exhibition areas, banks, concourse levels, mezzanines, commercial areas, cafeterias, restaurants, multipurpose rooms, health clubs, kitchens, employee facilities, and lower levels. Each area is served by separate supply and exhaust fans.

***Atria.*** Exhaust fans with exhaust inlets, located at the ceiling of the atrium, provide a minimum of six air changes per hour, or 19 $m^3$/sec, whichever is larger. When an atrium is over 17 m (56 ft) in height, atrium supply fans start to supply 100% outside air at the atrium floor level, with the capacity equivalent to 75% of the exhaust system. When an atrium is 17 m (56 ft) or less in height, supply air is provided through motorized window openings at the atrium floor level, or by supply fans with 100% outside air at the atrium floor level, with the capacity equivalent to 75% of the exhaust system.

***Closed Parking Garages.*** Parking garages are mechanically ventilated. This permits extraction of vehicle exhaust fumes as well as smoke generated by fires occurring within the sprinklered parking garage. Supply and exhaust fans are provided for each compartment of each level of closed (typically below ground) parking garages. Ventilation systems are designed to provide six air changes per hour. Supply fans will deliver 80% of the total exhaust to maintain negative pressure. Under normal operating conditions, the fans are controlled (on and off) by a carbon monoxide monitoring system. In emergency situations, a signal from the emergency command center overrides the normal operating system and operates the fans to extract the smoke from fires within the sprinklered parking garage.

Manual controls which override the normal operation of the parking garage ventilation system are also provided for the firefighters and the building operations personnel at the emergency command center located at ground-floor level. Carbon monoxide monitoring systems with a minimum of one sensor per 930 $m^2$ (10,011 $ft^2$), alarms, and controls are provided in the parking garages.

***Elevator Shafts.*** Ducted vent openings, 1.0 $m^2$ (9.5 $ft^2$) per elevator, with smoke dampers are provided to vent smoke to the atmosphere from each elevator shaft. The entire venting system is enclosed within a 2-hour fire rated enclosure. Dampers open upon

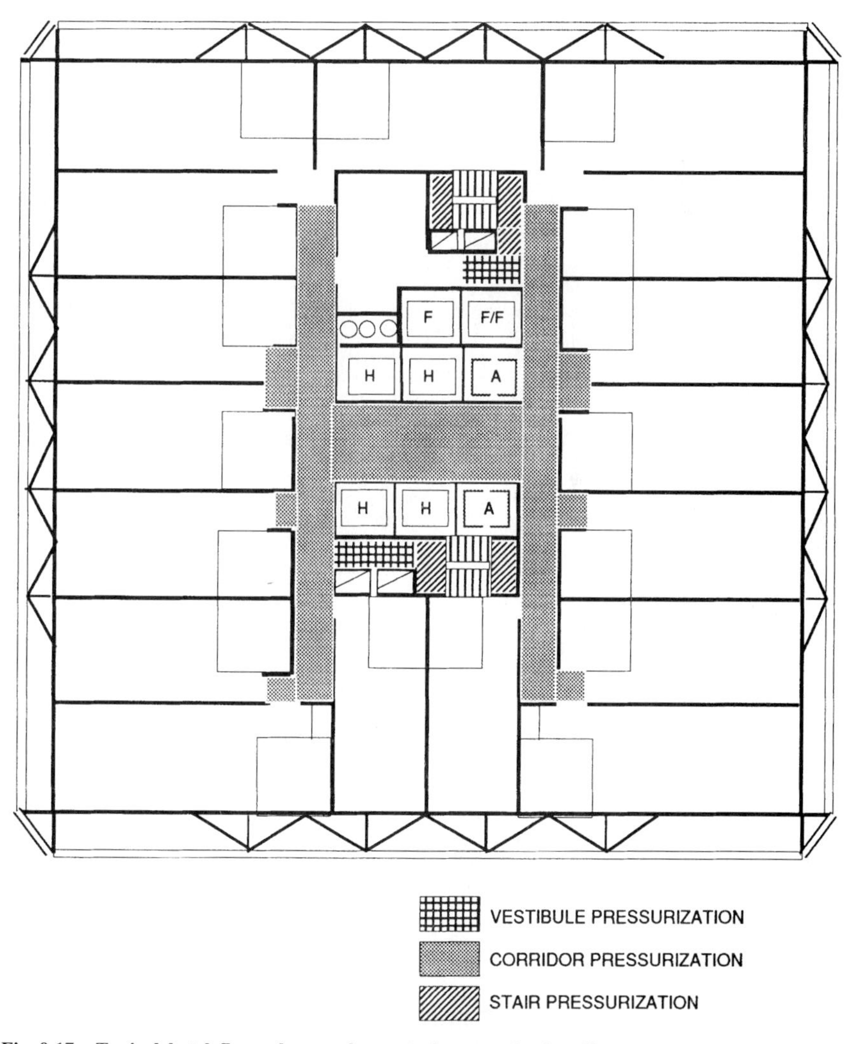


**Fig. 9.17 Typical hotel floor plan smoke control pressurization diagram.** (*Courtesy: Skidmore, Owings & Merrill.*)

a signal from a smoke detector located at the top of the shaft enclosure. Detectors are connected to both normal and emergency power.

## 5 Fire Detection Systems

Early fire detection and prompt alerting of the occupants of an emergency increase the available time to carry out orderly, staged building evacuation. Early detection coupled with prompt alarm transmission to the fire department via an alarm answering safety service, through decreased response time, aids the firefighters' ability to control and extinguish fires using the building's built-in fire protection systems before fire growth becomes life-threatening. A fire and life-safety matrix (Fig. 9.18) lists the placement of the fire detection devices throughout the building to provide early fire detection warning and alarm.

The fire detection, alarm, and communications systems are noncoded, zoned, and electrically supervised.

| | ELECTRICAL ROOMS | TRANSFORMER/ SWITCHGEAR ROOMS | MECHANICAL ROOMS | STAIR | ELEVATOR SHAFT | ELEVATOR CAB | ELEVATOR LOBBY | ELEVATOR MACHINE ROOMS | CORRIDORS | STORAGE ROOMS | TYPICAL OFFICE FLOORS | RETAIL | GARAGE | GENERATOR ROOM |
|---|---|---|---|---|---|---|---|---|---|---|---|---|---|---|
| DUCT DETECTOR | | | ● | | | | | | | | | | | |
| SMOKE DETECTOR | ● | ● | ● | ● | ● | | ● | ● | ● | ● | | ● | | ● |
| HEAT DETECTOR | | | | | | | | | | | | | | ● |
| FIRE FIGHTER PHONE | | | | | | ● | | ● | | | | | | |
| SPEAKER/HORN | | | | ● | | ● | ● | | ● | | ● | | ● | |
| SPRINKLER | | | | | | | | | | | | | | |
| TAMPER SWITCH | | | | ● | | | | | | | | | | |
| WATER FLOW SWITCH | | | | ● | | | | | | | | | | |
| WET STANDPIPE | | | | ● | | | | | | | | | | |
| EMERGENCY LIGHTING | ● | ● | ● | ● | | ● | ● | ● | ● | ● | ● | ● | ● | ● |
| EXIT SIGN | | | | ● | | | | | ● | | ● | ● | ● | |
| DOOR UNLOCK | | | | ● | | | | | | | | | | |
| SMOKE CONTROL | | | | | | | | | | | ● | | | |
| FIRE DEPT. VALVES | | | | ● | | | | | | | | | | |
| SPRINKLER CONTROL VALVES | | | | ● | | | | | | | | | ● | |

**Fig. 9.18 Fire and life-safety matrix.** (*Courtesy: Skidmore, Owings & Merrill.*)

***Heat Detectors.*** Heat detectors are provided in conjunction with special extinguishing systems used to protect specific hazards, such as computer rooms, the parking garage, and commercial cooking equipment. Power generation rooms are protected by both rate-of-rise heat detectors and automatic sprinklers. Detector coverage area in the parking garage is limited to a maximum of 30 $m^2$ (322.9 $ft^2$).

***Smoke Detectors.*** Area coverage addressable smoke detectors are located in all passenger elevator lobbies in conjunction with the elevator capture and recall system. Addressable smoke detectors are located at the top of all exit stairways and at the top of all elevator shafts.

Area-coverage smoke detectors of the addressable type provide immediate identification of the alarm location. Detector units are installed only in areas with ceiling heights of 10.5 m (34.5 ft) or less. Spacing between individual detectors must not exceed 7.5 m (24.5 ft). Area coverage must not exceed 70 $m^2$ (753.5 $ft^2$).

Area-coverage, system-connected addressable (maintenance-type) smoke detectors are installed in all major electrical transformer and switchgear rooms, elevator machine rooms, hotel rooms, apartments, and similar locations in conjunction with the special extinguishing systems protecting those spaces.

Area-coverage photoelectric smoke detectors are installed within the fan rooms to detect smoke in the exhaust air. The detectors are integrated with the controls for the smoke management system via the emergency command center.

Because of their large open layouts, atria are also equipped with projected linear-beam smoke detectors. The detectors are so positioned that the beams have an unobstructed projection range and provide complete coverage of the open areas.

***Manual Fire Alarm Systems (Pull Stations).*** Pull station locations must not exceed 1.5 m (5.0 ft) from exits and must be mounted not more than 1.25 m (4.0 ft) above the finished floor for exits from the building, exits to the roof, adjacent to exit stairways, and exits from parking garages. Maximum spacing for pull station locations in corridors is 61 m (22 ft). One zone per level for each office tower and multiple zones for lower-level structures and atria are provided. Zoning is limited to a maximum of 2000 $m^2$ (21,528 $ft^2$). Pull stations are positioned so that no person must travel more than 25 m (82 ft) in order to activate the system.

The following section describes the fire alarm, communications, and emergency systems for the three case study projects in general terms. The materials are applicable to other tall buildings as well.

## 9.4 FIRE ALARM, COMMUNICATIONS, AND EMERGENCY SYSTEMS

### 1 Fire Alarm Initiating and Alerting Devices

Operation of a manual pull station, a smoke detector, a heat detector, or a sprinkler water flow switch initiates direct alarm transmission to the central control center (Fig. 9.19) at ground-floor level and to a certified central station agency. Security and operations personnel are trained to relay all water flow alarms received immediately to the fire department.

Combined audible and visual fire-alerting devices are provided throughout all floors (see, for example, Figs. 9.20 and 9.21) and located such that alarm signals are clearly audible to occupants of every room or space. The speakers have a minimum sound level of 10 dB above ambient noise levels commonly found in the tenant spaces or 85 dB, whichever is greater.

Each floor of the building is zoned to such an extent that a reasonably quick determination of the precise location of the detector or pull station in alarm is possible. Fire alarm zones do not exceed 2000 $m^2$ (21,528 $ft^2$), except sprinkler water flow alarms, which are zoned on a per-floor basis.

Alarm speakers are not provided within staircase enclosures to minimize potential panic or confusion during an emergency.

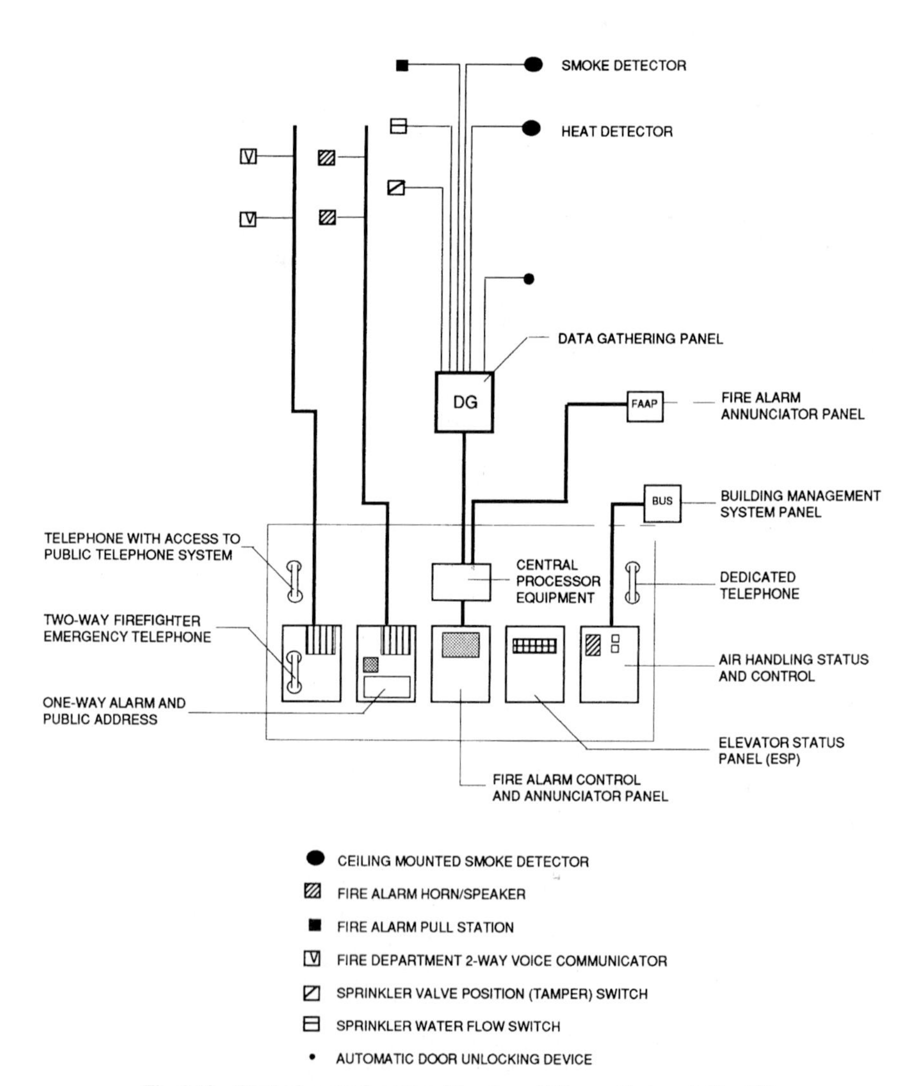


**Fig. 9.19 Central control station.** (*Courtesy: Skidmore, Owings & Merrill.*)

## 2 Fire Alarm Communications Systems

The fire department communications system consists of a two-way sound-powered telephone communications system between the emergency command center (fire command panel) and exit stairways on every fifth floor, at the exterior of any stairway exiting to the roof (located on the roof), in firefighters' service elevator cabs, fire pump rooms, the emergency generator room, elevator machine room, and major plant rooms. A fixed telephone handset is located in each designated area.

One-way public address speaker systems are provided from the emergency command center to the main lobby, elevator lobbies, elevators, protected stairway vestibules, corridors [spacing does not exceed 23 m (75.5 ft)], parking garages, all mechanical rooms, the emergency generator room, and service areas. Public address system controls enable the broadcast to be restricted to a single floor, to selected floors simultaneously, or to all floors. The emergency communications systems are electrically supervised and provided with secondary power from the emergency generator.

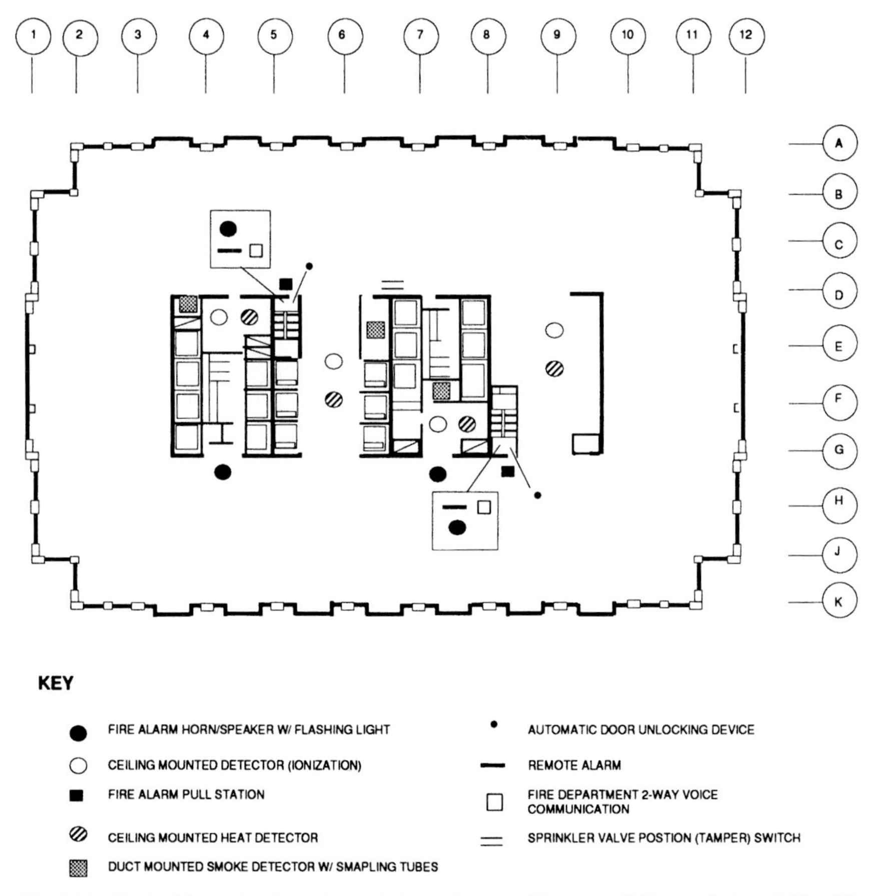


**Fig. 9.20 Typical floor plan detection and alarm diagram.** (*Courtesy: Skidmore, Owings & Merrill.*)

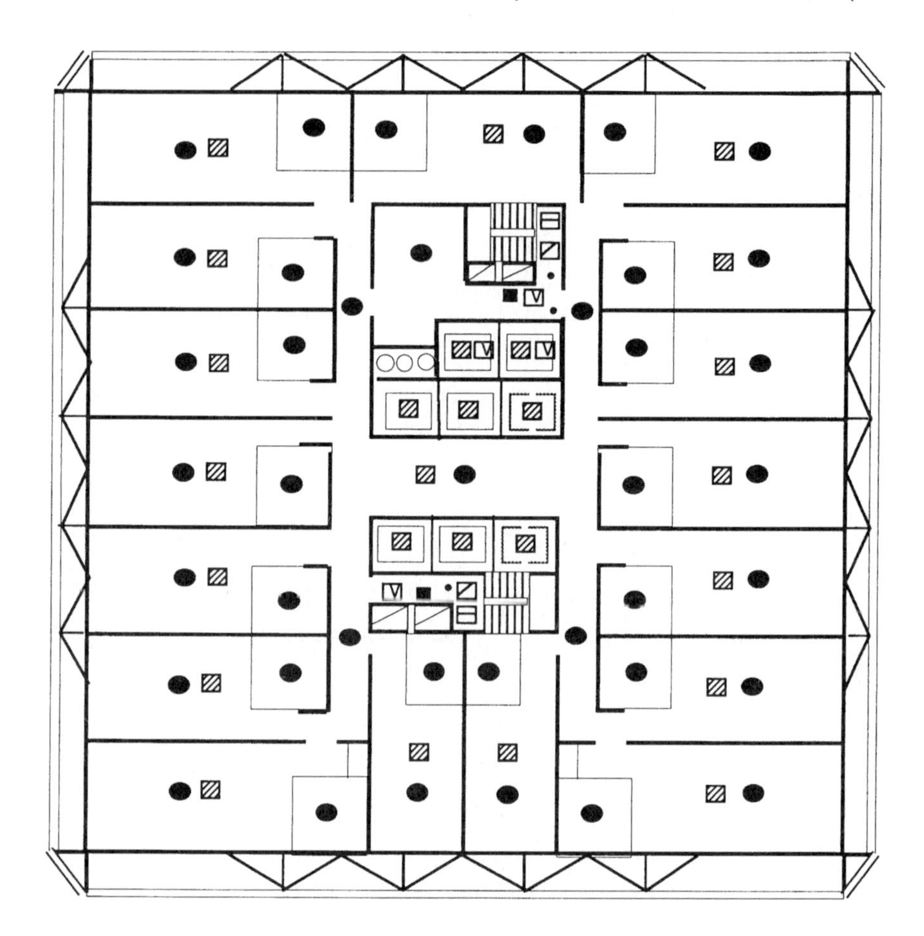



**Fig. 9.21 Typical hotel floor plan detection and alarm diagram.** (*Courtesy: Skidmore, Owings & Merrill.*)

### 3 Emergency Command Center

At the ground-floor level of the building complex, in proximity to the main entrance lobby, a dedicated space serves as the emergency command center for the fire department as well as the building's security and operations personnel. The firefighters' access to the command center as well as the firefighters' elevator is through the sprinkler-protected main lobby area.

The emergency command center contains the fire alarm indicating equipment, pump operating status indicators, manual controls for the smoke management systems, transmitters and receivers for communications systems, emergency generator status indicators, and elevator controls. The emergency command center is provided with emergency power and emergency lighting supplied from the emergency generator.

***Electric Standby Power System.*** A diesel-driven emergency generator with a fuel storage capacity for a minimum 24 hours of operation is provided and sized to operate lighting and ventilation at the emergency command center, escape lighting systems, exit signage, the fire detection alarm and emergency communications system, the stair pressurization fans, the smoke extract and floor pressurization fans, and the electric motor driven fire pumps.

Power from the emergency generator also serves the passenger elevators (one per bank at a time), the freight elevator, the fire-fighting elevators, lighting in the power generation rooms and other rooms or spaces housing control equipment for mechanical smoke extract systems. The transition from normal utility power to emergency generator power is designed to occur automatically upon loss of utility power.

The transfer switches are located in a 2-hour fire rated enclosure fully separated from both the main electrical switchgear and the power generation rooms. Mineral insulated (M.I.) power cable is used for all electric feeders and branch circuits serving the emergency power systems. The emergency risers are located within a separate room away from the normal power risers. The risers are enclosed in 2-hour fire rated walls. In addition, emergency exit signs and emergency exit lighting are provided with local battery backup.

## 9.5 MEANS OF EGRESS FOR FIRE SAFETY

Providing emergency egress from high-rise buildings is particularly difficult given their unusual heights. Elevators are generally unsafe in the event of a fire and will automatically shut down once they reach the ground floor of a high-rise building. Exit stairs can be used by the able-bodied, but in the case of the handicapped or the elderly, exit stairs can be difficult or impossible to use. Technology exists which can assist physically impaired individuals to safely exit buildings in the case of emergency. However, this technology is expensive and should be justified on the basis of its cost-benefit analysis. In the case of safety, benefits are based on both perceptions and demonstration of reduced risk. The architect ultimately has to deal with the monetary cost of reducing risk. Tall buildings typically should have places of refuge—floors or areas that may be sealed off and environmentally contained in the event of a fire. These places of refuge may provide safe haven for building occupants during fires or emergencies until rescue workers can arrive on the scene.

The following description applies to the high-rise hotel and apartment building in Barcelona, Spain (see Section 9.3).

### 1 Escape Routes within a Floor

***Hotel Floors.*** Travel distances within a guest room, which is separated from adjacent guest rooms by a 1-hour fire rated partition, are less than 11 m (36 ft). Exit access travel within the 1-hour fire rated corridor is approximately 18 m (59 ft) from the most remote room (Fig. 9.22). This is less than the 30 m (98.5 ft) allowed by the Barcelona code.

The small floor plate of the tower [approximately 1100 $m^2$ (11,840 $ft^2$)] and the limited number of occupants based upon the residential and hotel occupancy indicate that the two required exit stairs provide better than adequate egress capabilities and travel distances for the tower portion of the project. The hotel and apartment occupancies dictate a relatively even population distribution over all the floors of the tower. The even

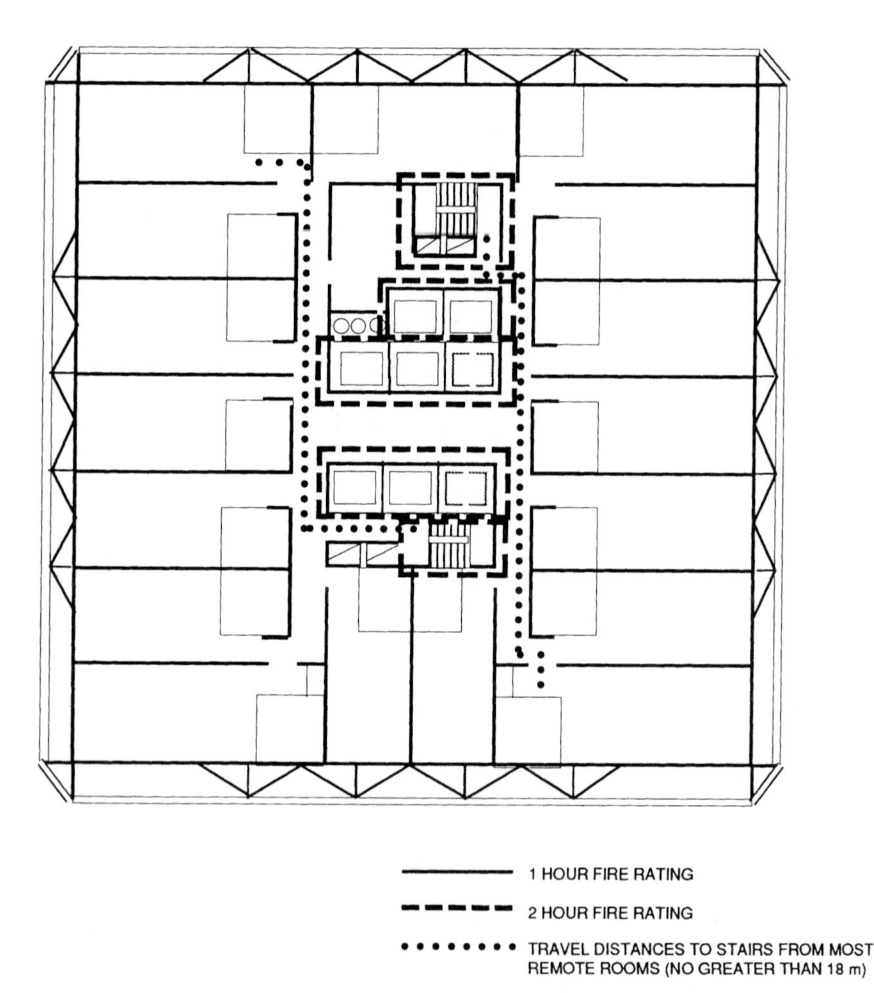


**Fig. 9.22 Typical hotel plan fire separation and exiting diagram.** (*Courtesy: Skidmore, Owings & Merrill.*)

distribution, combined with the low population per floor, will allow for a smooth flow of occupants into the egress stair system.

***Apartment Floors.*** Travel distances within any apartment, which is separated from adjacent apartments by a 1-hour fire rated partition, are less than 25 m (82 ft). Exit access travel within the 1-hour fire rated corridor is approximately 15 m (49 ft) from the most remote apartment. This is less than the 30 m (98 ft) allowed by the Barcelona code. All dead-end corridors in the tower are less than 15 m (49 ft) from an exit, as allowed by the Barcelona code. For the low-rise portions of the project, exit travel distances are limited to 30 m (98.5 ft) and dead-end corridors to 15 m (49 ft).

***Exits from a Floor.*** Exits from a story within the tower are within two internal stairways which are accessed through a vestibule. Both the stairways and the associated vestibules are enclosed by 2-hour fire rated construction (Fig. 9.23), and both are pressurized relative to the adjacent corridors and tenant space to prevent the passage of smoke into the stairways. Stairway shafts extend to the roof level and provide access to the roof for fire-fighting purposes. Each stair is sized to exit the entire population of the floor served.

All locking exit doors are equipped with a suitable fail-safe device which is automatically controlled via the emergency command center to provide free access to and from the stairways in emergency situations. All exit stairs from the low-rise portions of the project are enclosed in 2-hour fire rated construction.

***Final Exits and Discharge.*** One tower exit stair discharges directly to the exterior at ground-floor level. The second stair discharges through the sprinklered fire-resistant lobby at ground-floor level (Fig. 9.24). The stair doors are equal in width to the stairs that discharge through them.

***Means of Escape for the Disabled.*** Disabled persons who are unable to use the exit stairways would seek refuge in the 2-hour fire rated vestibule adjacent to the firefighters' elevator. They would then be assisted to the ground-floor level lobby via the firefighters' elevator.

Areas of refuge are provided in the 2-hour fire rated stair enclosures for the low-rise portions of the project. See *Building Design for Handicapped and Aged Persons* (CTBUH, Committee 56, 1992) for additional information on means of egress for the disabled.

## 2 Emergency Escape Lighting

An emergency lighting system is provided to define and illuminate the escape routes and to identify the location of fire alarm call points and fire-fighting equipment along such routes.

***Exit Signs.*** Exit signs are provided immediately below ceiling level, positioned to ensure that escape routes are clearly identified in an emergency. All exit, emergency exit, and escape route signs are illuminated so that they are legible at all times. Illumination may be accomplished by lamps external to the sign, lamps contained within the sign, a combination of internal and external lamps, or self-luminous signs. Escape routes are illuminated to achieve a horizontal illuminance on their centerline of not less than 0.2 lux (2 fc).

***Emergency Lighting.*** The emergency lighting system is designed to provide uniform illumination and is arranged to ensure that a failure in any one luminaire does not reduce the effectiveness of the system.

Emergency lighting luminaires are also provided in all passenger and firefighters' elevator cars (preferably battery powered), in toilet accommodations exceeding 8 $m^2$ (86.1 $ft^2$), and in all control and plant rooms.

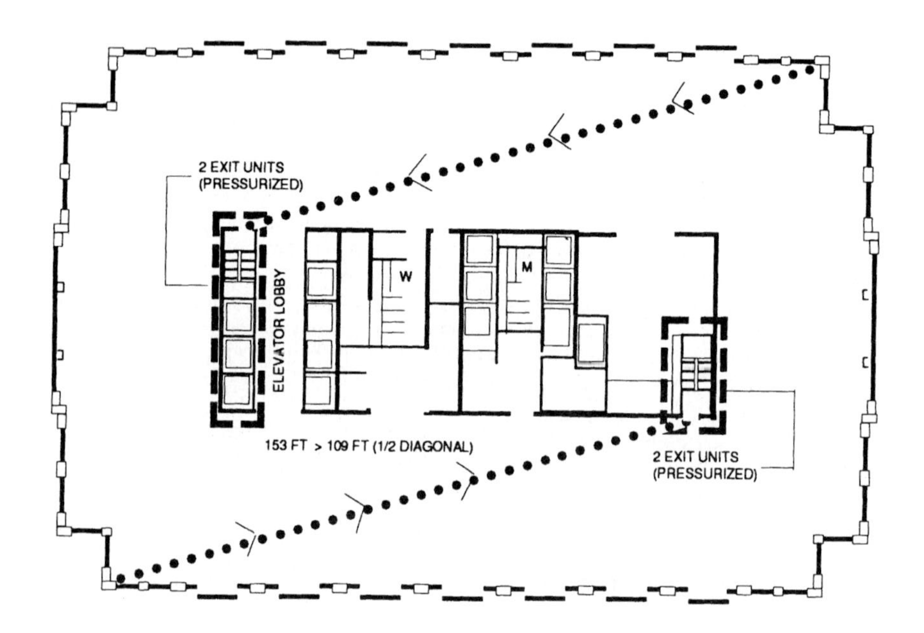


**KEY**

• • • • • • • • PATH OF EGRESS

▬ ▬ ▬ ▬ ▬ 2 HOUR RATED ENCLOSURE

**TYPICAL FLOOR LEVEL**

GROSS AREA: 28,975 $ft^2$

| OCCUPANT LOAD | GROSS AREA | $ft^2$ / PERSON | OCCUPANTS |
|---|---|---|---|
| OFFICE | 28,975 $ft^2$ | 100 | 290 |

EXIT UNITS REQUIRED: 3.87

EXIT UNITS PROVIDED: 4

**Fig. 9.23 Typical tower fire separation and exiting diagram.** (*Courtesy: Skidmore, Owings & Merrill.*)

The emergency lighting systems will reach their design level of illuminance within 5 seconds of the interruption of the normal power supply. The system is powered by the diesel-driven emergency generator. If the generator takes longer than 5 seconds to achieve the required output, a backup battery system is provided to supply the emergency lighting load until transfer is automatically accomplished.

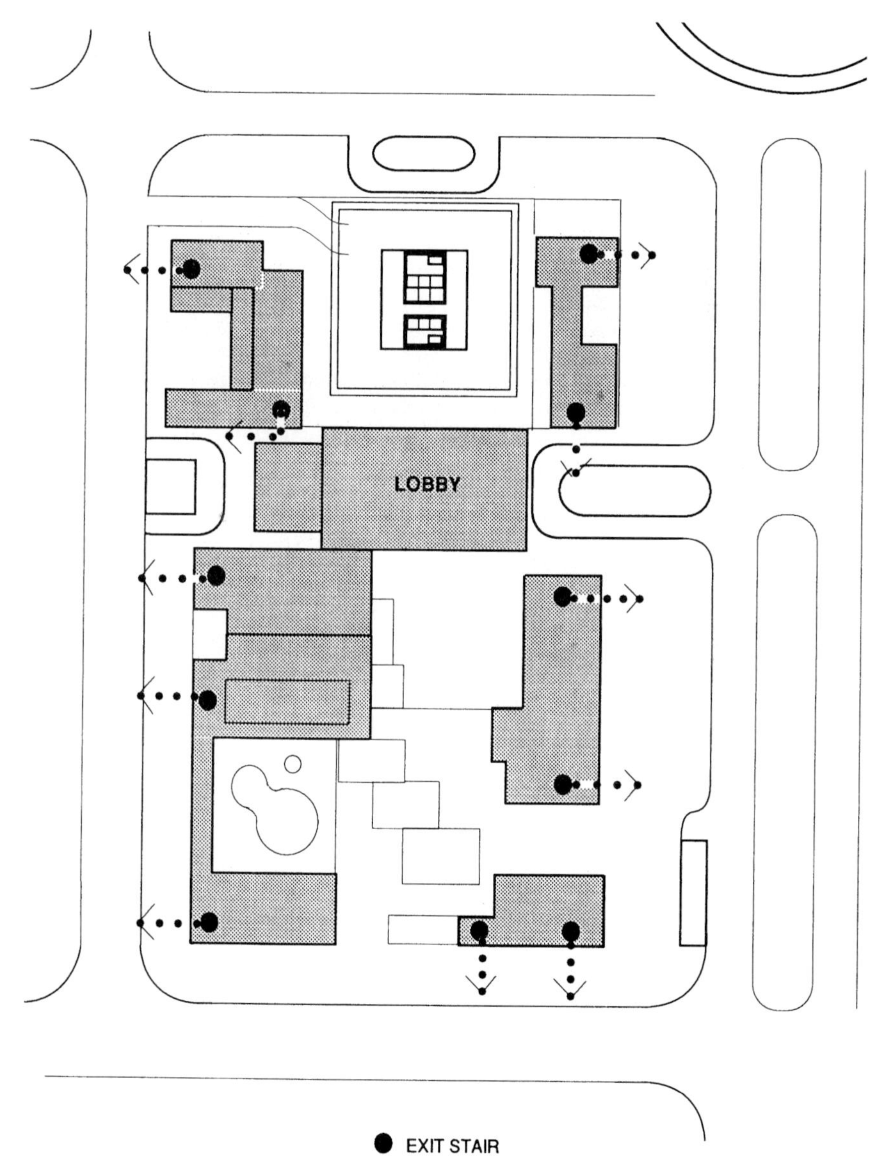


**Fig. 9.24 Ground floor plan exiting diagram.** (*Courtesy: Skidmore, Owings & Merrill.*)

## 9.6 STRUCTURAL DESIGN FOR FIRE PROTECTION

Fire engineering and protection methods for steel, concrete, and interior structures and finishes within the building must be specified by the architect and taken into consideration as part of the structural design of the building. Heights, areas, and cubic capacities of buildings play critical roles in the application of various regulatory requirements. In the case of an unusual building structure, care must be taken to ensure adequate fire protection of all the structural components of the building. The nature and extent of fire protection are also dictated for the most part by regulatory fire rating requirements.

### 1 Fire Resistance and Compartmentalization

The structural resistance to fire necessary to effectively resist partial or total collapse stemming from fire-induced stresses is part of the structural design. Horizontal fire separation in the form of fire rating is provided here for structural elements.

Electric and emergency power and communications closets on all levels are enclosed within 2-hour rated wall assemblies. All penetrations through the floor and ceiling assemblies are fire-stopped to maintain the required floor and ceiling assembly rating.

All three buildings discussed in the preceding sections are effectively subdivided against floor-to-floor fire extension and horizontal fire spread by means of the fire rated floor and ceiling assembly and the fire rated room and apartment partitions, as noted in the following.

Each floor is constructed as a compartment. Rated floor and ceiling assemblies extend from exterior wall to exterior wall, penetrated by rated vertical shafts only.

***Fire Separation (Fire Resistive Ratings)***

| | |
|---|---|
| Exterior bearing walls | 4 hours |
| Exterior nonbearing walls | Noncombustible |
| Exterior doors and windows | Noncombustible |
| Interior bearing walls | 4 hours |
| Interior nonbearing walls | Varies (see following) |
| Permanent partition, corridor partitions, and ceilings | Noncombustible |
| Shaft enclosures: | |
| stair | 2 hours |
| elevator shafts | 2 hours |
| pipe and duct shafts | 1 hour |
| Electrical equipment room(s), emergency standby power generation equipment room, electrical closets, elevator equipment rooms, fire pump and control rooms | 2 hours |
| Utility transformer vault | 3 hours |

| | |
|---|---|
| Driveways and loading spaces: | |
| Walls | Noncombustible |
| Floors above and below | 3 hours |
| Floor construction | 1 hour automatic (sprinkler system assumed) |
| Roof construction | 2 hours 0–4.3 m (0–14 ft) above floor<br>1 hour 4.3–6.1 m (14–20 ft) above floor<br>Noncombustible above 6.1 m (20 ft) |
| Structural frame: | |
| Beams, girders: | |
| Supporting roof only | 2 hours 0–4.3 m (0–14 ft) above floor<br>1 hour 4.3–6.1 m (14–20 ft) above floor<br>Noncombustible above 6.1 m (20 ft) |
| Supporting other floors | 3 hours |
| Exterior columns | 4 hours |
| Interior columns: | |
| Supporting roof only | 3 hours |
| Other floors | 4 hours |

## *2 Fire Engineering Concepts in Structural Design: A Case Study*

An engineered approach was used to evaluate the fire resistance and protection for the exposed steel elements of the Broadgate Building in London. All elements interior to the window wall were conventionally fire protected according to British standards, as for any building. In addition, the interior spaces were protected by an inspected sprinklered system with an emergency power hookup so as to improve reliability to a 99% level. With this degree of protection on the interior, the fire protection needs of the exterior elements in open air on a rational basis from fire loads in the building were reviewed.

The design for fire encompassed four stages. The first stage was to determine the nature of a fire that would occur in this particular building, taking into account the size and proportions of the floor plan, the amount of combustible material in the building, the total window area, and the air supply to the fire. The second stage in the design was to correlate the specific parameters of this building to established standards and full-scale tests already performed on a number of buildings in the United States and the United Kingdom. Through these correlations, the temperature and flame profile of an actual fire was determined. The third stage was to calculate the heat transfer into the steel structural elements, and determine the maximum steel temperature for the given fire. The calculations take into account the proximity of the steel to the fire, the shape of the elements, and the flame shields around the elements, if any, and are based upon the classical heat transfer theory. The final stage, after maximum steel temperatures were determined, was to perform a structural analysis of the entire three-dimensional structural system at these temperatures to determine the forces and building deformations caused by both thermal expansion and changing elastic properties. The purpose of this analysis was to ensure that steel temperatures were well below the critical level, that each individual element was adequate, and that the entire structural system was sound and stable and would support the building loads without imminent collapse.

Initial calculations were completed without any protection against fire. Here steel temperatures of 675°C (1247°F) and 704°C (1300°F) for arch and hanger/column members were found to be unacceptable.

One alternative was to develop an engineered flame shield to limit steel temperatures while preserving the shape characteristics. The flame shields are generally sheet steel attached to main members, but separated from them either by a small air gap or by a noncombustible insulation made of ceramic fibers. While the solution with flame shields was technically feasible, the resulting aesthetics when viewed from the interior were less desirable. Further, there were several drawbacks related to durability, corrosion protection, and long-term maintenance of the flame shield.

In lieu of flame shields, a solution involving a fire-resistant glass window wall was used. This is equivalent to having a fire rated barrier between the fire load and the exposed steel. The window wall also had the property of filtering the radiant heat from the fire source to a 50% value. The engineering verification then amounted to determining the steel temperatures for radiant heat in general and also for cases of isolated window breaks during a fire. A structural analysis of the entire system was performed for these two conditions to verify the structural integrity.

The temperatures of the steel with fire-resistant glass but without the aid of flame shields were considerably smaller and manageable, and led to solutions of a genuine exposed steel expression.

Two types of heat-resistant glass are available. In one, a layer of intumescent material is laminated between two layers of regular sheet glass. In the other, the chemistry of the glass is altered to toughen it against heat resistance, similar to Pyrex. The entire window wall system was fire-rated with tests according to British standards for the selected glass system.

Thermal stresses were calculated using maximum temperatures from the design fire on three floors of the building simultaneously. Stresses were also calculated using temperatures from the confined fire in various critical locations assuming localized window breaks. A reserve capacity of at least 25% was provided against the stresses induced by these temperatures generally, whereas the reserve capacity against the confined localized fire was 5%.

## 9.7 BIOMASS ENERGY SOURCE FOR AN URBAN MIXED-USE TOWER

Section 9.1 focused on the thermal comfort zone and the energy conserved through the alteration of this zone. Section 9.7 also addresses the issue of energy conservation by presenting an alternative source of energy for high-rise buildings. A completed research project is narrated in this section to illustrate the concepts involved.

### 1 Project Concept

As our understanding of the impact of fossil fuel usage on the global environment increases, it has become clear that the energy contribution from this source must be decreased. Alternative sources such as solar, wind, and biomass energy will undoubtedly become more attractive. This project proposes to use a locally available biomass fuel—waste paper—and a small-scale gas turbine power plant to provide electricity and steam for an urban high-rise building. The building is a 73-story mixed-use tower located on the near north side of Chicago.

When one considers the urban high-rise environment, there seems to be little opportunity to utilize solar and wind energy, for the density of buildings will block access to the sun and create turbulence in wind flow. However, substantial amounts of biomass are ubiquitous in the American city in the form of paper, most of which is used only briefly and then discarded. The vast quantities of waste paper generated by urban life include computer printouts, liner board and other packaging, photocopies, letterstock, catalogs, periodicals, newspapers, junk mail, tissues, and paper towels to mention only a few examples. Relatively little of this—about 20%—ever finds its way back to recycled paper. The rest is disposed of primarily in landfills, where it decomposes. It is feasible to collect and utilize this natural urban concentration of waste paper as an energy source. Moreover, using waste paper as a fuel does not contribute to global warming by greenhouse gases, because the trees grown to provide new paper absorb the $CO_2$ created by the paper burned as fuel.

This waste paper biomass resource has been available for some time, but there has been no efficient way to convert it to energy near the point of disposal. However, in the last 20 years, steam turbine systems using biomass fuels to cogenerate electricity and steam have been introduced in the sugar, paper-pulp, and forest products industries. A smaller and more sophisticated gas turbine system is used for this study. Since these cogeneration systems can produce substantial amounts of steam in addition to electricity, the choice of a mixed-use development offers more options for the use of steam, such as absorption chillers for summer cooling and winter heating of residential areas.

## 2 Building Description

As stated earlier, the project used in this study is a 73-story mixed-use building occupying an entire block on North Michigan Avenue in Chicago (Kao, 1991). It is similar in scope to a number of other mixed-use buildings built in this area in recent years. The building contains commercial retail space, office space, a hotel, and condominium apartments within a total gross area of 172,143 m$^2$ (1,853,000 ft$^2$). Below grade there are three levels of parking for 300 cars; the first level below grade contains a truck dock and mechanical space. The retail space begins at the Michigan Avenue level and includes 27,870 m$^2$ (300,000 ft$^2$) on floors 1 through 5. Separate entrance lobbies are provided at street level for the office, hotel, and apartment areas. The hotel lobby, restaurants, and function rooms are located on floor 6. The 18,580 m$^2$ (200,000 ft$^2$) of offices are placed on floors 7 through 14, which are reduced in size as the tower rises above the commercial base that occupies the entire site. After floor 15, which contains mechanical equipment, there is another reduction in floor size to accommodate the hotel guest rooms, which are placed on floors 16 through 29. Floors 30 and 31 contain the apartment sky lobby and various amenities, including shops and a health club. At floor 32 there is a final reduction in floor size with the beginning of the condominium apartments, which continue through floor 72. The building is terminated at floor 73 with mechanical equipment spaces.

The structure is enclosed with a curtain wall of clear glass and thin marble slabs of two colors set in white-painted aluminum frames. Some architectural illustrations for the building are shown in Figs. 9.25 through 9.30.

## 3 Energy Demands and Resources

***Peak Demand.*** Estimates of electric energy demands of the various building areas were made and are shown in Table 9.3; peak demands range from 54 W/m$^2$ (5.00

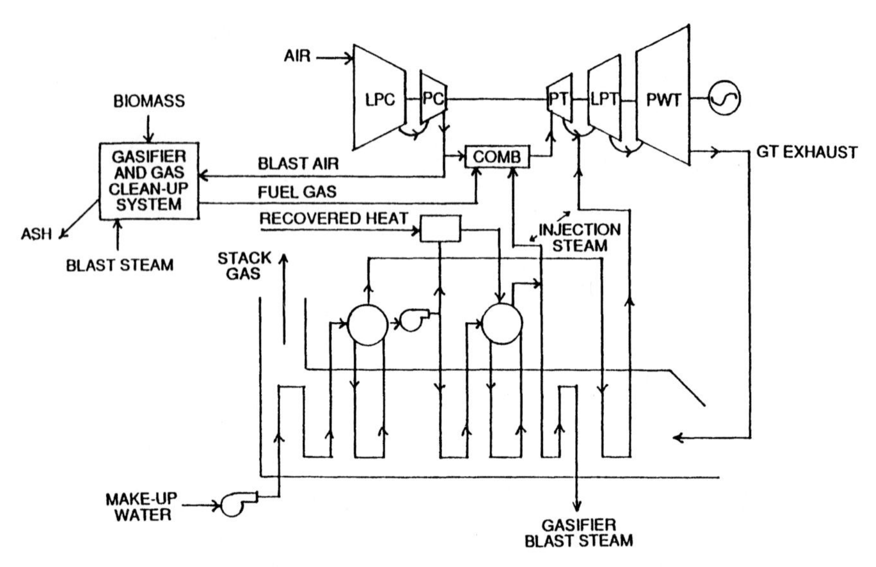


**Fig. 9.25 BIG/STIG—Biomass integrated gasifier/steam-injected gas turbine.** (*Courtesy: Illinois Institute of Technology, Department of Architecture.*)

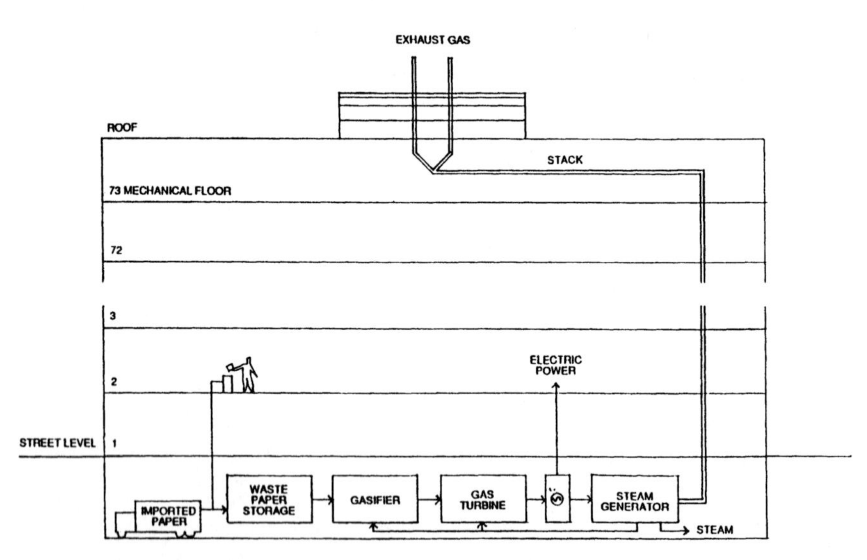


**Fig. 9.26 Diagram of building energy generation system.** (*Courtesy: Illinois Institute of Technology, Department of Architecture.*)

W/ft$^2$)for offices and commercial areas to 2.70 W/m$^2$ (0.25 W/ft$^2$) for parking areas. Office and retail demands are high during working hours on weekdays, but fall dramatically during evenings and weekends. Hotel and residential demands are lower and more constant, due to their larger proportions of heating and cooling. The total peak demand for the entire building is about 4.6 MW. Since the peak demand occurs only at infrequent intervals, it was decided to provide only 56% of the peak—2.7 MW—for 8 hours per day, 5 days per week. However, the smallest gas turbine that is available has an electric generating capacity of 5.4 MW. Therefore a small district generating system was to

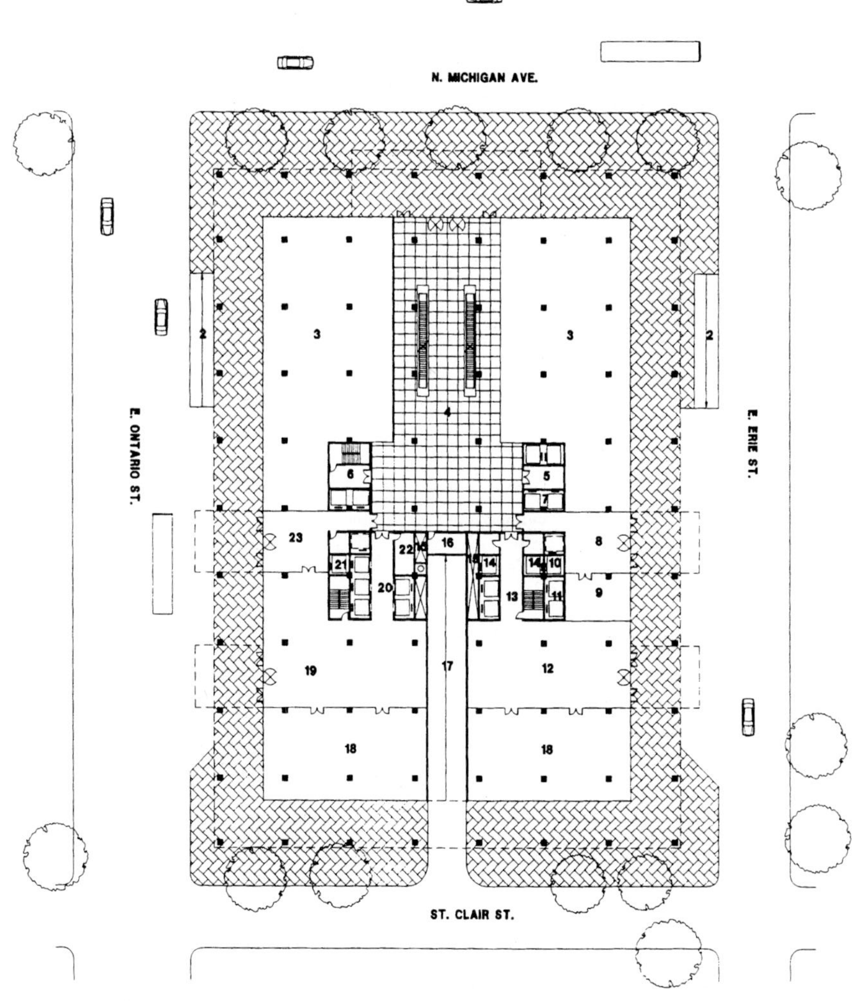


**Fig. 9.27 Multiuse high-rise building—ground floor plan.** (*Courtesy: Illinois Institute of Technology, Department of Architecture.*)

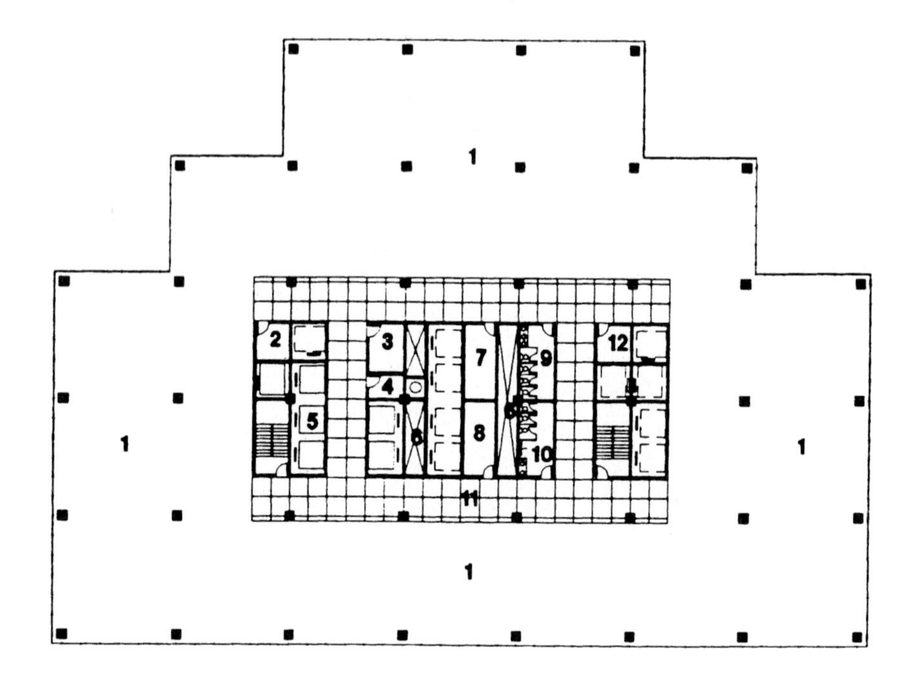


**TYPICAL OFFICE LEVEL**

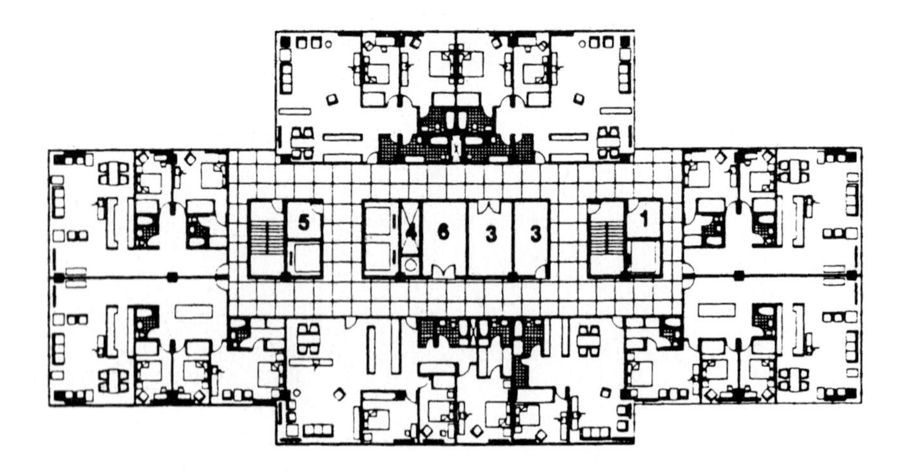


**APARTMENT LEVEL 52-72**

**Fig. 9.28 Multiuse high-rise building—floor plans.** (*Courtesy: Illinois Institute of Technology, Department of Architecture.*)

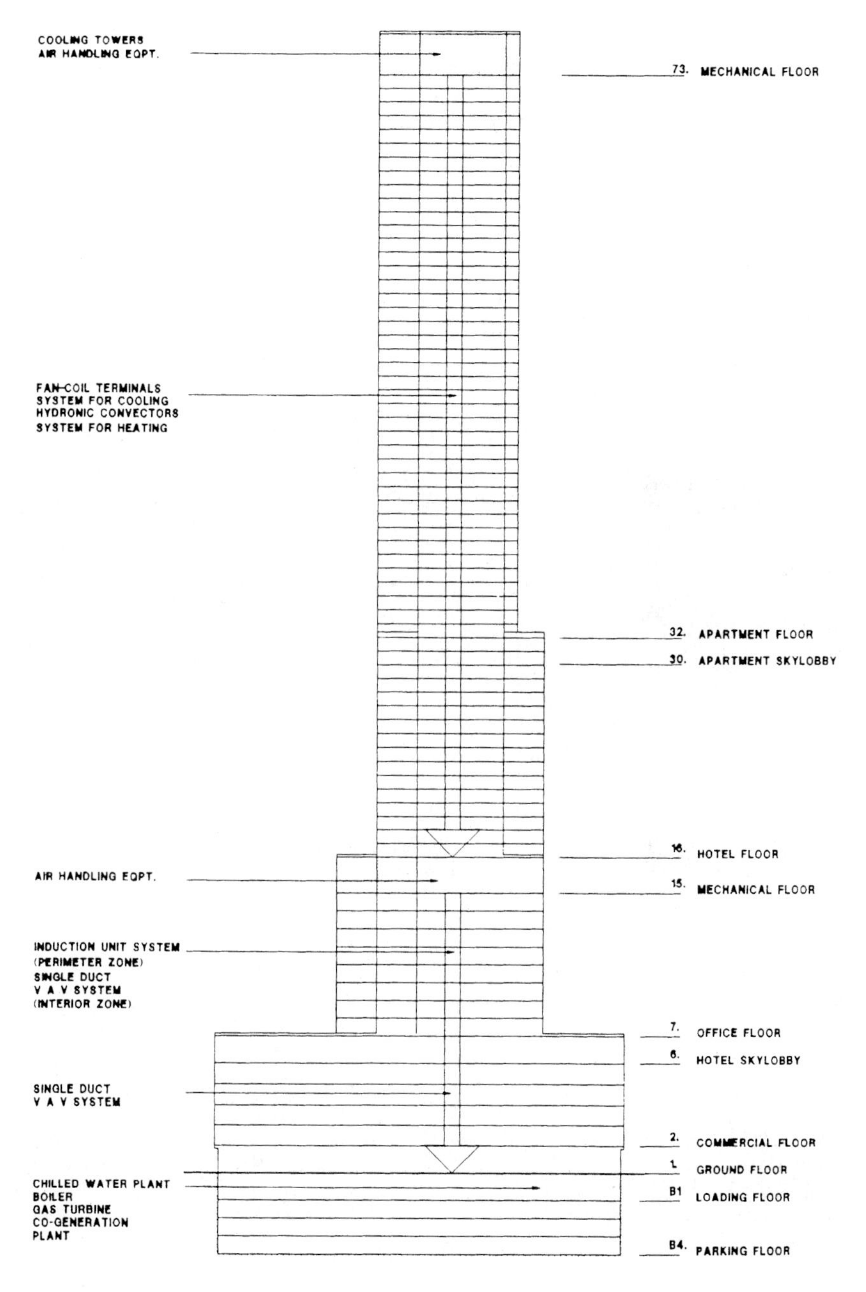


**Fig. 9.29 Multiuse high-rise building—section.** (*Courtesy: Illinois Institute of Technology, Department of Architecture.*)

**Fig. 9.30 Multiuse high-rise building—model.** (*Courtesy: Illinois Institute of Technology, Department of Architecture.*)

be created by offering to supply electric power to two existing office buildings on adjoining blocks at rates lower than the local utility. These buildings, which have a total gross floor area of 88,998 $m^2$ (958,000 $ft^2$), would also supply waste paper for fuel. They would receive another 2.7 MW of power, or about 60% of their peak demand, for the same weekly operating periods.

***Waste Paper Generation.*** To estimate the generation of waste paper, one large office building in Chicago was studied to obtain detailed information (Vanadol, 1991). This building produced an average of 430 $m^3$ (561 $yd^3$) of compacted waste paper per day. Assuming a density of 4005 kg/$m^3$ (250 lb/$ft^3$) this gave a total weight of about 623 kN (70 tons) per day, or 0.25 kg/$m^2$ (0.05 lb/$ft^2$) per day. This building contains a major corporate headquarters, a very large law firm, and several restaurants, so it may not be representative. Based on these data, we assumed a waste paper production of 0.110 kg/$m^2$ (0.022 lb/$ft^2$) per day for offices and commercial spaces.

The residential areas of hotel and apartments are estimated to produce about 0.011 kg/$m^2$ (0.0022 lb/$ft^2$) per day of paper for 7 days a week. In apartments this can vary widely from unit to unit. Also it may be more difficult to educate the residents to separate all types of paper from other kinds of trash. However, the separation of newspapers, junk mail, and periodicals into separate containers for collection can be accomplished with minimum education. The hotel restaurants also generate significant amounts of paper, primarily from containers for shipping food, which can be sorted for collection by kitchen personnel.

As shown in Table 9.4, the waste paper output of the three buildings will produce about 56% of the system fuel required each week; the rest must be imported from other nearby buildings.

### 4 Energy Conversion System

***Gas Turbines with Biomass Fuels.*** The use of gas turbines with biomass fuels has been investigated at the Princeton Center for Energy and Environmental Studies for applications in the sugar-cane industry (Larson and Williams, 1990). The authors' conclusion is that a biomass-integrated gasifier/steam-injected gas turbine (BIG/STIG) would be the most efficient system for using biomass fuels (Fig. 9.25). Their results are based on data collected from the actual operation of coal-gas and natural-gas fired gas

**Table 9.3 Project electric demand**

| Space or use | Floor area, $ft^2$ | Peak power, MW | Cogen power, MW | Cogen energy, kW-hr/wk |
|---|---|---|---|---|
| Parking | 313,000 | 0.08 | 0.08 | 3,130 |
| Commercial | 300,000 | 1.50 | 0.90 | 36,000 |
| Offices | 200,000 | 1.00 | 0.60 | 24,000 |
| Hotel | 340,000 | 0.68 | 0.34 | 13,600 |
| Apartments | 700,000 | 1.40 | 0.70 | 28,000 |
| Subtotal | 1,853,000 | 4.66 | 2.62 | 104,730 |
| 2 office bldgs. | 958,000 | 4.79 | 2.78 | 111,200 |
| Total | 2,811,000 | 9.45 | 5.40 | 215,930 |

turbines. In this system utilizing waste paper fuel, the paper would first be shredded and pressed into briquettes, which would then be fed into a Lurgi fixed-bed gasifier. The gas then goes to the combustion chamber, where it is ignited and fed into the turbine, which drives the electric generator. The exhaust gases are then conducted through a heat recovery system to produce steam, and finally exhausted through a stack. The steam can be either injected into the turbine to improve electric output efficiency, or piped away for use elsewhere.

***Gas Turbine Selection.*** For this project, the smallest available gas turbine was chosen to match the power level required by the three buildings in the system. This is the General Electric LM-38, based on the TF-34 jet engine developed for the A-10 attack aircraft. Using the LM-38 in the BIG/STIG configuration would produce 4 MW of electricity and 5715 kg/hr (12,600 lb/hr) of steam in the cogeneration mode at 56% efficiency, or 5.4 MW of electricity in the power-only mode at 30% efficiency, with a continuous range of intermediate outputs. In warmer months the system would be used in the electric-power-only mode. In winter, steam could be provided, as required, to the heating systems of the apartments and the hotel, which would have hot water fan-coil units, to reduce fuel consumption by the building boilers. However, this might not be cost-effective because the natural-gas fuel used for the boilers has a lower unit cost than electric power. As noted earlier, this cogeneration system would be operated only 8 hours per day, 5 days per week. This would allow ample downtime for maintenance of the turbine, and should permit an availability of 100% during scheduled operating times. Additional electric power will be supplied by the local utility.

***Energy System.*** The configuration of the complete building energy system is shown in Figure 9.26. Imported waste paper from the two nearby office buildings is received in the building truck dock, and together with waste paper collected in the building is moved by elevator to the lower-level mechanical space, where it is shredded and pressed into briquettes and stored in feed bins. During energy generation, the briquettes are fed into the gasifier, and the gas is piped to the combustion chamber at the turbine and generator, which are located on the lower level to provide maximum sound absorption for the operating noise. The turbine exhaust is then piped to the heat recovery system to produce steam, and exhausted out a stack which runs upward through the building to the roof.

**Table 9.4 Project biomass energy generation**

| Space or use | Floor area, $ft^2$ | Waste paper, $lb/ft^2$-day | Days-wk | Paper energy, kW-hr/wk | Electric energy, kW-hr/wk |
|---|---|---|---|---|---|
| Parking | 313,000 | 0 | 7 | 0 | 0 |
| Commercial | 300,000 | 0.022 | 6 | 87,120 | 26,217 |
| Offices | 200,000 | 0.022 | 5 | 48,400 | 14,568 |
| Hotel | 340,000 | 0.002 | 7 | 11,519 | 3,468 |
| Apartments | 700,000 | 0.002 | 7 | 23,716 | 7,138 |
| Subtotal | 1,853,000 | | | 170,755 | 51,391 |
| 2 office bldgs. | 958,000 | 0.022 | 5 | 231,836 | 69,782 |
| Subtotal | 2,811,000 | | | 402,591 | 121,173 |
| Imported paper | | 71.5 tons/wk | | 314,807 | 94,757 |
| Total | | | | 717,398 | 215,930 |

***Return Analysis.*** An internal rate of return analysis was made for the system. Chicago's high electric rates—$0.10/kW hr—may give this system an economic advantage here which might not be found in other urban areas. Larson and Williams (1990) give a cost of $1650/kW for the LM-38 BIG/STIG system, or a total cost of $8.91 million, and another $0.5 million will be needed for electrical connections to the two existing buildings. The income value of the electricity generated would have a value of about $1.12 million per year. Assuming a borrowing rate of 8%, this would result in a 7-year payback period and a rate of return on investment after payback of 9.25%.

It would be difficult to have a direct architectural expression related to the turbine system in the building, but a symbolic token of its presence is offered by the illumination of the bifurcated exhaust stack, which extends through the building roof enclosed by an abstract "powerhouse" of illuminated horizontal bands. For additional information on building economics and life-cycle costs, refer to Chapter PC-5 (CTBUH, Group PC, 1981).

### 5 Future Research

Further research needs to be done in this area, particularly on the actual quantities of waste paper produced by various building types. Nearly all the hardware proposed for the system exists, but these elements have never been used in this combination before, and this obviously is another subject for investigation. Finally, the economics of such systems must be studied in detail to establish their feasibility.

## 9.8 SUSTAINABLE ARCHITECTURE

### 1 Toward a Concept of Sustainable Architecture

Depending upon who defines the term or the context in which it is applied, *architectural sustainability* can mean different things. To many architects, sustainability is generally related to developing architecture which is sensitive to the environment, is energy conscious, promotes the conservation of natural resources, and promotes the recycled use of synthetic materials. Sometimes dubbed *green architecture,* this environmentally conscious architecture is but one facet of sustainable architecture which has emerged from the 1960s with the advent of solar panels and visionary creations, such as *Paolo Soleri's Arcology* (Arcidi, 1991a), and the ecoconsciousness of the 1970s.

The concept of sustainability in architecture is, in fact, an old idea that has been given new currency. Applications of energy efficient and environmentally compatible architecture have been around for a long time. Throughout the world there is ample evidence of architecture which has been developed with sensitivity to its locale and region. On the North American continent, indigenous cultures in the southwest, such as the Pueblo and Anastasi, used natural building sites to their greatest advantage, orienting buildings to take advantage of solar radiation during colder winter months, and deep shade and thermal cooling during hot summer months. In the modern era, architects such as Frank Lloyd Wright stressed that architecture must be compatible with its site. Wright sought innovative and expressive means to integrate buildings with their sites and contexts through both structural and material expression. In several of his Usonian houses, Wright experimented with passive solar radiation and thermal storage to heat and cool houses naturally many years before these concepts became fashionable. Further information on environmental influences such as thermal, lighting, noise, sun, shadow, and wind is available elsewhere (CTBUH, Group PC, 1981).

In addition to its environmental and ecological associations, sustainability has also taken on a broader meaning in terms of communities and cities. In an era of political and economic uncertainty in many major cities, sustaining urban infrastructures and redefining livability in both large and small communities have become issues with far-reaching architectural implications. The ultimate question is: How can we make communities, buildings, and landscapes more socially and environmentally responsive?

During the World Congress of Architects meeting in June 1993, the *Declaration of Interdependence for a Sustainable Future* was drafted. This declaration affirms sustainability as a holistic concept related to the natural, artificial, and social environments (UIA/AIA World Congress of Architects, 1993):

> A sustainable society restores, preserves, and enhances nature and culture for the benefit of life, present and future; a diverse and healthy environment is intrinsically valuable and essential to a healthy society; today's society is seriously degrading the environment and is not sustainable.
>
> We are ecologically interdependent with the whole natural environment; we are socially, culturally, and economically interdependent with all of humanity; sustainability, in the context of this interdependence, requires partnership, equity, and balance among all parties.
>
> Buildings and the built environment play a major role in the human impact on the natural environment and the quality of life; sustainable design integrates consideration of resource and energy efficiency, healthy buildings and materials, ecologically and socially sensitive land-use, and an esthetic sensitivity that inspires, affirms, and ennobles; sustainable design can significantly reduce adverse human impacts on the natural environment while simultaneously improving quality of life and economic well being.

Developing sustainability extends beyond the parameters of architecture, encompassing a wide range of social, economic, and political issues. "Today, our world view is undergoing a profound change. The threat of nuclear holocaust, ecological disasters, population explosion, and natural resource depletion have set in motion a truly radical paradigm shift" (Arcidi, 1991b). We are moving from a modern mechanistic view of the world to a new conception of reality—one that includes tradition and reintegrates the timeless principles of spiritual existence. In architecture, this translates into a new model of the way we see our world, its underlying form, and the forces that give it shape.

To a certain extent, the notion of sustainability is related to the concept of *sustainable development* in the fields of economics and environmental engineering. Sustainable development is essentially an endeavor to apply scientific, engineering, and economic knowledge to assist in rectifying the negative consequences that the unrestricted proliferation of technology helped create. The principal aspects of sustainable development are more efficient energy utilization through conservation and minimizing waste, greater recycling of materials, adoption of renewable energy sources, better management of resources, and more comprehensive economic and environmental appraisal employing life-cycle analyses. The explosive population growth throughout the world, particularly in developing nations, coupled with dramatic changes in technology and energy-intensive growth in the developed countries, increases the possibility of permanent damage to the global environment. Architectural design of tall buildings that accommodate thousands of people in a city must, therefore, account for such concerns.

## 2 Sustainability of Tall Buildings

Sustainable architecture can take many forms. At one end of the spectrum are the more proactive strategies, which focus on advances in high technology as solutions to energy conservation and environmental problems. Examples of these applications might include the use of photovoltaic panels, which were first developed to power satellites in the early 1960s. In early applications, the costs of photovoltaic panels and their installation, coupled with the costs of their physical support and power conversions, restricted their feasibility. However, advances in production and installation have made it possible to make these panels a part of the usual building materials specifications. As in the case of any new technology, the success of photovoltaic panels depends largely on their performance and the effectiveness of their integration with other building components and systems. In Chapter 7 the application of photovoltaic diaphragms in the facade of the World Arab Institute is discussed.

High-technology advances in other areas include supplementing electric energy through wind-powered generators, evaporative cooling systems for arid climates, and the use of geothermal energy and thermal storage. Adaptation of high technology to take advantage of renewable energy resources is promising but faces many economic and political obstacles. To date, these forms of technology have only been applied to isolated building examples and have yet to be applied to larger-scale enterprises, such as entire communities and cities.

Architects and developers are beginning to realize that many buildings must be designed so as to minimize the use of depletable resources or to revitalize existing structures. Using recycled building materials, as in the case of the Warsaw tower proposed for Poland (Arcidi, 1991b), exemplifies the attempts of architects to suggest alternative methods of creatively employing reusable materials to new uses. Adaptive reuse, which involves refurbishing and modernizing old buildings for new functions, is yet another example of creating sustainable architecture and, by extension, sustainable communities.

Environmental awareness extends to both the urban environment and the context in which the tall building is placed as well as to its interior environment. The issues of indoor and outdoor air quality and microclimate, and the potential toxicity of materials and chemicals used in building furnishings, components, and systems are also of concern to architects and building users. Developing truly sustainable architecture, therefore, requires a holistic view of the entire building enterprise, that is, consideration of the local and global environment, the availability of renewable and nonrenewable resources, community impact assessment, and the collaborative input of architects, engineers, planners, social scientists, behavioral scientists, and community-based groups.

Recent developments in public awareness have shifted the issue of sustainability from a peripheral concern to a global agenda. Although the effects of global warming, ozone depletion, acid rain, and deforestation have not been assessed, it is clear that these issues are already impacting the design and construction of tall buildings and the urban development. A new generation of architects must be trained to become aware of these concerns. Although the World Congress of Architects endorsed the notion of sustainability in architecture in principle, it must be noted that unless there is a genuine and comprehensive effort on the part of architects, developers, and politicians to develop and promulgate sustainable architecture and communities, the concept of sustainability will remain an unrealized goal.

### *3 Examples of Sustainable Architecture*

In 1978 the Netherlands International Bank (NMB) decided to change its image through a new headquarters in a suburb south east of Amsterdam. The building that resulted from the holistic consideration of all program needs was a direct result of the interdisciplinary team assembled. Representatives from the bank worked with architects, construction engineers, energy experts, interior design consultants, and others to formulate the final building design (Vale, 1991).

Energy efficiency is inherent throughout many aspects of the structure, starting with the construction. Mineral fiber insulation, a cavity of air, and an outer skin of brickwork encase the building. This outer skin becomes a heat sink, storing warmth from the sun and other internal sources, then releasing warmth when cooling occurs.

Daylighting was an important element of the design, which was influenced by the input of the people working in the building. Good natural lighting, natural ventilation, and exclusion of external noise had to be realized within the framework of the narrow European office floor plate.

Combining the building design, natural light sources, and alternative energy systems has resulted in a tall building prototype that may offer important lessons for the future. In a 3-month period it was possible to pay back construction and energy system costs as a direct result of the energy savings over that same period of time.

This example of sustainable tall building design not only encourages energy savings, alternative building design, and natural lighting, but also works of art, plants, and water. The new building affected employees in a positive manner by reducing absenteeism. Even the community regards the bank's image as more of a progressive, creative one. This tall building demonstrates that sustainable architecture is not a brake for architectural creativity—rather, it should serve to stimulate further creative investigation for any design team.

The Commerzbank headquarters in Frankfurt am Main, Germany, represents a decisive step in increasingly ecologically sophisticated buildings. The bank, like many other German businesses, embodies sustainable architecture wholeheartedly. The "garden in the sky" combined with a system of natural air-conditioning provides this bank with a creative twist to energy savings.

The overall form realized is a block of triangular "petals" with office floors spreading out from a core, which provides cooling to all floors. Large fans draw in additional air when needed. At every third floor, the offices are interspersed with gardens stag-

**Fig. 9.31 Menara Mesiniga (IBM tower).** (*Courtesy: Ken Yeang, T.R. Hamzah & Yeang.*)

gered around the height of the tower, so that offices view the gardens. Conventional opening windows were impractical, so windows were designed to open behind a windbreak layer of glazing.

All the gardens were made possible by a well-conceived structural form that provided clear-span spaces with six-story units suspended from the central core. The precedent-setting bank will certainly become one of the big sustainable tall buildings of the late twentieth century.

Menara Mesiniaga (IBM tower) in Malaysia represents yet another example of the sustainable tall building. Its program was designed to include spiraling vertical landscapes, recessed and shaded windows, single-core services on the hot side, naturally sunlit stairways and elevator lobbies, and spiraling balconies on the external wall with full-height sliding doors for natural ventilation.

The most powerful effects on the form of the building come from the sky courts and the sunshaded roof. These elements of the building not only shield the tower from tropical sun, but also provide natural ventilation flow.

### 4 Future Research

Sustainability refers to a broad spectrum of geopolitical issues pertaining to architecture, planning, energy, and the protection and enlightened use of the natural environment. The implications of sustainability—in both architecture and urban planning—are truly global in scope, extending beyond natural boundaries and continents. Therefore, to achieve a set of comprehensive goals and objectives relative to all aspects of sustainability governing architecture, the environment, and the use of synthetic and natural resources a united international effort is necessary.

Due to the magnitude and complexity of the issues affecting sustainability, future research must be multifaceted, yet coordinated among a diverse, interdisciplinary body of architects, planners, environmental and social scientists, as well as government, social, and civic agencies. For tall buildings, research must address the urban and economic impact of high-rise buildings on the environment and the urban context. The use of new and recycled materials must be addressed as well as development of innovative structural systems to help make tall buildings more efficient and cost-effective. Energy conservation, a significant issue in an era of limited natural energy resources, must be addressed through the development of more efficient and cost-effective heating and cooling systems for buildings in all climate extremes. Furthermore, research into more efficient means of heating and cooling buildings, such as passive and active solar systems, natural ventilation, thermal mass, and wind-powered electric generators, must continue to be promoted.

Finally, and perhaps most important, a concerted international effort must be mounted to educate people and raise their awareness of issues of sustainability at local, regional, and global scales. For it is fundamentally through education that architecture and the environment can be truly sustained from one generation to the next.

## 9.9 CONDENSED REFERENCES/BIBLIOGRAPHY

The following is a condensed bibliography for this chapter. Not only does it include all articles referred to or cited in the text, it also contains bibliography for further reading. The full citations will be found at the end of the volume. What is given here should be sufficient information to lead the reader to the correct article—author, date, and title. In case of multiple authors, only the first named is listed.

Arcidi 1991a, *Paolo Soleri's Arcology: Updating the Prognosis*
Arcidi 1991b, *Projects: Nine Proposals on Behalf of the Environment*
ASHRAE 1974, *Thermal Environment Conditions for Human Occupancy*
ASHRAE 1975, *Energy Conservation in New Building Design*
ASHRAE 1989, *ASHRAE Handbook, Fundamentals*
ASHRAE 1992, *ASHRAE Handbook, HVAC Systems and Equipment*
ASHRAE 1993, *ASHRAE Handbook 1993, Fundamentals*
Bedford 1964, *Basic Principles of Ventilation and Heating*
Branch 1993, *The State of Sustainability*
Brodsky 1991, *Calatrava to Design Cathedral Biosphere*
Browning 1992, *NMB Bank Headquarters*
Browning 1992a, *Megawatts for Buildings*
Browning 1992b, *Vaulting the Barriers to Green Architecture*
CTBUH, Committee 3, 1994, *Structural System for Tall Buildings*
CTBUH, Committee 8A, 1992, *Fire Safety in Tall Buildings*
CTBUH, Committee 56, 1992, *Building Design for Handicapped and Aged Persons*
CTBUH, Group PC 1981, *Planning and Environmental Criteria for Tall Buildings*
CTBUH, Group SC, 1980, *Tall Building Systems and Concepts*
Dubois 1915, *The Measurement of the Surface Area of Man*
Dufton 1932, *The Equivalent Temperature of a Room and Its Measurement*
Ellis 1953, *Thermal Comfort in Warm and Humid Atmospheres—Observations on Groups of Individuals in Singapore*
Fanger 1970, *Thermal Comfort*
Gunts 1993, *Blueprint for a Green Future*
Hee Sung Industry 1987, *Design and Construction of the Lucky-Goldstar Twin Towers*
Houghten 1923, *Determination of the Comfort Zone*
Hubka 1973, *Man as Criterion of Tall Buildings*
Iyengar 1989, *Exposed Steel Frames—A Unique Solution for Broadgate, London*
Jokl 1973, *Air-Conditioning of Tall Buildings and Its Influence on Layout and Structure*
Kao 1991, *A Multi-Use Highrise Building*
Kax 1991, *The Greening of Architecture*
Kohn Pederson Fox Associates 1985, *900 North Michigan Tower*
Larson 1990, *Biomass-Gasifier Steam Injected Gas Turbine Cogeneration*
McIntyre 1980, *Indoor Climate*
Missenard 1948, *Equivalence Thermique des Ambiences—Equivalence de Sejour*
Murphy/Jahn Architects 1985, *Oakbrook Terrace Tower*
Perkins & Will 1988, *100 North Riverside Plaza (Morton International Building)*
Rohles Jr. 1973, *The Revised Modal Comfort Envelope*
Rosiak 1972, *System of Natural Ventilation in Tall Buildings*
Skidmore, Owings & Merrill 1986, *AT&T Tower, Phase I*
Skidmore, Owings & Merrill 1990, *USG Tower*
Thomas 1991, *Mechanical Engineering Design Computer Programs for Buildings*
UIA/AIA World Congress of Architects 1993, *Declaration of Interdependence for a Sustainable Future*
Vale 1991, *Green Architecture: Design for a Sustainable Future*
Vanadol 1991, *Private Communication to Alfred Swenson*
Walker 1987, *Services and Structure—The Hong Kong and Shanghai Bank*

# 10

# Materials and Structures: Steel

Tall buildings employ several types of materials, both structural and nonstructural. Architects need to have a general understanding of the properties and applications of these materials so that they can select the most appropriate materials for a particular building project in collaboration with the engineer. Chapters 10 through 12 present a brief account of these structural materials—steel, concrete, and masonry—which are most commonly used for high-rise construction.

This chapter is intended to provide general information about structural steel, how it is produced, and how it can be selected and applied to tall building construction. Section 10.1 gives a brief history of the development of steel-frame construction. Sections 10.2 through 10.6 present the chemical composition and physical properties of steel, the fabrication process, and the application of steel. Structural systems employing steel as well as a number of case studies are then discussed in Sections 10.7 through 10.9. Some emerging or innovative trends and future research needs are presented in Section 10.10.

## *10.1 DEVELOPMENT OF STEEL-FRAME CONSTRUCTION*

The initial application of ferrous metal for structural building components was in the form of cast-iron columns and wrought-iron beams. The iron columns made it possible to reduce greatly the thickness of walls compared to earlier requirements for masonry construction. Thus the transition was made from bearing wall to skeleton frame construction, with the frame designed to resist lateral wind loads as well as gravity loads.

Structural steel shapes became available in 1885 and greatly accelerated the development of tall buildings (Grinter, 1949). Early applications were concentrated in the United States in Chicago and New York, with height records set and broken frequently. The high strength and modulus of elasticity of steel were advantageous from the viewpoints of both structural strength and stiffness. As a result, steel as a material was ideally suited for tall building structures.

Structural steel continues to play a major role in the construction of tall buildings. Most of the tallest buildings in the world are steel-framed, although concrete and steel-concrete composite construction is growing in use. The availability of steel of various strength levels, a wider selection of hot-rolled steel sections, innovative framing systems, and modern design methods, all have played a role in the continued use of steel for tall buildings.

## 1 Architecture as Structural Art

The notion of architecture as a structural art—the combined art and technique of designing, shaping, organizing, and decorating the stone, iron, wood, and glass of which a building is composed—was overwhelmed in the nineteenth century by the various classical revivals and a general rejection of the machine and industrialization on the part of many architects (Condit, 1964). Giedion (1941) refers to the "cultural schizophrenia" of the nineteenth century in its desire to hold on to tradition while engaging and utilizing science and technology.

It was the engineers who first pointed the way that a new structural art might profitably take. A small arch bridge, built between 1775 and 1779, over the River Severn at Coalbrookdale, England, was the first cast-iron structure. Between 1819 and 1826 Thomas Telford created the first suspension bridge over Menai Strait, leading to a cable suspension bridge in 1869–1883 by John and Washington Roebling—the Brooklyn Bridge. Gustave Eiffel demonstrated the structural principles of the two-hinged arch and the crescent-shaped arched truss in the Garabit Viaduct (1884), which would become prototypes for his famous tower in Paris (Fig. 10.1).

## 2 Cast-Iron Construction

More significant from an architectural perspective was the development of cast-iron construction technology and its application to conventional building types. An early and impressive use of cast iron for interior structural members was in St. Anne's Church in Liverpool, England (1770–1772). In the United States the new system of framing was first adopted by the architect-engineer William Strickland, who introduced cast-iron columns as balcony supports in the Chestnut Street Theater in Philadelphia (1820–1822). By the late 1840s, Daniel Badger and James Bogardus had established factories for prefabricated cast-iron components in New York (Condit, 1964; Giedion, 1941; Mumford, 1952).

Whereas many buildings, such as Lebrust's Biblioteque St. Geneviève in Paris, employed cast iron for the interior structure, the architects acquiesced to traditional means to express facades. Load-bearing stone masonry with classically inspired engaged columns and arches were typically used to express the building's "shell," whereas the more innovative and exciting use of cast iron was relegated to the expression of the interior structure and floor plates. Even when cast iron was employed directly on the facade, it was primarily relegated to use as ornamentation or as spandrels between floor levels.

The Crystal Palace, erected for the London International Exposition of 1851, demonstrated the full potential of the structural techniques of cast-iron construction and prefabricated assemblies. Conceived as a prefabricated "kit of parts," the Crystal Palace was one of the earliest prototypes of the curtain wall and skeleton. However, the significance of this achievement was largely lost on the architects of the nineteenth century since it was erected only for the purposes of the exposition (Condit, 1964).

An early example of large-scale iron framing that prefigured the dominant mode of the Chicago work was the John Schillito store in Cincinnati, Ohio (1876–1877). By combining interior iron columns and beams with narrow masonry piers in the exterior enclosure, all four elevations of the block-long six-story building were turned into the open cellular walls that the Chicago school later developed into forms of great power.

### 3 Steel-Frame Construction

In the late 1800s Chicago was the ideal crucible in which the development and expression of a new architectural typology could occur. The Great Chicago Fire of 1871 had laid to waste a vast portion of the city's original wood-frame structures, which meant the rebuilding of the entire commercial district with buildings incorporating fire-resistant materials. According to Condit (1964), Chicago's preeminence as a commercial center in the United States had been well established by the time of the fire. Its strategic location on Lake Michigan and its centralized geographic location provided access to worldwide

**Fig. 10.1 Eiffel Tower, Paris.** (*Courtesy: Paul Armstrong.*)

markets. Rapid agricultural expansion paralleled the growth of the railroad, and by the 1870s, Chicago had become a major land and waterway transportation hub.

With the railroads and the canals came the steel, coal, and lumber industries, all the raw materials essential to transforming Chicago into an industrial giant. Mumford (1952) characterized Chicago of the 1870s as "a brutal network of industrial necessities." The economic demands clearly established the need for a highly developed building art, but, as Condit writes, "it was the creative spirit and the stimulating intellectual climate of the city that made such an art possible."

Jenney, Sullivan, Root, and the other Chicago architects of the period recognized the architectural potential of the steel frame (Goldberger, 1981). The commercial steel-framed tall building was a new building type, which was the product of a new and innovative technology. The problem for the architect was how to give architectural expression to the frame without resorting to historic building styles. Buildings such as the Reliance (1894) in Chicago pointed the way to the future (Goldberger, 1981). Set within its terra-cotta-clad frame were huge sheets of glass, giving the entire building an extraordinary lightness. Although the architects were not yet able to fully express the vertical building structure, it represented a unique and unprecedented break with masonry-clad high-rise buildings. In many respects, the Reliance Building prefigured the modern glass-and-steel high-rise buildings in the latter part of this century.

The architectural and technical achievement of the Chicago school (see Chapters 1, 4, and 5) marked the establishment of a new style of architecture—and a new typology in the form of the steel high-rise office building. At the same time it represented the culmination of a structural evolution that extended over the century preceding it. According to Condit (1964), two parallel characteristics marked the development of the Chicago school: "One was highly utilitarian, marked by a strict adherence to function and structure derived from traditional vernacular prototypes. The other was formal and plastic, the product of a new theoretical spirit and the conscious determination to create rich, symbolic forms—and to create a new style expressive of contemporary American culture."

Today steel-frame construction is one of the most efficient and cost-effective construction types for tall buildings. It is used extensively throughout the world for midrise and high-rise building types and finds application for most supertall building structures. Steel framing is used widely because of its ductility in wind and seismic loading conditions, its relative strength to its light weight, its ease of assembly and field installation, and its economy in transport to the site. Typically, steel-framed buildings are clad with other sheathing materials such as glass, stone veneers, metal panels, or masonry veneers or panels. The frame can either be hidden by the cladding materials or it can be expressed as a component of the architectural aesthetics of the building. (For additional information on structure and its expression, refer to Chapters 2 and 4. Chapter 7 has information on steel facades.)

## 10.2 PROPERTIES AND COMPOSITION

### 1 Microstructure

Steel has many desirable properties that are suitable for tall building construction. Although corrosion resistance and fire resistance are two important properties that are lacking, the properties of high strength and stiffness generally far outweigh those deficiencies. The general properties of steel have been discussed extensively in the literature (Salmon and Johnson, 1990; Barsom, 1987).

At a very high temperature, steel turns into a liquid. As liquid steel cools, austenite grains are formed. These individual crystals, with the atoms arranged in repeating three-dimensional patterns, are essentially a solution of carbon and other elements in iron. If the temperature of the steel decreases slowly, the austenite undergoes a transformation in the temperature range of 700 to 540°C (1300 to 1000°F) to ferrite, which has a more limited solubility for carbon. To accommodate this limited solubility, the carbon precipitates as iron carbide (cementite). The alternating lamellae of ferrite and iron carbide are known as pearlite. Ferrite is a pure iron which is very ductile but has low tensile strength. Thus the steel becomes a mixture of a soft ferrite matrix and a hard pearlite matrix modifier. Examples of constructional steels having ferrite-pearlite microstructure are ASTM A36 carbon steel and A572 and A588 high-strength low-alloy steels (Lankford et al., 1985).

If the steel is cooled rapidly, the transformation of austenite to ferrite and pearlite is suppressed and, instead, very hard needlelike microstructures known as bainite and martensite are formed. Bainite forms in transformations above 230°C (450°F) and below about 540°C (1000°F). Martensite starts to form below about 260°C (500°F), with the transformation occurring almost instantly during rapid cooling. To develop a useful product, these microstructures must usually be tempered by reheating the steel and slowly cooling, thus improving the toughness and ductility of the material. Examples of quenched and tempered steels include ASTM A852, A514, and A517 (Lankford et al., 1985).

### 2 Chemical Composition

Steel is generally a mixture of iron and carbon with varying amounts of other elements, primarily sulfur, manganese, phosphorus, silicon, and aluminum. These and other elements are either unavoidably present or intentionally added in various combinations to achieve specific characteristics and properties of the finished steel products. The effects of some commonly used chemical elements on the properties of hot-rolled and heat-treated carbon and alloy steels are summarized in Table 10.1 (Lankford et al., 1985; ASM, 1978).

Carbon, the principal strengthening element in steel, increases hardness, tensile strength, and yield strength. On the other hand, increased amounts of carbon cause a decrease in ductility, toughness, and weldability. High-carbon steel with low ductility is not suitable for high-rise buildings in seismic zones.

Sulfur affects surface quality adversely, has a strong tendency to segregate, and decreases ductility, toughness, and weldability. It is generally considered an undesirable element except where machinability is important.

Manganese increases the hardness and strength of steel, although to a lesser degree than does carbon. Manganese combines with sulfur to form manganese sulfides, thus minimizing the harmful effects of sulfur.

Phosphorus strengthens carbon steel and increases atmospheric corrosion resistance. However, it causes a decrease in ductility and toughness.

Silicon and aluminum are the principal deoxidizers used in the manufacture of constructional steels. Aluminum is also used to control grain size.

## 10.3 FABRICATION OF STEEL PRODUCTS

In tall building construction, steel fabrication is a major item in the shop, from where the steel elements are transported to the construction site for erection. Steel products must be fabricated to create the desired shape and geometry. Fabrication operations typical in building construction include drilling or punching holes for bolts; shearing, gas cutting, or sawing components to the required width or length; and welding to develop the member cross section and its connections. Bending or straightening may also be required. Some aspects of these operations have been discussed by Brockenbrough and Barsom (1992) and are briefly reviewed in the following. In very tall buildings the loads become so large that standard rolled shapes for columns may not be adequate; in such instances, thick plates are welded together to fabricate special built-up columns. Similarly, large built-up girders may be required at progressively lower levels of buildings for attaining the required building stiffness.

***Shearing and Thermal Cutting.*** Thinner plates are typically cut at room temperature in a shear. The deformed metal on sheared edges is cold-worked and has higher strength and lower ductility and fracture toughness than the undeformed metal. In the extreme, this cold-worked surface layer may exhibit cracks perpendicular to the sheared edge. The difference in properties between the surface layer of sheared edges and the remainder of the component may be eliminated by machining the edge, although this is rarely needed.

Thicker material is generally gas cut instead of sheared. As the cutting torch moves along, the surrounding metal cools the cut surfaces rapidly and causes the formation of

**Table 10.1 Effects of alloying elements** *(Brockenbrough and Barsom, 1992)*

| Element | Effects |
|---|---|
| Aluminum (Al) | Deoxidizes or "kills" molten steel; refines grain size; increases strength and toughness. |
| Boron (B) | Small amounts (0.0005%) increase hardenability in quenched and tempered steels; used only in aluminum-killed steels; most effective at low-carbon levels. |
| Calcium (Ca) | Controls shape of nonmetallic inclusions. |
| Carbon (C) | Principal hardening element in steel; increases strength and hardness; decreases ductility, toughness, and weldability; moderate tendency to segregate. |
| Chromium (Cr) | Increases strength and atmospheric corrosion resistance. |
| Copper (Cu) | Increases atmospheric corrosion resistance. |
| Manganese (Mn) | Increases strength; controls harmful effects of sulfur. |
| Molybdenum (Mo) | Increases high-temperature tensile and creep strength. |
| Niobium (Nb) | Increases toughness and strength. |
| Nickel (Ni) | Increases strength and toughness. |
| Nitrogen (N) | Increases strength and hardness; decreases ductility and toughness. |
| Phosphorus (P) | Increases strength and hardness; decreases ductility and toughness; increases atmospheric corrosion resistance; strong tendency to segregate. |
| Silicon (Si) | Deoxidizes or "kills" molten steel. |
| Sulfur (S) | Considered undesirable except for machinability; decreases ductility, toughness, and weldability; adversely affects surface quality; strong tendency to segregate. |
| Titanium (Ti) | Increases creep and rupture strength and hardness. |
| Vanadium (V) and columbium (Nb) | Small additions increase strength. |

a heat-affected zone analogous to that of a weld. The depth of the heat-affected zone depends on the carbon and alloy content, thickness, preheat temperature, cutting speed, and postheat treatment. In addition to the microstructural changes that occur in the heat-affected zone, the cut surface may exhibit a slightly higher carbon content than the material below the surface.

***Bolting.*** Bolt holes may be formed by drilling, punching, or punching followed by reaming. All three operations cold-work the bolt hole surface. Drilling is preferable to punching because there is less cold working. Punching decreases the ductility and fracture toughness of the deformed surface layer, increases its strength, and, in the extreme, may cause the development of radial cracks in the deformed surface layer. Reaming the hole after punching eliminates any radial cracks and the severely deformed surface layer. Consequently, standard practice for building construction requires that holes be drilled when the thickness of the material is greater than the nominal bolt diameter plus 3 mm (⅛ in.); alternatively, such holes may be subpunched and reamed, provided the subpunched hole is at least 1.6 mm (1/16 in.) less than the nominal bolt diameter. However, holes in A514 steel must be drilled when the thickness exceeds 13 mm (½ in.).

***Welding.*** As stated before, welding is used to fabricate built-up members, such as I or box sections, and to join members, such as beam-to-column and truss connections. Generally it proves more economical to use welding for connections or attachments made in the shop, and to use bolting for field connections. Welds should be designed to transmit the required forces, but excessive welding should be avoided because it increases costs and may generate excessive residual stresses. Fillet welds are much more economical than groove welds and should be used where possible. Where groove welds are required, it should be indicated whether complete penetration or partial penetration joints are desired, as well as the effective throat that must be achieved with the latter. Details of the joint and selection of the welding process should be left to the fabricator to achieve the greatest economy. It is desirable to use standard prequalified joints whenever possible.

Designers should be alert to the potential problem of lamellar tearing in restrained welded joints. The elongation and reduction of area values of steel are usually much lower in the through-thickness direction than in the planar direction. This inherent directionality is of small consequence in many applications, but does become important in the design and fabrication of structures with highly restrained joints (AISC, 1973). It is reflected in a cracking phenomenon that starts underneath the surface of steel plates as a result of excessive through-thickness strain, usually associated with weld metal shrinkage in highly restrained joints. It has a steplike appearance consisting of a series of terraces parallel to the surface. The tearing may remain completely below the surface, or may emerge at the edges of plates or shapes or at weld toes. The incidence of lamellar tearing has increased with the trend toward heavy welded construction, particularly in joints with thick plates and heavy structural shapes. Careful selection of weld details, filler metal, and welding procedure can be effective in controlling this fracture mode.

***Cold Forming, Bending, and Straightening.*** Plates and shapes may be flattened and straightened during production, and often must be formed and bent during fabrication. Such operations are also used when steel members must be straightened after accidental impact, mishandling, or fire. Such operations induce inelastic deformations, and therefore the yield point, tensile strength, and ductility in the deformed regions tend to differ from those of the virgin steel.

Excessive strain in cold forming can exhaust ductility and cause cracking. Since the strain increases as the thickness increases and the bend radius decreases, it can be conveniently controlled by limiting the radius-to-thickness ratio. Table 10.2 gives some generally accepted minimum inside bend radii for forming plates of several steels. The values are for bend lines perpendicular to the direction of final rolling. When bend lines are parallel to the direction of final rolling, the radii must be increased (approximately doubled) because the ductility in the transverse direction is less than that in the longitudinal direction. Also, when bend lines are greater than about 900 mm (36 in.), the bend radius may have to be increased because of the additional restraint to plastic deformation.

***Heat Straightening.*** Members may be straightened by applying heat to selected locations. Dimensional changes are accomplished by restraining the free thermal expansion to cause plastic deformation, then allowing unrestrained thermal contraction during cooling. Often the unheated portion of the member provides sufficient natural restraint. Extensive dimensional changes can be achieved by controlling the restraint, temperature, heat pattern, and heating cycles. This method has been used successfully for decades to develop camber or sweep as well as to straighten, and it generally does not degrade steel properties significantly (Brockenbrough, 1970a; 1970b). Steel properties at room temperature have been discussed by Brockenbrough and Johnston (1981). Effects of elevated temperature on steel properties have been discussed by Brockenbrough (1970b).

**Table 10.2 Minimum radius for cold-bent plates with bend lines perpendicular to rolling direction** *(AISC, 1989)*

| | Thickness, mm (in.) | | | | |
|---|---|---|---|---|---|
| ASTM Designation | Up to 6.4 (up to ¼) | >6.4 to 13 (>¼ to ½) | >13 to 25 (>½ to 1) | >25 to 38 (>1 to 1½) | >38 to 51 (>1½ to 2) |
| A36 | 1½t* | 1½t | 2t | 3t | 4t |
| A242 | 2t | 3t | 5t | † | † |
| A441 | 2t | 3t | 5t | † | † |
| A529 | 2t | 2t | — | — | — |
| A572‡: | | | | | |
| Grade 42 | 2t | 2t | 3t | 4t | 5t |
| Grade 50 | 2½t | 2½t | 4t | † | † |
| Grade 60 | 3½t | 3½t | 6t | † | — |
| Grade 65 | 4t | 4t | † | † | — |
| A588 | 2t | 3t | 5t | † | † |
| A852§ | 2t | 2t | 3t | 3t | 3t |
| A514§ | 2t | 2t | 2t | 3t | 3t |

*t—plate thickness.

†It is recommended that steel in this thickness range be bent hot. Hot bending, however, may result in a decrease in the as-rolled mechanical properties.

‡Thickness may be restricted because of columbium content. Consult supplier.

§The mechanical properties of this steel result from a quench-and-temper operation. Hot bending may adversely affect these mechanical properties. If necessary to hot bend, fabricator should discuss procedure with steel supplier.

## 10.4 TYPES OF STEEL

### 1 Classification Based on Strength

Steel used for structural applications in tall buildings may be classified in four groups with progressively increasing strengths: carbon steels, high-strength low-alloy (HSLA) steels, heat-treated carbon and HSLA steels, and heat-treated alloy steels. Table 10.3 lists typical U.S. steels in these groups used in building construction and the specified yield and tensile strengths. Typical standards for HSLA plates and structural shapes in various countries are listed in Tables 10.4 and 10.5 (IISI, 1987).

Carbon steel (AISI, 1985) is technically defined as that in which:

1. No minimum content is specified for chromium, cobalt, columbium, molybdenum, nickel, titanium, tungsten, vanadium, zirconium, or any other element added to obtain a desired alloying effect.
2. The specified minimum content for copper does not exceed 0.40%.
3. The specified maximum content for any of the following elements does not exceed that noted: manganese 1.65%, silicon 0.60%, copper 0.60%.

Structural carbon steels are in the mild carbon category (0.15 to 0.29% carbon content). These steels exhibit definite yield points on the stress-strain curve for steel. Increased carbon content increases the yield strength but reduces ductility and weldability of the material. Carbon steels used for structural applications typically have a specified yield point of 200 to 275 MPa (30 to 40 ksi) in the as-rolled condition.

HSLA steels are those with chemical compositions developed to impart higher mechanical properties than can be obtained for carbon steels, but with alloying additions smaller than those for alloy steel. In some cases, greater atmospheric corrosion resistance is also obtained (for example, ASTM A242 type 1 and A588). HSLA steel has a specified minimum yield point in the range of 300 to 550 MPa (42 to 80 ksi) in the as-rolled condition.

Both carbon and HSLA steel can be heat-treated to obtain higher yield strengths. Such heat-treated carbon and HSLA steels have a specified minimum yield strength of 320 to 690 MPa (46 to 100 ksi).

In general, alloy steels are those with chemical compositions that exceed the limits for carbon steel and are not in the HSLA classification. Such steel used for structural applications is generally heat-treated to develop the desired strength and toughness and is thus designated as heat-treated alloy steel. The corresponding yield strength level is 620 to 690 MPa (90 to 100 ksi).

### 2 Heavy Structural Shapes

Designers and fabricators should be aware that most heavy structural wide-flange shapes used for tall buildings, and tees cut from them, have regions of low toughness, particularly at the flange-web intersection. This occurs because of the slower cooling at that location and (due to its geometry) the lower rolling pressures applied there in production. Special considerations must be given to prevent cracking. The heavy shapes most susceptible to this condition are those designated by ASTM as groups 4 and 5. AISC requires that when such shapes are used for members subject to primary tensile stress due to tension or flexure, the material toughness must be specified unless they are

**Table 10.3 Specified tensile properties of various steels** *(Brockenbrough, 1992)*

| ASTM designation | Plate thickness, mm (in.) | ANSI/ASTM group or weight/length for structural shapes* |
|---|---|---|
| Carbon steels | | |
| A36 | To 200 (8), incl. | To 634 kg/m, incl. |
| | | Over 634 kg/m |
| | Over 200 (8) | NC |
| A283, grade C | None specified | NC |
| A285, grade C | To 50 (2), incl. | NC |
| A515 or A516, grade 55 | To 305 (12), incl. | NC |
| A515 or A516, grade 60 | To 200 (8), incl. | NC |
| A515 or A516, grade 65 | To 200 (8), incl. | NC |
| A515 or A516, grade 70 | To 200 (8), incl. | NC |
| A573, grade 65 | To 40 (1½), incl. | NC |
| A573, grade 70 | To 40 (1½), incl. | NC |
| High-strength, low-alloy steels | | |
| A242 | To 20 (¾), incl. | Groups 1–2 |
| | Over 20 (¾) to 40 (1½), incl. | Group 3 |
| | Over 40 (1½) to 100 (4), incl. | Groups 4–5 |
| A588 | To 100 (4), incl. | Groups 1–5 |
| | Over 100 (4) to 125 (5), incl. | |
| | Over 125 (5) to 200 (8), incl. | |
| A572, grade 42 | To 150 (6), incl. | Groups 1–5 |
| A572, grade 50 | To 100 (4), incl. | Groups 1–5 |
| A572, grade 60 | To 32 (1¼), incl. | Groups 1–2 |
| A572, grade 65 | To 32 (1¼), incl. | Group 1 |
| Heat-treated carbon and high-strength, low-alloy steels | | |
| A633, grades C and D | To 65 (2½), incl. | NC |
| | Over 65 (2½) to 100 (4), incl. | |
| A633, grade E | To 100 (4), incl. | |
| | Over 100 (4) to 150 (6), incl. | |
| A678, grade C | To 20 (¾), incl. | NC |
| | Over 20 (¾) to 40 (1½), incl. | |
| | Over 40 (1½) to 50 (2), incl. | |
| A852 | To 100 (4), incl. | NC |
| Heat-treated alloy steel | | |
| A514 | To 65 (2½), incl. | NC |
| | Over 65 (2½) to 150 (6), incl. | |

*NC—not applicable.

| Specified yield stress, minimun | | Specified tensile stress, stress minimum or minimum–maximum | |
|---|---|---|---|
| MPa | ksi | MPa | ksi |
| | | Carbon steels | |
| 250 | 36 | 400–550 | 58–80 |
| 250 | 36 | 400 | 58 |
| 220 | 32 | 400–550 | 58–80 |
| 205 | 30 | 380–485 | 55–70 |
| 205 | 30 | 380–515 | 55–75 |
| 205 | 30 | 380–515 | 55–75 |
| 220 | 32 | 415–550 | 60–80 |
| 240 | 35 | 450–585 | 65–85 |
| 260 | 38 | 485–620 | 70–90 |
| 240 | 35 | 450–530 | 65–77 |
| 290 | 42 | 485–620 | 70–90 |
| | | High-strength, low-alloy steels | |
| 345 | 50 | 480 | 70 |
| 315 | 46 | 460 | 67 |
| 290 | 42 | 435 | 63 |
| 345 | 50 | 485 | 70 |
| 315 | 46 | 460 | 67 |
| 290 | 42 | 435 | 63 |
| 290 | 42 | 415 | 60 |
| 345 | 50 | 450 | 65 |
| 415 | 60 | 520 | 75 |
| 450 | 65 | 550 | 80 |
| | | Heat-treated carbon and high-strength, low-alloy steels | |
| 345 | 50 | 485–620 | 70–90 |
| 315 | 46 | 450–590 | 65–85 |
| 415 | 60 | 550–690 | 80–100 |
| 380 | 55 | 515–655 | 75–95 |
| 515 | 75 | 655–790 | 95–115 |
| 485 | 70 | 620–760 | 90–110 |
| 450 | 65 | 585–720 | 85–105 |
| 485 | 70 | 620–760 | 90–110 |
| | | Heat-treated alloy steel | |
| 690 | 100 | 760–895 | 110–130 |
| 620 | 90 | 760–895 | 100–130 |

joined by bolting (AISC, 1989). The same applies to members built up from plates exceeding 51 mm (2 in.) in thickness. Special fabrication procedures must also be followed for heavy structural shapes.

## 10.5 STEEL EXPOSED TO THE ATMOSPHERE

### 1 Weathering Steel

Architects have found that bare "weathering" steel provides an opportunity to create unique, attractive exposed structural frames. Under proper conditions, this steel weathers to a rich, dark, earthy color, with a natural even texture. The color varies somewhat with the type of atmosphere to which it is exposed, but is typically a deep, dark brown or black. A notable example of its use is the USX Tower in Pittsburgh, Pennsylvania. This steel has also been used to create attractive buildings of modest height.

Weathering steels are special HSLA steels with a resistance to atmospheric corrosion of at least four times that of carbon steel. Statistical methods have been developed for

**Table 10.4 Standards for HSLA steel plate in various countries** *(IISI, 1987)*

| | | |
|---|---|---|
| Germany | DIN 17100 | Steels for general structural purpose, quality specifications |
| | DIN 17102 | Weldable, fine-grain steels, normalized |
| France | NF A 35-501 | Structural steels |
| United Kingdom | BS 4360 | Weldable structural steels |
| United States | ASTM A242 | HSLA structural steels |
| | ASTM A441 | HSLA structural Mn V |
| | ASTM A514 | High-yield quenched and tempered alloy-steel plate suitable for welding |
| | ASTM A572 | Structural quality HSLA Nb V |
| | ASTM A588 | HSLA structural steel, 345-MPa (50 ksi) yield point up to 100 mm (4 in.) thick |
| | ASTM A633 | Normalized HSLA structural steel |
| | ASTM A656 | Hot-rolled structural steel, HSLA plate with improved formability |
| | ASTM A678 | Quenched and tempered C and HSLA plates for structural applications |
| | ASTM A710 | Low C age hardening Ni Cu Cr Mo Nb and Ni Cu Nb alloy steels |
| | ASTM A808 | Structural quality HSLA C Mn Nb V steel with improved notch toughness |
| | ASTM A852 | HSLA steel, quenched and tempered |
| Japan | JIS G3106 | Rolled steel for welded structures |
| | JIS G3114 | Hot-rolled atmospheric-corrosion-resisting steel for welded structure |
| | JIS G3128 | 700-MPa (101 ksi) high-yield plate for welded structures |

estimating the atmospheric corrosion resistance of weathering steels and for predicting their relative performance based on their chemical composition (Komp, 1987). Such steels first became available in 1933 for the construction of railroad hopper cars, and by the early 1960s they were being used on a widespread basis for heavy structures such as bridges, exposed building frames, and towers (Coburn, 1987).

When boldly exposed to the atmosphere, weathering steel builds up a tight protective oxide film that tends to resist further atmospheric corrosion. Alternate periods of dampness and dryness are necessary to develop the protective oxide film. In most of the United States it develops in 2 to 4 years, but in arid areas it may require more time. If scratched or marred, the oxide reforms.

Figure 10.2 compares the reduction in thickness with time for weathering steel (ASTM A242 type 1, A588, and A514) to that for carbon and copper bearing carbon steel, based on an industrial atmosphere in the United States. European performance varies somewhat from that in the United States because those countries are at a higher latitude, receive fewer hours of sunlight, and are at a lower angle. Thus the conditions for growth of the protective oxide film are less optimum and the corrosion rates are higher.

Certain guidelines must be followed in using weathering steels. When in a continually damp environment, or when buried in the ground, the performance of weathering steel is no better than that of carbon steel. Thus it is important not to allow debris to accumulate on ledges and not to permit vegetation to impinge on the structure. Crevices should be avoided. Design details should provide for drainage and not allow the accumulation of moisture. Flow of water containing salts, such as those used to control snow and ice, will accelerate corrosion and must be avoided. For example, leakage through expansion joints and improper drainage in parking structures have caused corrosion damage. Some coastal areas are not suitable for bare weathering steel because of the airborne salt. Surveys and experience with nearby structures can be used to determine the suitability of specific sites for proposed tall steel buildings employing steel.

**Table 10.5 Standards for HSLA structural steels in various countries** *(IISI, 1987)*

| | | |
|---|---|---|
| Germany | DIN 17100 | Steels for general structural purpose, quality specifications |
| | DIN 17102 | Weldable fine-grain steels, normalized |
| | DASt 011 | High-strength weldable fine-grain steels for structural purposes |
| France | NF A35-501 | Structural steels |
| | NA A35-504 | High-yield-stress steel beams and sections for welded constructions |
| United Kingdom | BS 4360 | Weldable structural steels |
| EEC | Euronorm 25 | Hot-rolled, nonalloy steels for general structural application |
| | Euronorm 113 | Hot-rolled, weldable fine-grain structural steels |
| United States | ASTM A572 | HSLA Nb V steels of structural quality |
| | ASTM A441 | HSLA structural Mn V steel |
| | ASTM A588 | HSLA structural steel, 345-MPa (50 ksi) minimum yield point up to 100 mm (4 in.) thick |
| Japan | JIS G3106 | Rolled steel for welded structures |
| International | ISO 630 | Structural steels |

Special considerations are required for bolted joints of weathering steel (Brockenbrough, 1983a). Basically, the details must be selected to achieve a tight joint that excludes moisture. Otherwise, formation of corrosion products within the joint can lead to expansive forces between the faying surfaces, which can severely bow the connected elements. Galvanized bolts and nuts should not be used with bare steel because, in the presence of moisture, the zinc coating will be sacrificed and leave the bolt unprotected. High-strength weathering steel bolts are available as ASTM A325 type 3 and A490 type 3.

Where weathering steel is used in the painted condition, the estimated paint life is about twice that of painted carbon steel.

## 2 Stainless Steel

There are many types of stainless steel available that readily withstand the effects of atmospheric corrosion and offer enhanced appearance. Well known as a material of enduring beauty, stainless steel provides strength and toughness over a wide temperature range and can be fabricated into a variety of shapes. Chromium is the primary alloying element that imparts corrosion resistance to this steel. Because of its high initial cost, the use of stainless steel is usually limited to curtain-wall components of tall buildings and other special items. The reader should refer to Chapter 1 for the case study on Mapfre office tower in Barcelona, Spain, in which stainless steel was chosen as the principal facade material. Another example of a stainless-steel facade is the Petronas Towers, now (1995) under construction, in Kuala Lumpur, Malaysia.

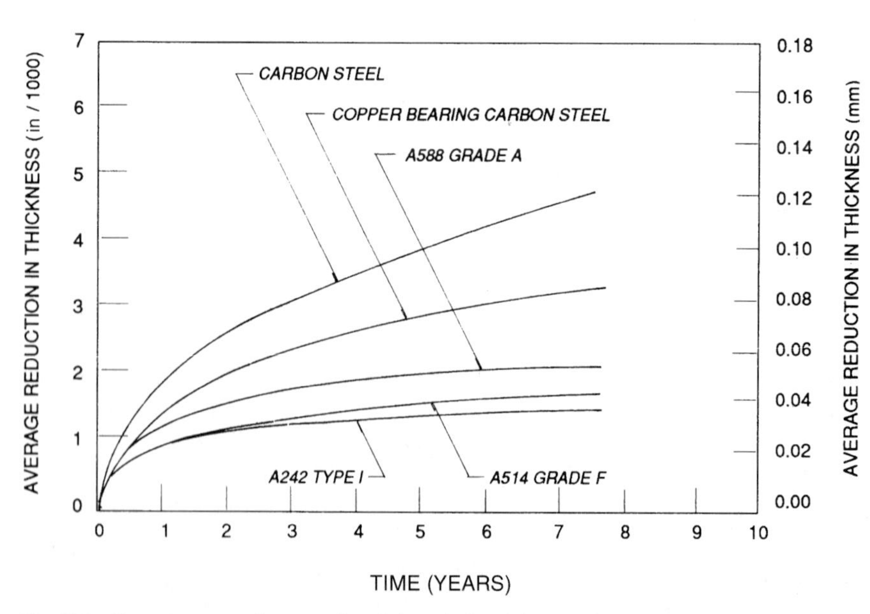


**Fig. 10.2 Corrosion curves for several steels in an industrial atmosphere.** (*Brockenbrough, 1992*).

## 10.6 APPLICATION OF STEEL

### 1 Product Forms

Steel is available in several product forms that are essential to the design and construction of tall buildings. The most familiar is the hot-rolled wide-flange section, which is used for columns and framing members, as well as for floor beams, bracing, and trusses. The wide-flange section has broad application because of its high structural efficiency in bending, its compactness that enables it to support high column loads with minimal sacrifice of the usable building areas, wide availability, and economical price. Hot-rolled angles and channels are also used as bracing and in other miscellaneous applications for tall buildings.

Hot-rolled plates are available for the fabrication of built-up sections, including box sections for columns, deep girders, and tapered members, as well as for connection plates for bracing members or trusses. They can also be used to fabricate shear wall systems that provide the major resistance to lateral loads. Sheet steel is used for the fabrication of decks, wall stud systems, curtain-wall systems, doors and door frames, and a variety of other components of tall steel buildings.

Structural steel tubing, including round, square, and rectangular sections, is being used increasingly in tall building applications, such as for lattice frames for tall atria. Advantages include structural efficiency in resisting compression, torsion, bending, and tension; clean uncluttered joints; and general aesthetic appeal. Although the unit material cost is typically higher than that of open sections, this tends to be offset by a lower total weight and a smaller painting area for exposed construction. One key to economical design with structural tubing is the design and detailing of joints. Information on this has been developed internationally (Wardenier et al., 1991).

Selecting the most economical steel for a given application will depend on a careful review of fabrication and erection costs, as well as the cost of material. Guidance can usually be obtained through local fabricators (Brockenbrough, 1983a).

### 2 Selection of Heavy Shapes

Numerous wide-flange sections have proliferated over the last decade or so to extend the normal range of application for this product. These "heavy shapes" have been used mainly in buildings of 30 or more stories. Applications include bending members in moment resisting frames, transfer girders to obtain wider column spacing at the base, interior long beam spans where column-free areas are desired, and heavy columns.

The heavy shapes offer several advantages. With a greater moment of inertia about the minor axis available for a given nominal section depth of columns, they can be used to reduce story height. This can be an important consideration where building height is limited by codes, and it may facilitate the inclusion of additional stories in the building. This may also reduce the amount of curtain wall required. Rolled sections tend to be more economical than equivalent built-up sections, and can usually be delivered quicker, unless special stocking programs for built-up sections have been adopted.

Although the use of heavy shapes can result in a more competitive system, the weight of the steel required for beams is likely to increase. Within limits, the deepest available beam will result in the least weight. For a given required section modulus, the depth of the beam can be reduced by 10 to 30% by using a heavy shape, but the weight will also increase by a similar percentage.

Heavy shapes may require special toughness considerations and fabrication practices as discussed previously.

## 3 Considerations in Seismic Zones

In seismic zones the record of steel in protecting life and minimizing property damage is historically good (EQE Engineering Inc., 1991). Indeed, either by itself or in conjunction with other materials, steel is the one material that is common to virtually all sound seismic design.

In general, resistance to significant earthquake motions must be achieved by energy dissipation in the nonlinear range of response. It would be uneconomical to design for an entirely elastic response. Because of the inherent ductility of steel, properly designed steel-framed buildings offer the best opportunity to withstand a seismic event without the need for extensive repair afterward. If repair is required, the cost is typically moderate.

Close attention must be paid to the design of ductile connections, because they are often the critical elements (Ali, 1986). Shop fabrication, with quality control provisions and inspection, can ensure that the desired performance level is achieved.

In addition to conventional systems, eccentric-braced frames (EBFs) in steel have been introduced in more recent years as economical systems to take advantage of the strength and ductility available (Popov, 1991; Roeder and Popov, 1978). Such frames typically have diagonal braces with at least one end connected to a beam a short distance from the beam-to-column connection. The short distance along the beam from the column to the brace connection, the link, is where most of the yielding and energy dissipation take place. The concept underlying EBFs is thus a major breakthrough for steel tall buildings in seismic zones. The benefits of wind bracing in increasing the building's lateral stiffness and the substantially increased ductility of the structural system can be readily realized in practice through the use of EBFs (Marsh, 1993).

Shear walls fabricated from steel plate for high-rise construction have been used in the United States and Japan and have demonstrated high strength and ductility in tests (Elgaaly and Caccese, 1990). This is yet another design choice that is likely to see expanded use in the future.

## 4 Considerations in Fire Protection

***Fire Resistance.*** The fire resistance of steel is particularly important for tall buildings. Structural steel does not support combustion and retains significant strength at elevated temperatures. Yet steel loses its strength under high temperature, and the necessity of fire protection for structural steel must be considered to protect life and property. Building codes usually specify fire resistance requirements based on such factors as height, floor area, and type of occupancy. Field-applied spray materials of cementitious or mineral fiber material are often the most economical choice. Gypsum wallboard or concrete encasement are other options. Although the approaches taken by the building codes in various countries differ somewhat, the focus is reportedly shifting from protection of the building structure and contents to provisions for safe egress for occupants and safe access for fire fighters (Almand, 1989). See Chapter CL-4 of *Tall Building Criteria and Loading* (CTBUH, Group CL, 1980) and *Fire Safety in Tall Buildings* (CTBUH, Committee 8A, 1992) for further information on fire resistance in tall buildings. Also, see Chapter 9 of this Monograph, in which the fire preventive design is treated from a mechanical system point of view.

In the past the resistance of an interior structural member to the effects of fire was usually determined in a standard fire test by observing the number of hours it successfully withstood exposure to a standard time-temperature curve. In more recent years, calculation methods have been developed and calibrated against the results of such tests. This makes it possible to determine the amount of fire protection material that must be applied to steel columns (AISI, 1980), beams (AISI, 1984), or trusses (AISI, 1981) to achieve a given required time of fire resistance. The calculation methods are available in a program for use on a personal computer (AISC, 1990b). The fire protection calculation relating to the practice in the United Kingdom has been presented by Law (1978). Design recommendations for European fire safety practice can be found in ECCS (1985).

Although the interior members of typical tall buildings require fire protection, steel may sometimes be used unprotected by other materials if a sprinkler system is included or if the application limits the proximity of combustible material.

When structural steel is located outside the building, the members are unlikely to reach the high temperatures that must be withstood by interior members in the event of a fire. Thus the location of members and the sizes of windows can often be selected to limit the temperature of unprotected members satisfactorily. Other alternatives that have been used successfully include liquid-filled (water and antifreeze) exterior columns of box or circular cross section, and flame shields to protect exterior spandrel beams from flames projecting through adjacent windows.

Because of the many variables involved, no standard fire test has been established for exterior structural members. However, rational analytical engineering methods have been established and confirmed by a large body of test data in various countries where exposed exterior steel structures have been built (AISI, 1979).

***Computer Programs.*** A more recent trend is the development of computer programs based on rigorous numerical simulation to calculate the structural behavior of steel or composite building frameworks at elevated temperatures (Schleich, 1987). The programs comprehend exposure to either standard or more realistic fire loadings based on the specific characteristics of the occupancy. This development includes full-scale fire tests to verify the simulation results. In addition, probability studies have been made to determine what structural loads should be considered to act in combination with the fire (Ellingwood and Corotis, 1990). Thus in the future, fire protection design is likely to become more closely integrated with the overall structural design of the tall building.

***Fire-Resistant Steel.*** Japan has introduced "fire-resistant" steels for building construction (Sakamoto et al., 1992). These are chrome-moly compositions, similar to those used for many years for process vessels, which maintain a higher percentage of yield strength at elevated temperature. They have been used in Japan to reduce or eliminate the amount of fire protection material required. However, the economics of using such steels depends greatly on the applicable fire codes, considering such things as the maximum temperature permitted and the extent to which a rational design method can be employed.

Additional information on steel tall buildings can be found in *Structural Design of Tall Steel Buildings* (CTBUH, Group SB, 1979) and in *Structural Systems for Tall Buildings* (CTBUH, Committee 3, 1994).

## 10.7 DESIGN CONSTRAINTS

During the late 1980s there were serious proposals to build the world's tallest building in the United States in New York, Los Angeles, Chicago, and Newark, New Jersey. (See also Chapter 5 of this Monograph.) While none of these came to fruition, construction began from 1970 to 1989 on more buildings over 50 stories than in all previous decades combined (CTBUH, 1986).

But around 1990, vacancy rates in the U.S. cities soared, the economy plummeted worldwide, and financing diminished. Except for China, Hong Kong, and the countries of Southeast Asia, it has been predicted that tall building design will not restart in large numbers until the existing surplus inventory of commercial building space is absorbed—and current estimates do not have that occurring until the late 1990s. However, if past trends hold true, the recovery of the construction industry will probably occur much sooner, particularly on a global scale. As a matter of fact, the next tallest building of the world, breaking the record of the Sears Tower, is under construction at the time of this writing. The 450-m (1476-ft)-high mixed steel and concrete building now under construction, known as Petronas Towers and located in Kuala Lumpur, Malaysia, is scheduled for completion in 1996.

As infrastructure costs are escalating rapidly worldwide, it is expected that construction recovery is likely to start primarily in urban areas. Because cities already have the foundation, albeit a decaying one, of extensive infrastructures, the development in the mid-1990s is expected to concentrate in urban areas, rather than suburban and rural areas. Since land costs are so high, developers may turn to high-rise buildings in urban areas by taking advantage of existing infrastructures in these areas.

When high-rise construction restarts, several trends that began in the 1980s are likely to continue. For the construction industry, the two most important ones are expected to be the move toward smaller floor plates, which directly affects the design of the building's structural system, and the increasingly complex designs mandated by unusual site constraints.

Two case studies of steel buildings located in two major U.S. cities, New York and Chicago, are presented here to illustrate the different design constraints for tall steel buildings that are faced by structural designers.

## 1 Morton International Building

***Site Constraints.*** For an architect or developer, the site of the Morton International Building shown in Fig. 10.3 is a dream come true. For an engineer, it could have been a nightmare. The building sits at the edge of Chicago's downtown area and is bordered by a roadway to the south, a cantilevered roadway to the west, and a city park and the Chicago River to the east. In addition it is an "air rights site" located above the active Metra and Amtrak railroad tracks, which lead to the nearby Union Station, a major commuter rail station. The air rights site lease included a caisson location diagram that depicted the only allowable locations for the building to "touch down." Still, despite the design challenges, it was a desirable site for its leasing advantages that included proximity to a major rail station, a prime location in an economically expanding area and an enviable view of the Chicago River.

***Program Requirements.*** General program requirements from the building's owner consisted of a 36-story building containing approximately 6503 $m^2$ (700,000 $ft^2$) of rentable office space and 1859 to 2322.5 $m^2$ (20,000 to 25,000 $ft^2$) of retail space with 2.7-m (9-ft)-high ceilings. In addition, the office space needed 4.5-m (15-ft) floor-to-ceiling heights to accommodate an anticipated 0.6-m (2-ft) raised-access computer floor, and several floors were designated for data center use and needed heavier than normal live-load capacity (Anderson and Karabatsos, 1991).

***Structural Constraints.*** The first problem for the project's architect and structural engineer was to deal with the railway passing beneath the site. Readily apparent from the caisson location diagram and the railroad track layout was the fact that the southwest corner of the site was inaccessible for foundation placements. At this location, the tracks curve and switch to make their final alignment to the station. In accordance with published railroad safety clearance requirements, there was no room for even the smallest foundation or supporting column in the southwest corner of the site.

A second foundation problem was that due to the multiple tracks, switches, and signal lines, a grade beam system tying the individual caissons together could not be installed. Since the building is on an air rights site, conditions precluded building a basement. This required the caissons to transmit all lateral wind loads from the structure to the soil.

**Fig. 10.3 Morton International Building, Chicago.** (*Courtesy: Gregory Murphey.*)

Due to the inability to place foundations in the southwest corner, initial schemes for the parking levels and the data center floor plates had a notch in the corner. However, this open area resulted in restricted and narrow floor plans, creating awkward space for planning a data center, and especially for planning a parking garage. Also, during this time, a prospective tenant for the data center indicated that he would only be interested in the building if the notch were filled. An interior truss system was considered to fill in the notch, but was eventually rejected due to the disadvantage in space planning for the parking facilities and data center, created by having large sloping columns running through the building. From these analyses emerged the concept for an exterior overhead truss.

The earliest designs for the overhead truss envisioned the use of cables similar to a suspension bridge. However, the general contractor and the architect and engineer concluded after studying this scheme that the connections of the cable members would be difficult.

It was then decided that the trusses should be built up from standard steel structural shapes and plates. Bolts were used for all tension connections and welding for compressive connections. The major truss members are built up from six 102- by 51-mm (4- by 2-in.) plates. The bolted connections are made with 35-mm (1⅜-in.)-diameter ASTM 490 bolts. Special paint was used to increase friction between the plates at the connections.

Typical connection knuckles were prefabricated and erected. These knuckles were used for both the overhead trusses and the lower trusses near level 1. The knuckles in compression were built up from four 152.5-mm (6-in.)-thick plates welded together. Vibratory stress relaxation techniques were used on the steel plates to minimize stress fractures and lamellar tearing of the plates during welding.

The overhead trusses cantilever 16.5 m (55 ft) and suspend 14 floors of data center and parking garage operations. Due to the unbalanced load condition created by the suspended floors, the vertical frames that support the overhead trusses were designed as cantilevered frames.

***Construction Sequence.*** The construction sequence of the framework was to build it 90 mm (3½ in.) out of plumb, toward the river. As the overhead trusses and floor structure were suspended on the west, the unbalanced loads caused the vertical frames to drift into a vertical plumb condition under dead load and partial live load.

During construction, column shafts were assembled on the ground and put into place by a crane. Once the hanging columns were in place, the floors were erected from the ground up—no different than a normal construction sequence. This construction sequence afforded safety for the construction crew by placing the floor decking below. Also, the trains were protected from potential falling objects.

***Structural Systems.*** Typically, an economical lateral system for a 36-story building would be a braced core. The railroad tracks, however, prohibited the use of that system. Further, there was no way to distribute the wind loads from a braced core to the caissons.

Instead, a tube structure, with the exterior columns spaced at 4.5 m (15 ft) on centers, was selected for the tower of the building. Exterior columns and spandrel beams were fabricated fully welded at the beam-column connection and were installed with a shear plate connection at the midspan of the beam. These "trees" were fabricated with two-story columns, as is common practice with tube structures.

Although not originally intended, the building's aesthetic expression leans heavily toward structuralist architecture because of the tubular structural skeleton and the large overhead trusses. (For additional information on the Morton building, refer to Chapter 7.)

## 2 750 Seventh Avenue

***Site Constraints.*** The site played a key role in the shape of the 750 Seventh Avenue building in New York City. The building is designed as a modern interpretation of the ziggurat, an ancient building form used in constructing temple towers. In form, it narrows as it increases in height, and is marked by spiraling outside staircases. The New York City structure is a glass-clad tower gradually stepping back along three sides, creating the illusion of giant steps ascending the structure's face (Fig. 10.4).

The building's design emerged at least in part as a result of the City's building code, which requires setbacks on all streets, and in this case that meant setbacks on three sides. The architect then took that requirement and added visual interest by designing them as sloping spirals.

***Structural Constraints.*** While visually fascinating, the unusual design, however, creates numerous structural challenges. On the lower levels the setbacks are as large as 3.1 m (10 ft), but become progressively smaller, reaching 1.1 m (3.5 ft) at the upper level. Adding to the complication was the owner's desire for column-free space. Had sloped columns been used, very large, unbalanced horizontal forces would have been introduced, especially at the lower floors.

The only way to satisfy the architectural design and the owner's requirements was to use a large number of transfer girders. Because the transfer girders are typically restricted to 1.1-m (3.5-ft) depth to fit within the ceiling construction, at the lower levels they have flanges up to 101.6 mm (4 in.) thick. The transfer girders had to be within the depth of the ceiling, which was typically 1.2 m (4 ft) from finished floor to finished ceiling.

Further complicating the design were wind tunnel tests indicating much larger overturning forces than the New York City code requires. As a result, the structure was designed as a "telescoping tube" with the girders getting heavier at the bottom. In plan, the tube is similar to an old sailor's telescope. As each section of the telescope resists the wind load, it transfers the load to the larger section below (AISC, 1990a).

***Structural System.*** The column spacing is 4.5 m (15 ft) on centers on the exterior. The closely spaced columns on the exterior combine with the spandrel beams to create a tube system. The braced core system helps transfer wind shear at the setbacks from one section to another. Because the architectural design did not allow corner columns, a modified tube system with cantilevers at each corner was designed.

Although there was a considerable penalty in steel tonnage due to column transfers, the structural engineers economized in the use of steel through efficient design of long-span composite beams with high-strength 275.8-MPa (40-ksi) composite metal deck. The beams were typically spaced 4.5 m (15 ft) on centers. This spacing and spans up to 14.6 m (48 ft) made it possible to take the maximum advantage of the live-load reduction allowed by the New York City code and minimized the number of steel pieces.

## 10.8 THE SPINE STRUCTURE

Today unusual architectural shapes are becoming the norm rather than the exception in tall building design. Further information about the form of tall buildings may be found in Chapter 5. Whereas the tall buildings of the past featured a wide variety of structural systems, they all shared a common regular, rectilinear form. When one considers tube

**Fig. 10.4 750 Seventh Avenue, New York.** (*Courtesy: Nathaniel Lieberman.*)

buildings such as the Sears Tower, the Amoco Building, and the John Hancock Center in Chicago or the World Trade Center in New York, one can observe that they all are essentially rectangles in plan. They were all built prior to the 1980s, when tall building architecture began featuring nonrectangular plans, multiple step backs, distinctive tops, tall and open entrance lobbies, and subterranean parking. These features often place a cost premium on closely spaced columns employed in tubular systems. As such, efforts were undertaken by structural designers to devise other solutions for tall building structural schemes.

An innovative steel structural system that emerged in the recent past is the so-called spine structure, which is felt to be an especially appropriate structural system for tall buildings in high seismic areas. In this type of structure, the spine is the lateral load resisting system that provides stiffness and stability. The spine consists of vertical or inclined elements, and shear membranes in the form of braced frames, walls, or Vierendeel girders. The vertical or inclined elements resist axial loads due to overturning moment, whereas the shear membranes resist shear forces due to lateral loads (Banavalkar, 1991). Two illustrative examples of spine structures are briefly reported in the following.

### *1 First Interstate World Center*

The 338-m (1108-ft)-tall First Interstate World Center (Library Square) in Los Angeles, shown in Fig. 10.5, is the world's tallest building in a UBC seismic zone 4 or equivalent at the time of this writing (UBC, 1991). The architects created a design with a number of step backs. While the simplest and most obvious structural system would have been a perimeter ductile tubular frame, this solution was very expensive and its dynamic characteristics were sensitive to wind-induced acceleration at the top floor.

Instead, the structural designer for the project examined the potential for designing a braced spine structure. This was rejected, however, because it was felt that the ductility of the structure to absorb energy in case of an unforeseen seismic event was below the desired level. Also, the diagonals crossing through the floor space presented a difficulty in interior space planning.

Instead, a design was developed featuring an uninterrupted braced square spine. The design of this spine was based on the deterministic approach for the ground motion caused by the maximum credible earthquake (100-year return period) on the San Andreas Fault assuming 5% damping.

The final design of the structure consists of a 22.5-m (73-ft 10-in.) square spine composed of a two-story-tall chevron-braced core with a perimeter ductile moment resisting frame having radially placed columns at 15° on centers.

The long-span floor framing coupled with a two-story-tall free spanning core loaded the corner core columns of the spine in such a way that the design was primarily governed by gravity load. To avoid any progressive catastrophic failure of the structure in case of buckling of a chevron-braced diagonal, two separate possible modes of failure were considered. In mode I the buckled diagonal assumes to have lost only axial load-carrying capacity, whereas in mode II the lower end of the buckled diagonal was physically removed. The safety of the structure was ensured by checking the remaining effective members and designing the connections for all possible load combinations.

An attempt was made to proportion the stiffness of the perimeter frame and core bracing in such a way as to initiate yielding in the ductile frame prior to buckling of the core diagonals. To establish the postyield behavior, a simplified nonlinear analysis was conducted by applying monotonically increasing load on the structure.

### 2 Figueroa at Wilshire Tower

Another spine structure in Los Angeles is the 53-story Figueroa at Wilshire tower. Three lateral load resisting systems were studied:

1. Perimeter ductile moment resisting frame
2. Dual system with braced (concentric or eccentric) core with perimeter ductile frame
3. Spine structure consisting of braced core and outrigger ductile frames with the long-span floor framing structured in such a way that the main columns participating in the lateral load resisting system are loaded heavily by gravity load

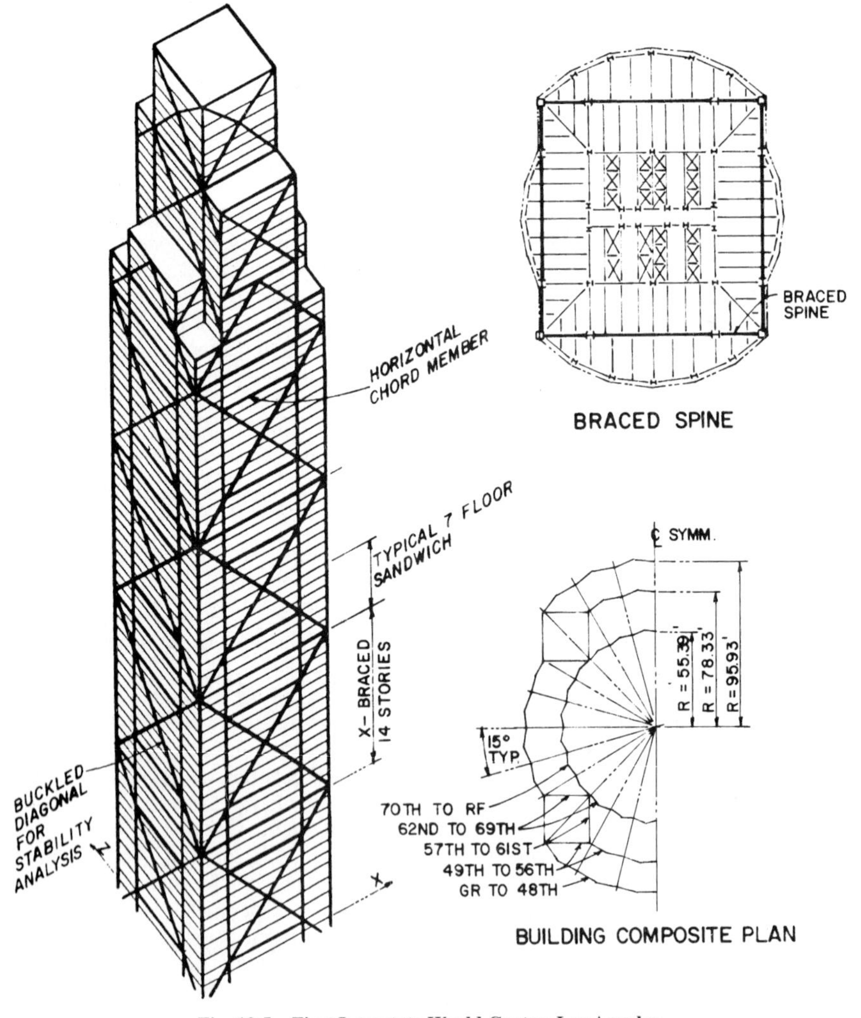


**Fig. 10.5 First Interstate World Center, Los Angeles.**

A cost comparison of these systems showed that the spine structure used the least amount of structural steel and provided an opportunity for column-free lease spaces. The interaction of the braced core and the outrigger ductile frames about both principal axes is essentially the same. In general, more than 65% of the shear is carried by the braced core, whereas almost 65% of the overturning moment is resisted by the perimeter columns in the outrigger frame.

Essentially the structure has three major components: interior core bracing, outrigger beams, and columns. The interior core is concentrically braced. The most important design criterion of the braces is that in the event of buckling of one of the compression diagonals, the horizontal members supporting the floor loads do not form a mechanism resulting in catastrophic failure of the floor.

The outrigger beams span approximately 12 m (40 ft) and have three major functions: (1) they support the design floor loads; (2) along with the core and the perimeter columns they act as a ductile moment resisting frame to carry a minimum of 25% of the design code level forces without the presence of interior core bracing; and (3) they have to possess enough stiffness to create effective linkage between interior core and perimeter columns to provide efficient overturning resistance to the seismic loads. To accomplish the latter function, 915-mm (36-in.)-deep outrigger beams were required.

Both perimeter and core columns were heavily loaded to provide overturning resistance to the entire structure. The perimeter columns were checked against full plastification of the outrigger beams.

## 10.9 COMPOSITE CONSTRUCTION

The term *composite construction* has assumed several meanings in recent years as a result of numerous creative combinations of steel and concrete in buildings, both for gravity systems (such as floor framing) and for lateral load resisting systems. The basic idea of using composite systems is to combine the benefits of both steel and concrete in a structure such that one complements the other in terms of strength, stiffness, member size, speed of construction, fire resistance, and other factors (Iyengar, 1977). In effect, such systems could be termed mixed or hybrid systems. In many instances, these systems are more economical than either an all concrete or an all steel system (Taranath, 1988).

In the following, three buildings that employed composite systems are presented as examples to demonstrate the application of such systems.

### 1 Home Savings Tower

Along with the move toward step backs, architects began to create buildings in the late 1970s and throughout the 1980s with greater sculptural elements. One of the best examples of the effect this can have on a building's structural form is the Home Savings tower in Los Angeles. The 24-story building, employing a composite system, is reminiscent of many older New York City buildings designed in the French chateau style. However, the building was designed so that the structural steel skeleton did not have to conform to the outside skin of the building.

Freeing the skin from the structure allowed the architect to get away from designing a glass box and instead allowed him to create a masonry appearance. Marble, precast concrete, and ceramic tile were hung on the steel frame to evoke an older architectural style. "Returning to masonry construction would be patently absurd, but adopting a masonry sensibility affords all the advantages of steel construction" (AISC, 1990a).

The building's office portion sits atop a below-ground Metro subway station and five levels of above-grade parking. The subway station is in concrete, with the steel frame beginning just below grade, instead of at the foundation. Seismic and wind shear forces were transferred at the plaza level to concrete ductile columns at the perimeter of the building. The perimeter steel beams, which are 0.9 m (3 ft) in depth, are welded to the steel erection columns.

The steel frame slopes inward at the sixth floor and again at the twenty-fourth. In addition to providing the aesthetic advantages that the architect desired, the slopes reflect the various functions of the building. Because the first five stories comprise a parking structure, the columns are pushed to the perimeter to minimize interference with vehicles. At the sixth floor there is a transition of these columns to fit the office space function. Rather than making an abrupt transition in the ductile frame, a decision was made to slope the frame and build up outriggers to pick up the vertical cladding—marble at the base and precast concrete above.

Complicating the project, however, was the need for the columns to slope in two directions, both a setback and a lateral slope. Because they slope in more than one direction, double-beveled cuts were required, which meant very tight controls during the fabrication stage.

The columns slope 0.7 m (2.5 ft) per story; some slope through just one story, whereas others slope through the entire two-story sky lobby. They form the new plate for the office space, and then they continue straight up to the penthouse level, where they sharply slope in to form the mansard roof. To create rounded corners, the columns above the sixth floor are moved 1.5 m (5 ft) back into the building.

### 2 Mellon Bank Center

The designers of the 55-story Mellon Bank Center in Philadelphia, Pennsylvania (Fig. 10.6) wanted the advantages inherent in a steel-framed structure, but they faced a problem. Wind tunnel tests revealed a vortex shedding problem with the crosswind structural response 50% larger than the forces stated in the building code (AISC, 1989). The structural engineer concluded that a composite structural system would be the most economical for this building. In addition to stiffening the frame, the composite design reduced the need for fire-proofing.

The 242-m (794-ft)-tall building is designed with tapering corners, which reduces the perceived mass of the structure. There are a six-story base and a tower portion with 47- by 47-m (154- by 154-ft) dimensions.

The building's lateral system is formed by the composite perimeter columns spaced 2.8 m (9.5 ft) on centers, forming a perimeter tube. Typical composite column schemes utilize the steel columns solely for erection purposes, with the bulk of the vertical load carried by the concrete. In this structure, however, restrictions on the overall size of the columns required the use of truly shared composite columns, with the concrete encasement and the steel columns both carrying significant portions of the vertical load.

Complicating the project was the fact that none of the 52 columns in the tower portion of the structure continued directly to the ground level. Instead, all of the perimeter columns were either sloped or picked up by trusses. The sloped column system enabled the columns to transfer into new positions, allowing for the enlargement of the lower floor sizes and still maintaining column-free lease space. Depending upon the different architectural constraints, groups of columns sloped at different floors. However, the sloped columns always formed a symmetrical system whereby sloped columns on opposite sides of a floor balanced out the overturning forces resulting from the slope.

In numerous cases, columns are terminated upon pick-up trusses that also are linked with their repositioned supporting columns. A unique sloped column system occurs between the tenth and thirteenth floors, where the four inside corner columns are picked up on a "two-legged tripod." Each two-legged tripod generates significant lateral forces that are all balanced out by again balancing one corner against the opposite end. The floor diaphragm, being the link between all columns, plays a key role in transferring these balancing forces across the floor. The most critical diaphragms are the fifth- and sixth-floor diaphragms, where in addition to supporting most of the sloped columns, the lateral wind forces are transferred from the perimeter to the core vertical truss.

**Fig. 10.6 Mellon Bank Center, Philadelphia, Pennsylvania.** (*Courtesy: B & H Photographics.*)

With some sloped columns generating 2002 kN (450 kip) in lateral force, the designers chose to place a 13.5-m (44-ft)-deep steel horizontal truss within the floor diagram. These trusses help transfer the wind forces to the core while passing the sloped column forces around the core to the opposite sloped columns.

Further complicating the design are sloped columns throughout most of the tower portion. Because the cut-out corners taper in above the tenth floor, 12 columns per floor (that is, three at each corner), sloped about 76 mm (3 in.) per floor.

Perimeter column sizes range from W14 × 400 wide-flange shapes (AISC, 1989) at the base to W24 × 76 wide-flange shapes near the top. To capitalize on the better bending capability of deeper shapes, beginning at the sixth floor, the engineer used WTM22 shapes. The shapes vary in weight, but average 0.6 m (24 in.) in depth. The WTM22 shapes averaged between 82.5 and 122 kg/m (182 and 269 lb/ft).

A vertical supertruss is located at the core that extends from the foundation up to the sixth floor. The supertruss is constructed of steel wide-flange shapes with the four corner columns being encased in 3- by 3- by 0.6-m (10- by 10- by 2-ft)-thick L-shaped concrete shear walls, thereby forming a composite steel and concrete supertruss. The supertruss is divided into two parts—a large 13.7-m (45-ft)-high truss between levels three and six, and a single cross truss on each face of the core, extending from the third level down to the foundation.

### 3 Proposed Miglin-Beitler Tower

The engineers for the Miglin-Beitler Tower also faced the problem of very small floors in their design of a proposed 610-m (2000-ft) structure in Chicago (see Chapter 2). They turned to composite design for the building.

The challenge to the design team was to create an economical and buildable structural frame capable of resisting vertical and lateral loads in a supertall building with a relatively small footprint. The stiffness of high-strength concrete was combined with the advantages of a steel floor system, including its inherent strength, speed of construction, and flexibility to allow tenant changes (Thornton et al., 1991).

The structural design of the building involved a cruciform tube system with six major components. The first is a 19.0-$m^2$ (62.5-$ft^2$) concrete core with walls of varying thickness. The interior cross walls of the core are generally not penetrated with openings. This contributes significantly to the lateral stiffness.

The second component is eight cast-in-place concrete fin columns located on the faces of the building and extending up 6 m (20 ft) beyond the 42.7- by 42.7-m (140- by 140-ft) tower footprint. They vary in dimension from 2 by 10 m (6.5 by 33 ft) at the base to 1.7 by 4.5 m (5.5 by 15 ft) at the middle to 1.4 by 4 m (4.5 by 13 ft) near the top.

The third is exterior steel Vierendeel trusses consisting of horizontal spandrels and two vertical columns located at each of the 18.6-m (61-ft)-wide faces on the four sides of the building between the fin columns. Exterior steel Vierendeel trusses are used to pick up each of the four cantilevered corners of the building. These Vierendeel trusses provide further resistance to lateral forces, in addition to improving the resistance of the entire structural system to torsion. Further, the trusses transfer dead load to the fin columns to eliminate tensile and uplift forces in the fin columns. All corner columns are eliminated for providing the corner offices with undisturbed views.

Eight link beams, the fourth component, connect the four corners of the core to the eight fin columns at every floor. These reinforced concrete beams are haunched at both ends for increased stiffness and reduced in depth at midspan to allow for the passage of mechanical ducts. By linking the fin columns and core they enable the full width of the building to act in resisting lateral forces. In addition to link beams at each floor, sets of two-story-deep outrigger walls are located at levels 16, 56, and 91. These outrigger walls enhance the interaction between exterior fin and columns and the core.

The fifth component is the conventional structural steel composite floor system having 0.5-m (18-in.)-deep rolled steel beams spaced at approximately 3 m (10 ft) on centers. A slab of stone concrete topping spans between the beams. The steel floor system is supported by the cast-in-place concrete elements.

The sixth and final component is a 183-m (600-ft)-tall steel-framed tower that tops the building. This braced frame will house observation levels, window washing equipment, mechanical equipment rooms, and an assortment of broadcasting equipment.

On each of the four faces of the building, steel Vierendeel trusses are employed to frame the 18.5-m (61-ft) clear opening between the fin columns. The trusses consist of a W36 horizontal beam at each level with two W36 verticals. To eliminate stresses produced by creep and shrinkage strains in the concrete fin columns, the verticals in the truss are provided with vertical slip connections. This has the added benefit of channeling all gravity loads on each of the building faces out to the fin columns to help eliminate uplift forces on the foundations.

The steel face Vierendeels are to be shop fabricated as horizontal trusses 3.8 m (12.5 ft) tall by 18.6 m (61 ft) long. Field connections are simple bolted connections. This system allows for all of the welded connections to be shop fabricated, which results in an economical and speedy solution.

At each of the corners of the Miglin-Beitler Tower, the floor slabs protrude beyond the fin columns by up to 8 m (26 ft). It was desirable to channel the gravity loads from these areas to the fin columns to help eliminate uplift in the foundations due to lateral loads.

An added challenge was to frame the corners without having a vertical element at the corner, thus allowing the corner offices unobstructed views. The solution to this problem was a Vierendeel truss, similar to the face Vierendeels. The corner Vierendeels are to consist of a horizontal steel beam at each level with a vertical steel beam at the centers of each face, thus allowing the corners of the building to be column-free. Unlike the face Vierendeels, all of the vertical connections are not slip connections. This is to allow the corner Vierendeels to resist unbalanced floor loads. (For another discussion of this building from the structural planning viewpoint, see Chapter 2.)

## 10.10 FUTURE RESEARCH

While the state of the art of tall steel building design is now well advanced and sophisticated, several unresolved structural issues and questions continue to attract the attention of research investigators and practitioners. Many are predominantly minor in nature, involving incremental refinements of existing quantitative procedures. Other topics are much broader in scope and reflect possible major innovations or trends in design technology. This latter category is certainly the more interesting and challenging, and, therefore, is the focus of the following general description of research needs.

### 1 Major Research Needs

***Serviceability.*** The primary load-carrying capabilities of structural components and assemblages have become well established and more accurately represented. This progress, coupled with the use of higher-strength materials, has resulted in lighter structural requirements for a minimum acceptable level of strength and safety. The advent of limit-states design (LSD), referred to as the load and resistance factor design (LRFD) in the United States (AISC, 1986), is another factor contributing to lighter member design. Consequently, serviceability has frequently become the governing limit for design, particularly in tall slender buildings. Controlling drift, wind or seismic accelerations, floor deflections, and vibrations are the main concerns relative to the owner's use and occupant satisfaction. Their design adequacy is thus more subjective.

It follows from the foregoing that the stiffness of the structure, its mass distribution, its damping characteristics, the method of analysis and its correlation to an acceptable response index, and selected recurrence intervals of the service loadings are all related variables. Various approaches can be utilized successfully to design a serviceable structure. These should be better documented for easy reference. In addition, new or special concepts for the mitigation of potential problems such as tuned mass or viscoelastic dampers, base isolation mechanisms, unique structural systems, and the inclusion of nonstructural elements (curtain walls and room partitions) in framing design can be further explored. Such work on serviceability issues is needed to keep pace with advances on the strength reliability of structural steel members and connections.

***Advanced Analysis and Computer-Aided Design.*** Even though these are two separate fields of expertise, they have become intimately linked because past exclusive reliance on manual calculations significantly constrains the potential of analysis and design techniques. Thus the pairing of improved but more complex structural models and solutions with the ever-increasing computer power can bring the simulation of structural response to more precise levels. Whereas linear elastic analysis had been the standard for years, second-order effects, inelasticity, stability, three-dimensional behavior, and semirigid connections are now being directly considered in several research studies, and are being at least partially implemented in some architectural-engineering (A-E) design offices. In the future, more application and acceptance of computer softwares incorporating such advanced analysis are expected.

Problems associated with combining such advanced technologies of structural design and computer processing can be imposing, such as verification of software accuracy for a variety of problems, user education and training, and the proper role of human engineering judgment, just to name a few. New analysis and design software needs to be foolproof, user friendly, and well documented for the average structural designer. As understanding and experience of the A-E profession grow in this area, the intended benefits can be realized in terms of optimized and more realistic designs.

***Earthquake-Resistant Design.*** Analysis and design of structures in earthquake zones is being largely performed through simplified models, assumptions, and coefficients. Some of these possess little or no rigorous justification, but represent consensus of perceptions on historical relative performance levels. With the anticipated progress in advanced analysis discussed previously, parallel advances in the more specialized seismic design field can be similarly attained. Specifically, such difficult topics as soil-structure interaction, time-history response (with nonlinearities), ductility under cyclic loads, and energy balance between earthquake motion record and the structure require detailed quantitative evaluation. The current seismic design limitations could thereby be overcome and replaced with more rational criteria.

## 2 Future Developments

Future innovations in steel design technology should focus on the development of the following items:

1. Simple production techniques for heavy structural shapes with improved toughness
2. Improved rational methods for rapid analytical determination of fire protection requirements
3. An economical system for shop-applied fire protection of structural members
4. Further research on fire-resistant steel for tall buildings
5. Better defined serviceability requirements to negate unneeded restrictions to the use of higher-strength steels
6. Economical and easily weldable steel with high strength, good ductility, and strain-hardening capacity for improved structures in seismic zones
7. Improved, economical, ready-to-use connection systems for steel-framed buildings (currently in progress at Lehigh University), including connections of tubular members
8. Devising new techniques for increasing building stiffness for the currently prevailing trend of small floor plans
9. Improved criteria for the design of steel-concrete composite framing systems, including connection systems
10. Improving the speed of construction by developing better field erection techniques

## 10.11 CONDENSED REFERENCES/BIBLIOGRAPHY

The following is a condensed bibliography for this chapter. Not only does it include all articles referred to or cited in the text, it also contains bibliography for further reading. The full citations will be found at the end of the volume. What is given here should be sufficient information to lead the reader to the correct article—author, date, and title. In case of multiple authors, only the first named is listed.

AISC 1973, *The Design, Fabrication, and Erection of Highly Restrained Connections to Minimize Lamellar Tearing*

AISC 1986, *Load and Resistance Factor Design Specification for Structural Steel Buildings*

AISC 1989, *Steel Construction Manual—ASD*

AISC 1990a, *Reinterpreting an Ancient Form*

AISC 1990b, *Stemfire: Steel Member Fire Protection Program*

AISI 1979, *Fire-Safe Structural Steel: A Design Guide*

AISI 1980, *Designing Fire Protection for Steel Columns*

AISI 1981, *Designing Fire Protection for Steel Trusses*

AISI 1984, *Designing Fire Protection for Steel Beams*

AISI 1985, *Steel Product Manual—Plates*

Ali 1986, *Seismic Design of Steel Tall Buildings*

Almand 1989, *Fire Resistance Requirements in Building Regulations: An International Comparison*

Anderson 1991, *Train Tracks Detour Building's Foundation*

ASCE 1988, *Wind Drift Design of Steel-Framed Buildings*

ASM 1978, *Properties and Selection: Irons and Steels*

ASM 1985, *Carbon and Alloy Steels*

Banavalkar 1991, *Spine Structures Provide Stability in Seismic Areas*

Barsom 1987, *Material Consideration in Structural Steel Design*

Bouwcentrum/Rotterdam 1963, *Modern Steel Construction in Europe*

Bresler 1968, *Design of Steel Structures*

Brockenbrough 1970a, *Experimental Stresses and Strains from Heat Curving*

Brockenbrough 1970b, *Theoretical Stresses and Strains from Heat Curving*

Brockenbrough 1981, *USS Steel Design Manual*

Brockenbrough 1983a, *Considerations in the Design of Bolted Joints for Weathering Steel*

Brockenbrough 1983b, *Structural Steel Design and Construction*

Brockenbrough 1992, *Material Properties*

Brockenbrough 1992, *Metallurgy*

Coburn 1987, *Theory and Practice in Use of Weathering Steels*

Colaco 1974, *Partial Tube Concept for Mid-Rise Structures*

Condit 1964, *The Chicago School of Architecture: A History of Commercial and Public Building in the Chicago Area, 1875–1925*

CTBUH 1986, *Tall Buildings of the World*

CTBUH, Committee 3, 1994, *Structural Systems for Tall Buildings*

CTBUH, Committee 8A, 1992, *Fire Safety in Tall Buildings*

CTBUH, Group CL, 1980, *Tall Building Criteria and Loading*

CTBUH, Group SB, 1979, *Structural Design of Tall Steel Buildings*

ECCS 1985, *Design Manual on the European Recommendations for the Fire Safety of Steel Structures*

Elgaaly 1990, *Steel Plate Shear Walls*

Ellingwood 1990, *Load Combinations for Buildings Exposed to Fire*

EQE Engineering Inc. 1991, *The Performance of Steel Buildings in Past Earthquakes*

Giedion 1941, *Space, Time, and Architecture*

Golderger 1981, *The Skyscraper*

Grinter 1949, *Theory of Modern Steel Structures*

Hart 1985, *Multi-Storey Buildings in Steel*

IISI 1987, *Pressure Vessel and Structural Steels*

Iyengar 1977, *Composite of Mixed Steel–Concrete Construction for Buildings*

Komp 1987, *Atmospheric Corrosion Ratings of Weathering Steels—Calculation and Significance*

Lankford 1985, *The Making, Shaping and Treating of Steel*

Law 1978, *Fire Safety of External Building Elements—The Design Approach*

Marsh 1993, *Earthquakes: Steel Structures Performance and Design Code Developments*

Mumford 1952, *Roots of Contemporary American Architecture: A Series of Thirty-Seven Essays Dating from the Mid-Nineteenth Century to the Present*

Ohashi 1990, *Development of Steel Plates for Building Structural Use*

Popov 1991, *U.S. Seismic Steel Codes*

Roeder 1978, *Eccentrically Braced Frames for Earthquakes*

Sakamoto 1992, *High-Temperature Properties of Fire Resistant Steel for Buildings*

Salmon 1990, *Steel Structures: Design and Behavior*

Schleich 1987, *Numerical Simulations, the Forthcoming Approach in Fire Engineering Design of Steel Structures*

Schueller 1977, *High-Rise Building Structures*

Structural Stability Research Council 1979a, *A Specification for the Design of Steel-Concrete Composite Columns*

Structural Stability Research Council 1979b, *Plastic Hinge Based Methods for Advanced Analysis and Design of Steel Frames*

Taranath 1988, *Structural Analysis and Design of Tall Buildings*

Thornton 1991, *Looking Down at the Sears Tower*

UBC 1991, *Earthquake Regulations*

Wardenier 1991, *Design Guide for CHS Joints under Predominantly Static Loading*

White 1987, *Building Structural Design Handbook*

# 11

# Materials and Structures: Concrete

The rapid development of concrete technology, methods of design, and means of construction during the recent past has facilitated the evolution of concrete into a viable structural material for tall buildings. This chapter primarily focuses on the new material and structural technology that has been developed and the unique characteristics of concrete favorable to the design of tall buildings. After providing a historical perspective, Sections 11.1 through 11.3 describe the material properties of concrete and the structural systems and concepts that have emerged in the recent past. In Section 11.4 information is presented on structural material selection, establishing the principal factors in structural materials that affect building costs and the total economy of the system. Sections 11.5 through 11.7 discuss new developments, particularly in high-strength concrete for seismic regions and high-performance concrete, whose superior strength and performance properties make it advantageous for tall buildings. Section 11.8 discusses reinforcing steel as a material for concrete tall buildings and introduces a twisted high-strength reinforcement bar. Section 11.9 offers some suggestions for future research.

## *11.1 HISTORICAL PERSPECTIVE*

Although concrete as a material has been known since ancient times, the practical use of reinforced concrete, which is a logical union of two materials (concrete and reinforcing steel), was only demonstrated in 1867 by Joseph Monier in Paris (Straub, 1964). Several others also demonstrated its practical use during the late nineteenth century. The invention of reinforced concrete considerably increased the importance and use of concrete in construction. Unlike steel, which was dominating the high-rise construction scene at that time (see Chapters 2, 5, and 10), concrete had the unique property of moldability, which allowed architects and engineers to shape the building and its elements in different elegant forms. It also had good fire resistance properties. However, it was yet no match for structural steel as far as strength was concerned.

## 1 Early Development of Reinforced Concrete

Extensive research on reinforced concrete in the United States and Germany in the early twentieth century established concrete as a material for building construction. Condit (1961) writes that most of the vital technical innovations in building construction since the development of iron framing have occurred in concrete construction. Primarily of French and German origin, these innovations have resulted in an increased efficiency of structures, greater variety of forms available to the designer, and, most significantly, a radical transformation of the whole empirical and aesthetic character of building forms. A comprehensive account of early developments of concrete high-rise buildings has been presented by Condit (1961) and Fintel (1985). The material presented here is largely based on Condit's and Fintel's accounts of the historical development of concrete.

Prior to World War I, concrete was used in the foundations of steel high-rise buildings and occasionally for floor slabs. However, after World War I, reinforced concrete multistory buildings were being built using flat slabs with column capitals. Apartment buildings up to 14 stories high were also being constructed. The structural principles of reinforced concrete were being explored only in the last quarter of the nineteenth century. Its low cost, its strength, durability, and continuity, and the practically unlimited range of forms in which it could be cast, made reinforced concrete economically feasible as a building material (Condit, 1961).

The first applications of reinforced concrete in building construction projects utilized reinforced concrete planks as if they were steel or wood supported by girders spanning from wall to wall and from column to column. In the beginning, concrete slabs were only used for small spans such as covering drains, projecting as balconies, and as infilling between joists (Bill, 1949). As early as 1909, the Swiss structural engineer Robert Maillart recognized the potential of reinforced concrete slabs in building construction: "The possibilities of reinforcing the concrete slab in a crosswise manner, however, introduced a new type of building element, capable of taking bending stresses in any direction, not only in the two directions of reinforcement..." (Bill, 1949).

## 2 Modern Developments

According to Maillart, the first innovations in reinforced concrete technology, which was primarily developed in France and Germany, were due to the French engineer Monier with his reinforced tubs, followed by arches and slabs that were able to span greater distances with relatively smaller thicknesses than stone arches and slabs. Another French engineer, François Hennebique, expanded on the then customary floor construction of steel sections, with concrete only considered as infilling (Bill, 1949). He concluded that the steel compression flanges were no longer necessary when concrete was considered as the body capable of withstanding compressive stresses. This formed the basis for his method of building.

In Paris, in the early 1900s, the architect Auguste Perret was experimenting with reinforced concrete in his buildings. Perret's aim was to blend the structural potentials of concrete with the logic of contemporary plans and the proportions and procedures of classical design (Curtis, 1986). Perret embraced the rationalist emphasis on the primacy of structure in the generation of architectural form, as initiated by Viollet-le-Duc, who had rejected the facile revivalism of the midnineteenth century and espoused a new style of architecture on the basis of "'truth' to structure and programme" (Curtis, 1986). For Perret, the concrete frame allowed open plans and wide apertures in the facades, as well as a flat roof that could be used as a garden. Restrained ornamentation, sober proportions, and an emphasis on the structural lines of force created an aesthetic that would become a precursor of modern architecture. The skeletal concrete structure was expressed at the top of the building, whereas at the lower floors, infill panels were employed. In Perret's architecture, structure was translated into art through an intuitive grasp of classical principles of organization.

Perret's most famous apprentice was Le Corbusier: "From Perret, Jeanneret [Le Corbusier] learned to think of concrete primarily in rectangular frames as if it were timber" (Curtis, 1986). Le Corbusier was not simply looking for style, but rather for guiding principles that might crystallize as forms later. Rationalism gave him a new perspective on tradition, as it was less concerned with ornamental details than with the anatomy underlying past forms. Perret's influence was evident in Le Corbusier's concept of the construction system he called the Domino, a name that invoked *domus* (Latin for house) and the game of dominoes. The Domino system was proposed as a rapid construction system for inexpensive and standardized housing based on a concrete skeleton, rubble infill for walls, and mass-produced windows, doors, and fixtures. The Domino structural system was essentially a series of reinforced concrete floor slabs supported by six columns, similar to the six points on a domino. The Domino system allowed for a "free plan" and a "free facade," and the houses constructed using this structural system could be laid out end to end in formal patterns, with some of them indented to form grassy communal areas (Curtis, 1986).

Le Corbusier also saw the potential of reinforced concrete being used as a structural system for high-rise housing. His 1922 proposal for the Ville Contemporaine was based on the concept of a series of 60-story cruciform towers arranged in a gridiron pattern separated by large areas of green spaces forming parks and gardens (Barnett, 1986). Le Corbusier went on to apply these ideas to central Paris in the Voisin Plan of 1925, in which he proposed to obliterate the central fabric of the city. Le Corbusier's objective was to relieve the congestion of the inner city and construct modern high-rise towers open to natural light, fresh air, and views of the city (Broadbent, 1990). The gridiron pattern of the towers was defined by a rectangular pattern of streets connected to high-speed highways. For high-rise office towers Le Corbusier proposed concrete structural frames with glass curtain-wall infill. However, for high-rise residential towers Le Corbusier saw the advantages of using reinforced concrete to create terraces and "balconies hollowed, cave-like, into the facades" of the apartment blocks (Broadbent, 1990). This was the modern city envisioned by Le Corbusier and the modern architects in the early part of the twentieth century. Although their ideas were implemented in many cities, the negative consequences of this strategy on the continuity and character of the existing cities would not be realized until later. (For more information on modern urban design principles, refer to Chapter 3.)

Although projects such as the Ville Contemporaine or the Voisin Plan were never realized, Le Corbusier's ideas about using reinforced concrete in high-rise building construction were eventually applied in the Unité d'Habitation in Marseilles (1947–1953). Le Corbusier compared the structural concept of the building to a bottle rack into which individual bottles are inserted (Curtis, 1986). Individual apartments were dry-assembled, lined with lead for sound insulation, and hoisted into place. The crude form of the concrete (*béton brute,* or *bare concrete*) is the result of Le Corbusier's decision to make roughness an aesthetic virtue due to the employment of different groups of contractors and a lack of quality control (Curtis, 1986). Raised on its large, tapered concrete columns (*pilotis*) like a credenza, the *machine à habiter* remains one of Le Corbusier's greatest architectural masterpieces—a consummate expression of brute structural strength combined with the refinement of machine-age materials. (See Chapter 7 for additional information on the Unité d'Habitation.)

## 3 *Reinforced Concrete Construction in the United States*

Condit (1961) writes that the pivotal figure in the development of reinforced concrete construction in the United States was Ernest L. Ransome. His early patents were issued in the 1880s, and by the turn of the century he was the leading designer of commercial and industrial buildings of reinforced concrete. His major contributions were advances in column-and-beam framing, which is commonly used in construction today. In the United Shoe Machinery Company at Beverly, Massachusetts (1903–1905), for example, the floor slab and joists were poured as a unit, the joists thus forming the parallel ribs of rectangular section, and the whole floor system resting on deep girders spanning between the columns in both directions. Beams and joists were reinforced near the undersurface—the area of maximum tension—and the number and size of bars were determined by the load on a particular member. Columns were reinforced by vertical bars and hoops or continuous helices to counteract the lateral deflection produced by wind loads and the tension at the surface resulting from the buckling tendency under high compressive loads. This principle was particularly important for a multistory building that housed heavy machinery. Although Ransome's technique represented a sound application of fundamental principles, it was based on an oversimplification of the physical properties and the behavior of concrete (Condit, 1961).

During and following World War II, reinforced concrete buildings were coming of age, and their styles and structural forms were gradually departing from those of the prevailing structural steel buildings. A group of reinforced concrete apartment buildings, known as the Clinton Homes, 12 to 14 stories high, were built in the early 1940s in Brooklyn, utilizing a flat-plate system (Fintel, 1985). Simultaneously, shear walls were employed as a cost-effective lateral load resisting system for high-rise buildings. The blend of these two systems to carry gravity and lateral loads became the most popular structural system for residential high-rise buildings.

The trend toward using more reinforced concrete in building construction in the United States was helped by the steel shortage during World War II (Condit, 1961). Following World War II, and particularly in the 1950s, concrete structures developed characteristics of their own, reflecting the inherent properties and performance characteristics of concrete as a material. Because of the ease of molding concrete into various shapes and forms, many reinforced concrete buildings of this period departed from the standard boxlike form. The Marina City twin towers complex in Chicago is a pioneer in the sculpturelike structural form of tall buildings. With the development of new materials and construction technology, reinforced concrete buildings reached heights of about 20 stories by the early 1950s. In 1958 this height climbed to 38 stories, and in 1962 the Americana Hotel in New York City attained a 50-story height, followed by the Marina City complex in Chicago, which rose to the 60-story mark (Fintel, 1985). The 70-story Lake Point Tower building in Chicago, employing the flat-plate shear wall-type construction, was completed in 1968.

Meanwhile the development of framed-tube buildings resulted in the 43-story DeWitt Chestnut apartment building in Chicago and the 38-story CBS Building in New York in 1965. This was succeeded by the 38-story Brunswick Building in Chicago (Fig. 11.1) and several other buildings. The 52-story One Shell Plaza was completed in Houston, Texas, in 1966 (Fig. 11.2).

Modified tubular forms were employed for several buildings in the 1980s. These include the bundled-tube form for the 58-story One Magnificent Mile building in Chicago and the diagonally braced tubes for the 50-story building at 780 Third Avenue in New York and the 60-story Onterie Center in Chicago (Fig. 11.3). Another development is the composite tube concept, where concrete is combined with structural steel to take advantage of the beneficial properties and negate the detrimental characteristics of the two materials (see Chapter 10). This concept combines a reinforced concrete peripheral framed tube, interior steel columns, and a steel slab system with a composite concrete topping. Some examples of this modified tubular system are the 24-story Gateway office building in Chicago, built in 1968, the 51-story One Shell Square in New Orleans constructed in 1972, the 74-story Water Tower Place in Chicago built in 1975, and the 55-story Southeast Financial Center in Miami constructed in 1986, to name a few. This concept was also used in the Far East in the 1980s (Sato, 1982). Fintel (1985) presents

**Fig. 11.1 Brunswick Building, Chicago.** (*Courtesy: Philip Turner.*)

a list of the world's tallest concrete buildings with heights of between about 30 and 80 stories. It is interesting to note that over 40% of these buildings were outside North America at the time the list was prepared. Several other tall concrete buildings have been built since then. Some notable examples are the 64-story Two Prudential Plaza and the 65-story 311 South Wacker Drive building in Chicago, both built in 1990. Other

**Fig. 11.2 One Shell Plaza, Houston, Texas.** (*Courtesy: Ezra Stoller.*)

supertall concrete buildings either recently completed or now under construction are the 105-story Ryugyong Hotel in Pyongyang, North Korea, the 104-story Metropolitan International building in Bangkok, Thailand, and the 78-story Central Plaza building in Hong Kong. The 374-m (1227-ft)-high Central Plaza building will be the world's tallest concrete building once it is completed, although it has fewer floors than the other two buildings.

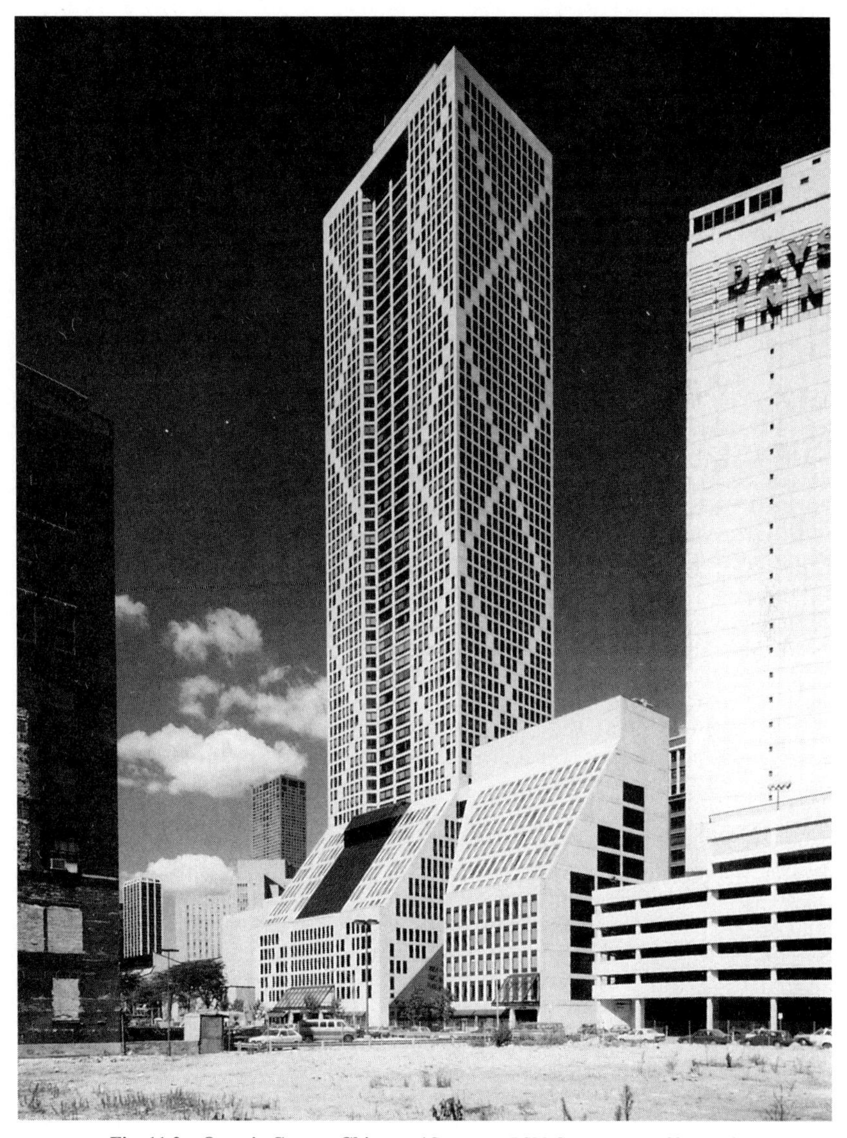

**Fig. 11.3 Onterie Center, Chicago.** (*Courtesy: PSM Corporation, Chicago.*)

## 11.2 PERFORMANCE CHARACTERISTICS

The large number of concrete tall buildings around the world suggests that there are several contributing factors that have accelerated the pace of concrete construction in tall buildings. Out of these factors, the performance of concrete as a structural material is significant. The performance of concrete is directly related to the inherent material properties and characteristics. The advantages of concrete as a material have been documented in a vast body of literature. Only the main points that are significant and relevant to performance are discussed here.

***Strength and Moldability.*** The development of high-strength concrete [up to 138 MPa (20 ksi)] permits smaller column sizes, resulting in a relative increase in valuable rentable floor space. Details of high-strength concrete in terms of design, material selection, proportioning, performance, and applications can be found in ACI (1990). In the past, low-strength concrete was a major disadvantage from this point of view. In addition, concrete permits substantial freedom in creative and aesthetic architectural expressions and engineering design because of its moldability. The introduction of lightweight concrete has helped circumvent some structural design difficulties in specific cases by reducing the dead load of the building. This also results in cost savings. The introduction of high-strength reinforcing bars has also resulted in considerable savings in the quantities of reinforcing steel per square unit of floor area.

***Fire Resistance.*** Structural concrete inherently provides adequate fire resistance and does not generally call for the additional fire protection measures that are required for structural steel. The satisfaction of specific requirements for beam width and spacing, restraint conditions, concrete aggregate type, slab thickness, column size, and concrete cover will maintain reinforcing or prestressing steel temperatures below prescribed limits for desired fire resistance ratings. These requirements are usually subject to statutory regulations.

***Durability.*** Concrete is a very durable material and normally gains strength with time. Usually the time-dependent creep phenomenon in concrete results in stress relaxation and consequent relief in peak stresses due to the redistribution of moments in the beams and columns. Of course, this is only qualitative and difficult to quantify, and it is not normally accounted for in design practice. Although concrete is vulnerable to chemical attacks, such incidences are generally rare and could be controlled by adopting preventive measures (Freedman, 1985a; Portland Cement Association, 1981).

***Building Stiffness.*** Concrete buildings have inherently good stiffness properties. Because of congestion in downtown areas, it is likely that in the future more and more slender buildings will be built in most large cities the world over. Slender concrete buildings have not only good stiffness properties, but also good damping characteristics under dynamic wind loads. The concrete floor systems can respond well to vibratory loads caused by heel impact and machinery. This is also true for elevator core areas, where precise tolerances are of utmost importance.

***Climatic Requirements.*** The climate and location of a city are important factors dictating the advantages of concrete as a structural material for tall buildings. In the Western countries, where temperatures go down to below freezing levels in the winter, artificial cost-intensive heating is required during concrete placement and subsequent protection

(ACI, 1991). In the hot and arid Middle Eastern countries, where temperatures rise to very high levels, on the other hand, special concreting techniques are necessary to protect the fresh concrete. Many countries that have moderate (for example, tropical or mediterranean) climates are exempted naturally from these requirements of cold- or hot-weather concreting during any time of the year. Some countries, such as those with the monsoon season, are vulnerable to frequent rains and, as such, are amenable to the disruption of concreting activities during periods of heavy rains. However, in many cases rains may be sporadic and of short duration. Also, countries with humid weather have less stringent requirements for continuous curing of concrete compared to areas with dry climate. Such humid weather provides a natural curing effect, resulting in durable concrete of good quality and strength.

The structural design of tall concrete buildings constructed in earthquake-prone areas requires special considerations. Such buildings call for special reinforcing steel detailing to provide adequate ductility to the structure (ACI, 1989).

***Building Acoustics.*** Acoustical control is an important item in tall buildings. However, there is usually no major concern for exterior noise control since most of the floors are located at elevated heights free from the street noise. The interior acoustical control through floors and walls becomes critical for tall buildings with diverse multipurpose occupancy. For commercial buildings, however, this is less critical than for residential buildings.

Although there are no set standard sound control criteria established at the present time, it is known that the softer and more porous the material, the better is its sound absorption performance, and the denser materials have better sound transmission resistance. Therefore sound absorption control will depend on the properties of the room surfaces in absorbing rather than reflecting sound waves. For concrete, porosity is the factor that will dictate the degree of sound absorption control.

Because of its weight and density characteristics, concrete is a good sound insulator. However, concrete is not a good material when impact noise is concerned, although such noises can be minimized by using resilient materials in combination with structural concrete floors, such as good carpeting on resilient padding.

## 11.3 CONCRETE BUILDING SYSTEMS

Since reinforced concrete is a relatively newer material for high-rise construction than structural steel, it was initially molded in close resemblance to its predecessor. However, because of the flexibility of concrete shapes and forms, new building structural systems and concepts have emerged in the recent past. The structural systems comprise the floor systems and the lateral load resisting systems. The floor systems usually represent the principal elements that carry the gravity loads applied in a building. These gravity loads are then transferred to the foundations via the columns and load-bearing walls. Floor systems could be either cast in place or precast. They could also be reinforced concrete or prestressed concrete (either pretensioned or posttensioned) systems. The importance of selecting a particular floor system is underlined by the fact that the floor is repeated many times in a high-rise building. [For additional information on structural framing and structural design of tall concrete buildings see *Structural Design of Tall Concrete and Masonry Buildings* (CTBUH, Group CB, 1978) and *Cast-in-Place Concrete in Tall Building Design and Construction* (CTBUH, Committee 21D, 1992). See also Chapter 4 for a description of the structural systems in a number of concrete tall buildings.]

## 1 Floor Systems

Detailed information on concrete floor systems may be found in Fintel and Ghosh (1983), Domel and Ghosh (1990), Schueller (1990), and Taranath (1988). The main systems are briefly discussed in the following.

***Flat Plates and Flat Slabs.*** Flat plates are horizontal two-way planar elements of solid concrete, usually of constant thickness without any drop panels or capitals at the columns. Flat slabs also constitute two-way beamless construction, except that they have drop panels or capitals at the columns. Flat plates and flat slabs allow maximum headroom and thereby reduce the total building height. The concrete soffit may be painted or plastered to achieve a desired architectural finish, thus eliminating the need for ceilings. This is particularly suitable for residential buildings. Some disadvantages of these systems are that they are usually subjected to large deflections, particularly due to long-term effects, and hence sagging of floors and other serviceability problems may take place. Also, floor openings for mechanical requirements are difficult to achieve because of the sensitivity of these systems to the size and location of such openings. An efficient span range for flat plates is 4.6 to 6.1 m (15 to 20 ft), and for flat slabs it is 6.1 to 7.6 m (20 to 25 ft).

***Concrete Joist Systems.*** Concrete joist floors consist of a one-way ribbed slab system (also called pan joist system). They are very popular for tall office building construction in North America. The corresponding two-way system, when the joists span in both directions, is called a waffle slab. In such systems, the voids between joists are formed with pans that are modular and reusable in nature with standard sizes. Joist construction is most suitable in office and commercial buildings. A commonly laid out office building has a square or rectangular configuration with a square or rectangular core at the center. One-way joists are typically used where the core walls parallel the exterior walls and the waffle slab system is used at the four corners.

Joist construction readily allows small penetrations to handle ductwork and pipe runs. It is a relatively lightweight system, particularly when lightweight concrete is used. Some major problems with such systems are that they are not readily adaptable to irregular plan configurations, the finished soffit is not generally suitable for exposed architectural application, and large floor openings cannot be easily framed. The economical span range for one-way concrete joists is 7.6 to 10.7 m (25 to 35 ft) and that for waffle slabs is 9.1 to 12.2 m (30 to 40 ft).

***Skip-Joist System.*** A modified version of the joist system is the skip-joist system, which is gaining more acceptance in the high-rise construction industry. Skip joists do not qualify as one-way joists by ACI 318-89 (1989) and are structurally designed as beams. They are usually spaced at 1.5 and 2 m (5 and 6.5 ft), employing 1346- and 1676-mm (53- and 66-in.) pans. The system is very economical as far as the volume of concrete and the number of shores required for forming are concerned (Taranath, 1988).

***Beam and Slab System.*** The beam and slab system consists of 102- to 152-mm (4- to 6-in.)-thick continuous slab poured integrally with the supporting beams usually spaced at a maximum of about 6.1 m (20 ft). This system is easily applicable to a variety of practical situations, the principal limitation being the depth available within the ceiling space. Usually such beams are placed over partition walls or at door heads. This system is very common in the developing countries because the construction technique has the least demand on technology. Also, this system is labor intensive and has its broadest applications in countries where the cost of labor is cheaper relative to the cost of materials. Some advantages of this system are that it readily adapts to large floor openings and to any plan configuration and that it is a relatively lightweight concrete system permitting relatively long spans.

## 2 Lateral Load Resisting Systems

For concrete high-rise buildings, three fundamental types of lateral load resisting systems are commonly employed:

1. Moment resisting frames
2. Shear walls
3. Combination systems

These three systems are briefly discussed in this section. Further details of lateral load resisting systems may be found in Fintel (1985), Khan (1966), Schueller (1977, 1990), and Taranath (1988).

***Moment Resisting Frames.*** Moment resisting frame structures consist of beams and columns connected together by moment resisting or rigid joints that are monolithic. When flat plates or slabs are used as floor systems, a portion of the flat plate or slab is utilized to act as shallow beams. Frames located inside a building have serious disadvantages because they result in a large number of columns, thereby drastically reducing valuable floor space and creating undesirable obstruction in an office or commercial-type occupancy, where openness is warranted. When exterior columns are closely spaced, it is possible to develop the perimeter of the building as a lateral load resisting system. Such a system is called a framed-tube system and is appropriate for very tall buildings. The Brunswick Building in Chicago (Fig. 11.1) and One Shell Plaza in Houston (Fig. 11.2) are examples of framed structures where columns are closely spaced. (See Section 11.1 for more examples of tube buildings.)

***Shear Walls.*** Shear walls are vertical planar diaphragms that provide rigidity to the building against lateral wind and seismic forces. They were first introduced in the 1940s. When two or more shear wall elements are connected together with relatively flexible members, they are called coupled shear walls. Shear walls are generally solid and occasionally have a limited number of openings to accommodate doors, windows, or mechanical duct runs. Shear walls may or may not carry large magnitudes of gravity loads, although some gravity loads should be deliberately applied in high-rise buildings to minimize effects of overturning as well as the development of tensile stresses due to lateral loads. Shear walls can be combined in plan to generate different forms such as box systems, angles, channels, and I-shaped cross sections. In general, shear walls are in the vertical plane and flat, but they also may be curved, folded, or in sloping planes.

Although the braced-frame concept that utilizes diagonal or K braces has been widely applied to steel structures, its application to concrete buildings is rather limited. The difficulty of making the connections in concrete discourages such applications. As a result, the shear wall is the naturally superior lateral load resisting element for concrete high-rise buildings. Shear walls can be located in a high-rise building in a variety of ways. They could be external, internal, or core walls. In commercial and office buildings, most partition walls are temporary or relocatable, and therefore it is undesirable to provide internal shear walls unless such walls are of permanent nature. The primary application of shear walls is in the core area, which often includes elevators, stairs, janitor closets, and mechanical shafts as well as some interior columns. Generally these core areas cover about 20 to 25% of the floor area and go through the full building height. In effect, the core walls result in vertical cantilever tubes that are efficient in lateral load resistance. In the same way, if the punched wall (for example, a wall with window openings) is placed on the exterior, the entire exterior skeleton can be utilized as a cantilever punched-tube system. The relative proportions of the members of the punched tube, the wall thickness, and the form of the wall may be explicitly controlled to meet any particular requirements of architectural form and structural action. This makes concrete a particularly suitable material for lateral load resisting systems (Iyengar, 1982, 1985).

The utility core requirements for residential high-rise buildings are less significant because of the architectural and mechanical demands on the building layout. Therefore the core wall concept is not practical for such buildings. However, the need for providing a fixed configuration of interior walls in apartment or hotel projects presents an opportunity to utilize some or all of these interior walls as shear walls. Because of their relative lack of ductility, shear walls are employed on a limited scale for high-rise buildings in seismic zones.

***Combination Systems.*** Floor systems are often a combination of beams, slabs, and joists. Similarly, columns and walls sometimes blend together. Lateral load resisting systems contain frames, walls, and bracing that are interactive. Because of the diversity of architectural functions and forms, as well as other utilitarian and aesthetic requirements in addition to cost and construction considerations coupled with technological advancement, building systems—in particular structural systems—have evolved rapidly into a variety of categories. It is natural then that to get optimum results, different systems should be investigated, both individually and in combination (Moreno and Zils, 1985). One such system is a combination of shear walls and rigid frames, commonly referred to as shear wall–frame interaction system in structural engineering terminology (Fintel, 1985; Taranath, 1988).

A basic requirement for such interaction is that the relative stiffness properties of each shear wall and each frame remain unchanged along the entire height of the building. In practice it is rarely possible to comply strictly with this requirement for various architectural and functional reasons. With the availability of sophisticated computer programs, which can simulate the actual interaction of the idealized elements, this is no longer a problem. However, since both gravity loads and wind shear increase downward in a tall building, the required member stiffnesses rise gradually toward the base of the building. Therefore frame–shear wall interaction within reasonable limits is expected to occur in most cases. A good example of a project where a shear wall and interactive frame structural system has been used for a 70-story concrete building is 311 South Wacker Drive in Chicago.

It is a known fact that the shape of a structure in plan has substantial influence on the lateral resistance of the building. This has been confirmed by the available computer analysis techniques applied to three-dimensional building models. Concrete as a material can be molded to any shape. A good example is the vertical cylindrical-shell-type Toronto City Hall building (Roy, 1965). Another application of the versatility of concrete as a material is seen in the exterior diagonalized tube buildings of the 1980s. An example of such a system is the Onterie Center in Chicago (Fig. 11.3). [Further information on this building is available in ENR (1985) and Gapp (1986).]

Other applications of concrete for lateral load resistance in high-rise buildings are the bundled-tube and composite-tube systems. In a bundled-tube system a number of individual tubes are "bundled" together to increase the tubular efficiency of the building. The concept of bundled tubes has led to the notion of free-form massing and its relevance to structural form and aesthetics of tall buildings (Ali, 1990). The emergence of a composite-tube system as a modification of the framed-tube system was accelerated to derive the benefits from both concrete and structural steel for tall buildings. Taranath (1988) has covered the topic on lateral systems for composite construction in considerable detail (see also Chapters 2 and 10).

### 3 Prestressed Concrete Systems

The application of prestressed concrete for high-rise buildings is a common practice worldwide. Prestressed concrete is most suited for long spans. For short spans it can be justified if the units are standardized and repeated many times in a project. In general, prestressing can be viewed as an aid to increase the span range of conventional reinforced concrete floor systems by about 30 to 40%. Prestressed concrete used in office and commercial buildings generally comprises shallow slabs and beams, one-way joists, skip joists, and waffle slabs for the floor systems. In residential buildings the more common type of prestressed element for floor systems is the flat plate.

For high-rise buildings, the principal justification for using prestressed concrete, from an engineering point of view, results from its following advantages:

1. Concrete members are generally crack-free and therefore stiffer and more durable than reinforced concrete members.
2. Smaller sections can be used.
3. Prestressed concrete is a resilient material in terms of recovery of deformation upon unloading. Also, prestressed concrete accommodates shrinkage and creep reasonably well.
4. The behavior of prestressed concrete is more predictable than that of reinforced concrete.
5. By adjusting the prestressing force, high-strength concrete may be effectively used for prestressed concrete members, and the design could be customized to meet specific requirements.
6. The prestressing operation by itself provides partial testing of steel and concrete.

Despite these advantages, there are certain limitations to prestressed concrete. Some of these limitations are:

1. Stronger materials are required for prestressed concrete, implying a higher cost.
2. The required formwork is complicated.
3. End anchorage and bearing plates are usually required.
4. Greater unit cost for skilled labor is required.
5. The structural design process is more elaborate since many design criteria must be satisfied.

Also, closer control is required of each construction phase. In general, prestressed concrete is more suitable to floor construction in conjunction with cast-in-place column and wall construction. The application of prestressed concrete for tall buildings in seismic regions has been discussed by Wakabayashi (1986).

Because of these limitations, prestressed concrete is not as popular in developing countries, where reinforced concrete is always preferred. However, this trend is changing. In industrialized countries, prestressing is very justifiable along with precasting for building construction. Industrialized prefabrication is about 30% faster than conventional methods, resulting in savings due to early completion.

A popular concept originating in Europe is partial prestressing, where the level of prestressing is adjusted to vary the cracking to a desired objective of stiffness or crack width.

For some high-rise apartment buildings in the West that used a novel system, the facade consists of prestressed precast concrete Vierendeel girders spanning the full face of the structure. The posttensioned facade is supported on four corner columns and provides vertical support to the edge of the floor slab, which is also prestressed. The facade provides lateral load resistance. This is a "total system" concept in prestressed concrete.

Prestressed concrete is particularly suitable for long spans in apartment buildings, provided the supporting walls are hollow concrete blocks or bricks with holes. Dowels could be employed between the precast prestressed floor planks, tees, or double tees and the wall to provide a positive connection between the two. The use of a posttensioned flat plate can yield relatively larger spans, and the spans could range from 4.9 to well over 11.6 m (16 to 38 ft). The plate is usually 127 to 254 mm (5 to 10 in.) thick. The maximum span-to-thickness ratio is about 45.

Basic codes and documents related to prestressed concrete are published by the American Concrete Institute (ACI) (1989) and the Prestressed Concrete Institute (1985). It may be noted here that unbonded posttensioned construction is popular in North American high-rise construction industry inasmuch as it eliminates the need for grouting, which is an expensive and time-consuming operation.

Worldwide, the concept of post-modernism in architecture and sophisticated computer-aided structural analysis techniques have made the building forms more varied. Different structural subsystems can be blended to formulate a total but varied system. With the aid of computers it is possible to analyze the structures in a three-dimensional envelope of the articulated structure and in mixed construction. This has paved the way for more and more prestressed concrete elements integrated in the tall building framework.

Even though developing countries may need to build within the realm of low technology, attention should not be confined within such a fixed boundary. Unless these countries start looking beyond this boundary, they will fall far behind the rest of the world in terms of technological development. Prestressed concrete has many positive potentials, and innovations in this area must be encouraged. As more high-rise buildings utilize prestressed concrete and the corresponding infrastructure is developed, it is expected that the architects and engineers in the developing nations will accept prestressed concrete more readily as a building material.

## 11.4 STRUCTURAL MATERIAL SELECTION

Whenever there is a proposal for building a tall building, the question arises: What is the most economical structural material for the building? In many countries where steel is an imported material and the infrastructure required for steel fabrication is absent, concrete is the predominant material for high-rise construction. Most of the developing countries fall into this category. The suitability of building either a steel or a reinforced concrete high-rise building at any location essentially depends upon a feasibility study that considered the various factors involved for the particular building. It is not always possible to quantify all factors in terms of monetary units, yet certain basic qualitative advantages and disadvantages of both steel and concrete as building materials are worth elaborating upon. The principal factors that affect the building cost are described in the following subsections.

### 1 Construction Techniques and Degree of Mechanization

A major factor in the decision on the type of building to be built is the relative availability of the latest construction techniques and the degree of mechanization. Before chalking out a definite construction program for a project, these factors need careful consideration.

The advancement of material technology goes hand in hand with the development of construction technology. Most developing countries are experimenting with different concrete construction techniques, including system formwork, use of precast elements, and concrete prestressing. In the international context, concrete as a material is advancing rapidly, from a technological viewpoint, relative to steel.

### 2 Local Infrastructure

The local infrastructure of a country determines to a large extent whether the construction industry will favor concrete or steel, or both. It usually includes skilled workers, professionals such as architects and engineers, plant facilities, equipment, and the like. Some countries in the Far East that in the past traditionally employed concrete for high-rise construction are going through a learning curve and trying to master the techniques and standards required for steel construction. The supply of labor is another significant factor to consider because it determines the labor cost.

### 3 Speed of Construction

A well-known fact is that a structural steel-framed building has definitely shown considerable advantage over other forms in the speed of construction. This is particularly so for high-rise buildings. If this vital aspect is not included when preliminary cost analysis and budgets are prepared for a building project, then a considerable cost penalty to the owner will result. This additional cost arises from such items as capital commitment over a longer period of time than necessary, extended interest payments for loans, and loss of rental. The speed of construction is generally correlated with the floor erection cycle time. It is commonly recognized that the typical floor cycle time for steel is substantially less than that for concrete. However, the speed of construction of steel structures is dependent on the availability of local infrastructure, methods of fire protection, and the pace of finishing and mechanical work to catch up with the building framing time. Another factor to consider is the size and complexity of the building. It must be borne in mind that the savings due to early completion of a steel building project may not be realized if the building is not rented within a reasonable period of time, and even the pace of construction of such a building may slow down if a forthcoming economic downturn in the rental market is perceived by the owner.

For both steel and concrete high-rise construction there exists generally an element of delay during the construction process which is reflected by a longer building framing time than originally estimated. For cost analysis, this additional time of construction delay may be spread uniformly over the total number of floors by an average lag time per floor (Ali and Ang, 1984a, 1984b). This lag time may result from a number of reasons, such as:

1. Equipment breakdown
2. Cash flow problems during construction
3. Unexpected labor shortage
4. Design modifications during construction

5. Drawing revisions and amendments
6. Construction errors and correction of field problems
7. Quality control related problems
8. Site accidents
9. Nontypical floors
10. Holidays
11. Disruption due to inclement weather
12. General mobilization and preparation delays
13. Coordination problems between trades and subtrades and between floor erection and finishing and mechanical items
14. Poor planning and construction management
15. Contract disputes

While items 4 and 5 are common to both concrete and steel, both have some particular significance for steel buildings. Design modification and changes can be accommodated relatively easily for concrete buildings whereas considerable time will be wasted if such changes are made at a belated stage of construction for steel buildings. It is important to note that the additional time needed for tall building construction basically arises from the delays in planning, execution, and lack of know-how. This time is in fact an overall time that occurs due to various unforeseen reasons at different stages of construction. The average lag time is thus a convenient measure of this overall delay applied to each floor. Consideration of such anticipated delay in construction is necessary to arrive at a realistic total completion time for a project. In the time and cost economics of tall buildings, the total completion time plays a crucial role. Since the delay time is often inevitable, it should be included in the time versus cost relationship in any economic analysis for financing high-rise buildings put up by speculative developers and private owners.

The process of steel construction has a long lead time for the various activities before actual construction can start, such as preparation of shop drawings, fabrication, and transportation to site, to name just a few. This lead time, however, is usually absorbed by preliminary site preparation, excavation, and foundation work. The frame erection starts when the foundation, or a major part of it, is completed. Evidently the lead time for steel does not in effect change the overall completion time. This lead time merely helps a steel building itself in the sense that it can be absorbed by the foundation work and hence does not cause any delay in construction. However, if due to poor project planning, the first batch of fabricated steel is not delivered on site when the foundation is completed, this will obviously delay the construction.

Concrete buildings, on the other hand, do not have the foregoing limitations. Building framing can continue immediately after the foundation construction. On the whole, steel structure construction requires more comprehensive planning and preparation at the front end and less flexibility for design modifications and amendments midstream. Despite these limitations, physical construction times for steel, that is, floor cycle time and total project completion time, are faster than for conventional cast-in-situ concrete.

As early as in the 1970s, Cowan (1976) pointed out that "twenty years ago it was axiomatic that reinforced concrete could not compete with steel in Australia for buildings taller than ten stories. At present, reinforced concrete is cheaper than structural steel for tall buildings, and most of the tall buildings in Australia are of reinforced concrete."

Likewise, Powys (1971) reported that from the point of view of the developer "one of the basic economic considerations is the selection of the material to be used in the structure. Steel is more expensive than concrete, but permits faster building time, less bulky walls and columns, and less weight. It should be more attractive in cities such as Sydney, where carrying costs arising from land tax and city rates assessed on unimproved values are especially onerous. However, most tall buildings in Australia are currently being built in concrete."

Based on an investigative study, Ali and Ang (1984a, 1984b, 1985) reported that in Singapore concrete buildings were more economical than steel buildings for up to about 60 stories. The issue of whether steel or concrete is the more economical material for high-rise construction in Singapore was being debated prior to this study. Although Singapore is a "concrete country" traditionally, the Japanese steel industry was trying to introduce steel in the local construction scene, which resulted in the controversy. Following the paper presentation by Ali and Ang (1984b) at the International Conference on Tall Buildings in Singapore in October, 1984, the debate was stirred up once again. A similar study for Chicago by Ali (1989) found the approximate break-even level to be at 35 stories, that is, concrete could be more economical than steel for high-rise buildings lower than 35 stories. It must be remembered that such break-even levels are not magic numbers and are valid within the framework of the assumptions involved. However, they give an approximate indication of height for typical high-rise buildings for which concrete could be a more cost-effective material. Further, such break-even levels are not static numbers since they may change very rapidly with the improvement of technology and the unpredictable play of the market forces.

## 4 High-Rise Concrete Construction in Seismic Zones

The following major characteristics should be provided in a building located in a seismic zone (Wakabayashi, 1986):

1. High strength-to-weight ratio of the material
2. High plastic deformation capacity
3. Low degradation of strength and stiffness
4. Continuity of structural elements
5. Cost-effectiveness

Different structural materials meet these requirements in different ways.

Because of the high strength and ductility characteristics, structural steel is recognized as the most efficient material for tall building construction in seismic zones. Cast-in-place concrete has the advantage of satisfying the requirements of items 4 and 5. To meet the disadvantages related to items 1 and 2, special detailing and concrete strength provisions are followed (ACI, 1989). With more research in this area, acceptability of concrete for seismic zones is gradually increasing.

It is not sufficient to make a structure "strong" in a seismic zone; it must also be provided with sufficient ductility. In addition, since tall buildings tend to be flexible, it is necessary to control the lateral drift due to earthquake loading. This necessitates an adequate amount of stiffness for the building. Excessive stiffness will, however, attract larger seismic inertia forces. The aim of earthquake-resistant design is thus to achieve an optimum stiffness of the building. Also, while designing a tall building, *flexibility* should not be confused with *ductility*.

Prestressing of concrete for a structural element adversely affects its performance in seismic zones because it results in a reduction of its deformability. The earthquake resistive properties of concrete high-rise buildings can be stated in terms of the overall performance of such buildings in the following decreasing order—nonprestressed cast-in-place concrete, prestressed cast-in-place concrete, nonprestressed precast concrete, prestressed precast concrete.

Composite structures combining structural steel and concrete often used for high-rise buildings exhibit characteristics that are intermediate between those of steel and concrete. In Japan, which is an earthquake-prone country, they have often been used (Wakabayashi, 1986).

Fintel (1991) suggests that for residential buildings, shear walls should be used to prevent collapse, even though shear walls have been known to be stiff and not sufficiently ductile. He reports:

> After observing the devastations and the resulting staggering loss of life in many earthquakes, particularly those in Managua in 1972, Mexico City in 1985, and America in 1988, the author believes it to be the responsibility of the engineering profession to make sure that residential buildings in particular be constructed with shear walls. Whether such walls are made of plain concrete, traditionally reinforced, or reinforced for ductility will depend upon the economic capacity of a given society and on engineering judgment; however, they all protect life and in most cases provide good protection of property.

For additional discussion on concrete high-rise buildings using high-strength concrete in seismic zones, see Section 11.6.

## 11.5 RECENT DEVELOPMENTS

### 1 Materials

***High-Strength Concrete.*** High-rise buildings in reinforced concrete have been built over the past few decades in large numbers all over the world. The principal factors contributing to this phenomenon are the improvement in the quality of materials and in construction techniques as well as the development of new structural systems.

Concrete of up to about 21-MPa (3000-psi) strength would prevail only up to about 30 years ago. Gradually the strength of concrete increased to 28 and 35 MPa (4000 and 5000 psi) and occasionally to 49 MPa (7000 psi). A strength of 76 MPa (11,000 psi) was utilized in 1981 on two Chicago buildings (Fintel, 1985). It is now possible to produce concrete with a strength of up to 138 MPa (20,000 psi). Such high strength is attainable because of the development of superplasticizers (ACI, 1979) and other admixtures. The need for high-strength concrete (HSC) arose in order to minimize the column size at lower levels of high-rise buildings. Such a reduction of column sizes is necessary to increase the valuable rentable floor space. A state-of-the-art report was presented by Russell (FIP/CEB, 1990). Researchers are now routinely making 172- to 242-MPa (25,000- to 35,000-psi) very high-strength concrete. Although for the time being such high-strength concrete is expensive and used primarily for specialized rehabilitation, it could be widespread by the turn of the century (Tarricone, 1991). A superhigh-strength 690-MPa (100,000-psi) concrete has recently been tested in the laboratory of the University of Illinois (Tarricone, 1991). The continuing development of higher-strength concrete is certainly going to influence the design and construction of concrete tall buildings of the future. Burnett (1991) reports that another special material called silica fume concrete was being produced in Australia since 1977 and has been used successfully for a number of projects. A very recent building project where high-strength concrete has been used is the 59-story GLG Center in Atlanta. High-strength concrete is discussed in further detail, with an emphasis on its application in high seismic areas, in Section 11.6.

***High-Performance Concrete.*** Another recent introduction into the concrete construction literature is high-performance concrete (HPC). Concrete in which improvements in qualities in addition to compressive strength are realized in justifying the cost of a building is called high-performance concrete. These qualities are durability, corrosion resistance, ductility, and energy absorption, to name just a few. A review of about 100 high-performance concrete structures throughout the world reveals that the use of high-performance concrete would be economically justifiable in only 15 to 25% of them if high compressive strength were the only criterion (Malier, 1991). A description of research needs is presented by Carino and Clifton (1991), and the application of high-performance concrete to tall buildings has been discussed by Moreno (1990). Section 11.7 details the use of high-performance concrete in tall buildings and current developments.

***Lightweight Concrete.*** Lightweight concrete is another material that has been developed to help circumvent some problems of high-rise structures. Savings due to the reduction of dead loads for high-rise buildings utilizing lightweight concrete could be substantial. Such savings are realized in the reduction of loads on columns and foundations, which will result in a reduction of material consumption. The higher cost of lightweight concrete relative to normal-weight concrete is thus compensated by the savings in the cost of columns and, particularly, foundations.

***High-Strength Reinforcing Steel.*** Another area of development is the production of high-strength reinforcing steel. Steels with yield strengths of 414 and 518 MPa (60 and 75 ksi) came into vogue with the introduction of the ultimate-strength-design (USD) method by the ACI building code. High-strength reinforcing steel results in reduced material consumption and hence more savings with respect to the working-stress-design (WSD) method. The application of bundled bars and the introduction of no. 14 and no. 18 bars also reduced the congestion of steel bars and proved particularly useful for columns in high-rise buildings. Another high-strength steel used in tall building construction in parts of Europe and Asia is cold-twisted Ribbed-Tor steel. Section 11.10 discusses the properties and characteristics of cold-twisted Ribbed-Tor steel and presents three case studies introducing the use of this steel in tall building construction.

## 2 Construction Techniques

Construction techniques for building concrete high-rise structures have evolved very rapidly during the last three decades. Developments in formwork, such as slip forms, flying forms, table forms, plastic-lined forms, and tunnel forms for modular construction (Ali and Napier, 1989; Jensen, 1986; Ceco Industries, Inc., 1985) have facilitated early construction of high-rise buildings. Other techniques are lift slab construction, precasting, and prefabrication of modular units. Factors affecting the selection of horizontal formwork systems have been reported by Hanna and Sanvido (1991). Other developments are improvements in construction equipment and methods, and better material handling techniques (Camellerie, 1985).

## 3 Design Methods

New structural design methods such as ultimate-strength design, currently known as strength design method, and the recognition of limit-strength design concepts that account for inelastic behavior of both the structure and its critical sections have helped in utilizing high-strength materials for reinforced concrete structures. Further, the development of flat-slab and flat-plate construction, the introduction of frame–shear wall interaction concepts for high-rise buildings, the employment of shear panels, and staggered wall-beam systems have also facilitated the rapid evolution of reinforced concrete buildings from being a mere imitation of steel buildings to more mature structural systems in their own right. More recently, the development of the framed-tube system and its modifications (tube in tube, bundled tube, composite tube, and braced tube) has made it possible to build concrete buildings to great heights that were unimaginable only a few decades ago (see Section 11.1). The development of new design methods and techniques for concrete buildings in seismic zones is another important research area that should be explored by researchers and investigators.

## 11.6 HIGH-STRENGTH CONCRETE FOR SEISMIC REGIONS

The application of high-strength concrete in highly seismic regions has lagged behind its application in regions of low seismicity. One of the primary reasons has been a concern with the inelastic deformability of high-strength concrete structural members under reversed cyclic loading of the type induced by earthquake excitation. This section discusses the current state of application of high-strength concrete [with a specified compression strength in excess of 41,370 kPa (6000 psi)] in buildings across the United States, including major West Coast cities, and then it will discuss the properties of high-strength concrete that are relevant to structural applications in high seismic regions. Particular attention is paid to the inelastic deformability of reinforced concrete structural members under reversed cyclic loading. The underlying concepts are applicable to other countries with regions of high seismicity (for example, Japan).

### 1 High-Strength Concrete in Buildings

In the early 1950s the tallest concrete buildings were in the 20-story height range. By 1975 the 74-story-high Water Tower Place, until recently the tallest concrete building in the world, had already been constructed. This virtual revolution within a very short time span was made possible by a number of factors, the most important of which were the availability of:

1. New, improved construction methods
2. Bigger cranes
3. High-strength materials
4. Innovative structural systems
5. High-storage, high-speed computer hardware and the corresponding software that gave the structural engineer unprecedented analytical capabilities

It is futile to speculate which factors were more or less important than the others; all of them contributed to the dramatic growth in the heights of concrete buildings.

Figure 11.4 shows a series of nine concrete buildings, each, with the exception of Two Prudential Plaza, having been the tallest concrete building in the world at the time of its completion. It is clear that the growth in the height of concrete buildings has gone hand in hand with the availability of higher- and higher-strength concrete.

Almost incredibly, seven of the nine record-setting buildings are located in Chicago, a city that in many ways has pioneered the evolution of high-strength concrete technology. However, very recently there has been an impressive spread in the availability of ultra-high-strength concrete [with a specified compression strength in excess of 68,950 kPa (10,000 psi)]. Figure 11.5 shows that 82,740-kPa (12,000-psi) or higher-strength concrete has been used in the last 3 or 4 years in the U.S. cities of Atlanta, Cleveland, Minneapolis, New York, and, most significantly, Seattle, which is in UBC seismic zone 3.

In fact, the highest known concrete strength ever used in a building has been 131,005 kPa (19,000 psi) in the composite columns of Seattle's 62-story 231.34-m (759-ft)-high Two Union Square. The strength was obtained by the use of:

1. What may be a record low water-to-cement ratio of 0.22 (this is the single most important factor in increasing strength and reducing shrinkage and creep)
2. The strongest of available cements
3. A superplasticizer that reduces the need for water and provides the necessary workability
4. A very strong, small 9.5-mm (⅜-in.) round glacial aggregate available locally
5. Silica fume (increasing the strength by about 25%)
6. A design strength obtained at 56 rather than the usual 28 days
7. An extraordinarily thorough quality assurance program

The 131,005-kPa (19,000-psi) strength was the by-product of the design requirement for an extremely high modulus of elasticity of 49.6 million kPa (7.2 million psi). The stiffness was desired in order to meet the occupant comfort criterion for the completed building. The same concrete strength was later used in the composite columns of the

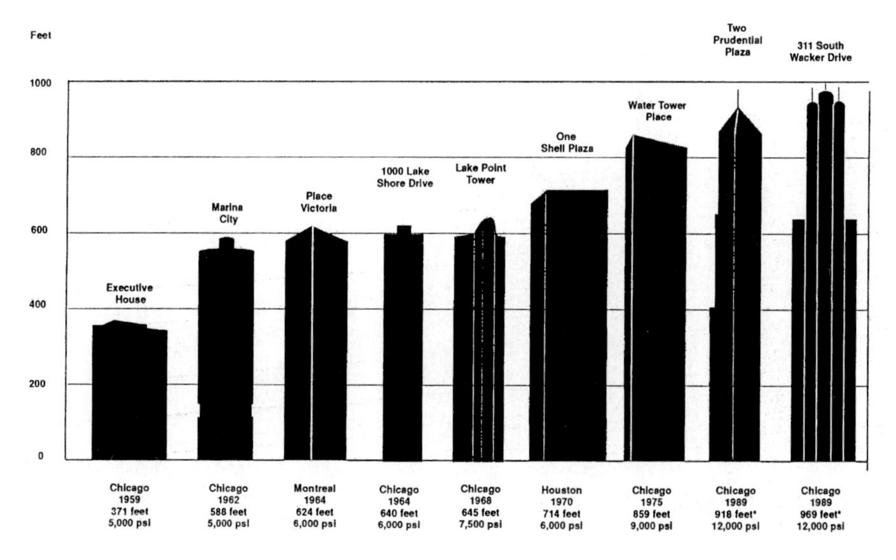


**Fig. 11.4 Tallest concrete buildings at time of completion.** (*Courtesy: Portland Cement Association.*)

shorter 44-story Pacific First Centre in Seattle, Washington (Fig. 11.6), and 65,502.5-kPa (9500-psi) concrete has been used at 600 California in San Francisco and at 1300 Clay in Oakland, both composite buildings. A 55,160-kPa (8000-psi) concrete has been used in several all-concrete Bay Area buildings, including the 19-story Fillmore Building (Fig. 11.7).

The spread in the use of high-strength concrete in Southern California has been hampered by the City of Los Angeles code provision restricting the strength of concrete to a maximum of 41,370 kPa (6000 psi). Even then, concrete strengths in excess of 41,370 kPa (6000 psi) have been used in the Great American Plaza office-hotel-garage complex in San Diego, a 14-story residential building at 5th and Ash in San Diego, and the 22-story Pacific Regent (senior citizen housing) at LaJolla.

## 2 Advantages and Properties of High-Strength Concrete

The three biggest advantages that high-strength concrete has over most other building materials, including normal-weight concrete, and which make its use attractive in high-rise buildings, are that it provides more:

1. Strength per unit cost
2. Strength per unit weight
3. Stiffness per unit cost

***Cost.*** Commercially available 96,530-kPa (14,000-psi) concrete costs significantly less in dollars per cubic yard than 3½ times the price of 27,580-kPa (4000-psi) concrete. In fact, the unit price goes up relatively little as the concrete strength increases from

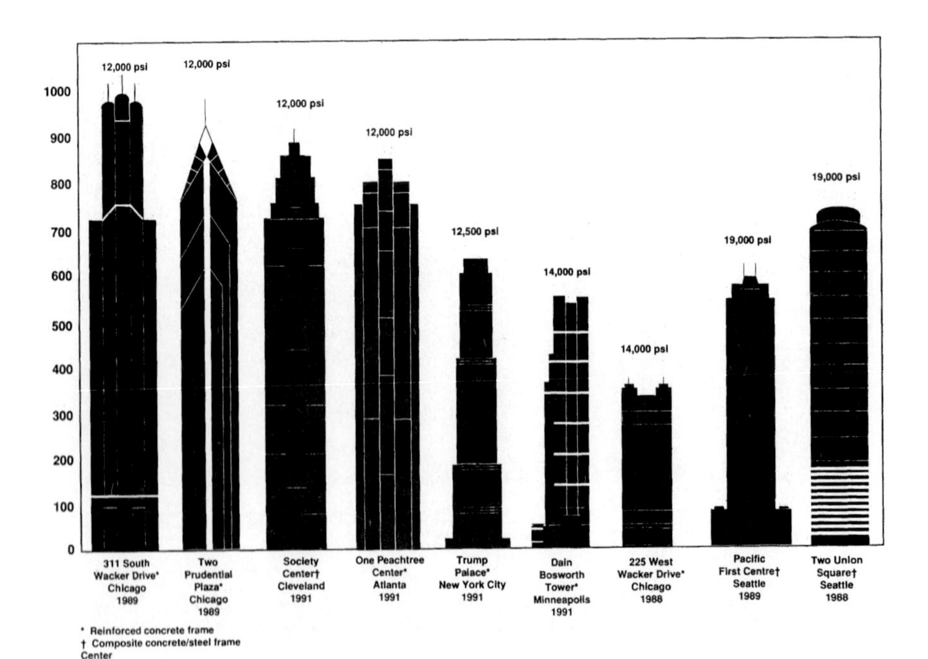


**Fig. 11.5 Higher-strength concrete used in tall buildings.** (*Courtesy: Portland Cement Association.*)

**Fig. 11.6 44-story Pacific First Centre, Seattle, Washington.** (*Courtesy: Skilling Ward, Magnusson, Barkshire.*)

**Fig. 11.7 19-story Fillmore Building, San Francisco, California.** (*Courtesy: Watry Design Group.*)

27,580 to 68,950 kPa (4000 to 10,000 psi). Thus high-strength concrete gives the user more strength per dollar.

***Unit Weight.*** The unit weight of concrete goes up only insignificantly as the concrete strength increases from moderate to very high levels. Thus more strength per unit weight is obtained, which can be a significant advantage for construction in high seismic regions, since earthquake-induced forces are directly proportional to mass.

***Modulus of Elasticity.*** The modulus of elasticity of concrete remains proportional to the square root of the compressive strength of concrete at the age of loading, even for high-strength concrete (Russell, 1990). The user thus obtains a higher stiffness per unit weight and unit cost. Indeed, it is quite common for a structural engineer to consider and specify high-strength concrete for its stiffness rather than for its strength.

***Creep.*** The specific creep (ultimate creep strain per unit of sustained stress) of concrete decreases significantly as the concrete strength increases. The most recent verification of this is available in *Deformation Properties and Ductility of High-Strength Concrete* (Bjerkeli et al., 1990). This is indeed a fortunate coincidence without which the application of high-strength concrete in high-rise buildings would have been seriously hampered. Because of the lower specific creep, high-strength concrete columns with their high stress levels suffer no more total shortening than normal-strength concrete columns with their lower strength levels. Otherwise the problem of differential shortening of vertical elements within high-rise buildings would have been aggravated by the use of high-strength concrete in columns.

### 3 Inelastic Deformability under Reversed Cyclic Loading

The inelastic deformability of high-strength concrete structural members under reversed cyclic loading of the type induced by earthquake excitation has been a concern that has slowed its application in high seismic regions.

The inelastic deformability of ultrahigh-strength concrete members [compressive strength ranging up to 107,562 kPa (15,600 psi)] under monotonic as well as reversed cyclic loading has been investigated through two series of tests (Shin et al., 1990). The specimens tested were columns under zero axial load. The principal test variables were the compressive strength of concrete, the percentage of longitudinal reinforcement, and the spacing of confinement reinforcement. The ratio $r/r_b$ turned out to be the most dominant factor influencing the magnitude of inelastic deformability, with the latter decreasing for increasing values of $r/r_b$. (Here $r$ is the tension reinforcement ratio and $r_b$ the reinforcement ratio producing balanced strain conditions.) However, even at very large $r/r_b$ ratios, substantial amounts of inelastic deformability were found to be available. Such deformability was generally found to increase with increasing concrete strengths. It was concluded that in the absence of axial loads acting simultaneously with flexure, high-strength reinforced concrete members subjected to reversed cyclic loading possess as much inelastic deformability as is likely to be required of them in practical situations.

In a very significant recent study, Muguruma and Watanabe (1990) have investigated the possibility of improving the ductility of high-strength concrete columns through the use of lateral reinforcement. Eight column specimens confined by lateral reinforcement having 328 and 793 MPa (47.6 and 115.0 ksi) yield strength were tested under reversed cyclic lateral loads with constant axial compressive load levels of 0.254 to 0.629 times the axial load-carrying capacity. The concrete compressive strengths were 86 and 116 MPa (12.4 and 16.8 ksi). The volumetric ratio of lateral reinforcement was 1.6% in all specimens. Test results indicated that very large ductilities could be achieved by using lateral reinforcement with high yield strength, even for such high-strength concrete columns.

In sum, the use of high-strength concrete [with specified compression strengths in excess of 41,370 kPa (6000 psi)] in multistory buildings is spreading across the United States, and is now not uncommon in major cities in UBC seismic zones 3 and 4. The big advantages that make such application attractive are:

1. More strength per unit cost and per unit weight
2. More stiffness per unit cost and per unit weight
3. Lower specific creep for high-strength concretes

One remaining concern that has somewhat slowed the spread of high-strength concrete in regions of high seismicity has been about the inelastic deformability of such structural members under reversed cyclic loading of the type caused by earthquakes. However, reassuring test results have begun to come in, indicating that even high-strength concrete columns under high levels of axial loading can be made ductile under reversed lateral loading, with proper confinement of the concrete.

## 11.7 HIGH-PERFORMANCE CONCRETE

The rediscovery of the importance of a low water-to-cement ratio for improving the strength in concrete started the revolution of using concrete for tall building structures. But low water-to-cement ratio concretes are difficult to mix. The use of superplasticizers which make the dry mixes more workable was the other necessary invention. These dry mixes that are characteristic of high-strength concrete were used in the 1970s and 1980s for the first crop of tall concrete buildings. This high-strength concrete was still too porous and therefore not durable enough over time. An improved material introduced in Section 11.5 is high-performance concrete (HPC), which is essentially the durable high-strength concrete with augmented properties obtained by the addition of small particles such as silica and fly ash to fill the void spaces.

The development of tall concrete buildings using high-strength concrete initially occurred mainly in Chicago (Moreno, 1990, 1992). The use of concrete for tall buildings was enhanced by the real-time optimization technique developed to give the owner smaller, much less expensive buildings with earlier occupancy dates. This was done not by specifying materials, but by adjusting the quality of the materials available, weather conditions, equipment available, and problems that developed primarily because of the smaller reinforced concrete column sizes in tall buildings. Many of these tall concrete buildings now stand in Chicago, Seattle, Hong Kong, and Bangkok (see Section 11.1).

### 1 Material Property Development

The question naturally arises: What material and design developments have occurred that allow concrete to partially displace steel in the market for vertical structural elements in high-rise construction? The answer comes partly from an understanding of the material properties and their relation to microstructure as well as the construction process. The major material performance deficits in ordinary-strength concrete are cracking (and related brittleness) and porosity (and related permeability). Both of these represent durability concerns. Pores and cracks in concrete allow intrusion of water, gases, and other agents of the outside, causing the durability problems of corrosion of reinforcing metals, freeze-thaw damage inside the concrete, and chemical deterioration.

In ordinary concrete both permeability and compressive strength are determined primarily by microstructure porosity. Increases in compressive strength have been made by an increase in close packing and by altering the microstructure in terms of pore-size distribution, total porosity, and nature of the solid phase (Mindess, 1984). The major developments in high-strength concrete can be viewed as attempts to reduce porosity, increase close packing of particles, and reduce brittleness. The main approaches to reducing brittleness and porosity were developed during the high-technology era in cements and concrete in the 1970s and 1980s. These main approaches are:

1. *Polymer impregnation:* A liquid-monomer pore infill which is later solidified in the pores
2. *Macro-defect-free (MDF) cement:* A kneaded and specially prepared mix to get rid of defects or pores, which contains a significant percentage of polymers
3. *Condensed silica fume (CSF) particle concrete:* An attempt to reduce open pore sizes by filling them with small silica fume particles, which chemically bond well to other aggregates
4. *Fiber-reinforced concrete:* Reduction of cracks by the addition of fibers, which can increase toughness, decrease crack size, or decrease stress intensity by interrupting crack energy (Hoff, 1984), or the use of discontinuous layers, laminated together, which also disrupt crack propagation

These approaches are not without their durability problems. Condensed silica fume tends to self-desiccate and has autogenous shrinkage cracking because of its low water-to-cement ratio (Paillere et al., 1989). Macro-defect-free cement can lose strength in moist conditions if the polymer used is water soluble. Polymer impregnation is often difficult to accomplish in situ, and the chemicals tend to be hazardous. Fiber-filled cements have to be mixed carefully (Hoff, 1984). Fibers pull out and laminates may delaminate, both of which can cause a weak zone in the material.

When high-performance cement was used for tall buildings, condensed silica fume concrete was used in the majority of cases. The first work on it was done in the early 1950s at the Norwegian Institute of Technology, and that effort continued to be centered in the Scandinavian countries until it was picked up for research in Canada in the 1970s (Sellevoid and Nilsen, 1987). However, there is much controversy associated with it, namely, the issue of quality control over time, that is, as it relates to durability. High condensed silica fume concrete with superplasticizers and a very low water-to-cement ratio is the general recipe for high-performance concrete.

## 2 Influence of Porosity and Microstructure on Properties

The relationship of strength to the lower water-to-cement ratio was explained by Feret and Abrams (*FIP/CEB,* 1990). Jambor and Zaitzen (*FIP/CEB,* 1990) showed that better pore-size distribution, reduced average pore size, and reduced maximum pore size lead to increased compressive strength. Later research shows that compressive strength of the cement paste itself appears to depend primarily upon total porosity, pore-size distribution, and the nature of the solid phase. These factors contribute to all the strength properties related to compression and bending as well as the elastic modulus. Using this information, lower porosity due to a reduced presence of unhydrated water, caused by the lower water-to-cement ratio, and greater solidity of the solid phase due to close packing of smaller particles and greater hydration can accomplish greater compressive-strength concrete (*FIP/CEB,* 1990). After that research development was achieved, the use of water reducers and superplasticizers further reduced the water-to-cement ratio and related total porosity and achieved a greater percentage of hydration. The use of fine-ground cement and fly ash achieved further close packing of the solid phase (*FIP/CEB,* 1990). An understanding of all three of these factors controlling strength has led to even further research exploration. Two specific types of major concrete and cement materials have emerged—dense silica fume concrete and macro-defect-free cement. Both of these materials use:

1. Pore fillers and binders such as silica fume or polymers
2. Low water-to-cement ratio
3. Plasticizers
4. Fine-ground portland cement with excellent hydration qualities
5. Materials with high strength potential such as polymers or pozzolans

Macro-defect-free cement has not been used in tall buildings due to the peculiar mixing requirements but "the use of CSF in concrete is the only way of producing concrete of normal workability with a strength level exceeding 80 N/mm$^2$ hence the development of HSC made of CSF" (*FIP/CEB,* 1990).

## 3 Current Trends

***Construction Practice.*** Keeping the needs of the construction practice in mind, Chicago's Material Service Corporation pioneered high-strength silica fume concrete (HSSFC). They used an optimization "process directed toward development of HSSFC that meets the needs of the entire construction team" (Detwiler, 1992). It is a performance concrete using the "end product approach to mix design" in which strength requirements are given but no specifications are enforced, and so adjustment to mix design is allowed during the project. Adjustments can be made for variabilities in raw materials, mechanical difficulties, varying delivery, and planning conditions. The HSSFC reduces the amount of steel and the volume of concrete in columns, has lower in-place costs, and results in an accelerated construction cycle (Detwiler, 1992). In high-rise construction, where the price is driven by the cost of columns and the speed of construction, the higher material cost of HSSFC is more than offset by the use of fewer materials.

***Performance Issues.*** The need for improved overall performance led to a broader definition of high-strength concrete. It includes high-performance concrete, which is a very workable concrete that yields superior strength, performance properties, and durability. The performance issues are ease of placement and early high strength—both for fast construction—and the material property issue of long-term maintenance of mechanical properties. Other issues are durability in all environments, toughness, stiffness, and volume stability (Carino and Clifton, 1991; Moreno, 1990). The ingredients for high-performance concrete, including a silica fume, as used in the Nova Scotia Plaza in Toronto, Ontario, Canada, are listed in Table 11.1.

The initial slightly higher overall cost of high-performance material compared to conventional concrete is expected to be offset by savings from (Moreno, 1990):

1. Enhanced workability and early high strength, which can reduce construction costs (especially finish and form removal)
2. Enhanced mechanical properties that will reduce the size of structural elements (especially columns or even beams and hence the overall height of the building and the amount of cladding)
3. Increased durability, which gives longer service life (Carino and Clifton, 1991)
4. High modulus of elasticity, which will yield less lateral sway of buildings

Aitcin and LaPlamte (1992) have discussed the developments of high-performance concrete in the North American context. In summary, the need was to resolve the problems of the high-strength concrete of the 1980s and to achieve greater overall performance characteristics and cost savings. These efforts transformed it to the high-performance concrete of the 1990s. The result is better performance over time with more predictable properties (Carino and Clifton, 1991). Future research includes reorganized quality control and assurance programs and enhanced construction methods carried out by a workforce of people skilled and knowledgeable in these new techniques and materials (Carino and Clifton, 1991; Moreno, 1990).

Various research organizations have developed initiatives to address these needs. The focus of their research interests has been well documented by Carino and Clifton (1991).

## 11.8 REINFORCING STEEL FOR CONCRETE

Tall buildings in reinforced concrete consist of structural elements such as beams, columns, and slabs, elements designed against flexure, shear, axial force, torsion, and their combinations. In any reinforced concrete structure the quality of the steel and concrete should be chosen in a manner that matches them well with each other, yielding the

**Table 11.1 Ingredients of high-performance concrete, Nova Scotia Plaza, Toronto, 1988** *(Courtesy: P.C. Aitcin and P. LaPlamte)*

| | |
|---|---|
| Water | 145 kg/m$^3$ |
| Cement (U.S. type I) | 315 kg/m$^3$ |
| Slag | 137 kg/m$^3$ |
| Silica fume | 36 kg/m$^3$ |
| Coarse aggregate | 1130 kg/m$^3$ |
| Fine aggregate | 745 kg/m$^3$ |
| Superplasticizer | 5.9 L |
| Water reducer | 0.9 L |
| Age, days: | |
| 7 | 66.9 MPa |
| 28 | 83.4 MPa |
| 56 | 89.0 MPa |
| 91 | 93.4 MPa |

required strength and resulting in economical returns. There was a time, a few decades ago, when mild-steel round bars were the only steel available for the construction of reinforced concrete structures. But the scenario has totally changed now and different types of steel, including deformed bars, are more commonly used to match the increased strengths of concrete. Other types of steel are Tor steel or twisted steel, cold-twisted Ribbed-Tor steel, hot-rolled ribbed bars, and similar other bars in addition to ordinary mild steel. Twisted bars are currently being used in Europe, particularly in Austria. One other type is I-steg steel bars, made by cold twisting two round mild-steel bars in special machines. Torsional steel bars are produced by cold twisting a deformed round mild-steel bar.

The application of deformed reinforcing bars is so widespread and well documented in the literature (Freedman, 1985a; ACI Committee 439, 1989) that it need not be repeated here. In Section 11.10, some details of cold-twisted Ribbed-Tor steel are presented. This twisted steel is currently being used in India for high-rise construction. Although this type of steel reinforcing is not known in the United States and Canada, it is gaining popularity in countries besides India. The properties of cold-twisted Ribbed-Tor steel, or simply Ribbed-Tor steel, that are somewhat different from those of other types of steel are discussed in Section 11.10, where three case studies are included to demonstrate the advantages of using Ribbed-Tor steel in moderately high buildings.

In sum, it may be said that Ribbed-Tor steel has several properties that could make it suitable for use in reinforced concrete structures. Compared to conventional mild-steel round bars used in India and many other developing countries, it brings almost 35 to 40% savings in steel for a slightly higher unit cost of material. On top of this, it has certain additional merits, such as effective continuous bond with concrete and higher bond resistance, which are not found in some other varieties of steel used in reinforced concrete. This gives promise for the wide use of this material for high-rise buildings. Since this is a high-strength steel, further research is needed to improve its ductility.

## 11.9 FUTURE RESEARCH

The application of concrete for tall buildings has now become universal, in both the developing and the industrialized countries. With economy and fire resistance, combined with the versatility of form, concrete has become a progressively more attractive material for tall building construction. To make concrete more acceptable to architects, engineers, and builders, research is required in the following areas:

1. Studies on integrating the functional and aesthetic design with structural, mechanical, and electrical systems. Better integration for concrete structures is required relative to steel, inasmuch as errors are difficult to correct after construction.
2. Further studies on the serviceability of concrete structures.
3. Research on the durability of concrete.
4. Research on the effects of movement due to temperature, creep, and shrinkage on the architectural, structural, and mechanical details.
5. Studies on natural and built-in architectural finishes in a variety of colors, tones, and textures without the need of additional treatment or painting.
6. Further investigations on high-strength and high-performance concrete, particularly when used in regions of high seismicity.

7. Research on improving the ductility of shear walls.
8. Studies on the damping characteristics of concrete tall buildings under dynamic loads.
9. Research on prefabricated capsules or modules for residential construction.
10. Development of design details for panel structures in seismic zones.
11. Studies on rational and practical design procedures to avoid progressive collapse, particularly due to blast loads.

## 11.10 APPENDIX: PROPERTIES AND APPLICATION OF RIBBED-TOR STEEL

### 1 General Properties of Ribbed-Tor Steel

Ribbed-Tor steel consists of a core of circular cross section with two continuous helical longitudinal ribs having a pitch of twist about 10 times the bar diameter formed on its periphery. Closely spaced short oblique ribs or lugs inclined at least at 35° with respect to the bar axis and having side faces gradually merging into the circumference of the core are made between these longitudinal ribs. A piece of Ribbed-Tor steel bar is shown in Fig. 11.8. Ribbed-Tor steel is manufactured from hot-rolled mild-steel ribbed bars by cold twisting. Twisting sets up a torsional stress in the steel, which is zero at the center and maximum at the circumference of the bar. The torsional stress passes from the elas-

**Fig. 11.8 Ribbed-Tor steel.** (*Courtesy: Tor-Isteg Steel Corporation.*)

tic range within the core gradually to the elastoplastic range at the periphery, and is the source of the additional tensile strength obtained in these bars when put to tension.

***Stress-Strain Characteristics.*** The cold-twisting process results in a substantial change in the properties of the material. In hot-rolled steel having $g/d = \infty$, where $g$ is the pitch of the twist and $d$ the diameter of the bar, the stress-strain curve has a pronounced yield range, as shown in Fig. 11.9, whereas the finished Ribbed-Tor steel having $g/d = 10$ has a continuous form without any yield range. In this case the stress corresponding to 0.2% strain is known as proof stress and is internationally adopted as a substitute for yield stress. The ultimate tensile stress of Ribbed-Tor steel is about 10 to 20% above the proof stress, and this is considered to be a sufficiently safe margin for the steel to be used in reinforced concrete construction.

Tests show that the proof stress as well as the ultimate stress increase with the decrease in pitch of twist, but the rupture elongation and the deformability decrease with a decrease in pitch. The values of proof stress and ultimate stress are found to coincide at about $g/d = 2$, which means that the deformability of steel is totally exhausted at this value. Ribbed-Tor steel with $g/d = 10$ is usually used in practice (see Fig. 11.10).

***Bendability.*** Ribbed-Tor steel can generally be bent through 180° around a pin of twice the bar diameter without fracture or crack formation on the surface (Fig. 11.11). This simple test may not be sufficient to prove that the bar is entirely acceptable as far as its bendability is concerned, since reverse bendability is sometimes needed. It may, for example, be necessary to straighten the bar and bend it at another place or change the bending angle. Also, the reverse bending test should be done to ascertain the bendability.

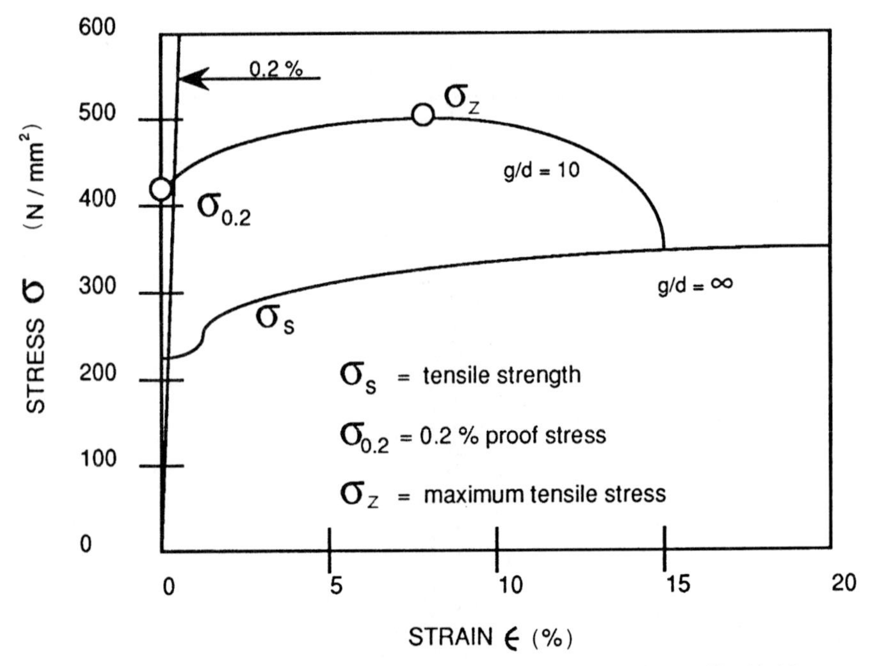


**Fig. 11.9 Stress-strain relation of Ribbed-Tor steel before and after twisting to *g/d* = 10.** (*Courtesy: Tor-Isteg Steel Corporation.*)

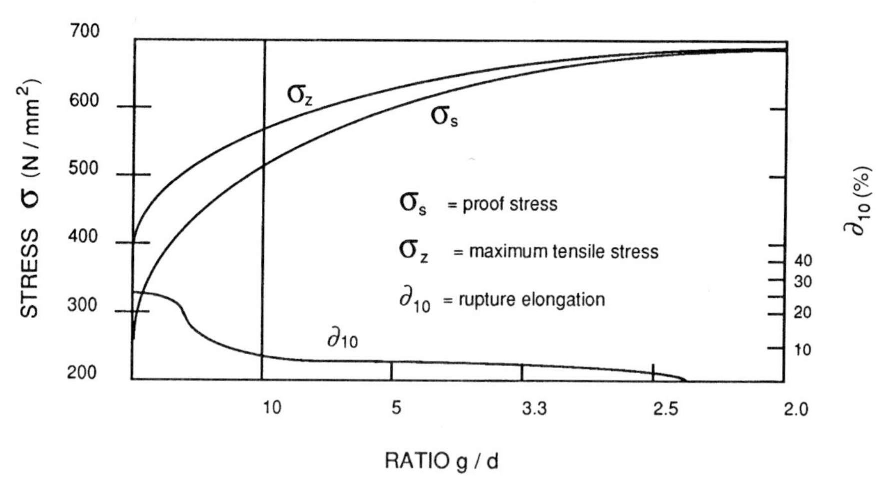


**Fig. 11.10 Relation between stress, *g/d* ratio, and rupture elongation.** (*Courtesy: Tor-Isteg Steel Corporation.*)

**Fig. 11.11 Bent specimen of Ribbed-Tor steel.** (*Courtesy: Tor-Isteg Steel Corporation.*)

Ribbed-Tor steel has an increased yield point (that is, proof stress) compared to mild-steel reinforcement, yet it has the required bendability. The lugs or the transverse ribs are rolled at an inclination of about 35° with the longitudinal axis.

***Response to Elevated Temperatures.*** Ribbed-Tor steel under the action of fire or during welding is subjected to considerable changes in temperature. Hence its response to a change in temperature is very important. For example, a fire will first heat the concrete surface in contact with the flames. The heat will penetrate gradually and cause high stress in the concrete. The resulting temperature gradient between outer and inner concrete may be as high as 1000°C (1832°F). The high stress thus generated will cause the concrete to spall off and expose the steel to the heat. In typical fire tests loaded concrete structures with Ribbed-Tor steel have been shown to stand up for 66 to 73 minutes. It is found that failure of Ribbed-Tor steel occurs generally at temperatures of 500 to 600°C (932 to 1112°F) whereas hot-rolled steel has shown distress at about 400°C (752°F). Thus a concrete structure reinforced with Ribbed-Tor steel has at least the same safety factor during and after the action of fire as a structure reinforced with hot-rolled steel.

***Weldability.*** For long spans, reinforcing bars are occasionally required to be joined, and for technical and economic reasons this is preferably done by welding. Ribbed-Tor steel has been satisfactorily welded using electric-flash butt welding, arc welding of double-V-butt joints, and welded lap seams.

Electric-flash butt welding is always preferable if a large number of welded joints are to be made at the same place. Arc-welded double-V-butt joints may be made in bars having a diameter of at least 20 mm (¾ in.). Mild-steel electrodes with acid rutile coating should be used for welding grade 400 Ribbed-Tor steel. Welding is done in a number of runs, following an approved sequence of runs. Lap joints may be welded in Ribbed-Tor steel of all diameters and grades. A run deposited in a continuous pass should not be longer than five times the bar diameter. Longer seams should be divided into sections, each of which should not be more than five times the bar diameter. Oxyacetylene welding is not permissible with Ribbed-Tor steel.

***Response to Impact or Explosion.*** In the design of reinforced concrete structures it may be necessary to take impact stress into account, which may be caused by impact of vehicles, derailment of trains, action of bombs, or explosion due to other causes. Under the circumstances the reinforcements are subjected to extra tensile stress, which has to be taken up by a large deformation of the steel since brittle fracture will cause disintegration of the structure. Impact tensile tests performed on cold-worked Ribbed-Tor steel specimens are found not to cause any loss in ductility or increase in brittleness in Ribbed-Tor steel. Tests further have shown that after an impact tensile stress due to an explosion, Ribbed-Tor steel remains in good condition and exhibits the same amount of necking as after a normal static tensile stress.

***Bond with Concrete.*** At the beginning Ribbed-Tor steel behaves similar to round bars in regard to its bond with concrete. As the bar tends to slip within the concrete, however, those sides of the ribs which face the ends where the force is applied are forced against the concrete locked in between the ribs, producing a new resistance to the initial slipping movement. A further increase in the tensile force will force the ribs increasingly into the concrete, stressing the concrete between the ribs in shear. At an even greater stress, fracture occurs by shearing of the concrete between the ribs. Hence round bars provide an adhesive bond whereas Ribbed-Tor steel provides a shear bond. In general, Ribbed-Tor steel has a bond resistance five times greater than that of plain round bars. This bond strength remains intact even after millions of load cycles.

## 2 Application Examples

To demonstrate applications of Ribbed-Tor steel in India, three moderately high buildings (by Indian standards) are selected. The structural planning aspects of these buildings are also presented. All three buildings are located in Calcutta and were designed with M20 concrete having a characteristic compressive strength of 20 $N/mm^2$ (2840 $lb/in.^2$), and Tor steel of quality Fe 415 having a characteristic tensile strength of 415 $N/mm^2$ (28,930 $lb/in.^2$). Here the characteristic strength is adopted as the strength below which no more than 5% of the test results may fall. The structural design for all these buildings was performed by the limit-states design method.

***Tower Block Building (Case 1).*** This is a 12-story commercial building in a fast-growing business center of Calcutta. Figures 11.12 and 11.13 present the plan and the elevation of the building. The covered area of each floor is 1080 $m^2$ (11,620 $ft^2$) and the total covered area of the building is 130,000 $m^2$ (1,398,600 $ft^2$). In addition there is a basement with a floor area of 1080 $m^2$ (11,620 $ft^2$) for parking. Other relevant dimensions are as follows:

| | |
|---|---|
| Length | 62.5 m (205 ft) |
| Breadth | 17.28 m (56.7 ft) |
| Height of each floor | 3.75 m (12.3 ft) |
| Total height above plinth level, including 1-m (3-ft) parapet at roof level | 46 m (151 ft) |
| Height of plinth above ground level | 0.75 m (2.5 ft) |
| Height of basement below ground level | 4.0 m (13 ft) |
| Spacing of columns along length of building | 5.75 m (19 ft) |
| Spacing of columns along width of building | 4.0 m (13 ft) for two end bays<br>8.0 m (26.2 ft) for central bay |

There is an expansion joint at the center of the building, which divides the superstructure into portions of 31.25 by 17.24 m (102.5 by 56.5 ft). The adopted live load is 4000 $N/mm^2$ (82 lb/ft). It is founded on reinforced concrete piles with a diameter of 500 mm (1.6 ft) and a length of 25 m (82 ft), having a capacity of 783 kN (176 kip). It is designed as a framed structural building with frame action in either direction. All exterior walls

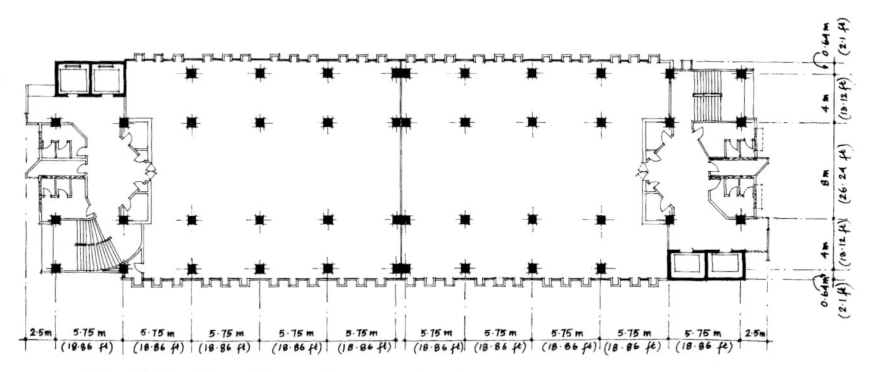


**Fig. 11.12 Plan of Tower Block building, Calcutta, India.** (*Courtesy: Lina Nath.*)

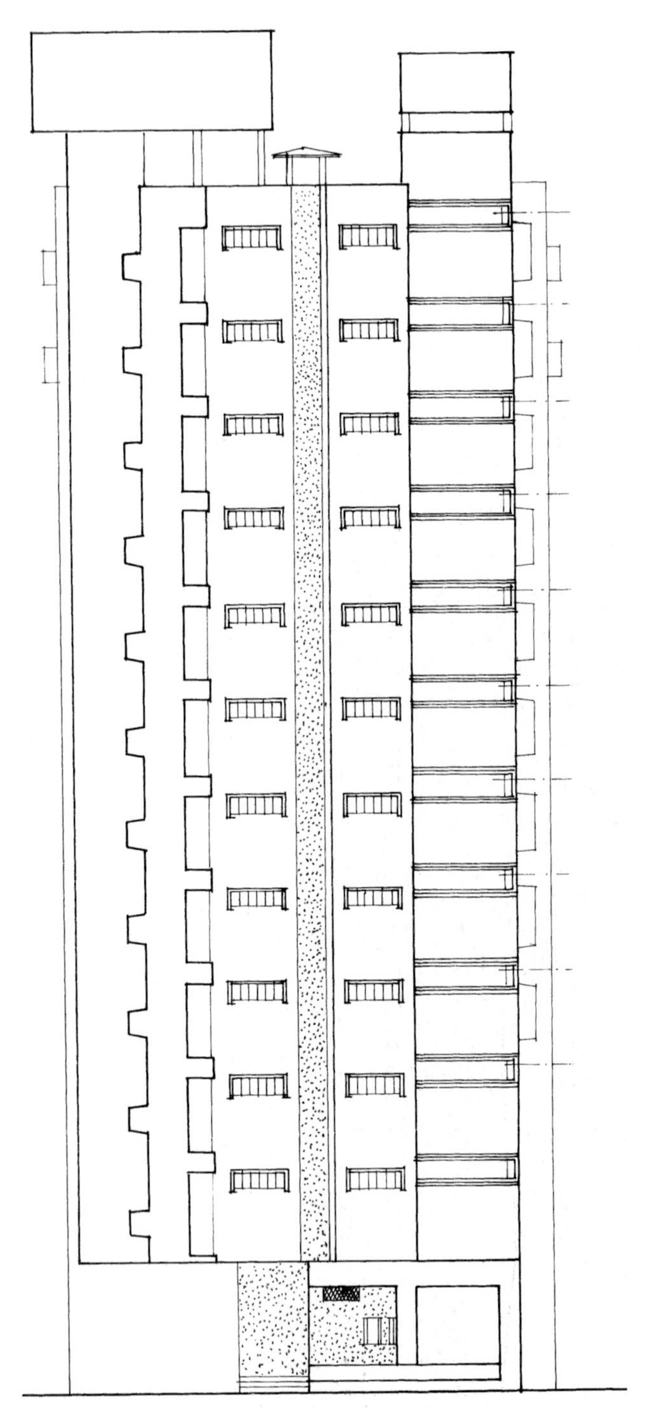

**Fig. 11.13 Elevation of Tower Block building, Calcutta, India.** (*Courtesy: Lina Nath.*)

are 250-mm (9.8-in.)-thick brick masonry and all interior walls are 125 mm (5 in.) thick, also of brick masonry. There is a provision of an equivalent dead load of 1000 N/m$^2$ (20.5 lb/ft$^2$) over the entire floor area to account for light timber partitions not shown in the plan. The live load in the stair landings and car parking areas is taken as 5000 N/m$^2$ (102 lb/ft$^2$). To determine the economy achieved with regard to the quantity of Ribbed-Tor steel of quality Fe 415 used in the building, the building was also designed with mild-steel round rebars (commonly used in India) of quality Fe 250 having a characteristic tensile strength of 250 N/mm$^2$ (35,500 lb/in.$^2$).

It was found that the steel requirement in slab panels is about 68% higher, in beams it is also about 68% higher, and in columns it is 46% higher when Fe 250 is used instead of Fe 415. The overall economy achieved by using Fe 415 steel is on the order of a net saving in steel of 38.3%. Since the price of cold-twisted Ribbed-Tor steel bars is about 10% higher than that of ordinary plain mild-steel round bars, the cost saving would be on the order of approximately 30%. The quantity of Tor steel used in the building comes to 72.8 kg/m$^2$ (15 lb/ft$^2$) based on the total built-up area of the building, whereas if Fe 250 steel were used instead, the corresponding quantity would be 116.9 kg/m$^2$ (24 lb/ft$^2$). Table 11.2 presents these figures.

***United Commercial Bank of India (Case 2).*** This is a proposed bank building having 16 floors and a basement, with a general live load of 4000 N/m$^2$ (82 lb/ft$^2$). But in specific areas, such as the vault room, cash room, book room, and other heavily loaded sections, the live load has been taken as 10,000 N/m$^2$ (205 lb/ft$^2$). A typical plan and a perspective view of the building are presented in Figs. 11.14 and 11.15. Referring to the plan, one can see that there are two reinforced concrete cores, each having three cells, at the two ends of the building. These cores accommodate the services, such as staircase, elevator, and toilet blocks, whereas the office area is located in between these two cores. Columns are spaced at 6.5 m (21 ft) along the length. The column spacing along the width, which has three bays for the office area, is 5.2 m (17 ft) for the end bays and 6.5 m (21 ft) for the central bay. The floor-to-floor height is 3.5 m (11 ft). The covered area in the plan increases uniformly from 646 m$^2$ (6950 ft$^2$) at the fourth-floor level to 813 m$^2$ (8747 ft$^2$) at the ground-floor level so that the building looks tapered on two sides, with additional inclined columns terminating with the end columns at the fourth-floor level. Beyond the fourth-floor level it continues with a constant covered area of 646 m$^2$ (6950 ft$^2$) for the remaining 12 floors. The basement parking garage has an area of 813 m$^2$ (8747 ft$^2$). Other specifications, such as external and internal walls, foundation, and light partition, are similar to those of the Tower Block building just discussed. Because

**Table 11.2 Comparative study of steel consumption for three example buildings**
*(Courtesy: Tor-Isteg Steel Corporation)*

| Case | Steel type | Steel consumption as a percentage of total steel in different elements: Slabs, raft, and pile caps | Beams, lintels, and cornices | Columns and walls | Total consumption kg/m$^2$ | (lb/ft$^2$) |
|---|---|---|---|---|---|---|
| 1 | Fe 415 | 23 | 44 | 33 | 72.8 | 15 |
| | Fe 250 | 24 | 46 | 30 | 116.9 | 24 |
| 2 | Fe 415 | 44 | 29 | 27 | 59 | 12 |
| | Fe 250 | 46 | 30 | 24 | 100 | 20 |
| 3 | Fe 415 | 51 | 31 | 18 | 57 | 12 |
| | Fe 250 | 53 | 31 | 16 | 91.8 | 19 |

of the relatively high stiffness of the cores, these are providing 95% resistance of the lateral loads due to wind or seismic forces, and only 5% resistance is developed in the reinforced concrete frames occupying the central region of the building. Thus these frames mainly carry dead and live loads of the building.

Of the total amount of steel consumed, the quantity of steel required using Tor steel of quality Fe 415 is as follows:

1. Slabs, including basement, 44%
2. Beams, including lintels and cornices, 29%
3. Columns, including basement walls and cores, 27%

The overall requirement of steel for the building is 59 kg/m$^2$ (12 lb/ft$^2$). If the structure is designed using Fe 250 steel, however, the overall steel requirement is found to be 100 kg/m$^2$ (20 lb/ft$^2$), that is, the consumption of steel of quality Fe 415 is about 40% less compared to Fe 250 steel (see Table 11.2).

***Staff Quarters, Reserve Bank of India (Case 3).*** This is a nine-story residential apartment building having an asymmetrical plan, as shown in Fig. 11.16. There are three apartments per floor, with a common staircase. The elevation of the building is shown in Fig. 11.17. The live load adopted for design is 2000 N/m$^2$ (41 lb/ft$^2$). Some other aspects of the building are:

1. Area of each floor, 358.5 m$^2$ (3857 ft$^2$); total area, 3226 m$^2$ (34,710 ft$^2$); and height of each floor, 3 m (10 ft).
2. The column width is restricted to only 250 mm (10 in.), as desired by the architect, and hence frame action could not be utilized effectively for resisting lateral loads applied in one direction, whereas frame action is found to be effective for lateral loads applied in the other direction.

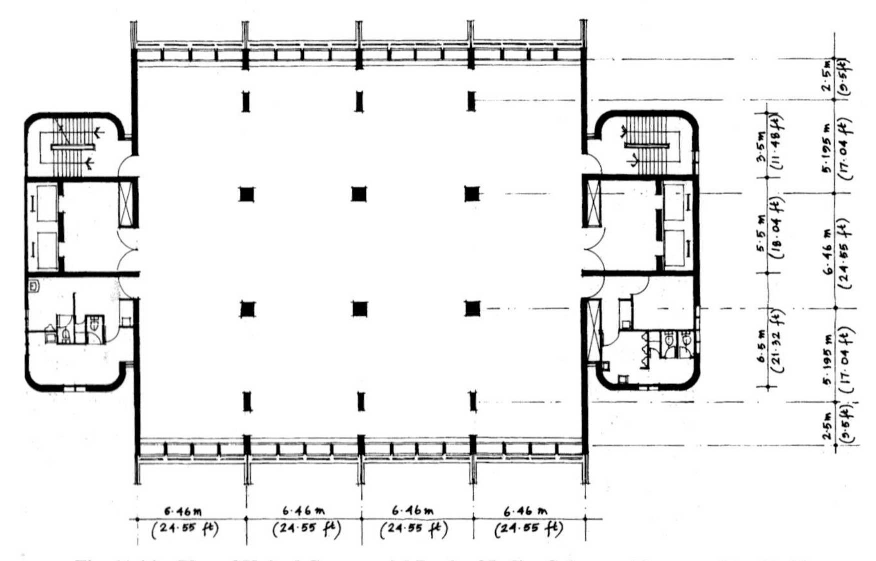


**Fig. 11.14 Plan of United Commercial Bank of India, Calcutta.** (*Courtesy: Lina Nath.*)

**Fig. 11.15 Perspective view of United Commercial Bank of India, Calcutta.** (*Courtesy: Lina Nath.*)

3. To cater to the lateral load requirements in the direction where the frames are found to be weaker, reinforced concrete shear walls are provided, which have replaced the brick masonry walls of the staircase and elevator shafts, as shown in the plan of Fig. 11.16. This has resulted in frame–shear wall interaction.
4. The effect of torsion due to noncoincidence of the center of gravity and the center of stiffness of the building has been accounted for in the design.
5. The column spacing has to be determined according to the dimensions of the rooms used for different purposes in a residential unit, as shown in Fig. 11.16.
6. The columns are founded on reinforced concrete piles with pile caps.
7. All exterior walls are of 250-mm (10-in.)-thick brick masonry and all interior walls of 125-mm (5-in.)-thick brick masonry.

Of the total quantity of steel of quality Fe 415 consumed in the building the quantities required for the individual components are as follows:

1. Slabs and pile caps, 51%, out of which pile caps consumed 34%
2. Beams, lintels, and cornices, 31%
3. Columns and reinforced concrete walls, 18%

The overall consumption of steel per unit of built-up area was found to be 57 kg/m$^2$ (12 lb/ft$^2$). Detailed results are given in Table 11.2.

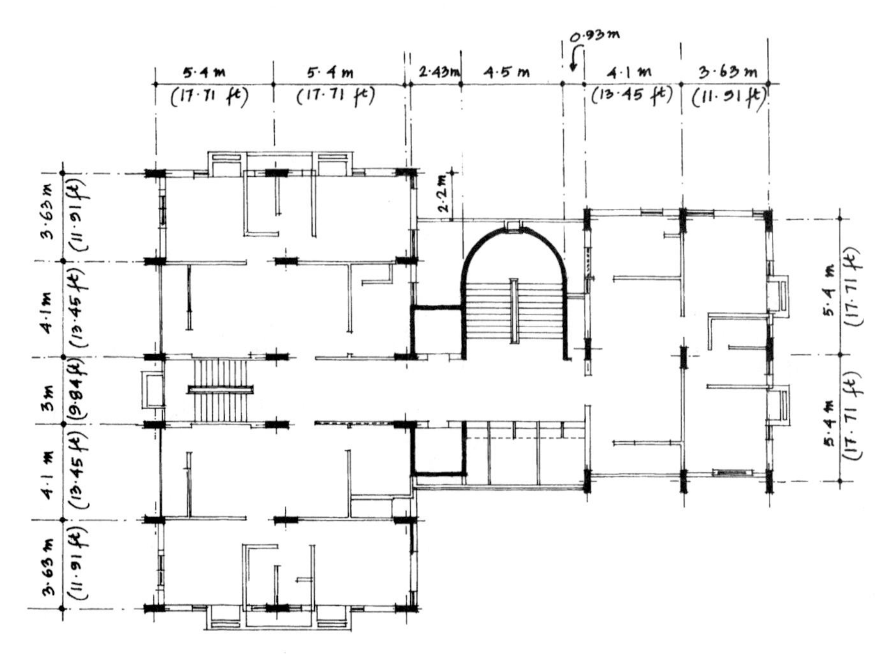


**Fig. 11.16 Plan of staff quarters, Reserve Bank of India.** (*Courtesy: Lina Nath.*)

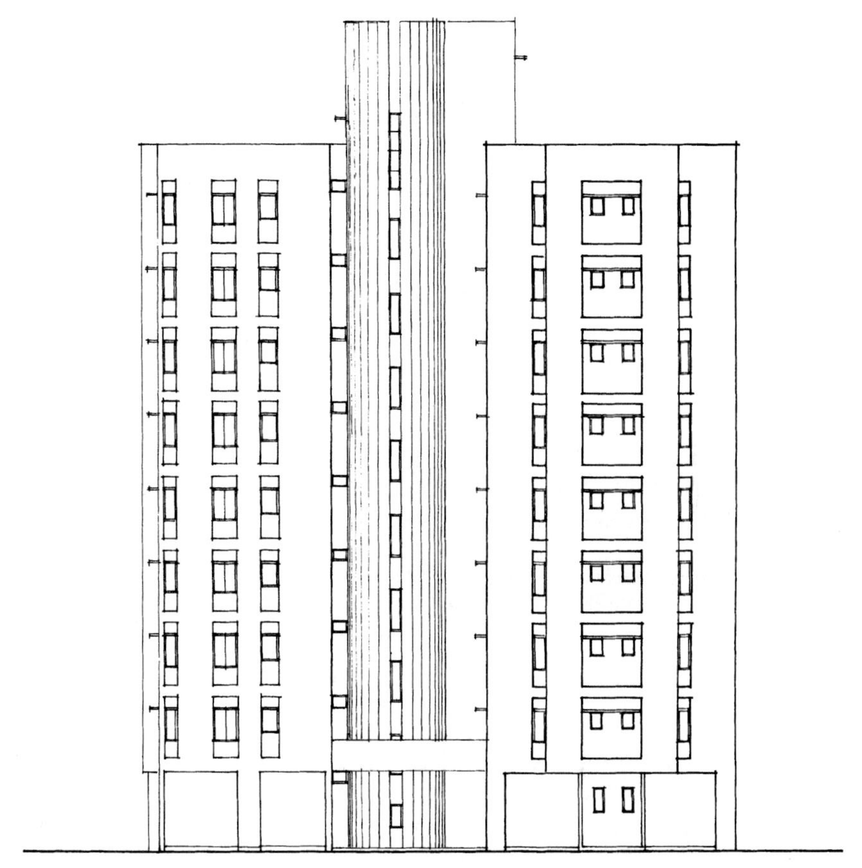

**Fig. 11.17 Elevation of staff quarters, Reserve Bank of India.** (*Courtesy: Lina Nath.*)

## 11.11 CONDENSED REFERENCES/BIBLIOGRAPHY

The following is a condensed bibliography for this chapter. Not only does it include all articles referred to or cited in the text, it also contains bibliography for further reading. The full citations will be found at the end of the volume. What is given here should be sufficient information to lead the reader to the correct article—author, date, and title. In case of multiple authors, only the first named is listed.

ACI 1979, *Super-Plasticizers in Concrete*

ACI 1982, *Recommendations for Concrete Members Prestressed with Unbonded Tendons*

ACI 318-89 1989, *Building Code Requirements for Reinforced Concrete*

ACI 1990, *High-Strength Concrete—Second International Symposium*

ACI 1991, *Manual of Concrete Practice*

ACI Committee 439 1989, *Steel Reinforcement—Physical Properties and U.S. Availability*

Aitcin 1992, *The Development of High Performance Concrete in North America*

Ali 1984a, *Comparative Evaluation of High-Rise Buildings in Singapore*

Ali 1984b, *Cost Effectiveness of Tall Buildings in Singapore—Structural Steel vs. Concrete*

Ali 1985, *Influence of Material Selection on Tall Building Economics*

Ali 1989, *Evaluation Technique for Concrete Structural Systems*

Ali 1989, *Structural System and Material Selection for a 40-Story Multi-Use Building in Chicago*

Ali 1990, *Integration of Structural Form and Esthetics in Tall Building Design: The Future Challenge*

Barnett 1986, *The Elusive City: Five Centuries of Design, Ambition, and Miscalculation*

Bill 1949, *Robert Maillart*

Bjerkeli 1990, *Deformation Properties and Ductility of High-Strength Concrete*

Broadbent 1990, *Emerging Concepts in Urban Space Design*

Burnett 1991, *Silica Fume Concrete in Melbourne, Australia*

Camellerie 1985, *Construction Methods and Equipment*

Carino 1991, *High-Performance Concrete: Research Needs to Enhance Its Use*

Ceco Industries, Inc. 1985, *Concrete Building—New Formwork Perspective*

Condit 1961, *American Building Art: The Twentieth Century*

Cowan 1976, *The Design of Reinforced Concrete in Accordance with the Metric SSA Concrete Structures Code*

CTBUH, Committee 21D, 1992, *Cast-in-Place Concrete in Tall Building Design and Construction*

CTBUH, Group CB, 1978, *Structural Design of Tall Concrete and Masonry Buildings*

CTBUH, Group CL, 1980, *Tall Building Criteria and Loading*

Curtis 1986, *Le Corbusier: Ideas and Forms*

Detwiler 1992, *High-Strength Silica Fume Concrete—Chicago Style*

Domel 1990, *Economical Floor Systems for Multi-Story Residential Buildings*

ENR 1985, *Mixed-Use Tower Will Show Off Concrete Braces*

Fintel 1983, *Economics of Long-Span Concrete Slab Systems for Office Buildings—A Survey*

Fintel 1985, *Multistory Structures*

Fintel 1991, *Shear Walls—An Answer for Seismic Resistance?*

FIP/CEB 1990, *High Strength Concrete State of the Art Report*

Freedman 1985a, *Properties of Materials for Reinforced Concrete*

Freedman 1985b, *Multistory Structures*

Gapp 1986, *Structurally Khan*

Hanna 1991, *Interactive Horizontal Formwork Selection System*

Hoff 1984, *Fiber Reinforced Concrete International Symposium*

Hurd 1990, *Concrete Giant Rises on Chicago Skyline*

Iyengar 1982, *Selected Works of Fazlur R. Khan (1929–1982)*

Iyengar 1985, *High-Rise System Developments in Concrete*

Jensen 1986, *Choosing a Forming System for Concrete Floors and Roofs*

Khan 1966, *Current Trends in Concrete High-Rise Buildings*

Malier 1991, *The French Approach to Using HPC*

Mindess 1984, *Relationships between Strength and Microstructure for Cement-Based Materials: An Overview*

Moreno 1985, *Optimization of High-Rise Concrete Buildings*

Moreno 1990, *High Performance Concrete for Tall Buildings in Chicago—Past, Present and Future*

Moreno 1992, *Concrete Technology for the 21st Century*

Muguruma 1990, *Ductility Improvement of High Strength Concrete Columns with Lateral Confinement*

Paillere 1989, *Effect of Fiber Additions on the Autogenous Shrinkage of Silica Fume Concrete*

Portland Cement Association 1981, *Effects of Various Substances on Concrete and Protective Treatments, where Required (IS001.05T)*

Powys 1971, *Economic Aspects of Planning and Design of Tall Buildings*

Prestressed Concrete Institute 1985, *PCI Design Handbook*

Roy 1965, *Toronto City Hall and Civic Square*

Russell 1990, *Shortening of High-Strength Concrete Members*

Saether 1985, *Prestressed Concrete in High-Rise Construction*

Sato 1982, *New Composite Structure System for High-Rise Buildings*

Schmidt 1966, *High-Rise Buildings of Reinforced Concrete—What Are the Limitations?*

Schueller 1977, *High-Rise Building Structures*

Schueller 1990, *The Vertical Building Structure*

Sellevoid 1987, *Condensed Silica Fume in Concrete: A World Review*

Shin 1990, *Flexural Ductility, Strength Prediction, and Hysteretic Behavior of Ultra-High-Strength Concrete Members*

Soretz 1965a, *Fatigue Behavior of High Yield Steel Reinforcement*

Soretz 1965b, *Tensile Stress on Reinforcing Steels under Sustained Load and Increasing Temperature*

Soretz 1974, *Wide Span Reinforced Concrete Floors*

Straub 1964, *A History of Civil Engineering*

Taranath 1988, *Structural Analysis and Design of Tall Buildings*

Tarricone 1991, *Out of the Lab...and into the Field?*

Wakabayashi 1986, *Design of Earthquake-Resistant Buildings*

Walsh 1990, *311 S. Wacker Stands Strong and Tall*

# 12

# Materials and Structures: Masonry

This chapter discusses the general design of masonry tall buildings, with a balanced emphasis on both architectural and technical issues. It completes the series that began with steel and concrete as materials employed for structural systems, presented in Chapters 10 and 11, respectively, by investigating masonry as a material utilized for tall buildings. In Section 12.1 the historical roots of masonry are traced from ancient to modern times. The use of this material is examined in Section 12.2 with regard to its inherent and traditional physical properties in conjunction with its design developments. Contemporary exterior skin applications are analyzed in Sections 12.3 and 12.4 in two general categories—nonstructural and structural masonry enclosures. Section 12.5 introduces possible future innovations for masonry design and construction and suggests several future research topics. [For further information on the subject, see Chapter CB-13 of *Structural Design of Tall Concrete and Masonry Buildings* (CTBUH, Group CB, 1978).]

## 12.1 HISTORICAL PERSPECTIVE

### 1 Ancient Civilization

The prime structural material for buildings until the latter half of the nineteenth century was masonry. Ancient civilizations considered masonry as the material of choice for monumental buildings, being valued for its aesthetics, durability, and strength. The stepped brick ziggurats of Mesopotamia, the stone pyramids of Egypt, the stepped platforms of the Mayan Yucatan, the ornate temple shafts of ancient India, and the spectacular African ruins from local granite rocks all suggest that humankind everywhere used masonry to scale the heights. The Hanging Gardens, one of the seven wonders of the ancient world, built at Babylon, are thought by some scholars to be the inspiration for the biblical story of the Tower of Babel. The sculptures on fine-grained sandstone are lavishly carved upon old Hindu temples. The largest masonry construction of ancient times is the tallest pyramid by Khufu at Giza at 230 m (756 ft) on a side and 147 m (481 ft) tall.

About 200 B.C., somewhat tall, articulated buildings began to appear, buildings not dependent on the massive pyramid form but similar to the wooden pillared structures of

China and Japan. This tradition of wooden buildings has persisted to the present time. Known as trabeated style, this was a system of horizontal beams raised to considerable heights by cantilevering and bracketing. Originally applied to command posts, gatehouses, watchtowers, and the like, the construction reached impressive heights in Ming Palaces and Buddhist pagodas by the year 1103. In that year Li Chich's *Methods and Designs of Architecture* was published, which established the modular system and essentially determined the uniform proportions of the size and spacing of pillars and beams used throughout the Chinese empire. Wooden pillars were replaced with stone masonry or with brick in towers scattered in the northern and eastern parts of China. The pagodas built from the eleventh to the thirteenth centuries, roofed with trabeated wooden beams and ceramic tile, reached heights of between 30 and 60 m (100 and 200 ft).

In the same period when the Asiatic Oriental culture was perfecting its carpentry technology with the pagoda form, European masonry technology was being perfected by French and German stone masons in Gothic cathedrals (Fig. 12.1). The nave vaults rose to incredible heights; for example, the one at Chartres (started in 1194) rose to 37 m (120 ft). The tallest Gothic cathedral of all was built in Beauvais, France, which collapsed in 1275 after rising to 48 m (157 ft). As is apparent, none of the pagoda or cathedral building technologies achieved the height of the tallest pyramid by Khufu at Giza.

## 2 Rise of Modern Tall Buildings

Most accounts of the tall building have emphasized the technological achievement of the steel frame over the economic and physical limitations of the structural use of masonry. In tall building architecture, load-bearing masonry walls have been supplanted by concrete and steel-frame construction. Masonry has been relegated to the role of veneer and cladding. As a result, the expression of the structural frame tends to be equated with architectural quality. However, as Scully (1988) points out, "technological determinism is always simplistic and, in the end, a dangerous evasion of all reality." As he states, the irony is that many masonry buildings, such as the Monadnock Building of 1889–1891 and the Rookery of 1886, have "outlasted the frame technique which was to outmode [them], since contemporary wall-bearing concrete construction can call forth a similar profile."

In 1891 Burnham and Root designed the 16-story tallest load-bearing masonry building in Chicago—the Monadnock Building. (See Chapters 5 and 7.) It was the tallest masonry building in the world at that time. With outer walls over 2 m (6 ft) thick, the design alternated projected windows and brick pilasters to produce a soaring excellence that has remained to this day a structural facade of classic beauty and simplicity. The invention of the steel frame, soon to be followed by reinforced concrete, relegated masonry to thick nonstructural shells, as in the Wainwright Building. Here terra-cotta spandrels, alternated with pilasters and ribbons of brick, were the forerunners of modern nonstructural masonry facades. The Woolworth Building was also clad in terra cotta backed by heavy thick common brick walls [up to 0.7 m (2.5 ft)]. When renovated in the 1970s, it was found by strain measurements that some gravity loads had been transferred to the masonry despite its intended nonstructural role as an enclosure of the steel frame. The period from the Reliance Building to the Woolworth Building represents only two decades, in which the use of nonstructural masonry facades was established. Forgotten was the structural masonry facade of the Monadnock. In fact, during an occasion to celebrate the modern exterior metal panel curtain wall in 1957, a plaque was placed upon the Monadnock Building to announce the demise of the load-bearing wall as a facade of tall buildings. With the advent of the subsequently published sophisticated structural masonry codes, this hasty recognition has proven to be somewhat premature.

**Fig. 12.1 Cologne Cathedral, detail of transept, Cologne, Germany.** (*Courtesy: Paul Armstrong.*)

## 3 Early Masonry High-Rise Facades

Before the widespread use of steel skeletal construction in the late nineteenth century, architects were restricted to using structural load-bearing masonry for buildings. Typically, these buildings were large and monumental in scale and were limited in height, especially by the standards of high-rise buildings that would follow. These buildings are significant in that they demonstrated how the architect can unify the vertical expression of the building. The classical tripartite organization of the high-rise building into base, shaft, and crown, so adeptly resolved by Adler and Sullivan in the Wainwright Building (1891; Fig. 12.2) in St. Louis and the Guaranty Building (1895) in Buffalo, was addressed in these early examples. Some other examples from the nineteenth century are the Marshall Field wholesale store, the Auditorium Building, and of course the Monadnock Building in Chicago.

***Marshall Field Wholesale Store.*** H. H. Richardson's Marshall Field wholesale store (1885–1887) in Chicago is generally referred to as establishing a significant turning point in the development of tall building form and the expression of the facade (Prescott, 1962). A load-bearing masonry structure, its outward expression is articulate, clear-cut, and clean. Classically derived rustication and articulation are carried over the entirety of its exterior walls. Goldberger (1981) writes that its tight Romanesque form was ordered in the same manner as a Renaissance palazzo (Fig. 12.3). The vestigial cornice is reduced to a rhythm of projecting "dots" and a terminating band, but the windows are grouped so that the expression of the basement or plinth, the *piano nobile,* subsidiary floors, and capping of an "attic," is retained. Three floors together form the *piano nobile,* in an attempt to reduce the apparent vertical scale of the building's seven floors.

Giedion (1964) speculates that the massive stone walls belong to another period, and that it showed Chicago architects how unobtrusively a great volume could be integrated. He also writes that Richardson injected into this building something of the vitality of the rising city, in a treatment redolent with dignity (Giedion, 1964).

Montgomery Schuyler, who knew Richardson personally, saw the Marshall Field store during his visit to Chicago in 1891. He correctly observed the significance of it as the inspiration for much that was relevant to the stripped style of Chicago commercial buildings of the 1880s and early 1890s (Charenbhak, 1981). In the essay *Glimpses: Chicago,* originally published in 1891, Schuyler describes the building as follows (Mumford, 1952):

> One of the earliest of the more modern and characteristic of commercial structures of Chicago, the Field Building is by Mr. Richardson also, a huge warehouse covering a whole square, and seven stories high....Its bigness is made apparent by the simplicity of its treatment and the absence of any lateral division whatever....The vast expanses of the fronts are unrelieved by any ornament except a leaf in the cornice, and a rudimentary capital in the piers and mullions of the colonnaded attic. The effect of the mass is due wholly to its magnitude, to the disposition of its openings, and to the emphatic exhibition of the masonic structure. The openings, except for an ample pier reserved at each corner, are equally spaced throughout. The vertical division is limited to a sharp separation from the intermediate wall veil of the basement on the one hand, and of the attic on the other. It must be that there is even a distinct infelicity in the arrangement of the five stories of this intermediate wall, the two superimposed arcades, the upper of which, by reason of its multiplied supports, is the more solid of aspect, and between which there is no harmonious relation, but contrariwise a competition. Nevertheless, the main division is so clear, and the handling throughout so vigorous, as to carry off even the more serious defect. Nothing of its kind could be more impressive than the rugged expanse of masonry, of which the building is expressed throughout, and which in the granite basement becomes Cyclopean in scale, and in the doorway Cyclopean in rude strength.

**Fig. 12.2 Wainwright Building (1891), St. Louis, Missouri.** (*Courtesy: Kevin Hinders.*)

**Fig. 12.3 Palazzo Strossi (1489), Florence, Italy.** (*Courtesy: Paul Armstrong.*)

***Auditorium Building.*** The Marshall Field wholesale store served as the prototype for Adler and Sullivan's Auditorium Building (1887–1889) in Chicago. Repeating the use of the rusticated plinth over the ground and the two mezzanine floors, Sullivan arcaded the next four as a *piano nobile* and terminated the whole block of 10 stories with an attic floor and a suggested cornice. "Confined within the limits of an architecture of tradition, Sullivan and Richardson were not yet able to accept the most direct solution of vertical expression" (Goldberger, 1981).

***Monadnock Building.*** Burnham and Root's Monadnock Building (1891) in Chicago, along with the Rookery (1886), are considered architectural masterpieces. Goldberger (1981) reports on the Monadnock Building:

> The last of the great solid masonry-walled structures, it rises to 16 stories. The smooth, clean brick walls are thick and heavy at the bottom and, then, as if to express the diminishing weight they bear as they rise, they taper to thinness. No ornamentation articulates the wall surfaces—only the gentle curve of the walls, and the shallow projections of the bay windows, which reinforce the verticality of the facade. The long, narrow site 62 m × 21 m (202 ft × 68 ft) results in the thin slab massing of the Monadnock Building giving it a sense of increased verticality and slenderness. However, the most significant element of the building are the walls, monumentally powerful, hand-edged in a way that calls to mind not only much of the industrial brickwork of the later decades of the nineteenth century, but is a harbinger of the abstract minimalism of the twentieth.

The austerity and straightforward expression of the facades of the Monadnock Building impressed Schuyler. In its battered and tapering walls and the architectural refinement of its facades, Schuyler saw the prototype of the Chicago office building (Jordy and Coe, 1961):

> The business buildings of Burnham & Root were the first tall buildings in which the conditions of both commercial architecture in general and of elevator architecture in particular were recognized and expressed....
>
> Here is a sixteen-story building without ornament, with one form of opening repeated from top to bottom and from end to end, with no relief to the expanse of blank wall and equal opening except what is furnished by the wide and shallow oriels, equally spaced, five on the long side and two on the short, which run through thirteen stories with only a string course to mark off the plinth and with not even a string course to mark off the cornice, or a single molding to relieve or to emphasize its outward curve. Nay, the entrance itself, in which some enrichment is allowed even in the severest buildings, is here only a rectangular hole in the wall like the rest. Add that the whole, excepting the granite lintels of the basement, is in a monochrome of dark brick and terra cotta.

### 4 Developments in Masonry Construction

By the 1950s the Swiss had demonstrated the potentialities of engineered brick masonry. In 1957 an 18-story load-bearing masonry apartment was built in Schwarmendingen, a suburb of Zurich, Switzerland, surpassing the height of the Monadnock Building in Chicago. At Biel-Met a cluster of load-bearing masonry high-rise buildings (ranging up to 25 stories) was built. The Swiss engineer P. T. Haller at the Federal Technical Institute in Zurich provided many of the test data to confirm the feasibility of high-rise load-bearing masonry construction (Haller, 1958). Thus after 66 years (1891–1957) structural skins once again became of interest, partly due to the traditional aesthetics of masonry, their possible economy, improved research knowledge of masonry behavior, and worldwide interest in structural masonry walls as an alternative to the structural frame and curtain-wall technology.

Since the late 1960s, buildings from 12 to 22 stories in height have been designed and constructed in both nonreinforced and reinforced masonry in Switzerland, central Europe, Scandinavia, and the British Isles. Several 25-story buildings in nonreinforced masonry have been designed for nonseismic areas on the basis of modern engineering methods and the recommendations of new, progressive building codes (CTBUH, Group CB, 1978). Structural masonry currently finds wide application in residential construction, such as multistory apartment buildings, student residences, and hotels. Also many office buildings and hospitals are being designed with masonry as the primary structural material.

The use of masonry in curtain-wall construction was severely challenged after World War II by the innovative use of glass, metals, thin stone veneers, and architectural concrete panels, all valued for their thinness, lightness, and aesthetic qualities. This gave rise to panel joint design in the form of plastic spines, organic caulks, and flexible gaskets, challenging the use of cement mortar in ornamental stone work. The demise of the terra-cotta industry by the 1950s was principally the result of the advent of international style architecture and less emphasis on traditional architectural ornamentation. However, in recent years, with the historic preservation and adaptive reuse of older buildings, terra cotta is once again being incorporated into high-rise architecture.

Some architects and engineers are critical of the common practice of employing brick veneer over steel stud framing in wall construction due to potential for corrosion (Mays, 1988). Some masonry specialists maintain that the brick veneer and steel stud wall system was introduced and promoted without adequate research. In 1987 the Brick Institute of America changed its recommended deflection criterion for such systems from $\frac{1}{360}$ of the wall height to about half that, with additional recommendations for a wider internal air space (Mays, 1988). This suggests the need for more research and careful attention to detailing.

For designers, the rise of contextualism and the interest in alternatives to the glass curtain wall have made masonry again a desirable material. As a veneer over structural steel or concrete, brick and other masonry materials perform just as well as glass, metal, or stone, if properly detailed. The modular size of unit masonry offers the potential for a wide range of formal expressions and ease of placement during construction. Furthermore, its flexibility allows it to be either prefabricated in panels or hand-laid on the site.

## 12.2 MASONRY MATERIALS AND DESIGN

### 1 Masonry Materials

The following are the most common masonry materials:

1. Stone masonry and clay brick masonry, which have a long record of use and have demonstrated a high degree of durability.
2. Concrete brick, which is used in a very similar way as clay brick, with some modifications in shape.
3. Concrete block, which is very popular because of its cost-effectiveness and efficiency, is used for both plain and reinforced masonry.

4. Hollow clay brick, which is similar to concrete block and is a relatively newer development.
5. Tile, which is a conventional material, provides an elegant, durable finish, and has good structural properties.

A joint material is required to provide the bonding between the masonry units so as to transform masonry into a composite material. The bonding agent is a portland cement type mortar or grout. The advantages of the composite nature of masonry lie in the economy and flexibility of fabrication of the structural units in that it is often cheaper to join together comparatively small elements (that is, the masonry units) than to create a single large element.

The properties of masonry are controlled by the individual properties of the component elements and their interactive characteristics (Francis et al., 1970; Hilsdorf, 1969; Shrive and Jessop, 1980; Ali, 1980, 1981; McNary and Abrams, 1985).

In view of the inherent weakness of unreinforced masonry, there is a severe limitation to its use for structures subjected to heavy loads. For tall buildings it is therefore imperative that steel reinforcing be added to increase the strength of the walls and to keep the wall thickness as slender as possible. To ensure adequate bond and interaction between the masonry unit and the reinforcement, as well as to eliminate the corrosion of the reinforcing bars, the reinforcement is placed in joints or cavities which are filled with grout or mortar. In reinforced concrete block masonry, steel bars are generally installed in the cores of the blocks. Some pilot studies have demonstrated that masonry could be used for up to 60-story-tall apartment buildings (CTBUH, Group CB, 1978). Further details on reinforced masonry can be found in Schneider and Dickey (1979), Dickey (1985), and Amrhein (1992).

Additional details on masonry materials and masonry assemblies can be found in Amrhein (1991), *Building News* (1981), Senbu et al. (1991), Redmond and Allen (1975), Sahlin (1971), Yao and Nathan (1989), and CTBUH, Group CB (1978). A narrative account of the masonry materials available and used in Canada can be found in Nicks (1980).

## 2 Design Considerations

Architectural function and general structural requirements, performance of materials, and methods of construction are the principal factors that determine the configuration and basic physical dimensions of a building. The configuration determines the shape and layout of the building in relation to the site conditions and space characteristics, whereas the physical dimensions constitute the logic behind the establishment of the size, order, and scale of the building. Different geometric forms of masonry units, wall panels made of these units for use in buildings, structural bays defined by a series of bearing walls in succession, and floor-to-floor heights offer a choice of geometric plans and elevations that form the basis of independent visual units or modules.

In order to obtain desirable aesthetic and architectural effects, these basic modules must be organized in a certain sequence. Further, these geometric forms may be combined in several ways to provide novel architectural expressions. The final physical form of a masonry building is essentially a logical consequence of the intimate arrangements of the masonry modules and is thus greatly influenced by the geometry of the selected basic modules. Thus the possibilities of building forms are almost limitless in that the forms could be quite extensive and intricate, depending upon the design objectives of the architect. A general guide in the search for simplicity, logic, and elegance in obtaining an acceptable design is the architect's discipline, which would aim at eliminating the arbitrary, erratic, and random elements in a building form. In 1938 Mies van der Rohe observed in this connection (Spaeth, 1985):

> Take a brick....how sensible is its small, hardy shape, so useful for every purpose. What logic in its bonding, pattern, and texture. What richness in the simplest wall surface, but what discipline this material imposes.

***Building Plan.*** The layout of horizontal elements in floors is dictated primarily by functional space requirements. The arrangement of vertical elements supporting the floors is a more complex problem since both the topology and the structure are involved. The principal types of structural masonry arrangements are:

1. Transverse wall plan, where the main walls are placed along the transverse direction of the building
2. Longitudinal wall plan, where the main walls are placed along the longitudinal direction of the building
3. Random or combination wall plan, where the main walls are placed in various combinations of transverse and longitudinal walls to create varied shapes

An important aspect to consider is the layout of masonry shear walls that carry the lateral wind or seismic loads. Appropriate consideration must be given to ensure that it is a wall of permanent nature and is designed and detailed by the structural engineer accordingly. Collaboration between the architect and the structural engineer at the initial design stage is most important to avoid misunderstanding and the need for design changes at a later stage. Another crucial aspect to consider is the possibility of future alterations in plan to suit changing needs. This is particularly true where structural interaction between floor diaphragms and walls has been used for lateral stability and load resistance, and where headroom, floor thickness, and architectural planning of the building preclude future addition of local framing elements. Architects and structural designers must discuss with the client the issue of possible future remodeling and any other restrictions to avoid future misunderstanding and expensive redesign estimates.

The building configuration in plan (and elevation) is another important issue to be considered, particularly for seismic codes. Many masonry buildings have collapsed during earthquakes either because their walls were not properly reinforced and tied together or because the wall configuration in plan and section did not provide adequate lateral stability and the seismic event resulted in an undesirable twisting motion of the buildings. Architects and engineers must pay careful attention to configuration principles when designing tall masonry buildings located in earthquake-prone regions (Ali, 1987). In addition to properly reinforcing masonry walls to withstand lateral loads, architects and engineers can collaborate to find wall configurations that will strengthen masonry walls and help them to withstand lateral forces. Designers also should account for movement in masonry walls with adequate and proper expansion joints, flexible anchorage, and bond breaks (Mays, 1988).

***Building Elevation.*** The design of building elevations involves the establishment of the basic geometric forms and their organization into a series of specific relationships. The design process requires a sequence of readjustments, with the end result being the sophistication of the original design concept in such a way that an elegant visual impact is created by the building's elevation. The geometric and aesthetic principles must be manifest in the form of several proportional systems or subsystems for the elevational treatment of tall masonry buildings. Aesthetically speaking, the disciplines imposed by masonry result in visual unity, a unity that involves several different materials. A masonry unit, whether brick or block, is large enough, when viewed from proximity, to form a benchmark pattern against which other objects in the surroundings can be scaled, yet small enough that it does not become overbearing in the building's visual impact. When viewed from afar, the masonry unit elegantly surrenders its individuality to the largest element or module to which it belongs, thereby developing harmony and rhythm within its own setting and in the greater context of its environment.

Windows and other openings in the building's exterior masonry walls play an important role in terms of a building's architectural expression. Their size, location, and shape are important from the visual and aesthetic viewpoints. The sizes of the openings in a way determine the massivity of a building. Large openings offer a feeling of lightness, resulting in an elevation with gridlike characteristics. Small openings on the other hand convey the impression of heavy mass. Similarly, if the openings are recessed into the elevational surface, an impression of massivity is conveyed. If the openings are projected beyond the elevational surface, they create a special visual impact complete with new patterns of light and shade. The shape of openings also influences the visual impression and must be given consideration during the design process. Openings should not be located in such a manner as to offer an impression of holes, but rather should be arranged as logical and natural interspace. Likewise, any discontinuity in walls should be located as a definite, clear division of space that is aesthetically consistent. A good example of the relationship of openings to the aesthetic effect of the exterior walls is the Monadnock Building in Chicago. (See Chapters 5 and 7.)

Whenever possible very large window and door openings adjacent to one another should be avoided in a tall masonry building since this may result in very narrow piers, which will act as slender columns with small cross-sectional areas, thereby increasing the complexity of the structural design. Large wall openings introduce a high degree of stress concentration at the reentrant corners of the openings. A wider pier is more efficient in carrying the large building loads transferred to it through the door and window lintels. This is particularly of significance at the lower levels of a tall building where the loads from the upper levels accumulate. When large windows and doors, and hence narrow piers, are unavoidable for architectural reasons, the structural designer must ensure that these piers are adequate to carry the applied loads.

The texture, color, and pattern of masonry components on the exterior surfaces of tall buildings can significantly impact the architectural expression of the facade. (See Chapters 5 and 7.) The rise in contextualism and an interest in alternatives to the glass curtain wall have once again made brick a desirable alternative to industrial materials (Mays, 1988). There are new uses for masonry in high-rise building construction, and developments in the production techniques of masonry, such as flash firing to alter the colors and sizes of masonry units and ceramic coatings, offer new possibilities in texture and color.

The aesthetic expression and design of facades are greatly influenced by the texture, color, and pattern of masonry (Fig. 12.4). Bond patterns, bull nosing, corbeling, and bearing arches provide a texture and variety to facades without requiring custom-made shapes. Adding certain chemicals, such as manganese to create "ironspot" patterns, to masonry mixtures prior to firing can create variations in color and texture (Mays, 1988). Corbeled and bull-nosed masonry can add to the three-dimensional appearance of the facade. Custom-produced masonry can produce a wide range of color and textural effects to meet almost any design need.

The joint patterns of masonry walls also must be given consideration due to their visibility. Masonry is desirable as a building material because each unit represents a modular size that is easy to handle during construction and adds a scale dimension to the entire building. Proper mixture of the mortar to produce the best bond, selection of color admixtures to the mortar, and careful tooling to produce pleasing aesthetic effects and prevent water infiltration are some of the most important considerations. Construction

**Fig. 12.4 Arched passageway, Cranbrook Academy of Art and Architecture, Bloomfield Hills, Michigan.** (*Courtesy: Paul Armstrong.*)

and control joints are very conspicuous. If they are not detailed properly, the effect can be disastrous from both a structural as well as an architectural point of view.

The location, size, and aesthetics of the wall joint also affect its performance. Aesthetics can be adversely effected when a joint is omitted or made very small for reasons of appearance, overstressing either the cladding or the sealant. The proper performance of a joint dictates certain parameters. For instance, the location of a joint usually occurs wherever dissimilar materials and discontinuous surfaces or structures meet or where continuous surfaces require expansion and contraction (Fisher, 1985). The dissimilar materials subject the joints to nonuniform movement. The selection of high-recovery sealants which can withstand the tendency of various materials to expand and contract at different, intermittent rates is essential (Fisher, 1985).

***Architectural Detailing.*** Masonry design consists of a very important aspect called detailing. An advantage of masonry construction is the flexibility of shape (Fig. 12.5). However, since unusual skill is necessary for masonry construction, careful attention must be given to proper installation and modular characteristics. Good detailing is an art by itself and requires a knowledge of theory and practical application.

Detailing involves the joints in terms of their thickness, shape, and movement. Joint sizes are controlled by the size of mortar sand aggregate, irregularities of the units, and workability of the mortar. Generally speaking, the joint thickness shall be more than twice the maximum particle diameter of the mortar sand aggregate (Monk, 1980). Larger units require greater tolerances to accommodate their inherent dimensional irregularities. Dimensionally dense stone can be much larger than unground clay or concrete units for a joint thickness. There are no rigorous rules for determining joint sizes. It is customary to use 9.5-mm (⅜-in.)-thick joints for most units using a cementitious mortar. Joints thicker than 9.5 mm (⅜ in.) are subjected to lateral flow from the weight of the freshly laid masonry. They are difficult to tool to produce concave joints against the unit edges to increase bond with the mortar and thus are prone to water penetration. From a structural point of view, thicker joints will result in reduced values in bearing, bending, and racking wall strengths. Thinner joints on the other hand are difficult to lay accurately to the mason's line, increasing the cost of labor.

Joints that are either recessed or projecting are potentially troublesome. They cannot be compacted with a concave tool. Their shape profiles permit the accumulation of ice and snow. Such joints should be avoided.

Detailing also involves the important aspect of waterproofing, flashings, and water drainage provisions. Flashings are provided to prevent migration of water once it enters into masonry construction. Since completely sealed masonry construction is not possible, flashings are provided to convey water back to the exterior. A weep drainage system is employed to send the water outside once it is collected by the flashing. Potential locations where water is likely to penetrate are parapet walls, chimney tops, garden walls, and window and door sills, to name a few. Locations where water will generally accumulate are shelf angles, lintel openings, and floor slabs. Eppel (1980) has discussed the mechanism of moisture penetration, summarized existing research efforts, and proposed a number of recommendations in this regard. Other detailing aspects of masonry are those of anchorage, ties, and shelf angles. These are generally required for nonstructural curtain walls. For wall ties, the following materials are commonly used: stainless steel, galvanized mild steel, brass, bronze, manganese bronze (high tensile brass), hard-drawn copper, aluminum, polypropylene, and gun metal. When selecting a particular tie or ties for a project, due consideration must be given to the aspects of corrosion, fire damage, and structural adequacy. Some of these issues are also discussed from another perspective in Sections 12.3 and 12.4.

**Fig. 12.5 Performing Arts Theater (1959–1973), Fort Wayne, Indiana.** (*Courtesy: Paul Armstrong.*)

Another important detailing issue (and which also relates to elevational design) is the location and detail of movement joints. Masonry can develop cracks due to variations in temperature and time-dependent effects as well as its exposure to rain and snow. Allowances for movement must be made by suitably designing the joints. It is impractical to recommend or advocate a comprehensive specification for the location and spacing of movement joints for all masonry structures. Each building must be considered individually and specifications prepared according to its particular condition and characteristics. Joints vary widely and are classified in diverse ways. Dynamic joints have more than 5% and static joints less than 5% total movement (Fisher, 1985). Butt joints occur in the same plane, lap joints between overlapping planes, and corner joints between angled planes. Further classifications include one-stage joints with their sealant at the exterior edge and two-stage or rain screen joints with the sealant at the back of the joint, concealed behind a vented air cavity and a surface rain deflector or an overlapping edge (Fisher, 1985). Control joints are located in concrete masonry walls to control the location of cracks that would otherwise occur because of shrinkage. They are intended to open and are filled with masonry or grout. Expansion joints are used in clay and brick masonry and must be compressible so that they may close.

The spacing of movement joints for masonry varies according to climate and material properties. Grimm (1992) provides general recommendations for vertical expansion and control joints in vertical walls. Vertical expansion joints in brick masonry walls should be placed at or near:

1. Changes in wall height or thickness
2. One side of pilasters, recesses, and chases
3. Wall intersections (corners)

Vertical control joints in concrete masonry walls should be placed at or near:

1. Changes in wall height or thickness
2. One side of pilasters, recesses, and chases
3. Wall intersections (corners)
4. One end of lintels and sills for wall openings less than 1.8 m (6 ft) wide and at both ends for openings 1.8 m (6 ft) or more in width

The maximum horizontal spacing between vertical control joints in concrete masonry walls is determined by:

1. Local average annual relative humidity
2. Type of concrete masonry units, that is, ASTM C 90, type I (moisture controlled) or type II (nonmoisture controlled)
3. Vertical spacing of bed joint reinforcement
4. Exposure to weather

When designing a movement joint, it is essential to consider what type of movement is anticipated. Movements in brickwork may be caused by expansion and contraction; hence a material that is easily compressible and resilient must be used. Certain types of fiberboard and similar materials are not suitable for expansion joints. Strips of V-shaped copper have been used successfully for a long time, but new materials such as premolded, extruded, closed-cell rubber or plastic sections are now often used. The sealing of movement joints has always been a problem. Sealants are chosen based upon their movement capacity. Manufacturers denote the movement capacity of a sealant as plus or minus a certain percentage of its width. For example, in a 25 mm (1-in.)-wide joint, a sealant with a capacity of ±25% means that the sealant can withstand the joint's compression to 19 mm (¾ in.) or its expansion to 32 mm (1¼ in.) without either adhesive failure of its bond or the cohesive (tearing) failure of the sealant (Fisher, 1985). When selecting a sealant for a particular situation, consideration must be given to its service life, hardness, curing time, adhesion to various substrates, weather and chemical resistance, and recovery (Fisher, 1985).

According to Fisher (1985), sealants fall into three general categories according to their cost and performance. The first category consists of low-grade caulking compounds, such as polybutene, polyisobutylene, and oil- and resin-based caulks, and has the lowest cost and performance capabilities. Intermediate-level sealants include acrylic latex, polyvinyl acetate latex, vulcanized butyl, Hypalon, and neoprene sealants. These sealants are more expensive than caulk, but offer a higher level of performance and longer life. The high-performance elastomeric sealants include solvent-based acrylics, polysulfides, polyurethane, and silicone. These sealants have a long service life (up to 20 years or more), a wide range of movement capabilities depending on sealant type, and short curing times. They also have minimum shrinkage, high strength, good resistance to weather and chemicals, and good resilience after compression.

Additional considerations for detailing the movement joints in seismic zones are necessary. Appropriate code requirements (for example, retaining the structural continuity of a wall) must be satisfied while designing movement joints for tall masonry buildings in seismic zones (ICBO, 1991). Provisions of other suitable connections and details for the building (that is, in addition to movement joints) are also necessary to make the design seismic-resistant.

Other detailing issues involve the connections between the floor system and the masonry walls. See Chapter CB-13 (CTBUH, Group CB, 1978) for a detailed treatment of this subject. Masonry detailing has also been discussed by Elmiger (1976) and Beall (1987).

***Structural Design.*** The acceptance of masonry as a structural material in recent years has increased its use for high-rise construction. Some early publications, such as Sahlin (1971), Sutherland (1969), and Hendry and Hasan (1976), provided some basic concepts of structural masonry design. Chapter CB-13 (CTBUH, Group CB, 1978) reported the developments in structural design up to that time. Further developments are reported by Hendry (1983). The latest building codes such as the *Basic Building Code* (BOCA, 1993) and the *Uniform Building Code* (ICBO, 1991) in the United States and similar codes in other countries, particularly in Europe, as well as research on slender walls (Amrhein, 1986) have resulted in substantial progress in structural masonry design. Current trends in structural wall design lean heavily toward thinness. For structural design of masonry, the working-stress design (WSD) method is the most common approach adopted in the United States, although the ultimate-strength design (USD) approach is also encouraged by the *Uniform Building Code* (ICBO, 1991).

## 12.3 NONSTRUCTURAL MASONRY ENCLOSURES

Between the Monadnock Building (1891) and the Woolworth Building (1913) the architectural details for nonstructural facades to enclose skeleton frames of steel and concrete were developed. On the Woolworth Building, the skin was covered with glazed terra cotta—a fired clay material that experienced a decline due to the extensive cracking of its glaze. By the 1950s nearly all the architectural terra-cotta producers seemed to find no demand for their product. Extensive research by ceramicists has shown that the low-fired clay bisque beneath the glaze underwent a gradual expansion caused by moisture. The fired surface glaze, not experiencing a similar volume change, cracked as the body to which it was bonded grew in volume. The condition may be likened to the cracking of hard-boiled egg shells from the expansion of the coagulated egg beneath the shell. Clay moisture expansion is due to moisture adsorption of water vapor onto the fired clay minerals. Unlike the moisture absorption of water from rain or condensing humidity in the porous voids, adsorbed moisture from atmospheric water vapor attaches itself by electrochemical means to the fired clay mineral surfaces. This condition causes deformational release of the strained mass of the fired hardened clay body, which generally will tend to shrink when cooled. This deformational condition is released by the adsorbed water, producing measurable expansion. Clay moisture expansion from adsorbed water vapor is quite opposite to concrete shrinkage from moisture evaporation from the hydrated gel of portland cement mortar systems. Glazed ceramic ware, either in the form of hand-shaped, sculptured, decorative terra-cotta bisque or of hydraulic pressed brick units of relative uniform dimensions, has not weathered well in many U.S. climates. The designer must consult each glazed-ware producer for knowledge from historical experience. Adequate ASTM specifications to assuredly prevent glazed-ware cracking are yet to be developed. There is, however, sufficient experience with the successful use of fired glazed clay ware to justify its continued use, but only upon close consultation with the manufacturers of the material who can cite successful field use.

Nonstructural exterior walls were as thick as 0.7 m (2.5 ft) in the Woolworth Building. In 1932, when the Empire State Building was built, masonry cladding was still 0.5 m (1.5 ft) thick. The traditional structural use of masonry was giving way to being an enclosing skin material over the frame (Monk, 1980). It was still cherished for its aesthetics, durability, and fire resistance, but by 1950 even this use was challenged by other types of modern curtain-wall construction employing glass, asbestos cement, slate, plastic-coated steel, anodized aluminum, and bronze. Thinness (to create maximum usable floor area) and lightweight (to minimize the size of the supporting frame members) were challenges that drove masonry into a considerably lesser market share in the palette of available cladding materials. Thin stone veneers were thus developed by the 1960s.

Figure 12.6 shows examples in Chicago of the evolution of structural and cladding systems. In the foreground is a detail of a stone load-bearing wall of a church. 900 North Michigan is in the middle ground. The granite panel veneer cladding is supported by the building's structural frame. The John Hancock Center looms in the background. It is the epitome of structural art with its large structural X braces expressed on its facades.

### 1 *Types of Nonstructural Enclosures*

Currently nonstructural masonry enclosures for high-rise buildings are of five general categories:

1. Veneers
2. Reinforced panel walls
3. Cavity walls
4. Thin stone assemblies
5. Solar screens

**Fig. 12.6** **Three Chicago buildings illustrating different structural and cladding systems. Foreground—stone church; middleground—900 North Michigan; background—John Hancock Center.** (*Courtesy: Bryan Richter.*)

Available to the designer is a considerable amount of published information, from which the following selected material has been extracted.

***Veneers.*** There are two kinds of veneers—hand-laid glazed ceramics and metal stud construction. Figure 12.7 shows standard detailing for terra-cotta that depends on mortar adhesion and metal ties to be secured to masonry backup construction. Fundamentally there is little difference between what is shown here and what was done on the Woolworth Building over 80 years ago. In the last 25 years there has been wide acceptance of metal stud construction wherein substantial economies are claimed by the elimination of the masonry backup.

Another classification of veneer is by the way it is attached to its backing. There are two types of veneer (ICBO, 1991; Dickey, 1985)—adhered veneer and anchored veneer. The requirements for the design are that the adhered veneer develop 345 N/mm$^2$ (50 psi) in shear and that the anchored veneer attachments develop twice the weight of the veneer being attached (Dickey, 1985).

***Reinforced Panel Walls.*** Reinforced panel walls were developed primarily where high lateral loads are required by the code, such as in seismic zones 2, 3, and 4 as designated on U.S. national seismology maps. Further, this technique is used to provide prefabricated panels. Frequently, when intended for prefabrication, flat panel walls must be reinforced to withstand gravity loads during lifting from the manufacturing beds and for building erection. Such loads frequently exceed design wind loads.

***Cavity Walls.*** Cavity walls were originally developed in England as a practical answer to leaky masonry in the rainy climate of the British Isles. Figure 12.8 illustrates innovative cavity wall construction details. Metal ties and other reinforcing are necessary to make the two wythes act together in resisting lateral loads. Rainwater is intercepted by the outer wythe, which drains on its cavity face downward to flashing and weep holes. Aided by flashing, drainage to the outside must be provided at the cavity air space at the bottom. Whereas air space alone will provide some degree of insulation, insulated cavity walls can be utilized where substantial resistance to heat flow is required. The insulation will cause greater differential movement between the inner and outer wythes due to greater thermal differences. This is especially so at corners, where vertical expansion joints must be provided each way from the corner. Despite their heavy weight and cost, cavity walls are widely used in maritime climates close to coastal boundaries of salt-water bodies where corrosion from salt aerosols is an ever-present danger to metal stud construction. Further, when cracking of the outside veneer is considered critical due to excessive flexibility of the metal stud, the designer may prefer the masonry cavity wall with the stiffer interior wythe backup in corrosive environments.

***Thin Stone Assemblies.*** In some ways thin stone assemblies are similar to veneer construction. However, the unique system assemblies and architectural detailing developed since 1950 are worthy of serious separate study. Figures 12.9 and 12.10 show thin stone veneer assemblies anchored to a structural steel frame and a concrete structural frame, respectively. These details and technology can be applied to a wide variety of stone materials, including marble, granite, sandstone, limestone, and slate. The panel system in these figures would apply to any material system. In the refacing of old construction the use of liner block supports is very traditional. In adhered thin stone assemblies the use of epoxy glue permits assemblage configurations which would otherwise not be possible.

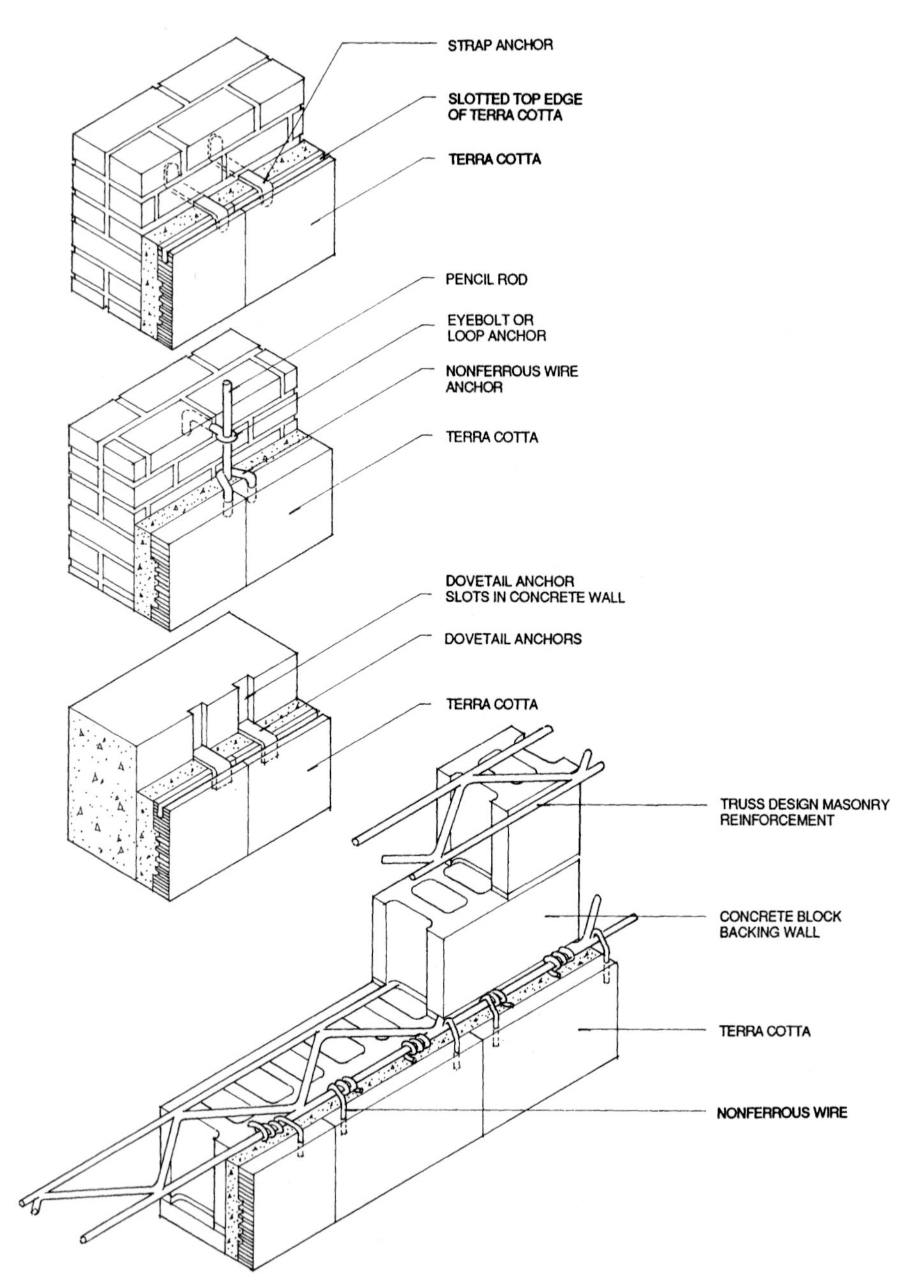


**Fig. 12.7 Terra-cotta detail.**

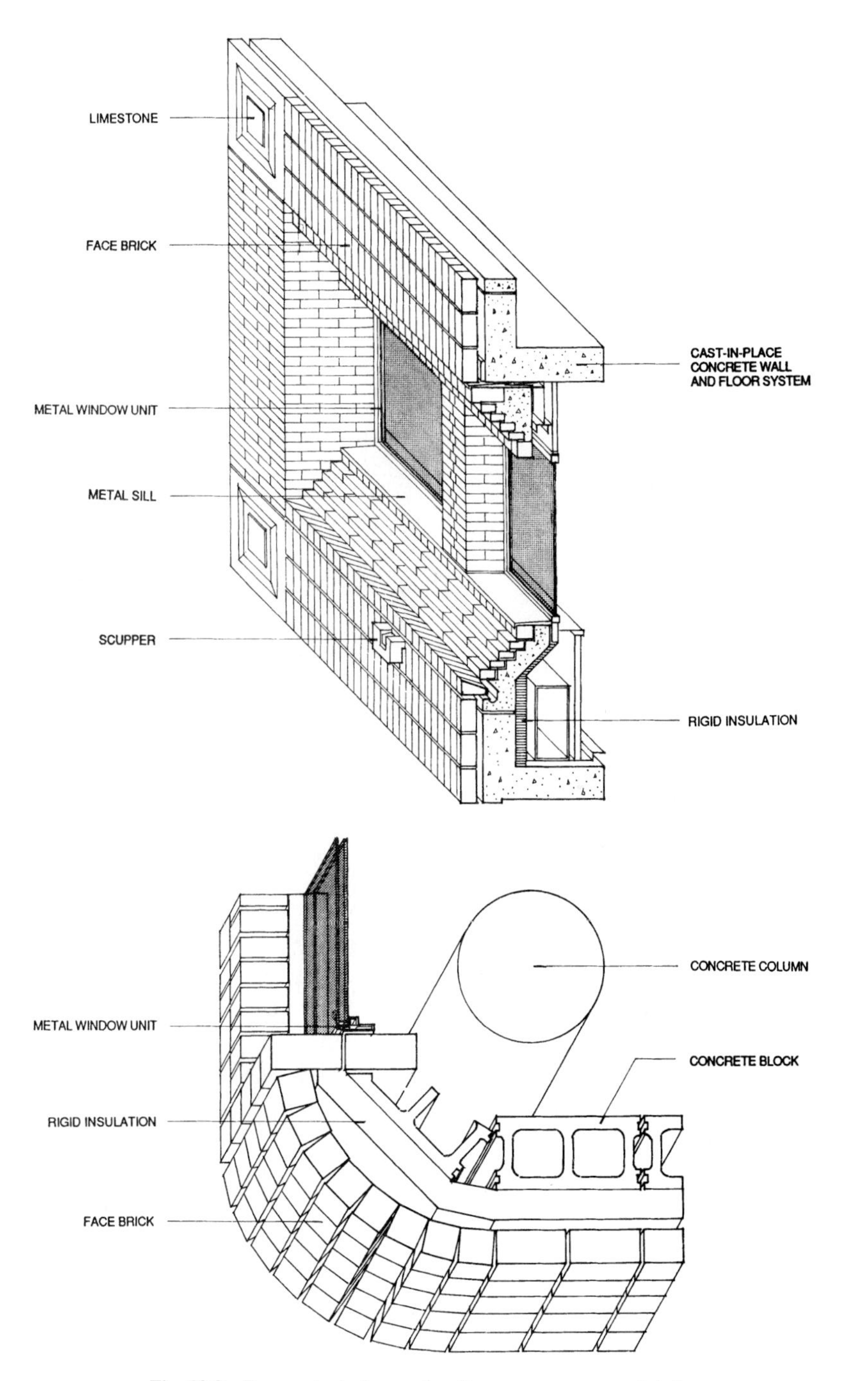


**Fig. 12.8 Recessed window and radius corner masonry details.**

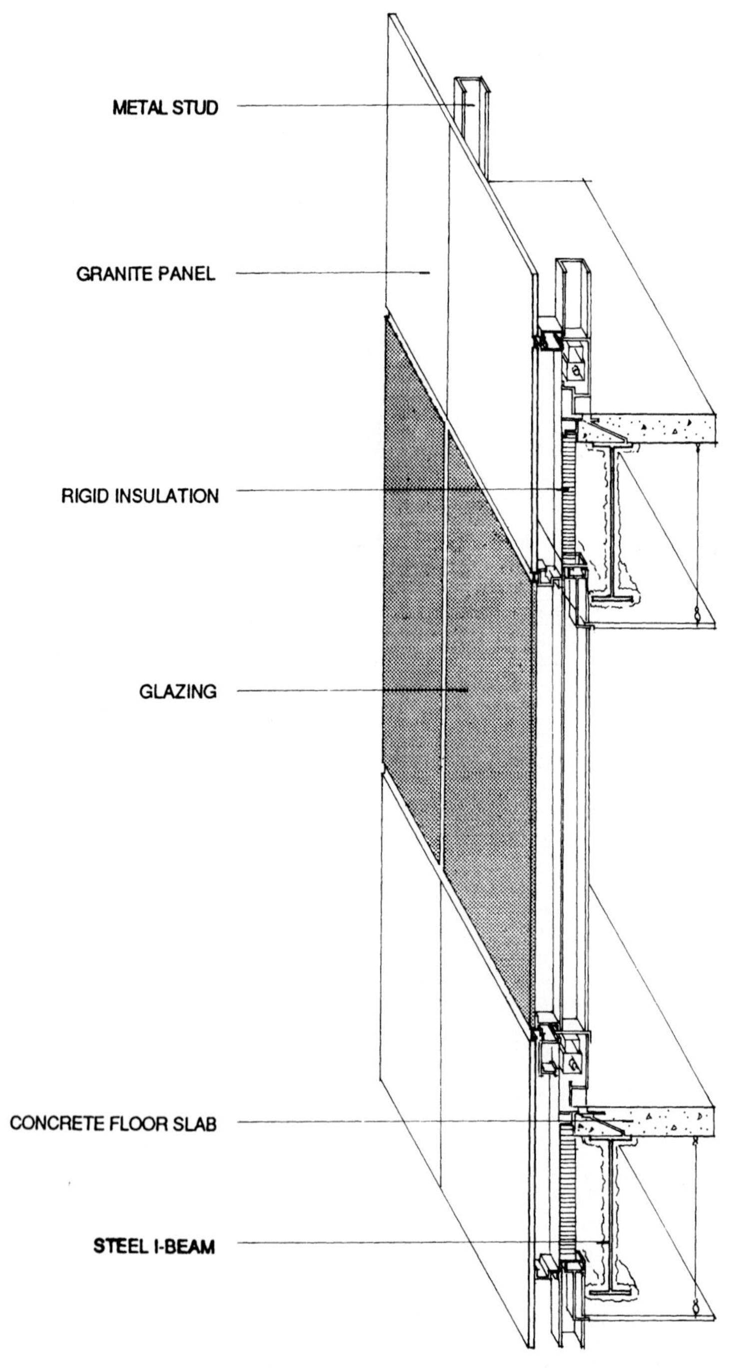


**Fig. 12.9 Thin stone veneer assembly anchored to a structural steel frame.**

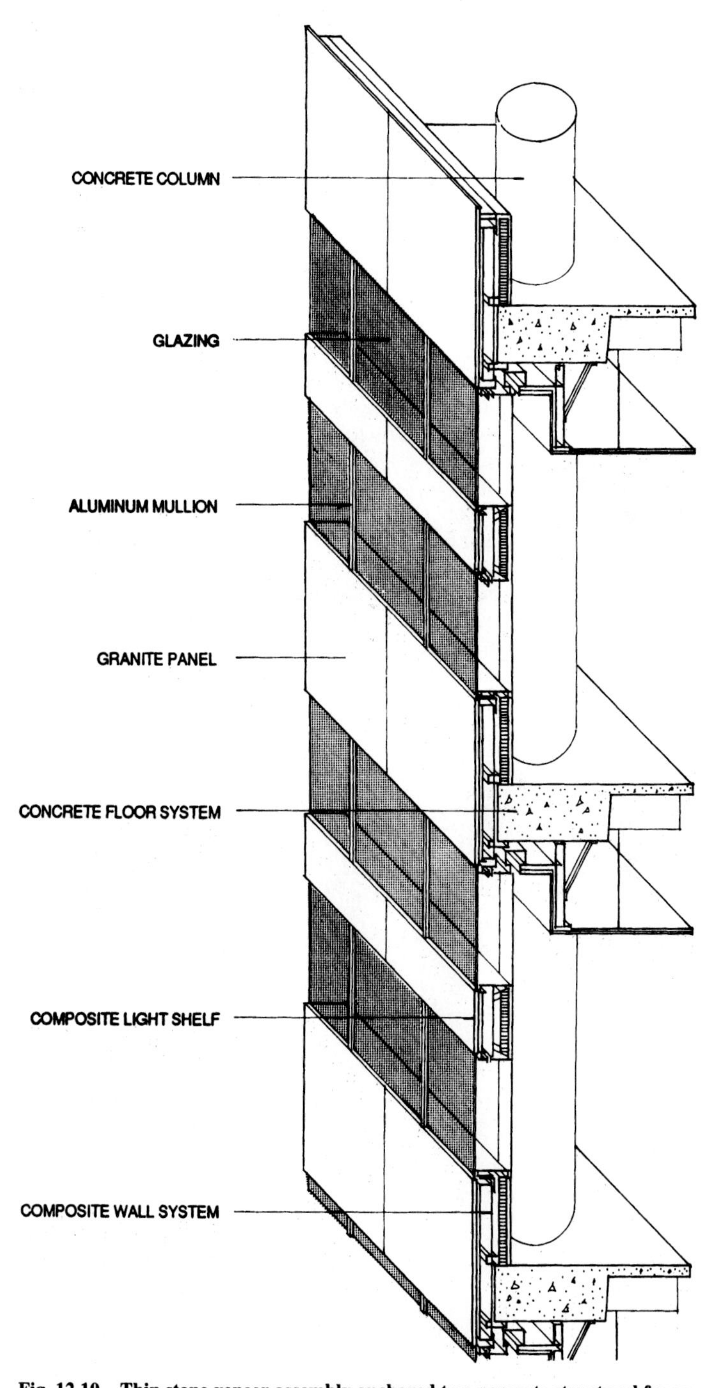


**Fig. 12.10 Thin stone veneer assembly anchored to a concrete structural frame.**

***Solar Screens.*** Solar screens are available both in clay tile and in patterned concrete designs (Fig. 12.11). The science of architectural climatology had been well developed by the mid-twentieth century. With the advent of large glass areas in high-rise buildings in 1960, the need to reduce solar heat gain was a necessity to reduce the operation costs of air-conditioning. Solar masonry screens were developed in response to this need. These seemingly nonstructural elements can be subject to structural forces for which their fragile forms are not suited. Two critical conditions have occurred that have led to unfortunate collapses with serious consequences when debris fell from many stories above. Both were due to building the screen wall tightly as infill panels between floors and between piers or columns. When such flexural panels are built as plates fixed on four edges, there is no built-in provision for relief due to differential thermal movements between screen and frame when the frame is exposed on the building skin.

Long-term creep in the columns of concrete frames or in the cantilever floor edge construction of concrete floors can result in loadings as if the screen wall were load-bearing. Unfortunately the shape of solar screen masonry units fails to provide resistance to the inadvertent loadings from unrelieved horizontal thermal movement, sideway building racking action from lateral wind, earthquake forces, stresses caused by vertical deflections of long cantilever overhangs, or vertical creep in concrete columns. Experience has shown that this problem is as likely to occur in steel frames as in concrete because steel frames are relatively slender and their facades experience substantial differential movement between the frame and the screen wall. Unlike concrete, steel does not display time-dependent creep effects, but it does have a slightly higher coefficient of thermal expansion (over 20% higher from available data).

Thorough attention to boundary conditions for both expansion joints and mechanical anchors is required to accommodate the differential movement described. Dependence

**Fig. 12.11 Interior view of patterned concrete solar screens, Concordia Theological Seminary, Fort Wayne, Indiana.** (*Courtesy: Paul Armstrong.*)

on the adhesion of mortar at screen wall boundaries is an undesirable design approach that has caused needless lawsuits as well as personal injury. Screen wall anchorage should therefore be carefully detailed.

### 2 Detailing

As nonstructural facades become thinner and thinner, greater attention must be given to the detailing of movement joints, flexible anchorage, and flashing. Much construction litigation has occurred over detailing issues. Masonry materials clad over skeleton frames can yield unsightly cracks in clay masonry due to failure to accommodate thermal and moisture movements by expansion joints and flexible anchorage. The same is true concerning shrinkage and use of control joints in concrete block construction. Note that in general, bricks expand and blocks contract when the net effects of all sources of differential movement are considered. The issue of detailing was discussed in depth earlier in Section 12.2.

### 3 Water Penetration

One of the greatest concerns of masonry nonstructural wall facades is the fact that they all tend to permit rainwater leakage. When masonry walls were 762 mm (30 in.) thick or more, water penetration was never a problem to building interiors. In tall buildings, the rainwater rundown accumulates to substantial thickness. This accumulation plus the lateral wind-driven rain in a given story merge into a continuous downward sheet of ever growing thickness. With the advent of glass curtain walls as in the Lever House in New York, the roofs at lower story setbacks were found to be inundated. Suddenly designers realized that rainwater descending over the face of highly absorptive brick wall accumulates at a reduced rate as water is temporarily held aloft within the porous voids of the wall until it is harmlessly evaporated into the atmosphere. While water penetrating into the interior was negligible in the older, thicker curtain walls, there was however a price to be paid for water held within the wall mass until evaporation took place. In 1977, when the terra-cotta skin of the Woolworth Building in New York was repaired, the 64-year-old building showed considerable rusting of the spandrel steel, although fortunately less so of the column steel. This was enhanced by the many crevices afforded by riveted built-up plate and steel angle construction that trapped moisture for long periods of time. While not fatal to the building's performance after 64 years, extensive and costly repairs were required. The older ungalvanized wall ties used to anchor the terra cotta had also rusted through in many places.

The observations of the Woolworth skin repair should put designers on notice about the careful attention that must be paid to wall flashings. Thorough articulation of all flashings should be made on the architectural details. Parapet copings, cornice projections, window sills, and wall belt courses present horizontal ledges on which descending rainwater can pool causing much damage, not only due to leakage to the interior but also due to localized freeze-thaw cycles. Projecting drip edge details, parapet copings, roof junctures, and spandrel or window openings are prime locations that require careful attention. Failure to take this matter seriously has resulted in much owner dissatisfaction. There is more construction litigation over the issue of water penetration than any other single cause. (The issue of waterproofing and drainage was also discussed in Section 12.2.)

***Rain Screen Walls.*** There are three traditional choices for masonry wall cladding—cavity walls with a concrete masonry unit backup, cavity walls with a metal stud backup, or precast concrete panels which are face-sealed at the joint (Keleher, 1993). These cladding systems rely entirely on a single barrier to shed water. Therefore it can be very difficult on site to ensure that every corner and crevice is sealed. The *rain screen* concept has been known for quite some time, but it is only recently that it is being applied to building construction. The forces that cause rain penetration to occur in walls include kinetic energy of the raindrop, gravity, capillarity, and air pressure (Labs, 1990). Of these forces, air pressure is now recognized as the major cause of water penetration problems. Most traditional methods of preventing rain penetration have concentrated on eliminating openings in the wall through which water can pass, that is, a face-sealed system.

There are generally two kinds of rain screens—drained back-ventilated and pressure-equalized compartmented (Keleher, 1993). Drained back-ventilated rain screens are a series of sheets, panels, or planks fixed to vertical support rails. The joints are designed to provide protection against the kinetic energy of wind-driven rain by incorporating baffles or by stringently controlling the width of narrow open joints in the face of the wall cladding. Relatively large quantities of rainwater are allowed to penetrate the joints and run down the reverse, hidden face of the cladding assembly. Drained back-ventilated walls require a water barrier at the back of the cavity. Therefore the cavity should be carefully and completely flashed and the run-off water brought to the outside of the wall surface frequently.

Pressure-equalized compartmented rain screens are similar in design to the drained back-ventilated rain screen wall, but they are specifically designed to control air pressure differentials by the introduction of a pressure-equalization (PE) chamber in the wall cavity. The pressure-equalization chamber in the wall cavity or at the joint nullifies the external force of wind-driven rain. An impermeable air barrier must be installed that will effectively seal the inner leaf of the wall so that pressure leakage will not compromise the positive air pressure built up in the cavity. Figure 12.12 illustrates a drained back-ventilated rain screen wall.

## 12.4 STRUCTURAL MASONRY ENCLOSURES

In many ways the structural design and the architectural detailing of structural masonry skins are similar to nonstructural skins. Unfortunately facades frequently receive less attention by structural engineers who are skilled in gravity and lateral force design and the analysis of differential building movements. Much construction litigation has shown a lack of collaboration of the design team on this issue. The nonstructural facade technology reviewed in the preceding section still applies to structural facades. However, the issue of thinness per se is no longer critical. Since structural enclosures are a major part of the load-resisting system for a masonry tall building, the structural design of these elements must comply with code requirements (Fig. 12.13). In the following, a discussion of load-bearing walls is presented following which the evolution of building codes for masonry is briefly discussed.

### 1 Load-Bearing Masonry Walls

With the advent of structural steel and subsequently reinforced concrete as materials for modern high-rise construction, masonry as a structural material for tall buildings was overshadowed. Although masonry has strength and fire-resistant properties, it had the drawback of weight. For tall structures, the mass of masonry became too great, thus adding excessive weight to the lower walls and foundation of the building. Much

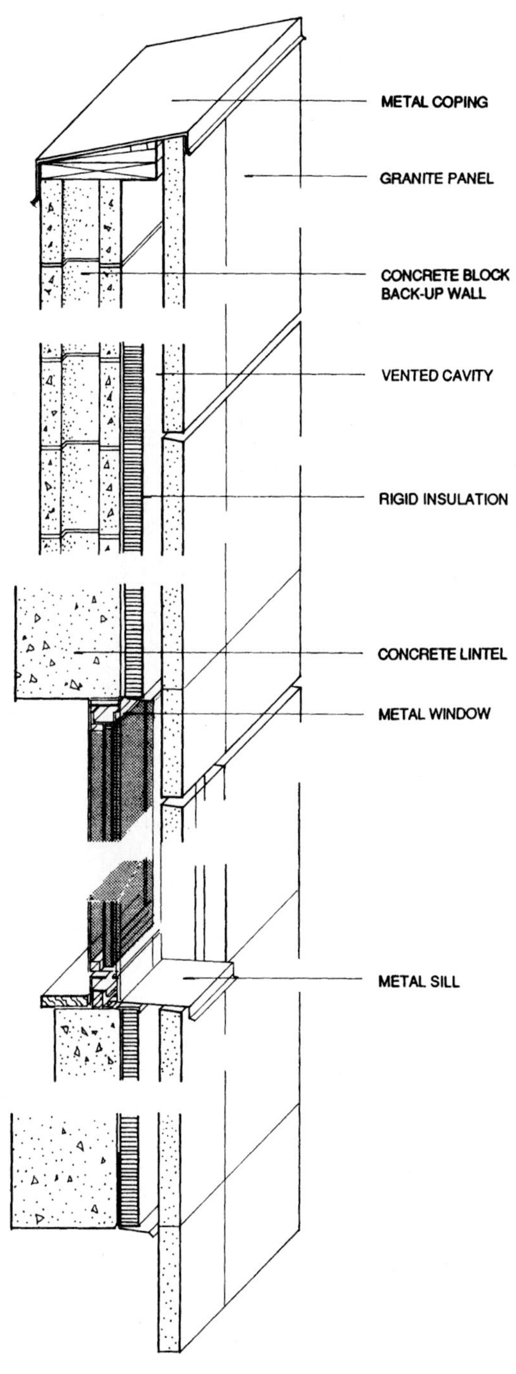


**Fig. 12.12 Rain screen wall section.**

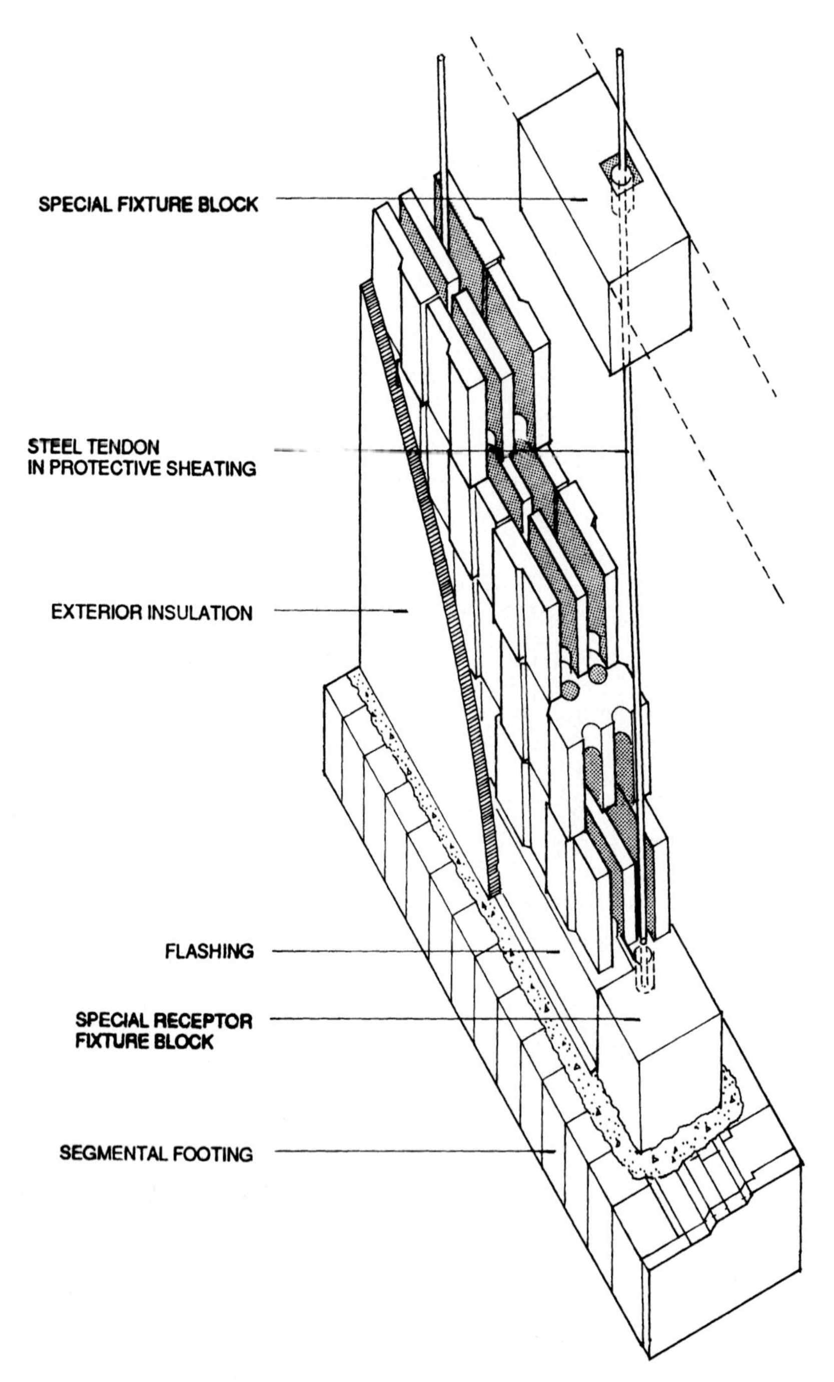


**Fig. 12.13 Posttensioned concrete block wall.**

research, both theoretical and experimental, was devoted to steel and concrete in terms of material technology and construction techniques. Refined methods of design resulting from this research brought about a major change in the architectural and structural disciplines and consequently the development of the tall building structural systems and architectural forms. Building codes were also refined and updated (particularly in seismic zones), resulting in wider applications of these two materials. A principal advantage of masonry construction is that the walls serve several important functions simultaneously. Some of these functions can be enumerated as follows:

1. The mass of the walls tends to be figural, unlike curtain walls, and better define geometric space.
2. They provide a total structural system supporting the loads and adequate lateral resistance to wind and earthquake forces.
3. They provide weather-resistant enclosure.
4. Superior thermal and acoustical insulation can be provided.
5. They provide excellent fire resistance.
6. A variety of architectural finishes with a wide choice of colors, textures, and patterns can be developed.
7. Techniques of wall construction are simple.
8. They reduce the number of materials to be handled and scheduled for a given project.

As discussed earlier, Haller's research into load-bearing masonry construction increased international awareness of masonry as a potential structural material for mid- to high-rise buildings. To increase their load-bearing capacity, Haller (1958) reconfigured the vertical masonry walls so that they were no longer simple lineal elements, but three-dimensional flat, bent, or curved wall plates. The wall elements could be hand-laid in place or brought to the site as prefabricated elements ready for erection. Sophisticated computer programs (as indicated in Section 12.5) using finite-element analysis technique are feasible for refined stress analysis of such elements, but they are beyond the scope of the material covered here.

Other pioneering construction methods have once again made load-bearing masonry feasible for mid- and high-rise buildings. High-lift grout techniques for load-bearing walls reduce the cost of construction. The wall is built as a cavity wall, using ties. Prior to grouting by pumping, the base is cleaned out of mortar droppings and debris through clean-outs at the base of the wall. The height of the pour is limited to the strength against pullout of the wall ties as the grouting hydrostatic pressure rises. Reinforcing is placed into the cavity as required by the structural drawings. Conceptually, reinforced brick masonry (RBM) may be regarded as a form of reinforced concrete wherein the masonry shells are finished forms. This analogy can be carried to the point of regarding brick masonry units as carefully hand-laid aggregate that doubles as permanent formwork.

Research in masonry has continued throughout Europe, Great Britain, and North America. The International Conference on Masonry Structural Systems held at the University of Texas at Austin in 1969 facilitated the international exchange of information by bringing together experts from all over the world who presented their research findings (Johnson, 1969). An account of high-rise residential buildings (17 and 18 stories high) was presented by Lechner (1969). Case studies of load-bearing mid- to high-rise brick buildings in Britain were presented by Hazeltine (1969) and Stockbridge and Hendry (1969). Several other conferences on masonry have been organized regionally and internationally since then, such as, the 5th International Brick Masonry Conference in Washington, D.C., in 1979, and later, the 4th North American Masonry Conference in

Los Angeles in 1987, the 5th North American Masonry Conference at Urbana-Champaign, Illinois, in 1990, and the 9th International Brick/Block Masonry Conference in Berlin, Germany, in 1991.

Developments also have taken place in the area of building codes and regulations for masonry in Europe, Great Britain, Canada, and the United States. These design guides and codes of practice of increasing sophistication provide the design rules for engineered brick and concrete masonry construction. Some of the most recent accomplishments in masonry construction include:

1. Development of general design principles and guidelines based on experimental evidence
2. Incorporation of new theoretical and experimental results in modern building codes
3. Improvements and progress in the manufacture of new masonry products and methods of construction
4. Developments of code regulations for buildings in seismic zones
5. Collaboration of progressive architects and engineers on developing innovative potentialities of modern structural masonry

Following serious failures of masonry structures during earthquakes (for example, at Napier, New Zealand, in 1931), masonry structures became unpopular in seismic regions (Holmes, 1967). Subsequent research work (Scrivener, 1972, 1976; Schneider, 1969; Anderson et al., 1980) and developments in construction techniques have overcome this limitation and also have paved the way for general acceptability of masonry as a structural material by practitioners and code-making bodies. In seismic zones it is now possible to adopt masonry as the material of construction for high-rise buildings with confidence by introducing reinforcing steel and thereby improving its ductility (ICBO, 1991; Paulay and Priestley, 1992).

Load-bearing masonry wall construction currently finds wide application in high-rise residential construction, such as multistory apartment buildings, student residences, and hotels in many countries of the world. Also, many high-rise office buildings and hospitals are being designed with masonry as the primary structural material.

## 2 Building Code Development

The history reviewed in Section 12.1 is appropriately expanded here where emphasis on structural use is made. Initially, during the earlier development of scientific structural analysis, masonry was the prime building material for nonhousing architecture. Yet today it is the least rationally engineered building material as compared to steel and concrete, though this is expected to change in the United States since the creation of the *Building Code Requirements for Masonry Structures* in 1988 by the joint ACI 530 and ASCE 5 committees. The publication was officially designated ACI 530-88/ASCE 5-88. The current publication is designated ACI 530-92/ASCE 5-92/TMS 402-92. Earlier the brick and block industries produced their own codes: *Recommended Practice for Engineered Brick Masonry* (BIA, 1969) and *Specification for the Design and Construction of Load-Bearing Concrete Masonry* (NCMA, 1970). For updated versions of these codes, see BIA (1990) and NCMA (1990). Other well-known building codes covering masonry in the United States are the *Uniform Building Code* (ICBO, 1991) and the *Basic Building Code* (BOCA, 1993). The UBC incorporates ultimate-strength design concepts for the structural design of slender walls and shear walls in addition to the working-stress design method.

The 1988 and 1992 joint committee work was an effort to bring these two masonry materials (that is, brick and concrete block) together into one code. Actually, this is very necessary as composite construction using both materials is in contemporary use and has been so for much of this century. However, even the design aids reviewed here have favored one material to the exclusion of the other. Each industry, brick or block, has its own technical representative, the Brick Institute of America (BIA) and the National Concrete Masonry Association (NCMA). Both are well-staffed with professionals with extensive experience. Both have produced technical notes and bulletins that have been received with wide recognition by professionals around the world. In addition, there are national trade associations that have been actively involved in ASTM committee work—ASTM C12 (mortar), ASTM CI5 (manufactured masonry units), and ASTM C18 (natural stone). Unfortunately there is no single, specific spokesperson for all masonry materials. An abortive attempt was made in the mid-1960s when the ASCE accepted a challenge to develop a masonry design manual. The lack of a unified professional leadership for structural use of masonry has been satisfactorily resolved with the creation of The Masonry Society (TMS) in 1967. Not only registered design engineers and architects are included, but also the contractor and labor segments of the industry are welcome to this organization. Membership includes a variety of others—economists, artists, building management officers, and building maintenance operatives (Borchelt and Wakefield, 1980).

In California, structural masonry became a matter of critical consideration after the Long Beach earthquake of 1933. Following this disaster, the California state legislature at Sacramento passed the so-called Field Act, which resulted in strict design engineering and construction inspection measures for all state school construction. From this impetus grew the art of reinforced brick masonry. The California experience was brought to the attention of designers by two publications—*Reinforced/Brick Masonry and Lateral Force Design* (Plummer and Blume, 1953) and *Building Code Requirements for Reinforced Masonry* (National Bureau of Standards, 1960). Structural masonry in the United States had a steady pace of development from its early beginnings in California in 1933 to the current joint code work and participation of many people and organizations.

In the United Kingdom a limit-states code for plain masonry was initially published in 1978 as BS 5628, Part 1, followed by Part 2 in a 1981 draft, which dealt with reinforced and prestressed masonry. A standard for engineered masonry construction was issued in Germany (DIN 1053, Part 2, 1981 draft), which was followed by a standard for reinforced masonry (CTBUH, Group CB, 1978). Both the British and German codes as well as other European codes are based on characteristic compressive strengths of masonry and specified safety factors. The U.S. ACI/ASCE/TMS code is based on allowable stresses.

Two international bodies, the International Council on Building Research (CIB) and the European Association of Brick and Tile Manufacturers (TBE), published their recommendations for masonry structures in 1981 and 1978, respectively. These documents were intended to be the bases for a subsequent enactment of national codes or revision of existing codes.

## 12.5 FUTURE INNOVATIONS IN MASONRY

### 1 Innovative Possibilities

Many innovative possibilities exist for the use of masonry for tall building construction. Masonry has not received the same attention from structural engineers in the past as has steel or concrete. A greater emphasis on masonry as a structural material will encourage designers to use it more often.

One such innovation is the use of composite masonry construction. A light conventional frame of steel designed to sustain the deadweight of the masonry is built and the composite action of the cured masonry built around the steel sustains the final design loads.

Thin, deep-beam applications may be suitable for spandrels, proscenium arches, decorative masonry expanses, and above-store windows, where thin deep walls requiring flexural action could be employed. The masonry is built atop an inverted steel T section to the web of which are welded steel stirrups. The steel section acts as a temporary support form. After curing, the system acts as a single unit in a composite manner.

Exterior structural masonry walls may be regarded as vertical plates (that is, as continuous media) into which an opening pattern of any form may be carved. Given computer analyses using finite-element theory and improved construction methods ranging from tower cranes to grout pumps, free forms can be produced at feasible costs. Arches, curved walls, and other shapes can now be added to buildings without great difficulty.

## 2 Research Needs

Research is required for masonry from both architectural and engineering viewpoints. The main hindrance to early engineering research in masonry was the fact that it was generally considered to be an architectural material. Progress in the construction of tall masonry buildings largely depends upon a better understanding of the mechanics of masonry from a behavioral point of view, innovative architectural concepts, improved construction techniques, and enhanced quality of materials. In the following, some future research needs are briefly stated.

1. Development of innovative masonry building plans where load-bearing walls may be dropped off at increasingly higher floor levels to permit larger spaces at those levels. This is to take advantage of the fact that at higher levels, loads are lighter, which makes it possible to reduce the number of load-bearing walls there.
2. In high-rise office and commercial buildings, large open spaces are required at the main floor level to accommodate escalators, open atria, wide entries, and for visibility from and to the street through large glass windows. Solid load-bearing masonry walls at the lower levels are antitheses to these requirements. Research is needed to create such large open wall areas at the base of tall masonry office buildings.
3. Development of other more efficient building shapes in plan rather than the common rectangular shape. Studies on various wall configurations in plan are required to assess the augmented lateral load carrying capacity for the shear walls due to such special configurations. Further, these shapes may create a greater amount of desirable leasable floor area.
4. Studies on new shapes of bricks and concrete blocks that will improve aesthetic quality and provide increased shear and compressive strength of masonry.
5. Improvements in placing techniques of masonry units to speed up the construction process are required. Moreover, studies on possible mechanization of the installation of masonry units or modules and on the related implications are essential.
6. Studies on structural form-producing potentials of modern masonry and on color, tone, and textural relationships of masonry materials.

7. Development of fast-setting, compatible bonding agents that possess all required functional and strength characteristics.
8. Improved architectural detailing of facades to eliminate the problem of water penetration. Further investigations into the applications of rain screen concepts are also necessary in this regard.
9. Theoretical studies to develop a constitutive law capable of accurately predicting the carrying capacity of masonry walls.
10. Studies on the composite behavior of masonry with other materials and on composite structural systems for tall buildings.
11. Studies on building movements to provide a more definitive understanding of the short-term and long-term movements and cracking characteristics and incorporation of research results into codes for the use of practitioners.
12. Research on innovative structural systems in masonry for lateral load resistance by exploring the possibility of analyzing by computers the structural shell by considering, for example, the building with perimeter load-bearing walls as a tubular structure for office buildings where large interior open space is desirable.
13. Studies on the improvement of ductility of masonry walls, particularly for tall buildings in seismic zones.
14. Studies on the interaction of masonry walls with floor slabs, and of nonstructural masonry skin with the structural framework of tall buildings.
15. Research on deep-beam behavior of masonry walls over large door or window openings.
16. Avoidance of progressive collapse of tall masonry buildings, particularly when subjected to explosive and blast loads.
17. Research on the prefabrication of masonry components to expedite the construction process.

## 12.6 CONDENSED REFERENCES/BIBLIOGRAPHY

The following is a condensed bibliography for this chapter. Not only does it include all articles referred to or cited in the text, it also contains bibliography for further reading. The full citations will be found at the end of the volume. What is given here should be sufficient information to lead the reader to the correct articles—author, date, and title. In case of multiple authors, only the first named is listed.

ACI 530-92/ASCE 5-92/TMS 402-92 Joint Committees 1992, *Building Code Requirements for Masonry Structures*

Ali 1980, *Failure Investigation of Masonry Structures*

Ali 1981, *Carrying Capacity of a Two-Phase Material in Axial Compression*

Ali 1987, *Influence of Architectural Configuration on Seismic Response of Buildings*

Amrhein 1986, *Slender Walls Research Program by California Structural Engineers*

Amrhein 1991, *Reinforced Grouted Brick Masonry*

Amrhein 1992, *Reinforced Masonry Engineering Handbook: Clay and Concrete Masonry*

Anderson 1980, *Seismic Design of Reinforced Concrete Masonry Walls*

Angerer 1969, *Appeal and Potential of the Polyfunctional Characteristics of Engineered Brickwork*

Beall 1987, *Masonry Design and Detailing*

BIA 1969, *Recommended Practice for Engineered Brick Masonry*

BIA 1990, *Building Code Requirements for Masonry Structures ACI 530/ASCE 5 and Specifications for Masonry Structures ACI 530/ASCE 6*

Blaser 1965, *Mies van der Rohe*

BOCA 1993, *Basic Building Code*

Borchelt 1980, *Writing a Design Standard: The Masonry Society's Building Standard*

Building News 1981, *Concrete Masonry Design Manual*

Charenbhak 1981, *Chicago School Architects and Their Critics*

CTBUH, Group CB, 1978, *Structural Design of Tall Concrete and Masonry Buildings*

CTBUH, Group CL, 1980, *Tall Building Criteria and Loading*

Dickey 1985, *Reinforced Concrete Masonry Construction*

Elmiger 1976, *Architectural and Engineering Concrete Masonry Details for Building Construction*

Eppel 1980, *State-of-the-Art Report: Rain Penetration in Masonry*

Esteva 1966, *Behavior under Alternating Loads of Masonry Diaphragms Framed by Reinforced Concrete Members*

Fisher 1985, *PIA Technics: Wall Joints*

Francis 1970, *The Effect of Joint Thickness and Other Factors on the Compressive Strength of Brickwork*

Giedion 1964, *Space, Time and Architecture: The Growth of a New Tradition*

Goldberger 1981, *The Skyscraper*

Grimm 1992, *Technics Topics: Masonry Crack Control Design Checklist*

Haller 1958, *Technical Properties of Masonry Made of Clay Bricks and Used for High-Rise Construction*

Hazeltine 1969, *Some Load-Bearing Brick Buildings in England*

Hendry 1976, *Effect of Slenderness and Eccentricity on the Compressive Strength of Walls*

Hendry 1983, *Recent Developments in the Design of Masonry Structures*

Hilsdorf 1969, *Investigation into the Failure Mechanism of Brick Masonry Loaded in Axial Compression*

Holmes 1967, *Masonry Buildings in High Intensity Seismic Zones*

ICBO 1991, *Uniform Building Code*

Johnson 1969, *Designing, Engineering and Constructing with Masonry Products*

Jordy 1961, *Montgomery Schuyler: American Architecture and Other Writings*

Keleher 1993, *Rain Screen Principles in Practice*

Labs 1990, *The Rain Screen Wall*

Lechner 1969, *Special Quality Brick Masonry Multistory Apartment Houses Built in Switzerland*

Mays 1988, *PIA Techniques: The Many Faces of Brick*

McNary 1985, *Mechanics of Masonry in Compression*

Mies van der Rohe 1938, *Inaugural Address*

Mikluchin 1969, *Morphotectonics of Masonry Structures*

Monk 1980, *Masonry Facade and Paving Failures*

Mumford 1952, *Roots of Contemporary American Architecture: A Series of Thirty-Seven Essays Dating from the Mid-Nineteenth Century to the Present*

National Bureau of Standards 1960, *Building Code Requirements for Reinforced Masonry*

NCMA 1970, *Specification for the Design and Construction of Load-Bearing Concrete Masonry*

NCMA 1990, *Building Code Requirements for Masonry Structures*

Nicks 1980, *Masonry Materials*

Paulay 1992, *Seismic Design of Reinforced Concrete and Masonry Buildings*

Plummer 1953, *Reinforced/Brick Masonry and Lateral Force Design*

Prescott 1962, *Formal Values and High Buildings*

Redmond 1975, *Compressive Strength of Composite Brick and Concrete Masonry Walls*

Sahlin 1971, *Structural Masonry*

Scully 1988, *American Architecture and Urbanism*

Schneider 1969, *Shear in Concrete Masonry Piers*

Schneider 1979, *Reinforced Masonry Design*

Scrivener 1972, *Reinforced Masonry—Seismic Behavior and Design*

Scrivener 1976, *Reinforced Masonry in a Seismic Area—Research and Construction Development in New Zealand*

Senbu 1991, *Effects of Admixtures on Compactibility and Properties of Grout*

Shrive 1980, *Anisotropy in Extruded Clay Units and Its Effect on Masonry Behavior*

Spaeth 1985, *Mies van der Rohe*

Stockbridge 1969, *Case Studies and Critical Evaluation of High-Rise Load-Bearing Brickwork in Britain*

Sutherland 1969, *Design Engineers' Approach to Masonry Construction*

Yao 1989, *Axial Capacity of Grouted Concrete Masonry*

# Current Questions, Problems, and Research Needs

This appendix identifies problem areas for further study and research. The issues listed represent a condensed version of the research needs summarized at the end of each chapter. Additional suggestions of study areas should be forwarded to the Council headquarters for transmission to the appropriate committee.

## 1 ARCHITECTURAL DESIGN: ISSUES AND CRITERIA

1. Effect of builtform density on the quality of urban space.
2. Establishment of scientifically based performance parameters as urban-control standards.
3. Tall building response to the quality of urban environmental balance.
4. Comprehensive documentation of design parameters and criteria intrinsic to tall buildings.
5. Effects of tall building mass on environmental conditions and microclimate.
6. Impact on tall building design of varying building codes and regulations in different countries.
7. Influence of the nature and context of the surrounding urban area on tall building form and design assumptions.
8. Development of design criteria for superlarge and superthin tall buildings.
9. Development of future tall building typologies.
10. Development of interdisciplinary approaches to resolving building planning, construction, and usage issues.
11. Identification of factors influencing tall building design and study of the interaction of these factors in determining the final outcome of the design process.
12. Tall building impact on society on a global scale.

## 2 BUILDING PLANNING AND DESIGN COLLABORATION

1. Integration of information about building science, practice, and technology into the architectural education system.

2. Research on priorities between open space and ground-floor commercial space.
3. Further studies on the future possibilities for mixed-use buildings as well as limitations and social implications of such buildings.
4. Optimizing the heights and floor areas of tall buildings.
5. Studying the best location for elevator banks and service cores.
6. Reducing building drift and minimizing building vibrations.
7. Application of structural innovations in other building types to the design of tall buildings.
8. Translating concepts and criteria developed in the academic setting to office practice.
9. Improvement of the interaction between academic and office practice at both university and professional levels.
10. Stimulation of the professional relationship between architect and engineer in an academic environment.

## 3 URBAN DESIGN AND DEVELOPMENT

1. How the impact of a tall building on a city's skyline affects the success of the builtform.
2. Examining the influence of culture on the form and character of a tall building.
3. Exploring changes in the growth and development of tall buildings when exposed to rapid population growth and land value escalation.
4. Investigating changes in zoning regulations during rapid population growth.
5. Influence of urban growth and rise of tall buildings on the surrounding rural and farm land.
6. Concerns of high-rise residential buildings in dense urban areas.
7. Tall building design in the context of a complete urban master plan.
8. Applicability of urban design and planning principles in industrialized countries to nonindustrialized cities in developing countries.
9. Interaction of public agencies and private developers in the preparation and implemention of viable and holistic masterplan concepts.
10. Relationships developing, or which are likely to develop, between private and public sectors in terms of addressing urban-scale design and planning issues.
11. Relationship of tall buildings with surface and underground transportation systems and with the general infrastructure.

## 4 STRUCTURAL EXPRESSION IN TALL BUILDINGS

1. Expression of new technology and implications on the building structure and aesthetics.
2. Comparison of construction and maintenance costs of tall buildings with structural expression with other types of tall buildings.

3. Maintaining the structural efficiency of tall buildings while limiting the total cost per square meter.
4. Structural innovations with the collaborative efforts of architects and engineers.
5. New structural materials and their roles in tall building expression.
6. Development of a broader public understanding of the contributions of the structural engineer to tall building design.
7. Impact of prefabrication and composite integration of building components and systems on tall building design.
8. Integration of structural form with architectural function for future tall buildings.

## 5 AESTHETICS AND FORM

1. New optimum form that will lead to an efficient structural system, resulting in substantial savings in cost while maintaining aesthetic integrity.
2. Development of building configurations, both horizontal and vertical, that will meet the basic needs such as providing lateral resistance to wind and seismic loads.
3. Human response to colors.
4. Building tops and their impact on the skyline.
5. Relationship at the base of high-rise buildings to the scale of the pedestrian, the street, and the overall building form.

## 6 PSYCHOLOGICAL ASPECTS

1. Symbolic aspects of a city form which can enhance legibility, making it easier for visitors and locals to find their way around.
2. Preparation of occupants of tall buildings for the physiological effects of height and building sway.
3. Further study on how tall building occupants experience building motion due to earthquakes and strong wind.
4. Design criteria for lobby and atrium spaces.
5. Determination of the optimal light in a high-rise work environment.
6. Role of the views from tall buildings on worker productivity.
7. Methods for improving the design of high-rise housing to make units more psychologically and socially satisfying to occupants, particularly families with young children.
8. Innovative ways to create more user-friendly parking garages, housing, office towers, and shopping centers.
9. Improvements in barrier-free design and safety.
10. Enhancing security and safety in tall buildings.
11. Postoccupancy evaluations of multiuse high-rise buildings.

## 7 FACADES: PRINCIPLES AND PARADIGMS

1. Research into facade types, materials, and details and their impact on the urban context.
2. Effects of macro- and microclimates on cladding systems.
3. Creation of innovative types of cladding systems that are durable, maintainable, and cost-efficient.
4. New developments in high-performance glass, low-emissivity glass, coated glass, and thermal glazing systems coupled with variably zoned and balanced HVAC systems.
5. Facade incorporation of mechanical air-handling systems and natural ventilation coupled with opaque, translucent, and transparent materials.
6. Economic, climatic, sound transmission properties, and life-cycle viability of tall building cladding systems.
7. Incorporation of computerized technology and intelligent building systems into facade design.
8. Interaction between nonstructural architectural elements of the facade and the structure.
9. Use of facade as a vehicle for structural and aesthetic innovations.
10. Studies on rain screen walls.

## 8 BUILDING SYSTEMS AUTOMATION AND INTEGRATION

1. Effects of professional specialization on computer integration of all related fields.
2. Development of standards for computer integration and an authority to do so.
3. Extension of computer automation and its integration into the academic environment.
4. Integration of techniques between different countries and languages.
5. Possible impact of altering the design or form of tall buildings as a reflection of the integration of intelligent building systems.
6. New roles that intelligent buildings will initiate and play in the building industry.
7. Systems development of specialized modules as integral components of a complete building.
8. Establishment of standardized interfaces and technical guidelines for computerized building systems.
9. Integration of AI-based systems in A-E education and research.
10. Development of object-oriented software applications that cross traditional professional barriers.
11. Holistic (macroanalysis) approach to the information flow processes throughout the complete building life cycle.

12. Defining and maintaining an electronic information highway system, including database structures, that will allow the direct exchange of information between specialized, independent, and proprietory software systems serving various professions and segments of the industry.

## 9 ENVIRONMENTAL SYSTEMS

1. Psychological and physiological effects of diurnal temperature variations on humans.
2. Studies on how much different the cost in energy due to operable window leakage in a mild climate would be in comparison with fixed windows, keeping in mind the goal of maximizing occupant comfort and productivity.
3. Examining factors related to a lessening of the abrupt transition between interior and exterior environments.
4. New sources of energy for use in tall buildings.
5. Health concerns about the effects of electromagnetic environments changing the design of environmental controls in tall buildings.
6. Hardware and material combinations to be used in a biomass energy system.
7. Economic feasibility of a biomass energy system.
8. Accurate estimate of the actual quantities of waste paper produced by various building types using biomass energy systems.
9. Real costs of sustainable design versus conventional design.
10. Efforts to educate the owners toward environmentally sensitive design solutions for their buildings.
11. Promotion of a change in the production industry to get more environmentally sensitive building product manufacturers.
12. Education of specifiers of building products toward environmentally sensitive building materials.
13. Accommodation of building codes, zoning, and other regulations for more environmentally conscious design.
14. Public participation in policy making that promotes sustainable development.
15. More efficient means of heating and cooling buildings.
16. Relationship of green architecture and aesthetics.

## 10 MATERIALS AND STRUCTURES: STEEL

1. Development of an economical, easily weldable steel with high strength, good ductility, and strain-hardening capacity for improved structures in seismic zones.
2. Better defined serviceability requirements to negate unneeded restrictions to the use of higher-strength steels.
3. Simple production techniques for heavy structural shapes with improved toughness.

4. More economical connection systems for steel-framed buildings, including connections of tubular members.
5. Development of inherently fire-resistive steel.
6. Generating improved criteria for the design of steel-concrete composite framing systems, including connection systems.
7. New concepts for mitigation of potential vibration problems, such as tuned mass or viscoelastic dampers.
8. Improving the speed of construction by developing better field erection techniques.

## 11 MATERIALS AND STRUCTURES: CONCRETE

1. How quality control can be reorganized to adjust to the new characteristics of high-strength reinforced concrete used for tall buildings.
2. Improving the integration of functional and aesthetic design with structural, mechanical, and electrical systems.
3. Durability and the effects of climate on concrete.
4. Effects of movement due to temperature, creep, and shrinkage on the architectural, structural, and mechanical details.
5. Development of natural and/or built-in architectural finishes in a variety of colors, tones, and textures.
6. Investigations on high-strength and high-performance concrete when used in regions of high seismicity.
7. Improvement of ductility of shear walls.
8. Increasing the damping characteristics of concrete tall buildings under dynamic loads.
9. Development of new aesthetics for concrete buildings by exploiting the moldability property of concrete.

## 12 MATERIALS AND STRUCTURES: MASONRY

1. Improvements in masonry construction methods.
2. Development and application of free forms in masonry at reasonable costs.
3. User education and training of masonry construction methods.
4. Applications of advanced analysis computer software to masonry construction.
5. Development of innovative masonry building plans where load-bearing walls may be dropped off at increasingly higher floor levels to permit larger spaces at those levels.
6. Exploring efficient building shapes in plan rather than the common rectangular shape.
7. Developing new shapes of bricks and concrete blocks that will improve aesthetic quality and provide increased shear and compressive strength.

8. Improving architectural detailing of facades to eliminate water penetration.
9. Development of a constitutive law capable of accurately predicting the carrying capacity of masonry walls.
10. Improving ductility of masonry walls, particularly in seismic zones.
11. Avoidance of progressive collapse of tall masonry buildings, particularly when subjected to explosive and blast loads.
12. Prefabrication of masonry components to expedite the construction process.

# Nomenclature

## *GLOSSARY*

**Absolute humidity.** Mass of water vapor per unit volume.

**Acid rain.** Rain that contains a high concentration of pollutants, chiefly sulfur dioxide and nitrogen oxide, released into the atmosphere by the burning of fossil fuels such as coal or oil.

**Acrophobia.** Fear of high places.

**Active solar system.** Building subsystem in which solar energy is collected and transferred predominantly by mechanical equipment powered by energy not derived from solar radiation.

**Adaptive behavior.** Processes of modifying one's response to an environmental stimulus in order to bring about an optimal level of stimulation. Over time, response to a particular stimulus generally will decrease.

**Adaptive reuse.** Conversion of an existing building to adapt to a different occupancy.

**Aesthetic field.** Exterior expanse of tall buildings within which aesthetic characteristics prevail.

**Addressable smoke detector.** Smoke detector, the location of which is known at a computerized remote panel.

**Agora.** Open space in a Greek or Roman town used as a marketplace or general meeting place, usually surrounded by porticos, as in a forum.

**Agoraphobia.** Fear of open spaces.

**Air changes.** Volume of air supplied to or exhausted from a building or room; usually expressed in terms of the number of complete changes of air per hour in the room or space under consideration.

**Air-handling unit.** Packaged assembly of air-conditioning components (coils, filters, fan, humidifier), which provides for the treatment of air before it is distributed to an air-conditioned space.

**Air rights site.** Location that has the legal property right for use of the space above a specified elevated plane; usually includes the right to ground support but excludes other rights to ground use; for example, the right to construct a building over a railroad track.

**Alarm indicator.** Alarm indicating system for the hearing and/or visually impaired and people with disabilities; for example, rotating beacons and strobe lights for the deaf.

**Alloy steel.** Steel with chemical composition that exceeds the limits for carbon steel.

**Anchorage.** Length of reinforcement, mechanical anchor, hook, or a combination thereof, beyond the point where the reinforcing is no longer required to resist calculated stresses.

**Anemometer.** Instrument for measuring air velocity.

**Annealing.** Treatment of steel by heating to near the critical range followed by slow cooling.

**Annunciation panel.** Fire and emergency control panel with alarm indicators that delineate the exact location of the fire or emergency.

**Anodized aluminum.** Aluminum provided with a hard, noncorrosive, electrolytic oxide film on the surface by electrolytic action.

**Aperture.** Opening in a facade. (*See also* Facade.)

**Arc weld.** Weld in which the heat of fusion is supplied by an electric arc.

**Arcade.** Covered passage with supporting arches, often associated with entrances to shops.

**Arch.** Curved structure in building, supporting itself over an open space, often by mutual pressure of the stones.

**Archetype.** Original pattern or model; Edmund Bacon's reference to a tall building's unique profile in relation to its configuration against the sky.

**Architectural detail.** Sketch of a building component or part showing every item minutely from an architectural point of view.

**Architectural differentiation.** Degree to which parts of the architectural environment are perceived as the same or different.

**Architectural expression.** Outward manifestation of building form by the exhibition of color, texture, substance, and decoration of the facade.

**Architectural style.** Common vocabulary of building forms, elements, and expression shared by a cultural group.

**Architexture.** Edmund Bacon's reference to a tall building's rhythmic stimuli where it meets the ground, and its ability to provide a harmonious and exciting experience for pedestrians walking around its base.

**Art deco.** Fashionable jazz age style concurrent with international modern in the 1920s and 1930s. The name derives from a Paris exhibition of decorative and industrial arts in 1925. It is characterized by nonfunctional "modernism" or streamlining motifs in architecture. It would begin the development of a new kind of "secular Gothic" vocabulary for the tall building, still keeping its verticality, with buttresslike setbacks and crowning spires. (*See also* Gothic; Modern architecture.)

**Art moderne.** *See* Art deco.

**Articulation.** Action or manner of jointing or interrelating architectural elements throughout a design or building.

**Artificial intelligence.** Collective term for a set of many specialities in which knowledge-based and expert techniques are generally used for design. (*See also* Expert system; Knowledge-based system.)

**Asbestos cement.** Dense, rigid board containing a high proportion of asbestos fibers bonded with portland cement; resistant to fire, flame, and weathering; has low resistance to heat flow; used as a building material in sheet form and corrugated sheeting.

**Atrium.** Derived from the central hall of a Roman house, it originally was a rectangular shaped courtyard around which the house was built. Atria are typically large interior climatically controlled multistory spaces in commercial and public buildings. Atria may be used as gathering spaces, often contain primary circulation elements, such as escalators and elevators, and are usually centrally located in a tall building adjacent to the main entrance.

**Austenite.** Solid solution of carbon and other elements in the gamma phase of iron.

**Automation.** Technique or system of operating and controlling a mechanical or productive process by highly automatic means, as by electronic devices with minimum or no human intervention.

**Axial force.** Resultant longitudinal internal component of force which acts perpendicular to the cross section of a structural member and at its centroid, producing uniform stress.

**Axiological.** Pertaining to the study of the nature, types, and criteria of values and value judgments, especially in ethics.

**Axis.** Imaginary straight line, indicating center of symmetry of a solid or plane figure or center of rotation of a solid body.

**BACNET.** Communication standard for building automation systems developed by ASHRAE. (*See also* Building automation systems.)

**Barrel vault.** Simplest form of vault, consisting of a continuous arched ceiling or roof of stone or brick of semicircular or pointed sections, unbroken in its length by cross vaults. (*See also* Vault.)

**Base.** Lowest (and often widest) visible part of a building, often distinctively treated. A base is distinguished from a foundation or footing in being visible rather than buried. (*See also* Foundation.)

**Base isolation.** Technique of isolating the base of a building from ground vibrations in a seismic zone.

**Batophobia.** Fear of high objects.

**Bay.** Spatial unit between windows, columns, or pilasters in a building or wall.

**Bay window.** Angular or curved projection of a house or building front filled by fenestration. (*See also* Fenestration.)

**Beam.** Structural member whose primary function it is to support loads transverse to its longitudinal axis.

**Bearing plate.** Steel slab, which is placed under a beam, column, girder, or truss to distribute the end reaction from the beam to its support.

**Beaux Arts style.** Style of architectural design associated with the teaching of the Ecole des Beaux Arts of Paris prior to the beginning of the modern movement.

**Bend radius.** Smallest radius of curvature into which a material can be bent without damage.

**Bendability.** Capacity of a bar to bend without formation of a crack.

**Bending stress.** Stress caused by bending of a structural member.

**Biomass.** Fuel material of natural origin, such as paper, wood chips, and sugar cane fiber.

**Blast furnace.** Refractory-lined vessel for producing iron from iron ore and other raw materials.

**Board (or board of directors).** Group selected to represent a larger group as a position of authority.

**Bond.** Adhesive force built up between steel bars and surrounding concrete at their interface.

**Borrowed light.** Daylight transmitted from one interior space to an adjacent interior space through a glazed opening.

**Box beam.** Hollow beam, usually rectangular in section. If fabricated of steel, the sides are steel plates welded together, or they may be riveted together by steel angles at the corners.

**Brace.** Secondary structural element provided to strengthen, stabilize, and/or stiffen the primary structural system against lateral loads.

**Braced frame.** Essentially vertical truss system for resisting lateral loads on the building system.

**Braced-tube system.** Tubular building in which braced frames have been incorporated into the structural system in mutually perpendicular directions.

**Bricolage.** A French term referring to the performance of a large number of tasks which are not subordinated to the availability of raw materials and tools conceived and procured for the purpose of the project. Rowe and Koetter in their book *Collage City* conceive bricolage as a technique of urban planning based on collaging building figures and set pieces forming regular and irregular urban spaces with specific references to site, context, and memory.

**Brise-soleil.** Sunbreak or check; now frequently an arrangement of horizontal or vertical fins, used in hot climates to shade window openings.

**Buckling.** Instability (loss of strength) of a member under compression with resultant deformations perpendicular to the member longitudinal axis.

**Buildable area.** Portion of a lot or land parcel on which a building can be constructed.

**Building automation systems.** Automatic (electronic, electrical, pneumatic) control systems used to operate and manage all equipment and processes in a building.

**Building code.** Mandatory method, procedure, and standard for developing and operating buildings.

**Building envelope (skin or enclosure).** Boundary of a building between its interior and exterior environments; essentially the outer walls or cladding, the roof, and the basement floor. In mechanical engineering, elements of a building which enclose conditioned spaces through which thermal energy may be transferred to or from the exterior. (*See also* Cladding.)

**Building footprint.** Area on a plane directly beneath a structure that has the same perimeter as the structure.

**Building management system.** *See* Building automation systems.

**Building plan.** Two-dimensional graphic representation of the design, location, and dimensions of a building, or parts thereof, seen in a horizontal plane viewed from above.

**Building services engineering.** Science and practice related to buildings that deal with heating, ventilation, air-conditioning, plumbing, fire protection, automatic control systems, lighting, electric power distribution, and telecommunications; also known as mechanical-electrical (M-E) systems and building environmental systems.

**Building system.** According to the National Electrical Code, plans, specifications, and documentation for a system of manufactured building or for a type of system of building components.

**Building system automation.** *See* Building automation systems.

**Built-up girder.** Girder of structural metal units such as plates and angles which are riveted, bolted, or welded together; girder of precast concrete units which are joined by shear connectors.

**Built-up plate construction.** Method of applying successive layers to form a thicker mass.

**Built-up section.** Section of a beam or column that is made up of component parts fastened together.

**Bundled-tube system.** Structural system in which structural framed tubes are arranged or bundled together in a way so that common walls of contiguous tubes are combined into single walls, thereby forcing total compatibility of stresses at the interface of such contiguous tubes. In a bundled tube, individual tube elements may be terminated at any appropriate level. (*See also* Framed tube.)

**Buttress.** Mass of masonry or brickwork projecting from or built against a wall to give additional strength, usually to counteract the lateral thrust of an arch, roof, or vault.

**CAD.** *See* Computer-aided design.

**CAE.** *See* Computer-aided engineering.

**CAM.** *See* Computer-aided manufacturing.

**Caisson.** Watertight chamber for conducting foundation work in wet ground.

**Camber.** Slight, usually upward curvature of a structural member or form to improve appearance or to compensate for anticipated deflections.

**Campanile.** Italian word for bell tower, usually separate from the main building.

**Candlepower.** Light intensity in any location where artificial lighting is used.

**Cantilever.** In construction, portion of a floor or deck extending beyond the vertical column; overhang.

**Capital.** Head or crowning feature of a column. (*See also* Column.)

**Carbon steel.** Steel with no or very low levels of alloying elements.

**Cardo.** Street used as a baseline to form a perpendicular grid pattern for Roman city planning.

**Cast-in-place.** Method of pouring concrete into a form fabricated on the construction site; also called in situ. (*See also* Concrete.)

**Cast iron.** Iron alloy, usually including carbon and silicon. A large range of building products are made of this material by pouring the molten metal into sand molds and then machining. It has high compressive strength, but low tensile strength.

**Catenary.** Curve formed by a flexible cord hung between two points of support.

**Cavity wall.** Wall containing continuous horizontal and vertical space(s) between wythes of masonry, with wythes tied together by metal ties.

**CD-ROM.** Compact disc—read-only memory. Compact disc that contains information that can only be read by a computer.

**Cellular floor deck.** Floor having hollow openings in it that provide ready-made raceways for distributing wiring for telecommunications and electric power.

**Cementite.** Iron carbide, the form in which carbon occurs in steel.

**Cementitious.** Having cementing properties.

**Characteristic strength.** Strength in compression or tension in a series of tests below which the results of a specified percentage will fall.

**Check valve.** Valve that allows flow in one direction.

**Chemical environment.** Surrounding in which the chemical constituent of the air that humans breathe is emphasized. (*See also* Environment.)

**Chiaroscuro.** Arrangement or treatment of light and dark in painting or architecture.

**Chord.** Principal member of a truss which extends from one end to the other, primarily to resist bending; usually one of a pair of such members. (*See also* Truss.)

**Circulation system.** Traffic flow pattern for an on-street system or in-parking facility, such as two-way or one-way.

**City.** Inhabited place of greater size, population, or importance than a town or village.

**Cityscape.** Stretch of scenery showing the arrangement of buildings in a city.

**Civil inattention.** Behavioral ritual that occurs when two or more people are mutually present but not involved in any form of interaction.

**Cladding.** External covering or skin applied to a structure for aesthetic or protective purposes.

**Classicism.** Revival of or return to the principles of Greek or Roman art and architecture.

**Classical conditioning.** Learning process in which a neutral stimulus comes to elicit a response after multiple pairings with another stimulus.

**Classical eclecticism.** Selection of elements from the architecture of Hellenic Greece or Imperial Rome for architectural decorative designs.

**Clo.** Unit for measuring the effect of clothing on thermal comfort and defined as the insulating value of a man's lightweight business suit and cotton underwear.

**Cold-twisting process.** Process by which rolled bars are twisted after cooling.

**Cold working.** Plastic deformation of metal at or near room temperature. This shaping is usually carried out by drawing, pressing, rolling, or stamping.

**Colonnade.** Row of columns spaced at regular intervals.

**Column.** Structural member whose primary function it is to support compressive loads parallel to its longitudinal axis.

**Column capital.** Enlargement of the end of a column designed and built to act as an integral unit with the column and flat slab and to increase the shearing resistance or reduce the moments, or both. (*See also* Shear.)

**Column spacing.** Spacing pattern between columns or rows of columns in any structure.

**Communication system.** Telephone or voice-activated system providing communications with police or security agencies, or internally within a facility.

**Compartmentalization.** Dividing a building or large space into two or more separate enclosures, each totally enclosed within fire separating barriers.

**Compatibility.** Ability to integrate with other systems or materials.

**Composite building.** Type of construction made up of different materials (such as concrete and structural steel) or of members produced by different methods (such as cast-in-place concrete and precast concrete). (*See also* Concrete.)

**Composite construction.** System in which the steel beams are fixed to the concrete slab above, so that the slab acts with the beams as a single unit to resist bending; also applies to other materials. (*See also* Slab.)

**Composite design.** Design employing the use of structural steel, concrete, and reinforcing designed in combination for structural interaction to achieve improved strength and performance of each constituent material.

**Composite tube concept.** Concept combining, but not limited to, a reinforced concrete peripheral framed tube, interior steel columns, and a steel slab system with a composite concrete topping. (*See also* Framed tube; Slab.)

**Compressive strength.** Measured maximum resistance of a concrete or mortar specimen to axial loading; expressed in stress or the specified resistance used in design calculations; in the United States expressed in pounds per square inch (psi).

**Compressive stress.** Stress caused by the shortening effect of an external compressive force.

**Computer-aided design (CAD).** Analysis and/or design, and/or modeling, and/or simulation, and/or layout of building design with the aid of a computer.

**Computer-aided engineering (CAE).** Engineering of a building, machine, or product encompassing all aspects of design, fabrication/manufacturing, installation/erection, and monitoring with the aid of a computer.

**Computer-aided manufacturing (CAM).** Manufacturing process for building components utilizing the analysis and/or design, and/or modeling, and/or simulation, and/or layout of manufacturing design with the aid of a computer.

**Computer networking.** Linking several independent computers so that they can communicate (exchange electronic information) with each other.

**Conceptual design.** Design phase in which the salient organizational characteristics of a building or architectural space are developed. It may incorporate rough study models, sketches, and diagrams.

**Concourse.** Open space where several roads or paths meet.

**Concrete.** Composite material that consists essentially of a binding medium within which are embedded particles or fragments of aggregate. In normal portland cement concrete, the binder is a mixture of portland cement and water and the aggregate is sand and gravel.

**Concrete block.** Hollow, prismatic concrete masonry unit.

**Concrete brick.** Solid concrete masonry unit, generally the shape of a rectangular prism.

**Concrete panel.** Rectangular piece of concrete, generally used for cladding.

**Condensation.** Formation of water on a surface because the air temperature falls below the dew point.

**Conduction.** Process of heat transfer through a substance in which heat is transferred from particle to particle, not as in convection by the movement of particles from one place to another, nor as in radiation.

**Confinement reinforcement.** Closely spaced special transverse reinforcement which restrains concrete in directions perpendicular to the applied stress.

**Connection.** Intersection between two or more members designed to transmit forces.

**Construction documents.** Architectural-engineering drawings, specifications, and schedules used to build and operate a building.

**Continuous casting.** Producing steel by pouring it in liquid form through a water-cooled mold in a continuous process.

**Contract documents.** Design plans and specifications for construction of a facility.

**Contraction.** Of concrete, the sum of volume changes due to shortening occurring as the result of all processes affecting the bulk volume of a mass of concrete.

**Control joint.** Formed, sawed, or tooled groove in a concrete structure to regulate the location and amount of cracking and separation resulting from the dimensional change of different parts of a structure so as to avoid the development of high stresses.

**Control valve.** Valve employed to control (shut off) a supply of water to a sprinkler system.

**Controlled rolling.** Thermomechanical treatment that controls the time, temperature, and deformation of steel plates during hot rolling.

**Convection.** Transmission of heat by natural or forced motion of the air, that is, by movement of the particles, as opposed to conduction or radiation.

**Corbel.** Projecting block, usually of stone, supporting a beam or other horizontal member. A series, each one projecting beyond the one below, can be used in constructing a vault or arch.

**Core.** Closed loop usually formed by a reinforced concrete wall designed primarily for carrying high lateral load in tall buildings; also used for housing mechanical services, pipes, elevators, and such. (*See also* Lateral loads.)

**Cornice.** Any projecting ornamental molding along the top of a building, wall, or arch, finishing or crowning it.

**Coupled shear wall.** Two or more reinforced concrete walls connected together with relatively flexible members designed primarily for carrying lateral loads and having a bending mode of displacement.

**CPM network diagram.** Schedule that shows a system of project planning, scheduling, and control which combines all relevant information into a single master plan, permitting the establishment of the optimum sequence and duration of operations. The interrelation of all efforts required to complete a construction project is shown.

**Creep.** Continuous deformation of concrete, masonry, or wood under a constant load.

**Critical-path method.** Project networking and planning technique with the deterministic mode.

**Cross-beam system.** System of construction employing large beams between two walls.

**Crowding.** Perception of the spatial restriction of an environment due to the density of individuals within it. Levels of crowding are perceived differently among individuals and across cultures.

**Crown.** Top part of an arch, including the keystone; also the uppermost portion of a tall building.

**Curtain wall.** Outside covering, usually of glass, on a building supported by steel or concrete structure.

**Cyclic loads.** Repeated loads occurring in cycles.

**Damping.** Resistance of a structure to displacement from an externally applied disturbance that is subsequently removed. The resistance may be provided by internal frictional resistance of the molecules of the structure, the drag effects of the surrounding medium, or contact with or connection to an adjacent structure.

**Database.** Computer information; data file.

**Daylighting.** Use of sky luminance, ground reflections, and sunlight to illuminate the interior of buildings. Often called natural lighting, but this includes moonlight.

**Dead load.** Permanent weight or load of a structure.

**Decumanus.** Street used as a baseline to form a perpendicular grid pattern for Roman city planning.

**Deep beam behavior.** Conditions of a beam in which the elastic distribution of stresses is nonlinear and there are significant compressive forces between loads and reactions.

**Deflection.** Horizontal or vertical movement of a point on a structure due to loads, creep, shrinkage, temperature changes, or settlements.

**Deformability.** Change of form, shape, or dimensions produced in a body by a stress or force without breach of the continuity of its parts.

**Demographics.** Size, distribution, and composition of, and changes within, a specified population group.

**Design-build.** Construction system usually employed when time is of the essence. Construction takes place as design plans are completed.

**Design standard.** Set of criteria established to define the design characteristics of a facility.

**Developer.** Individual or group that builds or causes to be built (develops) real estate.

**Dew point.** Temperature at which condensation of water vapor in the air takes place, that is, temperature at which the air is fully saturated.

**Diagonal bracing.** Inclined structural members carrying axial load employed to transform a structural frame into a truss to resist horizontal loads. (*See also* Truss.)

**Diaphragm action.** In-plane action of a floor system that maintains all columns framing into the floor in their same relative positions.

**Discharge type lamp.** Any lamp that produces light by means of phosphorus as a result of an electrical discharge through one or more gases or vapors within the lamp's envelope.

**Districts.** Areas within environments that have common characteristics or traits. Some urban examples include Paris' Left Bank, Chicago's Magnificent Mile, New York City's Wall Street, and San Francisco's Fisherman's Wharf.

**Diurnal change.** Change occurring during daytime.

**Domino construction system.** System of construction developed by Le Corbusier of columns supporting monolithic floor slabs; literal reference to the game of dominoes in which the six dots on the face of the domino represent a column grid. Floor slabs supported by perimeter columns allow spatial freedom at each floor level.

**Domus.** Latin word for house.

**Double cross-vaulted roof.** Vault formed by the intersection at right angles of two barrel vaults, consisting of an inner shell separated from a higher outer shell.

**Drawing.** Graphic (to scale) representation on paper of the building in the form of plans, sections, and elevations.

**Drift.** Lateral deflection of a structure due to load.

**Drift ratio.** Ratio of lateral deflection of a story to story height.

**Drop panel.** Thickened slab over a column in a two-way slab system to prevent slab failure from punching shear.

**Dry-bulb temperature.** Temperature shown on an ordinary thermometer, as opposed to wet-bulb temperature.

**Dry system.** System employing automatic sprinklers attached to a piping system containing air under atmospheric or higher pressures. Loss of pressure from the opening of a sprinkler or detection of a fire condition causes the release of water into the piping system and out the opened sprinkler.

**Ductility.** Property of a material by virtue of which the material may undergo large permanent deformation without rupture.

**Ductile moment resisting frame.** Moment resisting frame in which ductility is deliberately provided for increasing the energy-absorption capacity of buildings in seismic zones.

**Easement.** Right of accommodation (for a specific purpose) in land owned by another, such as right-of-way or free access to light and air.

**Eccentric-braced frame.** Frame in which the centerlines of braces are offset from the points of intersection of the centerlines of beams and columns.

**Eclecticism.** Style of architecture that borrows its elements of composition from various sources.

**Economic feasibility.** Indication of a project's feasibility in terms of costs and revenue. Excess revenue establishes the degree of feasibility.

**Edges.** Enclosing features of an environment, such as mountains, lakes, rivers, or major traffic ways.

**Edo period.** Time period in and around the city of Tokyo, Japan, that spans from 1614 to 1868.

**Effective temperature.** Most commonly used criterion for determining the thermal comfort zone, introduced by Yaglou in 1924 and adopted with slight modifications by ASHRAE. It takes account of temperature, humidity, and air movement, but ignores radiation.

**Egress.** Exit or means of exiting.

**Electric arc furnace.** Refractory-lined vessel for producing steel by means of an electric arc between carbon electrodes.

**Electronic data.** Information stored in computers and transmitted electronically between computers.

**Electronic drawing.** Graphic [architectural-engineering (A-E)] drawing stored in computer as electronic data; drawing on paper generated by computer.

**Elevation.** Drawing showing the vertical elements of a building, either exterior or interior, as a direct projection to a vertical plane.

**Elevator.** Device for transporting passengers or goods vertically from story to story in a building; also known as lift.

**Elevator shaft (elevator core).** Shaftway for the travel of one or more elevators, lifts, or dumbwaiters; includes the pit; terminates at the underside floor or grating of the overhead machinery space.

**Emergency lighting system.** Lighting designed to supply illumination that is essential to safety in the event of failure of the normal electric power supply.

**Engineered brick.** Brick having the nominal dimensions of 81.3 by 101.6 by 203.6 mm (3⅕ by 4 by 8 in.).

**Enthalpy.** Heat content per unit mass due to both latent heat and sensible heat.

**Environment.** Interrelated elements which constitute the area external to a systems model.

**Environmental impact.** Measurement of the environmental consequences of a facility in terms of such factors as air, noise, and water pollution levels that are generated by the facility.

**Environmental stimuli.** Sensations—auditory, visual, and the like—which emanate from the environment.

**Esplanade.** Level open space for walking or driving, often providing a view.

**Exhaust fan.** Fan that withdraws air from a localized area or a space in a building from which it is desired not to return the air to the central air-treatment system, as from a toilet.

**Expansion joint.** Joint between sections of a structure which allows expansion and contraction from temperature variations.

**Expert system.** Computer technique used for design forming a part or a set of many specialties; results in automatic design of components.

**Facade.** Face and, especially, the principal elevation of a building. (*See also* Cladding.)

**Fatigue.** Process of loss of strength of a material subjected to reversal of stress.

**Fenestration.** Windows and openings of a building envelope.

**Ferrite.** Pure iron constituent of steel which is very ductile but has low tensile strength.

**Fiber reinforced concrete.** Concrete containing asbestos, spun glass, or other fibers to reduce unit weight and improve tensile strength.

**Filigree.** Ornamental openwork of delicate or intricate design, typically related to cast-iron or steel elements on buildings.

**Fillet weld.** Weld of approximately triangular cross section joining two surfaces, approximately at right angles to each other, as in a lap joint.

**Financing.** Means of providing funds for a facility through private capital, public sale of general obligation or revenue bonds, special assessment or tax district funds, leases, nonprofit associations, or a combination of these sources.

**Finite-element analysis technique.** Examination method that investigates a single segment in an analogy developed for stress analysis, in which a continuous structure is subdivided into a number of discrete linear two-dimensional or three-dimensional segments.

**Fire damper.** Device used in air ducts to shut off or vary the volume of airflow through the duct. Fire dampers are used to isolate the smoke areas and prevent the flow of smoke through the duct system to smoke-free areas.

**Fire rating.** Fire-endurance rating for doors, shutters, walls, and the like, established by Underwriters' Laboratories, Inc., or another recognized and approved laboratory.

**Fire resistance.** Ability of a material or assembly to withstand fire or give protection from it; ability to confine a fire or to continue to perform a given structural function, or both.

**Fire separation.** Floor or wall (either without openings or with adequately protected openings) having the fire-endurance rating required by appropriate authorities; acts as a barrier against the spread of fire within a building.

**Flame shield.** Sheet steel attached to main members, but separated from them either by a small air gap or by a noncombustible insulation made of ceramic fibers.

**Flange.** Plate or element located at the top and bottom of a beam cross section.

**Flashing.** Device to prevent migration of water once it enters into masonry construction and convey water back to the building's exterior.

**Flat plate.** Two-way slab not supported by beams and without a drop panel or column capital.

**Flexure.** Bending of a member, as under a load.

**Floor area.** Area of a floor, measured by length times width; in some cases the total floor area of a facility.

**Floor slab.** Structural slab serving as a floor, usually of reinforced concrete.

**Flying form.** Large prefabricated unit of formwork designed for reuse.

**Formwork.** Total system of support for freshly placed concrete, including the mold or sheathing which contacts the concrete, as well as all supporting members, hardware, and necessary bracing. Falsework and shuttering are also used with essentially the same meaning.

**Foundation.** Lowest part of a structure, upon which the remainder of the structure rests.

**Framed tube.** Perimeter equivalent tube consisting of closely spaced columns and spandrels.

**Frame-shear wall (frame-shear truss).** Combined system in which a rigid frame and a shear wall (shear truss) act interactively. (*See also* Shear wall–frame interaction system.)

**Free facade.** Separation of load-bearing columns from the exterior walls enclosing interior spaces.

**Galleria.** Covered and lighted way for foot passengers, with booths and shops.

**Generators.** Activities that generate parking demand such as stores, office buildings, hospitals, and recreation facilities.

**Genius loci.** From Latin, literally the (guardian) spirit of a place; general atmosphere or distinctive character of a place.

**Girder.** Main horizontal supporting member or beam in a structure, usually supporting another beam or column.

**Glazed terra cotta.** Fired clay material that experienced a decline due to the extensive cracking of its glaze.

**Golden section.** Defined geometrically as a line that is divided such that the lesser portion is to the greater as the greater is to the whole. Algebraically, it can be expressed as the equation of two ratios: $a/b=b/(a+b)$. The Greeks recognized the dominating role the golden section played in the proportions of the human body and applied it to the proportions of temples. The Renaissance architects applied the principle of the golden section to buildings and public spaces. Le Corbusier used the golden section as a basis for his Modulor system.

**Gothic.** Architecture of the pointed arch, rib vault, flying buttress, and walls reduced to a minimum by spacious arcades, by gallery or triforium, and by spacious clerestory windows. These are not isolated motifs; they act together and represent a system of skeletal structure with active, slender, resilient members and membrane-thin infilling or no infilling at all.

**Grade.** Ground level.

**Green architecture.** Environmentally conscious architecture.

**Gridiron pattern.** Closely spaced lines which are perpendicular to each other.

**Gross area.** Entire floor area of a building, usually measured in square meters or square feet.

**Gross density.** Number of dwellings in relation to total land area, including all public rights-of-way and other uses.

**Grout.** Mortar containing a considerable amount of water so that it has the consistency of a viscous liquid, permitting it to be poured or pumped into joints, spaces, and cracks within masonry walls and floors, between pieces of ceramic clay, slate, and floor tile, and into the joints between preformed roof deck units.

**Gypsum wallboard.** Base for plaster; sheet having a gypsum core, faced with paper, which provides a good bond for plaster; usually manufactured in 406- by 1219-mm (16- by 48-in.) or 610 by 2438-mm (24- by 96-in.) panels, 9.5 or 12.7 mm (⅜ or ½ in.) thick with round or square edges; used primarily as an interior surfacing in a building.

**Halon system.** Fire suppression system that uses CFC refrigerants as the fire-extinguishing chemical. This is being phased out because of its impact on the earth's ozone layer.

**Haunched beam (slab).** Beam (slab) whose cross section thickens toward its supports.

**Heat detector.** Heat-detecting device that initiates an automatic signaling and alarm system (and other corrective actions) when the surrounding temperature exceeds its thermal setting.

**Heat exchanger.** Device designed to transfer heat between two physically separated fluids; generally consists of a cylindrical shell with longitudinal tubes. One fluid flows on the inside, the other on the outside.

**Heat pump.** Refrigeration system designed to utilize alternately or simultaneously the heat extracted at a low temperature and the heat rejected at a higher temperature for cooling and heating functions, respectively.

**Heat transfer.** Generic term for conduction, convection, and radiation.

**Heating, ventilating and air-conditioning (HVAC).** Includes the mechanical systems for heating, cooling, humidity control, and movement of air in buildings.

**Heliport.** Facility where helicopters land, take off, and are maintained or repaired.

**Heuristics.** Helping to discover or learn; specifically, designating a method of education in which the pupil proceeds along empirical lines, using rules of thumb, to find solutions or answers. Computer methods that do not necessarily lead to theoretically optimum solutions, but which are relatively straightforward to apply and lead to "good" solutions.

**High-lift grout technique.** Wall being built as a cavity wall using ties. Prior to grouting by pumping, the base is cleaned out of mortar droppings through the clean-outs shown. The height of pour is limited to the strength against pullout of the wall ties as the grouting hydrostatic pressure rises.

**High-performance concrete.** Concrete in which improvements in qualities such as durability, corrosion resistance, ductility, and energy absorption in addition to compressive strength are realized in justifying the cost of a building.

**High rise.** Multistory building, usually providing luxury or status.

**High-strength concrete.** Concrete that has a higher strength than normal of up to 138 MPa (20,000 psi). Such high strength is attainable because of the development of superplasticizers and other admixtures.

**High-strength low-alloy steel.** Steel with chemical composition developed to impart higher mechanical properties than can be obtained for carbon steels, but with alloying additions (and strength) less than those for alloy steels.

**High-strength reinforcing steel.** Steels that have a yield strength of between 414 MPa and 518 MPa (60 and 75 ksi).

**High-tech architecture.** Mode of architecture in which the building services are not only revealed, but emphasized. For example, ducts and pipes may be painted in bright colors to indicate their respective functions. An outstanding example is the Centre Pompidou in Paris.

**Historical database.** Computer information and data file that captures the knowledge and experience of experts and saves this information for future use.

**Holistic.** Emphasizing the importance of the whole and the interdependence of its parts. In urban design, viewing the city as an entity rather than a loose collection of individual buildings.

**Hollow clay brick.** Units in which the net area in every plane parallel to the bearing surface is not less than 60% of the gross area on the same plane.

**Holographic image, hologram.** Three-dimensional image produced by a technique of wavefront reconstruction using lasers to record diffraction patterns on photographic plates.

**Hot-rolled steel.** Steel obtained by rolling the metal while hot; results in a dark, oxidized, relatively rough surface.

**Hydration.** Formation of a compound by combining water with some other substance; in concrete, chemical reaction between cement and water; chemical reaction by which a substance combines with water, giving off heat to form a crystalline structure in its setting and hardening.

**Hygrometer.** Instrument for measuring the humidity in the air.

**Hyperbolic paraboloid.** Locus of a point whose difference in distances from two fixed points, or foci, is constant; curve formed by the section of a cone cut by a plane more steeply inclined to the base than to the side of the core.

**I beam.** Rolled or extruded structural beam having a cross section resembling the letter I.

**I-steg steel bar.** Steel bar made by cold twisting two round mild-steel bars in special machines.

**Imageability.** Ability of a building or environment to create a distinct mental representation or image. (*See also* Legibility.)

**Inelastic deformability.** Deformation of a material that does not disappear on removal of the force that produced it.

**Infrastructure.** Building components and systems that are fixed in location and which are fundamental to the consideration of project budget and development schedule.

**Ingot mold.** Cast-iron mold used to receive and solidify liquid steel.

**In-situ concrete.** *See* Cast-in-place.

**Integration.** Organization and coordination of systems so that they interact and operate harmoniously.

**Intelligent building.** Building with a fully integrated management and control system in which the services are monitored and controlled by a sophisticated computer-based management system.

**Interior diagonally braced truss.** System consisting of four megatrusses, each extending the entire width of the building. The chords of these trusses comprise the exterior columns of the building. (*See also* Truss.)

**International style.** Style stemming from the Bauhaus; characterized by purity, functionalism, and impersonality.

**Irradiance.** Flow of solar energy.

**Irradiation.** Total solar energy over a period of time, such as an hour or a day.

**Iterations.** Repetitions of a logical computer software process so that it ultimately converges to a solution.

**Joint.** Division between sections of a floor or slab which permits expansion and contraction of the floor or slab.

**Joist.** Structural element of a floor system which rests upon the beams or walls and supports the floor surface.

**Joule.** SI unit for energy, including heat energy.

**Knowledge-based system.** Relates to a knowledge base encompassing rules, facts, design heuristics, and procedural knowledge as well as an inference engine that contains reasoning and problem-solving strategies.

**Lamellar tearing.** Separation of highly restrained based metal caused by through-thickness strains induced by shrinkage of adjacent weld metal.

**Land-use plan.** Comprehensive plan classifying particular areas or plots by the manner in which they are being utilized, and their relationships with adjacent uses and integral components (such as transit systems).

**Landmark.** Distinctive feature of an environment used as reference point. Some urban examples include Stockholm's City Hall, Athens' Acropolis, Chicago's Water Tower, and New York City's Statue of Liberty.

**Late-modernism.** Architectural period following modernism in which seminal design issues introduced by modern architects are being explored.

**Latent heat.** Thermal energy expended in changing the state of a body without changing the temperature; for example, converting water to water vapor at the same temperature.

**Lateral drift.** Lateral deflection of a tall building.

**Lateral load.** Load on a building acting horizontally; for example, wind load.

**Lateral stiffness.** Resistance to bending forces from the side.

**Lease span.** Distance from a fixed interior element to the exterior window wall.

**Legibility.** Degree to which environmental features can be distinguished from each other. A legible environment is one that is easily learned and remembered. (*See also* Imageability.)

**Life cycle (of a building).** Analysis technique where optimum return on the project is evaluated on the total cost (cost of construction plus operating cost and maintenance) for the life of the building project.

**Lift slab construction.** Method of concrete construction in which floor and roof slabs are cast on or at ground level and hoisted into position by jacking; also slab that is a component of such construction.

**Limit state.** Condition in which a structure or component becomes unacceptable because it is no longer functional (serviceability) or is likely to collapse (strength).

**Limit-states design.** Design process that involves identification of all potential modes of failure (limit states) and maintaining an acceptable level of safety against their occurrence. The safety level is usually established on a probabilistic basis.

**Link beam.** Beam connecting two large vertical elements.

**Lintel.** Horizontal beam or stone bridging an opening.

**Live load.** Load which a structure must support beyond the actual weight of the buildings.

**Load bearing.** That which has to be carried or supported by a column, pier, or wall.

**Lobby.** Entrance area of a structure, which can contain elevators, stairs, offices, and other similar elements.

**Loggia.** Arcade or colonnade, frequently with an upper story.

**Low rise.** Multiunit residential dwelling which is usually no more than six or seven stories in height. "Low" is generally used to distinguish a building from a "high" rise.

**Luminous environment.** Visible band of radiation from any radiant source, including the sun and artificial light sources of various kinds.

**Macro-defect-free cement.** Kneaded and specially prepared mix of cement that gets rid of the defects or pores, containing a significant percentage of polymers.

**Manual pull station.** Fire or smoke alarm system that is activated manually.

**Masonry.** Construction composed of shaped or molded units, usually small enough to be handled by one person and composed of stone, ceramic brick or tile, concrete, glass, adobe, or the like.

**Mass.** Quantity of matter forming a body of indefinite shape, usually of relatively large size.

**Mass distribution.** To scatter or spread out a quantity of matter.

**Master plan.** Plan, usually graphic and drawn on a small scale, but often supplemented by written material, which depicts all the elements of a project or scheme.

**Means of egress.** Continuous path of travel from any point in the building to the ground level outside. Such paths must be readily available (clear of obstacles) at all times.

**Mechanical ventilation.** Process of supplying outdoor air to a building or removing air from it by mechanical means, such as with fans. The air that is supplied may not be heated, cooled, or air-conditioned.

**Megacity.** City that expands to such a large degree as to encompass other cities, towns, or communities, resulting in multiple urban centers.

**Megalithic tower.** Tall structure built of unusually large stones.

**Megalopolis.** City of enormous size, especially when thought of as the center of power or wealth.

**Megastructure.** Building structure, usually large, providing for multiple use, combining living, working, and service functions within the whole.

**Met.** Unit of measure (in $W/m^2$) of the metabolic rate of the body at which it produces energy. The metabolic rate increases proportionally with the body's activities.

**Metropolis.** Large urban center of culture or trade.

**Mezzanine.** Low-ceilinged story or extensive balcony, usually constructed above the ground floor.

**Microclimate.** Highly specific weather and growth conditions of a limited area, influenced by local geographic and atmospheric factors.

**Microstructure.** Structure, as of a metal or alloy, seen under a microscope.

**Mild steel.** Nearly pure iron having a very low carbon content, usually between 0.15 and 0.25%; ductile, rust-resistant material used in boilers, joists, pipes, and the like.

**Minaret.** Tall tower in, or contiguous to, a mosque with stairs leading up to one or more balconies from which the faithful are called to prayer.

**Minimalism.** Design style that makes the smallest or least possible use of any type of system.

**Mixed use.** Building that contains more than one type of occupancy. For example, a building with apartments on the upper floors and business establishments on the lower floors is a mixed-use building; also referred to as multiuse.

**Model.** Abstraction of reality. In evaluation technique, a program for evaluating, accounting, analyzing, or forecasting. A model may be manual or computerized; it may be explanatory or prescriptive; it can be symbolic (verbal or mathematical) or iconic (providing a physical resemblance). Two or more interactive models will form a linked system.

**Modeling.** Process of developing logical or mathematical models which attempt to describe the behavior of a system.

**Modern architecture.** Largely a reaction against the Beaux Arts eclecticism and historically based architectural models, its revolutionary approach sought to redefine architecture from the ground up. Architects rejected ornamentation and historical reference and, instead, embraced a more technological and rational expression of building form.

**Modified tube system.** System involving a modification of the basic framed-tube system.

**Modular ceiling system.** Ceiling units of standardized sizes or designs that can be arranged or fitted together in a variety of ways.

**Modulus of elasticity.** Ratio of stress to strain below the proportional limit.

**Molding.** Member of construction or decoration so treated as to introduce varieties of outline or contour in edges or surfaces, whether on projections or cavities, as on cornices, capitals, bases, door and window jambs, and heads; may be of any building material, but almost all derive at least in part from wood or stone prototypes.

**Moment.** Property by which a force tends to cause a body, to which it is applied, to rotate about a point or line; equal in magnitude to the product of the force and the perpendicular distance of the point from the line of action of the force.

**Moment frame.** Building frame system in which the lateral loads are primarily resisted by shear and flexure in the members and joints.

**Moment of inertia.** About an axis in the plane of a section, it is equal to the sum of the products of all the elementary areas times their distances squared from the axis.

**Moment resisting frame.** Structures consisting of beams and columns connected together by rigid joints or moment resisting connections that are monolithic.

**Monotonically increasing load.** Continuous load increases without interruption or occasional unloading.

**Morphology.** Study of the structure, classification, and relationships of living organisms.

**Morphosis.** Order or development of an organism; in this context, evolving form and structure of urban development.

**Movement joint.** Joint or gap between adjacent parts of a building, structure, or concrete work which permits their relative movement due to temperature changes (or other conditions) without rupture or damage.

**Mullion.** Vertical member separating (and often supporting) windows, doors, or panels set in series.

**Multiuse structure.** *See* Mixed use.

**Natural ventilation.** Ventilation by air movement caused by natural forces, rather than by fans.

**Nave.** Central and principal aisle in a longitudinal church.

**Neoplatonism.** Any of various schools of philosophy based on a modified Platonism, postulating a single source from which all forms of existence emanate and with which the soul seeks mystical union.

**Net area.** Net usable floor area of a building, that is, gross area minus area of floor openings service cores, and utility cores.

**Nodes.** Points where behavior is focused, usually at the intersection of paths or where paths are broken. Some urban examples include Paris' Arch of Triumph, Athens' Constitution Square (Syndagma) and Omonia, and New York City's Times Square.

**Noncomposite construction.** System not employing composite construction.

**Normalizing.** Treatment of steel by heating above the critical range followed by air cooling.

**Obelisk.** Monumental, four-sided stone shaft, usually monolithic and tapering to a pyramidal tip.

**Office landscape.** Method of interior office design whereby a large space is subdivided by such devices as low partitions, furnishings, and planters.

**One-way joist.** Type of framing system for floors or roofs in a concrete building; consists of a series of parallel joists which are supported by girders (perpendicular to the joists) between columns.

**One-way-slab system.** Structural system where the arrangement of reinforcement is intended to resist stresses due to bending in one direction only.

**Open-floor system.** Building plan with a minimum of internal subdivision between spaces designed for different usage.

**Operant conditioning.** Learning process that emphasizes the rewarding or punishing consequences of some activity through positive or negative reinforcement.

**Optimized building design.** Design aimed at minimizing cost, materials, and such without sacrificing performance.

**Oriel.** In medieval English architecture, bay window corbeled out from the wall of an upper story; bay projecting inside or out, extending a room; windowed bay or porch at the top of exterior stairs.

**Ornamentation.** In architecture, every detail of shape, texture, and color that is deliberately exploited or added to attract an observer.

**Orthogonal truncation.** To cut short at right angles.

**Outrigger beam.** Beam that projects beyond the walls of a building to serve as a support, as for hoisting tackle.

**Outrigger wall.** Temporary support extending out from a main structure.

**Overturning moment.** Moment causing overturning of a building or shear wall.

**Pagoda.** Tall pyramidal temple building, found in China, Japan, and Southern Asia.

**Palazzo.** Palace or large city residence.

**Pan joist system.** Method of construction utilizing horizontal structural concrete members usually supported by beams constructed by the aid of prefabricated form units (pans).

**Parabolic arch.** Arch similar to a three-centered arch, but whose radius is parabolic, with a vertical axis.

**Parapet.** Low guarding wall at any point of sudden drop, as at the edge of a terrace, roof, battlement, or balcony.

**Parking level.** Floor or level within a multistory parking facility.

**Parking lot.** Surface area for parking, off the street or beyond the right-of-way.

**Parti.** Fundamental concept of a design; simplified diagram representing a design concept.

**Passive solar system.** Building subsystem in which solar energy is collected and transferred predominantly by natural means; uses natural convection, conduction, or radiation to distribute thermal energy through a structure, within the limits of the indoor design temperature conditions.

**Pastiche.** Literary, artistic, or musical composition made up of bits from various sources; potpourri.

**Peak.** Period of maximum activity; can be by the hour, day of week, or seasonal.

**Peak demand.** Maximum electric energy load; depends on occupancy and climate.

**Peak-load limiting.** Automatic electric monitor or controller which can be used to limit maximum power demands of a building.

**Pearlite.** Alternating lamellae of ferrite and cementite.

**Pedestrianism.** Practice of walking, especially in preference to using vehicular transportation.

**Penthouse.** Structure occupying usually less than half the roof area of a flat-roofed building.

**Perception.** Processing of environmental stimuli. Environments can be perceived through sights, sounds, tastes, smells, and touch.

**Perceptual access.** Perceived ability to see through or out of a building in order to find a destination.

**Performance specification.** Statement of criteria for the design parameters whereby the performance of the activities and facilities will meet their intended goals. Assessment and test procedures are established to assure compliance.

**Personal space.** Portable "bubble" of space around each person which is used to regulate interpersonal distance. The size of this bubble varies widely, depending upon factors such as culture, age, gender, and level of intimacy between people.

**Perspective vista.** Defined or refined scenic view; focused view. Vistas may be closed or open (stretching to distant views).

**Piano nobile.** In Renaissance architecture and derivatives, a floor with formal reception and dining rooms; principal story in a house, usually one flight above the ground.

**Pier.** Column designed to support concentrated load.

**Pilaster.** Engaged pier or pillar, often with capital and base.

**Pile.** Concrete, steel, or wood column, usually less than 0.6 m (2 ft) in diameter, which is driven or otherwise introduced into the soil, usually to carry a vertical load or to provide lateral support.

**Pile cap.** Slab or connecting beam which covers the heads of a group of piles, tying them together so that the structural load is distributed and they act as a single unit.

**Pillar.** Pier; post or column.

**Pilotis.** A French term for pillars or stilts that carry a building, thereby raising it to an upper floor level and leaving the ground floor open.

**Pitch.** To incline downward, dip.

**Plain masonry.** Masonry without reinforcement, or reinforced only for shrinkage or temperature change.

**Plane truss.** Planar triangulated arrangement of members of a framework that are primarily in axial tension and compression. (*See also* Truss.)

**Plastic deformation.** Deformation beyond the point of elastic recovery, accompanied by continuing deformation with no further increase in stress; results in a permanent change in shape.

**Plasticizer.** Any of various substances added to a plastic material to keep it soft and viscous.

**Plastification.** To make or become plastic.

**Plenum.** In suspended ceiling construction, space between the suspended ceiling and the main structure above.

**Plinth.** Square or rectangular base for wall, column, pilaster, or door framing.

**Pneumatic.** Control system that uses compressed air to transmit information.

**Point building.** High-rise buildings with a comparatively small floor area and corridors of apartments that surround a central community space or central core.

**Poisson's ratio.** Ratio of transverse to longitudinal strain.

**Polymer.** One of a group of high-molecular-weight resinlike organic compounds whose structures usually can be represented by repeated small units. Some polymers are elastomers, some are plastics, and some are fibers.

**Polymer impregnation.** Liquid-monomer pore infill which later solidifies in the pores of concrete.

**Population density.** Number of people in a specified area of space; sometimes referred to as social density.

**Porosity.** Ratio, usually expressed as a percentage, of the volume of voids in a material to the total volume of the material, including the voids. The voids permit gases or liquids to pass through the material.

**Post-modernism.** In architecture, break with the canons of international style modernism. Functionalism and emphasis on the expression of structure are rejected in favor of a greater freedom of design. There is a new interplay of contemporary forms and materials, with frequent historic allusions, often ironic. Post-modernism also accepts the manifestations of commercial mass culture—bright colors, neon lights, and advertising signs.

**Posttensioned.** Poured-in-place concrete which is subsequently strengthened by the tightening of cables extended through the slab.

**Postyield behavior.** Material behavior beyond the yield point.

**Poured-in-place concrete.** Concrete poured into forms erected on the job site, as opposed to precast concrete.

**Power.** Flow of energy or rate of doing work; energy over time.

**Pozzolan.** Siliceous or siliceous and aluminous material, which in itself possesses little or no cementitious value but will, in finely divided form and in the presence of moisture, chemically react with calcium hydroxide at ordinary temperatures to form compounds possessing cementitious properties.

**Praxis.** Established practice, custom.

**Precast concrete.** Concrete building components fabricated at a plant site and shipped to the site of construction.

**Prestressed concrete.** Concrete in which internal stresses of such magnitude and distribution are introduced that the tensile stresses resulting from the service loads are counteracted to a desired degree. In reinforced concrete the prestress is commonly introduced by tensioning the tendons.

**Prestressed concrete system.** System that employs prestressed concrete; most suited for long spans as it increases the span range of conventional reinforced concrete floor systems by about 30 to 40% (*See also* Prestressed concrete.).

**Project team.** Team comprising owner, architect/engineer, and project manager.

**Proof stress.** Stress read from the intersection of a line drawn parallel to the initial straight portion of the stress-strain diagram from the point of 0.2% strain and the original stress-strain curve. This is used for metals having no sharp yield point to specify the permissible stress.

**Proscenium arch.** Arch or any equivalent opening in the proscenium wall through which the stage is seen by the audience.

**Psychrometric chart.** Graphic determination of certain thermodynamic relations, used for design of thermal environment. The horizontal axis gives the dry-bulb temperature; the vertical axis gives the water content of the atmosphere. Lines are drawn for relative humidity, wet-bulb temperature, and enthalpy.

**Psychrometrics.** Study of the properties of a mixture of air and water vapor.

**Public emergency reporting system (PERS).** System of several emergency reporting stations located in key egress and public gathering areas.

**Punched tube.** Tube building in which the window openings are as if "punched" in solid walls of the tube.

**Punched wall.** Wall with window openings.

**Quenching and tempering.** Treatment of steel by heating and holding at the austenitizing temperature, then quenching by immersion in water or oil; subsequently the steel is reheated below the critical temperature, held at that temperature, then cooled.

**Racking.** Stepping back successive courses of masonry in an unfinished wall.

**Radiation.** Energy transmitted by electromagnetic waves.

**Rain screen wall.** Recent technical development in which two exterior walls are constructed with a cavity left between them. It is similar to cavity wall construction except that rain is allowed to penetrate the outermost permeable wall. The inner wall is protected with a waterproof membrane, preventing water from penetrating the interior of the building.

**Raison d'être.** Reason for being; justification for existence.

**Rationalism.** Principle or practice of accepting reason as the only authority in determining one's opinions or course of action.

**Rebar.** Metal bars, wires, or other slender members which are embedded in concrete in such a manner that the metal and the concrete act together in resisting forces.

**Reentrant corner.** Internal or inside corner; usually used to describe angles less than 90°.

**Refuge area.** Safe area (usually the perimeter or top of a building) from where the building occupants can be rescued.

**Reinforced concrete.** Concrete containing reinforcement designed on the assumption that the concrete and the reinforcement act together in resisting forces.

**Reinforced concrete plank.** Precast, prestressed, hollow-core reinforced concrete plank, usually relatively lightweight; used for floor and roof decking; may carry a structural topping.

**Relative humidity.** Ratio of the quantity of water vapor actually present in the air to that present at the same temperature in a water-saturated atmosphere.

**Renaissance.** Architectural style developed in early fifteenth-century Italy during the rebirth (rinascimento) of classical art and learning; initially characterized by the use of the classical orders, round arches, and symmetrical composition.

**Renewable energy resource.** Source of energy that may be replenished.

**Responsible architecture.** Architecture that fulfills specific social, environmental, and contextual needs; that is responsive to the requirements of users, place, and climate.

**Reticulated.** Covered with netted lines; netted; having distinct lines crossing in a network.

**Reveal.** Side of an opening for a door, window, doorway, or the like, between the door or window frame and the outer surface of the wall; where the opening is not filled with the door or window, the whole thickness of the wall.

**Reversed cyclic loading.** Cyclic loading applied alternately in reversed directions.

**Ribbed-Tor steel.** Mild-steel round bars that are ribbed during rolling and then twisted in a twisting machine to improve the strength of steel.

**Rigid frame.** Structure made up of beam and column members joined together at their intersections in such a manner that there is no relative rotation between the intersecting members at a joint under applied load or deformation.

**Rolled beam.** Metal beam fabricated of steel made in a rolling mill.

**Romanesque.** Style emerging in western Europe in the early eleventh century, based on Roman and Byzantine elements, characterized by massive articulated wall structures, round arches, and powerful vaults, and lasting until the advent of Gothic architecture in the middle of the twelfth century.

**Rusticated.** Said of cut stone having strongly emphasized recessed joints and smooth or roughly textured block faces; used to create an appearance of impregnability in banks, palaces, courthouses, and the like. The border of each block may be rebated, chamfered, or beveled on all four sides, at top and bottom only, or on two adjacent sides; the face of the brick may be flat, pitched, or diamond-point, and if smooth may be hand- or machine-tooled.

**Schedule.** Table or itemized list of products and services required for construction, showing performance information.

**Section modulus.** Moment of inertia of the area of the cross section of a structural member divided by the distance of the neutral axis from the farthest point of the section; measure of the flexural strength of the beam.

**Seismic load.** Force produced on a structural mass owing to its acceleration, induced by an earthquake.

**Sensible heat.** Heat absorbed or emitted when the temperature changes without a change of state, as distinct from latent heat.

**Serial vision.** Sequence of images revealing what is viewed by pedestrians walking at a uniform pace from one end of an urban space to another; a method for analyzing the effects of archetype and architexture. (*See also* Architexture.)

**Setback.** Reduction in the horizontal cross section of a building, causing the upper levels of the building to be narrower than the lower levels.

**Shaft.** Portion of a column, colonnette, or pilaster between base and capital.

**Shear.** Internal force tangential to the plane on which it acts.

**Shear deformation.** Change in form due to shear.

**Shear-lag effect.** Consequence occurring in bending members with wide thin flanges, causing the axial flange stresses remote from the web to be less, and those close to the web more, than would be calculated on the basis of the proportionate distance from the neutral axis. The effect is due to in-plane shearing deformations of the flange. In a tube building, difference in axial stress in perimeter columns with maximum at the corner columns and minimum at the columns at the center.

**Shear wall.** Reinforced concrete wall designed primarily for carrying lateral loads and having a bending mode of displacement.

**Shear wall–frame interaction system.** Combined system utilizing the interactive role of shear walls and rigid frames to resist lateral loads. (*See also* Frame-shear wall.)

**Shelf angle.** Angle iron that is fixed to a girder to carry the ends of joists; formwork or hollow tiles of a concrete slab; also, angle iron attached to a wall to carry nonstructural panels or veneers.

**Siamese connection.** Twin (double) water supply connections to sprinkler piping and standpipe systems located outside building for use by the fire department.

**Sill.** Horizontal timber, at the bottom of the frame of a wood structure, which rests on the foundation.

**Simulation.** Imitation of reality, often carried out using analog or digital computers. It enables studies to be carried out on "paper" which may be impractical to perform on the real system.

**Simulation model.** Mathematical model developed to imitate the behavior of a given improvement.

**Single-use structure.** Building that contains only one type of occupancy, such as only residential or only commercial. (*See also* Mixed use.)

**Site.** Area on which a facility or other improvement is constructed.

**Site characteristics.** Physical features of a site, such as shape, area, topography, soil conditions, and access.

**Skeleton structure.** Supporting structural framework of a building consisting of thin, discrete members.

**Skin.** Outer veneer or face ply of a laminated construction.

**Skip-joist system.** Reinforced concrete slab system in which the joists are spaced wider than the typical 0.9 m (3 ft) between joists, employing special pan forms to improve economy.

**Skyscraper.** Name originally applied in the United States to tall, multistory buildings built entirely with an iron skeleton, later with steel or reinforced concrete, or both.

**Slab.** Exposed wearing surface laid over the beams of a structure.

**Slab building.** High-rise building with a corridor layout floor plan.

**Slenderness ratio.** Ratio of height to width of building; for columns, ratio of effective length to radius of gyration.

**Slip connection.** Bearing-type connection in steel construction allowing movement.

**Slip form.** Form that moves, usually continuously, during placing of the concrete. Movement may be either horizontal or vertical. Slip forming is like an extrusion process with the forms acting as moving dies to shape the concrete.

**Slip-formed concrete.** Form designed to move upward slowly (usually by means of hydraulic jacks or screw jacks), supported by the hardened concrete of the wall section which was poured previously.

**Smoke damper.** Damper arranged to seal off airflow automatically through part of an air duct system so as to restrict passage of smoke.

**Smoke detector.** Smoke detecting device that initiates an automatic signaling and alarm system (and other corrective actions) when smoke is detected in the space.

**Smoke management system.** Mechanical-electrical system used to control and remove smoke from a building and to provide safe areas and escape routes for building occupants.

**Software.** Logical procedures in electronic coded format in the computer that process information.

**Solar envelope.** Way in which a building affects its surroundings in terms of its shadows.

**Solar screen.** Nonstructural openwork or louvered panel of a building arranged so as to act as a sunshading device. (*See also* Brise-soleil.)

**Sonic environment.** Vibration of air at the audible range of frequency.

**Space frame.** Three-dimensional framework, as contrasted to a plane frame.

**Space truss.** Three-dimensional truss.

**Spall.** Small fragment or chip removed from the face of a stone or masonry unit by a blow or by action of the elements.

**Span.** Length of a beam; distance between columns.

**Spandrel.** Wall between the head of a window and the sill of the window above it. (An upturned spandrel continues above the roof or floor line.)

**Spandrel beam.** Floor-level beams in the faces of a building, usually supporting the edges of the floor slabs.

**Spandrel glass.** Opaque glass used in windows and curtain walls to conceal spandrel beams, columns, or other internal construction.

**Specific volume.** Volume of a unit mass; the reciprocal of density.

**Specifications.** Construction information (descriptions, procedures, conditions, regulations) presented in a text format.

**Spine structure.** Structural system in which the spine, consisting of vertical or inclined elements and braced frames, walls, or Vierendeel girders, resists lateral loads.

**Splay.** Sloped surface, or surface that makes an oblique angle with another, especially at the sides of a door, window, or proscenium, so that the opening is larger on one side than on the other; large chamfer; reveal at an oblique angle to the exterior face of the wall.

**Sprinkler, automatic.** Fire suppression device which operates automatically when its heat-actuated element is heated to or above its thermal rating, allowing water to discharge over a specific area.

**Sprinkler head.** Water outlet device in the sprinkler system that opens when a fire is detected.

**Sprinkler system.** Integrated system of piping connected to a water supply, with sprinklers that will automatically initiate water discharge over a fire area. When required, the sprinkler system also includes a control valve and a device for actuating an alarm when the system opens.

**Squatter.** Settler without authority.

**Stainless steel.** Steel with at least 4% chromium, which offers superior resistance to corrosion.

**Standards.** Uniform methods for testing, design, and practice. They are developed by consensus procedures that include public review and comments.

**Standpipe.** Vertical pipe system used to secure a uniform water pressure.

**Standpipe system.** Pipes with connections for water supply provided for the manual application of water to fires in buildings. Vertical standpipes with hose connections on each floor are typically located in stairwells.

**Steel.** Mixture of iron and carbon with varying amounts of other elements, primarily manganese, phosphorus, sulfur, and silicon.

**Steel frame.** Structure with a framework composed of steel columns and beams.

**Steel reinforcement.** Process that increases the strength of walls and keeps the wall thickness as slender as possible within practical limits.

**Stereometry.** Art of masonry. (*See also* Masonry.)

**Stick building.** Eclectic American style, mainly of cottage architecture, in the second half of the nineteenth century, predominately in wood, characterized by jagged, angular elements expressing exposed frame construction.

**Stiffness.** Structural property of resistance to bending.

**Stimulus.** Physical or environmental event that excites a receptor or sense organ.

**Street level.** That portion of a facility which is located at the same level as the adjacent street.

**Strength design method.** *See* Ultimate-strength design.

**Stress analysis.** In structural engineering, analytical determination of the stresses in the elements of a structure resulting from an applied load.

**String course.** Horizontal band of masonry, generally narrower than other courses, extending across the facade of a structure and in some instances encircling such decorative features as pillars or engaged columns; may be flush or projecting, and flat-surfaced, molded, or richly carved. (*See also* Facade.)

**Structural.** Of a load-bearing member or element of a building.

**Structural expression.** Outward manifestation of a tall building in which the logical nature of load transfer in the structural system is demonstrated.

**Structural form.** Shape or contour of the structural frame of tall buildings as distinguished from the color, texture, substance, and other visual patterns of the exterior skin.

**Structural masonry.** Masonry used for structural reasons, that is, to carry applied loads.

**Structural optimization.** Minimization of cost or quantity of material in a structure.

**Structural steel.** Steel, rolled in a variety of shapes (such as beams, angles, bars, plates, sheets, or strips) and fabricated for use as load-bearing structural members or elements.

**Structural system.** Three-dimensional complex assemblage of various combinations of interconnected structural elements.

**Structuralism.** Name often given to the expression of structure and space together, that integration which Louis Kahn defined as "Structure is the space."

**Strut.** Brace or any piece of a frame which resists thrusts in the direction of its own length; may be upright, diagonal, or horizontal.

**Subsystem.** Part of a larger system, but also being a system in itself.

**Superframe.** Structural frame used in tall building construction that resists all of the gravity loads and wind loads at the same time. The gravity loads are not directly supported by the frame, but are transferred through floor trusses located at approximately every 12 stories.

**Superstructure.** That part of a building or structure which is above the level of the adjoining ground or the level of the foundation.

**Supertall building.** Tall building over 70 to 80 stories high.

**Sustainable development.** Endeavor to apply scientific, engineering, and economic knowledge to assist in rectifying the negative consequences that the unrestricted proliferation of technology helped create.

**Sway.** Translational lateral motion of a building.

**Sweep.** Characteristic of any large, curving form, mass, or shape; that is, sweep of a curved wall.

**Symbolism.** Hidden meanings and associations attached to buildings, building types, or environments by individuals, groups, or cultures.

**Synergy.** Action of two or more substances, organs, or organisms to achieve an enhanced overall effect of which each is individually capable.

**System.** Collective description of a set of interconnected things, parts, or activities; a combination of mechanical-electrical products linked together to function as a single unit.

**Systems approach.** In building construction, prefabricated assemblies, components, and parts which are combined into single integrated units utilizing industrialized production techniques.

**Table form.** Slab set horizontally and carried on supports.

**Tabula rasa.** Literally, "erased plane." A concept of designing and building in which the ground plane or site is conceived as a "new" condition without regard to past or memory.

**Tee.** Beam composed of a stem and a flange in the form of a T.

**Tektonik.** Art of cutting and joining timber.

**Telecommunications.** Science or process of exchanging information by voice or electronic data using telephones, including wireless, and other electronic equipment.

**Tensile strength.** Maximum stress reached in a tensile test.

**Terra cotta.** Hard, unglazed fired clay; used for ornamental work and roof and floor tile; weighs approximately 1920 $kg/m^3$ (120 $lb/ft^3$).

**Terrace.** Raised level space or platform supported by a wall of masonry in a garden or park, or extension from a building, often used as a promenade provided with seats. (*See also* Masonry.)

**Territoriality.** Behavior based on perceived ownership of an environment or part of an environment. Territorial markers are most visible in personal, private environments such as homes; however, they can often be seen, albeit in more subtle forms, in public places such as park benches or train stations.

**Thermal environment.** Ambient heat in the air and radiant heat from any hot object such as the sun, fireplace, or radiator.

**Thermal regulator.** For a motor or motor compressor, a device that protects the motor against dangerous overheating due to either failure to start or overload.

**Thermocouple.** Electric thermometer consisting of a pair of suitable wires of different materials joined together so that an electric current is produced with a change of temperature.

**Tie.** Closed loop of reinforcing bars encircling the longitudinal steel in columns; also, a member in tension.

**Tile.** Ceramic surfacing unit, usually thin in relation to the facial area; made from clay or a mixture of clay and other ceramic materials; has either a glazed or an unglazed face.

**Tool.** In stonework a fluted, flat surface that usually carries 5 to 30 concave grooves per centimeter (2 to 12 per inch).

**Torsion.** Twisting of a structural member about its longitudinal axis by two equal and opposite torques, one at one end and the other at the opposite end.

**Transfer girder.** Beam designed to transfer column loads due to a difference in column spacing in the building grid.

**Travertine.** Type of striated marble, usually buff and white colored.

**Triangulation.** Process by which some external stimulus provides a link between people and prompts strangers to talk to each other.

**Tripartite.** Divided into three parts.

**Truss.** Subassemblage of structural members that is intended to resist the applied loads primarily through axial tension and compression.

**Tube-in-tube system.** Building with an inner core tube system and an exterior perimeter tube system.

**Tubular building.** Building employing a tubular structural concept.

**Tuned mass damper.** Damping system in which a large mass placed at the top of the building is tuned to have a period equal to the building's natural period and tends to oscillate with the same period as the building's, but in the opposite direction.

**Twisted high-strength reinforcement.** Steel bar used in concrete construction to provide additional strength.

**Typology.** Study of types, symbols, or symbolism.

**Ultimate-strength design.** In concrete design, method of proportioning structures or members to have failure capacities equal to or greater than the elastically computed moments, shears, and axial forces corresponding to a specified multiple of the working loads and assuming a nonlinear distribution of flexural stresses.

**Ultimate stress.** Maximum stress in a material under load.

**Urban.** Constituting or comprising a city or town.

**Urban conservation.** Protective measures designed to maintain the function and quality of an area by requiring adequate maintenance and preventing inappropriate developments or changes in the use of land and buildings for the purpose of preventing deterioration and blight.

**Urban core.** Central business district or downtown of an urban area.

**Urban design.** Plans for the structures, facilities, and services of a city (or a significant portion of it) that will meet the human activity needs thereof, including industrial, governmental, economic, social, cultural, and aesthetic; to fashion a city.

**Urban planning.** Science that deals with the establishment of guidelines, policies, and implementation measures designed to rationalize the use of natural, human, economic, and technical resources in order to assure an acceptable quality of life for the inhabitants of urban settlements.

**Urban redevelopment.** Clearance and rebuilding of structures that are deteriorated or obsolete, or both, in which the original function may be maintained, a new more appropriate use assigned, or a mixed use established.

**Urban rehabilitation.** Measures designed to improve a blighted or deteriorated area without total demolition, normally including the administrative control of land and building use for the purpose of restoring the area to its original function, bringing about a new, more appropriate use, or a combination of the two. Typical measures include installation of amenities in individual buildings, introduction of neighborhood buildings, completion, restoration, or replacement of building facades, and the like.

**Urban renewal.** Coordinated urban programs intended to improve blighted, deficient, and obsolete areas and their structures (especially housing), inadequate transportation, sanitation, and other services and facilities, nonconforming land use, traffic congestion, and the like through redevelopment, rehabilitation, and conservation. (*See* Urban redevelopment; Urban rehabilitation.)

**Urban (United States).** Communities with 2500 or more persons.

**Urban zone.** Generally, an area of concentrated building development and/or population density. It may be an area defined by a common system of transportation services and public amenities.

**Usonian.** Usonia, a term used frequently by Frank Lloyd Wright, was Samuel Butler's name for the United States. Wright referred to ideal, affordable houses and utopian villages as Usonian.

**Variable.** Factor that can vary in quantity or degree of impact. Variables may be dependent or independent.

**Variable air volume.** Constant-temperature HVAC system; differentiates airflow or volume rate.

**Vault.** Masonry covering over an area that uses the principle of the arch.

**Veneer.** Thin sheet of wood that has been sliced, rotary-cut, or sawn from a log; used as one of several plies in plywood for added strength or as a facing material on a less attractive wood. An outside wall facing of brick, stone, and such provides a decorative, durable surface, but is not load-bearing.

**Veranda.** Covered porch or balcony, extending along the outside of a building, planned for summer leisure.

**Vernacular.** Architectural style, language, or idiom of one's own country, city, or area.

**Vestibule.** Anteroom or small foyer leading into a larger space.

**Vibratory stress.** Strain caused by an oscillating, power-operated machine used to agitate fresh concrete so as to eliminate gross voids including entrapped air and to produce intimate contact with form surfaces and embedded materials.

**Vierendeel girder.** *See* Vierendeel truss.

**Vierendeel truss.** Frame or beam having rigid joints between verticals and chords and no diagonal member; also called Vierendeel girder. (*See also* Joint.)

**Virtual reality.** Computer simulation where perception of real-world experience is augmented by enhancing graphics and sound so as to portray a visually compelling sense of reality.

**Viscoelastic.** Having or exhibiting viscous and elastic properties.

**Vision glass.** Translucent sheet glass, usually having one face roughened.

**Visual impact.** Effect of scale, profile, and aesthetics of buildings on the sensibilities of a viewer.

**Visual perception.** Manner by which the physical world is understood visually by the viewer or user; intellectual understanding of the world based on the sensory acquisition of physical information through sight and vision.

**Vortex shedding.** Production of vortices (eddies) on alternate sides of slender exposed structures, caused by wind loads and leading to large vibrations.

**Waffle slab.** Two-way concrete joist floor construction.

**Wall tie.** In masonry, type of anchor used to secure facing to a backup wall; mortared into joints during setting.

**Water flow alarm.** Sounding device activated by a water flow detector or alarm check valve.

**Water flow detector.** Electric signaling indicator or alarm check valve actuated by water flow in one direction only.

**Way-finding.** Process of navigating and finding one's way around an environment.

**Weathering steel.** Steel with a chemical composition such that, when boldly exposed to the atmosphere, builds up a tight protective oxide film that resists further atmospheric corrosion.

**Web.** Vertical plate, or its equivalent, that joins the top and bottom flanges in a beam.

**Weep drainage system.** Small opening in a wall or window member, through which accumulated condensation or water may drain to the building exterior, as from the base of a cavity wall, a wall flashing, or a skylight.

**Well-point method.** Hollow rods with perforated intakes at their pointed lower ends, driven into the ground and connected to a pump to remove water at an excavation site.

**Wet-bulb temperature.** Temperature indicated by a mercury-in-glass thermometer wrapped in a damp wick whose far end is dipped in water. This is lower than the dry-bulb temperature of a thermometer without a wick, because the evaporation of the water on the wick cools the mercury bulb. The difference depends on the relative humidity, which can be determined by reading both the dry-bulb temperature and the wet-bulb temperature of two identical thermometers mounted side by side.

**Wet system.** System employing automatic sprinklers attached to a piping system containing water and connected to a water supply so that water discharges immediately from sprinklers opened by a fire.

**Wide-flange section.** Structural beam of rolled steel or concrete having a shape whose cross section resembles the letter H; has wider flanges than an I-beam. (*See also* Rolled beam; Concrete.)

**Wind bracing.** Any brace, such as a strut, which strengthens a structure or framework against the wind.

**Wind drag.** To pull or draw with force as a result of wind.

**Wind load.** Total force exerted by the wind on a structure or part of a structure.

**Working-stress design method.** Method of design in which structures or members are proportioned for prescribed working loads at stresses that are well below their ultimate values. A linear distribution of flexural stresses is assumed.

**Wrought iron.** Commercially pure iron of fibrous nature; valued for its corrosion resistance and ductility; used for water pipes, water tank plates, rivets, stay bolts, and forged work.

**Wythe.** Each continuous vertical section of a wall one masonry unit in thickness.

**X brace.** Braces which cross each other diagonally to form the letter X. (*See* Brace.)

**Yield point.** Point during increasing stress at which the proportion of stress to strain becomes substantially less than it has been at smaller values of stress. Below this point the stress-strain curve may be assumed to be linear, above it is usually nonlinear.

**Yield stress.** Stress at the yield point of a material, that is, at which an increase in strain occurs without a significant increase in stress.

**Ziggurat.** Ancient Assyrian or Mesopotamian pyramidal temple in which the stories become progressively smaller with height. Some versions had a continuous ramp to the top.

**Zone.** In an air-conditioning, heating, or lighting system, a space, group of spaces, or circuit served by the system, whose actions are regulated by a single control. In urban planning, a region with a specific use. (*See* Zoning; Urban zone.)

**Zoning.** Legal regulation of the use of land and buildings, in which the density of population and the height, bulk, and spacing of structures are also specified.

**Zoning ordinance.** Legal instrument that establishes regulations governing the use of land and the location, height, use, and land coverage of buildings in the various use districts.

**Zoning variance.** Permission granted as relief from some specific and unusual hardship imposed by the strict interpretation of the zoning ordinance.

## ABBREVIATIONS

| | |
|---|---|
| ACI | American Concrete Institute |
| ADA | Americans with Disabilities Act |
| A-E | Architect-engineer |
| AES | *Architecture and Engineering* series |
| AI | Artificial intelligence |
| AIA | American Institute of Architects |
| AISC | American Institute of Steel Construction |
| AISE | Association of Iron and Steel Engineers |
| AISI | American Iron and Steel Institute |
| ANSI | American National Standards Institute |
| ASC | Access-control system |
| ASCE | American Society of Civil Engineers |
| ASHRAE | American Society of Heating, Refrigerating and Air Conditioning Engineers |
| ASM | American Society of Metals |
| ASTM | American Society for Testing and Materials |
| BAS | Building automation systems |
| BIA | Brick Institute of America |
| BIG/STIG | Biomass-integrated gasifier/steam injected gas turbine |
| BMS | Building management systems |
| BOCA | Building Officials Conference of America |
| BOD | Biological oxygen demand |
| BSE | Building services engineering |
| CAD | Computer-aided design |
| CAE | Computer-aided engineering |
| CAM | Computer-aided manufacturing |
| CCTV | Closed-circuit television system |
| CD-ROM | Compact disc–read-only memory |
| CIAM | Congrès International d'Architecture Moderne |
| CIB | International Council on Building Research |
| CNR | National Council for Research |
| CPM | Critical-path method |
| CRT | Cathode-ray tube |
| CSF | Condensed silica fume particle concrete |
| DBT | Dry-bulb temperature |
| DDC | Direct digital controllers |
| DIT | Dacca (Dhaka) Improvement Trust |

| | |
|---|---|
| EBF | Eccentric braced frame |
| ECCS | European Convention for Constructional Steelwork |
| ENR | *Engineering News Record* |
| EPIDC | East Pakistan Industrial Development Corporation |
| FAR | Floor area ratio |
| FCC | Fire control center |
| FIP/CEB | Federation Internationale de la Précontrainte (International Federation for Prestressing) Comité Européen du Béton (European Concrete Committee), since 1976 Comité Euro-International du Béton |
| HOK | Hellmuth, Obata & Kassabaum, Inc. |
| HPC | High-performance concrete |
| HSC | High-strength concrete |
| HSLA | High-strength low-alloy steel |
| HSSFC | High-strength silica fume concrete |
| HUD | Housing and Urban Development (U.S.) |
| HVAC | Heating, ventilating, and air-conditioning |
| IBA | International Building Exhibition |
| IBC | International Business Center |
| IBST | Intelligent building systems technology |
| ICBO | International Conference of Building Officials |
| IISI | International Iron and Steel Institute |
| IIT | Illinois Institute of Technology |
| IMI | International Masonry Institute |
| IREM | Institute of Real Estate Management |
| LAN | Local-area network |
| LDC | Lucky Development Company |
| LRFD | Load and resistance factor design |
| LSD | Limit-states design |
| MCA | Motijheel commercial area |
| MCAA | Masonry Contractors Association of America |
| MDF | Macro-defect-free cement |
| M-E | Mechanical-electrical |
| MRC | Masonry Research Committee |
| MRT | Mean radiant temperature |
| NBC | National Building Code of Canada |
| NBS | National Bureau of Standards |
| NCMA | National Concrete Masonry Association |
| NFPA | National Fire Protection Association |
| NIST | National Institute of Standards and Technology |
| PPS | Project for Public Spaces |

| | |
|---|---|
| PWD | Public Works Department |
| RAJUK | RAJdhani Unayan Kartiphakkha |
| RBM | Reinforced brick masonry |
| RH | Relative humidity |
| S&H | Syska & Hennessy |
| SCPRF | Structural Clay Products Research Foundation |
| SOM | Skidmore, Owings & Merrill |
| SV | Specific volume |
| TBE | European Association of Brick and Tile Manufacturers |
| TMS | The Masonry Society |
| UBC | Uniform Building Code |
| UFAS | Uniform Federal Accessibility Standards |
| URA | Urban Redevelopment Authority |
| USD | Ultimate-strength design |
| VAV | Variable air volume |
| WAPDA | Water and Power Development Authority |
| WBT | Wet-bulb temperature |
| WSD | Working-stress design |

## UNITS

In the table below are given conversion factors for commonly used units. The numerical values have been rounded off to the values shown. The British (Imperial) System of unites is the same as the American System except where noted. Le Système International d'Unités (abbreviated "SI") is the name formally given in 1960 to the system of units partly derived from, and replacing, the old metric system.

| SI | American | Old metric |
|---|---|---|
| | Length | |
| 1 mm | 0.03937 in. | 1 mm |
| 1 m | 3.28083 ft | 1 m |
| | 1.093613 yd | |
| 1 km | 0.62137 mile | 1 km |
| | Area | |
| 1 mm$^2$ | 0.00155 in.$^2$ | 1 mm$^2$ |
| 1 m$^2$ | 10.76392 ft$^2$ | 1 m$^2$ |
| | 1.19599 yd$^2$ | |
| 1 km$^2$ | 247.1043 acres | 1 km$^2$ |
| 1 hectare | 2.471 acres(1) | 1 hectare |

| | Volume | |
|---|---|---|
| 1 $cm^3$ | 0.061023 $in.^3$ | 1 cc |
| | | 1 ml |
| 1 $m^3$ | 35.3147 $ft^3$ | 1 $m^3$ |
| | 1.30795 $yd^3$ | |
| | 264.172 gal(2) liquid | |

| | Velocity | |
|---|---|---|
| 1 m/sec | 3.28084 ft/sec | 1 m/sec |
| 1 km/hr | 0.62137 mi/hr | 1 km/hr |

| | Acceleration | |
|---|---|---|
| 1 $m/sec^2$ | 3.28084 $ft/sec^2$ | 1 $m/sec^2$ |

| | Mass | |
|---|---|---|
| 1 g | 0.035274 oz | 1 g |
| 1 kg | 2.2046216 lb(3) | 1 kg |

| | Density | |
|---|---|---|
| 1 $kg/m^3$ | 0.062428 $lb/ft^3$ | 1 $kg/m^3$ |

| | Force, weight | |
|---|---|---|
| 1 N | 0.224809 lbf | 0.101972 kgf |
| 1 kN | 0.1124045 tons(4) | |
| 1 MN | 224.809 kips | |
| 1 kN/m | 0.06853 kips/ft | |
| 1 $kN/m^2$ | 20.9 $lbf/ft^2$ | |

| | Torque, bending moment | |
|---|---|---|
| 1 N-m | 0.73756 lbf-ft | 0.101972 kgf-m |
| 1 kN-m | 0.73756 kip-ft | 101.972 kgf-m |

| | Pressure, stress | |
|---|---|---|
| 1 $N/m^2$ = 1 Pa | 0.000145038 psi | 0.101972 $kgf/m^2$ |
| 1 $kN/m^2$ = 1 kPa | 20.8855 psf | |
| 1 $MN/m^2$ = 1 MPa | 0.145038 ksi | |

| | Viscosity (dynamic) | |
|---|---|---|
| 1 $N\text{-}sec/m^2$ | 0.0208854 $lbf\text{-}sec/ft^2$ | 0.101972 $kgf\text{-}sec/m^2$ |

| | Viscosity (kinematic) | |
|---|---|---|
| 1 $m^2/sec$ | 10.7639 $ft^2/sec$ | 1 $m^2/sec$ |

| | Energy, work | |
|---|---|---|
| 1 J = 1 N-m | 0.737562 lbf-ft | 0.00027778 W-hr |
| 1 MJ | 0.37251 hp-hr | 0.27778 kW-hr |
| | **Power** | |
| 1 W = 1 J/sec | 0.737562 lbf-ft/sec | 1 W |
| 1 kW | 1.34102 hp | 1 kW |
| | **Temperature** | |
| K = 273.15 + °C | °F = (°C × 1.8) + 32 | °C = (°F − 32)/1.8 |
| K = 273.15 + 5/9 (°F − 32) | | |
| K = 273.15 + 5/9 (°R − 491.69) | | |

(1) Hectare as an alternative for $km^2$ is restricted to land and water areas.
(2) 1 $m^3$ = 219.9693 Imperial gallons.
(3) 1 kg = 0.068522 slugs.
(4) 1 American ton = 2000 lb; 1 kN = 0.1003612 Imperial ton; 1 Imperial ton = 2240 lb.

## Abbreviations for Units

| | | | |
|---|---|---|---|
| Btu | British thermal unit | kW | kilowatt |
| °C | degree Celsius (centigrade) | lb | pound |
| $cm^3$ | cubic centimeter | lbf | pound force |
| cm | centimeter | $lb_m$ | pound mass |
| °F | degree Fahrenheit | MJ | megajoule |
| ft | foot | MPa | megapascal |
| g | gram | m | meter |
| gal | gallon | mi | mile |
| hp | horsepower | ml | milliliter |
| hr | hour | mm | millimeter |
| Imp | British Imperial | MN | meganewton |
| in. | inch | N | newton |
| J | joule | oz | ounce |
| K | kelvin | Pa | pascal |
| kg | kilogram | psf | pound per square foot |
| kgf | kilogram-force | psi | pound per square inch |
| kip | 1000 pound force | °R | degree Rankine |
| km | kilometer | sec | second |
| kN | kilonewton | slug | 14.594 kg |
| kPa | kilopascal | W | watt |
| ksi | kips per square inch | yd | yard |

# References/Bibliography


Abelson, P. H., and Dorfman, M. (Eds.), 1982
COMPUTERS AND ELECTRONICS, *Science,* vol. 215, no. 4534.

ACI, 1979
SUPER-PLASTICIZERS IN CONCRETE, Special Publication SP-62, American Concrete Institute, Detroit, Mich.

ACI 423.3R/ASCE Committee, 1982
RECOMMENDATIONS FOR CONCRETE MEMBERS PRESTRESSED WITH UNBONDED TENDONS, American Concrete Institute, Detroit, Mich.

ACI 318-89, 1989
BUILDING CODE REQUIREMENTS FOR REINFORCED CONCRETE, American Concrete Institute, Detroit, Mich.

ACI, 1990
HIGH-STRENGTH CONCRETE—SECOND INTERNATIONAL SYMPOSIUM, Special Publication SP-121, American Concrete Institute, Detroit, Mich.

ACI, 1991
MANUAL OF CONCRETE PRACTICE, American Concrete Institute, Detroit, Mich.

ACI 530-92/ASCE 5-92/TMS 402-92 Joint Committees, 1992
BUILDING CODE REQUIREMENTS FOR MASONRY STRUCTURES, American Concrete Institute, Detroit, Mich.

ACE Committee 439, 1989
STEEL REINFORCEMENT—PHYSICAL PROPERTIES AND U.S. AVAILABILITY, *ACI Materials Journal,* vol. 86, January-February.

Adler, D., 1893
THE ARCHITECTURE OF JOHN WELLBORN ROOT, Chicago Tribune LII, October 29; quoted in D. Hoffman, Johns Hopkins University Press, Baltimore, Md., 1967.

AISC, 1973
THE DESIGN, FABRICATION, AND ERECTION OF HIGHLY RESTRAINED CONNECTIONS TO MINIMIZE LAMELLAR TEARING, *Engineering Journal,* vol. 10, no. 3, American Institute of Steel Construction, Chicago, Ill.

AISC, 1978
SPECIFICATIONS FOR THE DESIGN, FABRICATION AND ERECTION OF STRUCTURAL STEEL FOR BUILDINGS, American Institute of Steel Construction, Chicago, Ill.

AISC, 1986
LOAD AND RESISTANCE FACTOR DESIGN SPECIFICATION FOR STRUCTURAL STEEL BUILDINGS, American Institute of Steel Construction, Chicago, Ill., September 1.

AISC, 1989
STEEL CONSTRUCTION MANUAL—ASD, American Institute of Steel Construction, Chicago, Ill.

AISC, 1990a
REINTERPRETING AN ANCIENT FORM, *Modern Steel Construction,* vol. 30, no. 5, American Institute of Steel Construction, Chicago, Ill., pp. 24–28.

AISC, 1990b
STEMFIRE: STEEL MEMBER FIRE PROTECTION PROGRAM, American Institute of Steel Construction, Chicago, Ill.

AISI, 1979
FIRE-SAFE STRUCTURAL STEEL: A DESIGN GUIDE, American Iron and Steel Institute, Washington, D.C.

AISI, 1980
DESIGNING FIRE PROTECTION FOR STEEL COLUMNS, American Iron and Steel Institute, Washington, D.C.

AISI, 1981
DESIGNING FIRE PROTECTION FOR STEEL TRUSSES, American Iron and Steel Institute, Washington, D.C.

AISI, 1984
DESIGNING FIRE PROTECTION FOR STEEL BEAMS, American Iron and Steel Institute, Washington, D.C.

AISI, 1985
STEEL PRODUCT MANUAL—PLATES, American Iron and Steel Institute, Washington, D.C.

Aitcin, P. C., and LaPlamte, P., 1992
THE DEVELOPMENT OF HIGH PERFORMANCE CONCRETE IN NORTH AMERICA, in *High Performance Concrete from Material to Structures,* Proceedings of Conference, Cachan, France, September; Chapman and Hall, London, England, pp. 412–420.

Alam, M. A., 1991
DHAKA: PLANNING HOUSING AND THE ENVIRONMENT. SHELTER AND THE LIVING ENVIRONMENT, Ministry of Public Works, Government of Bangladesh, October.

Ali, M. M., 1980
FAILURE INVESTIGATION OF MASONRY STRUCTURES, Proceedings of 2d Canadian Masonry Symposium, Carleton University, Ottawa, Ont., Canada.

Ali, M. M., 1981
CARRYING CAPACITY OF A TWO-PHASE MATERIAL IN AXIAL COMPRESSION, Proceedings of Canadian Congress of Applied Mechanics (CANCAM-81), Moncton, N.B., Canada.

Ali, M. M., 1986a
SEISMIC DESIGN OF STEEL TALL BUILDINGS, Proceedings of 3d U.S. National Conference on Earthquake Engineering, Charleston, S.C., August.

Ali, M. M., 1986b
THE FORGOTTEN HALF OF DESIGN: STRUCTURAL ART FORM, Proceedings of Association of Collegiate Schools of Architecture (ACSA), West Central Regional Meeting, Champaign, Ill., September.

Ali, M. M., 1987a
ACADEMIC VS. INDUSTRIAL RESEARCH IN ARCHITECTURAL ENGINEERING, Proceedings of American Society for Engineering Education (ASEE) Annual Conference, Reno, Nev., pp. 239–242.

Ali, M. M., 1987b
INFLUENCE OF ARCHITECTURAL CONFIGURATION ON SEISMIC RESPONSE OF BUILDINGS, Proceedings of 5th Canadian Conference on Earthquake Engineering, Ottawa, Ont., Canada.

Ali, M. M., 1988a
CURRENT TRENDS IN THE USE OF ADVANCED COMPUTER TECHNOLOGY FOR BUILDING DESIGN AND CONSTRUCTION, Proceedings of American Society for Engineering Education (ASEE) Annual Conference, Portland, Ore.

Ali, M. M., 1988b
IMPACT OF COMPUTERS ON ARCHITECTURAL ENGINEERING RESEARCH AND EDUCATION, *International Journal of Applied Engineering Education,* vol. 4, no. 5, pp. 461–465.

Ali, M. M., 1989
STRUCTURAL SYSTEM AND MATERIAL SELECTION FOR A 40-STORY MULTI-USE BUILDING IN CHICAGO, Proceedings of International Conference on High-Rise Buildings, Nanjing, China, March 25–27.

Ali, M. M., 1990
INTEGRATION OF STRUCTURAL FORM AND ESTHETICS IN TALL BUILDING DESIGN: THE FUTURE CHALLENGE, Proceedings of the World Congress on Tall Buildings, Hong Kong, November.

Ali, M. M., 1993a
DEVELOPMENTS IN THE APPLICATION OF COMPUTER TECHNOLOGY TO BUILDING ENGINEERING, Proceedings of Symposium on Computer Integration of the Building Industry, Dallas, Tex., June 10–12.

Ali, M. M., 1993b
MASS RAPID TRANSIT SYSTEM FOR DHAKA CITY, *The Bangladesh Observer,* Dhaka, Bangladesh, August 30.

Ali, M. M., and Ang, T., 1984a
COMPARATIVE EVALUATION OF HIGH-RISE BUILDINGS IN SINGAPORE, Technical Report, National Iron & Steel Mills Ltd., Singapore.

Ali, M. M., and Ang, T., 1984b
COST EFFECTIVENESS OF TALL BUILDINGS IN SINGAPORE—STRUCTURAL STEEL VS. CONCRETE, Proceedings of International Conference on Tall Buildings, Singapore, October.

Ali, M. M., and Ang, T., 1985
INFLUENCE OF MATERIAL SELECTION ON TALL BUILDING ECONOMICS, *Construction Industry Development Board Review Journal,* vol. 1, no. 2, Singapore, October-December.

Ali, M. M., and Anway, R. R., 1988
BUILDING TECHNOLOGY FORECAST AND EVALUATION OF STRUCTURAL SYSTEMS, Technical Report, Contract RS DACA 88-86-D-0014-35, U.S. Army Construction Engineering Research Laboratory (CERL), Champaign, Ill., May.

Ali, M. M., and Napier, T., 1989
EVALUATION TECHNIQUE FOR CONCRETE STRUCTURAL SYSTEMS, *ACI Concrete International Magazine,* July.

Almand, K., 1989
FIRE RESISTANCE REQUIREMENTS IN BUILDING REGULATIONS: AN INTERNATIONAL COMPARISON, Melbourne Research Laboratories, Broken Hill Proprietary Co. Ltd., Australia.

Altman, I., 1975
THE ENVIRONMENT AND SOCIAL BEHAVIOR, Brooks/Cole Publishing, Monterey, Calif.

Amrhein, J. E., 1986
SLENDER WALLS RESEARCH PROGRAM BY CALIFORNIA STRUCTURAL ENGINEERS, in *Advances in Tall Buildings,* L. S. Beedle (Ed.), Council on Tall Buildings and Urban Habitat, Lehigh University, Bethlehem, Pa.

Amrhein, J. E., 1991
REINFORCED GROUTED BRICK MASONRY, 13th edition, Masonry Institute of America, Los Angeles, Calif.

Amrhein, J. E., 1992
REINFORCED MASONRY ENGINEERING HANDBOOK: CLAY AND CONCRETE MASONRY, Masonry Institute of America, Los Angeles, Calif.

Anderson, C., and Karabatsos, J., 1991
TRAIN TRACKS DETOUR BUILDING'S FOUNDATION, *Modern Steel Construction,* vol. 31, no. 4, American Institute of Steel Construction, Chicago, Ill., pp. 32–38.

Anderson, D. L., Nathan, N. D., Cherry, S., and Gajer, R. B., 1980
SEISMIC DESIGN OF REINFORCED CONCRETE MASONRY WALLS, Proceedings of 2d Canadian Masonry Symposium, Carleton University, Ottawa, Ont., Canada.

Angerer, E. W., 1969
APPEAL AND POTENTIAL OF THE POLYFUNCTIONAL CHARACTERISTICS OF ENGINEERED BRICKWORK, in *Design Engineering and Construction with Masonry Products,* F. B. Johnson (Ed.), Gulf Publishing, Houston, Tex.

Anonymous 1930
SKYSCRAPERS: PROPHECY IN STEEL, Fortune, Time, Inc., American Institute of Steel Construction, New York.

ANSI A117.1, 1980
AMERICAN NATIONAL STANDARD SPECIFICATIONS FOR MAKING BUILDINGS AND FACILITIES ACCESSIBLE TO AND USABLE BY PHYSICALLY HANDICAPPED PEOPLE, American National Standards Institute, New York, N.Y.

Anthony, K. H., 1981
INTERNATIONAL HOUSE: HOME AWAY FROM HOME?, Ph.D. Dissertation, Department of Architecture, University of California, Berkeley.

Anthony, K. H., 1985a
PUBLIC PERCEPTIONS OF RECENT PROJECTS: A BERKELEY CLASS CONDUCTS EVALUATIONS OF FIVE BUILDINGS AND SPACES, *Architecture,* March, pp. 93–99.

Anthony, K. H., 1985b
THE SHOPPING MALL: A TEENAGE HANGOUT, *Adolescence,* vol. 20, no. 78, Summer, pp. 307–312.

Appleyard, D., 1969
WHY BUILDINGS ARE KNOWN, *Environment and Behavior,* vol. 1, no. 2, December, pp. 131–156.

Appleyard, D., 1976
PLANNING A PLURALIST CITY, MIT Press, Cambridge, Mass.

Appleyard, D., 1979
THE ENVIRONMENT AS A SOCIAL SYMBOL, *Ekistics,* vol. 278, September/October, pp. 272–281.

Appleyard, D., and Fishman, L., 1977
HIGH-RISE BUILDINGS VERSUS SAN FRANCISCO: MEASURING VISUAL AND SYMBOLIC IMPACTS, in *Human Response to Tall Buildings,* D. Conway (Ed.), Dowden, Hutchinson & Ross, Stroudsburg, Pa., pp. 81–100.

Appleyard, D., Lynch, K., and Myer, J. R., 1964
THE VIEW FROM THE ROAD, MIT Press, Cambridge, Mass.

Architectural Book Publishing, 1974
ARCHITECTURE OF SKIDMORE, OWINGS & MERRILL—1963–1973.

*Architectural Forum,* 1958
INLAND'S STEEL SHOWCASE: THE FIRST MODERN OFFICE BUILDINGS IN THE CHICAGO LOOP, vol. 108, April, pp. 88–93.

*Architectural Forum,* 1968
THINK MIES: IBM'S NEW CHICAGO HEADQUARTERS, vol. 129, September, p. 97.

*Architectural Record,* 1958
INLAND STEEL BUILDING, CHICAGO, vol. 123, April, pp. 69–178.

*Architectural Record,* 1974
THE HIGH RISE OFFICE BUILDINGS—THE PUBLIC SPACES THEY MAKE, vol. 155, no. 3, March.

*Architectural Record,* 1985
THE BANK OF CHINA, HONG KONG, September, p. 136.

Archival File of Press Releases, 1988–1992
WATTERSON TOWERS, Illinois State University, Normal.

Arcidi, 1991a
PAOLO SOLERI'S ARCOLOGY: UPDATING THE PROGNOSIS, *Progressive Architecture,* March, pp. 95–100.

Arcidi, 1991b
PROJECTS: NINE PROPOSALS ON BEHALF OF THE ENVIRONMENT, *Progressive Architecture,* March, pp. 76–78.

Aregger, H., and Glaus, O., 1975
HIGHRISE BUILDING AND URBAN DESIGN, Frederick A. Praeger, New York, N.Y.

Arnold, C., 1986
A GUIDE TO EARTHQUAKE MOTION AND ITS EFFECT ON BUILDINGS, *Journal of Architectural and Planning Research,* vol. 3, pp. 305–313.

Arnold, C., and Reitherman, R., 1982
BUILDING CONFIGURATIONS AND SEISMIC DESIGN, John Wiley, New York, N.Y.

ASCE, 1988
WIND DRIFT DESIGN OF STEEL-FRAMED BUILDINGS, ASCE Task Committee on Drift Control of Steel Building Structures of the Committee on Design of Steel Building Structures, *ASCE Journal of Structural Engineering,* vol. 114, no. 9, September 1.

ASHRAE 55-74, 1974
THERMAL ENVIRONMENT CONDITIONS FOR HUMAN OCCUPANCY, American Society of Heating, Refrigerating, and Air Conditioning Engineers, New York, N.Y.

ASHRAE 90-75, 1975
ENERGY CONSERVATION IN NEW BUILDING DESIGN, American Society of Heating, Refrigerating, and Air Conditioning Engineers, New York, N.Y.

ASHRAE, 1977
ASHRAE HANDBOOK 1977, FUNDAMENTALS, American Society of Heating, Refrigerating, and Air Conditioning Engineers, New York, N.Y.

ASHRAE, 1989
ASHRAE HANDBOOK, FUNDAMENTALS, American Society of Heating, Refrigerating, and Air Conditioning Engineers, New York, N.Y.

ASHRAE, 1992
ASHRAE HANDBOOK, HVAC SYSTEMS AND EQUIPMENT, American Society of Heating, Refrigerating, and Air Conditioning Engineers, New York, N.Y.

ASM, 1978
PROPERTIES AND SELECTION: IRONS AND STEELS, in *Metals Handbook,* vol. 1, American Society of Metals, Metals Park, Ohio.

ASM, 1985
CARBON AND ALLOY STEELS, in *Metals Handbook,* vol. 1, American Society of Metals, Metals Park, Ohio.

Attoe, W., 1981
SKYLINES: UNDERSTANDING AND MOLDING URBAN SILHOUETTES, John Wiley, New York, N.Y.

Aynsley, R. M., 1973
THE ENVIRONMENT AROUND TALL BUILDINGS, Proceedings of 12th National/Regional Conference, Sydney, Australia, August; Lehigh University, Bethlehem, Pa., pp. 256–274.

Bacon, E. N., 1976
DESIGN OF CITIES, Viking Penguin, New York, N.Y.

Bain, B. A., 1989
AN APPROACH TO BUILDINGS: THE ENTRY SEQUENCE, Masters Thesis, School of Architecture, University of Illinois at Urbana-Champaign, Champaign.

Banavalkar, P. V., 1991
SPINE STRUCTURES PROVIDE STABILITY IN SEISMIC AREAS, *Modern Steel Construction,* vol. 31, no. 1, American Institute of Steel Construction, Chicago, Ill., January.

Bangladesh Bureau of Statistics, 1986
1986 STATISTICAL YEARBOOK OF BANGLADESH, December.

Banham, R., 1960
THEORY AND DESIGN IN THE FIRST MACHINE AGE, MIT Press, Cambridge, Mass.

Barnett, J., 1973
THE FUTURE OF TALL BUILDINGS: SYSTEMS AND CONCEPTS, in *Planning and Design of Tall Buildings,* Proceedings of Conference, Lehigh University, August 1972, vol. Ia; American Society of Civil Engineers, New York, N.Y., pp. 85–90.

Barnett, J., 1986
THE ELUSIVE CITY: FIVE CENTURIES OF DESIGN, AMBITION AND MISCALCULATION, Harper & Row, New York, N.Y.

Barsom, J. M., 1987
MATERIAL CONSIDERATION IN STRUCTURAL STEEL DESIGN, *Engineering Journal,* vol. 24, no. 3, American Institute of Steel Construction, Chicago, Ill.

Bartone, C., 1991
ENVIRONMENTAL CHALLENGE IN THIRD WORLD CITIES, *Journal of the American Planning Association,* vol. 57, no. 4, Autumn, American Planning Association.

Beall, C., 1987
MASONRY DESIGN AND DETAILING, McGraw-Hill, New York, N.Y.

Bechtel Civil Inc., 1989
LUCKY-GOLDSTAR INTERNATIONAL BUSINESS CENTER: CONCEPTUAL DESIGN REPORT, San Francisco, Calif., May.

Becker, F., 1990
THE TOTAL WORKPLACE: FACILITIES MANAGEMENT AND THE ELASTIC ORGANIZATION, Van Nostrand Reinhold, New York, N.Y.

Bedford, T., 1964
BASIC PRINCIPLES OF VENTILATION AND HEATING, H. K. Lewis, London, England.

Bednar, M. J., 1986
THE NEW ATRIUM, McGraw-Hill, New York, N.Y.

Beedle, L. S., 1971
WHAT'S A TALL BUILDING? Preprint no. 1553 (M20), ASCE Annual and Environmental Engineering Meeting, St. Louis, Mo., October; American Society of Civil Engineering, New York, N.Y.

Beedle, L. S., 1974
ON TALL BUILDINGS AND THE ESTHETIC ENVIRONMENT, Proceedings of Conference on Tall Buildings, Kuala Lumpur, Malaysia, December 2–5; Institution of Engineers, Kuala Lumpur, Malaysia, pp. 1-1–1-6.

Beedle, L. S. (Ed.), 1992
TALL BUILDINGS AROUND THE WORLD, Keynote Paper, International Conference on Tall Buildings, Kuala Lumpur, Malaysia, July 28–30.

Beedle, L. S., and Iyengar, H., 1982
SELECTED WORKS OF FAZLUR R. KHAN: FIRST WISCONSIN CENTER, IABSE Structures C-23, pp. 72–73.

*Beijing Review,* 1985
HIGH-RISES SPELL ISOLATION FOR RESIDENTS, vol. 28, August 19, p. 27.

Benson, R., 1987
THE STATE OF SPACE, *Inland Architect,* March/April, pp. 55–61.

Benton, T., 1987
LE CORBUSIER: ARCHITECT OF THE CENTURY, Arts Council of Great Britain and authors, Balding and Mansell, UK Ltd., Great Britain.

Berger, P., 1993
LAST OF THE TOWERS, *Inland Architect*, vol. 37, no. 3, May/June, pp. 48–53.

Berman, M., 1982
ALL THAT IS SOLID MELTS INTO AIR: THE EXPERIENCE OF MODERNITY, Simon and Schuster, New York, N.Y.

BIA, 1969
RECOMMENDED PRACTICE FOR ENGINEERED BRICK MASONRY, Brick Institute of America, McLean, Va.

BIA, 1990
BUILDING CODE REQUIREMENTS FOR MASONRY STRUCTURES ACI 530/ASCE 5 AND SPECIFICATIONS FOR MASONRY STRUCTURES ACI 530/ASCE 6, Technical Notes on Brick Construction, no. 3, Brick Institute of America, McLean, Va.

Bigart, H., 1965
A NIGHT OF CONFUSION, FRUSTRATION AND ADVENTURE, *The New York Times*, Thursday, November 11, pp. 1, 37.

Bill, M., 1949
ROBERT MAILLART, Les Editions d'Architecture SA., Erlanbach-Zurich, Switzerland.

Billington, D. P., 1985
THE TOWER AND THE BRIDGE: THE NEW ART OF STRUCTURAL ENGINEERING, Princeton University Press, Princeton, N.J.

Billington, D. P., 1986
ENGINEER AS ARTIST—FROM ROEBLING TO KHAN, in *Technique and Aesthetics in the Design of Tall Buildings*, Institute for the Study of the High-Rise Habitat, Lehigh University, Bethlehem, Pa.

Billington, D. P., and Goldsmith, M. (Eds.), 1986
TECHNIQUE AND AESTHETICS IN THE DESIGN OF TALL BUILDINGS, Institute for the Study of the High-Rise Habitat, Lehigh University, Bethlehem, Pa.

Bjerkeli, L., Tomaszewicz, A., and Jensen, J. J., 1990
DEFORMATION PROPERTIES AND DUCTILITY OF HIGH STRENGTH CONCRETE, in *High-Strength Concrete—2d International Symposium*, Publication SP-121, American Concrete Institute, Detroit, Mich., pp. 215–238.

Blake, P., 1974
THE RISE OF A MODERN ARCHITECTURE, *Dialogue*, vol. 7, no. 3.

Blakeslee, S., 1992
NAVIGATING LIFE'S MAZE: STYLES SPLIT THE SEXES, *The New York Times*, Tuesday, May 26, pp. B1, B8.

Blaser, W., 1965
MIES VAN DER ROHE, Fredrick A. Praeger, Inc., New York.

Block, A. B., 1985
CASUALTY REPORTS, *Forbes*, vol. 135, May 20, p. 170.

BOCA, 1978
THE BOCA BASIC BUILDING CODE, 7th edition, Building Officials and Code Administrators International, Chicago, Ill.

BOCA, 1993
BASIC BUILDING CODE, Building Officials and Code Administrators International, Chicago, Ill.

Boles, D. D., 1988
FRANKFURT MIXED-USE COMPLEX, *Progressive Architecture*, Reinhold Publishing, Cleveland, Ohio, January, pp. 96–98.

Boles, D. D., 1989
MADE IN MINNEAPOLIS, *Progressive Architecture*, Reinhold Publishing, Cleveland, Ohio, March.

Boles, D. D., and Murphy, J., 1985
CINCINNATI CENTERPIECE, *Progressive Architecture,* Reinhold Publishing, Cleveland, Ohio, October.

Bonavia, 1986
THE HONG KONG BANK: INTRODUCTION, *Progressive Architecture,* vol. 67, no. 3, Reinhold Publishing, Cleveland, Ohio, March, pp. 67–74.

Borchelt, J. G., and Wakefield, D., 1980
WRITING A DESIGN STANDARD: THE MASONRY SOCIETY'S BUILDING STANDARD, Proceedings of 2d Canadian Masonry Symposium, Ottawa, Ont., Canada.

Bosselman, P., 1983
SHADOWBOXING—KEEPING THE SUNLIGHT ON CHINATOWN'S KIDS, *Landscape Architecture,* vol. 73, July, pp. 74–76.

Bosselman, P., Flores, J., and O'Hare, T., 1983
SUN AND LIGHT FOR DOWNTOWN SAN FRANCISCO, Environmental Simulation Laboratory, Institute of Urban and Regional Development, College of Environmental Design, University of California, Berkeley.

Boubekri, M., Hulliv, R. B., and Boyer, L. L., 1991
IMPACT OF WINDOW SIZE AND SUNLIGHT PENETRATION ON OFFICE WORKERS' MOOD AND SATISFACTION, *Environment and Behavior,* vol. 23, no. 4, July, pp. 474–493.

Bouwcentrum/Rotterdam, 1963
MODERN STEEL CONSTRUCTION IN EUROPE, Elsevier, Amsterdam, Netherlands.

Brady, D., and English, M., 1993
COLOR, *Inland Architect,* vol. 37, no. 2, March/April, pp. 42–45.

Branch, M., 1987
STATE OF ILLINOIS CENTER UPDATE, *Progressive Architecture,* November, pp. 27–28.

Branch, M. A., 1993
THE STATE OF SUSTAINABILITY, *Progressive Architecture,* vol. 74, no. 6, Reinhold Publishing, Cleveland, Ohio, June, pp. 47–51.

Bresler, B. L., and Scalzi, J. B., 1968
DESIGN OF STEEL STRUCTURES, John Wiley, New York, N.Y.

Brill, M., with Margulis, S. T., Konar, E., and BOSTI in association with Westinghouse Furniture Systems, 1984
USING OFFICE DESIGN TO INCREASE PRODUCTIVITY, vols. 1 and 2, Workplace Design and Productivity, Buffalo, N.Y.

Bristol, K. G., 1991
THE PRUITT-IGOE MYTH, *Journal of Architectural Education,* vol. 44, no. 3, May, pp. 163–171.

Broadbent, G., 1990
EMERGING CONCEPTS IN URBAN SPACE DESIGN, Van Nostrand Reinhold (International), New York, N.Y.

Brockenbrough, R. L., 1970a
EXPERIMENTAL STRESSES AND STRAINS FROM HEAT CURVING, *ASCE Journal of the Structural Division,* vol. 96, no. ST7, New York, N.Y.

Brockenbrough, R. L., 1970b
THEORETICAL STRESSES AND STRAINS FROM HEAT CURVING, *ASCE Journal of the Structural Division,* vol. 96, no. ST7, New York, N.Y.

Brockenbrough, R. L., 1983a
CONSIDERATIONS IN THE DESIGN OF BOLTED JOINTS FOR WEATHERING STEEL, *Engineering Journal,* American Institute of Steel Construction, Chicago, Ill., vol. 20, no 1.

Brockenbrough, R. L., 1983b
STRUCTURAL STEEL DESIGN AND CONSTRUCTION, in *Standard Handbook for Civil Engineers,* McGraw-Hill, New York, N.Y.

Brockenbrough, R. L., 1992
MATERIAL PROPERTIES, in *Constructional Steel Design: An International Guide,* Elsevier, Barking, Essex, England.

Brockenbrough, R. L., and Barsom, J. M., 1992
METALLURGY, in *Constructional Steel Design: An International Guide,* Elsevier, Barking, Essex, England.

Brockenbrough, R. L., and Johnston, B. G., 1981
USS STEEL DESIGN MANUAL, USX Corp., Pittsburgh, Pa.

Brodsky, S., 1991
CALATRAVA TO DESIGN CATHEDRAL BIOSPHERE, *Progressive Architecture,* Reinhold Publishing, Cleveland, Ohio, p. 23.

Brotchie, J. F., 1969
A GENERAL PLANNING MODEL, *Management Science Theory,* vol. 10.

Brotchie, J. F., 1974
SYSTEMATIC DESIGN OF MULTISTORY BUILDINGS, *ASCE Journal of the Structural Division,* vol. 100, no. ST7, July.

Brotchie, J. F., and Linzey, M. P. T., 1971
A MODEL FOR INTEGRATED BUILDING DESIGN, *Building Science,* vol. 6, no. 3, September.

Brown, D., Sijpkes, P., and Maclean, M., 1986
THE COMMUNITY ROLE OF PUBLIC INDOOR SPACE, *Journal of Architectural and Planning Research,* vol. 3, pp. 161–172.

Brown, J. S., 1982
COMPUTERS IN DESIGN AND CONSTRUCTION: EVERYONE READS THE SAME MUSIC, *Modern Steel Construction,* 3d Quarter, American Institute of Steel Construction, Chicago, Ill., pp. 18–19.

Browning, W. D., 1992
NMB BANK HEADQUARTERS, *Urban Land,* vol. 51, no. 7, July, pp. 23–25.

Browning, W. D., and Lovins, A., 1992a
MEGAWATTS FOR BUILDINGS, *Urban Land,* vol. 51, no. 7, July, pp. 26–29.

Browning, W. D., and Lovins, A., 1992b
VAULTING THE BARRIERS TO GREEN ARCHITECTURE, *Architectural Record,* vol. 180, December, p. 16.

Bruegmann, R., 1991
LOCAL ASYMMETRIES, *Inland Architect,* vol. 35, no. 3, March/April, pp. 43–49.

Brugmann, B., and Sletteland, G. (Eds.), 1971
THE ULTIMATE HIGHRISE: SAN FRANCISCO'S MAD RUSH TOWARD THE SKY, San Francisco Bay Guardian Books, San Francisco, Calif.

BTM, 1974
AWARD WINNING BUILDING SYSTEM, *Building Technology and Management,* vol. 12, October.

Building News, Inc., 1981
CONCRETE MASONRY DESIGN MANUAL, 4th edition, Los Angeles, Calif.

Burnett, I., 1991
SILICA FUME CONCRETE IN MELBOURNE, AUSTRALIA, *ACI Concrete International Magazine,* August.

Bush-Brown, A., 1984
SKIDMORE, OWINGS & MERRILL ARCHITECTURE AND URBANISM—1973–1983, Van Nostrand Reinhold, New York, N.Y.

Camellerie, J. F., 1985
CONSTRUCTION METHODS AND EQUIPMENT, in *Handbook of Concrete Engineering,* M. Fintel (Ed.), Van Nostrand Reinhold, New York, N.Y.

Carino, N. J., and Clifton, J. R., 1991
HIGH-PERFORMANCE CONCRETE: RESEARCH NEEDS TO ENHANCE ITS USE, *Concrete International Magazine,* September, pp. 70–72.

Ceco Industries, Inc., 1985
CONCRETE BUILDING—NEW FORMWORK PERSPECTIVE, Chicago, Ill.

Charenbhak, W., 1981
CHICAGO SCHOOL ARCHITECTS AND THEIR CRITICS, UMI Research Press, Ann Arbor, Mich.

Chowdhury, K., 1985
LAND USE PLANNING IN BANGLADESH, National Institute of Local Government, Dhaka, Bangladesh.

Clark, W. C., and Kingston, J. L., 1930
THE SKYSCRAPER: A STUDY IN THE ECONOMIC HEIGHT OF MODERN OFFICE BUILDINGS, American Institute of Steel Construction, New York, N.Y.

Coburn, S., 1987
THEORY AND PRACTICE IN USE OF WEATHERING STEELS, Proceedings of National Engineering Conference, American Institute of Steel Construction, Chicago, Ill.

Codella, F., 1983
WHERE DO WE GO FROM HERE ARCHITECTURALLY, in *Developments in Tall Buildings,* Council on Tall Buildings and Urban Habitat, Hutchinson Ross, Stroudsburg, Pa.

Codella, F. L., and McArthur, A. J., 1973
ARCHITECTURE OF TALL BUILDINGS, in *Planning and Design of Tall Buildings,* Proceedings of ASCE-IABSE International Conference, New York, vol. 1a, no. 1–8.

Cohen, M. A., 1991
URBAN POLICY AND ECONOMIC DEVELOPMENT—AN AGENDA FOR THE 1990s, A World Bank Policy Paper, April.

Colaco, J. P., 1974
PARTIAL TUBE CONCEPT FOR MID-RISE STRUCTURES, *Engineering Journal,* 4th Quarter, American Institute of Steel Construction, Chicago, Ill.

Condit, C. W., 1961
AMERICAN BUILDING ART: THE TWENTIETH CENTURY, Oxford University Press, New York, N.Y.

Condit, C. W., 1964
THE CHICAGO SCHOOL OF ARCHITECTURE: A HISTORY OF COMMERCIAL AND PUBLIC BUILDING IN THE CHICAGO AREA, 1875–1925, University of Chicago Press, Chicago, Ill.

Contini, E., 1973
TALL BUILDINGS: URBAN PLANNING FACTORS, Fritz Engineering Laboratory Report 369-56C, Lehigh University, Bethlehem, Pa.

Conway, D. (Ed.), 1977
HUMAN RESPONSE TO TALL BUILDINGS, Dowden, Hutchinson and Ross, Stroudsburg, Pa.

Cooper Marcus, C., and Francis, C. (Eds.), 1990
PEOPLE PLACES: DESIGN GUIDELINES FOR URBAN OPEN SPACE, Van Nostrand Reinhold, New York, N.Y.

Cooper Marcus, C., and Hogue, L., 1977
DESIGN GUIDELINES FOR HIGH-RISE FAMILY HOUSING, in *Human Response to Tall Buildings,* D. J. Conway (Ed.), Dowden, Hutchinson and Ross, Stroudsburg, Pa., pp. 240–277.

Council on Tall Buildings and Urban Habitat (CTBUH) 1978–1981
PLANNING AND DESIGN OF TALL BUILDINGS, a Monograph, 5 vols., American Society of Civil Engineers, New York, N.Y.

CTBUH, 1978
PHILOSOPHY OF TALL BUILDINGS, Sections 1 and 2, M. M. Moser (Ed.), American Society of Civil Engineers, New York, N.Y.

CTBUH, 1986a
ADVANCES IN TALL BUILDINGS, Van Nostrand Reinhold, New York, N.Y.
CTBUH, 1986b
THE ONE-HUNDRED TALLEST BUILDINGS IN THE WORLD, *Advances in Tall Buildings,* Van Nostrand Reinhold, New York, N.Y., Table 1.
CTBUH, 1986c
HIGH-RISE BUILDINGS: RECENT PROGRESS, Council on Tall Buildings and Urban Habitat, Lehigh University, Bethlehem, Pa.
CTBUH, 1986d
TALL BUILDINGS OF THE WORLD, Council on Tall Buildings and Urban Habitat, Lehigh University, Bethlehem, Pa.
CTBUH, 1988
SECOND CENTURY OF THE SKYSCRAPER, Van Nostrand Reinhold, New York, N.Y.
CTBUH, 1990
TALL BUILDINGS: 2000 AND BEYOND, Council on Tall Buildings and Urban Habitat, Lehigh University, Bethlehem, Pa.
CTBUH, Committee 8A, 1992
FIRE SAFETY IN TALL BUILDINGS, *Tall Buildings and Urban Environment* Series, McGraw-Hill, New York, N.Y.
CTBUH, Committee 12A, 1992
CLADDING, *Tall Buildings and Urban Environment* Series, McGraw-Hill, New York, N.Y.
CTBUH, Committee 21D, 1992
CAST-IN-PLACE CONCRETE IN TALL BUILDING DESIGN AND CONSTRUCTION, *Tall Buildings and Urban Environment* Series, McGraw-Hill, New York, N.Y.
CTBUH, Committee 56, 1992
BUILDING DESIGN FOR HANDICAPPED AND AGED PERSONS, *Tall Buildings and Urban Environment* Series, McGraw-Hill, New York, N.Y.
CTBUH, Committee 3, 1994
STRUCTURAL SYSTEMS FOR TALL BUILDINGS, *Tall Buildings and Urban Environment* Series, McGraw-Hill, New York, N.Y.
CTBUH, Group CB, 1978
STRUCTURAL DESIGN OF TALL CONCRETE AND MASONRY BUILDINGS, vol. CB, *Planning and Design of Tall Buildings,* American Society of Civil Engineers, New York, N.Y.
CTBUH, Group CL, 1980
TALL BUILDING CRITERIA AND LOADING, vol. CL, *Planning and Design of Tall Buildings,* American Society of Civil Engineers, New York, N.Y.
CTBUH, Group PC, 1981
PLANNING AND ENVIRONMENTAL CRITERIA FOR TALL BUILDINGS, vol. PC, *Planning and Design of Tall Buildings,* American Society of Civil Engineers, New York, N.Y.
CTBUH, Group SB, 1979
STRUCTURAL DESIGN OF TALL STEEL BUILDINGS, vol. SB, *Planning and Design of Tall Buildings,* American Society of Civil Engineers, New York, N.Y.
CTBUH, Group SC, 1980
TALL BUILDING SYSTEMS AND CONCEPTS, vol. SC, *Planning and Design of Tall Buildings,* American Society of Civil Engineers, New York, N.Y.
Cowan, H. J., 1976
THE DESIGN OF REINFORCED CONCRETE IN ACCORDANCE WITH THE METRIC SSA CONCRETE STRUCTURES CODE, Sydney University Press, Sydney, Australia.
Cowan, R., 1974
TALL BUILDINGS FOR PEOPLE—AESTHETICS AND AMENITY, in *Tall Buildings and People?* Proceedings of Conference, Great Britain, September 17–19; Institution of Structural Engineers, London, England.

Crosbie, M. J., 1992
STEEL REFLECTIONS, *Architecture,* vol. 81, no. 11, December, pp. 85–90.

Cullen, G., 1961
TOWNSCAPES, Reinhold Publishing, New York, N.Y.

Cullen, G., 1971
THE CONCISE TOWNSCAPE, Van Nostrand Reinhold, New York, N.Y.

Curtis, W. J. R., 1986
LE CORBUSIER: IDEAS AND FORMS, Rizzoli International, New York, N.Y.

Deasy, C. M., 1985
DESIGNING PLACES FOR PEOPLE, Watson Guptill, New York, N.Y.

*Design News,* 1986
HIGH-TECH ELEVATOR BUTTONS DETER VANDALS, vol. 42, May 19, pp. 18–19.

Detwiler, G., 1992
HIGH-STRENGTH SILICA FUME CONCRETE—CHICAGO STYLE, *Concrete International,* October, pp. 32–36.

Development—Population Research and Review, 1989
UNDERSTANDING THE DYNAMICS OF POPULATION—PROJECTIONS AND PREDICTIONS, Special Issue, Canadian International Development Agency, February, p. 9.

Devlin, K., 1990
AN EXAMINATION OF ARCHITECTURAL INTERPRETATION: ARCHITECTS VERSUS NON-ARCHITECTS, *Journal of Architectural and Planning Research,* vol. 7, no. 3, pp. 235–244.

DIAGNOSTIC AND STATISTICAL MANUAL OF MENTAL DISORDERS, 1987
3d edition, revised, American Psychiatric Association, Washington, D.C.

Dickey, W. L., 1985
REINFORCED CONCRETE MASONRY CONSTRUCTION, in *Handbook of Concrete Engineering,* M. Fintel (Ed.), Van Nostrand Reinhold, New York, N.Y.

Dixon, J. M., 1993
AMERICAN SALUTE TO A EUROPEAN CITY, *Progressive Architecture,* Reinhold Publishing, Cleveland, Ohio, September, pp. 52–61.

Doctor, R. M., and Kahn, A. P., 1989
THE ENCYCLOPEDIA OF PHOBIAS, FEARS AND ANXIETIES, Facts on File, Inc., New York, N.Y.

Dorgan, C. E., 1991
BUILDING SYSTEMS AUTOMATION—INTEGRATION, Proceedings of Conference, University of Wisconsin, Madison, June.

Dorgan, C. E., 1992
BUILDING SYSTEMS AUTOMATION—INTEGRATION, Proceedings of Conference, Dallas, Texas, June.

Domel, A., and Ghosh, S., 1990
ECONOMICAL FLOOR SYSTEMS FOR MULTI-STORY RESIDENTIAL BUILDINGS, *ACI Concrete Internal Magazine,* September, pp. 37–40.

Doubliet, S., and Fisher, T., 1986
THE HONGKONGBANK, *Progressive Architecture,* Reinhold Publishing, Cleveland, Ohio, March, pp. 67–74.

Dovey, K., 1991
TALL STOREYS: CORPORATE TOWERS AND SYMBOLIC CAPITAL, Proceedings of Environmental Design Research Association (EDRA) Conference, Oaxtepec, Mexico.

DuBois, D., and DuBois, E. F., 1915
THE MEASUREMENT OF THE SURFACE AREA OF MAN, *Archives of Internal Medicine,* vol. 15, pp. 868–881.

Dufton, A. F., 1932
THE EQUIVALENT TEMPERATURE OF A ROOM AND ITS MEASUREMENT, Building Research Technical Paper 13, H. M. Stationery Office, London, England.

Durand, J. N. L., 1836
RECUEIL ET PARALLELE DES EDIFICES DE TOUT GENRE, ANCIENS ET MODERNES, Vincent, Freal & Cie, Paris, France.

ECCS, 1985
DESIGN MANUAL ON THE EUROPEAN RECOMMENDATIONS FOR THE FIRE SAFETY OF STEEL STRUCTURES, European Convention for Constructional Steelwork, Publication 35.

Eckbo, G., 1964
URBAN LANDSCAPE DESIGN, McGraw-Hill, New York, N.Y.

Edgell, 1928
THE AMERICAN ARCHITECTURE OF TO-DAY, Scribner, New York, N.Y.

Edmonson, K. W., 1991
EYESORES OR ASSETS? HOW LANDSCAPE ARCHITECTS, DEVELOPERS AND USERS PERCEIVE LARGE PARKING LOTS, Masters Thesis, Department of Landscape Architecture, University of Illinois at Urbana-Champaign, Urbana.

Elgaaly, M., and Caccese, V., 1990
STEEL PLATE SHEAR WALLS, Proceedings of AISC National Engineering Conference, Chicago, Ill.

Ellingwood, B., and Corotis, R., 1990
LOAD COMBINATIONS FOR BUILDINGS EXPOSED TO FIRE, Proceedings of AISC National Engineering Conference, Chicago, Ill.

Ellis, F. P., 1953
THERMAL COMFORT IN WARM AND HUMID ATMOSPHERES—OBSERVATIONS ON GROUPS OF INDIVIDUALS IN SINGAPORE, *Journal of Hygiene,* vol. 51, pp. 386–404.

Elmiger, A., 1976
ARCHITECTURAL AND ENGINEERING CONCRETE MASONRY DETAILS FOR BUILDING CONSTRUCTION, National Concrete Masonry Association.

*Engineering News Record,* 1971
FAZLUR R. KHAN: AVANT GARDE DESIGNER OF HIGH-RISES, vol. 182, August, pp. 16–18.

*Engineering News Record,* 1988
SOM/IBM FIELD QUERIES ON JOINT CADD SYSTEM, May 26, p. 13.

ENR, 1985
MIXED-USE TOWER WILL SHOW OFF CONCRETE BRACES, *McGraw-Hill Construction Weekly,* January 31.

Eppel, F., 1980
STATE-OF-THE-ART REPORT: RAIN PENETRATION IN MASONRY, Proceedings of 2d Canadian Masonry Symposium, Ottawa, Ont., Canada.

EQE Engineering Inc., 1991
THE PERFORMANCE OF STEEL BUILDINGS IN PAST EARTHQUAKES, American Iron and Steel Institute, Washington, D.C.

Esteva, L., 1966
BEHAVIOR UNDER ALTERNATING LOADS OF MASONRY DIAPHRAGMS FRAMED BY REINFORCED CONCRETE MEMBERS, RILEM International Symposium on the Effects of Repeated Loading of Materials and Structural Elements, Mexico.

Evans, G. W., Smith, C., and Pezdek, K., 1982
COGNITIVE MAPS AND URBAN FORM, *Journal of the American Planning Association,* vol. 48, no. 2, Spring, pp. 232–244.

Fanger, P. O., 1970
THERMAL COMFORT, Danish Technical Press, Copenhagen, Denmark.

Fenves, S. J., Maher, M. L., and Sriram, D., 1984
EXPERT SYSTEMS: C.E. POTENTIAL, *Civil Engineering, ASCE,* vol. 54, no. 10, pp. 44–47.

Fintel, M., 1985
MULTISTORY STRUCTURES, in *Handbook of Concrete Engineering,* 2d edition, Van Nostrand Reinhold, New York, N.Y.

Fintel, M., 1986
NEW FORMS IN CONCRETE, in *Technique and Aesthetics in the Design of Tall Buildings,* Institute for the Study of the High-Rise Habitat, Lehigh University, Bethlehem, Pa.

Fintel, M., 1991
SHEAR WALLS—AN ANSWER FOR SEISMIC RESISTANCE?, *ACI Concrete International Magazine,* July.

Fintel, M., and Ghosh, S., 1983
ECONOMICS OF LONG-SPAN CONCRETE SLAB SYSTEMS FOR OFFICE BUILDINGS—A SURVEY, *ACI Concrete International Magazine,* February, pp. 21–34.

FIP/CEB, 1990
HIGH STRENGTH CONCRETE STATE OF THE ART REPORT, Prepared by Joint Working Group of Fédération Internationale de la Précontrainte (FIP) and the Comité Euro-International du Béton (CEB).

Fischer, R., 1962
DEFLECTION OF TALL BUILDINGS, *Consulting Engineer,* vol. 18, no. 2, February.

Fischer, R. E., 1972
OPTIMIZING THE STRUCTURE OF THE SKYSCRAPER, *Architectural Record,* Architectural Engineering: Skyscrapers, McGraw-Hill, New York, N.Y., October.

Fisher, T., 1985
P/A TECHNICS: WALL JOINTS, *Progressive Architecture,* Reinhold Publishing, Cleveland, Ohio, February, pp. 105–110.

Fisher, T., 1988
A BOW TO BAHRAIN, *Progressive Architecture,* Reinhold Publishing, Cleveland, Ohio, May, pp. 65–73.

Fishman, R., 1977
URBAN UTOPIAS IN THE TWENTIETH CENTURY, Basic Books, New York, N.Y.

Fox, A. J., Jr. (Ed.), 1971
BANK BUILDING, SUSPENDED 30 FT IN AIR, SPANS 273 FT, *Engineering News Record,* vol. 187, no. 19, November 4, p. 30.

Fox, A. J., Jr. (Ed.), 1975
TOWER'S CORNER BRACING CUTS STEEL WEIGHT BY 14%, *Engineering News Record,* vol. 195, no. 16, October 6, p. 32.

Frampton, K., 1992
MODERN ARCHITECTURE: A CRITICAL HISTORY, Thames and Hudson, London, England.

Francescato, G., Weidemann, S., Anderson, J., and Chenoweth, R., 1979
RESIDENTS' SATISFACTION IN HUD-ASSISTED HOUSING: DESIGN AND MANAGEMENT FACTORS, U.S. Department of Housing and Urban Development, Washington, D.C.

Francis, A. J., Horman, C. B., and Jerrems, L. E., 1970
THE EFFECT OF JOINT THICKNESS AND OTHER FACTORS ON THE COMPRESSIVE STRENGTH OF BRICKWORK, Proceedings 2nd International Conference, Stoke-on-Trent, England.

Frederking, M. A., and Penz, A. J., 1973
AN INTEGRATED METHODOLOGY FOR OFFICE BUILDING ELEVATOR DESIGN, in *Environmental Design Research,* vol. 1, Selected Papers, 4th International Environmental Design Research Association (EDRA) Conference, W. F. E. Preiser (Ed.), Dowden, Hutchinson and Ross, Stroudsburg, Pa., pp. 448–460.

Freedman, S., 1985a
PROPERTIES OF MATERIALS FOR REINFORCED CONCRETE, *Handbook of Concrete Engineering,* M. Fintel (Ed.), Van Nostrand Reinhold, New York, N.Y.

Freedman, S., 1985b
MULTISTORY STRUCTURES, in *Handbook of Concrete Engineering,* M. Fintel (Ed.), Van Nostrand Reinhold, New York, N.Y.

Fuerst, J. S., 1985
HIGH-RISE LIVING: WHAT TENANTS SAY, *Journal of Housing,* vol. 42, May/June, pp. 88–90.

Fujisawa, H., 1987
1400 FOOT HIGH STEEL OFFICE BUILDING, Masters Thesis, Illinois Institute of Technology, Department of Architecture, Chicago.

Gan, R., 1974
TALL BUILDINGS AND URBAN PLANNING, Proceedings of Conference on Tall Buildings, Kuala Lumpur, Malaysia, December 2–5; Institution of Engineers, pp. 2-1–2-10.

Gapp, P., 1986
STRUCTURALLY KHAN, *Chicago Tribune,* Chicago, Ill., October 12.

Gero, J. S., and Julian, W. G., 1975
INTERACTION IN THE PLANNING OF BUILDINGS, in *Spatial Synthesis in Computer Aided Building Design,* C. M. Eastman (Ed.), Applied Sciences, London, England.

Giangreco, E., 1973
GENERAL REPORT, Proceedings of Italian National Conference on Tall Buildings, Sorrento, Italy, October 31; Collegio dei Tecnici dell'Acciaio, Milan, Italy, pp. 11–50.

Gibbs, K. T., 1984
BUSINESS ARCHITECTURAL IMAGERY IN AMERICA, 1870–1930, UMI Research Press, Ann Arbor, Mich.

Giedion, S., 1967
SPACE, TIME, AND ARCHITECTURE: THE GROWTH OF A NEW TRADITION, Harvard University Press, Cambridge, Mass.

Ginsberg, Y., and Churchman, A., 1985
THE PATTERN AND MEANING OF NEIGHBOR RELATIONS IN HIGH-RISE HOUSING IN ISRAEL, *Human Ecology,* vol. 13, no. 4, December, pp. 467–484.

Goffman, E., 1963
BEHAVIOR IN PUBLIC PLACES, Free Press, New York, N.Y.

Goldberger, P., 1981
THE SKYSCRAPER, Alfred A. Knopf, New York, N.Y.

Goldberger, P., 1983
ON THE RISE: ARCHITECTURE AND DESIGN IN A POSTMODERN AGE, Times Books, New York, N.Y.

Goldman, J., 1980
THE EMPIRE STATE BUILDING BOOK, St. Martin's Press, New York, N.Y.

Goldsmith, M., 1986
FAZLUR KHAN'S CONTRIBUTIONS IN EDUCATION, in *Techniques and Aesthetics in the Design of Tall Buildings,* Institute for the Study of the High-Rise Habitat, Lehigh University, Bethlehem, Pa.

Goldsmith, M., and Blaser, W. (Ed.), 1987
BUILDINGS AND CONCEPTS, Rizzoli International, New York, N.Y.

Gordon, D. E., 1990
DESIGN FOR THE BIG ONE, *Architecture,* vol. 79, no. 2, February.

Graham, B., 1986
COLLABORATION IN PRACTICE BETWEEN ARCHITECT AND ENGINEER, in *Technique and Aesthetics in the Design of Tall Buildings,* Institute for the Study of the High-Rise Habitat, Lehigh University, Bethlehem, Pa., pp. 1–5.

Grant, A., 1990
BEAUTY AND BRIDGES, *Esthetics in Concrete Bridge Design,* S. C. Watson and M. K. Hurd (Eds.), American Concrete Institute Publication MP-1, Detroit, Mich.

Greene, B., 1983
BUT SHOPPING IS FOR WOMEN, *The San Diego Union,* Saturday, February 19, p. B-15.

Gregory, W. G., 1962
THOUGHTS ON THE ARCHITECTURE OF HIGH BUILDINGS, in *Thoughts on the Architecture of High Buildings,* S. Mackey (Ed.), Symposium on the Design of High Buildings: Proceedings of the Golden Jubilee Congress of the University of Hong Kong, Hong Kong University Press.

Grimm, C. T., 1992
TECHNICS TOPICS: MASONRY CRACK CONTROL DESIGN CHECK LIST, *Progressive Architecture,* Reinhold Publishing, Cleveland, Ohio, December, pp. 34–35.

Grinter, L. E., 1949
THEORY OF MODERN STEEL STRUCTURES, vol. 1, Macmillan, New York, N.Y.

Groat, L., 1982
MEANING IN POST-MODERN ARCHITECTURE: AN EXAMINATION USING THE MULTIPLE SORTING TASK, *Journal of Environmental Psychology,* vol. 2, pp. 3–22.

Groat, L. N., 1984
CONTEXTUAL COMPATIBILITY IN ARCHITECTURE: AN INVESTIGATION OF NON-DESIGNERS' CONCEPTUALIZATIONS, Report R84-3, Publications in Architecture and Urban Planning, Center for Architecture and Urban Planning Research, University of Wisconsin-Milwaukee.

Groat, L., and Canter, D., 1979
DOES POST-MODERNISM COMMUNICATE?, *Progressive Architecture,* vol. 60, Reinhold Publishing, Cleveland, Ohio, December, pp. 84–87.

Grossman, J. S., 1985
780 THIRD AVENUE—FIRST HIGH RISE DIAGONALLY BRACED CONCRETE STRUCTURE, *Concrete International,* vol. 7, no. 2, February, pp. 53–56.

Grube, O. W., Pran, P., and Schulze, F., 1977
100 YEARS OF ARCHITECTURE IN CHICAGO, Chicago, Ill., pp. 76–79.

Guise, D., 1985
DESIGN AND TECHNOLOGY IN ARCHITECTURE, John Wiley, New York, N.Y., pp. 7–8.

Gunts, E., 1993
BLUEPRINT FOR A GREEN FUTURE, *Architecture,* vol. 82, no. 6, June, pp. 47–51.

Haber, G. M., 1977
THE IMPACT OF TALL BUILDINGS ON USERS AND NEIGHBORS, in *Human Response to Tall Buildings,* D. J. Conway (Ed.), Dowden, Hutchinson and Ross, Stroudsburg, Pa., pp. 45–57.

Haber, G. M., 1986
RESPONSE OF THE ELDERLY AND HANDICAPPED TO LIVING IN A BARRIER-FREE TALL BUILDING, in *Advances in Tall Buildings,* L. S. Beedle (Ed.), Van Nostrand Reinhold, New York, N.Y., pp. 115–129.

Hackett, G., and Gonzalez, D. L., 1987
SAN JUAN'S TOWERING INFERNO: A SUSPICIOUS HOTEL FIRE LEAVES CLOSE TO 100 DEAD, *Newsweek,* vol. 109, January 12, p. 24.

Hall, E. T., 1966
THE HIDDEN DIMENSION, Doubleday, New York, N.Y.

Haller, J. P., 1958
TECHNICAL PROPERTIES OF MASONRY MADE OF CLAY BRICKS AND USED FOR HIGH-RISE CONSTRUCTION, *Schweizerische Bauzeitung,* vol. 76, no. 28; Canadian National Research Council Technical Translation, TT-792, 1959.

Halpern, K., 1978
DOWNTOWN USA: URBAN DESIGN IN NINE AMERICAN CITIES, Whitney Library of Design, New York, N.Y.

Halprin, L., 1963
CITIES, Van Nostrand Reinhold, New York, N.Y.

Handlin, D., 1985
AMERICAN ARCHITECTURE, Thames & Hudson, London, England, pp. 167–197.

Hanna, A., and Sanvido, V., 1991
INTERACTIVE HORIZONTAL FORMWORK SELECTION SYSTEM, *ACI Concrete International Magazine,* August.

Harriman, M. S., 1991
SET IN STONE: TECHNICAL DETAILS DISTINGUISH WELL DESIGNED STONE CLADDING FROM DISTRESSED VENEERS, *Architecture,* vol. 80, no. 2, American Institute of Architects, February, pp. 79–86.

Harriman, M. S., 1992
FULL METAL JACKET, *Architecture,* vol. 81, no. 3, American Institute of Architects, March, pp. 103–110.

Harris, A. J., 1961
ARCHITECTURAL MISCONCEPTIONS OF ENGINEERING, *Royal Institute of British Architects Journal,* vol. 68, February, pp. 130–133.

Hart, F., 1985
ONE HUNDRED YEARS OF STEEL FRAMED STRUCTURES 1872–1972, in *Multi-Story Buildings in Steel,* 2d edition, G. B. Godfrey (Ed.), Nicholas Publishing, New York.

Hart, F., Henn, W., and Sontag, H., 1985
MULTI-STOREY BUILDINGS IN STEEL, Nichols Publishing, New York, N.Y.

Hayden, D., 1977
SKYSCRAPER SEDUCTION, SKYSCRAPER RAPE, *Heresies,* May, pp. 108–115.

Hayek, F. A., 1973
LAW, LEGISLATION AND LIBERTY, vol. 1: RULES AND ORDER, University of Chicago Press, Chicago, Ill.

Hayes-Roth, F., Waterman, D. A., and Lenat, D. B. (Eds.), 1983
BUILDING EXPERT SYSTEMS, Addison-Wesley, Reading, Mass.

Hazeltine, B. A., 1969
SOME LOAD-BEARING BRICK BUILDINGS IN ENGLAND, in *Designing, Engineering and Construction with Masonry Products,* F. B. Johnson (Ed.), Gulf Publishing, Houston, Tex.

Heckscher, A., 1977
OPEN SPACES: THE LIFE OF AMERICAN CITIES, Harper & Row, New York, N.Y.

Hedge, A., 1986
OPEN VERSUS ENCLOSED WORKSPACES: THE IMPACT OF DESIGN ON EMPLOYEE REACTIONS TO THEIR OFFICES, in *Behavioral Issues in Office Design,* J. Wineman (Ed.), Van Nostrand Reinhold, New York, N.Y., pp. 139–176.

Hee Sung Industry, 1987
DESIGN AND CONSTRUCTION OF THE LUCKY-GOLDSTAR TWIN TOWERS, Seoul, Korea.

Hendry, A. W., 1983
RECENT DEVELOPMENTS IN THE DESIGN OF MASONRY STRUCTURES, in *Developments in Tall Buildings,* L. S. Beedle (Ed.), Council on Tall Buildings and Urban Habitat, Lehigh University, Bethlehem, Pa.

Hendry, A. W., and Hasan, S. S., 1976
EFFECT OF SLENDERNESS AND ECCENTRICITY ON THE COMPRESSIVE STRENGTH OF WALLS, Proceedings of 4th International Brick Masonry Conference, Brussels, Belgium.

FIP/CEB, 1990
STATE OF THE ART REPORT, SR-90-1, Bulletin d'Information no. 197, Fédération Internationale de la Précontrainte, London, Spring, pp. 7, 48, 66.

Hilsdorf, H. K., 1969
INVESTIGATION INTO THE FAILURE MECHANISM OF BRICK MASONRY LOADED IN AXIAL COMPRESSION, Proceedings of International Conference on Masonry Structural Systems, University of Texas, Austin; Gulf Publishing, Houston, Tex.

Hodgkison, R. L., 1968
AN ULTRA HIGH-RISE CONCRETE OFFICE BUILDING, Masters Thesis, Illinois Institute of Technology, Department of Architecture, Chicago.

Hoff, G. (Ed.), 1984
FIBER REINFORCED CONCRETE INTERNATIONAL SYMPOSIUM, American Concrete Institute, SP-81, pp. 414, 419.

Hoffman, D., 1967
THE MEANINGS OF ARCHITECTURE: BUILDINGS AND WRITINGS OF JOHN WELLBORN ROOT, Horizon Press, New York, N.Y.

Holmes, I. L., 1967
MASONRY BUILDINGS IN HIGH INTENSITY SEISMIC ZONES, in *Designing, Engineering and Constructing with Masonry Products,* F. B. Johnson (Ed.), Gulf Publishing, Houston, Tex.

Hood, R., 1925
EXTERIOR ARCHITECTURE OF OFFICE BUILDINGS, *Architectural Form,* September, p. 97.

Houghten, F. C., and Yaglou, C. P., 1923
DETERMINATION OF THE COMFORT ZONE, *Transactions of the American Society of Heating and Ventilating Engineers,* vol. 29, p. 361.

Hubka, J., 1973
MAN AS CRITERION OF TALL BUILDINGS, Proceedings of 10th Regional Conference, April; CSVTA—Czechoslovak Scientific and Technical Association, Bratislava, Czechoslovakia (two editions: English and Czech).

Hurd, M., 1990
CONCRETE GIANT RISES ON CHICAGO SKYLINE, *Concrete Construction,* vol. 35, January.

Hurt, H., 1988
HOTEL FIRE: CAGE OF HORROR, *Reader's Digest,* vol. 132, January, pp. 39–46, 181–202.

Husaini, B., Moore, S. T., and Castor, R. S., 1991
SOCIAL AND PSYCHOLOGICAL WELL-BEING OF BLACK ELDERLY LIVING IN HIGH-RISES FOR THE ELDERLY, *Journal of Gerontological Social Work,* vol. 16, no. 3/4, pp. 57–78.

Huxtable, A. L., 1984
THE TALL BUILDING ARTISTICALLY RECONSIDERED: THE SEARCH FOR A SKYSCRAPER STYLE, Pantheon Books, New York, N.Y.

ICBO, 1988
UNIFORM BUILDING CODE, International Conference of Building Officials, Whittier, Calif.

ICBO, 1991
UNIFORM BUILDING CODE, International Conference of Building Officials, Whittier, Calif.

IISI, 1987
PRESSURE VESSEL AND STRUCTURAL STEELS, in *High-Strength Low Alloy Steels,* International Iron and Steel Institute, Brussels, Belgium.

IREM, 1981
MANAGING THE OFFICE BUILDING, Institute of Real Estate Management, Chicago, Ill.

Islam, N., 1990
DHAKA METROPOLITAN FRINGE, LAND AND HOUSING DEVELOPMENT, Dhaka City Museum on behalf of Center for Urban Studies, December.

Iyengar, H., 1977
COMPOSITE OF MIXED STEEL—CONCRETE CONSTRUCTION FOR BUILDINGS, American Society of Civil Engineers, New York, N.Y

Iyengar, H., 1982
SELECTED WORKS OF FAZLUR R. KHAN (1929-1982), IABSE Structures, C-23/82.

Iyengar, H., 1985
HIGH-RISE SYSTEM DEVELOPMENTS IN CONCRETE, ACI Fall Convention, Chicago, Ill.

Iyengar, H., 1986
STRUCTURAL AND STEEL SYSTEMS, in *Techniques and Aesthetics in the Design of Tall Buildings,* Institute for the Study of the High-Rise Habitat, Lehigh University, Bethlehem, Pa.

Iyengar, H., 1989
EXPOSED STEEL FRAMES—A UNIQUE SOLUTION FOR BROADGATE, LONDON, Proceedings of National Steel Construction Conference, Nashville, Tenn., pp. 15-1–15-25.

Iyengar, H., Zils, J., and Shinn, R., 1993
EXPOSED STEEL FRAME CREATES ARCHITECTURAL EXCITEMENT, *Modern Steel Construction,* March, pp. 18–28.

Jacobs, J., 1984
THE MALL: AN ATTEMPTED ESCAPE FROM EVERYDAY LIFE, Waveland Press, Prospect Heights, Ill.

Jencks, C., 1973
LE CORBUSIER AND THE TRAGIC VIEW OF ARCHITECTURE, Harvard University Press, Cambridge, Mass.

Jencks, C., 1980
SKYSCRAPERS—SKYPRICKERS—SKYCITIES, Rizzoli International, New York, N.Y.

Jencks, C., 1984
THE LANGUAGE OF POST-MODERN ARCHITECTURE, Rizzoli International, New York, N.Y.

Jencks, C., 1988
ARCHITECTURE TODAY, Harry N. Abrams, New York, N.Y.

Jensen, D., 1986
CHOOSING A FORMING SYSTEM FOR CONCRETE FLOORS AND ROOFS, *Concrete Construction,* vol. 31, no. 1, January.

John, W. J., and Shaw, M. R., 1989
MAKING INTELLIGENT BUILDINGS A REALITY, Proceedings of High Tech Buildings 89 Conference, London, England, June, pp. 133–143.

Johnson, F. B. (Ed.), 1969
DESIGNING, ENGINEERING AND CONSTRUCTION WITH MASONRY PRODUCTS, Gulf Publishing, Houston, Tex.

Johnson, P., 1978
MIES VAN DER ROHE, 3d edition, Museum of Modern Art, New York, N.Y.

Jokl, M., 1973
AIR-CONDITIONING OF TALL BUILDINGS AND ITS INFLUENCE ON LAYOUT AND STRUCTURE, Proceedings of 10th Regional Conference, April; CSVTA—Czechoslovak Scientific and Technical Association, Bratislava, Czechoslovakia (two editions: English and Czech).

Jordy, W. H., and Coe, R. (Eds.), 1961
MONTGOMERY SCHUYLER: AMERICAN ARCHITECTURE AND OTHER WRITINGS, vol. II, Belknap Press of the Harvard University Press, Cambridge, Mass.

Kao, Chi-Hwa, 1991
A MULTI-USE HIGHRISE BUILDING, Masters Thesis, Illinois Institute of Technology, Chicago.

Kaplan/McLaughlin/Diaz, 1985
CELEBRATING URBAN GATHERING PLACES, *Urban Land,* May, pp. 10–14.

Kasuda, T., 1970
USE OF COMPUTERS FOR ENVIRONMENTAL ENGINEERING RELATED TO BUILDINGS, Proceedings of Symposium, National Bureau of Standards, Gaithersburg, Md., November/December.

Kavanagh, T. C., 1972
ENVIRONMENTAL SYSTEMS, Theme Report, Planning and Design of Tall Buildings, Proceedings of Conference, Lehigh University, Bethlehem, Pa., August, vol. 1a; American Society of Civil Engineers, New York, N.Y., pp. 7–40.

Kax, J. H., 1991
THE GREENING OF ARCHITECTURE, *Architecture,* vol. 80, no. 5, May, pp. 61–63.

Keegan, E., 1990
ILLUSIONS IN CHICAGO, *Inland Architect,* vol. 34, no. 1, January/February.

Keegan, P., 1991
BLOWING THE ZOO TO KINGDOM COME, *Lingua Franca,* December, pp. 15–21.

Keleher, R., 1993
RAIN SCREEN PRINCIPLES IN PRACTICE, *Progressive Architecture,* Reinhold Publishing, Cleveland, Ohio, August, pp. 33–39.

Kennett, E., 1993
CD-ROM: THE FUTURE IS NOW, *A/E/C Systems Computer Solutions,* vol. 2, no. 5, September-October.

Kerr, D. D., 1990
THE RAIN SCREEN WALL, *Progressive Architecture,* Reinhold Publishing, Cleveland, Ohio, August, pp. 47–54.

Khan, F., 1966
CURRENT TRENDS IN CONCRETE HIGH-RISE BUILDINGS, Proceedings of Symposium on Tall Buildings, University of Southampton, England.

Khan, F. R., 1967
THE JOHN HANCOCK CENTER, *ASCE Civil Engineering Magazine,* October.

Khan, F. R., 1971
THE JOHN HANCOCK CENTER, *Civil Engineering,* vol. 37, October, p. 40.

Khan, F. R., 1972a
PHILOSOPHICAL COMPARISON BETWEEN MAILLART'S BRIDGES AND SOME RECENT CONCRETE BUILDINGS, in *Civil Engineering: History, Heritage and the Humanities,* Princeton University, Princeton, N.J., pp. 14–17.

Khan, F. R., 1972b
THE FUTURE OF HIGHRISE STRUCTURES, *Progressive Architecture,* Reinhold Publishing, Cleveland, Ohio, October.

Khan, F. R., 1973a
EVOLUTION OF STRUCTURAL SYSTEMS FOR HIGH-RISE BUILDINGS IN STEEL AND CONCRETE, Regional Conference on Tall Buildings, Bratislava, Czechoslovakia.

Khan, F. R., 1973b
NEWER STRUCTURAL SYSTEMS AND THEIR EFFECT ON THE CHANGING SCALE OF CITIES, Proceedings of Conference on Tall Buildings, Zurich, Switzerland, October; SIA-Fachgruppen für Brückenbau und Hochbau (FBH) und für Architektur (FGA), Zurich, Switzerland.

Khan, F. R., 1974
A CRISIS IN DESIGN—THE NEW ROLE OF THE STRUCTURAL ENGINEER, Proceedings of National Conference on Tall Buildings, Kuala Lumpur, Malaysia, December; Institution of Engineers, Kuala Lumpur, Malaysia.

Khan, F. R., 1978
THE ROLE OF TALL BUILDINGS IN URBAN SPACE, 2001, in *Urban Space for Life and Work,* Proceedings of Conference, Paris, France, November 1977, bol. I; CTICM, Paris, France, pp. 141–148 (in English and French).

Khan, F. R., 1982a
100-STORY JOHN HANCOCK CENTER IN CHICAGO—A CASE STUDY OF THE DESIGN PROCESS, *IABSE Journal,* J-16/82, August.

Khan, F. R., 1982b
THE RISE AND FALL OF STRUCTURAL LOGIC IN ARCHITECTURE, *Chicago Architectural Journal,* vol. 2.

Khan, F. R., and Elnimieri, M. M., 1983
STRUCTURAL SYSTEMS FOR MULTI-USE HIGH-RISE BUILDINGS, in *Developments in Tall Buildings,* Council on Tall Buildings and Urban Habitat, Hutchinson Ross, Stroudsburg, Pa.

Khan, F. R., and Nassetta, A. F., 1970
TEMPERATURE EFFECTS ON TALL STEEL FRAMED BUILDINGS: PART 3—DESIGN CONSIDERATIONS, *Engineering Journal,* American Institute of Steel Construction, Chicago, Ill., vol. 7, no. 4, October, p. 122.

Kleihus, J. P., and Klotz, H., 1986
INTERNATIONAL BUILDING EXHIBITION BERLIN 1987, DAM, Frankfurt am Main, Germany.

Knight, J. L., Jr., Howells, R. H., and Weiss, S., 1979
CROWDING IN ELEVATORS: DURATION OF POST-EXPOSURE EFFECTS, in *Proceedings of Environmental Design: Research, Theory, Application,* A. Seidel and S. Danford (Eds.), Conference held in Buffalo, N.Y., May. Environmental Design Research Association, Washington, D.C., p. 68.

Knowles, R. L., 1981
SUN, RHYTHM, FORM, MIT Press, Cambridge, Mass.

Kohn, E., 1990
THE TALL BUILDING, Proceedings of 4th World Congress on Tall Buildings: 2000 and Beyond, Hong Kong, November.

Kohn Pedersen Fox Associates, 1985
900 NORTH MICHIGAN TOWER, Original Construction Documents, New York, N.Y.

Komp, M. E., 1987
ATMOSPHERIC CORROSION RATINGS OF WEATHERING STEELS—CALCULATION AND SIGNIFICANCE, *Material Performance,* vol. 26, no. 7.

Kotela, C., 1972
TRENDS IN THE DEVELOPMENT OF TALL BUILDINGS IN POLAND, in *Planning and Design of Tall Buildings,* Proceedings of Regional Conference, Warsaw, Poland, November, vol. 1; Warsaw Technical University, Polish Group of IABSE, Warsaw, Poland, pp. 22–29.

Kowinski, W. S., 1985
THE MALLING OF AMERICA: AN INSIDE LOOK AT THE GREAT CONSUMER PARADISE, William Morrow, New York, N.Y.

Kozloff, M., 1990
SKYSCRAPERS, THE LATE IMPERIAL MOB, *Art Forum,* vol. 29, December, pp. 96–101.

Kreimer, A., 1973
BUILDING THE IMAGERY OF SAN FRANCISCO: AN ANALYSIS OF CONTROVERSY OVER HIGH-RISE DEVELOPMENT, 1970–1971, Environmental Design Research, EDRA IV, Conference Proceedings, 1970–1971, vol. 2, W. F. E. Preiser (Ed.), Dowden, Hutchinson and Ross, Stroudsburg, Pa., pp. 221–231.

Kridler, C., and Stewart, R. K., 1992
ACCESS FOR THE DISABLED: PART I, *Progressive Architecture,* Reinhold Publishing, Cleveland, Ohio, July, pp. 41–42.

Krier, R., 1979
URBAN SPACE, Rizzoli International, New York, N.Y.

Krier, R., 1983
ELEMENTS OF ARCHITECTURE, Architectural Design Publications, London, England.

Labs, K., 1990
THE RAIN SCREEN WALL, *Progressive Architecture,* Reinhold Publishing, Cleveland, Ohio, August, pp. 47–52.

Lamela, A., 1973
TOWN PLANNING AND VERTICAL ARCHITECTURE, THE PROBLEMS RAISED THEREBY AND THE NECESSITY OF A NEW OUTLOOK, Proceedings of Regional Conference on Tall Buildings, Madrid, Spain, September 17–19; Tipografia Artistica, Madrid, Spain, pp. 187–206 (two editions: English and Spanish).

Lankford, W. T., Jr., et al., 1985
THE MAKING, SHAPING AND TREATING OF STEEL, Association of Iron and Steel Engineers, Pittsburgh, Pa.

Larson, E. D., and Williams, R. H., 1990
BIOMASS-GASIFIER STEAM INJECTED GAS TURBINE COGENERATION, *Journal of Engineering for Gas Turbines and Power,* American Society of Mechanical Engineers, vol. 112, pp. 157–163.

Law, M., 1978
FIRE SAFETY OF EXTERNAL BUILDING ELEMENTS—THE DESIGN APPROACH, *Steel Construction Engineering Journal,* vol. 15, no. 8, American Institute of Steel Construction, Chicago, Ill.

Le Corbusier, 1927
TOWARDS A NEW ARCHITECTURE, Translation by F. Etchells, Architectural Press, London, England.

Lechner, H., 1969
SPECIAL QUALITY BRICK MASONRY MULTISTORY APARTMENT HOUSES BUILT IN SWITZERLAND, in *Designing, Engineering and Constructing with Masonry Products,* F. B. Johnson (Ed.), Gulf Publishing, Houston, Tex.

Lee, D. J., 1985
THE FINE ART OF STRUCTURAL ENGINEERING, *Structural Engineer,* vol. 63A, no. 11, November.

Levin, B., Gordon, B., Bliss, P., Finlayson, A., Nichols, M., Diotte, K., and Wallace, B., 1987
DEATH IN A TOWERING INFERNO, *MacLean's,* vol. 100, January 12, pp. 16–19.

Liang, C., 1973
URBAN LAND USE ANALYSIS—A CASE STUDY OF HONG KONG, Ernest Publications, Hong Kong.

Liebenberg, A. C., 1983
THE AESTHETIC EVALUATION OF BRIDGES, in *Handbook of Structural Concrete,* F. K. Kong et al. (Eds.), McGraw-Hill, New York, N.Y.

Lim, B., 1992
THE ARCHITECTURAL CHARACTERISTICS OF SOME TALL BUILDINGS SINCE SEARS TOWER, Proceedings of International Conference on Tall Buildings, Kuala Lumpur, Malaysia, July 28–30.

Lim, H. J., 1992
A SUPER TALL THIN OFFICE BUILDING, Masters Thesis, Illinois Institute of Technology, Department of Architecture, Chicago, Ill.

Loos, A., 1928
THE PRINCIPLE OF CLADDING, in *Ins Leere Gesprochen,* Georges Crès & Cie, Paris, France; reprinted in RAUMPLAN VERSUS PLAN LIBRE: ADOLF LOOS AND LE CORBUSIER, 1919–1930, M. Risselada (Ed.), Rizzoli International, New York, N.Y., 1989.

Loriers, M. C., 1988
THROUGH THE LOOKING GLASS, *Progressive Architecture,* Reinhold Publishing, Cleveland, Ohio, May, pp. 94–97.

Lynch, K., 1960, 1969
THE IMAGE OF THE CITY, MIT Press, Cambridge, Mass.

Lynch, K., 1981
A THEORY OF GOOD CITY FORM, MIT Press, Cambridge, Mass.

Lynch, K., 1984
GOOD CITY FORM, MIT Press, Cambridge, Mass.

Lynch, K., 1990
CITY SENSE AND CITY DESIGN: WRITINGS AND PROJECTS OF KEVIN LYNCH, MIT Press, Cambridge, Mass., pp. 351–378.

Lynch, K., and Hack, G., 1984
SITE PLANNING, MIT Press, Cambridge, Mass.

MacMillan, A., and Metzstein, I., 1974
AMENITY AND AESTHETICS OF TALL BUILDINGS, in *Tall Buildings and People?* Proceedings of Conference, Great Britain, September 17–19; Institution of Structural Engineers, London, England, pp. 91–96.

Maki, F., 1972
THE MENACE OF TALL BUILDINGS, in *Planning and Design of Tall Buildings,* Proceedings of Conference on Tall Buildings, Lehigh University, Bethlehem, Pa., August, vol. Ia; American Society of Civil Engineers, New York, N.Y., pp. 91–92.

Malier, Y., 1991
THE FRENCH APPROACH TO USING HPC, *ACI Concrete International Magazine,* July.

Mankowski, T., 1972
DISCUSSION: THE PROBLEMS OF TALL BUILDING SPATIAL FUNCTIONS, in *Planning and Design of Tall Buildings,* Proceedings of Regional Conference, Warsaw, Poland, November, vol. 1; Warsaw Technical University, Polish Group of IABSE, Warsaw, Poland, pp. 43–44.

Marks, I. M., 1969
FEARS AND PHOBIAS, Academic Press, New York, N.Y.

Marksbury, D. L., 1984
ARTIFICIAL INTELLIGENCE: CAE'S OUTER LIMITS, *Consulting Engineer,* December, pp. 39–41.

Marsh, J. W., 1993
EARTHQUAKES: STEEL STRUCTURES PERFORMANCE AND DESIGN CODE DEVELOPMENTS, *Engineering Journal,* 2d Quarter, American Institute of Steel Construction, Chicago, Ill.

Martin, R., 1987
CENTER THROWS CHICAGO FOR A LOOP, *Insight,* September 7, pp. 62–63.

Marx, K., 1848
THE COMMUNIST MANIFESTO, quoted in Berman, 1982.

Mays, V., 1988
P/A TECHNICS: THE MANY FACES OF BRICKS, *Progressive Architecture,* Reinhold Publishing, Cleveland, Ohio, July, pp. 102–107.

Mays, V., 1989
THE USES OF GLASS, *Progressive Architecture,* Reinhold Publishing, Cleveland, Ohio, March, pp. 106–111.

McHarg, I., 1992
DESIGN WITH NATURE, John Wiley, New York, N.Y.

McIntyre, D. A., 1980
INDOOR CLIMATE, Applied Science Publishers, London, England.

McNary, W. S., and Abrams, D. P., 1985
MECHANICS OF MASONRY IN COMPRESSION, *ASCE Journal of Structural Engineering,* April.

Menon, A. L. K., 1966
A NINETY STORY APARTMENT BUILDING USING AN OPTIMIZED CONCRETE STRUCTURE, Masters Thesis, Illinois Institute of Technology, Department of Architecture, Chicago.

Mies van der Rohe, L., 1922
TWO GLASS SKYSCRAPERS, originally published in *Früucht,* vol. 1, pp. 122–124, reprinted in P. Johnson, *Mies van der Rohe,* 1978, p. 187.

Mies van der Rohe, L., 1924
ARCHITECTURE AND THE TIMES, originally published in *Der Querschnitt,* vol. 4, pp. 31–32, reprinted in P. Johnson, *Mies van der Rohe,* 1978, pp. 191–192.

Mies van der Rohe, L., 1938
INAUGURAL ADDRESS, Illinois Institute of Technology, Chicago.

Mies van der Rohe, L., 1950
ARCHITECTURE AND TECHNOLOGY, Address at the Union of the Institute of Design and the Illinois Institute of Technology, Chicago; reprinted in P. Johnson, *Mies van der Rohe,* 1978, pp. 203–204.

Mies van der Rohe, L., 1960
ARCHITECTURE AND CIVILIZATION, Address given on the Occasion of Receiving the Gold Medal of the American Institute of Architects; reprinted in P. Carter, *Mies van der Rohe at Work,* Thames & Hudson, London, England, 1974.

Mikluchin, P. T., 1969
MORPHOTECTONICS OF MASONRY STRUCTURES, in *Masonry Structural Systems,* Proceedings of International Conference, University of Texas, Austin.

Mindess, S., 1984
RELATIONSHIPS BETWEEN STRENGTH AND MICROSTRUCTURE FOR CEMENT-BASED MATERIALS: AN OVERVIEW, Proceedings of Materials Research Society Symposia, vol. 42, Pittsburgh, Pa.

Ministry of National Development, 1985
DESIGN GUIDELINES ON ACCESSIBILITY FOR THE DISABLED IN BUILDINGS, Development and Building Control Division, Public Works Department, Singapore.

Missenard, A., 1948
EQUIVALENCE THERMIQUE DES AMBIENCES—EQUIVALENCES DE SEJOUR, *Chaleur et Industrie,* vol. 29, pp. 159–172, 189–198.

Monk, C. B., 1980
MASONRY FACADES AND PAVING FAILURES, Proceedings of 2d Canadian Masonry Symposium, Ottawa, Ont., Canada.

Mönninger, M., 1993
CROWNED SKYSCRAPER, *Progressive Architecture,* Reinhold Publishing, Cleveland, Ohio, September, pp. 62–65.

Moreno, J., 1990
HIGH PERFORMANCE CONCRETE FOR TALL BUILDINGS IN CHICAGO—PAST, PRESENT AND FUTURE, Proceedings of 4th World Congress on Tall Buildings, Council on Tall Buildings and Urban Habitat, Hong Kong, November.

Moreno, J., 1992
CONCRETE TECHNOLOGY FOR THE 21ST CENTURY, Proceedings of Conference on Byproducts Utilization, June 11; Canmet & University of Wisconsin Publishers.

Moreno, J., and Zils, J., 1985
OPTIMIZATION OF HIGH-RISE CONCRETE BUILDINGS, ACI Fall Convention, Chicago, Ill.

Morris, A. E. J., 1974
HISTORY OF URBAN FORM: PREHISTORY TO THE RENAISSANCE, John Wiley, New York, N.Y.

Mozingo, L., 1989
WOMEN AND DOWNTOWN OPEN SPACES, *Places,* vol. 6, no. 1, pp. 38–47.

Muguruma, H., and Watanabe, F., 1990
DUCTILITY IMPROVEMENT OF HIGH STRENGTH CONCRETE COLUMNS WITH LATERAL CONFINEMENT, in *High-Strength Concrete—2d International Symposium,* Publication SP-121, American Concrete Institute, Detroit, Mich., pp. 47–60.

Mumford, L., 1938
THE CULTURE OF CITIES, Harcourt, Brace, New York, N.Y.

Mumford, L., 1952
ROOTS OF CONTEMPORARY AMERICAN ARCHITECTURE: A SERIES OF THIRTY-SEVEN ESSAYS DATING FROM THE MID-NINETEENTH CENTURY TO THE PRESENT, Reinhold Publishing, New York, N.Y.

Mumford, L., 1961
THE HISTORY OF CITIES, Harcourt, Brace, and World, New York, N.Y.

Murphy, J., 1985
2000 AND BEYOND: THE NEW STATE OF ILLINOIS CENTER: INFAMOUS, A NOBLE EFFORT, OR BOTH?, *Progressive Architecture,* Reinhold Publishing, Cleveland, Ohio, December, pp. 72–79.

Murphy, J., 1990
TO BE CONTINUED, *Progressive Architecture,* Reinhold Publishing, Cleveland, Ohio, March, pp. 102–112.

Murphy/Jahn Architects, 1985
OAKBROOK TERRACE TOWER, Original Construction Documents, Chicago, Ill.

Nahemow, L., Lawton, M. P., and Howell, S. C., 1977
ELDERLY PEOPLE IN TALL BUILDINGS: A NATIONWIDE STUDY, in *Human Response to Tall Buildings,* D. J. Conway (Ed.), Dowden, Hutchinson and Ross, Stroudsburg, Pa., pp. 175–181.

Napier, T., and Golish, M., 1982
A SYSTEMS APPROACH TO MILITARY CONSTRUCTION, U.S. Army Construction Engineering Research Laboratory (CERL), Report P-132, Champaign, Ill.

National Bureau of Standards, 1960
BUILDING CODE REQUIREMENTS FOR REINFORCED MASONRY, American Standard A41.2-1960, Handbook 74, October 21.

National Bureau of Standards, 1970
USE OF COMPUTERS FOR ENVIRONMENTAL ENGINEERING RELATED TO BUILDINGS, American Standard, National Institute of Standards and Technology, Department of Commerce, Gaithersburg, Md., p. 84.

*National Geographic,* 1989
SKYSCRAPERS, February, pp. 149–173.

National Organization of the Handicapped, 1983
REQUIREMENTS FOR ACCESS: MANUAL FOR BUILDING ENTRANCIBLE AND USABLE FOR HANDICAPPED PEOPLE (Geboden toegang: handboek voor het toegangkelijk en bruikbaar ontwerpen en bouwen voor gehandicapten mensen), 7th edition, National Orgaan Gehandicaptenbeleid; in samenwerking met de Gemeenschappelijke Medische Dienst a.o., Utrecht, Netherlands.

Navier, L. M. H., 1826
LECTURE NOTES, Ecole Polytechnique, *Strength of Materials* textbook.

Nawaz, T., 1991
DHAKA: 2000 A.D. AND BEYOND, AN URBAN BLACK HOLE OR A LIVABLE WORLD CITY, Seminar by Canada-Bangladesh Forum held at Carleton University, April 22; *Daily Star,* Dhaka, June 28.

Nawaz, T., 1992
A PROPOSAL FOR FUTURE ALTERNATE GROWTH OPTIONS FOR DHAKA, BANGLADESH, Proposal for Alternate Development Plan submitted to Ford Foundation in Bangladesh, June.

NCMA, 1970
SPECIFICATION FOR THE DESIGN AND CONSTRUCTION OF LOAD-BEARING CONCRETE MASONRY, National Concrete Masonry Association, Arlington, Va.

NCMA, 1990
BUILDING CODE REQUIREMENTS FOR MASONRY STRUCTURES, NCMA TEK Notes, no. 113A, National Concrete Masonry Association, Arlington, Va.

Nemiah, J. C., 1985
PHOBIC DISORDERS (PHOBIC NEUROSES), in *Comprehensive Textbook of Psychiatry/IV,* H. I. Kaplan and B. J. Sadock (Eds.), vol. 1, 4th edition, pp. 894–904.

Newman, O., 1972
DEFENSIBLE SPACE: CRIME PREVENTION THROUGH URBAN DESIGN, Macmillan, New York, N.Y.

Newman, O., 1973
DEFENSIBLE SPACE: CRIME PREVENTION THROUGH URBAN DESIGN, Collier Books, Toronto, Ont., Canada.

*Newsweek,* 1965
THE LONGEST NIGHT, vol. 60, no. 92, November 22, pp. 27–33.

*Newsweek,* 1993
TERROR HITS HOME: THE HUNT FOR THE BOMBERS, vol. 121, no. 10, March 8.

NFPA, 1978
LIFE SAFETY CODE, National Fire Protection Association, Boston, Mass.

Nicholas, L. H., 1987
MEGA-CITIES OF THE FUTURE, in *Development—Cities,* Canadian International Development Agency (CIDA), Winter.

Nicholson, P., 1989
HIGH TECH BUILDINGS: PRODUCT FROM PROCESS, Proceedings of Conference on High Tech Buildings, London, England, June; Blenheim Online Publications.

Nicks, J. F., 1980
MASONRY MATERIALS, Proceedings of 2d Canadian Masonry Symposium, Ottawa, Ont., Canada.

Norberg-Schulz, C., 1974, 1988
ROOTS OF MODERN ARCHITECTURE, A. D. A. EDITA, Tokyo Co. Ltd., Japan.

Norberg-Schulz, C., 1979, 1980
GENIUS LOCI: TOWARDS A PHENOMENOLOGY OF ARCHITECTURE, Rizzoli International, New York, N.Y.

Norberg-Schulz, C., 1980
MEANING IN WESTERN ARCHITECTURE, Rizzoli International, New York, N.Y.

Normoyle, J. B., and Foley, J. M., 1988
THE DEFENSIBLE SPACE MODEL OF FEAR AND ELDERLY PUBLIC HOUSING RESIDENTS, *Environment and Behavior,* vol. 20, no. 1, January, pp. 50–74.

Oh, S., 1992
DEVELOPMENT OF HIGH-RISE MIXED-USE BUILDING UTILIZING HIGH STRENGTH CONCRETE, Masters Thesis, Illinois Institute of Technology, Department of Architecture, Chicago.

Ohashi, M., et al., 1990
DEVELOPMENT OF STEEL PLATES FOR BUILDING STRUCTURAL USE, Nippon-Steel Technical Report 44, Tokyo, Japan.

Oldenburg, R., and Brisset, D., 1980
THE ESSENTIAL HANGOUT, *Psychology Today,* vol. 13, no. 11, April, pp. 82–84.

Pagani, C., Hoepli, U., and Milano, E., 1955
ITALY'S ARCHITECTURE TODAY, Ulrico Hoepli, Milan, Italy.

Paillere, A. M., Build, M., and Serramo, J. J., 1989
EFFECT OF FIBER ADDITIONS ON THE AUTOGENOUS SHRINKAGE OF SILICA FUME CONCRETE, *ACI Materials Journal,* March.

Passini, R., 1984
WAYFINDING IN ARCHITECTURE, Van Nostrand Reinhold, New York, N.Y.

Paulay, T., and Priestley, M. J. M., 1992
SEISMIC DESIGN OF REINFORCED CONCRETE AND MASONRY BUILDINGS, John Wiley, New York, N.Y.

Peets, E., 1968
ON THE ART OF DESIGNING CITIES, MIT Press, Cambridge, Mass.

Pelli, C., 1982
SKYSCRAPERS, Perspecta 18: Yale Papers in Architecture, MIT, Cambridge, Mass.

Perkins & Will, 1988
100 NORTH RIVERSIDE PLAZA (MORTON INTERNATIONAL BUILDING), Original Construction Document.

Petrie, W. J., 1981
A HIGH-RISE OFFICE BUILDING USING LONG-SPAN TUBE STRUCTURE, Masters Thesis, Illinois Institute of Technology, Department of Architecture, Chicago.

Phillips, M., 1965
BLACKOUT VIGNETTES ARE EVERYWHERE YOU LOOK, *The New York Times,* Thursday, November 11, p. 37.

Planning Commission, Ministry of Planning, 1985
THE THIRD FIVE YEAR PLAN, Government of the People's Republic of Bangladesh, December.

Plummer, H., and Blume, J., 1953
REINFORCED/BRICK MASONRY AND LATERAL FORCE DESIGN, November.

Polshek, J., 1988
CONTEXT AND RESPONSIBILITY, Rizzoli, New York, N.Y.

Popov, E. P., 1991
U.S. SEISMIC STEEL CODES, *AISC Engineering Journal,* vol. 28, no. 3.

Portland Cement Association, 1981
EFFECTS OF VARIOUS SUBSTANCES ON CONCRETE AND PROTECTIVE TREATMENTS, WHERE REQUIRED (IS001.05T), Portland Cement Association, Skokie, Ill.

Posokhin, M. V., 1973
DESIGN AND TRENDS IN THE DEVELOPMENT OF TALL BUILDINGS IN MOSCOW, in *Planning and Design of Tall Buildings* Proceedings of Conference, Lehigh University, Bethlehem, Pa., August 1972, vol. Ia; American Society of Civil Engineers, New York, N.Y., pp. 183–198.

Powell, R., 1989
KEN YEANG: RETHINKING THE ENVIRONMENTAL FILTER, Landmark Books, Singapore.

Powys, R. O., 1971
ECONOMIC ASPECTS OF PLANNING AND DESIGN OF TALL BUILDINGS, Australian Report on Environmental Aspects of Planning and Design of Tall Buildings, Fritz Engineering Laboratory Report 369.32, Lehigh University, Bethlehem, Pa.

Praege, F. A., 1963
ARCHITECTURE OF SKIDMORE, OWINGS AND MERRILL—1950–1962, New York.

Prager, F., and Scaglia, G., 1970
BRUNELLESCHI: STUDIES OF HIS TECHNOLOGY AND INVENTIONS, MIT Press, Cambridge, Mass.

Prescott, J. A., 1962
FORMAL VALUES AND HIGH BUILDINGS, Symposium on the Design of High Buildings, Proceedings of the Golden Jubilee Congress of the University of Hong Kong, S. Mackey (Ed.); Hong Kong University Press, 1962.

Prestressed Concrete Institute, 1985
PCI DESIGN HANDBOOK, 3d edition, Chicago, Ill.

*Progressive Architecture,* 1968
CHICAGO TO HAVE ANOTHER MIES OFFICE BUILDING BY 1969, vol. 49, June, pp. 48–49.

*Progressive Architecture,* 1980a
FACADE AT RIGHT ANGLES, December.

*Progressive Architecture,* 1980b
THINKING TALL, December.

*Progressive Architecture,* 1990a
BUILDING TOPS, February.

*Progressive Architecture,* 1990b
FOSTER'S JAPANESE NEEDLE, October.

Prokesch, S., and Meier, B., 1993
INCHING BACK TO LIFE, TRADE CENTER TALLIES THE COST: BUSINESSES CONCERNED WITH HOW FAR INSURANCE WILL COVER THEM, *The New York Times,* Sunday, March 7, pp. 1, 19.

Public Works Department, 1990
CODE ON BARRIER-FREE ACCESSIBILITY IN BUILDINGS, Singapore.

Rachman, S. J., 1990
FEAR AND COURAGE, W. H. Freeman, New York, N.Y., chap. 6, "Claustrophobia and the Fear of Suffocation."

Raschko, B. B., 1982
HOUSING INTERIORS FOR THE DISABLED AND ELDERLY, Van Nostrand Reinhold, New York, N.Y.

Rasmussen, S. E., 1962
EXPERIENCING ARCHITECTURE, MIT Press, Cambridge, Mass.

Redmond, T., and Allen, M., 1975
COMPRESSIVE STRENGTH OF COMPOSITE BRICK AND CONCRETE MASONRY WALLS, ASTM STP589, American Society for Testing and Materials, Philadelphia, Pa.

Reinhardt, R., 1971
ON THE WATERFRONT: THE GREAT WALL OF MAGNIN, in *The Ultimate Highrise: San Francisco's Mad Rush Toward the Sky,* B. Brugmann and G. Sletteland (Eds.), San Francisco Bay Guardian Books, San Francisco, Calif., pp. 92–137.

Renn, J. J., 1990
WOLF POINT CENTER, A MULTIFUNCTIONAL TALL BUILDING, Illinois Institute of Technology, Department of Architecture, Chicago.

Risselada, M., 1989
RAUMPLAN VERSUS PLAN LIBRE, Rizzoli International, New York, N.Y.

Robbie, R. G., 1972
THE PERFORMANCE CONCEPT IN BUILDING: THE WORKING APPLICATION OF THE SYSTEMS APPROACH TO BUILDING, Joint RILEM-ASTM-CIB Symposium, Proceedings of Symposium, Philadelphia, Pa., pp. 201–206.

Robertson, L. E., 1973
HEIGHTS WE CAN REACH, *AIA Journal,* vol. 59, no. 1, January, pp. 25–30.

Robins, A. W., 1987
OBSERVATIONS, *Architectural Record,* January.

Rodriquez, A. M., 1970
A FORM STIFFENED HIGH-RISE APARTMENT BUILDING, Masters Thesis, Illinois Institute of Technology, Department of Architecture, Chicago.

Roeder, C. W., and Popov, E. P., 1978
ECCENTRICALLY BRACED FRAMES FOR EARTHQUAKES, *ASCE Journal of the Structural Division,* vol. 104, no. 3, March.

Rohles, F. H., Jr., 1973
THE REVISED MODAL COMFORT ENVELOPE, *ASHRAE Transactions,* vol. 79, part II, p. 52.

Root, J. W., 1883
THE VALUE OF TYPE IN ART, originally published in *Inland Architect and Builder,* vol. 2, November, p. 132; reprinted in Hoffman, 1967, pp. 169–171.

Root, J. W., 1890
A GREAT ARCHITECTURAL PROBLEM, originally published in *Inland Architect and News Record,* vol. 15, June, p. 68; reprinted in Hoffman, 1967, pp. 130–142.

Rosiak, B., 1972
SYSTEM OF NATURAL VENTILATION IN TALL BUILDINGS, Proceedings of Regional Conference on the Planning and Design of Tall Buildings, Warsaw, Poland, November, vol. II; Warsaw Technical University, Polish Group of IABSE, Warsaw, Poland.

Ross, W. D., 1955
ARISTOTLE: SELECTIONS, Charles Scribners, Lyceum Editions, Philosophy Series, New York, N.Y.

Rossi, A., 1982
ARCHITECTURE OF THE CITY, MIT Press, Cambridge, Mass.

Rowe, C., and Koetter, F., 1978
COLLAGE CITY, MIT Press, Cambridge, Mass.

Roy, H. E., 1965
TORONTO CITY HALL AND CIVIC SQUARE, *Journal of the American Concrete Institute,* December.

Rubanenko, B., 1973
DESIGN AND CONSTRUCTION OF TALL BUILDINGS, in *Planning and Design of Tall Buildings,* Proceedings of Conference, Lehigh University, Bethlehem, Pa., August 1972, vol. Ia; American Society of Civil Engineers, New York, N.Y., pp. 231–247.

Ruchelman, L., 1978
PLANNING URBAN SERVICES IN SUPPORT OF HIGH RISE BUILDINGS, *2001: Urban Space for Life and Work,* Proceedings of Conference, Paris, France, November 1977, vol. II; CTICM, Paris, France (in English and French).

Russell, H. G., 1990
SHORTENING OF HIGH-STRENGTH CONCRETE MEMBERS, in *High Strength Concrete—Second International Symposium,* Publication SP-121, American Concrete Institute, Detroit, Mich., pp. 1–20.

Russell, J. S., 1990
DESIGNING THE SUPER-THIN NEW BUILDINGS, *Architectural Record,* vol. 178, October, pp. 105–109.

Sabbagh, K., 1989
SKYSCRAPER, Viking, New York, N.Y.

Saether, K., 1985
PRESTRESSED CONCRETE IN HIGH-RISE CONSTRUCTION, ACI Fall Convention, Chicago, Ill.

Sahlin, S., 1971
STRUCTURAL MASONRY, Prentice-Hall, Englewood Cliffs, N.J.

Sakamoto, Y., Yamaguchi, T., Ohashi, M., and Saito, H., 1992
HIGH-TEMPERATURE PROPERTIES OF FIRE RESISTANT STEEL FOR BUILDINGS, American Society of Civil Engineers, *Structural Engineering Journal,* February, pp. 392–407.

Saliga, P. (Ed.), 1990
THE SKY'S THE LIMIT: A CENTURY OF CHICAGO SKYSCRAPERS, Rizzoli, New York, N.Y.

Salmon, C. G., and Johnson, J. E., 1990
STEEL STRUCTURES: DESIGN AND BEHAVIOR, Harper and Row, New York, N.Y.

Sasaki, M., 1964
A TALL OFFICE BUILDING, Masters Thesis, Illinois Institute of Technology, Department of Architecture, Chicago.

Sato, K., 1982
NEW COMPOSITE STRUCTURE SYSTEM FOR HIGH-RISE BUILDINGS, Proceedings of the Technical Forum on High-Rise Steel Structures, Singapore, March.

Schleich, J. B., 1987
NUMERICAL SIMULATIONS, THE FORTHCOMING APPROACH IN FIRE ENGINEERING DESIGN OF STEEL STRUCTURES, *Revue Technique Luxembourgeoise.*

Schmertz, M., 1975
OFFICE BUILDING DESIGN, *Architectural Record,* pp. 87–192.

Schmidt, W., 1966
HIGH-RISE BUILDINGS OF REINFORCED CONCRETE—WHAT ARE THE LIMITATIONS?, *Journal of the American Concrete Institute,* December.

Schmitt, R., 1991
SCHUYLKILL RIVEREDGE PROJECT, PHILADELPHIA, PENNSYLVANIA, in *Urban Synergy: Process, Projects, and Projections,* School of Architecture, University of Illinois at Urbana-Champaign, Champaign.

Schneider, R. R., 1969
SHEAR IN CONCRETE MASONRY PIERS, Reports on Tests at California State Polytechnic University, Pomona, Calif.

Schneider, R. R., and Dickey, W. L., 1979
REINFORCED MASONRY DESIGN, Prentice-Hall, Englewood Cliffs, N.J.

Schueller, W., 1977
HIGH-RISE BUILDING STRUCTURES, John Wiley, New York, N.Y.

Schueller, W., 1990
THE VERTICAL BUILDING STRUCTURE, Chap. 7: "Building Structure Systems," Van Nostrand Reinhold, New York, N.Y.

Schulze, F., and Harrington, K., 1993
CHICAGO'S FAMOUS BUILDINGS, University of Chicago Press, Chicago, Ill.

Scrivener, J. C., 1972
REINFORCED MASONRY—SEISMIC BEHAVIOR AND DESIGN, Bulletin, N.Z. Society of Earthquake Engineering, vol. 5, no. 4.

Scrivener, J. C., 1976
REINFORCED MASONRY IN A SEISMIC AREA—RESEARCH AND CONSTRUCTION DEVELOPMENT IN NEW ZEALAND, Proceedings of 1st Canadian Masonry Symposium, E. L. Jessop and M. A. Ward (Eds.), University of Calgary, Alta., Canada.

Scully, V., 1988
AMERICAN ARCHITECTURE AND URBANISM, Henry Holt, New York, N.Y.

Scuri, P., 1990
LATE-TWENTIETH CENTURY SKYSCRAPERS, Van Nostrand Reinhold, New York, N.Y.

Selby, R., 1993
URBAN SYNERGY: PROCESS, PROJECTS, AND PROJECTIONS, School of Architecture, University of Illinois at Urbana-Champaign, Champaign.

Selby, R., and Warfield, J., 1992
BEYOND EDMUND BACON'S PHILADELPHIA: A CASE STUDY OF URBAN METAMORPHOSES, Proceedings of International Association of People-Environment Studies International Conference: Socio-Environment Studies (IAPS 12), Thessaloniki, Greece.

Sellevoid, E. J., and Nilsen, T., 1987
CONDENSED SILICA FUME IN CONCRETE: A WORLD REVIEW, in *Supplementary Cementing Materials,* Canadian Government Printing Center, Ottawa, Ont., Canada, pp. 164, 168.

Senbu, O., Abe, M., Matsushima, Y., Baba, A., and Sugiyama, M., 1991
EFFECTS OF ADMIXTURES ON COMPACTIBILITY AND PROPERTIES OF GROUT, Proceedings of 9th International Brick/Block Masonry Conference, Berlin, Germany, vol. 1.

Sharp, D., 1989
TROPICAL HEIGHTS, *RIBA* (*Royal Institute of British Architects*) *Journal,* August, pp. 42–46.

Sheehan, W. J., 1971
SECURITY, Australian Report on Environmental Aspects of the Design of Tall Buildings, Fritz Engineering Laboratory Report 369.32, Lehigh University, Bethlehem, Pa.

Shin, S. W., Kamara, M., and Ghosh, S. K., 1990
FLEXURAL DUCTILITY, STRENGTH PREDICTION, AND HYSTERETIC BEHAVIOR OF ULTRA-HIGH-STRENGTH CONCRETE MEMBERS, in *High-Strength Concrete—2d International Symposium,* Publication SP-121, American Concrete Institute, Detroit, Mich., pp. 239–264.

Shrive, N. G., and Jessop, E. L., 1980
ANISOTROPY IN EXTRUDED CLAY UNITS AND ITS EFFECT ON MASONRY BEHAVIOR, Proceedings of 2d Canadian Masonry Symposium, Ottawa, Ont., Canada.

Siddiqui, K., Qadir, S. R., Alamgir, S., and Huq, S., 1990
SOCIAL FORMATION IN DHAKA CITY, University Press Ltd., Dhaka, Bangladesh.

Siegel, C., 1962
STRUCTURE AND FORM IN MODERN ARCHITECTURE, Reinhold Publishing, New York, N.Y.

Siegel, C., 1975
STRUCTURE AND FORM IN MODERN ARCHITECTURE, Translation by T. E. Burton, Robert E. Krieger, Huntington, N.Y.

Sitte, C., 1889
CITY PLANNING ACCORDING TO ARTISTIC PRINCIPLES, Phaidon, London, England.

Skidmore, Owings & Merrill, 1986
AT&T TOWER, PHASE I, Original Construction Documents, Chicago, Ill.

Skidmore, Owings & Merrill, 1988a
ENERGY APPLICATION: GUIDE TO THE ENERGY PROGRAM, IBM Architecture and Engineering Series, SC33-8161.

Skidmore, Owings & Merrill, 1988b
HVAC APPLICATION: LOADS REFERENCE GUIDE, IBM Architecture and Engineering Series, SC33-8162.

Skidmore, Owings & Merrill, 1988c
HVAC APPLICATION: DUCTS REFERENCE GUIDE, IBM Architecture and Engineering Series, SC33-8164.

Skidmore, Owings & Merrill, 1988d
LIGHTING APPLICATION: LIGHTS REFERENCE GUIDE, IBM Architecture and Engineering Series, SC33-8168.

Skidmore, Owings & Merrill, and SKM Systems Analysis Inc., 1988e
POWER APPLICATION: POWER REFERENCE GUIDE, IBM Architecture and Engineering Series, SC33-8170.

Skidmore, Owings & Merrill, 1989
PIPING APPLICATION: PIPES REFERENCE GUIDE, IBM Architecture and Engineering Series, SC33-8166.

Skidmore, Owings & Merrill, 1990
USG TOWER, Original Construction Documents, Chicago, Ill.

Sletteland, G., 1971
A CITY OF SKYSCRAPERS, in *The Ultimate Highrise: San Francisco's Mad Rush Toward the Sky,* B. Brugmann and G. Sletteland (Eds.), San Francisco Bay Guardian Books, San Francisco, Calif., pp. 14–29.

Smith, G. E. K., 1955
ITALY BUILDS, Reinhold Publishing, New York, N.Y.

Smith, G. H., and Slack, W. J., 1990
TECHNICS: CURTAIN WALLS—OPTIONS AND ISSUES, *Progressive Architecture,* Reinhold Publishing, Cleveland, Ohio, April, pp. 45–62.

Smith, P. C., 1990
THE EXPRESSIVE ROLE OF THE TALL BUILDING IN HONG KONG, in *Tall Buildings: 2000 and Beyond,* Proceedings of 4th World Congress on Tall Buildings, Hong Kong, November.

Smith, W., 1991
FAULTY TOWERS: STUDENTS AT ILLINOIS STATE U. LEARN TO LIVE WITH HIGH ANXIETY IN THE WORLD'S LARGEST COLLEGE DORM, *Chicago Tribune,* Sunday, September 1, Tempo Section, pp. 1, 5.

Soleri, P., 1973
TALL BUILDINGS AND GIGANTISM, Mexican Conference on Tall Buildings, Mexico City, Mexico, March 7–9.

Sommer, R., 1986
STUDENT REACTIONS TO FOUR TYPES OF RESIDENCE HALLS, *Journal of College Student Personnel,* vol. 9, pp. 232–237.

Sommer, R., 1987
CRIME AND VANDALISM IN UNIVERSITY RESIDENCE HALLS: A CONFIRMATION OF DEFENSIBLE SPACE THEORY, *Journal of Environmental Psychology,* vol. 7, pp. 1–12.

Soretz, S., 1965a
FATIGUE BEHAVIOR OF HIGH YIELD STEEL REINFORCEMENT, Tor-Isteg Steel Corporation, vol. 26, Luxembourg.

Soretz, S., 1965b
TENSILE STRESS ON REINFORCING STEELS UNDER SUSTAINED LOAD AND INCREASING TEMPERATURE, Tor-Isteg Steel Corporation, vol. 31, Luxembourg.

Soretz, S., 1974
WIDE SPAN REINFORCED CONCRETE FLOORS, Tor-Isteg Steel Corporation, vol. 42, Luxembourg.

Spaeth, D., 1985
MIES VAN DER ROHE, Rizzoli International, New York, N.Y.

Spreiregen, P. D., 1965
URBAN DESIGN: THE ARCHITECTURE OF TOWNS AND CITIES, The American Institute of Architects, McGraw-Hill, New York, N.Y.

Stackhouse, J., 1992
FOREIGN-AID ARMY REGROUPING IN FACE OF FAILURE, *The Globe and Mail,* Toronto, Ont., Canada, April 4.

Stamps, A. E., 1991
PUBLIC PREFERENCES FOR HIGH RISE BUILDINGS: STYLISTIC AND DEMOGRAPHIC EFFECTS, *Perceptual and Motor Skills,* vol. 72, pp. 839–844.

Stern, R. A. M., 1986
PRIDE OF PLACE: BUILDING THE AMERICAN DREAM, Houghton Mifflin, Boston, American Heritage, New York, N.Y.

Stern, R. A. M., Gilmartin, G., and Mellins, T., 1978
NEW YORK 1930: ARCHITECTURE AND URBANISM BETWEEN THE TWO WARS, Rizzoli International, New York, N.Y.

Stockbridge, J. G., and Hendry, A. W., 1969
CASE STUDIES AND CRITICAL EVALUATION OF HIGH-RISE LOAD-BEARING BRICKWORK IN BRITAIN, in *Designing, Engineering and Construction with Masonry Products,* F. B. Johnson (Ed.), Gulf Publishing, Houston, Tex.

Straub, H., 1964
A HISTORY OF CIVIL ENGINEERING, MIT Press, Cambridge, Mass.

Strauss, A. L., 1976
IMAGES OF THE AMERICAN CITY, Transaction Books, New Brunswick, N.J.

Structural Stability Research Council, 1979a
A SPECIFICATION FOR THE DESIGN OF STEEL-CONCRETE COMPOSITE COLUMNS, Task Group 20, *AISC Engineering Journal,* vol. 16, no. 4.

Structural Stability Research Council, 1979b
PLASTIC HINGE BASED METHODS FOR ADVANCED ANALYSIS AND DESIGN OF STEEL FRAMES, Compendium of Papers for Task Group 29 Workshop, April 5.

Sullivan, L. H., 1896
THE TALL OFFICE BUILDING ARTISTICALLY CONSIDERED, first published in Lippincott's 57, March; rpt. in Sullivan, *Kindergarten Chats and Other Writings,* George Wittenborn, New York, N.Y., 1947.

Sundstrom, E., 1987
WORK ENVIRONMENTS: OFFICES AND FACTORIES, in *Handbook of Environmental Psychology,* D. Stokols and I. Altman (Eds.), vol. 1, John Wiley, New York, N.Y., pp. 733–782.

Sundstrom, E., and Sundstrom, M. G., 1986
WORK PLACES: THE PSYCHOLOGY OF THE PHYSICAL ENVIRONMENT IN OFFICES AND FACTORIES, Cambridge University Press, New York, N.Y.

Sutherland, R. J. M., 1969
DESIGN ENGINEERS' APPROACH TO MASONRY CONSTRUCTION, in *Design Engineering and Construction with Masonry Products,* F. Johnson (Ed.), Gulf Publishing, Houston, Tex.

Syrovy, P., 1973
DEVELOPMENT, ARCHITECTURE AND TOWN-PLANNING CONDITIONS OF DESIGN AND CONSTRUCTION OF TALL BUILDINGS, Proceedings of 10th Regional Conference, Bratislava, Czechoslovakia, April; CSVTA—Czechoslovak Scientific and Technical Association, Bratislava, Czechoslovakia (two editions: English and Czech), pp. 65–72.

Taranath, B., 1988
STRUCTURAL ANALYSIS AND DESIGN OF TALL BUILDINGS, McGraw-Hill, New York, N.Y.

Tarricone, P., 1991
OUT OF THE LAB...AND INTO THE FIELD? *ASCE Civil Engineering,* July.

Taylor, W. R., 1992
IN PURSUIT OF GOTHAM: CULTURE AND COMMERCE IN NEW YORK, Oxford University Press, New York, N.Y.

Telford, T., 1811
FIFTH REPORT OF THE COMMISSIONERS FOR ROADS AND BRIDGES IN THE HIGHLANDS OF SCOTLAND, Report of January 3, 1811, House of Commons, April 8.

Telford, T., 1838
LIFE OF THOMAS TELFORD, CIVIL ENGINEER, J. Rickman (Ed.), Jaine and Luke G. Hansard and Sons, London, England.

*The New York Times,* 1993
BLAST HITS TRADE CENTER, BOMB SUSPECTED; 5 KILLED, THOUSANDS FLEE SMOKE IN TOWERS, vol. 142, February 27.

Thomas, V. C., 1990
INTELLIGENT COMPUTER PROGRAMS FOR BUILDING SERVICES ENGINEERING DESIGN, *ASHRAE Transactions,* vol. 96, part 2.

Thomas, V. C., 1991
MECHANICAL ENGINEERING DESIGN COMPUTER PROGRAMS FOR BUILDINGS, *ASHRAE Transactions,* vol. 97, part 2.

Thornton, C. H., Hungspruke, U., and Descenza, R. P., 1991
LOOKING DOWN AT THE SEARS TOWER, *Modern Steel Construction,* American Institute of Steel Construction, Chicago, Ill., August.

*Time,* 1965
THE NORTHEAST: THE DISASTER THAT WASN'T, vol. 86, November 19, pp. 36–43.

*Time,* 1993
TOWER TERROR, vol. 141, no. 10, March 8.

Toong, H. D., and Gapta, A., 1982
PERSONAL COMPUTERS, *Scientific American,* pp. 87–107.

Torroja, E., 1958
THE BEAUTY OF STRUCTURES, in *Philosophy of Structures,* University of California Press, Berkeley, Calif.

Tuchman, J., 1989
LESLIE E. ROBERTSON: ENR MAN OF THE YEAR, *Engineering News Record,* February 23, pp. 38–44.

Tucker, J. B., 1985
SUPER SKYSCRAPERS: AIMING FOR 200 STORIES, *High Technology,* January.

UBC, 1991
EARTHQUAKE REGULATIONS, Section 2312, Uniform Building Code 1991, International Conference of Building Officials, Whittier, Calif.

UFAS, 1984
UNIFORM FEDERAL ACCESSIBILITY STANDARDS, *Federal Register* (49FR31528), Washington, D.C., August 7.

UIA/AIA, World Congress of Architects, 1993
DECLARATION OF INTERDEPENDENCE FOR A SUSTAINABLE FUTURE, June 18–21, Chicago, Ill.

Vale, B., 1991
GREEN ARCHITECTURE: DESIGN FOR A SUSTAINABLE FUTURE, Thames and Hudson, London, England.

Van der Ryn, S., and Silverstein, M., 1967
DORMS AT BERKELEY: AN ENVIRONMENTAL ANALYSIS, Center for Planning and Development Research, Berkeley, Calif.

Vanadol, T., 1991
PRIVATE COMMUNICATION TO ALFRED SWENSON, Illinois Institute of Technology, Chicago.

Venturi, R., 1966
COMPLEXITY AND CONTRADICTION IN ARCHITECTURE, Museum of Modern Art, New York, N.Y.

Vergara, C. J., 1989
HELL IN A VERY TALL PLACE, *The Atlantic Monthly,* vol. 264, September, pp. 72–78.

Vierk, K. E., 1986
CONTINUING THE EVOLUTION OF THE SKYSCRAPER: A 142 STORY STEEL K-BRACED MULTI-USE BUILDING, Masters Thesis, Illinois Institute of Technology, Department of Architecture, Chicago.

Vilar, A. (Ed.), 1970
ADVANCED STEEL CONSTRUCTION IMPLEMENTS BELLUSCHI-ROTH DESIGN FOR A 41-STORY OFFICE BUILDING, *Building Design & Construction,* vol. 11, no. 1, January, pp. 66–74.

Vuyosevich, R. D., 1991
SEMPER AND TWO AMERICAN GLASS HOUSES, Reflections 8: *Journal of the School of Architecture,* University of Illinois at Urbana-Champaign, Champaign, Spring, pp. 4–11.

Wakabayashi, M., 1986
DESIGN OF EARTHQUAKE-RESISTANT BUILDINGS, McGraw-Hill, New York, N.Y.

Walker, D., 1987
SERVICES AND STRUCTURE—THE HONG KONG AND SHANGHAI BANK, in *Great Engineers,* St. Martins Press, New York, N.Y., pp. 178–191.

Walsh, M. B., 1990
311 S. WACKER STANDS STRONG AND TALL, *Dodge Construction News,* vol. 44, no. 166, August, pp. 4–12.

Wardenier, J., et al., 1991
DESIGN GUIDE FOR CHS JOINTS UNDER PREDOMINANTLY STATIC LOADING, CIDECT, Verlag TUV Rheinland GmbH, Cologne, Germany.

Warfield, J., 1991
EDMUND BACON: 1991–92 RECIPIENT OF THE PLYM DISTINGUISHED PROFESSORSHIP IN ARCHITECTURE, School of Architecture, University of Illinois at Urbana-Champaign, Champaign.

Warfield, J., and Selby, R., 1992
TOMORROW'S SKYLINE: A SEARCH FOR THE ROLE OF TALL BUILDINGS IN URBAN DESIGN, Proceedings of International Conference on Tall Buildings: Reach for the Sky, Kuala Lumpur, Malaysia.

Warfield, J., Zheng, S., and Hammond, J., 1989
SHANGHAI: URBAN DIRECTIONS IN NEW CHINA, Proceedings of 2d International Convention on Urban Planning, Housing and Design, Singapore, July.

Weisman, G., 1979
WAY-FINDING IN THE BUILT ENVIRONMENT: A STUDY IN ARCHITECTURAL LEGIBILITY, Ph.D. Dissertation, Department of Architecture, University of Michigan.

Weisman, G., 1981
EVALUATING ARCHITECTURAL LEGIBILITY: WAY-FINDING IN THE BUILT ENVIRONMENT, *Environment and Behavior,* vol. 13, no. 2, March, pp. 189–204.

Weisman, L. K., 1992
DISCRIMINATION BY DESIGN: A FEMINIST CRITIQUE OF THE MAN-MADE ENVIRONMENT, University of Illinois Press, Urbana.

Wener, R., 1988
DOING IT RIGHT: EXAMPLES OF SUCCESSFUL APPLICATION OF ENVIRONMENT-BEHAVIOR RESEARCH, *Journal of Architectural and Planning Research,* vol. 5, no. 4, Winter, pp. 284–300.

White, R. N., and Salmon, C. G., 1987
BUILDING STRUCTURAL DESIGN HANDBOOK, John Wiley, New York, N.Y.

White, T. H., 1965
WHAT WENT WRONG? SOMETHING CALLED 345 KV, *Life,* vol. 59, November 19, pp. 46–52.

Whyte, W. H., 1980
THE SOCIAL LIFE OF SMALL URBAN SPACES, The Conservation Foundation, Washington, D.C.

Whyte, W. H., 1988
CITY: REDISCOVERING THE CENTER, Doubleday, New York, N.Y.

Williamson, R. C., 1981
ADJUSTMENT TO THE HIGHRISE: VARIABLES IN A GERMAN SAMPLE, *Environment and Behavior,* vol. 13, no. 3, pp. 289–310.

Wineman, J. D., 1982
THE OFFICE ENVIRONMENT AS A SOURCE OF STRESS, in *Environmental Stress,* G. W. Evans (Ed.), Cambridge University Press, New York, N.Y., pp. 256–285.

Wineman, J. D. (Ed.), 1986
BEHAVIORAL ISSUES IN OFFICE DESIGN, Van Nostrand Reinhold, New York, N.Y.

Wolner, E. W., 1989
URBAN ICON, *Inland Architect,* vol. 34, no. 1, January/February, pp. 44–48.

Woodward, L. N., 1978
MODERN SHOPPING CENTER DESIGN: PSYCHOLOGY MADE CONCRETE, *Real Estate Review,* vol. 8, no. 2, Summer, pp. 52–55.

World Health Organization, 1980
CLASSIFICATIONS OF IMPAIRMENTS, DISABILITIES AND HANDICAPS, Geneva, Switzerland.

Xu, P., 1991
MULTI-USE HIGH-RISE CONCRETE BUILDING, Masters Thesis, Illinois Institute of Technology, Department of Architecture, Chicago.

Yao, C., and Nathan, N. D., 1989
AXIAL CAPACITY OF GROUTED CONCRETE MASONRY, Proceedings of 5th Canadian Masonry Symposium, vol. 1, University of British Columbia, Vancouver, B.C., Canada.

Yeang, K., 1990
HIGH-RISE DESIGN FOR HOT HUMID PLACES, Manuscript available from T. R. Hamzah and Yeang, Kuala Lumpur, Malaysia, August.

Youssef, N., 1991
MEGASTRUCTURE: A NEW CONCEPT FOR SUPERTALL BUILDINGS, *Modern Steel Construction,* August.

Zalcik, T., and Franco, D., 1973
TOWN-PLANNING AND SOCIOLOGICAL CONSIDERATIONS OF TALL BUILDING CONSTRUCTION IN SLOVAKIA, Proceedings of 10th Regional Conference, Bratislava, Czechoslovakia, April; CSVTA—Czechoslovak Scientific and Technical Association, Bratislava, Czechoslovakia (two editions: English and Czech), pp. 34–39.

Zelanski, P., 1987
SHAPING SPACE, Holt, Rinehart and Winston, New York, N.Y., pp. 104–112.

Zils, J. J., and Clark, R. S., 1986
THE CONCRETE DIAGONAL, *ASCE Civil Engineering,* October, pp. 48–50.

Zoll, S., 1974
THE UTILITY OF VERY HIGH BULK, in *Tall Buildings and People?* Proceedings of Conference, Great Britain, September 17–19; Institution of Structural Engineers, London, England, pp. 36–41.

Zotti, E., 1987
STACKING THE DECKS FOR BETTER DESIGN, *Planning,* vol. 53, no. 10, October, pp. 19–22.

Zuckerman, M., Miserandino, M., and Bernieri, F., 1983
CIVIL INATTENTION EXISTS—IN ELEVATORS, *Personality and Social Psychology Bulletin,* vol. 9, no. 4, December, pp. 578–586.

# Contributors

The following is a list of those who have contributed to the manuscript for this Monograph. The names, affiliations, and countries of each contributor are given.

**Ali, M. M.,** University of Illinois, Urbana-Champaign, Illinois, USA
**Anthony, K. H.,** University of Illinois, Urbana-Champaign, Illinois, USA
**Armstrong, P. J.,** University of Illinois, Urbana-Champaign, Illinois, USA
**Attia, E.,** Eli Attia Architects, New York, New York, USA
**Babu, A.,** University of Illinois, Urbana-Champaign, Illinois, USA
**Billington, D. P.,** Princeton University, Princeton, New Jersey, USA
**Billington, S. L.,** University of Texas, Austin, Texas, USA
**Brockenbrough, R.,** R. L. Brockenbrough & Associates, Inc., Pittsburgh, Pennsylvania, USA
**Chang, P. C.,** Illinois Institute of Technology, Chicago, Illinois, USA
**Ciamarra, M. P.,** Pica Ciamarra Associati, Naples, Italy
**Clark, R. J.,** Skidmore, Owings and Merrill, Chicago, Illinois, USA
**Cowan, H. J.,** University of Sydney, Sydney, Australia
**Dolan, P.,** University of Illinois, Urbana-Champaign, Illinois, USA
**Dry, C.,** University of Illinois, Urbana-Champaign, Illinois, USA
**Dutta, C. R.,** Jadavpur University, Calcutta, India
**Eilken, D.,** University of Illinois, Urbana-Champaign, Illinois, USA
**Elnimeiri, M.,** Illinois Institute of Technology, Chicago, Illinois, USA
**Eyres, J.,** University of Illinois, Urbana-Champaign, Illinois, USA
**Ghosh, S. K.,** Portland Cement Association, Skokie, Illinois, USA
**Gibbon, M. E.,** University of Illinois, Urbana-Champaign, Illinois, USA
**Goldsmith, M.,** Illinois Institute of Technology, Chicago, Illinois, USA
**Gray, M.,** Illinois Institute of Technology, Chicago, Illinois, USA
**Haines, L. W.,** Bechtel, San Francisco, California, USA
**Iwankiw, N.,** American Institute of Steel Construction, Chicago, Illinois, USA
**Jones, D. S.,** Bechtel, San Francisco, California, USA
**Jordon, R. S.,** Bechtel, San Francisco, California, USA
**Kao, C. H.,** Illinois Institute of Technology, Chicago, Illinois, USA
**Kelly, M.,** John & Associates, Chicago. Illinois, USA
**Kim, M. K.,** University of Illinois, Urbana-Champaign, Illinois, USA
**McDermott, A.,** University of Illinois, Urbana-Champaign, Illinois, USA
**Melnick, S.,** American Institute of Steel Construction, Chicago, Illinois, USA
**Monk, C.,** Wiss, Janney, Elstner Associates, Inc., Northbrook, Illinois, USA
**Nath, L.,** Jadavpur University, Calcutta, India
**Nawaz, T.,** T. Nawaz Architects and Planners, Nepean, Ontario, Canada

**Obata, G.,** Hellmuth, Obata & Kassabaum, Inc., St. Louis, Missouri, USA
**Ortiz, I.,** Ortiz Leon Arquitectos, Madrid, Spain
**Patel, R.,** University of Illinois, Urbana-Champaign, Illinois, USA
**Pennypacker, R.,** Bechtel, San Francisco, California, USA
**Salen, J. S.,** University of Illinois, Urbana-Champaign, Illinois, USA
**Selby, R. I.,** University of Illinois, Urbana-Champaign, Illinois, USA
**Sengupta, B.,** Jadavpur University, Calcutta, India
**Sharpe, D. C.,** Illinois Institute of Technology, Chicago, Illinois, USA
**Shoemaker, D. W.,** University of Illinois, Urbana-Champaign, Illinois, USA
**Slowik, L. J.,** Bechtel, San Francisco, California, USA
**Smith, R. J.,** Bechtel, San Francisco, California, USA
**Swenson, A.,** Illinois Institute of Technology, Chicago, Illinois, USA
**Tange, K.,** Kenzo Tange Associates, Tokyo, Japan
**Thomas, V.,** University of Oklahoma, Norman, Oklahoma, USA
**Yeang, K.,** T. R. Hamzah & Yeang Sdn. Bhd., Selangor, Malaysia
**Warfield, J. P.,** University of Illinois, Urbana-Champaign, Illinois, USA
**Wong, A.,** Bechtel, San Francisco, California, USA

# Building Appendix

### 10 Columbus Circle, 1989[2]
**San Francisco, California**
Architect: Eli Attia Associates

### 101 California Street, 1982
**San Francisco, California**
*Architect:* Burgee/Johnson Associates
*Developer:* Gerald D. Hines Interests
*Structural Engineer:* CBM Engineers
*Building Services Engineer:* Cygma Energy Services

### 101 Park Avenue, 1985
**New York, New York**
*Architect:* Burgee/Johnson Associates
*Developer:* H.J. Kalikow & Company, Inc.
*Structural Engineer:* CBM Engineers
*Building Services Engineer:* Cosentini Associates

### 123 North Wacker, 1987[1]*
**Chicago, Illinois**
*Architect:* Perkins & Will
*Developer:* Rubloff, Inc.
*Structural Engineer:* Perkins & Will
*Building Services Engineer:*

### 1300 Clay, 1989
**Oakland, California**
*Architect:* IDG Architects/Kendall Heton
*Developer:* Bramalea Pacific
*Structural Engineer:* Skilling Ward Magnusson Barkshire
*Building Services Engineer:* SJ Engineering (mech), Silverman Light (elec)

### 30th Street Station, 19341
**Philadelphia, Pennsylvania**
*Architect:* Graham, Anderson, Probst & White

### 311 South Wacker, 1990
**Chicago, Illinois**
*Architect:* Kohn Pedersen Fox Associates
*Developer:* Lincoln Property Company
*Structural Engineer:* Brockette/Davis/Drake, Inc.
*Building Services Engineer:* BL & P Engineers, Inc.

### 333 West Wacker, 1983
**Chicago, Illinois**
*Architect:* Kohn Pedersen Fox Associates
*Developer:* Urban Investment & Development Co.
*Structural Engineer:* GCE of Illinois
*Building Services Engineer:* Environmental Systems Design

### 600 California Street, 1991
**San Francisco, California**
*Architect:* Kohn Pedersen Fox Associates
*Developer:* Lincoln Property Company
*Structural Engineer:* Skilling Ward Magnusson Barkshire
*Building Services Engineer:* Flack & Kurtz

### 712 Fifth Avenue
**New York, New York**
*Architect:* Burgee/Johnson Associates
*Developer:* Soloman Equities, Inc.
*Structural Engineer:* Weiskopf & Pickworth
*Building Services Engineer:* Jaros, Baum & Bolles

*Numbered footnotes appear at the end of the Appendix.

### 750 Seventh Avenue, 1990
**New York, New York**
*Architect:* Kevin Roche, John Dinkeloo and Associates
*Developer:* Soloman Equities, Inc.
*Structural Engineer:* Weiskopf & Pickworth
*Building Services Engineer:* Jaros, Baum & Bolles

### 780 Third Avenue, 1983
**New York, New York**
*Architect:* Skidmore, Owings & Merrill
*Developer:* Cohen Brothers Realty & Construction
*Structural Engineer:* Skidmore, Owings & Merrill
*Building Services Engineer:* Skidmore, Owings & Merrill

### 801 Grand Avenue, 19901
**Des Moines, Iowa**
*Architect:* Hellmuth, Obata & Kassabaum, Inc./Brooks, Borg & Skiles
*Developer:* Principal Financial Group

### 860-880 Lakeshore Drive, 19521
**Chicago, Illinois**
*Architect:* Mies Van Der Rohe

### 9 West 57th Street, 1973
**New York, New York**
*Architect:* Skidmore, Owings & Merrill
*Developer:* Solow Development Corporation
*Structural Engineer:* Weidlinger Association
*Building Services Engineer:* Cosentini Associates

### 900 North Michigan Avenue, 1988
**Chicago, Illinois**
*Architect:* Kohn Pedersen Fox Associates
*Developer:* Urban Investment & Development Co.
*Structural Engineer:* Alfred Benesch Company
*Building Services Engineer:* Environmental Systems Design

### Adamjee Court Building, 1959[1]
**Dhaka, Bangladesh**
*Architect:* Tajuddin Bhamani
*Developer:* Adamjee Jute Mills Corporation

### Albuquerque Plaza, 19901
**Albuquerque, New Mexico**
*Architect:* Hellmuth, Obata & Kassabaum, Inc.
*Developer:* Beta West Properties, Inc.

### Allied Bank Tower, 1986
**Dallas, Texas**
*Architect:* I.M. Pei & Partners
*Developer:* Criswell Development Company
*Structural Engineer:* CBM Engineers
*Building Services Engineer:* Cosentini Associates

### American Standard Bldg. (Radiator Bldg.), 1924[1]
**New York, New York**
*Architect:* Hood & Foulifax

### American Surety Building, 1896[1]
**New York, New York**
*Architect:* Bruce Price

### Americana Hotel, 1962[1]
**New York, New York**
*Architect:* Morris Lapidus & Associates
*Developer:* Lawrence-Tisch Company

### Amin Court Building, 1961[1]
**Dhaka, Bangladesh**
*Architect:* Tajuddin Bhamani
*Developer:* Amin Jute Mills Corporation

### AMOCO Building, 1973[1]
**Chicago, Illinois**
*Architect:* Perkins & Will
*Developer:* Standard Oil Realty
*Structural Engineer:* Perkins & Will

### AmSouth/Harbert Plaza, 1989
**Birmingham, Alabama**
*Architect:* Hellmuth, Obata & Kassabaum, Inc.
*Developer:* Amsouth Bank & Harbert International, Inc
*Structural Engineer:* Daturn Moore Partnerships Association
*Building Services Engineer:* Miller-Weaver Associates, Inc.

**AT&T Corporate Center, 1989**
**Chicago, Illinois**
*Architect:* Skidmore, Owings & Merrill
*Developer:* Stein & Company
*Structural Engineer:* Skidmore, Owings & Merrill
*Building Services Engineer:* Skidmore, Owings & Merrill

**AT&T Headquarters, 1984**
**New York, New York**
*Architect:* Johnson/Burgee Associates
*Developer:* 195 Broadway Corporation
*Structural Engineer:* Robertson, Fowler and Associates
*Building Services Engineer:* Cosentini Associates

**Auditorium Building, 1889[1]**
**Chicago, Illinois**
*Architect:* Adler & Sullivan

**Bangladesh Bank Building, 1963**
**Dhaka, Bangladesh**
*Architect:* Thariani & Company
*Developer:* Government of Pakistan
*Structural Engineer:* Thariani & Company
*Building Services Engineer:* Thariani & Company

**Bank of America Building, 1969[1]**
**San Francisco, California**
*Architect:* Skidmore, Owings & Merrill—San Francisco/Pietro Belluschi
*Developer:* Bank of America
*Structural Engineer:* H.J. Brunnier & Associates

**Bank of Asia, 1983**
**Bangkok, Thailand**
*Architect:* Sumet Jumsai Associates
*Developer:* Asia Properties Corporation Ltd.
*Structural Engineer:* Tawatchai Nakata
*Building Services Engineer:* Rak Sumanmongko

**Bank of China, 1988**
**Hong Kong, China**
*Architect:* I.M. Pei & Partners
*Developer:* Bank of China/Hong Kong
*Structural Engineer:* Robertson, Fowler & Associates
*Building Services Engineer:* Jaros, Baum & Bolles

**Bank of Manhattan, 1929[1]**
**New York, New York**
*Architect:* H. Craig Severance/Yasou Matsui

**Bank of the Southwest Tower, 1982[2]**
**Houston, Texas**
*Architect:* Murphy/Jahn (completed building designed by Kohn Pedersen Fox Associates)
*Developer:* Century Development Corporation
*Structural Engineer:* Le Messurier Associates/SCI

**Bayard Building, 1898[1]**
**New York, New York**
*Architect:* Louis Sullivan

**Biblioteque St. Genevieve, 1850[1]**
**Paris, France**
*Architect:* Henri Labrouste

**Bima Bhaban[1]**
**Dhaka, Bangladesh**
*Architect:* Vastukalavid Architects

**Board of Trade, 1930[1]**
**Chicago, Illinois**
*Architect:* Holabird & Root

**BP America, 1986**
**Cleveland, Ohio**
*Architect:* Hellmuth, Obata & Kassabaum, Inc.
*Developer:* Original Standard Oil of Ohio
*Structural Engineer:* William O'Neal of KKBNA
*Building Services Engineer:* Jaros, Baum & Bolles

**Broadacre City, 1936[2]**
**Oak Park, Illinois**
*Architect:* Frank Lloyd Wright

**Broadgate Building, 1990**
**London, England**
*Architect:* Skidmore, Owings & Merrill
*Developer:* Rosehaugh Stanhope Development
*Structural Engineer:* Skidmore, Owings & Merrill
*Building Services Engineer:* Jaros, Baum & Bolles

**Brunswick Building, 1965**[1]
**Chicago, Illinois**
*Architect:* Skidmore, Owings & Merrill
*Developer:* Washington-Dearborn Property
*Structural Engineer:* Skidmore, Owings & Merrill

**Building B, World Financial Center, 1988**
**New York, New York**
*Architect:* Cesar Pelli & Associates
*Developer:* Olympia & York Equity Corporation
*Structural Engineer:* Lev Zetlin Associates, Inc.
*Building Services Engineer:* Flack & Kurtz

**Business Men's Assurance Tower, 1964**
**Kansas City, Missouri**
*Architect:* Skidmore, Owings & Merrill
*Developer:* Business Men's Assurance Company of America
*Structural Engineer:* Skidmore, Owings & Merrill
*Building Services Engineer:* Black & Veatch

**Carbide and Carbon Building, 1929**[1]
**Chicago, Illinois**
*Architect:* Burnham Brothers

**Carson Pirie Scott Department Store, 19041**
**Chicago, Illinois**
*Architect:* Louis Sullivan
*Developer:* Schlesinger & Mayer

**Casa a Milano, ca.1950**[1]
**Milan, Italy**
*Architect:* Luigi Figini & Gino Pollini

**CBS Building, 1965**[1]
**New York, New York**
*Architect:* Eero Saarinen
*Developer:* CBS Studios
*Structural Engineer:* Paul Weidlinger

**Central Plaza, 1992**
**Hong Kong, China**
*Architect:* Nu Chun Man & Associates
*Developer:* Sun Hong Kai Properties
*Structural Engineer:* Ove Arup & Partners
*Building Services Engineer:* Ove Arup & Partners

**Chanin Building, 1929**[1]
**New York, New York**
*Architect:* Sloan & Robertson
*Developer:* Chanin Real Estate

**Chicago Place, 1990**
**Chicago, Illinois**
*Architect:* Skidmore, Owings & Merrill
*Developer:* Brookfield Developments
*Structural Engineer:* Skidmore, Owings & Merrill
*Building Services Engineer:* Skidmore, Owings & Merrill

**Chicago Stock Exchange, 1894**[1]
**Chicago, Illinois**
*Architect:* Adler & Sullivan

**Chicago World Trade Center, ca.1982**[2]
**Chicago, Illinois**
*Architect:* Skidmore, Owings & Merrill
*Structural Engineer:* Skidmore, Owings & Merrill

**Christian Science Center, 1973**
**Boston, Massachusetts**
*Architect:* I.M. Pei & Partners
*Developer:* First Church of Christ Scientist
*Structural Engineer:* Weiskopf & Pickworth
*Building Services Engineer:* Syska & Hennessey

**Chrysler Corporation Building, 1930**
**New York, New York**
*Architect:* William Van Allen
*Developer:* Chrysler Corporation
*Structural Engineer:* Ralph Squire & Sons
*Building Services Engineer:* Lewis T.M. Ralston

**Cite de Refuge (Salvation Army Building), 1933**[1]
**Paris, France**
*Architect:* Le Corbusier

**Citicorp Center, 1977**
**New York, New York**
*Architect:* Hugh Stubbins & Associates
*Developer:* Citicorp Corporation
*Structural Engineer:* Le Messurier Associates/SCI
*Building Services Engineer:* Joseph R. Loring & Associates

**City Investing Company Building, 1908[1]**
**New York, New York**
*Architect:* Frances H. Kimball
*Developer:* Robert E. Dowling/City Investing Company

**Civic Center/Daley Center, 1965[1]**
**Chicago, Illinois**
*Architect:* C.F. Murphy/Skidmore, Owings & Merrill/Loebl, Schlossman, Bennett & Dart
*Developer:* Chicago Public Building Commission

**Clinton Homes[1]**
**Brooklyn, New York**
*Architect:* Grueben Partnership
*Developer:* Glick Organization

**Commerce Square, 1993**
**Philadelphia, Pennsylvania**
*Architect:* Pei, Cobb, Freed & Partners
*Developer:* Maguire-Thomas Partners/IBM Real Estate and Construction Department
*Structural Engineer:* CBM Engineers
*Building Services Engineer:* Cosentini Associates

**Commerzbank Headquarters, 1991**
**Frankfurt, Germany**
*Architect:* Norman Foster & Associates
*Developer:* Commerzbank, Inc.
*Structural Engineer:* Ove Arup & Partners
*Building Services Engineer:* J. Roger Preston & Associates

**Complesso Immobiliare, ca.1950[1]**
**Milan, Italy**
*Architect:* Luigi Morfetti

**Concordia Theological Seminary, 1958[1]**
**Fort Wayne, Indiana**
*Architect:* Eero Saarinen

**Crystal Palace, 1851[1]**
**London, England**
*Architect:* Joseph Paxton

**Daily News Building, 1930[1]**
**New York, New York**
*Architect:* Hood and Howells

**Dewitt Chestnut Apartment Tower, 1963[1]**
**Chicago, Illinois**
*Architect:* Skidmore, Owings & Merrill
*Developer:* Metropolitan Structures, Inc.
*Structural Engineer:* Skidmore, Owings & Merrill

**DG Bank, 1992**
**Frankfurt, Germany**
*Architect:* Kohn Pedersen Fox Associates
*Developer:* PGGM Group & DG Bank Group
*Structural Engineer:* Ingenieursozietat BGS
*Building Services Engineer:* Ingenieursozietat Reuter Ruhgartner

**Eastern Federal Union Insurance Building, 1965**
**Dhaka, Bangladesh**
*Architect:* Vastukalavid Architects
*Developer:* Eastern Federal Life Insurance Company
*Structural Engineer:* M. Shahidullah

**Empire State Building, 1931[1]**
**New York, New York**
*Architect:* Shreve, Lamb, and Harmon
*Developer:* John Jacob Raskob
*Structural Engineer:* H.G. Balcom

**Equitable Building, 1915[1]**
**New York, New York**
*Architect:* Ernest Graham

**Federal Center, 1964**
**Chicago, Illinois**
*Architect:* Mies Van Der Rohe/C.F. Murphy
*Developer:* General Services Administration/Public Building Service
*Structural Engineer:* A. Epstein & Sons
*Building Services Engineer:* Schmidt, Gardener & Erickson

**Federal Reserve Bank, 1973**
**Minneapolis, Minnesota**
*Architect:* Gunnar Birkerts & Associates
*Developer:* Federal Reserve Bank
*Structural Engineer:* Skilling Helle Christiansen Robertson
*Building Services Engineer:* Jaros, Baum & Bolles

**Field Buildings, 1893[1]**
**Chicago, Illinois**
*Architect:* Graham, Anderson, Probst & White

**Figueroa at Wilshire Building (Wilshire Financial Building), 1990**
**Los Angeles, California**
*Architect:* Albert C. Martin & Associates
*Developer:* Mitsui Fodosan
*Structural Engineer:* CBM Engineers, Inc.
*Building Services Engineer:* I.A. Naman & Associates

**Fillmore Building[1]**
**San Francisco, California**
*Architect:* The Watry Design Group

**First Boston Corporation Building, 1970[1]**
**Boston, Massachusetts**
*Architect:* Pietro Belluschi
*Structural Engineer:* Office of James Ruderman

**First Interstate World Center, 1989**
**Los Angeles, California**
*Architect:* I.M. Pei & Partners
*Developer:* Maguire Thomas Partners/Pacific Enterprises
*Structural Engineer:* CBM Engineers, Inc.
*Building Services Engineer:* James A. Knowles & Associates

**First National Plaza, 1969[1]**
**Chicago, Illinois**
*Architect:* Perkins & Will/C.F. Murphy
*Developer:* First National Bank of Chicago
*Structural Engineer:* Perkins & Will

**First Wisconsin Bank, 1973**
**Milwaukee, Wisconsin**
*Architect:* Skidmore, Owings & Merrill
*Developer:* First Wisconsin Development Corporation
*Structural Engineer:* Skidmore, Owings & Merrill
*Building Services Engineer:* Cosentini Associates

**Flatiron Building, 1902[1]**
**New York, New York**
*Architect:* Daniel H. Burnham & Company

**Foley Square Federal Office, 1994**
**New York, New York**
*Architect:* Hellmuth, Obata & Kassabaum, Inc.
*Developer:* General Services Administration
*Structural Engineer:* Ysrael A. Sienrick
*Building Services Engineer:* Flack & Kurtz

**Fountain Plaza, 1984**
**Portland, Oregon**
*Architect:* Zimmer Gunsul Frasca Partnership
*Developer:* Olympia & York
*Structural Engineer:* KPFF Consulting Engineers
*Building Services Engineer:* PAE Consulting Engineers

**Gateway Office Building, 1984**
**Chicago, Illinois**
*Architect:* Skidmore, Owings & Merrill
*Developer:* Tishman-Speyer
*Structural Engineer:* Skidmore, Owings & Merrill
*Building Services Engineer:* Skidmore, Owings & Merrill

**Genoa Building, ca.1950[1]**
**Genoa, Italy**
*Architect:* Daneri

**GLG Tower, 1993**
**Atlanta, Georgia**
*Architect:* Rabun Hogan Ota Rasche Architects
*Developer:* Gullstedt Gruppen
*Structural Engineer:* Harald Neilsen Engineers
*Building Services Engineer:* Hudson, Everett, Simonson & Mullis Associates

**Grattacielo Pirelli, 1959[1]**
**Milano, Italy**
*Architect:* Gio Ponti
*Developer:* Pirelli
*Structural Engineer:* Pierluigi Nervi

**Griffin Office Tower, ca.1992[2]**
**Dallas, Texas**
*Architect:* Hellmuth, Obata & Kassabaum, Inc.

**Guaranty Building, 1895**[1]
**Buffalo, New York**
*Architect:* Adler & Sullivan

**Habib Bank Building**[1]
**Dhaka, Bangladesh**
*Architect:* Thariani & Company
*Developer:* Habib Bank Ltd.
*Structural Engineer:* Thariani & Company
*Building Services Engineer:* Thariani & Company

**Hartford Company Buildings, 1961**
**Chicago, Illinois**
*Architect:* Skidmore, Owings & Merrill
*Developer:* Hartford Fire Insurance Company
*Structural Engineer:* Skidmore, Owings & Merrill
*Building Services Engineer:* Skidmore, Owings & Merrill

**Home Insurance Building, 1885**[1]
**Chicago, Illinois**
*Architect:* William Le Baron Jenney

**Hong Kong and Shanghai Bank, 1986**
**Hong Kong, China**
*Architect:* Norman Foster & Associates
*Developer:* HS Property Management Company, Ltd.
*Structural Engineer:* Ove Arup & Partners
*Building Services Engineer:* J. Roger Preston & Associates

**Horton Plaza, 1986**
**San Diego, California**
*Architect:* John Jerde
*Developer:* Ernest Hahn, Inc.
*Structural Engineer:* Robert Englekirk, Inc.
*Building Services Engineer:* David Chen & Associates

**Hotel, Chiavari, ca.1950**[2]
**Chiavari, Italy**
*Architect:* Nardi Greco

**Hotel, Sauce D'Ulzio, ca.1950**[1]
**Sauce D'Ulzio, Italy**
*Architect:* Bonade Bottino

**Hotel, Seistriele, ca.1950**[1]
**Sestriele, Italy**
*Architect:* Bonade Bottino

**Humana Headquarters Building, 1985**
**Louisville, Kentucky**
*Architect:* Michael Graves
*Developer:* Humana, Inc.
*Structural Engineer:* DeSimone, Chaplin & Associates
*Building Services Engineer:* Carestsky & Associates

**Hyatt Regency Hotel at Peachtree Center, 1967**[1]
**Atlanta, Georgia**
*Architect:* John Portman & Associates
*Developer:* The Portman Company

**Hyatt Regency at San Francisco, 1973**[1]
**San Francisco, California**
*Architect:* John Portman & Associates
*Developer:* The Portman Company

**IBM Building, 1973**[1]
**Chicago, Illinois**
*Architect:* Mies Van Der Rohe/C.F. Murphy
*Developer:* International Business Machines Corporation

**IBM Headquarters, 1983**
**New York, New York**
*Architect:* Edward Larrabee Barnes
*Developer:* International Business Machines Corporation
*Structural Engineer:* Office of James Ruderman
*Building Services Engineer:* Joseph Loring & Associates

**IDS Center, 1972**[1]
**Minneapolis, Minnesota**
*Architect:* Johnson/Burgee Associates
*Developer:* Investors Diversified Services
*Building Services Engineer:* Clint Hedsten

**Illinois Tower (Illinois Center), ca.1970**
**Chicago, Illinois**
*Architect:* Fujikawa, Johnson & Associates
*Developer:* Metropolitan Structures, Inc.
*Structural Engineer:* Alfred Benesch Company
*Building Services Engineer:* Cosentini Associates

### Inland Steel Building, 1958
**Chicago, Illinois**
*Architect:* Skidmore, Owings & Merrill
*Developer:* Inland Steel
*Structural Engineer:* Skidmore, Owings & Merrill
*Building Services Engineer:* Skidmore, Owings & Merrill

### International House, University of California, 1930[1]
**Berkeley, California**
*Architect:* George W. Kelham
*Developer:* John D. Rockefeller
*Structural Engineer:* H. J. Brunnier
*Building Services Engineer:* Hunter D. Hudson

### Janata Bank Building, 1985[1]
**Dhaka, Bangladesh**
*Architect:* DDC/BCL
*Developer:* Janata Bank (Concord Dev. Builder)

### Jin Mao Building [3] (Projected completion 1998)
**Shanghai, China**
*Architect:* Skidmore, Owings & Merrill

### John Hancock Center, 1968
**Chicago, Illinois**
*Architect:* Skidmore, Owings & Merrill
*Developer:* John Hancock Mutual Life Insurance Co.
*Structural Engineer:* Skidmore, Owings & Merrill
*Building Services Engineer:* Skidmore, Owings & Merrill

### John Hancock Tower, 1973
**Boston, Massachusetts**
*Architect:* I.M. Pei & Partners
*Developer:* John Hancock Mutual Life Insurance Co.
*Structural Engineer:* Office of James Ruderman
*Building Services Engineer:* Cosentini Associates

### Koln Cathedral, 1248[1]
**Koln, Germany**
*Architect:* Master Gerhard

### Kostabi World Tower, 1994[2]
**New York, New York**
*Architect:* Eli Attia
*Developer:* Mark Kostabi

### La Citta Nuova, 1914[2]
**Milan, Italy**
*Architect:* Antonio Sant' Elia

### La Grande Arche, 1989[1]
**Paris, France**
*Architect:* Johan Otto von Spreckelsen
*Developer:* Societe d'Economie Mixte Nationale Tete Defense

### Lake Point Tower, 1968[1]
**Chicago, Illinois**
*Architect:* Schipporeit-Henrich Architects/Graham, Anderson, Probst & White

### Larkin Office Tower, 1904[1]
**Buffalo, New York**
*Architect:* Frank Lloyd Wright
*Developer:* Larkin Company
*Structural Engineer:* Paul Mueller

### Law Courts Building, 1994
**Naples, Italy**
*Architect:* Beguinot, Capobianco, Pica Ciamarra & Zagaria
*Developer:* Ministero di Grazie e Guistizia
*Structural Engineer:* Giangreco/Giordano
*Building Services Engineer:* Parolini

### Leiter Building I, 1879[1]
**Chicago, Illinois**
*Architect:* William Le Baron Jenney

### Leiter Building II, 1891[1]
**Chicago, Illinois**
*Architect:* William Le Baron Jenney

### Leo Burnett Building, 1989
**Chicago, Illinois**
*Architect:* Kevin Roche John Dinkeloo and Associates
*Developer:* John Buck Co.
*Structural Engineer:* Cohen Baretto Marchettos Engineers
*Building Services Engineer:* Environmental Systems Design

## Lever House, 1952
**New York, New York**
*Architect:* Skidmore, Owings & Merrill
*Developer:* Lever Brothers
*Structural Engineer:* Skidmore, Owings & Merrill
*Building Services Engineer:* Skidmore, Owings & Merrill

## L'Institut du Monde Arabe, 1987
**Paris, France**
*Architect:* Jean Nouvel
*Developer:* Arab World Institute
*Structural Engineer:* Louis Gruittet
*Building Services Engineer:* SETEC

## Lloyd's of London, 1981
**London, England**
*Architect:* Sir Richard Rogers & Partners
*Developer:* Lloyd's of London Insurance
*Structural Engineer:* Ove Arup & Partners
*Building Services Engineer:* Graham Anthony

## Lucky-Goldstar International Business Center, 1988
**Seoul, South Korea**
*Architect:* Bechtel Civil, Inc.
*Developer:* Lucky-Goldstar Development Co., Ltd.
*Structural Engineer:* Bechtel Civil, Inc.
*Building Services Engineer:* Bechtel Civil, Inc.

## Lucky-Goldstar Twin Towers, 1987
**Seoul, South Korea**
*Architect:* Skidmore, Owings & Merrill
*Developer:* Lucky-Goldstar Development Co., Ltd.
*Structural Engineer:* Skidmore, Owings & Merrill
*Building Services Engineer:* Skidmore, Owings & Merrill

## Maiden Law Tower[1]
**New York, New York**
*Architect:* Johnson/Burgee Associates

## Manhattan Building, 1891[1]
**Chicago, Illinois**
*Architect:* William Le Baron Jenney

## Mapfre Tower, 1992
**Barcelona, Spain**
*Architect:* Inigo Ortiz & Enrique Leon Arquitectos
*Developer:* Mapfre Mutual Life Insurance Company
*Structural Engineer:* OTEP International Associates
*Building Services Engineer:* Antonio Carrion Associates

## Marina City Twin Towers, 1964[1]
**Chicago, Illinois**
*Architect:* Bertrand Goldberg & Associates

## Marine Midland Bank, 1970
**Rochester, New York**
*Architect:* Skidmore, Owings & Merrill
*Developer:* Marine Midland Grace Trust Company
*Structural Engineer:* Skidmore, Owings & Merrill
*Building Services Engineer:* Jaros, Baum & Bolles

## Marshall Field Wholesale Store, 1887[1]
**Chicago, Illinois**
*Architect:* Henry Hobson Richardson
*Developer:* Marshall Field

## Masonic Temple, 1892[1]
**Chicago, Illinois**
*Architect:* Burnham & Root

## Mellon Bank Center, 1990
**Philadelphia, Pennsylvania**
*Architect:* Kohn Pedersen Fox Associates
*Developer:* Richard I. Rubin & Company, Inc.
*Structural Engineer:* Office of Irwin G. Cantor
*Building Services Engineer:* Flack & Kurtz

## Menara Mesiniaga/IBM Building, 1992
**Selangor, Malaysia**
*Architect:* T.R. Hamzah & Ken Yeang SDN BHP
*Developer:* Mesiniaga SDN BDH
*Structural Engineer:* Reka Perunding
*Building Services Engineer:* Norman Disney & Young

**Messe Tower, 1990**
**Frankfurt, Germany**
*Architect:* Murphy/Jahn
*Developer:* ECE Project Management GMBLT
*Structural Engineer:* Dr. Fritz Noetzold, Inc.
*Building Services Engineer:* Brendel

**Metropolitian International Building[3] (Projected completion 1995)**
**Bangkok, Thailand**
*Architect:* Design + Develop
*Developer:* Maung Roong/Pornpat
*Structural Engineer:* ACT/Skilling Ward Magnusson Barkshire
*Building Services Engineer:* ACT/Flack & Kurtz

**Metropolitan Square, 1988[1]**
**St. Louis, Missouri**
*Architect:* Hellmuth, Obata & Kassabaum, Inc.
*Developer:* Metropolitan Life Insurance Company
*Structural Engineer:* KKBNA

**Miglin-Beitler Tower, 1993[2]**
**Chicago, Illinois**
*Architect:* Cesar Pelli & Associates
*Developer:* Miglin-Beitler Development
*Structural Engineer:* Dr. Charles Thorton

**Millennium Tower, 1994[2]**
**Tokyo, Japan**
*Architect:* Norman Foster & Associates

**Monadnock Building, 1891[1]**
**Chicago, Illinois**
*Architect:* Burnham & Root

**Montauk Building, 1882[1]**
**Chicago, Illinois**
*Architect:* Burnham & Root

**Morton International Building, 1990**
**Chicago, Illinois**
*Architect:* Perkins & Will
*Developer:* Morton Salt Company
*Structural Engineer:* Perkins & Will
*Building Services Engineer:* Environmental Systems Design

**Museum of Art, 1928[1]**
**Philadelphia, Pennsylvania**
*Architect:* Horace Trambauer & C. Clark Zantzinger

**Netherlands International Bank (NMB) Headquarters, 1987[1]**
**Amsterdam, Netherlands**
*Architect:* Alberts & Van Haut
*Developer:* Maatshappi Voor Bedrufsobjecten

**New Tokyo City Hall Project, 1991**
**Tokyo, Japan**
*Architect:* Kenzo Tange Associates
*Developer:* Municipal of Tokyo
*Structural Engineer:* Muto Associates
*Building Services Engineer:* Inuzuka Engineering Consultants

**New York Coliseum Building, 1990[2]**
**New York, New York**
*Architect:* Skidmore, Owings & Merrill
*Developer:* Galbreath-Ruffin Corporation
*Structural Engineer:* Skidmore, Owings & Merrill

**Northwestern Atrium Center, 1987**
**Chicago, Illinois**
*Architect:* Murphy/Jahn
*Developer:* Tishman Midwest Management
*Structural Engineer:* Lev Zetlin Associates, Inc.
*Building Services Engineer:* Mike Oppenheimer Associates

**Norwest Center, 1988**
**Minneapolis, Minnesota**
*Architect:* Cesar Pelli & Associates
*Developer:* Gerald D. Hines Interests
*Structural Engineer:* Cohen Baretto Marchettos Engineers
*Building Services Engineer:* I.A. Naman & Associates

**Oakbrook Terrace Tower, 1987**
**Oakbrook, Illinois**
*Architect:* Murphy/Jahn
*Developer:* Miglin-Beitler Development Company
*Structural Engineer:* CBM Engineers
*Building Services Engineer:* Environmental Systems Design

**Olivetti Factory Building, 1954[1]**
**Milano, Italy**
*Architect:* Luigini & Gino Pollini

### One Liberty Place, 1987
**Philadelphia, Pennsylvania**
*Architect:* Murphy/Jahn
*Developer:* Rouse Associates
*Structural Engineer:* Lev Zetlin Associates, Inc.
*Building Services Engineer:* Flack & Kurtz

### One Magnificent Mile, 1983
**Chicago, Illinois**
*Architect:* Skidmore, Owings & Merrill
*Developer:* The Levy Organization
*Structural Engineer:* Skidmore, Owings & Merrill
*Building Services Engineer:* Skidmore, Owings & Merrill

### One Shell Plaza, 1971
**Houston, Texas**
*Architect:* Skidmore, Owings & Merrill
*Developer:* Gerald D. Hines Interests
*Structural Engineer:* Skidmore, Owings & Merrill
*Building Services Engineer:* Chenault & Brady

### One Shell Square, 1972
**New Orleans, Louisiana**
*Architect:* Skidmore, Owings & Merrill
*Developer:* Gerald D. Hines Interests
*Structural Engineer:* Skidmore, Owings & Merrill
*Building Services Engineer:* Chenault & Brady

### One South Wacker Building, 1983
**Chicago, Illinois**
*Architect:* Murphy/Jahn
*Developer:* Harvey Walker & Company
*Structural Engineer:* Alfred Benesch Company
*Building Services Engineer:* Cosentini Associates

### Onterie Center, 1985
**Chicago, Illinois**
*Architect:* Skidmore, Owings & Merrill
*Developer:* PSM International Corporation
*Structural Engineer:* Skidmore, Owings & Merrill
*Building Services Engineer:* Skidmore, Owings & Merrill

### Oryx Energy Company Corporate Headquarters, 1991[1]
**Dallas, Texas**
*Architect:* Hellmuth, Obata & Kassabaum, Inc.
*Developer:* ORYX Energy Company

### Overseas Chinese Bank Center, 1976
**Singapore, Singapore**
*Architect:* I.M. Pei & Partners
*Developer:* Overseas Chinese Bank Corporation
*Structural Engineer:* Ove Arup & Partners
*Building Services Engineer:* Cosentini Associates

### Pacific First Centre, 1989[1]
**Seattle, Washington**
*Architect:* W.A. Callison Partnership
*Developer:* Prescott
*Structural Engineer:* Skilling Ward Magnusson Barkshire

### Palazzo Strozzi, 14901
**Florence, Italy**
*Architect:* Guiliano da Sangallo

### Palmolive Building, 1930[1]
**Chicago, Illinois**
*Architect:* Holabird & Root

### Paramount Building, 1927[1]
**New York, New York**
*Architect:* Rapp & Rapp

### Pennzoil Building, 1976
**Houston, Texas**
*Architect:* Burgee/Johnson Associates
*Developer:* Gerald D. Hines Interests
*Structural Engineer:* Ellisor Engineers
*Building Services Engineer:* I.A. Naman & Associates

### Performing Arts Building, 1973[1]
**Fort Wayne, Indiana**
*Architect:* Louis Kahn

### Petronas Towers[3] (Projected completion 1996)
**Kuala Lumpur, Malaysia**
*Architect:* Cesar Pelli & Associates
*Developer:* Kuala Lumpur City Center
*Structural Engineer:* Thorton-Thomasetti Engineers
*Building Services Engineer:* Ranhill Bersekutu SDN. BND.

**Pittsburgh Plate Glass Industries Headquarters, 1983**
**Pittsburgh, Pennsylvania**
*Architect:* Burgee/Johnson Associates
*Developer:* PPG Industries, Inc.
*Structural Engineer:* Robertson, Fowler & Associates
*Building Services Engineer:* Cosentini Associates

**Pontiac & Champlain Building, 1891[1]**
**Chicago, Illinois**
*Architect:* Holabird & Roche

**Procter & Gamble Headquarters, 1985**
**Cincinnati, Ohio**
*Architect:* Kohn Pedersen Fox Associates
*Developer:* Procter & Gamble
*Structural Engineer:* Weiskopf & Pickworth
*Building Services Engineer:* Syska & Hennessey

**Promontory Apartments, 1949[1]**
**Chicago, Illinois**
*Architect:* Mies Van Der Rohe

**Pruitt-Igoe Housing, 1955[1]**
**St. Louis, Missouri**
*Architect:* Hellmuth, Yamasaki & Leinweber
*Developer:* St. Louis Housing Authority

**RCA Building, 1933[1]**
**New York, New York**
*Architect:* Cross & Cross
*Developer:* RCA

**Reliance Building, 1895[1]**
**Chicago, Illinois**
*Architect:* Daniel Burnham & Company

**Republic Bank Center, 1984**
**Houston, Texas**
*Architect:* Burgee/Johnson Associates
*Developer:* Gerald D. Hines Interests
*Structural Engineer:* CBM Engineers
*Building Services Engineer:* I.A. Naman & Associates

**Reserve Bank Quarters, 1984**
**Calcutta, India**
*Architect:* Architects Collaborated
*Developer:* Reserve Bank of India
*Structural Engineer:* Dr. Bratish Sengupta
*Building Services Engineer:* Architects Collaborated

**Restaurant Tower, ca.1940[2]**
**Italy**
*Architect:* Mario Ridolfi

**Rockefeller Center, 1933[1]**
**New York, New York**
*Architect:* Harrison, Abramovitz & Harris
*Developer:* John D. Rockefeller

**The Rookery, 1888[1]**
**Chicago, Illinois**
*Architect:* Burnham & Root

**R.R. Donnelley Building, 1912[1]**
**Chicago, Illinois**
*Architect:* Howard Van Doren Shaw
*Developer:* R.R. Donnelley and Sons

**Seagram Building, 1959[1]**
**New York, New York**
*Architect:* Mies Van Der Rohe/Phillip Johnson
*Developer:* Seagram Corporation

**Sears Tower, 1974**
**Chicago, Illinois**
*Architect:* Skidmore, Owings & Merrill
*Developer:* Sears Roebuck & Company
*Structural Engineer:* Skidmore, Owings & Merrill
*Building Services Engineer:* Skidmore, Owings & Merrill

**Sena Bhaban[1]**
**Dhaka, Bangladesh**
*Architect:* Pratsthapana
*Developer:* Nirman International

**Shelton Hotel, 1930[1]**
**New York, New York**
*Architect:* Arthur L. Harmon

**Singer Building, 1908[1]**
**New York, New York**
*Architect:* Ernest Flagg
*Developer:* Singer Company

**Southeast Financial Center, 1983**
**Miami, Florida**
*Architect:* Skidmore, Owings & Merrill
*Developer:* Gerald D. Hines Interests
*Structural Engineer:* Skidmore, Owings & Merrill
*Building Services Engineer:* I.A. Naman & Associates

### Standard Oil Building, 1926[1]
**New York, New York**
*Architect:* Carrere & Hastings
*Developer:* Standard Oil Company (New Jersey)

### State of Illinois Center, 1985
**Chicago, Illinois**
*Architect:* Murphy/Jahn
*Developer:* Illinois Capital Development Board
*Structural Engineer:* Kolbjorn Saether
*Building Services Engineer:* Shepard, Eisenburg Inc. (mech), John Mohan (elec)

### Texas Commerce Tower, 1982
**Houston, Texas**
*Architect:* I.M. Pei & Partners
*Developer:* Gerald D. Hines Interests
*Structural Engineer:* CBM Engineers
*Building Services Engineer:* I.A. Naman & Associates

### Times Building, 1904[1]
**New York, New York**
*Architect:* Eidlitz & Mac Kenzie

### Tokyo Telecom Center, 1994[1]
**Tokyo, Japan**
*Architect:* Nippon-Sogo/Hellmuth, Obata & Kassabaum, Inc.
*Developer:* Tokyo Teleport Center

### Toronto City Hall, 1965[1]
**Toronto, Canada**
*Architect:* Viljo Revell/John B. Parkins & Associates
*Developer:* City of Toronto
*Structural Engineer:* Severud Eistad Krueger Associates

### Toronto-Dominion Centre, 1967[1]
**Toronto, Canada**
*Architect:* Mies Van Der Rohe/John B. Parkins & Asscciates
*Developer:* The Toronto Dominion Bank and CEMP Investments Ltd.
*Structural Engineer:* C. D. Carruthers & Wallace, Inc.
*Building Services Engineer:* H.H. Angus & Associates

### Torre Velasca, 1958[1]
**Milano, Italy**
*Architect:* L. Barbiano di Belgiojoso, E. Peresuitti & E.N. Rogers

### Tower Block Building, 1986
**Calcutta, India**
*Architect:* ACME Consultants Pvt. Ltd.
*Developer:* The Calcutta Municipal Corporation
*Structural Engineer:* ACME Consultants Pvt. Ltd.
*Building Services Engineer:* ACME Consultants Pvt., Ltd.

### Transamerica Pyramid, 1976[1]
**San Francisco, California**
*Architect:* William Pereira & Associates

### Transco Building, 1983
**Houston, Texas**
*Architect:* Johnson/Burgee Associates
*Developer:* Gerald D. Hines Interests
*Structural Engineer:* CBM Engineers
*Building Services Engineer:* I.A. Naman & Associates

### Tribune Tower, 1925[1]
**Chicago, Illinois**
*Architect:* Hood & Howells
*Developer:* Chicago Tribune Company

### Trump Tower, 1982
**New York, New York**
*Architect:* Swanke, Hayden & Connell
*Developer:* Trump Organization
*Structural Engineer:* Office of Irwin G. Cantor
*Building Services Engineer:* Digiacomo Associates, W.A., Inc.

### Twin Towers on the Green Axis, 1994
**Naples, Italy**
*Architect:* Pica Ciamarra Associati
*Developer:* ICLA/Pizzarotti
*Structural Engineer:* A. Grimaldi/R. Sparacio
*Building Services Engineer:* Sogi s.r.l/Guareschi s.r.l/Technitalia s.r.l.

### Two Prudential Plaza, 1990
**Chicago, Illinois**
*Architect:* Loebl, Schlossman & Hackl
*Developer:* Prudential Plaza Associates/NLI Insurance
*Structural Engineer:* CBM Engineers
*Building Services Engineer:* Environmental Systems Design

### Two Shell Plaza, 1972
**Houston, Texas**
*Architect:* Skidmore, Owings & Merrill
*Developer:* Gerald D. Hines Interests
*Structural Engineer:* Skidmore, Owings & Merrill
*Building Services Engineer:* Chenault & Brady

### Two Union Square, 1990
**Seattle, Washington**
*Architect:* NBBJ Group
*Developer:* UNICO Properties, Inc.
*Structural Engineer:* Skilling Ward Magnusson Barkshire
*Building Services Engineer:* Boullion, Christofferson & Schairer, Inc.

### Unite d' Habitation, 1952[1]
**Marseilles, France**
*Architect:* Le Corbusier

### United Commercial Bank of India[3]
**Calcutta, India**
*Architect:* ACME Consultants Pvt. Ltd.
*Developer:* United Commercial Bank of India
*Structural Engineer:* ACME Consultants Pvt. Ltd.
*Building Services Engineer:* ACME Consultants Pvt. Ltd.

### United Gulf Bank, 1987
**Manama, Bahrain**
*Architect:* Skidmore, Owings & Merrill
*Developer:* United Gulf Bank
*Structural Engineer:* Skidmore, Owings & Merrill
*Building Services Engineer:* Pan Arab Consulting Engineering

### United Nations Plaza, 1974
**New York, New York**
*Architect:* Kevin Roche John Dinkeloo and Associates
*Developer:* United Nations Development Corporation
*Structural Engineer:* Weiskopf & Pickworth/Irwin Cantor
*Building Services Engineer:* Cosentini Associates/Jaros, Baum & Bolles

### USG Building, 1990
**Chicago, Illinois**
*Architect:* Skidmore, Owings & Merrill
*Developer:* Stein & Company
*Structural Engineer:* Skidmore, Owings & Merrill
*Building Services Engineer:* Skidmore, Owings & Merrill

### USX Tower, 1970
**Pittsburgh, Pennsylvania**
*Architect:* Harrison and Abramovitz & Abbe
*Developer:* United States Steel Corporation
*Structural Engineer:* Skilling Helle Christiansen Robertson
*Building Services Engineer:* Jaros, Baum & Bolles

### Vila Olimpica-Hotel de la Artes, 1992
**Barcelona, Spain**
*Architect:* Skidmore, Owings & Merrill
*Developer:* The Travelstead Group
*Structural Engineer:* Skidmore, Owings & Merrill
*Building Services Engineer:* Skidmore, Owings & Merrill

### Villa Contemporaine, 1922[2]
**Paris, France**
*Architect:* Le Corbusier

### Villa Radieuse, 1933[2]
**Paris, France**
*Architect:* Le Corbusier

### Voisin Plan, 1925[2]
**Paris, France**
*Architect:* Le Corbusier

### Wainwright Building, 1891[1]
**St. Louis, Missouri**
*Architect:* Adler & Sullivan

**Waldorf-Astoria Hotel, 1931**[1]
**New York, New York**
*Architect:* Schultze & Weaver

**Warsaw Tower, 1990**[2]
**Warsaw, Poland**
*Architect:* William McDonough

**Water Tower Place, 1976**
**Chicago, Illinois**
*Architect:* Loebl, Schlossman, Dart & Hackl
*Developer:* Urban Investment & Development Company
*Structural Engineer:* C.F. Murphy
*Building Services Engineer:* C.F. Murphy

**Waterside Plaza, 1974**[1]
**New York, New York**
*Architect:* Lewis Davis/Samuel Brody & Associates
*Developer:* Waterside Redevelopment Company

**Watterson Towers, Illinois State University, 1968**
**Normal, Illinois**
*Architect:* Fridstein & Fitch
*Developer:* Illinois State University
*Structural Engineer:* Fridstein & Fitch
*Building Services Engineer:* Fridstein & Fitch

**Weld Atrium, 1985**
**Kuala Lumpur, Malaysia**
*Architect:* T. R. Hamzah & Yeang Sdn. Bhd.
*Developer:* Ban Seng Sdn. Bhd.
*Structural Engineer:* Reka Perunding Sdn. Bhd
*Building Services Engineer:* Jurutera Perunding LC Sdn. Bhd.

**Woolworth Building, 1913**[1]
**New York, New York**
*Architect:* Cass Gilbert
*Developer:* Woolworth Company

**World Trade Center Towers, 1973**
**New York, New York**
*Architect:* Minoru Yamasaki & Associates
*Developer:* Port Authority of New York & New Jersey
*Structural Engineer:* Skilling Helle Christiansen Robertson
*Building Services Engineer:* Jaros, Baum & Bolles

**Worldwide Plaza, 1989**
**New York, New York**
*Architect:* Skidmore, Owings & Merrill
*Developer:* Zeckendorf Corporation
*Structural Engineer:* Skidmore, Owings & Merrill
*Building Services Engineer:* Cosentini Associates

**Wrigley Building, 1924**[1]
**Chicago, Illinois**
*Architect:* Graham, Anderson, Probst & White
*Developer:* Wrigley Company

**Xerox Building, 1980**
**Chicago, Illinois**
*Architect:* Murphy/Jahn
*Developer:* Romanek Golub & Company
*Structural Engineer:* CBM Engineers
*Building Services Engineer:* CFMA Engineers

## *Notes:*

- The material represented in this appendix was researched thoroughly. However, the Editorial Group is not responsible for inaccuracies.
- For buildings constructed before 1960, records were generally not found regarding the developers and engineers.
- Some buildings listed in the text may not have appeared in this appendix due to lack of information.
- Year of completion was not available for some buildings.

[1]Full information not available.
[2]Project proposal (project never built).
[3]Building under construction.

# Building Index

The page numbers in italics indicate pictures of buildings.

# Name Index

# Subject Index